AF353147

Testing of Materials and Elements in Civil Engineering

Testing of Materials and Elements in Civil Engineering

Volume 1

Editor

Krzysztof Schabowicz

MDPI • Basel • Beijing • Wuhan • Barcelona • Belgrade • Manchester • Tokyo • Cluj • Tianjin

Editor
Krzysztof Schabowicz
Wroclaw University of Science
and Technology
Poland

Editorial Office
MDPI
St. Alban-Anlage 66
4052 Basel, Switzerland

This is a reprint of articles from the Special Issue published online in the open access journal *Materials* (ISSN 1996-1944) (available at: https://www.mdpi.com/journal/materials/special_issues/ testing_civil_engineering).

For citation purposes, cite each article independently as indicated on the article page online and as indicated below:

LastName, A.A.; LastName, B.B.; LastName, C.C. Article Title. *Journal Name* **Year**, *Volume Number*, Page Range.

Volume 1
ISBN 978-3-0365-1888-6 (Hbk)
ISBN 978-3-0365-1889-3 (PDF)

Volume 1-2
ISBN 978-3-0365-1577-9 (Hbk)
ISBN 978-3-0365-1578-6 (PDF)

© 2021 by the authors. Articles in this book are Open Access and distributed under the Creative Commons Attribution (CC BY) license, which allows users to download, copy and build upon published articles, as long as the author and publisher are properly credited, which ensures maximum dissemination and a wider impact of our publications.
The book as a whole is distributed by MDPI under the terms and conditions of the Creative Commons license CC BY-NC-ND.

Contents

About the Editor

Krzysztof Schabowicz

A specialist in the field of building engineering, building law, diagnostics and maintenance of engineering structures. He deals with ventilated fac¸ades, in particular production technology and testing of external fiber cement cladding in the field of detection, identification and classification of degradation and damage processes, as well as the methodology of these tests. He conducts scientific research and development works related to the implementation of non-destructive diagnostics devices and technologies in construction works, including the use of artificial intelligence. Author and co-author of 5 books, over 200 publications and 9 patents. Has more than 500 citations in Web of Science. He serves as an Editor of Materials (MDPI) and Editorial Board member of Civil Engineering and Architecture (HRPUB), and Nondestructive Testing and Diagnostics (SIMP). He developed more than 200 reviews of journal and conference articles.

Editorial

Testing of Materials and Elements in Civil Engineering

Krzysztof Schabowicz

Faculty of Civil Engineering, Wrocław University of Science and Technology, Wybrzeże Wyspiańskiego 27, 50-370 Wrocław, Poland; krzysztof.schabowicz@pwr.edu.pl

Abstract: This issue is proposed and organized as a means to present recent developments in the field of testing of materials in civil engineering. For this reason, the articles highlighted in this issue should relate to different aspects of testing of different materials in civil engineering, from building materials and elements to building structures. The current trend in the development of materials testing in civil engineering is mainly concerned with the detection of flaws and defects in elements and structures using destructive, semi-destructive, and nondestructive testing. The trend, as in medicine, is toward designing test equipment that allows one to obtain a picture of the inside of the tested element and materials. Very interesting results with significance for building practices of testing of materials and elements in civil engineering were obtained.

Keywords: testing; diagnostic; building materials; elements; civil engineering

Citation: Schabowicz, K. Testing of Materials and Elements in Civil Engineering. *Materials* **2021**, *14*, 3412. https://doi.org/10.3390/ma14123412

Received: 29 May 2021
Accepted: 18 June 2021
Published: 20 June 2021

Publisher's Note: MDPI stays neutral with regard to jurisdictional claims in published maps and institutional affiliations.

Copyright: © 2021 by the author. Licensee MDPI, Basel, Switzerland. This article is an open access article distributed under the terms and conditions of the Creative Commons Attribution (CC BY) license (https://creativecommons.org/licenses/by/4.0/).

1. Introduction

The field of testing of materials in civil engineering is very wide, and is interesting from an engineering and scientific point of view [1–3]. This issue is proposed and organized as a means to present recent developments in the field of testing of materials in civil engineering. For this reason, the articles highlighted in this issue should relate to different aspects of testing of different materials in civil engineering, from building materials and elements to building structures [4–6]. The current trend in the development of materials testing in civil engineering is mainly concerned with the detection of flaws and defects in elements and structures using destructive, semi-destructive, and nondestructive testing.

This issue mainly focuses on different novel testing approaches, the development of single and hybrid measurement techniques, and advanced signal analysis. The topics of interest include but are not limited to the testing of materials and elements in civil engineering, testing of structures made of novel materials [7–9], condition assessment of civil materials and elements, detecting defects invisible on the surface, damage detection and damage imaging, diagnostics of cultural heritage monuments, structural health monitoring systems, modeling and numerical analyses, nondestructive testing methods, and advanced signal processing for nondestructive testing [10,11].

2. Description of the Articles Presented in the Issue

Grouted rock bolts represent one of the most used elements for rock mass stabilization, as analyzed by [12], and reinforcement and the grouting quality have a crucial role in the load transfer mechanism. At the same time, the grouting quality, as well as the grouting procedures, are the least controlled in practice. This paper deals with the non-destructive investigation of grouting percentage through an analysis of the rock bolt's natural frequencies after applying an artificial longitudinal impulse to its head by using a soft-steel hammer as a generator. A series of laboratory models, with different positions and percentages of the grouted section, simulating grouting defects, were tested. A comprehensive statistical analysis was conducted and a high correlation between the grouting percentage and the first three natural frequencies of rock bolt models has been established. After validation of FEM numerical models based on experimentally obtained values, a further analysis includes consideration of grout stiffness variation and its impact on the natural frequencies of rock bolts [12].

In their paper [13], in order to create and make available design guidelines, recommendations for energy audits, data for analysis and simulation of the condition of masonry walls susceptible to biological corrosion, deterioration of comfort parameters in rooms, or deterioration of thermal resistance, were given. The paper analyzes various types of masonry wall structures occurring in and commonly used in historical buildings over the last 200 years. The summary is a list of results of particular types of masonry walls and their mutual comparison. On this basis, a procedure path has been proposed which is useful for monitoring heat loss, monitoring the moisture content of building partitions, and improving the hygrothermal comfort of rooms. The durability of such constructions has also been estimated, and the impact on the condition of the buildings that have been preserved and are still in use today was assessed [13].

In [14], the laboratory testing of the construction materials and elements is a subset of activities inherent in sustainable building materials engineering. Two questions arise regarding the test methods used: the relation between test results and material behavior in actual conditions on the one hand, and the variability of results related to uncertainty on the other. The paper presents the analysis of the results and uncertainties of the two simple independent test examples (bond strength and tensile strength) in order to demonstrate discrepancies related to the ambiguous methods of estimating uncertainty, and the consequences of using test methods when method suitability for conformity assessment has not been properly verified. These examples are the basis for opening a discussion on the test methods development direction which makes it possible to consider them as "sustainable". The paper addresses the negative impact of the lack of complete test models, taking into account proceeding with the uncertainty regarding erroneous assessment risks. Adverse effects can be minimized by creating test methods appropriate for the test's purpose (e.g., initial or routine tests) and handling uncontrolled uncertainty components. Sustainable test methods should ensure a balance between widely defined tests and evaluation costs and the material's or building's safety, reliability, and stability [14].

The article by [15] presents the possibilities of using foamed asphalt in the recycling process to produce the base layer of road pavement constructions, in Polish conditions. Foamed asphalt was combined with reclaimed asphalt pavement (RAP) and hydraulic binder (cement). Foamed asphalt mixtures with cement (FAC) were made, based on these ingredients. To reduce stiffness and cracking in the base layer, foamed asphalt (FA) was additionally used in the analyzed mixes containing cement. The laboratory analyses allowed estimating the stiffness and fatigue durability of the conglomerate. In the experimental section, measurements of deflections are made, modules of pavement layers are calculated, and their fatigue durability is determined. As a result of the research, new fatigue criteria for FAC mixtures and the correlation factors of stiffness modulus and fatigue durability in situ with the results of laboratory tests were developed. It is anticipated that FAC recycling technology will provide durable and safe road pavements [15].

The article by [16] presents experimental tests of a new type of composite bar that has been used as shear reinforcement for concrete beams. In the case of shearing concrete beams reinforced with steel stirrups, according to the theory of plasticity, the plastic deformation of stirrups, and stress redistribution in stirrups cut by a diagonal crack, are permitted. Tensile composite reinforcement is characterized by linear-elastic behavior throughout the entire strength range. The most popular type of shear reinforcement is closed-frame stirrups, and this type of fiber-reinforced polymer (FRP) shear reinforcement was the subject of research by other authors. In the case of FRP stirrups, rupture occurs rapidly, without the shear reinforcement being able to redistribute stress. An attempt was made to introduce a quasi-plastic character into the mechanisms transferring shear by appropriately shaping the shear reinforcement. Experimental material tests covered the determination of the strength and deformability of straight glass fiber-reinforced polymer (GFRP) bars and GFRP headed bars. Experimental studies of shear-reinforced beams with GFRP stirrups and

GFRP headed bars were carried out. This allowed a direct comparison of the shear behavior of beams reinforced with standard GFRP stirrups and a new type of shear reinforcement: GFRP headed bars. Experimental studies demonstrated that GFRP headed bars could be used as shear reinforcement in concrete beams. Unlike GFRP stirrups, these bars allow stress redistribution in bars cut by a diagonal crack [16].

The paper by [17] examines the effect of PBO (P-phenylene benzobisoxazole)–FRCM (fabric-reinforced cementitious matrix) reinforcement on the stiffness of eccentrically compressed reinforced concrete columns. Reinforcement with FRCM consists of bonding composite meshes to the concrete substrate by means of mineral mortar. Longitudinal and/or transverse reinforcements made of PBO (P-phenylene benzobisoxazole) mesh were applied to the analyzed column specimens. When assessing the stiffness of the columns, the focus was on the effect of the composite reinforcement itself, the value and eccentricity of the longitudinal force, and the decrease in the modulus of elasticity of the concrete, with increasing stress intensity in the latter. Dependencies between the change in the elasticity modulus of the concrete and the change in the stiffness of the tested specimens were examined. The relevant standards, providing methods of calculating the stiffness of composite columns, were used in the analysis. Regarding columns, which were strengthened only transversely with PBO mesh, reinforcement increases their load capacity, and at the same time, the stiffness of the columns increases due to the confinement of the cross-section. The stiffness depends on the destruction of the concrete core inside its composite jacket. In the case of columns with transverse and longitudinal reinforcement, the presence of longitudinal reinforcement reduces longitudinal deformations. The columns failed at higher stiffness values in the whole range of the eccentricities [17].

The paper by [18] presents the results of an experimental investigation into stop-splayed scarf joints which was carried out as part of a research program at the Wroclaw University of Science and Technology. A brief description of the characteristics of scarf and splice joints appearing in historical buildings is provided, with special reference to stop-splayed scarf joints (so-called "bolt of lightning" joints) which were widely used, for example, in Italian Renaissance architecture. Analyses and studies of scarf and splice joints in bent elements as presented in the literature are reviewed, along with selected examples of analyses and research on tensile joints. It is worth noting that the authors in practically all the cited literature draw attention to the need for further research in this area. Next, the results of the authors' own research on beams with stop-splayed scarf joints, strengthened using various methods, e.g., by means of drawbolts (metal screws), steel clamps and steel clamps with wooden pegs, which were subjected to four-point bending tests, are presented. Load-deflection plots were obtained for load-bearing to bending of each beam in relation to the load-bearing of a continuous reference beam. A comparative analysis of the results obtained for each beam series is presented, along with conclusions and directions for further research [18].

Non-destructive testing of concrete for defects detection, using acoustic techniques, is currently performed mainly by human inspection of recorded images [19]. The images consist of the inside of the examined elements, obtained from testing devices such as the ultrasonic tomograph. However, such an automatic inspection is time-consuming, expensive, and prone to errors. To address some of these problems, this paper aims to evaluate a convolutional neural network (CNN) toward an automated detection of flaws in concrete elements using ultrasonic tomography. There are two main stages in the proposed methodology. In the first stage, an image of the inside of the examined structure is obtained and recorded by performing ultrasonic tomography-based testing. In the second stage, a convolutional neural network model is used for the automatic detection of defects and flaws in the recorded image. In this work, a large and pre-trained CNN is used. It was fine-tuned on a small set of images collected during laboratory tests. Lastly, the prepared model was applied for detecting flaws. The obtained model has proven to be able to accurately detect defects in examined concrete elements. The presented approach for automatic detection of flaws is being developed with the potential to not only detect defects of one type but also to classify various types of defects in concrete elements [19].

The reliability and safety of power transmission depend first and foremost on the state of the power grid, and mainly on the state of the high-voltage power line towers [20]. The steel structures of existing power line supports (towers) have been in use for many years. Their in-service time, the variability in structural, thermal and environmental loads, the state of foundations (displacement and degradation), the corrosion of supporting structures and lack of technical documentation are essential factors that have an impact on the operating safety of the towers. The tower state assessment used to date, consisting of finding the deviation in the supporting structure apex, is insufficient because it omits the other necessary condition, the stress criterion, which is not to exceed allowable stress values. Moreover, in difficult terrain conditions, the measurement of the tower deviation is very troublesome, and for this reason, it is often not performed. This paper presents a stress-and-strain analysis of the legs of 110 kV power line truss towers with a height of 32 m. They have been in use for over 70 years and are located in especially difficult geotechnical conditions—one of them is in a gravel mine on an island surrounded by water, and the other stands on a steep, wet slope. Purpose-designed fiber Bragg grating (FBG) sensors were proposed for strain measurements. Real values of stresses arising in the tower legs were observed and determined over a period of one year. Validation was also carried out based on geodetic measurements of the tower apex deviation, and a residual magnetic field (RMF) analysis was performed to assess the occurrence of cracks and stress concentration zones [20].

The paper by [21] explores the microstructural evolution characteristics of tailings sand samples from different types of infiltration failure during the infiltration failure process. A homemade small infiltration deformation instrument is used to test the infiltration failure characteristics of the tailings sand during the infiltration failure process. Evolutionary characteristics of the internal microstructure pores and particle distribution were also studied. Using CT (computerized tomography) technology to establish digital image information, the distribution of the microscopic characteristics of the particle distribution and pore structure after tailing sand infiltration were studied. Microscopic analysis was also performed to analyze the microscopic process of infiltration and destruction, as well as to see the microscopic structural characteristics of the infiltration and destruction of the total tailings. The test results show that there are obvious differences in the microstructure characterization of fluid soil and piping-type infiltration failures. Microstructure parameters have a certain functional relationship with macro factors. Combining the relationship between macrophysical and mechanical parameters and microstructural parameters, new ideas for future research and the prevention of tailings sand infiltration and failure mechanisms are provided [21].

The paper by [22] presents the results of tests for flexural tensile strength ($f_{ct,fl}$) and fracture energy (G_f) in a three-point bending test of prismatic beams with notches, which were made from steel fiber-reinforced high-strength concrete (SFRHSC). The registration of the conventional force–displacement (F–δ) relationship and unconventional force-crack tip opening displacement (CTOD) relationship was made. On the basis of the obtained test results, estimations of the parameters $f_{ct,fl}$ and G_f in the function of the fiber-reinforcement ratio were carried out. The obtained results were applied to building and validating a numerical model with the use of the finite element method (FEM). A non-linear concrete damaged plasticity model CDP was used for the description of the concrete. The obtained FEM results were compared with the experimental ones that were based on the assumed criteria. The usefulness of the flexural tensile strength and fracture energy parameters for defining the linear form of weakening of the SFRHSC material under tension was confirmed. The author's own equations for estimating the flexural tensile strength and fracture energy of SFRHSC, as well as for approximating deflections (δ) of SFRHSC beams as the function of crack tip opening displacement (CTOD) instead of crack mouth opening displacement (CMOD), were proposed [22].

The accepted methods for testing concrete are not favorable for determining its heterogeneity [23]. The interpretation of the compressive strength result as a product of destructive force and cross-section area is burdened with significant understatements. It is assumed erroneously that this is the lowest value of strength at the height of the tested sample. The top layer of concrete floors often crumble, and the strength tested using sclerometric methods does not confirm the concrete class determined using control samples. That is why it is important to test the distribution of compressive strength in a cross-section of concrete industrial floors with special attention to surface top layers. This study presents strength tests of borehole material taken from industrial floors using the ultrasonic method, with exponential spot heads with a contact surface area of 0.8 mm^2 and a frequency of 40 kHz. The presented research project anticipated the determination of strength for samples in various cross-sections at the height of elements and destructive strength in the strength testing machine. It was confirmed that for standard and big borehole samples, it is not possible to test the strength of concrete in the top layer of the floor by destructive methods. This can be achieved using the ultrasonic method. After the analysis, certain types of distributions of strength across concrete floor thickness were chosen from the completed research program. The gradient and anti-gradient of strength were proposed as new parameters for the evaluation of floor concrete quality [23].

Ventilated facades are becoming an increasingly popular solution for the external part of walls in buildings [24]. They may differ in many elements, among others, cladding (fiber cement boards, HPL plates, large-slab ceramic tiles, ACM panels, stone cladding), types of substructures, console supports, etc. The main element that characterizes ventilated facades is the use of an air cavity between the cladding and thermal insulation. Unfortunately, in some respects, they are not yet standardized and tested. Above all, the requirements for the falling-off of elements from ventilated facades during a fire are not precisely defined by, among other things, the lack of clearly specified requirements and testing. This is undoubtedly a major problem, as it significantly affects the safety of evacuation during a fire emergency. For the purposes of this article, experimental tests were carried out on a large-scale facade model, with two types of external facade cladding. The materials used as external cladding were fiber cement boards and large-slab ceramic tiles. The model of the large-scale test was 3.95 m × 3.95 m; the burning gas released from the burner was used as the source of fire. The facade model was equipped with thermocouples. The test lasted one hour, and the cladding materials showed different behavior during the test. Large-slab ceramic tiles seemed to be a safer form of external cladding for ventilated facades. Unfortunately, they were destroyed much faster, by about 6 min. Large-slab ceramic tiles were destroyed within the first dozen or so minutes, then their destruction did not proceed or was minimal. In the case of fiber cement boards, the destruction started from the eleventh minute and increased until the end of the test. The author referred the results of the large-scale test to testing on samples carried out by other authors. The results presented the convergence of the large-scale test with samples. External claddings were equipped with additional mechanical protection. The use of additional mechanical protection to maintain external cladding elements increases their safety but does not completely eliminate the problem of the falling-off of parts of the facade. As research on fiber cement boards and large-slab ceramic tiles has suggested, these claddings were a major hazard due to fall-off from the facade [24].

The aim of this study [25] was to investigate the effect of plasterboards' humidity absorption on their performance. The specimens' hydration procedure consisted of consecutive immersing in water and subsequent drying at room temperature. Such a procedure was performed to increase the moisture content within the material volume. The microstructural observations of five different plasterboard types were performed through optical and scanning electron microscopy. The deterioration of their properties was evaluated using a three-point bending test and a subsequent ultrasonic (ultrasound testing (UT)) longitudinal wave velocity measurement. Depending on the material porosity, a loss of UT wave velocity from 6% to 35% and a considerable decrease in material strength from 70% to 80% were observed. Four types of approximated formulae were proposed to describe the dependence of UT wave velocity on the board moisture content. It was found that the proposed UT method could be successfully used for the on-site monitoring of plasterboards' hydration processes [25].

Concrete structure joints are filled in mainly in the course of sealing works ensuring protection against the influence of water. This paper by [26] presents the methodology for testing the mechanical properties of ESD pseudoplastic resins (E—elastic deformation, S—strengthening control, D—deflection control) recommended for concrete structure joint fillers. The existing standards and papers concerning quasi-brittle cement composites do not provide an adequate point of reference for the tested resins. The lack of a standardized testing method hampers the development of materials universally used in expansion joint fillers in reinforced concrete structures, as well as the assessment of their properties and durability. An assessment of the obtained results by referring to the reference sample has been suggested in the article. A test stand and a method of assessing the mechanical properties results (including adhesion to the concrete surface) of pseudoplastic resins in the axial tensile test have been presented [26].

The compaction index is one of the most important technological parameters during asphalt pavement construction, which may be negatively affected by the wrong asphalt paving machine setting, weather conditions, or the mix temperature. Presented in [27], this laboratory study analyzes the asphalt mix properties in case of inappropriate compaction. The reference mix was designed for an AC 11 S wearing layer (asphalt concrete for a wearing layer with maximum grading of 11 mm). Asphalt mix samples used in the tests were prepared using a Marshall device with the compaction energy of $2 \times 20, 2 \times 35, 2 \times 50,$ and 2×75 blows, as well as in a roller compactor where the slabs were compacted to various heights: 69.3 mm (+10% of nominal height), 66.2 mm (+5%), 63 mm (nominal), and 59.9 mm (−5%), which resulted in different compaction indexes. Afterward, the samples were cored from the slabs. Both Marshall samples and cores were tested for air void content, stiffness modulus in three temperatures, indirect tensile strength, and resistance to water and frost indicated by the ITSR value. It was found that either an insufficient or excessive level of compaction can cause a negative effect on the road surface performance [27].

The importance of surface roughness and its non-destructive examination has often been emphasized in structural rehabilitation. The innovative procedure presented in [28] enables the estimation of concrete-to-concrete strength, based on a combination of low-cost, area-limited tests and geostatistical methods. The new method removes the shortcomings of the existing one, i.e., it is neither qualitative nor subjective. The interface strength factors, cohesion and friction, can be estimated accurately based on the collected data on surface texture. The data acquisition needed to create digital models of the concrete surface can be performed by terrestrial close-range photogrammetry or other methods. In the presented procedure, limitations to the availability of concrete surfaces are overcome by the generation of subsequential Gaussian random fields (via height profiles) based on the semivariograms fitted to the digital surface models. In this way, the randomness of the surface texture is reproduced. The selected roughness parameters, such as mean valley depth and, most importantly, the geostatistical semivariogram parameter sill, were transformed into contact bond strength parameters based on the available strength tests. The proposed procedure estimates the interface bond strength based on the geostatistical methods applied to the numerical surface model and can be used in practical and theoretical applications [28].

The core part of a hybrid truss bridge is the connection joint that combines the concrete chord and steel truss-web members [29]. To study the mechanical behavior and failure mode of steel–concrete connection joints in a hybrid truss bridge, static model tests were carried out on two connection joints at the scale of 1:3, under a horizontal load that was provided by a loading jack mounted on the vertical reaction wall. The specimen design, experimental setup and testing procedure were introduced. In the experiment, the displacement, strain level, concrete crack and experimental phenomena were factually recorded. Compared with the previous study results, the experimental results in this study demonstrated that the connection joints had an excellent bearing capacity and deformability. The minimum ultimate load and displacement of the two connection joints were 5200 kN and 59.01 mm, respectively. Moreover, the connection joints exhibited multiple failure modes, including the fracture of gusset plates, the slippage of high-strength bolts, the local buckling of compressive splice plates, the fracture of tensile splice plates and concrete cracking. Additionally, the strain distribution of the steel–concrete connection joints followed certain rules. It is expected that the findings from this paper may provide a reference for the design and construction of steel–concrete connection joints in hybrid truss bridges [29].

The static elastic modulus (Ec) and compressive strength (fc) are critical properties of concrete [30]. When determining Ec and fc, concrete cores are collected and subjected to destructive tests. However, destructive tests require certain test permissions and large sample sizes. Hence, it is preferable to predict Ec using the dynamic elastic modulus (Ed), through non-destructive evaluations. A resonance frequency test performed according to ASTM C215-14, and a pressure wave (P-wave) measurement conducted according to ASTM C597M-16, are typically used to determine Ed. Recently, developments in transducers have enabled the measurement of shear wave (S-wave) velocities in concrete. Although various equations have been proposed for estimating Ec and fc from Ed, their results deviate from experimental values. Thus, it is necessary to obtain a reliable Ed value for accurately predicting Ec and fc. In this study, Ed values were experimentally obtained from P-wave and S-wave velocities in the longitudinal and transverse modes; Ec and fc values were predicted using these Ed values through four machine learning (ML) methods: support vector machine, artificial neural networks, ensembles, and linear regression. Using ML, the prediction accuracy of Ec and fc was improved by 2.5–5% and 7–9%, respectively, compared with the accuracy obtained using classical or normal-regression equations. By combining ML methods, the accuracy of the predicted Ec and fc was improved by 0.5% and 1.5%, respectively, compared with the optimal single variable results [30].

The paper by [31] describes tests conducted to identify the mechanisms occurring during the fracture of single-edge notches loaded in three-point bending (SENB) specimens made from an Al–Ti laminate. The experimental tests were complemented with microstructural analyses of the specimens' fracture surfaces and an in-depth analysis of acoustic emission (AE) signals. The paper presents the application of the AE method to identify fracture processes in the layered Al–Ti composite, using a non-hierarchical method for clustering AE signals (k-means) and analyses using waveform time domain, fast Fourier transform (FFT Real) and waveform continuous wavelet, based on the Morlet wavelet. These analyses made it possible to identify different fracture mechanisms in Al–Ti composites, which is very significant for the assessment of the safety of structures made from this material [31].

The aspects regarding the stiffness of the connections between the beams that support the storage pallets and the uprights are very important in the analysis of the displacements and stresses in the storage racking systems. The main purpose of the paper by [32] is to study the effects of both upright thickness and tab connector types on rotational stiffness and on the capable bending moment of the connection. For this purpose, 18 different groups of beam-connector-upright assemblies are prepared by combining three types of beams (different sizes of the box cross-section), three kinds of upright profiles (with a different thickness of the section walls), and two types of connectors (four-tab connectors and five-tab connectors). Flexural tests were carried out on 101 assemblies. For the assemblies containing the uprights with a thickness of 1.5 mm, the five-tab connector leads to a higher value of the capable moment and higher rotational stiffness than similar assemblies with four-tab connectors. A contrary phenomenon happens in the case of the assemblies containing those upright profiles having a thickness of 2.0 mm, regarding the capable design moment. It is shown how the safety coefficient of connection depends on both the rotational stiffness and capable bending moment [32].

Concrete shrinkage is a phenomenon that results in a decrease in volume in the composite material during the curing period. The method for determining the effects of restrained shrinkage is described in Standard ASTM C 1581/C 1581M–09a. This article [33] shows the calibration of measuring rings with respect to the theory of elasticity, and the analysis of the relationship of steel ring deformation to high-performance concrete tensile stress as a function of time. Steel rings equipped with strain gauges are used for the measurement of strain during the compression of the samples. The strain is caused by the shrinkage of the concrete ring specimen that tightens around steel rings. The method allows registering the changes to the shrinkage process over time and evaluating the susceptibility of concrete to cracking. However, the standard does not focus on the details of the mechanical design of the test bench. To acquire accurate measurements, the test bench needs to be calibrated. Measurement errors may be caused by an improper, uneven installation of strain gauges, imprecise geometry of the steel measuring rings, or incorrect equipment settings. The calibration method makes it possible to determine the stress in a concrete sample, leading to its cracking at the specific deformation of the steel ring [33].

The article by [34] proposes using the acoustic emission (AE) method to evaluate the degree of change in the mechanical parameters of fiber cement boards. The research was undertaken after a literature review, due to the lack of a methodology that would allow nondestructive assessment of the strength of cement–fiber elements. The tests covered the components cut out from a popular type of board available on the construction market. The samples were subjected to environmental (soaking in water, cyclic freezing–thawing) and exceptional (burning with fire and exposure to high temperature) factors, and then to three-point bending strength tests. The adopted conditions correspond to the actual working environment of the boards. When applying the external load, AE signals were generated which were then grouped into classes, and initially assigned to specific processes occurring in the material. The frequencies occurring over time for the tested samples were also analyzed, and microscopic observations were made to confirm the suppositions based on the first part of the tests. Comparing the results obtained from a group of samples subjected to environmental and exceptional actions, significant differences were noted between them, which included the types of recorded signal class, the frequency of events, and the construction of the microstructure. The degradation of the structure, associated with damage to the fibers or their complete destruction, results in the generation under load of AE signals that indicate the uncontrolled development of scratches, and a decrease in the frequency of these events. According to the authors, the methodology used allows the control of cement–fiber boards in use. The registration and analysis of active processes under the effect of payloads makes it possible to distinguish mechanisms occurring inside the structure of the elements, and to formulate a quick response to the situation when the signals indicate a decrease in the strength of the boards [34].

The article by [35] discusses one of the methods of dielectric constant determination in a continuous way, which is the determination of its value based on the amplitude of the wave reflected from the surface. Based on tests performed on model asphalt slabs, the research presented how the value of the dielectric constant changes, depending on the atmospheric conditions of the measured surface (dry, covered with water film, covered with ice, covered with snow, covered with de-icing salt). Coefficients correcting dielectric constants of hot mix asphalt (HMA) determined in various surface atmospheric conditions were introduced. It was proposed to determine the atmospheric conditions of the pavement with the use of wavelet analysis in order to choose the proper dielectric constant correction coefficient and, therefore, improve the accuracy of the pavement layer thickness estimation based on the ground-penetrating radar (GPR) method [35].

Phenomena occurring during the curing of concrete can decrease its mechanical properties, specifically its strength and serviceability, even before it is placed [36]. This is due to excessive stresses caused by temperature gradients, moisture changes, and chemical processes arising during the concreting and in hardened concrete. At stress concentration sites, microcracks form in the interfacial transition zones (ITZ) in the early phase and propagate deeper into the cement paste or to the surface of the element. Microcracks can contribute to the development of larger cracks, reduce the durability of structures, limit their serviceability, and, in rare cases, lead to their failure. It is thus important to search for a tool that allows objective assessment of damage initiation and development in concrete. The objectivity of the assessment lies in it being independent of the constituents and additives used in the concrete or of external influences. The acoustic emission-based method presented in this paper allows damage detection and identification in the early age of concrete (before loading) for different concrete compositions, curing conditions, temperature variations, and in reinforced concrete. As such, this method is an objective and effective tool for damage process detection [36].

The authors of [37] suggest a wire-mesh method to classify the particle shape of large amounts of aggregate. This method is controlled by the tilting angle and opening size of the wire mesh. The more rounded the aggregate particles, the more they roll on the tilted wire mesh. Three different sizes of aggregate, 11–15, 17–32, and 33–51 mm, were used for assessing their roundness after classification, using the sphericity index to sort them into rounded, sub-rounded/sub-angular, and angular. The aggregate particles with different sphericities were colored differently and then used for classification via the wire-mesh method. The opening sizes of the wire mesh were 6, 11, and 17 mm, and its frame was 0.5 m wide and 1.8 m long. The ratio of aggregate size to mesh-opening size was between 0.6 and 8.5. The wire mesh was inclined at various angles of 10°, 15°, 20°, 25°, and 30° to evaluate the rolling degree of the aggregates. The aggregates were rolled and remained on the wire mesh between 0.0–0.6, 0.6–1.2, and 1.2–1.8 m, depending on their sphericity. A tilting angle of 25° was the most suitable angle for classifying aggregate size ranging from 11–15 mm, while the most suitable angle for aggregate sizes of 17–32 and 33–51 mm was 20°. The best ratio for the average aggregate size to mesh-opening size for the aggregate roundness classification was 2 [37].

Taking into account the possibilities offered by two imaging methods, X-ray micro-computed tomography (μCT) and two-dimensional optical scanning, this article by [38] discusses the possibility of using these methods to assess the internal structure of spun concrete, particularly its composition after hardening (Michałek 2020). To demonstrate the performance of the approach based on imaging, laboratory techniques based on physical and chemical methods were used as verification. A comparison of the obtained results of applied research methods was carried out on samples of spun concrete, characterized by the layered structure of the annular cross-section. Samples were taken from the power pole E10.5/6c (Strunobet-Migacz, Lewin Brzeski, Poland) made by one of the Polish manufacturers of prestressed concrete E-poles precast in steel molds. The validation shows that optical scanning followed by appropriate image analysis is an effective method for evaluation of the spun concrete internal structure. In addition, such analysis can significantly complement the results of the laboratory methods used so far. In a fairly simple way, through the porosity image, it can reveal improperly selected parameters of concrete spinning, such as speed and time, and, through the distribution of cement content in the cross-section of the element, it can indicate compliance with the requirement for the corrosion durability of spun concrete. The research methodology presented in the paper can be used to improve the production process of poles made of spun concrete; it can be an effective tool for verifying concrete structure [38].

In the study by [39], the effects of the mixing conditions of waste-paper sludge ash (WPSA) on the strength and bearing capacity of controlled low-strength material (CLSM) were evaluated, and the optimal mixing conditions were used to evaluate the strength characteristics of CLSM with recyclable WPSA. The strength and bearing capacity of CLSM with WPSA were evaluated using unconfined compressive strength tests and plate bearing tests, respectively. The unconfined compressive strength test results show that the optimal mixing conditions for securing 0.8–1.2 MPa of target strength under 5% of cement content conditions can be obtained when both WPSA and fly ash are used. This is because WPSA and fly ash, which act as binders, have a significant impact on overall strength when the cement content is low. The bearing capacity of weathered soil increased from 550 to 575 kPa over time, and CLSM with WPSA increased significantly, from 560 to 730 kPa. This means that the bearing capacity of CLSM with WPSA was 2.0% higher than that of weathered soil immediately after construction; furthermore, it was 27% higher at 60 days of age. In addition, the allowable bearing capacity of CLSM corresponding to the optimal mixing conditions was evaluated, and it was found that this value increased by 30.4% until 60 days of age. This increase rate was 6.7 times larger than that of weathered soil (4.5%). Therefore, based on the allowable bearing capacity calculation results, CLSM with WPSA was applied as a sewage pipe backfill material. It was found that CLSM with WPSA performed better as backfill and was more stable than soil immediately after construction. The results of this study confirm that CLSM with WPSA can be utilized as sewage-pipe backfill material [39].

Non-destructive testing (NDT) methods are an important means to detect and assess rock damage [40]. To better understand the accuracy of NDT methods for measuring damage in sandstone, this study compared three NDT methods, including ultrasonic testing, electrical impedance spectroscopy (EIS) testing, computed tomography (CT) scan testing, and a destructive test method, elastic modulus testing. Sandstone specimens were subjected to different levels of damage through cyclic loading, and different damage variables derived from five different measured parameters—longitudinal wave (P-wave) velocity, first-wave amplitude attenuation, resistivity, effective bearing area and the elastic modulus—were compared. The results show that the NDT methods all reflect the damage levels for sandstone accurately. The damage variable derived from the P-wave velocity is more consistent with the other damage variables, and the amplitude attenuation is more sensitive to damage. The damage variable derived from the effective bearing area is smaller than that derived from the other NDT measurement parameters. Resistivity provides a more stable measure of damage, and damage derived from the acoustic parameters is less stable. By developing P-wave velocity-to-resistivity models based on theoretical and empirical relationships, it was found that differences between these two damage parameters can be explained by differences between the mechanisms through which they respond to porosity, since the resistivity reflects pore structure, while the P-wave velocity reflects the extent of the continuous medium within the sandstone [40].

The H spin-lattice relaxometry (T_1, longitudinal) of cement pastes with 0 to 0.18 wt % polycarboxylate superplasticizers (PCEs) at intervals of 0.06 wt % from 10 min to 1210 min was investigated in [41]. Results showed that the main peak in T_1 relaxometry of cement pastes was shorter and lower along with the hydration times. PCEs delayed and lowered this main peak in the T_1 relaxometry of cement pastes at 10 min, 605 min and 1210 min, which was highly correlated to its dosages. In contrast, PCEs increased the total signal intensity of T_1 of cement pastes at these three times, which still correlated to its dosages. Both changes of the main peak in T_1 relaxometry and the total signal intensity of T_1 revealed interferences in evaporable water during cement hydration by the dispersion mechanisms of PCEs. The time-dependent evolution of the weighted average T_1 of cement pastes with different PCEs between 10 min and 1210 min was found to be regular to the four-stage hydration mechanism of tricalcium silicate [41].

Control of technical parameters obtained by ready-mixed concrete may be carried out at different stages of the development of concrete properties, and by different participants involved in the construction investment process [42]. According to the European Standard EN 206 "Concrete—specification, performance, production and conformity", mandatory control of concrete conformity is conducted by the producer during production. As shown by the subject literature, statistical criteria set out in the standard, including the method for concrete quality assessment based on the concept of concrete family, continue to evoke discussions and raise doubts. This justifies seeking alternative methods for concrete quality assessment. This paper presents a novel approach to quality control and the classification of concrete based on combining statistical and fuzzy theories as a means of representation of two types of uncertainty: random uncertainty and information uncertainty. In concrete production, a typical situation when fuzzy uncertainty can be taken into consideration is the conformity control of concrete compressive strength, which is conducted to confirm the declared concrete class. The proposed procedure for quality assessment of a concrete batch is based on defining the membership function for the considered concrete classes and establishing the degree of belonging to the considered concrete class. It was found that concrete classification set out by the standard includes too many concrete classes of overlapping probability density distributions, and the proposed solution was to limit the scope of compressive strength to every second class so as to ensure the efficacy of conformity assessment conducted for concrete classes and concrete families. The proposed procedures can lead to two types of decisions: non-fuzzy (crisp) or fuzzy, which point to possible solutions and their corresponding preferences. The suggested procedure for quality assessment allows researchers to classify a concrete batch in a fuzzy way with the degree of certainty less than or equal to 1. The results obtained confirm the possibility of employing the proposed method for quality assessment in the production process of ready-mixed concrete [42].

In recent years, the application of fiber-reinforced plastics (FRPs) as structural members has been promoted [43]. Metallic bolts and rivets are often used for the connection of FRP structures, but there are some problems caused by corrosion and stress concentration at the bearing position. Fiber-reinforced thermoplastics (FRTPs) have attracted attention in composite material fields because they can be remolded by heating and manufactured at excellent speed compared with thermosetting plastics. In this paper, we propose and evaluate the connection method using rivets produced from FRTPs for FRP members. It was confirmed through material tests that an FRTP rivet provides stable tensile, shear, and bending strength. Then, it was clarified that a non-clearance connection could be achieved by the proposed connection method, so initial sliding was not observed, and connection strength linearly increased as the number of FRTP rivets increased through the double-lapped tensile shear tests. Furthermore, the joint strength of the beam using FRTP rivets could be calculated with high accuracy, using the method for bolt joints in steel structures through a four-point beam bending test [43].

Polymer pipes are used in the construction of underground gas, water, and sewage networks [44]. During exploitation, various external forces work on the pipeline which cause its deformation. In this paper, numerical analysis and experimental investigations of polyethylene pipe deformation at different external load values (500, 1000, 1500, and 2000 N) were performed. The authors measured the strains of the lower and upper surfaces of the pipe during its loading moment using resistance strain gauges, which were located on the pipe at equal intervals. The results obtained from computer simulation and experimental studies were comparable. An innovative element of the research presented in the article is the recognition of the impact of the proposed values of the load of polyethylene pipe on the change in its deformation [44].

The shear and particle-crushing characteristics of the failure plane (or shear surface) in catastrophic mass movements are examined with a ring shear apparatus, which is generally employed owing to its suitability for large deformations [45]. Based on the results of previous experiments on waste materials from abandoned mine deposits, we employed a simple numerical model based on ring shear testing using the particle flow code (PFC^{2D}). We examined drainage, normal stress, and the shear velocity-dependent shear characteristics of landslide materials. For shear velocities of 0.1 and 100 mm/s and normal stress (NS) of 25 kPa, the numerical results are in good agreement with those obtained from experimental results. The difference between the experimental and numerical results of the residual shear stress was approximately 0.4 kPa, for NS equal to 25 kPa and 0.9 kPa for NS equal to 100 kPa, for both drained and undrained conditions. In addition, we examined the particle-crushing effect during shearing using the frictional work concept in PFC. We calculated the work done by friction at both peak and residual shear stresses, and then used the results as crushing criteria in the numerical analysis. The frictional work at peak and the residual shear stresses ranged from 303 kPa·s to 2579 kPa·s for the given drainage and normal stress conditions. These results showed that clump particles were partially crushed at peak shear stress [45].

The main objective of the research presented in [46] was to develop a solution to the global problem of using steel waste obtained during rubber recovery during tire recycling. A detailed comparative analysis of the mechanical and physical features of concrete composite with the addition of recycled steel fibers (RSF) in relation to the steel fiber concrete commonly used for industrial floors was conducted. A study was carried out using micro-computed tomography and the scanning electron microscope to determine the fibers' characteristics, including the EDS spectrum. In order to designate full performance of the physical and mechanical features of the novel composite, a wide range of tests was performed, with particular emphasis on the determination of the tensile strength of the composite. This parameter, appointed by tensile strength testing for splitting, residual tensile strength testing (3-point test), and a wedge splitting test (WST), demonstrated the increase of tensile strength (vs. unmodified concrete) by 43%, 30%, and 70% respectively to the method. The indication of the reinforced composite's fracture characteristics using the digital image correlation (DIC) method allowed the researchers to illustrate the map of deformation of the samples during WST. The novel composite was tested in reference to the circular economy concept and showed 31.3% lower energy consumption and 30.8% lower CO_2 emissions than a commonly used fiber concrete [46].

Existing buildings, especially historical buildings, require periodic or situational diagnostic tests [47]. If a building is in use, advanced non-destructive or semi-destructive methods should be used. In the diagnosis of reinforced concrete structures, tests allowing assessment of the condition of the reinforcement and concrete cover are particularly important. The article presents the non-destructive and semi-destructive research methods that are used for such tests, as well as the results of tests performed for selected elements of a historic water tower structure. The assessment of the corrosion risk of the reinforcement was carried out with the use of a semi-destructive galvanostatic pulse method. The protective properties of the concrete cover were checked by the carbonation test and the phase analysis of the concrete, for which X-ray diffractometry and thermal analysis methods were used. In order to determine X the position of the reinforcement and to estimate the concrete cover thickness distribution, a ferromagnetic detection system was used. The comprehensive application of several test methods allowed mutual verification of the results and the drawing of reliable conclusions. The results indicated a very poor state of reinforcement, loss in the depth of cover, and sulfate corrosion [47].

The paper by [48] presents the possibility of using low-module polypropylene dispersed reinforcement (E = 4.9 GPa) to influence the load-deflection correlation of cement composites. Problems have been indicated regarding the improvement of elastic range when using that type of fiber as compared with a composite without reinforcement. It was demonstrated that it was possible to increase the ability to carry stress in the Hooke's law proportionality range in the mortar and paste types of composites reinforced with low-module fibers, i.e., $V_f = 3\%$ (in contrast to concrete composites). The possibility of having good strengthening and deflection control in order to limit the catastrophic destruction process was confirmed. This paper identifies the problem of deformation assessment in composites with significant deformation capacity. Determining the effects of reinforcement based on a comparison with a composite without fibers is suggested as a reasonable approach, as it enables the comparison of results obtained by various universities under different research conditions [48].

Arcan shear tests with digital image correlation were used to evaluate the shear modulus and shear stress–strain diagrams in the plane defined by two principal axes of the material orthotropy [49]. Two different orientations of the grain direction, as compared to the direction of the shear force in specimens, were considered: perpendicular and parallel shear. Two different ways were used to obtain the elastic properties based on the digital image correlation (DIC) results from the full-field measurement and from the virtual strain gauges with the linear strains: perpendicular to each other and directed at an angle of $\pi/4$

to the shearing load. In addition, their own continuum structural model for the failure analysis in the experimental tests was used. The constitutive relationships of the model were established in the framework of the mathematical multi-surface elastoplasticity for the plane stress state. The numerical simulations performed by the finite element program after implementation of the model demonstrated the failure mechanisms from the experimental tests [49].

After obtaining the value of shear wave velocity (V_S) from the bender elements test (BET), the shear modulus of soils at small strains (G_{max}) can be estimated [50]. Shear wave velocity is an important parameter in the design of geo-structures subjected to static and dynamic loading. While bender elements are increasingly used in both academic and commercial laboratory test systems, there remains a lack of agreement when interpreting the shear wave travel time from these tests. Based on the test data of 12 Warsaw glacial quartz samples of sand, two different approaches were primarily examined for determining V_S. They are both related to the observation of the source and received *BE* signal, namely, the first time of arrival and the peak-to-peak method. These methods were performed through visual analysis of BET data by the authors, so that subjective travel time estimates were produced. Subsequently, automated analysis methods from the GDS Bender Element Analysis Tool (BEAT) were applied. Here, three techniques in the time-domain (TD) were selected, namely, the peak-to-peak, the zero-crossing, and the cross-correlation functions. Additionally, a cross-power spectrum calculation of the signals was completed, viewed as a frequency-domain (FD) method. Final comparisons between subjective observational analyses and automated interpretations of BET results showed good agreement. There is compatibility, especially between the two methods of the first time of arrival and cross-correlation, which the authors considered the best interpreting techniques for their soils. Moreover, the laboratory tests were performed on compact, medium, and well-grained sand samples with different curvature coefficients and mean grain sizes [50].

The reduction in natural resources and aspects of environmental protection necessitate alternative uses for waste materials in the area of construction [51]. Recycling is also observed in road construction, where mineral–cement emulsion (MCE) mixtures are applied. The MCE mix is a conglomerate that can be used to make the base layer in road pavement structures. MCE mixes contain reclaimed asphalt from old, degraded road surfaces, aggregate for improving the gradation, asphalt emulsion, and cement as a binder. The use of these ingredients, especially cement, can cause shrinkage and cracks in road layers. The article presents selected issues related to the problem of cracking in MCE mixtures. The authors of the study focused on reducing the cracking phenomenon in MCE mixes by using an innovative cement binder with recycled materials. The innovative cement binder, based on dusty by-products from cement plants, also contributes to the optimization of the recycling process in road surfaces. The research was carried out in the fields of stiffness, fatigue life, crack resistance, and shrinkage analysis of mineral–cement emulsion mixes. It was found that it was possible to reduce the stiffness and cracking in MCE mixes. The use of innovative binders will positively affect the durability of road pavements [51].

The windblown sand-induced degradation of glass panels influences the serviceability and safety of these panels. In this study, the degradation of glass panels subject to windblown sand at different impact velocities and impact angles was studied based on a sandblasting test simulating a sandstorm [52]. After the glass panels were degraded by windblown sand, the surface morphology of the damaged glass panels was observed using scanning electron microscopy, and three damage modes were found: a cutting mode, smash mode, and plastic deformation mode. The mass loss, visible light transmittance, and effective area ratio values of the glass samples were then measured to evaluate the effects of the windblown sand on the panels. The results indicate that, at high abrasive feed rates, the relative mass loss of the glass samples decreases initially and then remains steady with increases in impact time, whereas it increases first and then decreases with an increase in impact angle, such as that for ductile materials. Both the visible light transmittance and effective area ratio decrease with increases in the impact time and velocities. There exists a positive linear relationship between the visible light transmittance and effective area ratio [52].

Durability tests against fungal action for wood-plastic composites are carried out in accordance with European standard ENV 12038, but the authors of the manuscript try to prove that the assessment of the results performed according to these methods is imprecise and suffers from a significant error [53]. Fungal exposure is always accompanied by high humidity, so the result of tests made by such a method is always burdened with the influence of moisture, which can lead to a wrong assessment of the negative effects of the action of the fungus itself. The paper (Wiejak 2021) has shown a modification of such a method that separates the destructive effect of fungi from moisture accompanying the test's destructive effect. The functional properties selected to prove the proposed modification are changes in the mass and bending strength after subsequent environmental exposure. It was found that the intensive action of moisture measured in the culture chamber at about (70 ± 5)%, i.e., for 16 weeks, at (22 ± 2) °C, which was the fungi culture in the accompanying period, led to changes in the mass of the wood-plastic composites, amounting to 50% of the final result of the fungi resistance test, and changes in the bending strength amounting to 30–46% of the final test result. As a result of this research, the correction for assessing the durability of wood–polymer composites against biological corrosion has been proposed. The laboratory tests were compared with the products' test results following three years of exposure to the natural environment [53].

Reduced maintenance costs of concrete structures can be ensured by efficient and comprehensive condition assessment [54]. Ground-penetrating radar (GPR) has been widely used in the condition assessment of reinforced concrete structures and it provides completely non-destructive results in real-time. It is mainly used for locating reinforcement and determining concrete cover thickness. More recently, research has focused on the possibility of using GPR for reinforcement corrosion assessment. In this paper, an overview of the application of GPR in the corrosion assessment of concrete is presented. A literature search and study selection methodology were used to identify the relevant studies. First, the laboratory studies are shown. After that, the studies for application on real structures are presented. The results have shown that the laboratory studies have not fully illuminated the influence of the corrosion process on the GPR signal. In addition, no clear relationship was reported between the results of the laboratory studies and the on-site inspection. Although GPR has a long history in the condition assessment of structures, it needs more laboratory investigations to clarify the influence of the corrosion process on the GPR signal [54].

The paper by [55] attempts to compare three methods of testing floor slip resistance and the resulting classifications. Polished, flamed, brushed, and grained granite slabs were tested. The acceptance angle values (α_{ob}) obtained through the shod ramp test, slip resistance value (SRV), and sliding friction coefficient (μ) were compared in terms of the correlation between the series, the precision of each method, and the classification results assigned to each of the three obtained indices. It was found that the evaluation of a product for slip resistance was strongly related to the test method used and the resulting classification method. This influence was particularly pronounced for low-roughness slabs. This would result in risks associated with inadequate assessments that could affect the safe use of buildings and facilities [55].

Standard sensors for the measurement and monitoring of temperature in civil structures are liable to mechanical damage and electromagnetic interference [56]. A system of purpose-designed fiber-optic FBG sensors offers a more suitable and reliable solution—the sensors can be directly integrated with the load-bearing structure during construction, and it is possible to create a network of fiber-optic sensors to ensure not only temperature measurements but also measurements of strain and of the moisture content in the building envelope. The paper describes the results of temperature measurements of a building's two-layer wall using optical fiber Bragg grating (FBG) sensors, and of a three-layer wall using equivalent classical temperature sensors. The testing results can be transmitted remotely. In the first stage, the sensors were tested in a climatic test chamber to determine their characteristics. The paper describes the test results of temperature measurements carried out in the winter season for two multilayer external walls of a building, in relation to the environmental conditions recorded at that time, i.e., outdoor temperature, relative humidity, and wind speed. Cases are considered with the biggest difference in the level of the relative humidity of air recorded in the observation period. It was found that there is greater convergence between the theoretical and the real temperature distribution in the wall for high levels (~84%) of the outdoor air's relative humidity, whereas, at the humidity level of ~49%, the difference between theoretical and real temperature histories is substantial and totals up to 20%. A correction factor is proposed for the theoretical temperature distribution [56].

The paper by [57] deals with a complex analysis of acoustic emission signals that were recorded during freeze-thaw cycles in test specimens produced from air-entrained concrete. An assessment of the resistance of concrete to the effects of freezing and thawing was conducted on the basis of signal analysis. Since the experiment simulated the testing of concrete in a structure, a concrete block with a height of 2.4 m and width of 1.8 m was produced to represent a real structure. When the age of the concrete was two months, samples were obtained from the block by core drilling and were subsequently used to produce test specimens. Testing of the freeze-thaw resistance of concrete employed both destructive and non-destructive methods including the measurement of acoustic emission, which took place directly during the freeze-thaw cycles. The recorded acoustic emission signals were then meticulously analyzed. The aim of the conducted experiments was to verify whether measurement using the acoustic emission method during freeze-thaw (F-T) cycles is more sensitive to the degree of damage of concrete than the more commonly employed construction testing methods. The results clearly demonstrate that the acoustic emission method can reveal changes (e.g., minor cracks) in the internal structure of concrete, unlike other commonly used methods. The analysis of the acoustic emission signals using a fast Fourier transform revealed a significant shift of the dominant frequency toward lower values when the concrete was subjected to freeze-thaw cycling [57].

An original experimental method was used to investigate the influence of water and road salt with an anti-caking agent on the material used in pavement construction layers [58]. This method allowed the monitoring of material changes resulting from the influence of water and road salt with an anti-caking agent over time. The experiment used five different mineral road mixes, which were soaked separately in water and brine for two time intervals (2 days and 21 days). Then, each sample of the mix was subjected to tests of the complex module using the four-point bending (4PB-PR) method. The increase in mass of the soaked samples and the change in value of the stiffness modulus were analyzed. Exemplary tomographic (X-ray) imaging was performed to confirm the reaction of the road salt and anti-caking agent (lead agent) with the material. Based on the measurements of the stiffness modulus and absorption, the correlations of the mass change and the value of the stiffness modulus were determined, which may be useful in estimating the sensitivity of mixes to the use of winter maintenance agents, e.g., road salt with an anti-caking agent (sodium chloride). It was found that the greatest changes occur with mixes intended for base course layers (mineral cement mix with foamed asphalt (MCAS) and mineral-cement-emulsion mixes (MCE)), and that the smallest changes occur for mixes containing highly modified asphalt (HIMA) [58].

Cracking in non-load-bearing internal partition walls is a serious problem that frequently occurs in new buildings in the short term after putting them into service, or even before completion of construction [59]. Sometimes, it is so considerable that it cannot be accepted by the occupiers. The article presents tests of cracking in ceramic walls with a door opening connected in a rigid and flexible way along vertical edges. The first analyses were conducted using the finite element method (FEM), and afterward, the measurements of deformations and stresses in walls on deflecting floors were performed at full scale in the actual building structure. The measurements enabled the authors to determine floor deformations leading to the cracking of walls and to establish a dependency between the values of tensile stresses within the area of the door opening corners and their location along the length of walls, and the type of vertical connection with the structure [59].

The paper by [60] discusses the problems connected with the long-term exploitation of reinforced concrete post-tensioned girders. The scale of problems in the world related to the number of cable post-tensioned concrete girders built in the 1950s and still in operation is very large and possibly has very serious consequences. The paper presents an analysis and evaluation of the results of measurements of the deflection and strength and homogeneity of concrete in cable–concrete roof girders of selected industrial halls located in Poland, exploited for over 50 years. On the basis of the results of displacement monitoring in the years 2009 to 2020, the maximum increments of deflection of the analyzed girders were determined. Non-destructive, destructive, and indirect evaluation methods were used to determine the compressive strength of concrete. Within the framework of the indirect method recommended in standard PN-EN 13791, a procedure was proposed by the authors to modify the so-called base curve for determining compressive strength. Due to the age of the analyzed structural elements, a correction factor for the age of the concrete was taken into account in the strength assessment. The typical value of the characteristic compressive strength was within the range of 20.3–28.4 MPa. As a result of the conducted tests, the concrete class assumed in the design was not confirmed, and its classification depended on the applied test method. The analyzed girders, despite their long-term exploitation, can still be used for years, on the condition that regular periodical inspections of their technical condition are carried out. The authors emphasize the necessity for a permanent and cyclic diagnostic process and monitoring of the geometry of girders, as they are expected to operate much longer than was assumed by their designers [60].

The paper by [61] presents an implementation of purpose-designed optical fiber Bragg grating (FBG) sensors intended for the monitoring of real values of strain in reinforced road structures in areas of mining activity. Two field test stations are described. The first enables analysis of the geogrid on concrete and ground subgrades. The second models the situation of subsoil deformation due to mining activity at different external loads. The paper presents a system of optical fiber sensors registering strain and temperature dedicated to the investigated concrete mattress. Laboratory tests were performed to determine the strain characteristic of the FBG sensor–geogrid system with respect to standard load. As a result, it was possible to establish the dependence of the geogrid strain on the forces occurring within it. This may be the basis for an analysis of the mining activity effect on right-of-way structures during precise strain measurements of a geogrid using FBG sensors embedded within it. The analysis of the results of measurements in the aspect of forecasted and actual static and dynamic effects of mining on the stability of a reinforced road structure is of key importance for detailed management of road investment, and for the appropriate repair and modernization management of the road structure [61].

Geopolymer concrete (GPC) offers a potential solution for sustainable construction by utilizing waste materials [62]. However, the production and testing procedures for GPC are quite cumbersome and expensive, which can slow down the development of mix design and the implementation of GPC. The basic characteristics of GPC depend on numerous factors such as the type of precursor material, type of alkali activators and their concentration, and liquid to solid (precursor material) ratio. To optimize time and cost, artificial neural networks (ANN) can be a lucrative technique for exploring and predicting GPC characteristics. In this study, the compressive strength of fly-ash-based GPC, with bottom ash as a replacement for fine aggregates, as well as fly ash, is predicted using a machine learning-based ANN model. The data inputs are taken from the literature as well as in-house lab scale testing of GPC. The specifications of GPC specimens act as input features of the ANN model to predict compressive strength as the output, while minimizing error. Fourteen ANN models are designed which differ in the backpropagation training algorithm, the number of hidden layers, and the neurons in each layer. The performance analysis and comparison of these models in terms of mean squared error (MSE) and coefficient of correlation (R) resulted in a Bayesian regularized ANN (BRANN) model for the effective prediction of compressive strength of fly-ash and bottom-ash based geopolymer concrete [62].

The paper by [63] analyzes the issue of reduction of load capacity in fiber cement board during a fire. Fiber cement boards were put under the influence of fire by using a large-scale facade model. Such a model is a reliable source of knowledge regarding the behavior of facade cladding and the way fire spreads. One technical solution for external walls—a ventilated facade—is gaining popularity and is used more and more often. However, the problem of the destruction during a fire of a range of different materials used in external facade cladding is insufficiently recognized. For this study, the authors used fiber cement boards as the facade cladding. Fiber cement boards are fiber-reinforced composite materials, mainly used for facade cladding, but are also used as roof cladding, drywall, drywall ceilings and floorboards. This paper analyzes the effect of fire temperatures on facade cladding using a large-scale facade model. Samples were taken from external facade cladding materials that were mounted on the model at specific locations above the combustion chamber. Subsequently, three-point bending flexural tests were performed, and the effects of temperature and the integrals of temperature and time functions on the samples were evaluated. The three-point bending flexural test was chosen because it is a universal method for assessing fiber cement boards, as cited in Standard EN 12467. The test also allows easy reference to results in other literature [63].

The paper by [64] presents the possibilities of determining the range of stresses preceding the critical destruction process in cement composites, with the use of micro-events identified by means of a sound spectrum. The presented test results refer to the earlier papers in which micro-events (destruction processes) were identified, but without determining the stress level of their occurrence. This paper indicates a correlation of the stress level corresponding to the elastic range with the occurrence of micro-events in traditional and quasi-brittle composites. Tests were carried out on beams (with and without reinforcement) subjected to four-point bending. In summary, it is suggested that the conclusions can be extended to other test cases (e.g., compression strength), which should be confirmed by the appropriate tests. The paper also indicates a need for further research to identify micro-events. The correct recognition of micro-events is important for the safety and durability of traditional and quasi-brittle cement composites [64].

The paper by [65] contains the results of a newly developed residual-state creep test, performed to determine the behavior of a selected geomaterial in the context of reactivated landslides. Soil and rock creep is a time-dependent phenomenon in which deformation occurs under constant stress. Based on the examination results, it was found that the tested clayey material (from Kobe, Japan) shows tertiary creep behavior only under shear stress higher than the residual strength condition, and primary and secondary creep behavior under shear stress lower than or equal to the residual strength condition. Based on the data, a model for predicting the critical or failure time is introduced. The time until the occurrence of the conditions necessary for unlimited creep on the surface is estimated. As long-term precipitation and infiltrating water in the area of the landslides are identified as the key phenomena initiating collapse, the work focuses on the prediction of landslides, with identified surfaces of potential damage as a result of changes in the saturation state. The procedure outlined is applied to a case study, and considerations as to when the necessary safety work should be carried out are presented [65].

The article by [66] is focused on the medium-term negative effect of groundwater on the underground grout elements. This is the physical–mechanical effect of groundwater, which is known as erosion. We conducted a laboratory verification of the erosional resistance of grout mixtures. A new test apparatus was designed and developed, since there is no standardized method for testing at present. An erosion stability test of grout mixtures and the technical solutions of the apparatus for the test's implementation are described. This apparatus was subsequently used for the experimental evaluation of the erosional stability of silicate grout mixtures. Grout mixtures with activated and non-activated bentonite are tested. The stabilizing effect of cellulose relative to erosion stability has also been investigated. The specimens of grout mixtures were exposed to flowing water stress for a set period of time. The erosional stabilities of the grout mixtures were assessed on the basis of weight loss (WL) as a percentage of initial specimen weight. The lower the grout mixture weight loss, the higher its erosional stability, and vice versa [66].

3. Conclusions

As mentioned at the beginning, this issue was proposed and organized as a means of presenting recent developments in the field of non-destructive testing of materials in civil engineering. For this reason, the articles highlighted in this issue relate to different aspects of the testing of different materials in civil engineering, from building materials and elements to building structures. Interesting results, with significance for the materials, were obtained, and all of the papers have been precisely described.

Funding: This research received no external funding.

Institutional Review Board Statement: Not applicable.

Informed Consent Statement: Not applicable.

Conflicts of Interest: The authors declare no conflict of interest.

References

1. Schabowicz, K. Modern acoustic techniques for testing concrete structures accessible from one side only. *Arch. Civ. Mech. Eng.* **2015**, *15*, 1149–1159. [CrossRef]
2. Hoła, J.; Schabowicz, K. State-of-the-art non-destructive methods for diagnostic testing of building structures—Anticipated development trends. *Arch. Civ. Mech. Eng.* **2010**, *10*, 5–18. [CrossRef]
3. Hoła, J.; Schabowicz, K. Non-destructive diagnostics for building structures: Survey of selected state-of-the-art methods with application examples. In Proceedings of the 56th Scientific Conference of PANCivil Engineering Committee and PZITB Science Committee, Krynica, Poland, 19–24 September 2010. (In Polish).
4. Schabowicz, K.; Gorzelanczyk, T. Fabrication of fibre cement boards. In *The Fabrication, Testing and Application of Fibre Cement Boards*, 1st ed.; Ranachowski, Z., Schabowicz, K., Eds.; Cambridge Scholars Publishing: Newcastle upon Tyne, UK, 2018; pp. 7–39. ISBN 978-1-5276-6.
5. Drelich, R.; Gorzelanczyk, T.; Pakuła, M.; Schabowicz, K. Automated control of cellulose fiber cement boards with a non-contact ultrasound scanner. *Autom. Constr.* **2015**, *57*, 55–63. [CrossRef]
6. Chady, T.; Schabowicz, K.; Szymków, M. Automated multisource electromagnetic inspection of fibre-cement boards. *Autom. Constr.* **2018**, *94*, 383–394. [CrossRef]
7. Schabowicz, K.; Józwiak-Niedzwiedzka, D.; Ranachowski, Z.; Kudela, S.; Dvorak, T. Microstructural characterization of cellulose fibres in reinforced cement boards. *Arch. Civ. Mech. Eng.* **2018**, *4*, 1068–1078. [CrossRef]
8. Schabowicz, K.; Gorzelanczyk, T.; Szymków, M. Identification of the degree of fibre-cement boards degradation under the influence of high temperature. *Autom. Constr.* **2019**, *101*, 190–198. [CrossRef]
9. Schabowicz, K.; Gorzelanczyk, T. A non-destructive methodology for the testing of fibre cement boards by means of a non-contact ultrasound scanner. *Constr. Build. Mater.* **2016**, *102*, 200–207. [CrossRef]
10. Schabowicz, K.; Ranachowski, Z.; Józwiak-Niedzwiedzka, D.; Radzik, Ł.; Kudela, S.; Dvorak, T. Application of X-ray microtomography to quality assessment of fibre cement boards. *Constr. Build. Mater.* **2016**, *110*, 182–188. [CrossRef]
11. Ranachowski, Z.; Schabowicz, K. The contribution of fibre reinforcement system to the overall toughness of cellulose fibre concrete panels. *Constr. Build. Mater.* **2017**, *156*, 1028–1034. [CrossRef]
12. Bačić, M.; Kovačević, M.; Jurić Kaćunić, D. Non-Destructive Evaluation of Rock Bolt Grouting Quality by Analysis of Its Natural Frequencies. *Materials* **2020**, *13*, 282. [CrossRef]
13. Bajno, D.; Bednarz, L.; Matkowski, Z.; Raszczuk, K. Monitoring of Thermal and Moisture Processes in Various Types of External Historical Walls. *Materials* **2020**, *13*, 505. [CrossRef]
14. Szewczak, E.; Winkler-Skalna, A.; Czarnecki, L. Sustainable Test Methods for Construction Materials and Elements. *Materials* **2020**, *13*, 606. [CrossRef]
15. Skotnicki, L.; Kuźniewski, J.; Szydlo, A. Stiffness Identification of Foamed Asphalt Mixtures with Cement, Evaluated in Laboratory and In Situ in Road Pavements. *Materials* **2020**, *13*, 1128. [CrossRef] [PubMed]
16. Bywalski, C.; Drzazga, M.; Kaźmierowski, M.; Kamiński, M. Shear Behavior of Concrete Beams Reinforced with a New Type of Glass Fiber Reinforced Polymer Reinforcement: Experimental Study. *Materials* **2020**, *13*, 1159. [CrossRef]
17. Trapko, T.; Musiał, M. Effect of PBO–FRCM Reinforcement on Stiffness of Eccentrically Compressed Reinforced Concrete Columns. *Materials* **2020**, *13*, 1221. [CrossRef]

18. Karolak, A.; Jasieńko, J.; Nowak, T.; Raszczuk, K. Experimental Investigations of Timber Beams with Stop-Splayed Scarf Carpentry Joints. *Materials* **2020**, *13*, 1435. [CrossRef] [PubMed]
19. Słoński, M.; Schabowicz, K.; Krawczyk, E. Detection of Flaws in Concrete Using Ultrasonic Tomography and Convolutional Neural Networks. *Materials* **2020**, *13*, 1557. [CrossRef] [PubMed]
20. Juraszek, J. Fiber Bragg Sensors on Strain Analysis of Power Transmission Lines. *Materials* **2020**, *13*, 1559. [CrossRef]
21. Shi, Y.; Li, C.; Long, D. Study of the Microstructure Characteristics of Three Different Fine-grained Tailings Sand Samples during Penetration. *Materials* **2020**, *13*, 1585. [CrossRef]
22. Bywalski, C.; Kaźmierowski, M.; Kamiński, M.; Drzazga, M. Material Analysis of Steel Fibre Reinforced High-Strength Concrete in Terms of Flexural Behaviour. Experimental and Numerical Investigation. Experimental and Numerical Investigation. *Materials* **2020**, *13*, 1631. [CrossRef]
23. Stawiski, B.; Kania, T. Tests of Concrete Strength across the Thickness of Industrial Floor Using the Ultrasonic Method with Exponential Spot Heads. *Materials* **2020**, *13*, 2118. [CrossRef]
24. Schabowicz, K.; Sulik, P.; Zawiślak, Ł. Identification of the Destruction Model of Ventilated Facade under the Influence of Fire. *Materials* **2020**, *13*, 2387. [CrossRef] [PubMed]
25. Ranachowski, Z.; Ranachowski, P.; Dębowski, T.; Brodecki, A.; Kopec, M.; Roskosz, M.; Fryczowski, K.; Szymków, M.; Krawczyk, E.; Schabowicz, K. Mechanical and Non-Destructive Testing of Plasterboards Subjected to a Hydration Process. *Materials* **2020**, *13*, 2405. [CrossRef] [PubMed]
26. Logoń, D.; Schabowicz, K.; Wróblewski, K. Assessment of the Mechanical Properties of ESD Pseudoplastic Resins for Joints in Working Elements of Concrete Structures. *Materials* **2020**, *13*, 2426. [CrossRef] [PubMed]
27. Wróbel, M.; Woszuk, A.; Franus, W. Laboratory Methods for Assessing the Influence of Improper Asphalt Mix Compaction on Its Performance. *Materials* **2020**, *13*, 2476. [CrossRef] [PubMed]
28. Kozubal, J.; Wróblewski, R.; Muszyński, Z.; Wyjadłowski, M.; Stróżyk, J. Non-Deterministic Assessment of Surface Roughness as Bond Strength Parameters between Concrete Layers Cast at Different Ages. *Materials* **2020**, *13*, 2542. [CrossRef]
29. Tan, Y.; Zhu, B.; Qi, L.; Yan, T.; Wan, T.; Yang, W. Mechanical Behavior and Failure Mode of Steel–Concrete Connection Joints in a Hybrid Truss Bridge: Experimental Investigation. *Materials* **2020**, *13*, 2549. [CrossRef]
30. Park, J.; Sim, S.; Yoon, Y.; Oh, T. Prediction of Static Modulus and Compressive Strength of Concrete from Dynamic Modulus Associated with Wave Velocity and Resonance Frequency Using Machine Learning Techniques. *Materials* **2020**, *13*, 2886. [CrossRef]
31. Świt, G.; Krampikowska, A.; Pała, T.; Lipiec, S.; Dzioba, I. Using AE Signals to Investigate the Fracture Process in an Al–Ti Laminate. *Materials* **2020**, *13*, 2909. [CrossRef]
32. Dumbrava, F.; Cerbu, C. Experimental Study on the Stiffness of Steel Beam-to-Upright Connections for Storage Racking Systems. *Materials* **2020**, *13*, 2949. [CrossRef]
33. Zieliński, A.; Kaszyńska, M. Calibration of Steel Rings for the Measurement of Strain and Shrinkage Stress for Cement-Based Composites. *Materials* **2020**, *13*, 2963. [CrossRef]
34. Adamczak-Bugno, A.; Krampikowska, A. The Acoustic Emission Method Implementation Proposition to Confirm the Presence and Assessment of Reinforcement Quality and Strength of Fiber–Cement Composites. *Materials* **2020**, *13*, 2966. [CrossRef] [PubMed]
35. Wutke, M.; Lejzerowicz, A.; Garbacz, A. The Use of Wavelet Analysis to Improve the Accuracy of Pavement Layer Thickness Estimation Based on Amplitudes of Electromagnetic Waves. *Materials* **2020**, *13*, 3214. [CrossRef]
36. Trąmpczyński, W.; Goszczyńska, B.; Bacharz, M. Acoustic Emission for Determining Early Age Concrete Damage as an Important Indicator of Concrete Quality/Condition before Loading. *Materials* **2020**, *13*, 3523. [CrossRef]
37. Park, S.; Lee, J.; Lee, D. Aggregate Roundness Classification Using a Wire Mesh Method. *Materials* **2020**, *13*, 3682. [CrossRef] [PubMed]
38. Michałek, J.; Sobótka, M. Assessment of Internal Structure of Spun Concrete Using Image Analysis and Physicochemical Methods. *Materials* **2020**, *13*, 3987. [CrossRef] [PubMed]
39. Park, J.; Hong, G. Strength Characteristics of Controlled Low-Strength Materials with Waste Paper Sludge Ash (WPSA) for Prevention of Sewage Pipe Damage. *Materials* **2020**, *13*, 4238. [CrossRef]
40. Yin, D.; Xu, Q. Comparison of Sandstone Damage Measurements Based on Non-Destructive Testing. *Materials* **2020**, *13*, 5154. [CrossRef] [PubMed]
41. Pang, M.; Sun, Z.; Li, Q.; Ji, Y. 1H NMR Spin-Lattice Relaxometry of Cement Pastes with Polycarboxylate Superplasticizers. *Materials* **2020**, *13*, 5626. [CrossRef]
42. Skrzypczak, I.; Kokoszka, W.; Zięba, J.; Leśniak, A.; Bajno, D.; Bednarz, L. A Proposal of a Method for Ready-Mixed Concrete Quality Assessment Based on Statistical-Fuzzy Approach. *Materials* **2020**, *13*, 5674. [CrossRef]
43. Matsui, T.; Matsushita, Y.; Matsumoto, Y. Mechanical Behavior of GFRP Connection Using FRTP Rivets. *Materials* **2021**, *14*, 7. [CrossRef]
44. Gnatowski, A.; Kijo-Kleczkowska, A.; Chyra, M.; Kwiatkowski, D. Numerical–Experimental Analysis of Polyethylene Pipe Deformation at Different Load Values. *Materials* **2021**, *14*, 160. [CrossRef] [PubMed]
45. Jeong, S.; Kighuta, K.; Lee, D.; Park, S. Numerical Analysis of Shear and Particle Crushing Characteristics in Ring Shear System Using the PFC2D. *Materials* **2021**, *14*, 229. [CrossRef] [PubMed]

46. Pawelska-Mazur, M.; Kaszynska, M. Mechanical Performance and Environmental Assessment of Sustainable Concrete Reinforced with Recycled End-of-Life Tyre Fibres. *Materials* **2021**, *14*, 256. [CrossRef]
47. Tworzewski, P.; Raczkiewicz, W.; Czapik, P.; Tworzewska, J. Diagnostics of Concrete and Steel in Elements of an Historic Reinforced Concrete Structure. *Materials* **2021**, *14*, 306. [CrossRef] [PubMed]
48. Logoń, D.; Schabowicz, K.; Roskosz, M.; Fryczowski, K. The Increase in the Elastic Range and Strengthening Control of Quasi Brittle Cement Composites by Low-Module Dispersed Reinforcement: An Assessment of Reinforcement Effects. *Materials* **2021**, *14*, 341. [CrossRef]
49. Bilko, P.; Skoratko, A.; Rutkiewicz, A.; Małyszko, L. Determination of the Shear Modulus of Pine Wood with the Arcan Test and Digital Image Correlation. *Materials* **2021**, *14*, 468. [CrossRef] [PubMed]
50. Gabryś, K.; Soból, E.; Sas, W.; Šadzevičius, R.; Skominas, R. Warsaw Glacial Quartz Sand with Different Grain-Size Characteristics and Its Shear Wave Velocity from Various Interpretation Methods of BET. *Materials* **2021**, *14*, 544. [CrossRef] [PubMed]
51. Skotnicki, Ł.; Kuźniewski, J.; Szydło, A. Research on the Properties of mineral–Cement Emulsion Mixtures Using Recycled Road Pavement Materials. *Materials* **2021**, *14*, 563. [CrossRef]
52. Zhao, Y.; Liu, R.; Yan, F.; Zhang, D.; Liu, J. Windblown Sand-Induced Degradation of Glass Panels in Curtain Walls. *Materials* **2021**, *14*, 607. [CrossRef]
53. Wiejak, A.; Francke, B. Testing and Assessing Method for the Resistance of Wood-Plastic Composites to the Action of Destroying Fungi. *Materials* **2021**, *14*, 697. [CrossRef]
54. Tešić, K.; Baričević, A.; Serdar, M. Non-Destructive Corrosion Inspection of Reinforced Concrete Using Ground-Penetrating Radar: A Review. *Materials* **2021**, *14*, 975. [CrossRef] [PubMed]
55. Sudoł, E.; Szewczak, E.; Małek, M. Comparative Analysis of Slip Resistance Test Methods for Granite Floors. *Materials* **2021**, *14*, 1108. [CrossRef] [PubMed]
56. Juraszek, J.; Antonik-Popiołek, P. Fibre Optic FBG Sensors for Monitoring of the Temperature of the Building Envelope. *Materials* **2021**, *14*, 1207. [CrossRef] [PubMed]
57. Topolář, L.; Kocáb, D.; Pazdera, L.; Vymazal, T. Analysis of Acoustic Emission Signals Recorded during Freeze-Thaw Cycling of Concrete. *Materials* **2021**, *14*, 1230. [CrossRef]
58. Mackiewicz, P.; Mączka, E. The Impact of Water and Road Salt with Anti-Caking Agent on the Stiffness of Select Mixes Used for the Road Surface. *Materials* **2021**, *14*, 1345. [CrossRef]
59. Kania, T.; Derkach, V.; Nowak, R. Testing Crack Resistance of Non-Load-Bearing Ceramic Walls with Door Openings. *Materials* **2021**, *14*, 1379. [CrossRef] [PubMed]
60. Bednarz, L.; Bajno, D.; Matkowski, Z.; Skrzypczak, I.; Leśniak, A. Elements of Pathway for Quick and Reliable Health Monitoring of Concrete Behavior in Cable Post-Tensioned Concrete Girders. *Materials* **2021**, *14*, 1503. [CrossRef]
61. Juraszek, J.; Gwóźdź-Lasoń, M.; Logoń, D. FBG Strain Monitoring of a Road Structure Reinforced with a Geosynthetic Mattress in Cases of Subsoil Deformation in mining Activity Areas. *Materials* **2021**, *14*, 1709. [CrossRef]
62. Aneja, S.; Sharma, A.; Gupta, R.; Yoo, D. Bayesian Regularized Artificial Neural Network Model to Predict Strength Characteristics of Fly-Ash and Bottom-Ash Based Geopolymer Concrete. *Materials* **2021**, *14*, 1729. [CrossRef]
63. Schabowicz, K.; Sulik, P.; Zawiślak, Ł. Reduction of Load Capacity of Fiber Cement Board Facade Cladding under the Influence of Fire. *Materials* **2021**, *14*, 1769. [CrossRef] [PubMed]
64. Logoń, D.; Juraszek, J.; Keršner, Z.; Frantík, P. Identifying the Range of Micro-Events Preceding the Critical Point in the Destruction Process in Traditional and Quasi-Brittle Cement Composites with the Use of a Sound Spectrum. *Materials* **2021**, *14*, 1809. [CrossRef]
65. Bhat, D.; Kozubal, J.; Tankiewicz, M. Extended Residual-State Creep Test and Its Application for Landslide Stability Assessment. *Materials* **2021**, *14*, 1968. [CrossRef] [PubMed]
66. Boštík, J.; Miča, L.; Terzijski, I.; Džaferagić, M.; Leiter, A. Grouting below Subterranean Water: Erosional Stability Test. *Materials* **2021**, *14*, 2333. [CrossRef] [PubMed]

Article

Grouting below Subterranean Water: Erosional Stability Test

Jiří Boštík *, Lumír Miča, Ivailo Terzijski, Mirnela Džaferagić and Augustin Leiter

Faculty of Civil Engineering, Brno University of Technology, Veveří 331/95, 60200 Brno, Czech Republic; mica.l@fce.vutbr.cz (L.M.); terzijski.i@fce.vutbr.cz (I.T.); karzicm@study.fce.vutbr.cz (M.D.); leiter.a@fce.vutbr.cz (A.L.)
* Correspondence: bostik.j@fce.vutbr.cz

Abstract: The article is focused on the medium-term negative effect of groundwater on the underground grout elements. This is the physical–mechanical effect of groundwater, which is known as erosion. We conduct a laboratory verification of the erosional resistance of grout mixtures. A new test apparatus was designed and developed, since there is no standardized method for testing at present. An erosion stability test of grout mixtures and the technical solutions of the apparatus for the test's implementation are described. This apparatus was subsequently used for the experimental evaluation of the erosional stability of silicate grout mixtures. Grout mixtures with activated and non-activated bentonite are tested. The stabilizing effect of cellulose relative to erosion stability has been also investigated. The specimens of grout mixtures are exposed to flowing water stress for a certain period of time. The erosional stabilities of the grout mixtures are assessed on the basis of weight loss (WL) as a percentage of initial specimen weight. The lower the grout mixture weight loss, the higher its erosional stability and vice versa.

Keywords: erosional stability; laboratory testing; grout mixtures; groundwater; test apparatus

Citation: Boštík, J.; Miča, L.; Terzijski, I.; Džaferagić, M.; Leiter, A. Grouting below Subterranean Water: Erosional Stability Test. *Materials* **2021**, *14*, 2333. https://doi.org/10.3390/ma14092333

Academic Editor: Krzysztof Schabowicz

Received: 3 February 2021
Accepted: 27 April 2021
Published: 30 April 2021

Publisher's Note: MDPI stays neutral with regard to jurisdictional claims in published maps and institutional affiliations.

Copyright: © 2021 by the authors. Licensee MDPI, Basel, Switzerland. This article is an open access article distributed under the terms and conditions of the Creative Commons Attribution (CC BY) license (https://creativecommons.org/licenses/by/4.0/).

1. Introduction

The objective of grouting a soil and rock mass is to create an "underground grout element" (UGE). This may have the function of reinforcement (support), sealing, or both. Possible complications arise from this basic objective of grouting works during grouting below the groundwater table. Simplifying somewhat, the results of grouting under the groundwater level are:

- UGE may not even be able to form;
- UGE may be able to form, but only with limited functionality;
- UGE may form with good functionality, but due to the subsequent action of groundwater, degrades gradually (to limited or zero functionality).

The reasons for the failed creation of an UGE (or insufficient functionality) can be different: for example, inappropriately selected type or character of the grout mixture, due to the porosity of soil, or inappropriately selected grouting procedure (e.g., low or high grouting pressure). These reasons and the contexts are usually discussed in the literature dealing with grouting, e.g., [1,2]. The implied reasons are usually also unambiguously unrelated to the presence or activity of groundwater.

Focusing only on the problems arising from grouting under the groundwater level, these problems logically result from the presence of underground water and its effect on the formation of or already-formed UGE. The negative effects of groundwater can be divided in terms of time to (this is a sub-division of the terms used by the authors of this article within their research work):

- short-term effects;
- medium-term effects; and
- long-term effects.

21

The short-term effects of groundwater (especially flow) on formed UGE are usually physical–mechanical; the grout mixture has not solidified enough. Most commonly, it is the segregation and/or flushing out grout mixtures from the UGE space. For example, this impact (authors label it as dispersion) is included in the research of Baluch et al. [3]. In this case, the dispersiveness of grout mixtures was tested by pouring grout mixture into a water-filled beaker.

The medium-term effects of groundwater on UGE are usually physical–mechanical as well; however, compared to the short-term effects, there is no grout mixture dilution or flushing, but rather it is erosion. The formation of the UGE is mechanically disrupted and its function is negated as a result of the erosion process. The described state can occur by using grout mixture with a low erosional resistance (erosional stability) or by using a grout mixture in which the final erosional resistance is achieved very slowly (i.e., slow setting of the mixture). This process is in analogy with the internal erosion of soil by piping. This phenomenon has been investigated by many authors who studied it on soil [4] or soil treated by chemical stabilizers [5]. Research of silicate grout mixture erosion has not been found.

In the course of short-term and medium-term effects of groundwater, chemical degradation processes occur. In these cases, they are not of great significance as the grout mixture has been flushed out or was already eroded.

The long-term activity of groundwater includes, once again, erosion. The possibility of suffusion cannot be excluded either. However, chemical degradation processes, which are commonly known as "corrosion", may have the greatest importance. This is only applied when water contains corrosive substances. Wang et al. [6] described the damage of adding solids after seawater's long-term corrosion, as well as analyses of its stress–strain curve and the relationship between damage variables and grouting solids.

The line between the time span of the above-mentioned categories is not fixed, which obviously arises from the above-mentioned facts. It depends, for example, on the speed of the solidifying and hardening processes of the grout mixture, on the hydraulic gradient or on the types and amounts of chemical substances present in the water. An order-of-magnitude estimate of the time span is:

- short-term effect—from tens of minutes to a low number of hours after the application of the grout mixture;
- medium-term effect—from a low number of hours to tens of hours after the application of the grout mixture;
- long-term effect—more than hundreds or thousands of hours after the application of the grout mixture.

For example, in the experiments described below, the line between the short-term and medium-term effects can be estimated to be between 5 and 6 hours after the production of the grout mixture. Within 5 h of its production, the grout mixture was flushed out as a result of the effect of flowing water.

The danger arising from physical–mechanical processes is the largest for grouting of highly permeable and water-bearing horizons. This is where a significant negative underground grout element process of disruption by ground water may be expected. For low permeability horizons (though water-bearing), the intensity of groundwater flow is, from the given perspective, naturally insignificant.

Erosion is caused by the mechanical effects of the surrounding substance's movement (in this case flowing water); therefore, the erosional resistance of grout mixtures can be deduced from their strengths. However, there is no general relation between strength and erosional resistance. This is understandable, since there is no "standard" intensity of erosive action, nor are there "standard" conditions. In addition to measuring the strength, there are non-standard tests of erosional resistance that attempt to directly simulate the process of erosion. However, these are single-purpose devices to a limited extent.

For laboratory testing of grout mixture resistances, it is possible to use both mentioned principles. The first is to monitor the strength of grout mixtures. Applied shear strength measurements (e.g., the penetration method) are the best for the monitoring of relatively low-strength "young" grout mixtures. The second method for determining the erosional resistance is a test of erosional stability. This test measures the resistance of the grout mixture (grouted element) to mechanical deterioration caused by flowing water.

The description of the erosional stability test, technical solutions and implementation of test apparatus is the main subject of this article. This test was subsequently used for the experimental evaluation of the erosional stability of silicate grout mixtures.

2. Overview of Laboratory Test Apparatus for Erosional Stability Testing of Geomaterials

Studies on internal erosion originally focused on a mechanical principle, where particle and opening sizes in the soil that allow particle movement were primarily investigated [7,8].

Over the years, a number of different laboratory methods have been developed in order to test soils according to their susceptibility to erosion, e.g., the pinhole test, flume tests, the jet erosion test and the rotation cylinder test. These tests have been summarized by various authors, e.g., Wan and Fell [9,10].

The rotating cylinder test (RCT) was developed by Moore and Masch [11]. The RCT uses a block of soil, suspended and submerged, inside a rotating cylindrical chamber, where the rotation of the cylinder induces a flow around the specimen which causes erosion. Torque is applied to the specimen and the erosion rates are measured. The results are used to estimate applied stress and erodibility parameters.

Arulanandan et al. [12] extended previous research by performing erosion tests on samples by circulating collected water on undisturbed samples inserted into a hydraulic flume from the bottom. The flume test is used for modeling erosion mechanism, where water is flowing parallel to the soil surface at a certain speed and depth. The erosion rate is only visually observed and described in most cases, while hydraulic shear stress on the soil surface is deduced from the flow velocity and water depth.

Lefebvre et al. [13] and Rohan et al. [14] developed the drill hole test (DHT). The DHT is conducted by applying a flow rate-controlled pressure flow to a cylindrical clay sample using a predrilled axial hole. The friction head loss along the specimen is measured to determine shear stress and the collected eroded material is used to calculate the erosion rate.

A special type of flow-over-surface test is the erosion function apparatus (EFA) developed by Briaud et al. [15]. The EFA test uses site-specific soil samples acquired via thin-walled tubes to generate an erosion rate and shear stress. During the EFA test, a data acquisition system records the velocity and amount of soil eroded. These data are used to calculate the erosion rate and shear stress.

The jet erosion test (JET) was developed at the Agricultural Research Service Hydraulic Engineering Research Unit, Stillwater, Oklahoma [16]. This method is applicable to a wide range of soils. JET can be performed in situ [17,18] or in a laboratory [19] using tube samples or remolded samples in compaction molds. Testing has been successfully carried out on specimens as small as 75 mm (3 inches) in diameter and uses a submerged hydraulic jet to produce scour erosion. JET estimates the critical shear stress needed to initiate erosion.

The hole erosion test (HET) was developed by Wan and Fell [10] to measure the erosional properties of soils. HET involves measuring an accelerating flow rate through an eroding pre-formed hole in a test specimen. HET enables the estimation of the critical shear stress and erosion rate coefficient. All test results of the hole erosion tests give an interesting relationship between the critical stress and the erosion rate coefficient. A greater critical stress implies a greater erosion rate index (i.e., slower erosion).

Sanchez et al. [20] were the first to develop an internal erosion test within a modified triaxial apparatus to evaluate the erosion of core embankment materials. Recent experiments by Bendahmane et al. [21] revealed the complex effects of confining pressure on internal erosion. The triaxial erosion test was conducted by Bendahmane et al. [21,22] to measure the effects of internal flows on sand/kaolin samples.

Richards and Reddy [23] developed a true triaxial system to investigate the piping potential of both cohesive and non-cohesive soils. A computer-controlled triaxial testing apparatus was modified to allow the independent control of hydraulic gradient and stress state for investigating the initiation and development of soil internal erosion, and the stress–strain behavior of soil subjected to internal erosion.

One of the most recently presented pieces of experimental apparatus is the cross erosion test (CET) [24,25], which is devoted to the measurement of the initiation of suffusion. The test consists of a clear water injection into a first drill hole, and the recovery of washed-out particles suspended in water in another drill hole. This technique can be portable, in order to consider the in situ risks of internal erosion at dams and dikes.

All the mentioned methods, in most cases, include a circular cross-sectional shape of the specimen. The position of the specimen during the test with regard to the direction of water flow is mainly vertical, except for in the HET test, where it is horizontal. Monitoring of piezometric height takes place using a piezometer or manometer.

The evaluation of erosional resistance is mainly based on quantifying the amount of eroded material; specifically, the amount of washed-out material in the outflow and hydraulic gradient. Internal erosion in an internally unstable soil will occur when the hydraulic gradient exceeds a certain critical value.

As stated earlier, a summary of the findings from the literature indicates that quite considerable attention has been devoted to experimental soil erosion modeling. However, experimental modeling of erosion, especially of grout mixtures, is not often mentioned in the literature. One method of laboratory testing of erosional resistance is the principle described by Verfel [1]. The test principle is basically the same as that for determining soil erosion, i.e., the grout mixture (partially stiffened) is exposed to water flow stress. Here, the erosional resistance is tested on a specimen of grout mixture with a height of 70 mm, which is cast into a cylindrical vessel (the test cell). A hole is formed in the specimen by inserting a glass stick with a diameter of 8 mm during preparation of the specimen. The glass stick is removed after 18 h and water is fed into the upper cell space (above the upper specimen surface) under such a pressure as to form a gradient that causes water flow through the coaxial hole in the specimen at a velocity of about 2 ms^{-1}. The result of this laboratory test is weight loss of the grout mixture, which is eroded by the water flowing through a hole in the specimen for one hour.

The erosion tests are also mentioned in Austrian standard ÖNORM B 4452 [26]. The laboratory testing process of the erosion test and schematic drawings and the dimensions of the test cell (specimen) are described in the appendix of this standard. The principle of the test is also based on investigation of hole erosion behaviors. A cylindrical hole is formed in the axis of the cylindrical test specimen from the cut-off wall material. The hole's diameter in this case varies from 1–10 mm. The specimen is placed into a test cell, which is connected to a water-pressure system during the test.

3. Experimental Verification of Erosional Stability of Grout Mixtures

The development of the research framework carried out at the Faculty of Civil Engineering, Brno University of Technology, included the method for testing the erosional stability, according to Verfel [1]. We decided to design and develop our own apparatus, since there is no standardized method for testing at present, and it was not possible to purchase an appropriate laboratory apparatus.

3.1. Apparatus and Test Process

The test apparatus for determining erosion stability consists of several basic structural units (Figure 1):

- storage (recycling) vessel;
- flow controller;
- test cell;
- filter;

- pump; and
- supporting structure.

The storage vessel accumulates a test fluid (water) and enables its recycling. There is a submersible pump placed in the vessel that is connected to the flow controller via a pipe. The flow controller is used to maintain constant hydraulic conditions (i.e., constant volume flow rate of 0.1 L/s at a water flow velocity of 2 ms^{-1}) through the test cell. If needed, it is possible to increase or decrease the volume flow rate using the control button, which is on the flow controller. The solid particle filter is part of the test circuit. The filter captures particles of the tested matter that erode during the test, and any other impurities that may occur in the circuit.

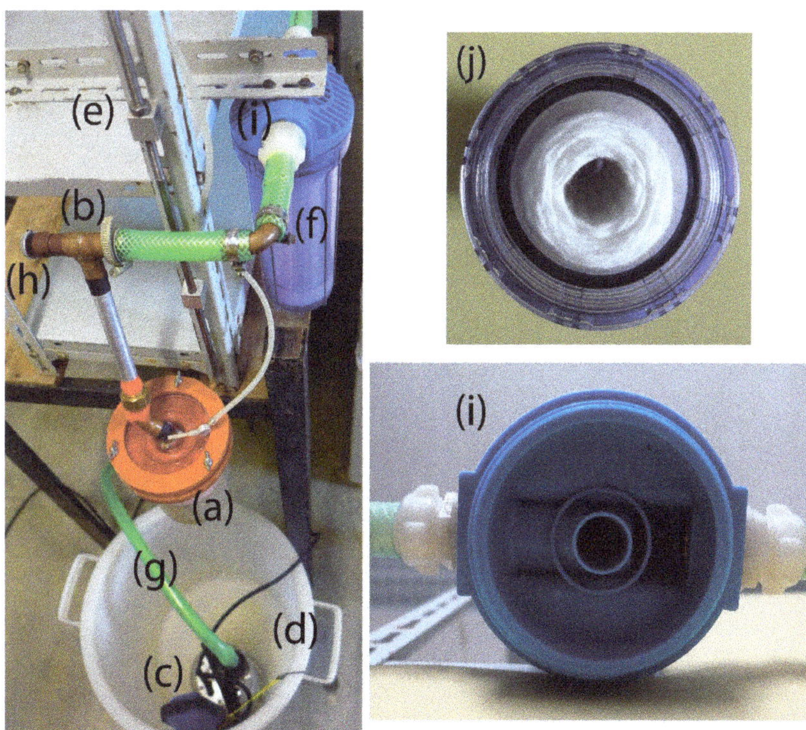

Figure 1. Test apparatus for determining erosional stability: (**a**) test cell; (**b**) flow controller; (c) pump; (**d**) storage/recycling vessel; (**e**) supporting structure; (**f**) filter; (**g**) supply pipe; (**h**) control button; (**i**) cap of the filter; (**j**) opened filter view.

The test cell contains a specimen of the grout mixture. The cell has two parts. The grout mixture specimen is placed in the lower part of the cell. There is an 8 mm round hole in the bottom. The upper part is a pressure cap and both parts can be sealed together. The cap is equipped with an inlet fitting, flow straightener and air vent. The diameter of the test cell in the bottom is about 100 mm; the cell slightly widens upwards conically. A test cell holder enables the cell position to be adjusted.

The test specimen is prepared by casting grout mixture into the test cell. A glass (or plastic) stick with a diameter of 8 mm is inserted into the hole at the bottom of the test cell (prior to casting). The specimen height in the cell should be about 70 mm at the time of starting the test (Figure 2). The glass stick is removed before starting the test, and there is an 8 mm round hole that remains in the bottom of the specimen. The hole in the bottom of the test cell is temporarily sealed (glued) in the next step, and the upper space of the grout mixture is flooded with water. After the cell is closed, it is placed in the cell holder and connected to the flow controller via a pipe. The flow controller must be adjusted so that the water flow velocity through the non-eroded hole in the axis of the specimen is about $2\,\mathrm{ms}^{-1}$.

According to the standard procedure described in [1], tests should be initiated after 18 h of grout mixture production. However, within the demonstration examples shown in Sections 3.2 and 4, during the erosional stability test in compliance with the prescribed time delay (18 h) from the grout mixture preparation to the beginning of the test, there was no specimen disruption caused by flowing water. Since this is a purely comparative test of grout mixture, this step was necessary to change the conditions of the test to detect potential differences between suspensions (i.e., to capture the potential improvement expected for modified grout mixtures). It turned out that, technically, the easiest way was to change the time delay. This was based on the fact that the age of a specimen must allow for the creation of the desired hole in the specimen, and the measurement of the weight loss after the test. After several experiments, the time delay was set to six hours. This kind of test can only be taken as characteristic for properties of UGE if the grout mixture is appropriately stable. A stability limit of 2% for increasing the grout mixture's specific weight between its casting and the beginning of the water flow was used (as recommended).

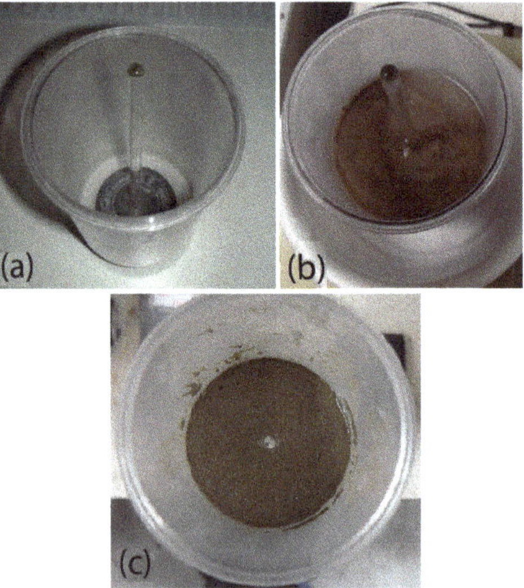

Figure 2. Specimen preparation: (**a**) empty test cell; (**b**) test cell with grout mixture; (**c**) specimen ready for testing.

The water flows into the test cell (Figure 3) through the filter and the flow controller from the storage/recycling vessel during the test. The velocity of water flowing during the tests was kept constant at the value of $2\,\mathrm{ms}^{-1}$. Drinkable water from the Brno aqueduct was used for these experiments.

The test cell is equipped with a flow straightener. This is a specially shaped inlet that forces the inflowing water jet through a large number of small holes. Such a small diameter of the outflowing water jet formed flooded stream and also ensured good venting. Air bubbles can be purged through the air vent in the cap of the test cell. Undisturbed water then flowed into the hole in the specimen. The water (with possible eroded particles of the specimen) falls into the storage/recycling vessel. The height of the test cell above the storage/recycling vessel was set to allow flow rate monitoring using the volumetric method. If the erosional effect is negligible and therefore the water is clear, it can be reused. Water reuse is permitted by a submersible pump in the storage/recycling vessel. This eliminates the use of tap water. The test was terminated after one hour of water flowing through the specimen.

The erosional stability test was evaluated by recording/verifying the following: specimen weight before and after the test, the water flow rate, and the elapsed time. The erosional stability of the grout mixture was assessed on the basis of weight loss (WL) as a percentage of the initial specimen weight. The lower the grout mixture weight loss (WL), the higher its erosional stability (erosion resistance) and vice versa. A WL value equal to 0, or very close to 0 (cases in which only the inlet edge of the hole in the specimen is impacted), means that failure of the grout mixture in the allotted time did not happen.

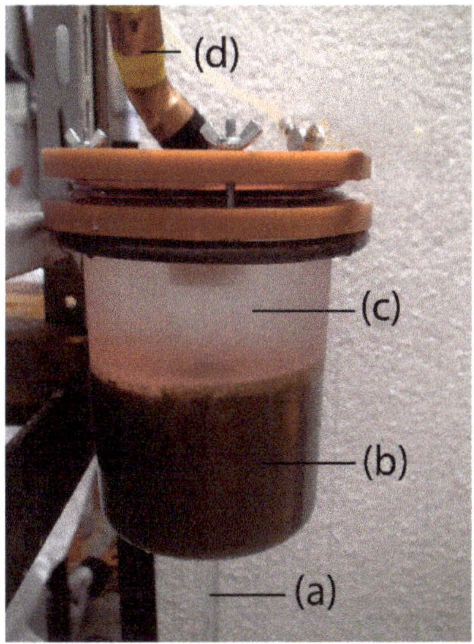

Figure 3. Grout mixture specimen during testing: (**a**) effluent water; (**b**) specimen of grout mixture; (**c**) test cell; (**d**) supply pipe.

3.2. Application of the Erosional Stability Test—Examples of Experiments

The test described in the previous section and the created laboratory apparatus were used for subsequent tests. These activities were primarily focused on finding ways to reduce the negative physical–mechanical effects of flowing groundwater on the underground grout elements created by grouting. Here, further focus was only on the testing of selected options, leading to an increased erosional stability of grout mixtures.

Grout mixtures based on silicate composites were the only subjects tested. From a broader set of these grout mixtures, clay–cement sealing mixtures were selected. The grout mixtures were prepared on the basis of Na^+ ion-activated bentonite, commercially designated as Bentovet K (Gemerská nerudná spoločnosť, Hnúšťa, Slovakia) and also on non-activated bentonite, commercially designated as Bentonit 75 (manufacturer: Keramost, a.s., Most, Czech Republic).

For determination of the mineralogical compositions, bentonite samples were subjected to X-ray diffraction analyses. The specimens were crushed prior to analysis in isopropanol using the McCrone Micronising Mill with the addition of 20 wt.% of an internal standard (i.e., after addition: 80% of specimen + 20% of standard). The used standard was fluorite (CaF_2). The addition of the internal standard allowed the quantification of the amorphous phase. X-ray diffraction (XRD) analyses were performed on a Rigaku Smartlab apparatus with a Cu-anode ($\lambda K\alpha = 0.15418$ nm), 2D position-sensitive detector used in the 1D mode and fixed divergence screens in conventional Bragg-Brentano parafocusing Θ-Θ reflection geometry. The measured angular area was 5–80 °2θ. The step size was 0.013°

2θ and the scan step time was 255 s. Total measurement time was 5769 s. The data were processed using Panalytical HihgScore 3 plus software. Quantitative phase analyses were performed using the Rietveld method. The structural patterns from the database of ICSD 2012 were used to refine the structural specification and for quantification. The results of the quantitative phase analyses are listed in Table 1. The overall compositions of the specimens, including the content of the amorphous phase, were quantified, as was the composition of the isolated crystalline part. Diffractograms are shown in Figures 4 and 5.

Portland composite cement CEM II/BM (S-LL) 32.5 R (manufacturer: Českomoravský cement, a.s., Mokrá, Czech Republic) was used for the laboratory preparation of the grout mixtures.

The laboratory production of the grout mixture followed common preparation methods used in practice. A two-stage process, as well as a single-stage process, of grout mixture preparation was used. The two-stage process of preparation was only used for grout mixtures based on Bentovet K. The single-stage process of preparation was used for the grout mixtures based on Bentovet K and for the grout mixtures based on Bentonit 75. The two-stage process for grout mixture preparation was as follows:

- The water bentonite mixture (BM) was prepared, mixing time was 10 min.
- The BM was left to cure in a closed vessel for approximately 24 h at a temperature of $20 \pm 2\,°C$.

- The grout mixture itself was prepared by adding cement to the cured BM while stirring continuously. The stirring time was 5 min in this case.

The single-stage process of the grout mixture preparation was as follows:

- The water bentonite mixture was prepared—it was mixed for 10 min (grout mixtures on the basis of Bentovet K) or for 5 min (grout mixtures on the basis of Bentonit 75).
- Cement was added gradually to the BM while mixing continuously. It was mixed for 5 min in this case.

Mixing of the grout (both two-stage process and single-stage process) was accomplished through the use of a laboratory stirrer (Heidolph RZR 2020). A dissolver stirrer with a diameter of 100 mm and revolutions of 1300 min^{-1} was used.

After cellulose was added to the finished mixture, stirrer revolutions were reduced to 400 min^{-1} and the process of stirring took place for a further 1 min.

Table 1. Results of quantitative phase analysis.

Specimen	Bentonit 75		Bentovet K	
Mineral	Specimen Composition Including Amorphous Phase [%]	Specimen Composition of Crystalline Part Only [%]	Specimen Composition Including Amorphous Phase [%]	Specimen Composition of Crystalline Part Only [%]
Montmorillonite	50.8	63.5	51.1	67.1
Quartz	8.3	10.4	7.6	10.0
Illite	5.1	6.3	8.3	10.9
Kaolinite	1.2	1.5	-	-
Calcite	0.2	0.2	-	-
Siderite	10.6	13.2	-	-
Anatase	4.0	4.9	-	-
Feldspar (plagioclase)	-	-	1.3	1.7
Chabazite	-	-	1.3	1.7
Chlorite	-	-	6.5	8.6
Amorphous phase	19.8	-	23.9	-

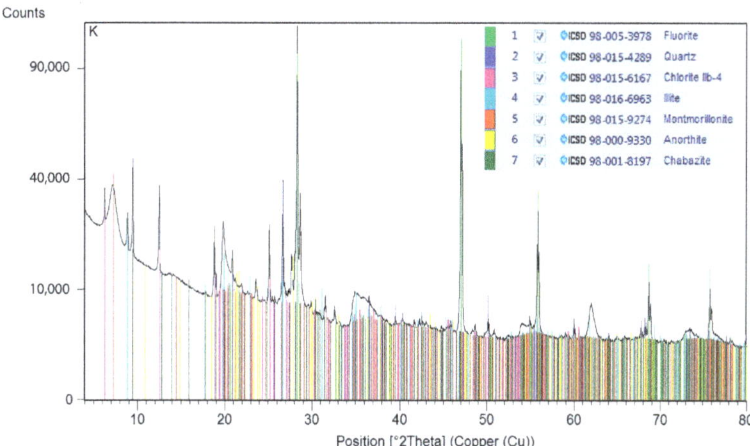

Figure 4. Diffractogram of Bentovet K.

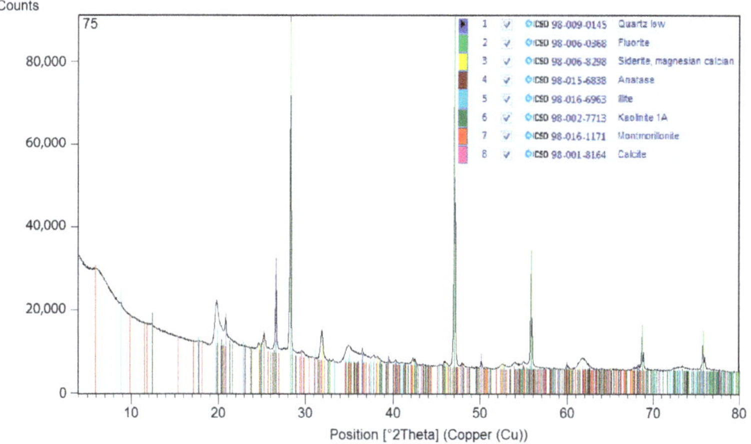

Figure 5. Diffractogram of Bentonit 75.

In the first stage of the experimental work, the so-called "standard grout mixture" was designed and tested. The composition of the grout mixture (1 m^3) was as follows: 42.6 kg of activated bentonite (dry weight), 350 kg of cement, and water for the remainder. This was a grout mixture recipe that was created at the Brno University of Technology previously, having been actually utilized for sealing works by the BauGeo company. In the next phase of the experimental work, design and testing of the modified grout mixtures were carried out. The compositions of the modified grout mixtures were modified with the aim of increasing erosional stability.

Some designed and subsequently tested modifications included, e.g., an increase in bentonite content in the grout mixture, using a different type of bentonite in the grout mixture, or the use of special additives to enhance grout mixture thixotrophy.

4. Results and Discussion

4.1. Increase in Bentonite Content in the Grout Mixture (Example One)

The standard dose of Bentovet K was 42.6 kg (dry) per 1 m^3 of initial bentonite mixture (i.e., the standard grout mixture with Bentovet K labeled as K1). In an effort to improve the grout mixture's erosional stability, the Bentovet K dose was gradually increased to 50, 57.5 and 65 kg per 1 m^3 of bentonite mixture. The dosages used during the two-stage process of grout mixture preparation are shown in Table 2.

The used Bentovet K concentration range covered the entire useful range of this grout mixture type. If the minimum concentration was used, an unstable grout mixture is expected. If greater than the maximum concentration was used, the grout mixture is difficult to pump and inject. The cement dosage in all tested mixtures was constant: 350 kg/m^3 of grout mixture. The impact of these changes on the grouts' erosional stability was observed.

Table 2. Composition and WL of grout mixtures based on Bentovet K.

Indication	Bentonite Mixture (BM) *		Grout Mixture *		Bentonite Concentration [% of the Grout Mixture Weight]	Weight Loss (WL) [%]
	Bentovet K (Dry Weight) [kg]	Water [kg]	BM [kg]	Cement CEM II/B-M (S-L)/32,5R [kg]		
K1	42.6	984.2	907.8		3.0	4.7
K2	50.0	981.5	910.7	350.0	3.5	11.4 [+]
K3	57.5	978.7	915.1		4.0	2.1
K4	65.0	975.9	919.6		4.5	5.0

* Bentonite (grout) mixture composition is specified per m^3. [+] Arithmetical average from two tests (12.5% and 10.2%).

An example of erosional damage can be seen in Figure 6. This represents the inlet side of the specimens before (a) and after the tests (b, c, d, e), and the confirmed percent value of weight loss of the grout mixture (WL). The contour of the hole in the specimen before (green highlighted) and after (red highlighted) the tests is marked in Figure 6. In Figure 7, the specimen's cross section after the test of K2 grout mixture with the lowest erosional stability is presented. Again, the location of the hole contour in the specimen and the inlet side before the test is highlighted in green. The edge of the failure area after the test is highlighted in red. It is obvious from the figure that the decreasing erosional stability of the grout mixtures causes significant distortion of the specimen's inlet edges. This phenomenon is accompanied by a widening of the hole in the specimen. It is clear that the extent of the widening depends on the value of WL. The higher the WL value, the wider the hole, and vice versa. Concurrently, the hole is widened along the full height of the specimen and it can be noticeably asymmetric (with regard to the longitudinal axis of the specimen), see Figure 7.

The erosional stabilities of these grouts appear to be uncorrelated with the used bentonite dose (within the given range). This is not surprising, since in the case of erosional stability, the most significant factor is the strength of solidified grout, which is in the test time of 6 h, determined by the amount and type of cement, respectively, using the water/cement ratio. These parameters were constant in a given series of grout mixtures (cement dose), or changed insignificantly (water/cement ratio), see Table 2.

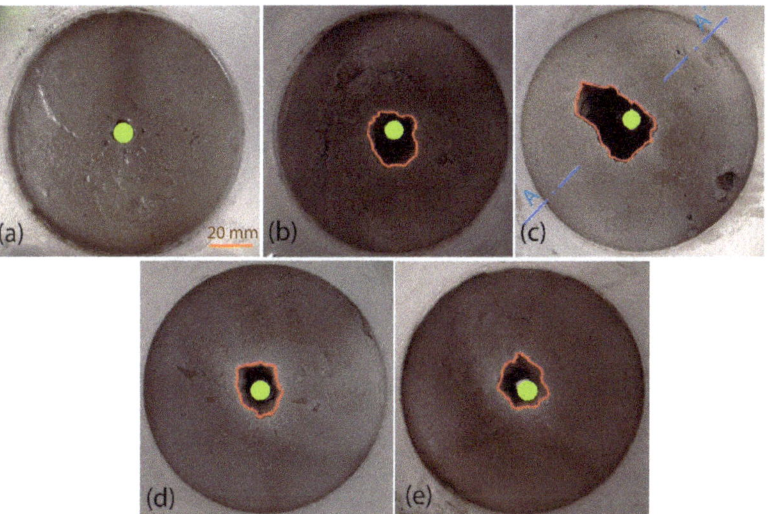

Figure 6. Inlet side of specimens with Bentovet K: (**a**) standard grout mixture K1 before the test; (**b**) grout mixture K1 after the test—WL = 4.7%; (**c**) grout mixture K2 after the test—WL = 11.4%; (**d**) grout mixture K3 after the test—WL = 2.1%; (**e**) grout mixture K4 after the test—WL = 5.0% (the edge of the failure area in red, the initial hole in green).

Figure 7. Specimen cross section A–A' after the test (see Figure 6c)—grout mixture K2 based on Bentovet K (the edge of the failure area in red, the initial hole and inlet side in green).

4.2. Use of Different Bentonite Type in the Grout Mixture (Example Two)

Chemically activated (natrificated) bentonites are commonly used in grout mixtures (Bentovet K was used in this study—see example one). This means that the natural Ca form of the main bentonite mineral, montmorillonite, is artificially converted to the Na form during the natrification process. This form has significantly improved solvating ability. If the suspension of natrificated bentonite is mixed with cement, the hydration of cement releases Ca-ions that, in turn, convert the suspended Na-bentonite (montmorillonite) back to the Ca form. The stabilizing ability of bentonite, therefore, decreases over time. However, simultaneously, the gel phase and later the crystalline phase of Al, Ca and Fe hydrosilicates from the cement are formed. Therefore, the mixture gradually solidifies and hardens. An ordinary well-designed grout mixture maintains its fluidity for a relatively long time period before strengthening processes gradually take over.

The aforementioned processing scheme is not suitable for most mixtures with increased erosional stability. If we replace the natrified bentonite in the grout mixture with Ca-bentonite, the dose must be, given its lower solvating capacity, significantly higher (see Tables 2 and 3). Generally, depending on the parameters of both forms of bentonite, the dose may increase by a factor of three to five. The advantage for such a suspension is that the Na and Ca ions are not reversely interchanged, e.g., the strengthening processes start significantly faster. The disadvantage is the higher price of the total doses of Ca-bentonite in grout mixtures.

Table 3. Composition * and WL of grout mixtures based on Bentonit 75.

Indication	Bentonit 75 (Dry Weight) [kg]	Cement CEM II/B-M (S-L)/32,5R [kg]	Water [kg]	Bentonite Concentration [% of the Grout Mixture Weight]	Weight Loss (WL) [%]
B1	130.0	350.0	824.9	10.0	1.4
B5	200.0		793.5	14.9	0.2

* Composition is specified per m^3.

The two variants of grout mixtures based on chemically non-activated (i.e., in the Ca-form) bentonite (Bentonit 75) were prepared and tested. The used bentonite concentration range was again chosen to cover the entire useful range of this type of grout mixture (from instability to difficult pumping and injection). Due to the bentonite characteristics, considerably higher doses of bentonite were used. The particular dosages used during a single-stage preparation process are shown in Table 3.

The erosional stability in the case of a grout mixture based on non-activated bentonite (see Table 3 and Figure 8) appears to be significantly higher (i.e., the WL is lower) than that of a grout mixture based on activated bentonite. Using non-activated bentonite or increasing its concentration in the grout mixture is a simple solution that can be easily implemented in practical applications.

The inlet side of the specimens before (a) and after the test (b and c) are shown in Figure 8. The vertical cross section of the grout mixture B5 specimen after the test is shown in Figure 9. The contour of the hole before and after the test is marked as in the case of the grout mixture based on Bentovet K. It is apparent that erosion only occurs at the inlet edge of the hole of the specimen. However, in the case of grout mixtures based on activated bentonite (see, e.g., Figure 7), there is a significant distortion of the hole along the entire specimen length.

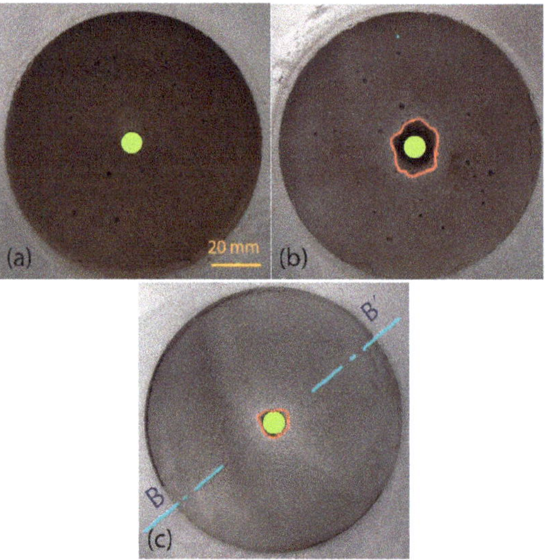

Figure 8. Inlet side of specimens with Bentonit 75: (**a**) grout mixture B1 before the test; (**b**) grout mixture B1 after the test—WL = 1.4%; (**c**) grout mixture B5 after the test—WL = 0.2% (the edge of the failure area in red, the initial hole in green).

This can be explained by the fact that the suspension has significantly higher strength after six hours. This is due to the significantly lower water/cement ratio compared to the grout mixtures with activated bentonite (e.g., 2.35 for grout mixture B1 or 2.48 for grout mixtures K3; see also composition of grout mixtures in Tables 2 and 3).

Figure 9. Specimen cross section B–B' after the test (see Figure 8c)—grout mixture B5 based on Bentonit 75 (the edge of the failure area in red, the initial hole and inlet side in green).

4.3. Use of Additives to Enhance Grout Mixtures' Thixotropy (Example Three)

To increase the yield stress and thixotropy of the grouts, cellulose derivatives can be used. In addition to the formerly obligatory carboxy-methyl-cellulose (CMC), a number of modified derivatives are available today that have improved dispersibility and solubility in water, with varying efficiencies. We used Culminal C9133 cellulose, which is a methyl-hydroxy-propylene cellulose with a viscosity of the standard (2%) solution in water of 4500–6500 mPa.s. It is suitable for application in systems with a large number of dispersed substances.

Several variants of grout mixtures based on Bentovet K and Bentonit 75 were prepared with different concentrations of cellulose in the grout mixture (0.5% and 0.75% of the cement weight). The cement dosage in all tested grout mixtures was constant; again, 350 kg/m^3 of the grout mixture.

When cellulose is added to the grout mixture, there is an expected increase in the yield stress and thixotropy. It has a positive effect on the erosional stability of the grout mixture as it increases substantially (Figures 10 and 11). Concerning grout mixture on the basis of activated bentonite, there was a verified increase by a factor of 1.7 in the erosional stability after 0.5% of cellulose was added (a decrease in the WL value from 4.7% to 2.7% for the K1 grout mixture) and an increase by a factor of 2.1 in the erosional stability (a decrease in the WL value from 7.1% to 3.4%) for the K2 grout mixture (single-stage process of the grout mixture preparation, designated as Kb in Figure 10).

Regarding grout mixtures on the basis of non-activated bentonite, there is no such increase in erosional stability as that in grout mixtures based on activated bentonite. There was an increase by a factor of 1.3 in the erosional stability of the grout mixture after 0.5% cellulose was added to the B1 grout mixture. It corresponded to a decrease in the WL value from 1.4% to 1.1%.

There was an increase by a factor of 2.0 in the erosional stability of the grout mixture after 0.75% cellulose was added to the B1 grout mixture. This corresponded to the decrease in the WL value from 1.4% to 0.7%.

The stabilizing effect of these molecular chains is temporary; therefore, these measures should be combined with those ensuring rapid solidification of grout mixtures (e.g., the use of non-activated instead of activated bentonite).

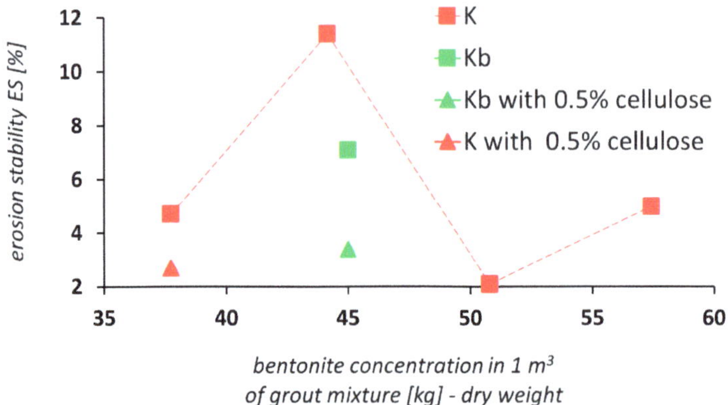

Figure 10. Weight loss of grout mixture (WL) vs. bentonite concentration: grout mixtures based on Bentovet K (K—two-stage process of grout mixture preparation, Kb—single-stage process of grout mixture preparation).

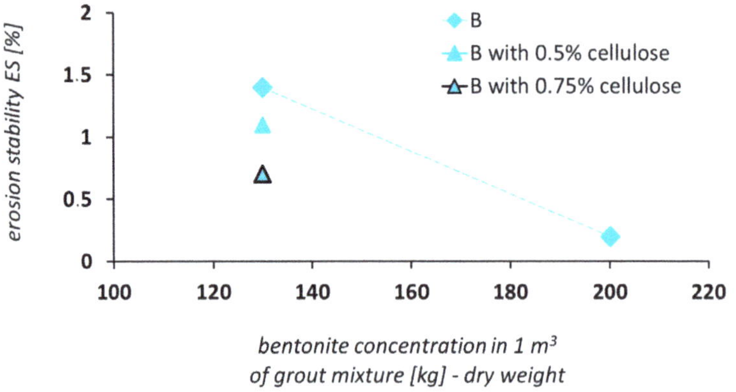

Figure 11. Grout mixture based on Bentonit 75: weight loss of grout mixture (WL) vs. bentonite concentration.

5. Conclusions

The erosional stability test is one way to allow for the evaluation of medium-term resistance of a grout mixture to the mechanical influence of flowing water. The authors are not aware of any published papers and results on erosion tests of silicate grout mixtures. From this point of view, we can consider these experimental works as the first of their kind.

A new laboratory apparatus for performing the test was presented here. The apparatus was designed for the purpose of comparatively evaluating grouts' erosional resistances:

- The principal and subsequent evaluation of the erosional resistance was based on the study of Verfel (1992);
- Erosional stability was quantified by the relative weight loss of a grout (WL) during the test. Grout mixture, which is resistant to erosion, has a WL value equal to 0. This is a condition for real practical use;
- The apparatus is useful for practical use in ordinary laboratory conditions due to its operational and structural simplicity;
- This also allows a visual assessment of the extent of the erosional impact on a specimen. For example, only the inlet edge of the hole in the specimen is impacted, etc.;
- The enhanced apparatus design allows, if necessary, for erosion stability determination for different boundary conditions than those specified by Verfel (1992), or those that

selected during the performed tests. In particular, there is the possibility of water flow control and an option to choose the grout mixture specimen height in the test cell. Alternatively, a simple irreversible modification (enlargement) of the hole in the test cell (grout specimen) is possible.

In addition, the use of this apparatus was demonstrated in some tests. The erosional stabilities of the grout mixtures were evaluated 6 h after production. It was confirmed that the choice of bentonite type, adjusting a bigger dose of non-activated bentonite, or the use of cellulose in the grout mixture may lead to a change in the grout mixture erosional stability. Based on the results presented in the paper, it can be stated that:

- A substantial increase in the erosional stability of grout mixture was reached by replacing activated bentonite with non-activated bentonite. It appeared that there was a decrease in WL value even to 0.2, i.e., the value was very close to 0;
- By increasing the dose of non-activated bentonite in the grout mixture, the erosional resistance of the grout mixture was increased. A 54% increase in the bentonite dose led to decrease of WL of seven times. Using non-activated bentonite (alternatively an increase of its concentration in the grout mixture) is a measure, which does not lead to a more complex composition of the grout mixture (or it does not require more complex preparation process of the grout mixture). Such a solution can be easily implemented in practical applications;
- Concerning the grout mixture on the basis of activated bentonite, the erosional stability seems not to correlate with the bentonite dose used (within the tested range). The value gains were in the range between 2.1 and 11.4;
- The addition of cellulose to the prepared grout mixture results in a positive effect on the erosional stability, i.e., there was a decrease in WL. The 0.5% cellulose concentration in the grout mixture led to a decrease of WL of 2.1 times. However, the stabilizing effect of cellulose was temporary; therefore, these measures should be combined with those ensuring rapid solidification of grout mixtures.

Author Contributions: Conceptualization, I.T., J.B. and L.M.; methodology, I.T. and J.B.; investigation, J.B., I.T., L.M., M.D. and A.L.; writing—original draft preparation, J.B., M.D., and I.T.; supervision, I.T., J.B. and L.M. All authors have read and agreed to the published version of the manuscript.

Funding: The article was written with the financial support of the state budget through the Ministry of Industry and Trade of the Czech Republic within the project No. FR-TI4/261: "Research into the behavior of grout mixtures when applied below the subterranean water, development of new grout mixture, the methodology of its design, implementation and monitoring method" and under the project No. LO1408: "AdMaS UP—Advanced Materials, Structures and Technologies", which is supported by the Ministry of Education, Youth and Sports of the Czech Republic under the "National Sustainability Programme I".

Institutional Review Board: Not applicable.

Informed Consent Statement: Not applicable.

Data Availability Statement: Data are contained within the article.

Conflicts of Interest: The authors declare no conflict of interest.

References

1. Verfel, J. *Injektování Hornin a Výstavba Podzemních Stěn [Rock Grouting and Diaphragm Wall Construction]*; MÚS Bradlo: Bratislava, Czech Republic, 1992; p. 551, ISBN 80-7127-043-1. (In Czech)
2. Warner, J. *Practical Handbook of Grouting: Soil, Rock, and Structures*; John Wiley & Sons: Hoboken, NJ, USA, 2004.
3. Baluch, K.; Baluch, S.; Yang, H.-S.; Kim, J.-G.; Kim, J.-G.; Qaisrani, S. Non-Dispersive Anti-Washout Grout Design Based on Geotechnical Experimentation for Application in Subsidence-Prone Underwater Karstic Formations. *Materials* **2021**, *14*, 1587. [CrossRef] [PubMed]
4. Benahmed, N.; Bonelli, S. Investigating Concentrated Leak Erosion Behaviour of Cohesive Soils by Performing Hole Erosion Tests. *Eur. J. Environ. Civ. Eng.* **2012**, *16*, 43–58. [CrossRef]

5. Liang, Y.; Yeh, T.-C.J.; Ma, C.; Zhang, J.; Xu, W.; Yang, D.; Hao, Y. Experimental Investigation on Hole Erosion Behaviors of Chemical Stabilizer Treated Soil. *J. Hydrol.* **2020**, 125647. [CrossRef]
6. Wang, H.; Liu, Q.; Sun, S.; Zhang, Q.; Li, Z.; Zhang, P. Damage Model and Experimental Study of a Sand Grouting-Reinforced Body in a Seawater Environment. *Water* **2020**, *12*, 2495. [CrossRef]
7. Arulanandan, K.; Perry, E.B. Erosion in Relation to Filter Design Criteria in Earth Dams. *J. Geotech. Eng.* **1983**, *109*, 682–698. [CrossRef]
8. Terzaghi, K.; Peck, R.B.; Mesri, G. *Soil Mechanics in Engineering Practice*, 3rd ed.; John Wiley & Sons, Inc.: New York, NY, USA; Chichester, UK, 1996; pp. 549, 592. ISBN 978-0-471-08658-1.
9. Suits, L.D.; Sheahan, T.; Wan, C.; Fell, R. Laboratory Tests on the Rate of Piping Erosion of Soils in Embankment Dams. *Geotech. Test. J.* **2004**, *27*, 1–9. [CrossRef]
10. Wan, C.F.; Fell, R. Investigation of Rate of Erosion of Soils in Embankment Dams. *J. Geotech. Geoenviron. Eng.* **2004**, *130*, 373–380. [CrossRef]
11. Moore, W.L.; Masch, F.D. Experiments on the Scour Resistance of Cohesive Sediments. *J. Geophys. Res. Space Phys.* **1962**, *67*, 1437–1446. [CrossRef]
12. Arulanandan, K.; Gillogley, E.; Tully, R. *Development of a Quantitative Method to Predict Critical Shear Stress and Rate of Erosion of Natural Undisturbed Cohesive Soils (Technical Report GL-80-5)*; University of California: Davis, CA, USA, 1980; p. 99.
13. Lefebvre, G.; Rohan, K.; Douville, S. Erosivity of Natural Intact Structured Clay: Evaluation. *Can. Geotech. J.* **1985**, *22*, 508–517. [CrossRef]
14. Drnevich, V.; Rohan, K.; Lefèbvre, G.; Douville, S.; Milette, J.-P. A New Technique to Evaluate Erosivity of Cohesive Material. *Geotech. Test. J.* **1986**, *9*, 87. [CrossRef]
15. Briaud, J.L.; Ting, F.C.K.; Chen, H.C.; Cao, Y.; Han, S.W.; Kwak, K.W. Erosion Function Apparatus for Scour Rate Predictions. *J. Geotech. Geoenviron. Eng.* **2001**, *127*, 105–113. [CrossRef]
16. Hanson, G.J.; Cook, K.R. Apparatus, Test Procedures, and Analytical Methods to Measure Soil Erodibility in Situ. *Appl. Eng. Agric.* **2004**, *20*, 455–462. [CrossRef]
17. Shugar, D.; Kostaschuk, R.; Ashmore, P.; Desloges, J.; Burge, L. In Situ Jet-Testing of the Erosional Resistance of Cohesive Streambeds. *Can. J. Civ. Eng.* **2007**, *34*, 1192–1195. [CrossRef]
18. Mahalder, B.; Schwartz, J.S.; Palomino, A.M.; Zirkle, J. Estimating Erodibility Parameters for Streambanks with Cohesive Soils Using the Mini Jet Test Device: A Comparison of Field and Computational Methods. *Water* **2018**, *10*, 304. [CrossRef]
19. Regazzoni, P.-L.; Marot, D. Investigation of Interface Erosion Rate by Jet Erosion Test and Statistical Analysis. *Eur. J. Environ. Civ. Eng.* **2011**, *15*, 1167–1185. [CrossRef]
20. Sanchez, R.L.; Strutynsky, A.I.; Silver, M.L. *Evaluation of the Erosion Potential of Embankment Core Materials using the Laboratory Tri-Axial Erosion Test Procedure (Technical Report GL-83-4)*; Army Engineers Waterways Experiment Station: Viksburg, MS, USA, 1983; p. 335.
21. Bendahmane, F.; Marot, D.; Rosquoet, F.; Alexis, A. Characterization of Internal Erosion in Sand Kaolin Soils. *Revue Eur. De Génie Civ.* **2006**, *10*, 505–520. [CrossRef]
22. Bendahmane, F.; Marot, D.; Alexis, A. Experimental Parametric Study of Suffusion and Backward Erosion. *J. Geotech. Geoenviron. Eng.* **2008**, *134*, 57–67. [CrossRef]
23. Suits, L.D.; Sheahan, T.C.; Richards, K.S.; Reddy, K.R. True Triaxial Piping Test Apparatus for Evaluation of Piping Potential in Earth Structures. *Geotech. Test. J.* **2010**, *33*, 1–13. [CrossRef]
24. Monnet, J.; Plé, O.; Nguyen, D.M.; Plotto, P. A New Test for the Characterization of Suffusion into Embankment and Dam. In Proceedings of the 6th International Conference on Scour and Erosion (ICSE-6), Paris, France, 27–31 August 2012; pp. 1073–1080.
25. Monnet, J.; Plé, O.; Nguyen, D.M.; Plotto, P. *Characterization of the Susceptibility of the Soils to Internal Erosion, In Geotechnical and Geophysical Site Characterization 4*, 1st ed.; Coutinho, R.Q., Mayne, P.W., Eds.; Taylor & Francis Group: London, UK, 2013; pp. 765–770, ISBN 978-0-415-62136-6.
26. ÖNORM B 4452. *Erd-Und Grundbau Dichtwände im Untergrund [Geotechnical Engineering/Foundation Engineering—Cut-off Walls]*; Österreichisches Normungsinstitut: Wien, Austria, 1998. (In German)

 *materials*

Article

Extended Residual-State Creep Test and Its Application for Landslide Stability Assessment

Deepak R. Bhat [1], Janusz V. Kozubal [2] and Matylda Tankiewicz [3,*]

[1]   Engineering Division, Okuyama Boring Co. Ltd., Tokyo 103-0004, Japan; deepakrajbhat@gmail.com
[2]   Faculty of Civil Engineering, Wrocław University of Science and Technology, 50-370 Wrocław, Poland; janusz.kozubal@pwr.edu.pl
[3]   Department of Building Engineering, Wrocław University of Environmental and Life Sciences, 50-375 Wrocław, Poland
*   Correspondence: matylda.tankiewicz@upwr.edu.pl

Abstract: This paper contains the results of a newly developed residual-state creep test performed to determine the behavior of a selected geomaterial in the context of reactivated landslides. Soil and rock creep is a time-dependent phenomenon in which a deformation occurs under constant stress. Based on the examination results, it was found that the tested clayey material (from Kobe, Japan) shows tertiary creep behavior only under shear stress higher than the residual strength condition and primary and secondary creep behavior under shear stress lower or equal to the residual strength condition. Based on the data, a model for predicting the critical or failure time is introduced. The study traces the development of the limit state based on the contact model corresponding to Blair's body. The time to occurrence of the conditions necessary for unlimited creep on the surface is estimated. As long-term precipitation and infiltrating water in the area of the landslides are identified as the key phenomena initiating collapse, the work focuses on the prediction of landslides with identified surfaces of potential damage as a result of changes in the saturation state. The procedure outlined is applied to a case study and considerations as to when the necessary safety work should be carried out are presented.

Keywords: residual-state creep; saturation front; landslides

Citation: Bhat, D.R.; Kozubal, J.V.; Tankiewicz, M. Extended Residual-State Creep Test and Its Application for Landslide Stability Assessment. *Materials* **2021**, *14*, 1968. https://doi.org/10.3390/ma14081968

Academic Editor: Angelo Marcello Tarantino

Received: 27 January 2021
Accepted: 12 April 2021
Published: 14 April 2021

Publisher's Note: MDPI stays neutral with regard to jurisdictional claims in published maps and institutional affiliations.

Copyright: © 2021 by the authors. Licensee MDPI, Basel, Switzerland. This article is an open access article distributed under the terms and conditions of the Creative Commons Attribution (CC BY) license (https://creativecommons.org/licenses/by/4.0/).

1. Introduction

Landslide processes pose a real threat to engineering structures. Reasonably accurate prediction of failure allows the application of protective actions, thus avoiding human casualties and reduction in property damage. Many methods have been developed for spatial analysis of the phenomenon from a mechanical point of view, supported by a regional reliability approach [1–3]. The transfer of risk management to the level of geographic information system (GIS) tools has been accompanied by methods of building risk maps, supporting the assessment and management of spatial phenomena [4,5]. A different direction of research focused on describing the dependence of slope stability over time. In such a case, the mechanical properties of the substrate or contact layer were related to the time of action of forces causing slippage [6] or environmental factors, especially, for example, rainfall [7]. The selected aspect of rainfall action and stable infiltration conditions in previously partially saturated soil and its influence on landslide initiation have been analyzed by many authors [8–10]. The regional general scale of the issue based on the Italian experience is analyzed in Ref. [11]. A comprehensive review of different methods for predicting landslide failure time can be found in Ref. [12]. The connection of physical phenomena with prediction of stability changes and possible threat gives a chance to create early warning systems for rapture damage occurrence [13,14]. The methodology used in warning systems includes echoes of acoustic waves, statistical observations and spatial and temporal variability of regional landslides, as well as changes in pore pressure measured by sensors.

In issues of slope stability, the creep of geomaterials is especially important [15]. In general, landslides are created more or less rapidly depending on the conditions. Creep is one of various types of landslides, characterized by an imperceptibly slow, steady, downward movement and is found mainly in clayey soils. Soil creep as a physical phenomenon is a time-dependent effect in a which deformation occurs under constant stress. Theoretically, an ideal creep curve consists of three stages that have different deformation properties according to the shape of the strain–time curve. Primary creep, the so-called transient or fading phase, can be defined as a creep deformation during which the strain rate decreases continuously with time (i.e., decreasing strain rate). Secondary creep, the secondary phase, consists of deformations at a constant rate (i.e., constant strain rate), which is sometimes also called the non-fading phase. With tertiary creep, the deformation is continuously increasing and leads to creep failure (i.e., increasing and accelerating strain rate). A widely acknowledged concept of creep distinguishing between the different phases of creep movement was discussed in Refs. [16,17]. In the primary and secondary stages of creep, soil materials are in a stable condition, but they may collapse when reaching the tertiary stage of creep. Many extensive laboratory investigations have been conducted with triaxial apparatus and oedometers to examine the creep behavior of various kinds of soils [18–20], but almost all of them focus only on the pre-peak-state creep behavior of soil materials. Numerous theories, equations and models have been proposed to account for the creep behavior of clayey soils [21–23] and various geotechnical issues [24,25]. However, they have also focused on the pre-peak state of shear stress. Nevertheless, since the long-term strength of the shear zone soil of a reactivated slow-moving landslide is nearly equivalent to its residual strength [26], a creep test should be conducted at the residual state of shear stress for understanding the actual creep characteristics of clay soil materials.

The issue of slope stability can be analyzed in both ultimate and serviceability states [27]. In the case of a material with rheological features and the existence of a clear yield point, correct description of the phenomena requires spatial recognition of the variability of soil features. Many researchers have been interested in predicting creep failure. However, capturing the extremely slow movement of large-scale landslides through instrumentation is very difficult. Moreover, if a small number of field monitoring data are available, the actual trend of the movement variation with time will not be clear. On the other hand, standard laboratory tests are not feasible for representing long-term behavior, and long-term tests are not easy to perform. An estimation of damage time based on a clay creep study was introduced in Ref. [28]. Afterwards, it was extended in Ref. [29] to all subsequent areas of material work in the creep regime. The first successful attempts to describe the phenomena of tertiary creep in stability applications were introduced in Refs. [30,31] and later developed in Ref. [32]. There have been also attempts to perform a spatial description of creep phenomena affecting landslides by means of numerical models [33,34].

The main objective of this study was to deal with the above mentioned issues by means of an extended residual-state creep test. Based on this, an attempt to develop a method of predicting landslide displacement was made. The work focuses on the problem of landslides with identified surfaces of potential damage. Long-term precipitation and infiltrating water in the vicinity of the landslide area are identified as the key initiating phenomena for the considered type of landslides (creep). Based on the residual-state creep test results of typical clayey soils, prediction curves were proposed, and the time to occurrence of the conditions necessary for unlimited creep on this surface was estimated. In the description of the material, the fractional derivative theory has been used [35–37]. It is a form of a well-functioning fit to test results involving a two-layered model based on Blair-type elements. It gives good fitting results for a small amount of experimental data, while reducing the number of initial assumptions. Numerous articles describing fractional derivatives [38,39] and a wider interest in their possibilities in mechanics [40–43] have been introduced. The presented concept of the critical time related to the extended laboratory tests is not deterministic. The approach allows for estimating the stability of a rock or soil block with the defined slip surface, taking into account the expected time of its operation and the change in the indicator with the climate's influence. Loss of equilibrium was connected with changes in the movement of the saturation front (negative pressure in the profile).

2. Materials and Methods

2.1. Creep Tests—Creep Test Apparatus and Test Methods

The standard ring shear machine allows for continuous shearing of soil samples up to very large shear deformations in one direction without a change in the geometry of the specimens [44,45]. It is used mainly to investigate the residual strength of soils [46–48]. However, it is not suitable for creep tests at the residual state of shear. In order to identify creep behavior, it is necessary to modify the standard version of the apparatus that was introduced by Bhat et al. [49–51]. The modification is mainly based on the transitional change of strain-controlled shear into creep load shearing without completely releasing the applied shear stress, as shown in Figure 1. In other words, the machine has been modified in such way that it can shear clayey soil material in strain-controlled as well as stress-controlled patterns under drained conditions. In the beginning, the material is sheared under a strain-controlled pattern, and after the specimen reaches its residual state of shear, different sets of constant creep loads are applied until it fails repeatedly. The modified ring shear machine enables measurement of the shear deformation (i.e., displacement) with respect to time under the applied constant creep stress [52,53].

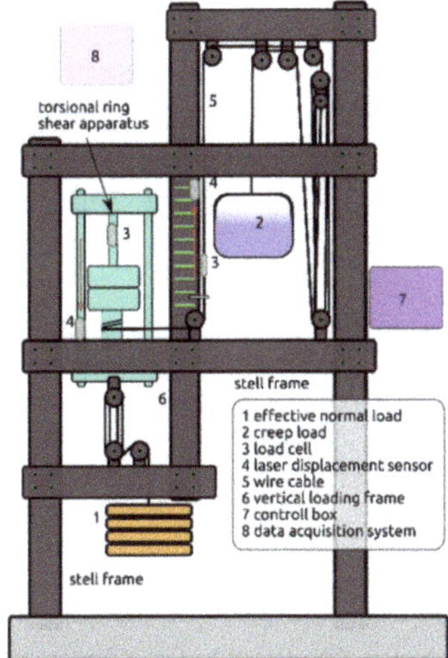

Figure 1. Structural features of creep test apparatus.

In the residual-state creep test, there are two main steps: (1) the ring shear test and (2) the residual-state creep test. The test scheme is shown in Figure 2. The ring shear test is performed to obtain the residual state of shear of fully saturated specimens. This state is confirmed when constant values of readings for the load cell and displacement sensors after a large displacement are achieved. The details of the ring shear test that is related to this study have been discussed by Bhat et al. [54,55]. Next, in the residual-state creep test, the creep stress is initially applied at a certain value of the residual creep stress ratio (RCSR). The value of RCSR is the ratio of the applied constant creep stress to the residual strength of soil. The specimen is maintained for several hours in the same condition to determine whether the effect of the creep behaviors is significant or not. Similarly, the creep load is applied accordingly to the subsequent load factor values until the specimen fails.

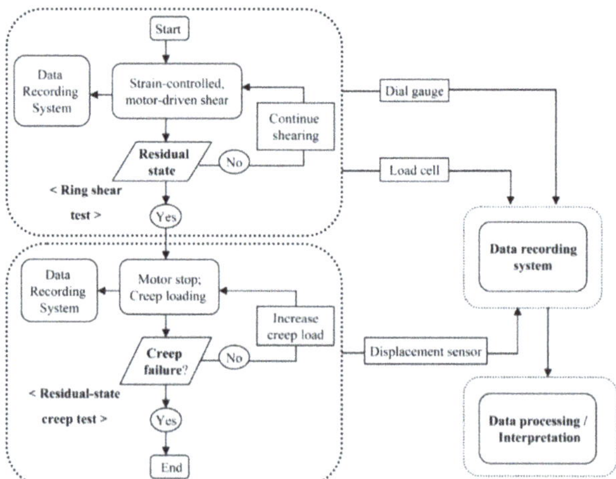

Figure 2. Overall experimental flow of the residual-state creep test.

2.2. Creep Test on Clayey Soils

In this study, a clayey soil sample collected from the Toyooka-kita landslide area is considered. This is one of the major reactivated landslides, which is located at Yokawa-cho of the city Miki in Hyogo Prefecture, Japan. The details of the study area, such as the topographic map, the geological map, location sampling points, etc., are presented in Ref. [55]. A series of laboratory tests were performed to estimate the physical properties and mechanical properties according to the Laboratory Testing Standards of Geomaterials [56–59]. The specific gravity of the tested soil was 2.67 g/cm^3. A plastic limit of 47%, liquid of 81% and, consequently, a plasticity index of 34% were obtained. The specimen consisted of a 21.3% clay fraction (<2 μm), 52.8% silt fraction (2–75 μm) and 25.9% sand fraction (75–425 μm). X-ray diffraction tests confirmed the presence of clay minerals in the tested samples and revealed that smectite is a major clay mineral.

The samples formed in the laboratory were used in examinations, and the precise procedure is described in Ref. [52]. Specimens were tested in accordance with the methodology described in Section 2.1. First, the ring shear test was performed to obtain the residual-state of shear. In each case, it was achieved after 10 cm of shear displacement. For certainty, the ring shear tests were conducted up to 15 cm of shear displacement. The value of the effective stress was fixed at a constant (i.e., 98.10 kPa) for both the ring shear test and the creep tests. With this load, the soil was in an over-consolidated state with an overconsolidation ratio (OCR) value of 2. The residual shear strength of the tested soil was 8.86 kPa. After achieving residual strength, the constant creep stresses were applied step by step until the specimen reached failure. In this work, the sequence of RCSR values from 0.9000 to 1.0300 was used. The concept and overall experimental procedure for one complete test pattern (Test I) are presented in Figure 3, and the results are summarized in Table 1. In this study, different stages of creep were defined based on the change in displacements. In the primary stage of creep, the change in displacement decreases. When the change in the displacement remains constant, it is called the secondary stage of creep. In the tertiary stage of creep, the change in displacement is suddenly increased, which leads to failure. In Table 1, the value t_1 represents the total time at the end of the primary stage of creep (i.e., beginning of secondary creep) and δ_1 is the corresponding displacement. Similarly, the values t_f and δ_c represent the total time at the end of secondary creep (i.e., beginning of tertiary creep) and the total displacement, respectively (as shown in Figure 3c). These are also called the failure time and the critical displacement. A "failure" indicates that the specimen has reached tertiary creep, and "no failure" represents secondary creep. In the case of Test I, the sample reached failure at RCSR 1.0025. In subsequent tests, the sample was subjected to subsequent loads up to failure. A compilation of the procedures for Tests II-X is shown in Figure 4. A summary of the creep test of a clay soil is presented in Figure 5. The test results confirmed that the tested soils exhibit the secondary stage of creep when RCSR $\leq$ 1.0 and the tertiary stage of creep for RCSR > 1.0 (Figure 5).

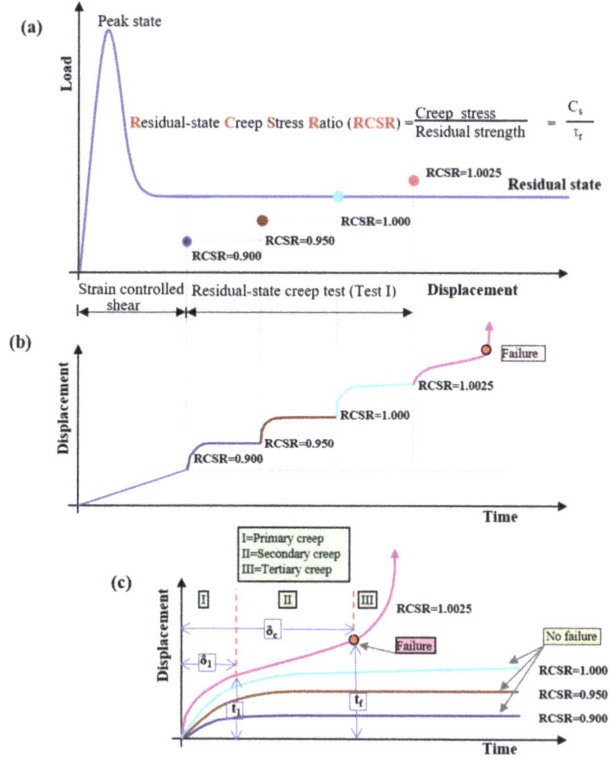

Figure 3. Overall experimental procedure for Test I: (**a**) application of constant creep load for residual-state creep stress ratio (RCSR) values from 0.9000 to 1.0025; (**b**) automatically recorded data from the creep test; (**c**) analysis results for the interpretation of different creep stages.

Table 1. Residual state creep tests summary for Test I.

Test No.	RCSR	t_1 (s)	δ_1 (mm)	t_f (s)	δ_c (mm)	Remarks
I (1)	0.9000	3526	2.3672	25,692	2.3045	No failure
I (2)	0.9500	6872	2.8793	26,570	2.8840	No failure
I (3)	1.0000	8338	3.2688	28,844	3.3599	No failure
I (4)	1.0025	18,184	3.7340	54,308	3.8970	Failure

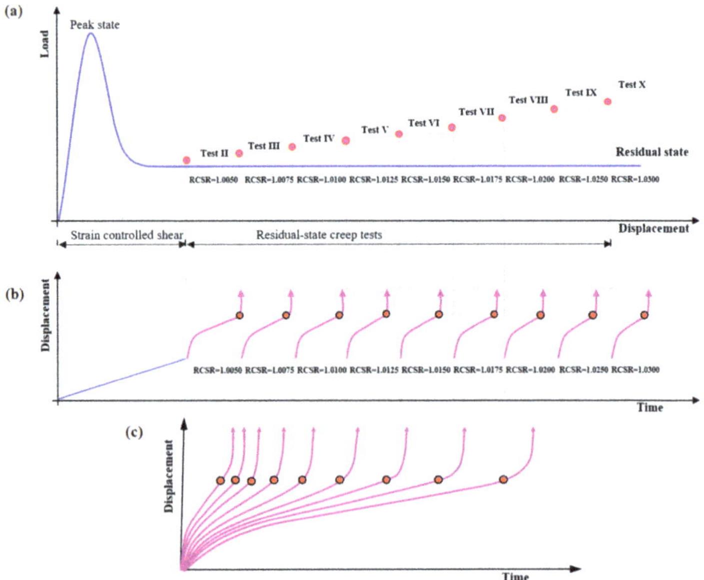

Figure 4. Overall experimental procedure for Tests II-X: (**a**) application of constant creep load for RCSR values from 1.0050 to 1.0300; (**b**) automatically recorded data from the creep tests; (**c**) analysis results for the interpretation of different creep stages.

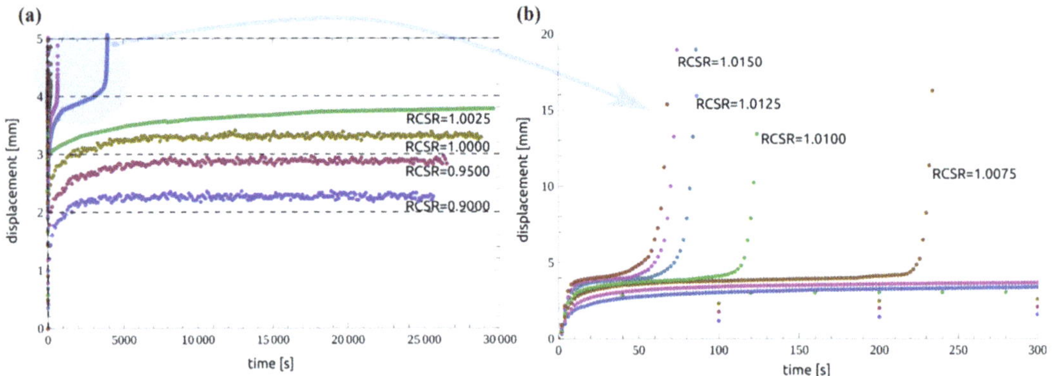

Figure 5. Summary of creep tests for RCSR up to 1.0150: (**a**) complete time range of 0–30,000 s; (**b**) initial time of experiment of 0–300 s.

2.3. Fractional Integrals and Derivatives

To describe the creep clay behavior, a rheological model based on fractional derivatives was used. The model contains three material parameters. A brief introduction to the fractional derivative account was limited to the definition of Caputo for the sake of clarity. The necessary prerequisite is the gamma function:

$$\Gamma(z) = \int_0^\infty e^{-t} t^{z-1} dt, \tag{1}$$

where the variable z belongs to real numbers. Selected values of this function are as follows (where n is a natural):

$$\Gamma\left(\frac{1}{2}\right) = \sqrt{\pi}, \Gamma\left(\frac{1}{2}+n\right) = \frac{(2n)!\sqrt{\pi}}{4^n n!}. \tag{2}$$

Another important dependence is the Laplace representation of a function derivative:

$$\mathcal{L}\left\{\frac{d^n y(t)}{dt^n}, t, s\right\} = s^n Y(s) - \sum_{k=1}^{n} s^{n-k} y^{(k-1)}. \tag{3}$$

The fractional derivative Caputo function $y(t) : [0, T] \to R$ for the physical case in $[t_1, t_2, t_3]$ when $\alpha_1 \in [0, 1]$ and $n_c = \text{ceil}(\alpha_1)$ has the following form:

$$^{C}D_a{}^{\alpha_1}y(t) = \frac{1}{\Gamma(n_c - \alpha_1)} \int_a^t (t - \tau)^{n_c - \alpha_1 - 1} y^{n_c}(\tau)d\tau, \tag{4}$$

In the case of fractional derivatives (also called differintegral), integration and differentiation are inverse operations:

$$^{C}_{\alpha_1}D^{\alpha_1}y(t) = {}_{\alpha_1}I^{n-\alpha_1}y^{(n)}(t) \tag{5}$$

Solving the differential equation of the model with the operator method (Laplace transform $\mathcal{L}\{\cdot\}$), it is necessary to determine the fractional derivative transform:

$$\mathcal{L}\left\{ {}^{C}D_a{}^{\alpha_1}y(t), t, s\right\} = s^{\alpha_1}Y(s) - \sum_{\kappa=1}^{n_c} s^{\alpha_1 - \kappa} y^{\kappa-1}(0). \tag{6}$$

The classical approach to modeling real bodies is based on using the basic elements as for a perfectly elastic body (Hooke's law), where the relationship between strains and stresses is described by the following:

$$\sigma(t) = E \in (t), \tag{7}$$

and for Newtonian liquid by the following:

$$\sigma(t) = \eta d \in (t), \tag{8}$$

where E and η are material constants. The physical body is characterized by both elastic and viscosity features. The combination of these can be modeled by means of basic element systems, e.g., as commonly used Maxwell's fluid, Voight's body and standard models. A number of variants have been developed and described in the literature. The proposed model has an element that mixes both relations of pure elasticity with viscosity. The introduced concept of fractional derivatives allows for simple description of a combination of these two features in the basic Blair element as follows:

$$\sigma(t) = X_{\alpha_1} \cdot {}^{C}D^{\alpha_1} \in (t), \tag{9}$$

where $^{c}D^{\alpha_1}$ is a fractional derivative of Caputo for $a = 0$. The value of X_{α_1} is the constant coefficient accounting for the dimensional relationship between the material constant X and the order of the derivative α_1. Equation (9) gives a perfectly elastic body for $\alpha_1 = 0$:

$$\sigma(t) = X_{\alpha_1} \cdot {}^{C}D^0 \in (t), \tag{10}$$

where $X_{\alpha_1} = E$; and a Newtonian liquid for $\alpha_1 = 1$:

$$\sigma(t) = X_{\alpha_1} \cdot {}^{C}D^1 \in (t), \tag{11}$$

where $X_{\alpha 1} = \eta$. For $\alpha_1 \in \left(0, \frac{1}{2}\right)$, the body has predominantly elastic properties, and when $\alpha_1 \in \left[\frac{1}{2}, 1\right)$, it has viscous properties. Hence, the relaxation module for the single time step based on springpot body has the following form:

$$G(t) = \frac{E}{\Gamma(1 - \alpha_1)} \left(\frac{t}{\tau}\right)^{-\alpha_1}, \tag{12}$$

where $\tau = \frac{\eta}{E}$, and the creep module is its reverse:

$$J(t) = G(t)^{-1}. \tag{13}$$

The adjustment of creep function parameters to experimental data involves determining the mechanical parameters E and η and the fractional derivative value α_1 with the least square method.

2.4. Filtering and Fitting Calculation Results

In order to establish the time to reach the third stage of creep, the data from the experiment were analyzed using the signal filtering method. When evaluating the measurements in the second creep phase, discrete values of displacement readings were found as noise around local average values. The reason is the limited resolution of the AC/DC sensor. In order to normalize the data and to eliminate reading errors, a number of filtering tests were carried out. For further calculations, the Wiener filter was used. The results obtained are presented in the following figures. Figure 6 shows datasets (RCSR = 1.000, 0.9500, 0.9000) with mean, median and variance values and filtered values. Figure 7 compares the raw data with the filtered ones, and Figure 8 presents a compilation of these values with the model. As can be seen, the approach presented is adequate for describing the residual-state creep test results presented in the article.

The maximum displacement depends on RCSR, where for RCSR > 1, the value x_{lim} means the appearance of a sudden non-linear displacement increase. The formula was fitted in a nonlinear procedure, resulting in the following forms:

$$x_{lim} = -0.1079 + 0.1079 \, e^{3.4996 \, \text{RCSR}} \quad \text{RCSR} \leq 1, \tag{14a}$$

$$x_{lim} = -11.59 + 15.38 \, \text{RCSR} \quad \text{RCSR} > 1. \tag{14b}$$

Hence, the time to reach the state of tertiary stage is presented in the form of an exponential function:

$$t_c = 101.768 + 0.0001088 \, e^{-1144.32(-1.02 + \text{RCSR})} \tag{15}$$

which is presented together with the laboratory results in Figure 9.

The results obtained led to the description of the creep phenomenon for the investigated soil. A representation of the material as a contour map of the creep zones is shown in Figure 10, where the zones are separated from each other by colors. Zone III was classified as a prohibited area due to the very short duration to material rupture.

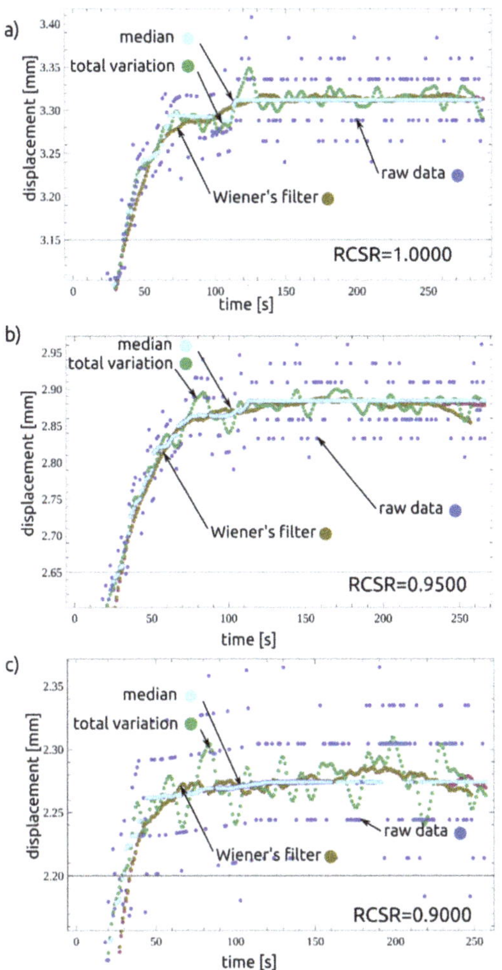

Figure 6. Effect comparison of filters, prepared for (**a**) RCSR = 1.0000, (**b**) RCSR = 0.9500 and (**c**) RCSR = 0.9000.

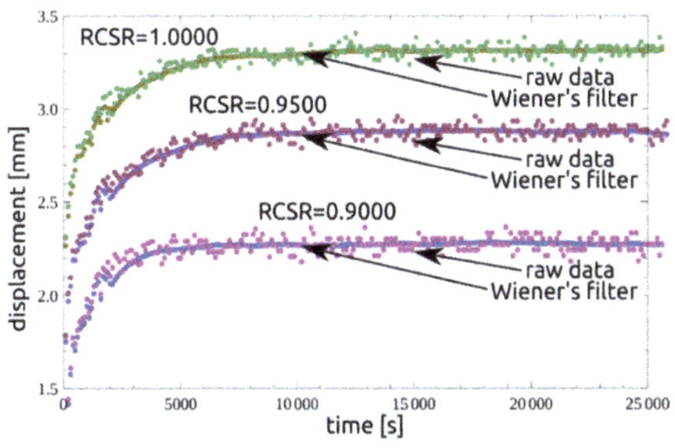

Figure 7. Wiener's filter based on raw data for RCSR = {0.9000, 0.9500, 1.0000}.

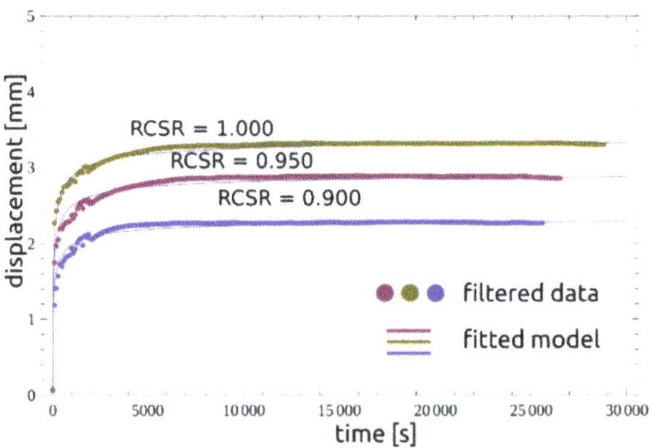

Figure 8. Comparison between the fitted Blair model and filtered data for RCSR = {0.9000, 0.9500, 1.0000}.

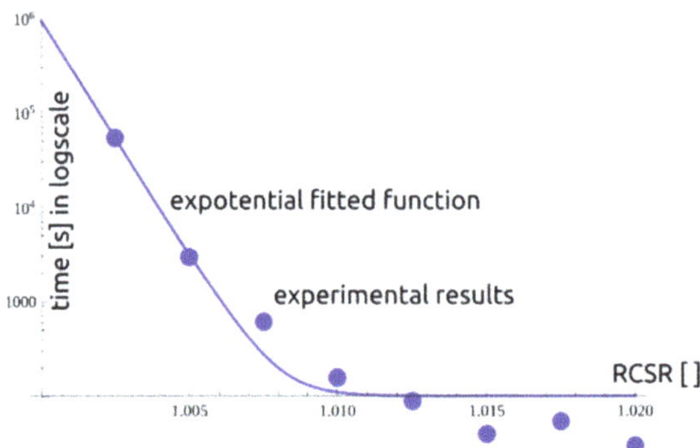

Figure 9. The dependency of RCSR vs. time.

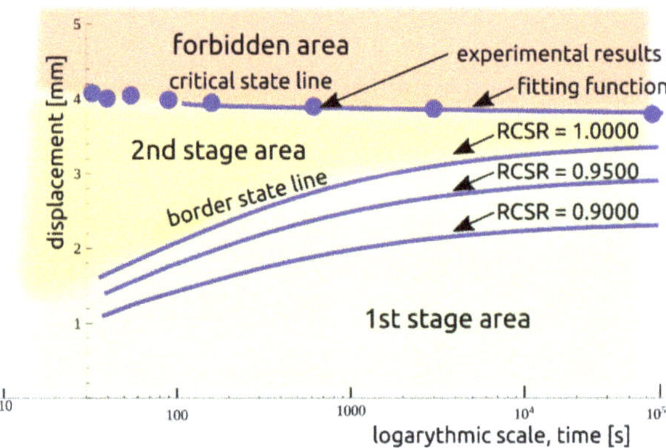

Figure 10. The contour map of creep phenomenon areas.

2.5. Water Flow in Unsaturated Substrate

The outcomes from the previous sections were used to estimate the time needed for the landslide to lose stability as a result of changes in the water content of the identified layer, e.g., due to long-term precipitation. As the water content increases, the cohesion of the partially saturated material decreases exponentially, following the concept of Matsushi and Matsukury [60]. In the paper, an abrupt loss of cohesion was assumed with the contact surface reaching full saturation θ_s. In this case, the strength parameters are reduced to the minimum values, corresponding to the saturation state as in the experimental procedure described in the previous subsections.

The work adopts the description of water flow in unsaturated soil according to the Buckingham Darcy law, implemented in Richard's equation:

$$\frac{\partial \theta}{\partial t} = \frac{\partial}{\partial z}\left[k\left(\frac{\partial h}{\partial z} + 1\right)\right], \tag{16}$$

where h is the pressure expressed in meters of water height, θ is the actual water content, t is time and z is a vertical coordinate corresponding to the influence of potential energy. In Equation (16), k describes the permeability of soil under conditions of variable pore saturation with water:

$$k = k_s k_r, \tag{17}$$

where k_r is a dimensionless coefficient, with values from the range (0, 1] modifying the full saturation water permeability k_s. Richard's Equation (16) describes the flow in a porous material with a number of simplifying assumptions. It does not take into account the full balance of masses, including gas phase, and the influence of temperature and takes into account only the flow of liquid phase in open pores. There are many variants of solving this one-dimensional flow problem in unsaturated soils, starting from many approximate analytical solutions describing the time of saturation movement and direct empirical solutions via the Talbot–Ogden method or percolation to numerical methods. These latter methods can be applied in the classical version, represented by, e.g., FlexPDE, or taking into account the specificity of the nonlinearity of the problem and optimizing for solution stability and time, executed by Hydrus-1D solution [61–63], where the k_r value is described with Van Genuchten's [64] model:

$$\theta = \theta_r + \frac{\theta_s - \theta_r}{\left(1 + |\alpha_2 h|^{n_1}\right)^m}, \quad h > 0, \tag{18a}$$

$$\theta = \theta_s, \qquad h \le 0, \tag{18b}$$

$$k_r = \Theta^{\frac{1}{2}}\left[1 - \left(1 - \Theta^{\frac{1}{m}}\right)^m\right]^2, \tag{19}$$

$$\Theta = \frac{\theta - \theta_r}{\theta_s - \theta_r}, \tag{20}$$

where $m = 1 - \frac{1}{n_1}$, n_1 for the Mualem model; n_1 is the exponent in the soil water retention function; Θ is effective saturation; θ_s is the saturated soil water content and θ_r is the residual soil water content. The typical soil water characteristic curve [65,66] was described using the parameters collected in Table 2 for different types of soils.

Table 2. Hydraulic properties of various soils.

Soil	n_1 (-)	α_2 (m^{-1})	k_s (m/s)
sand	4–8.5	1–5	10^{-2}–10^{-5}
silt	2–4	0.1–1	10^{-6}–10^{-9}
clay	1.1–2.5	0.01–0.1	10^{-9}–10^{-13}

2.6. Application of the Proposed Model—Case Study

The solution presented can be applied in specific geotechnical cases. In the paper, a slip over the identified surface within a potential landslide as a result of precipitation and change in the saturation state of the layer is considered. The scheme of the task is presented in Figure 11. A one-way downward flow forced by a constant value of pore overpressure as the upper boundary condition and free drainage as the downward boundary condition were assumed. Calculations were made with use of the Hydrus-1D program for all combinations of variables: h (m) = {2.00, 3.00, ... , 6.00}; α_2 (cm^{-1}) = {0.0100, 0.0133, ... , 0.0200}; and k_s (cm/h)= {0.100, 0.150, ... , 0.300}. The complete set of assumptions for the issue was defined as follows:

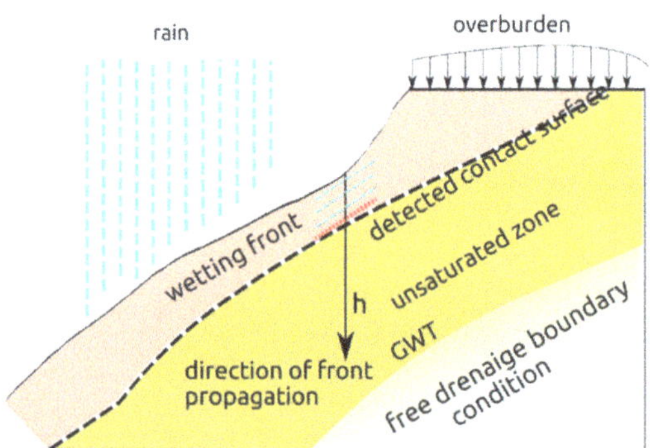

Figure 11. The task scheme (GTW—ground water table).

- Duration of the experiment: 1000 h;
- Minimum value of time increment: 0.01 h;
- The maximum value of time increment: 1.0 h;
- Maximum of 200 iterations;
- Soil water characteristic curve (SWCC) without hysteresis;
- Upper bound condition (along z axis) constant pressure head;
- Lower bound condition (along z axis) free drainage;
- Initial condition as pressure heads.

3. Results and Discussion

Different times of water flow in the pores cause different times for the front to moisturize the discontinuity area—the condition for layer slippage. In order to illustrate the concepts presented, calculations of the time to achieve failure in the identified surface were made. In the case study, a layer of homogeneous intact material with a set of variables, {h, α_2 = 0.0100, k_s}, was assumed. All variables were treated as deterministic, and the wavefront velocity v_F was an interpolating function. The results of saturation front velocity are presented in Figure 12, where a wide range of k_s and h values are considered for four different values of α_2.

Critical failure time t_{cft} is equal to the sum of Equation (15) and the time of wave transmission through a layer with thickness h:

$$t_{cft} = \frac{3600 \cdot 24 \cdot h}{v_F} + t_c \qquad (21)$$

The magnitude of t_{cft} is important for planning the management of a landslide-prone slope. The result of the time is illustrated in Figure 13, where four RCSR variants are shown for a fixed α_2 value equal to 0.010. A general assumption of a constant value of the RCSR load factor is made for illustration.

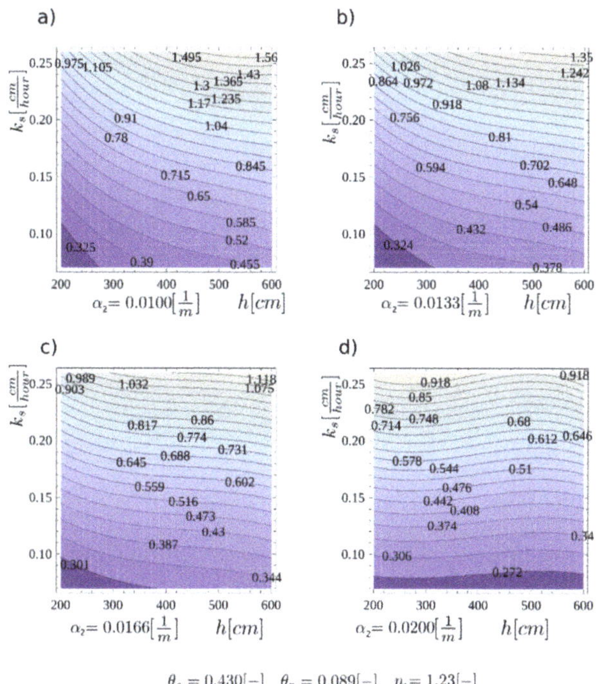

Figure 12. The velocity of the wetting head v_F for four cases of the soil water retention function parameter: (**a**) $\alpha_2 = 0.0100$ cm^{-1}; (**b**) $\alpha_2 = 0.0133$ cm^{-1}; (**c**) $\alpha_2 = 0.0166$ cm^{-1}; (**d**) $\alpha_2 = 0.0200$ cm^{-1}.

In the approach, a method for combining tertiary rheological effects with the concept of filtration in partially saturated soils is presented. The effect of wetting the material layer while it is subjected to shear forces at the contact surface leads to destruction. The main research problem was to estimate the time from the onset of surface wetting (rain, suspended or standing water) to the loss of slope stability t_{cft}. With the initial identification of the surface location of the existing potential failure and the determination of the t_{cft}, an important aspect is also the determination of methods and treatments to increase the critical failure time—especially for the purpose of evacuation or protection works. The obtained results presented in Figure 14 allow for comparing the influence of a change in thickness and a change in water permeability coefficient. For RCSR 1.0000, the change in thickness from 2 to 3 m resulted in a t_{cft} change within the range of 5–10% (for the range of k_s {0.10–0.25 cm/s}, the change in k_s from 0.10 to 0.20 cm/s results in a time change of 105% (for constant thickness 2 m)). This indicates that it is essential to determine the permeability of soils in the area of the projected landslide. It also highlights potential methods to improve stability and increase critical failure time.

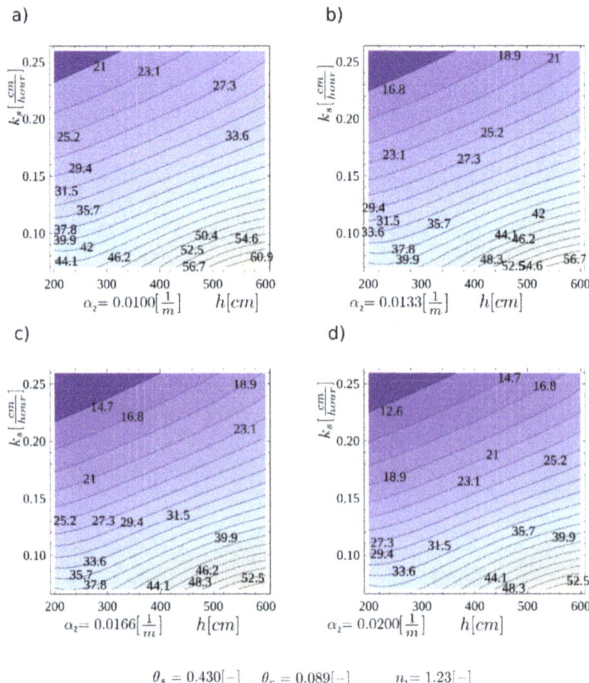

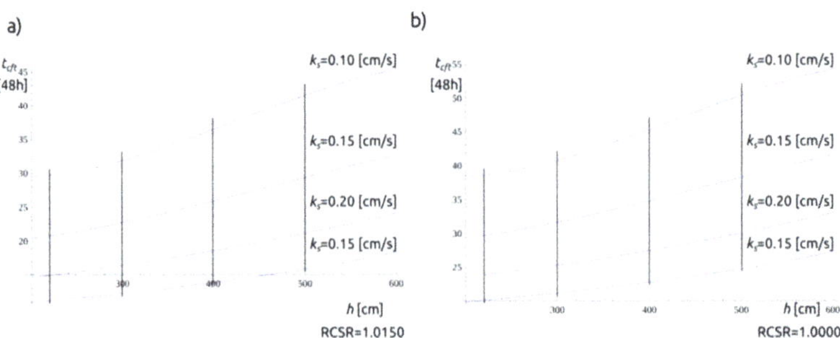

Figure 13. The differences (as in Equations (15) and (21)) in time to reach the limit t_{cft} state for the four load factor RCSR values: (**a**) RCSR = 1.0000; (**b**) RCSR = 1.0050; (**c**) RCSR = 1.0100; (**d**) RCSR = 1.0150 (time is in 24 h).

Figure 14. Values of t_{cft} at selected RCSR values and profiles (**a**) for RCSR = 1.0150 and (**b**) RCSR = 1.0000.

4. Conclusions

Situations resulting in the loss of slope stability are characterized by a pattern in which significant roles are played by terrain, pre-existing landslides and environmental factors, such as continuous precipitation. A method for determining the critical time required to initiate the landslide process as a consequence of moisture change is presented. The contact feature observed and examined in the modified ring shear apparatus, also taking into account the overburden factor, is relevant here. Clayey soil samples were subjected to residual-state creep tests. The objective was to obtain the results for the contact surface in the state of full soil saturation. The values from the laboratory investigations were filtered in order to remove measurement errors and noise. A Wiener-type denoising model was selected. To describe the rheological side of the phenomenon, a rheological model based on fractional derivatives was used. This model, requiring only few parameters, allowed to obtain correct matches to the results of the experiment. In the article, the complex model was introduced to assess the stability of the joint in originally unsaturated material subjected to hydraulic load. The initiated filtration process resulted in a front passage through the layer. The change in the contact parameters to fully saturated led to the start of the destruction process. The overload parameter has been introduced as part of the estimation of the survival time of objects in a hazardous state, associated with changes leading to an increase in saturation, such as sudden rainfall, flooding or failure of the drainage system. A calculation example is presented to illustrate the methodology for dealing with the post-continuous states. The main conclusion is the demonstration of the influence of the permeability and thickness of the layer on critical time in the selected case. The application of the above methodology to specific geotechnical cases can significantly improve the estimation of the time at which action should be taken. Additionally, it indicates potential directions for the development of methods to increase this interval. A future direction for the development of the approach given is the use of reliability analysis, which would allow the formulation of a comprehensive operating system.

Author Contributions: Conceptualization, D.R.B. and J.V.K.; methodology, D.R.B. and J.V.K.; software, J.V.K.; validation, D.R.B., J.V.K. and M.T.; formal analysis, D.R.B., J.V.K. and M.T.; investigation, D.R.B.; resources, D.R.B.; data curation, D.R.B., J.V.K. and M.T.; writing—original draft preparation, D.R.B., J.V.K. and M.T.; writing—review and editing, D.R.B., J.V.K. and M.T.; visualization, J.V.K.; supervision, J.V.K.; project administration, J.V.K.; funding acquisition, M.T. All authors have read and agreed to the published version of the manuscript.

Funding: This research received no external funding.

Institutional Review Board Statement: Not applicable.

Informed Consent Statement: Not applicable.

Data Availability Statement: Data available on request due to restrictions eg privacy or ethical.

Acknowledgments: The first author would like to acknowledge Ryuichi Yatabe (Emeritus Professor, Ehime University, Japan) and N. P. Bhandary (Professor, Ehime University, Japan) for giving a series of advice on the residual-state creep test during his PhD (2011–2014) at Ehime University, Japan.

Conflicts of Interest: The authors declare no conflict of interest.

References

1. Fell, R.; Corominas, J.; Bonnard, C.; Cascini, L.; Leroi, E.; Savage, W. On Behalf of the JTC-1 Joint Technical Committee on Landslides and Engineered Slopes (2008) Guidelines for Landslide Susceptibility, Hazard and Risk Zoning for Land Use Planning. *Eng. Geol.* **2008**, *102*, 85–98. [CrossRef]
2. Van Westen, C.J.; Castellanos, E.; Kuriakose, S.L. Spatial Data for Landslide Susceptibility, Hazard, and Vulnerability Assessment: An Overview. *Eng. Geol.* **2008**, *102*, 112–131. [CrossRef]
3. Le, Q.H.; Nguyen, T.H.V.; Do, M.D.; Le, T.C.H.; Pham, V.S.; Nguyen, H.K.; Luu, T.B. TXT-Tool 1.084-3.1: Landslide Susceptibility Mapping at a Regional Scale in Vietnam. In *Landslide Dynamics: ISDR-ICL Landslide Interactive Teaching Tools: Volume 1: Fundamentals, Mapping and Monitoring*; Sassa, K., Guzzetti, F., Yamagishi, H., Arbanas, Ž., Casagli, N., McSaveney, M., Dang, K., Eds.; Springer International Publishing: Cham, Switzerland, 2018; pp. 161–174. [CrossRef]
4. Dietrich, W.; Montgomery, D. A Digital Terrain Model for Mapping Shallow Landslide Potential. *Univ. Calif.-Univ. Wash.* **1998**. Available online: http://calm.geo.berkeley.edu/geomorph/shalstab/index.htm (accessed on 12 April 2021).
5. Rossi, G.; Catani, F.; Leoni, L.; Segoni, S.; Tofani, V. Hiresss: A Physically Based Slope Stability Simulator for HPC Applications. *Nat. Hazards Earth Syst. Sci.* **2013**, *13*, 151–166. [CrossRef]
6. Tianbin, L.; Mingdong, C. Time Prediction of Landslide Using Verhulst Inverse-Function Model. *J. Geol. Hazards Enveronment Preserv.* **1996**, *3*, 13–17.
7. Stedinger, J.R.; Vogel, R.M.; Lee, S.U.; Batchelder, R. Appraisal of the Generalized Likelihood Uncertainty Estimation (GLUE) Method. *Water Resour. Res.* **2008**, *44*, 12. [CrossRef]
8. Kim, J.; Jeong, S.; Park, S.; Sharma, J. Influence of Rainfall-Induced Wetting on the Stability of Slopes in Weathered Soils. *Eng. Geol.* **2004**, *75*, 251–262. [CrossRef]

9. Jeong, S.; Lee, K.; Kim, J.; Kim, Y. Analysis of Rainfall-Induced Landslide on Unsaturated Soil Slopes. *Sustainability* **2017**, *9*, 1280. [CrossRef]

10. Chen, X.; Guo, H.; Song, E. Analysis Method for Slope Stability under Rainfall Action. *Landslides Eng. Slopes* **2008**, *7*, 1507–1515.

11. Ponziani, F.; Pandolfo, C.; Stelluti, M.; Berni, N.; Brocca, L.; Moramarco, T. Assessment of Rainfall Thresholds and Soil Moisture Modeling for Operational Hydrogeological Risk Prevention in the Umbria Region (Central Italy). *Landslides* **2012**, *9*, 229–237. [CrossRef]

12. Intrieri, E.; Carlà, T.; Gigli, G. Forecasting the Time of Failure of Landslides at Slope-Scale: A Literature Review. *Earth-Sci. Rev.* **2019**, *193*, 333–349. [CrossRef]

13. Rosser, N.; Lim, M.; Petley, D.; Dunning, S.; Allison, R. Patterns of Precursory Rockfall Prior to Slope Failure. *J. Geophys. Res. Earth Surf.* **2007**, *112*. [CrossRef]

14. Vilhelm, J.; Rudajev, V.; Lokajíček, T.; Živor, R. Application of Autocorrelation Analysis for Interpreting Acoustic Emission in Rock. *Int. J. Rock Mech. Min. Sci.* **2008**, *45*, 1068–1081. [CrossRef]

15. Singh, D.P. A Study of Creep of Rocks. *Int. J. Rock Mech. Min. Sci. Geomech. Abstr.* **1975**, *12*, 271–276. [CrossRef]

16. Hunter, G.J.; Khalili, N. A Simple Criterion for Creep Induced Failure of Over-Consolidated Clays. In Proceedings of the ISRM International Symposium. International Society for Rock Mechanics and Rock Engineering, Melbourne, Australia, 19 November 2000.

17. Petley, D.; Mantovani, F.; Bulmer, M.; Zannoni, F. The Interpretation of Landslide Monitoring Data for Movement Forecasting. *Geomorphology* **2005**, *66*, 133–147. [CrossRef]

18. Feda, J. Interpretation of Creep of Soils by Rate Process Theory. *Géotechnique* **1989**, *39*, 667–677. [CrossRef]

19. Augustesen, A.; Liingaard, M.; Lade, P.V. Evaluation of Time-Dependent Behavior of Soils. *Int. J. Geomech.* **2004**, *4*, 137–156. [CrossRef]

20. Brandes, H.G.; Nakayama, D.D. Creep, Strength and Other Characteristics of Hawaiian Volcanic Soils. *Géotechnique* **2010**, *60*, 235–245. [CrossRef]

21. Ter-Stepanian, G. Creep of a Clay during Shear and Its Rheological Model. *Géotechnique* **1975**, *25*, 299–320. [CrossRef]

22. Leoni, M.; Karstunen, M.; Vermeer, P.A. Anisotropic Creep Model for Soft Soils. *Géotechnique* **2008**, *58*, 215–226. [CrossRef]

23. Yin, Z.-Y.; Chang, C.S.; Karstunen, M.; Hicher, P.-Y. An Anisotropic Elastic–Viscoplastic Model for Soft Clays. *Int. J. Solids Struct.* **2010**, *47*, 665–677. [CrossRef]

24. Cascini, L.; Calvello, M.; Grimaldi, G.M. Groundwater Modeling for the Analysis of Active Slow-Moving Landslides. *J. Geotech. Geoenviron. Eng.* **2010**, *136*, 1220–1230. [CrossRef]

25. Bhat, D.R.; Wakai, A. Numerical Simulation of a Creeping Landslide Induced by a Snow Melt Water. *Tech. J.* **2019**, *1*, 71–78. [CrossRef]

26. Wang, S.; Wu, W.; Wang, J.; Yin, Z.; Cui, D.; Xiang, W. Residual-State Creep of Clastic Soil in a Reactivated Slow-Moving Landslide in the Three Gorges Reservoir Region, China. *Landslides* **2018**, *15*, 2413–2422. [CrossRef]

27. Sharifzadeh, M.; Tarifard, A.; Moridi, M.A. Time-Dependent Behavior of Tunnel Lining in Weak Rock Mass Based on Displacement Back Analysis Method. *Tunn. Undergr. Space Technol.* **2013**, *38*, 348–356. [CrossRef]

28. Tavenas, F.; Leroueil, S. Creep and Failure of Slopes in Clays. *Can. Geotech. J.* **1981**, *18*, 106–120. [CrossRef]

29. Amitrano, D.; Helmstetter, A. Brittle Creep, Damage, and Time to Failure in Rocks. *J. Geophys. Res. Solid Earth* **2006**, *111*. [CrossRef]

30. Saito, M. Forecasting the Time of Occurrence of a Slope Failure. In Proceedings of the 6th International Conference Soil Mechanics and Foundation Engineering, Montreal, QC, Canada, 8–15 September 1965; pp. 537–541.

31. Saito, M. Forecasting Time of Slope Failure by Tertiary Creep. In *Proceedings of the 7th International Conference on Soil Mechanics and Foundation Engineering*; Sociedad Mexicana de Mecánica de Suelos: Mexico City, Mexico, 1969; Volume 2, pp. 677–683. Available online: https://citeseerx.ist.psu.edu/viewdoc/download?doi=10.1.1.612.2877&rep=rep1&type=pdf (accessed on 12 April 2021).

32. Federico, A.; Popescu, M.; Murianni, A. Temporal Prediction of Landslide Occurrence: A Possibility or a Challenge. *Ital. J. Eng. Geol. Environ.* **2015**, *1*, 41–60.

33. Chen, H.; Lee, C.F. A Dynamic Model for Rainfall-Induced Landslides on Natural Slopes. *Geomorphology* **2003**, *51*, 269–288. [CrossRef]

34. Antolini, F.; Barla, M.; Gigli, G.; Giorgetti, A.; Intrieri, E.; Casagli, N. Combined Finite–Discrete Numerical Modeling of Runout of the Torgiovannetto Di Assisi Rockslide in Central Italy. *Int. J. Geomech.* **2016**, *16*, 04016019. [CrossRef]

35. Atanacković, T.M.; Pilipović, S.; Stanković, B.; Zorica, D. *Fractional Calculus with Applications in Mechanics*; Wiley Online Library: Hoboken, NJ, USA, 2014.

36. Diethelm, K. *The Analysis of Fractional Differential Equations: An Application-Oriented Exposition Using Differential Operators of Caputo Type*; Springer Science & Business Media: Berlin/Heidelberg, Germany, 2010.

37. Kilbas, A.A.; Srivastava, H.M.; Trujillo, J.J. *Theory and Applications of Fractional Differential Equations*; Elsevier: Amsterdam, The Netherlands, 2006.

38. Podlubny, I. *Fractional Differential Equations: An Introduction to Fractional Derivatives, Fractional Differential Equations, to Methods of Their Solution and Some of Their Applications*; Elsevier: Amsterdam, The Netherlands, 1998.

39. Kilbas, A.A.; Marichev, O.; Samko, S. *Fractional Integrals and Derivatives (Theory and Applications)*; Gordon and Breach Science Publishers: Yverdon, Switzerland, 1993.

40. Paola, M.D.; Zingales, M. Exact Mechanical Models of Fractional Hereditary Materials. *J. Rheol.* **2012**, *56*, 983–1004. [CrossRef]
41. Di Paola, M.; Pirrotta, A.; Valenza, A. Visco-Elastic Behavior through Fractional Calculus: An Easier Method for Best Fitting Experimental Results. *Mech. Mater.* **2011**, *43*, 799–806. [CrossRef]
42. Di Paola, M.; Pinnola, F.P.; Zingales, M. A Discrete Mechanical Model of Fractional Hereditary Materials. *Meccanica* **2013**, *48*, 1573–1586. [CrossRef]
43. Di Paola, M.; Pinnola, F.P.; Zingales, M. Fractional Differential Equations and Related Exact Mechanical Models. *Comput. Math. Appl.* **2013**, *66*, 608–620. [CrossRef]
44. Bishop, A.W.; Green, G.E.; Garga, V.K.; Andresen, A.; Brown, J.D. A New Ring Shear Apparatus and Its Application to the Measurement of Residual Strength. *Géotechnique* **1971**, *21*, 273–328. [CrossRef]
45. Stark, T.D.; Eid, H.T. Modified Bromhead Ring Shear Apparatus. Geotech. *Test. J.* **1993**, *16*, 100–107. [CrossRef]
46. Bromhead, E.N.; Dixon, N. The Field Residual Strength of London Clay and Its Correlation with Loboratory Measurements, Especially Ring Shear Tests. *Géotechnique* **1986**, *36*, 449–452. [CrossRef]
47. Tiwari, B.; Brandon, T.L.; Marui, H.; Tuladhar, G.R. Comparison of Residual Shear Strengths from Back Analysis and Ring Shear Tests on Undisturbed and Remolded Specimens. *J. Geotech. Geoenviron. Eng.* **2005**, *131*, 1071–1079. [CrossRef]
48. Hong, Y.; Yu, G.; Wu, Y.; Zheng, X. Effect of Cyclic Loading on the Residual Strength of Over-Consolidated Silty Clay in a Ring Shear Test. *Landslides* **2011**, *8*, 233–240. [CrossRef]
49. Bhat, D.R.; Bhandary, N.P.; Yatabe, R. Method of Residual-State Creep Test to Understand the Creeping Behaviour of Landslide Soils. In *Landslide Science and Practice: Volume 2: Early Warning, Instrumentation and Monitoring*; Margottini, C., Canuti, P., Sassa, K., Eds.; Springer: Berlin/Heidelberg, Germany, 2013; pp. 635–642. [CrossRef]
50. Bhat, D.R.; Bhandary, N.P.; Yatabe, R.; Tiwari, R.C. A New Concept of Residual-State Creep Test to Understand the Creeping Behavior of Clayey Soils. In Proceedings of the GeoCongress 2012: State of the Art and Practice in Geotechnical Engineering, Oakland, CA, USA, 25–29 March 2012; pp. 683–692. [CrossRef]
51. Bhat, D.R. Residual-State Creep Test in Modified Torsional Ring Shear Machine: Methods and Implications. *Int. J. Geomate* **2011**, *1*, 39–43. [CrossRef]
52. Bhat, D.R.; Yatabe, R.; Bhandary, N.P. Creeping Displacement Behavior of Clayey Soils in A New Creep Test Apparatus. In Proceedings of the Soil Behavior and Geomechanics, Shanghai, China, 26–28 March 2014; pp. 275–285. [CrossRef]
53. Bhat, D.R.; Bhandary, N.P.; Yatabe, R. Residual-State Creep Behavior of Typical Clayey Soils. *Nat. Hazards* **2013**, *69*, 2161–2178. [CrossRef]
54. Bhat, D.R.; Yatabe, R. Effect of Shearing Rate on Residual Strength of Landslide Soils. In *Engineering Geology for Society and Territory*; Lollino, G., Giordan, D., Crosta, G.B., Corominas, J., Azzam, R., Wasowski, J., Sciarra, N., Eds.; Springer International Publishing: Cham, Switzerland, 2015; Volume 2, pp. 1211–1215. [CrossRef]
55. Bhat, D.R.; Yatabe, R.; Bhandary, N.P. Study of Preexisting Shear Surfaces of Reactivated Landslides from a Strength Recovery Perspective. *J. Asian Earth Sci.* **2013**, *77*, 243–253. [CrossRef]
56. JIS. *A 1202:2009 Test Method for Density of Soil Particles*; JIS: Tokyo, Japan, 2009.
57. JIS. *A 1203:2009 Test Method for Water Content of Soils*; JIS: Tokyo, Japan, 2009.
58. JIS. *A 1204:2009 Test Method for Particle Size Distribution of Soils*; JIS: Tokyo, Japan, 2009.
59. JIS. *A 1205:2009 Test Method for Liquid Limit and Plastic Limit of Soils*; JIS: Tokyo, Japan, 2009.
60. Matsushi, Y.; Matsukura, Y. Cohesion of Unsaturated Residual Soils as a Function of Volumetric Water Content. *Bull. Eng. Geol. Environ.* **2006**, *65*, 449. [CrossRef]
61. Simunek, J.; Van Genuchten, M.T.; Sejna, M. The HYDRUS-1D Software Package for Simulating the One-Dimensional Movement of Water, Heat, and Multiple Solutes in Variably-Saturated Media. *Univ. Calif.-Riverside Res. Rep.* **2005**, *3*, 1–240.
62. Simunek, J.; Jacques, D.; van Genuchten, M.T.; Mallants, D. Multicomponent Geochemical Transport Modeling Using HYDRUS-1 D and HP 1. *J. Am. Water Resour. Assoc.* **2006**, *42*, 1537–1547. [CrossRef]
63. Šimůnek, J.; Van Genuchten, M.T.; Šejna, M. The HYDRUS Software Package for Simulating Two-and Three-Dimensional Movement of Water, Heat, and Multiple Solutes in Variably-Saturated Media. *Tech. Man. Version* **2006**, *1*, 241.
64. Van Genuchten, M.T. A Closed-Form Equation for Predicting the Hydraulic Conductivity of Unsaturated Soils. *Soil Sci. Soc. Am. J.* **1980**, *44*, 892–898. [CrossRef]
65. Mualem, Y. A New Model for Predicting the Hydraulic Conductivity of Unsaturated Porous Media. *Water Resour. Res.* **1976**, *12*, 513–522. [CrossRef]
66. Singh, V.P. *Kinematic Wave Modeling in Water Resources: Environmental Hydrology*; John Wiley & Sons: Hoboken, NJ, USA, 1997.

 *materials*

Article

Identifying the Range of Micro-Events Preceding the Critical Point in the Destruction Process in Traditional and Quasi-Brittle Cement Composites with the Use of a Sound Spectrum

Dominik Logoń [1,*], Janusz Juraszek [2], Zbynek Keršner [3] and Petr Frantík [3]

[1] Faculty of Civil Engineering, Wrocław University of Science and Technology, Wybrzeże Wyspiańskiego 27, 50-370 Wrocław, Poland
[2] Faculty of Materials, Civil and Environmental Engineering, University of Technology and Humanities in Bielsko-Biała, Willowa 2, Building C, 43-309 Bielsko-Biała, Poland; jjuraszek@ath.bielsko.pl
[3] Faculty of Civil Engineering, Brno University of Technology, Veveří 331/95, 602 00 Brno, Czech Republic; kersner.z@fce.vutbr.cz (Z.K.); frantik.p@fce.vutbr.cz (P.F.)
* Correspondence: dominik.logon@pwr.edu.pl

Abstract: This paper presents the possibilities of determining the range of stresses preceding the critical destruction process in cement composites with the use of micro-events identified by means of a sound spectrum. The presented test results refer to the earlier papers in which micro-events (destruction processes) were identified but without determining the stress level of their occurrence. This paper indicates a correlation of 2/3 of the stress level corresponding to the elastic range with the occurrence of micro-events in traditional and quasi-brittle composites. Tests were carried out on beams (with and without reinforcement) subjected to four-point bending. In summary, it is suggested that the conclusions can be extended to other test cases (e.g., compression strength), which should be confirmed by the appropriate tests. The paper also indicates a need for further research to identify micro-events. The correct recognition of micro-events is important for the safety and durability of traditional and quasi-brittle cement composites.

Keywords: AE acoustic emission; micro-events; sound spectrum; traditional and quasi-brittle cement composites

Citation: Logoń, D.; Juraszek, J.; Keršner, Z.; Frantík, P. Identifying the Range of Micro-Events Preceding the Critical Point in the Destruction Process in Traditional and Quasi-Brittle Cement Composites with the Use of a Sound Spectrum. *Materials* **2021**, *14*, 1809. https://doi.org/10.3390/ma14071809

Academic Editor: Leif Kari

Received: 31 January 2021
Accepted: 29 March 2021
Published: 6 April 2021

Publisher's Note: MDPI stays neutral with regard to jurisdictional claims in published maps and institutional affiliations.

Copyright: © 2021 by the authors. Licensee MDPI, Basel, Switzerland. This article is an open access article distributed under the terms and conditions of the Creative Commons Attribution (CC BY) license (https://creativecommons.org/licenses/by/4.0/).

1. Introduction

Acoustic Emission (AE) is a well-researched method for testing building materials and structures, which has been used for a very long time in civil engineering [1–6]. As a review of the literature shows, papers concentrate on determining the destruction process of materials. AE has been used in monitoring early-age and early hydration acoustic emission of cement paste [7,8]. This method was used to determine the Hooke's law range and identify the destruction process in cement composites in various cases of loading [9–12]. AE (events sum AE) has been applied to determine the first crack [13–15], micro- and macro-cracks and their propagation in the fracture process in traditional and quasi-brittle cement composites with and without reinforcement [16–23]. Previous tests show the possibility of using AE in various materials, including traditional and high-strength [24–27] cement composites. The effectiveness of the acoustic emission measurements has been used to monitor structures [28–30]. Diagnostics and structure monitoring are still being modified with this method [31–33].The majority of papers concerning the destruction process identification are based on the measurement of the AE and AE sum as well as on the spectrograms of the AE signals. Research papers have also demonstrated the possibility of recording the destruction stages with the use of spectrogram images [4].

Various methods of stress determination are still being improved in the process of identification of failure processes in different materials and structures [34–39], but the correlation of the stress level with AE in terms of Hooke's law has not been applied.

Previous research work focused on the correlation between AE and the failure processes of each of the composite components based on the sound spectrum [40–45]. It has been found that, for the accurate recognition of composite failure processes, the AE recording should be expanded to include the analysis of each sound separately (as well as a single signal in a very small range of frequencies) and the analysis of the range of sounds corresponding to a given mechanical effect with the use of an acoustic spectrum. It was noticed that the 2D and 3D acoustic spectra should be correlated with the load–deflection curve and with other acoustic effects, which enables the identification of the failure process.

Previous papers [46,47] have focused on the possibility of identifying AE micro-events in the area preceding the occurrence of a critical crack initiating the destruction process in cement composites. It has been indicated that there is a possibility of predicting the occurrence of the f_{cr} (first crack) based on an analysis of the 2D and 3D sound spectra corresponding to the occurring groups of micro-events that precede the end of the load–deflection approximate proportionality area. Tests were conducted on a small-size (40 mm × 40 mm × 160 mm) specimen of a paste with dispersed reinforcement.

This paper indicates the occurrence of micro-events in a wide range of large cement composite beams (paste, mortar and concrete) with traditional and dispersed reinforcement. Before a decision was made on including examples of specimens in this paper, a wide range of composites were analysed and extremely different specimens were selected. The identification of micro-events was carried out by recording the AE on beams with a dimension of 150 mm × 150 mm × 600 mm. Conclusions were drawn based on the analysis of the sound spectrum and the corresponding amplitudes. The repeatability of the results indicates the possibility of considering the conclusions presented in this paper as more generally applicable. The main goal of this paper is the assessment of the level of stress at which micro-events appear that precede the occurrence of the critical crack f_{cr} (LOP—the limit of proportionality) and the flexural strength at bending f_{max} (MOR—the modulus of rupture).

The originality of the presented paper relates to the determination of the stress level (2/3 of the elastic range) for which there occurs the grouping of micro-events in the frequency range and the increase in relative amplitudes of the sound spectrum that precede the destruction process of cement composites. The paper does not identify the types and causes of the micro-events, indicating the need for such research in the future.

2. Materials and Methods

The AE tests were carried out on the specimens that were presented in previous papers. The specimens were selected in a manner ensuring that they differ significantly in terms of reinforcement and deformation capacity [43–49]. Acoustic measurements concentrated on the recording of micro-events. The micro-events were recognized by the space sound spectrum in the range of 0.2–20 kHz.

2.1. Materials Used for Tests

Materials for the preparation of the cement composites: c—Portland cement CEM I (class 42.5R) produced by "Górażdże" cement plant in Górażdże (Poland), silica fume (10% of cement mass), fly ash (20% of cement mass), sand of 0–2 mm, superplasticizer (SP) and tap water (w), with the w/b = 0.35 (b = cement + fly ash + silica fume).

Composites:

(1) Mortar without reinforcement—cement:sand (volume) = 1:4.5.
(2) The paste composite was reinforced with dispersed synthetic structural polypropylene fibres (compliance with ASTM C-1116)—specific weight 0.91 kg/dm^3, flexural strength f_t = 620–758 MPa, E = 4.9 GPa, l = 54 mm, equivalent diameter 0.48 mm, l/d = 113 and V_f = 6%.
(3) The concrete composite was reinforced with traditional continuous ST500-b reinforcing bars (ArcelorMittal, Warsaw, Poland) with a diameter (d) = 10 mm. Four

continuous structural bars in the corners of the beam were placed with stirrups of d = 6 mm, positioned every 150 mm.

2.2. Preparation of Specimens for Tests

The ingredients were mixed in the concrete mixer and then used to cast the samples. Beams (150 mm × 150 mm × 600 mm) were cast in slabs and then cured in water at 20 ± 2 °C. After 180 days of ageing, the beams were prepared for the bending tests.

The concrete specimens with traditional reinforcement were not notched. The paste and mortar samples were notched. These beams were turned 90° and cut to the depth of 30 mm. The width of the cut was 3 mm.

2.3. Description of the Test Stand

Four-point bending tests were carried out in the testing machine with closed-loop servo control displacement, see Figure 1. The load–deflection curves were obtained according to ASTM C 1018 but the tests were based on the measurement of the continuous and constant displacement of crosshead and the rate was 1 mm/min. The following data were obtained:

-   the flexural strength at bending f_{max} (MOR—the modulus of rupture), and the flexural strength at the first crack f_{cr} (LOP—the limit of proportionality);
-   the characteristic points on the load–deflection curve, f_x (F_x-load; ε_x-deflection; W_x-energy);
-   energy (work) as proportional to the area under the load–deflection curve up to the characteristic point.

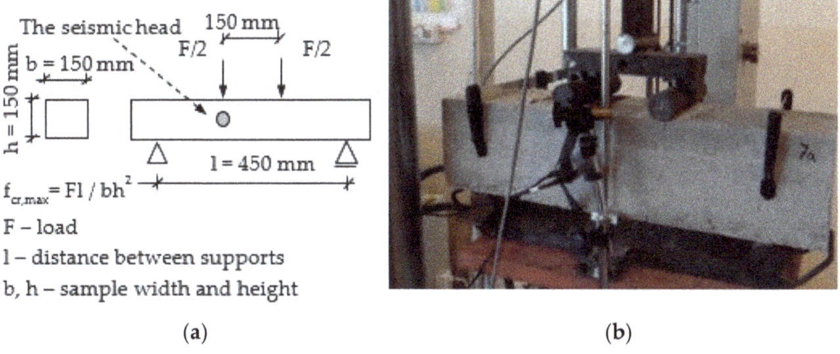

Figure 1. Four-point bending test: the specimen before the test in the test machine: (**a**) diagram of the test [44], (**b**) the specimen before test.

Acoustic emission effects (AE) were registered and recorded in order to monitor the progress of the cracking process during the monitoring of the load–deflection curve. A seismic head HY919 was used to record the acoustic emission effects in the range from 0.2 to 20 kHz. The sampling rate of the recorded waveforms was 44.1 kHz. The head was placed on the side in the central part of the loaded beams.

The mechanical effects of the composites were correlated with the recorded acoustic spectrum effects. The 3D and 2D sound spectra were measured with the use of the Spectra PLUS-SC program (Pioneer Hill Software LLC, Sequim, WA, USA). The measured AE effects were presented as 3D and 2D acoustic spectra.

The load–deflection curves of quasi-brittle cement composites with the corresponding acoustic effects (including sound spectra) are presented. Based on the results included in this paper (and [44]), Figure 2 has been modified by adding micro-events. The spectrum of micro-events was placed directly above the background noise spectrum and below the spectrum corresponding to the micro-crack. The 2D spectra (Figure 2c—amplitude and corresponding frequencies) were determined in relation to an established test duration point correlated with the destruction process, see Figure 2a. The determination of 2D spectra in the time range of the occurrence of the same destruction phenomena increases the range of amplitudes and should not be compared with the spectrum referred to the established test duration point. Infrasound, low- and very high-frequency sounds were not measured; this frequency range is also intended to be recorded in future tests, but it requires heads with a different measurement range.

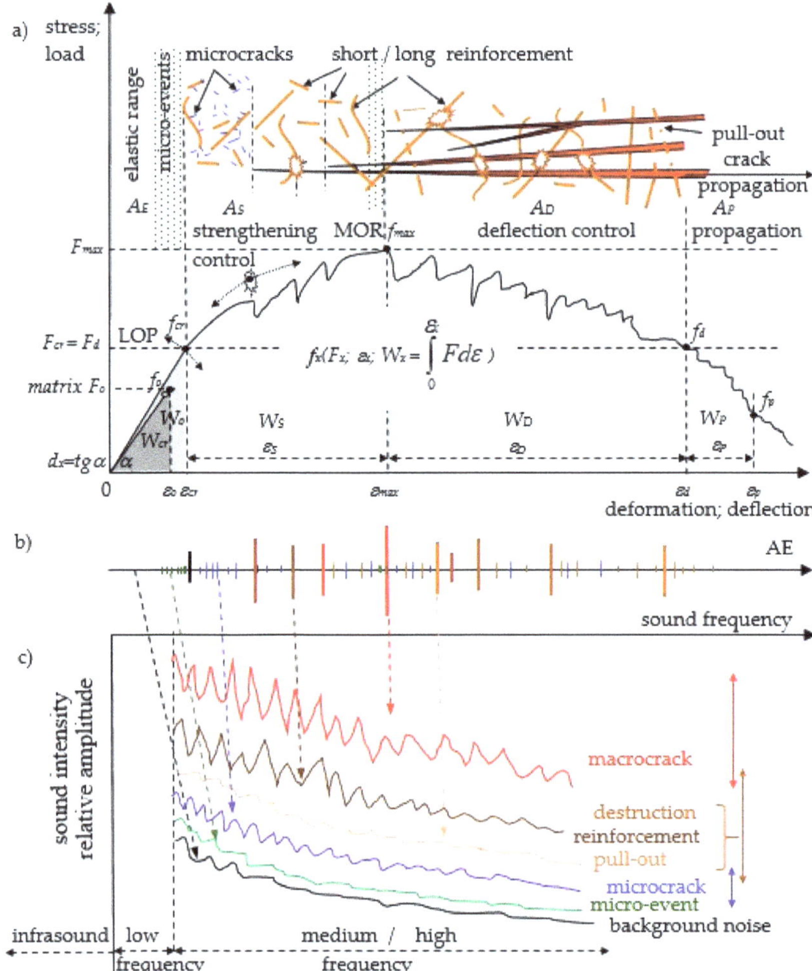

Figure 2. The quasi-brittle composite: (**a**) load–deflection curve; (**b**) AE—acoustic emission effects; (**c**) 2D acoustic spectrum (amplitude depending on sound intensity and frequency).

The ESD reinforcement effect (Eng. elastic range, strengthening control and deflection control) is presented by characteristic points f_x (load F_x; deflection ε_x; absorbed energy W_x). The absorbed energy is determined as the area under the load–deflection curve. The following symbols are used: A_E, W_E (the area and absorbed energy corresponding to the elastic range); A_S, W_S (the area and absorbed energy corresponding to the strengthening control); A_D, W_D (the area and absorbed energy corresponding to the deflection control); and A_P, W_P (the area and absorbed energy corresponding to the propagation). Deformation ability was determined as $d_x = \text{load}/\text{deflection}$.

3. Results

Figure 3 presents the examples of the tested specimens in the bending test. A specimen of mortar without reinforcement demonstrates the characteristic catastrophic brittle destruction process after achieving $f_{cr} = f_{max}$. At the critical point f_x (load; deflection; energy), the following results were obtained: $f_{cr} = f_{max}(13.0 \text{ kN}; 0.71 \text{ mm}; 4.658 \text{ J})$. The deformation ability is $tg\alpha = 18.2 \text{ kN/mm}$.

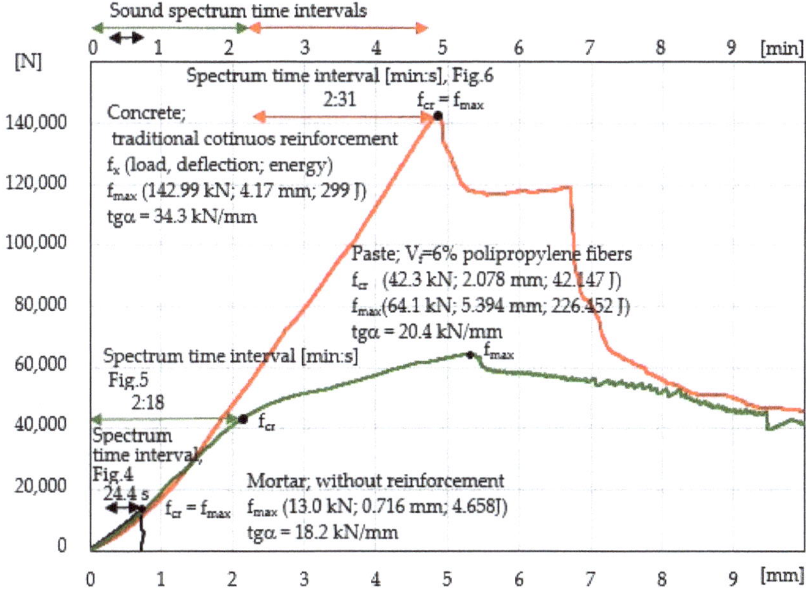

Figure 3. Load–deflection curves of the specimens during the bending test with the analysed sound spectrum time intervals.

A cement paste specimen with the maximum possible volume of polypropylene fibres $V_f = 6\%$ shows a significant improvement in the ability to carry stress in the Hooke's law range, i.e., $f_{cr}(42.3 \text{ kN}; 2.078 \text{ mm}; 42.147 \text{ J})$, and $tg\alpha = 20.4 \text{ kN/mm}$. It is a typical ESD composite with a significant strengthening control range A_s. The multicracking effects occur in that area. At point f_{max}, the following results were obtained: $f_{max}(64.1 \text{ kN}; 5.394 \text{ mm}; 226.452 \text{ J})$, and $tg\alpha = 20.4 \text{ kN/mm}$. After that point, further destruction processes based on the sound spectrum in the deflection control area A_D and the destruction propagation area A_P were not been analysed in this paper.

A concrete specimen with traditional continuous reinforcement demonstrates a strong ability to carry stress in the elastic range A_E. After critical stress at the characteristic point $f_{cr} = f_{max}$, a macrocrack occurs with $f_{max}(142.99 \text{ kN}; 4.17 \text{ mm}; 299 \text{ J})$ and $tg\alpha = 34.3 \text{ KN/mm}$. A significant drop in force is visible after exceeding f_{max} by ca. 16%, but a further destruction process is limited by the continuous reinforcement's ability to carry stress to the deflection of 6.7 mm. Another deformation results in the catastrophic destruction process.

The results of the acoustic emission measurement for all specimens are presented in the same manner (in the same amplitude and frequency range), see Figures 4–6. The measurement range of the head limits the AE measurement to the range of 0.2–20 kHz. The lower diagrams present the 3D sound spectrum. Below the 3D spectrum, the 2D spectrum amplitude in the range from −10 to −130 dB was placed. Due to the lack of possibility of presenting all the results from the entire destruction process, only the most characteristic ranges of the sound spectrum are presented, mainly those connected with micro-events preceding f_{cr} and f_{max}. The load-deflection test and the AE measurement (mortar and concrete) did not begin at the same time. The time ranges for mortar and concrete precede the occurrence of f_{cr}.

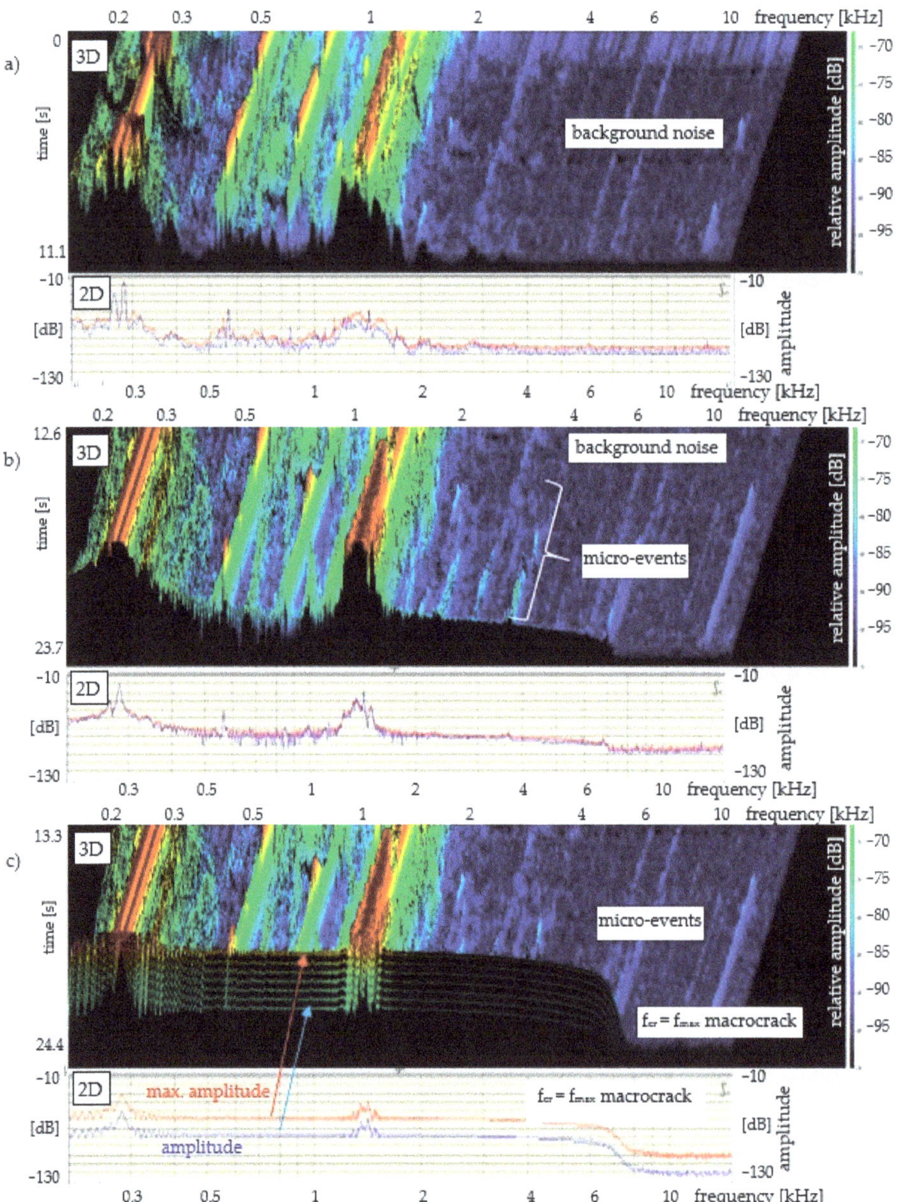

Figure 4. Mortar without reinforcement: bending test—AE; sound spectrum 3D and 2D: (**a**) background noise; (**b**) micro-events; (**c**) micro-events and macrocrack $f_{cr} = f_{max}$.

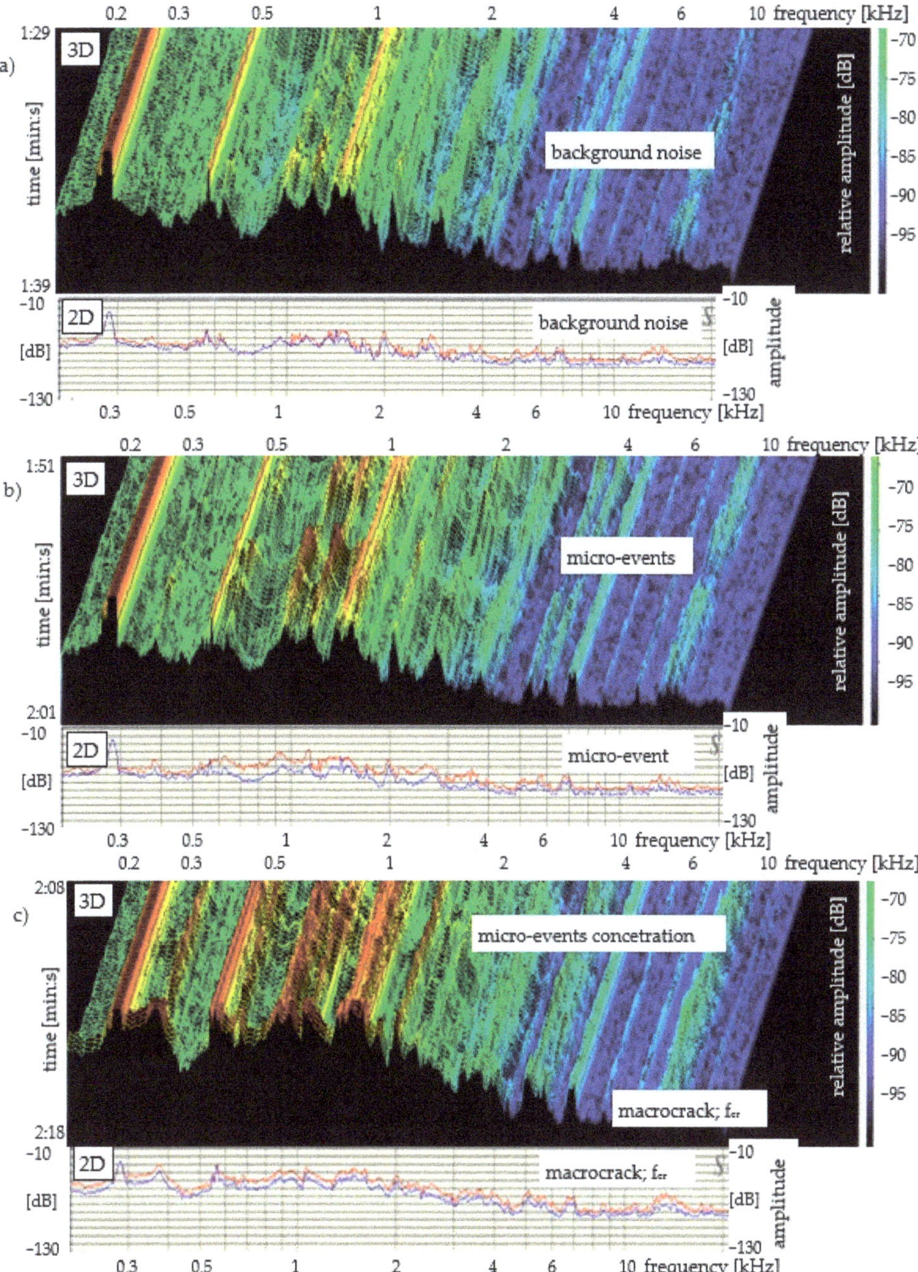

Figure 5. Cement paste with $V_f = 6\%$ in the bending test and the AE, with 3D and 2D sound spectra: (**a**) background noise; (**b**) micro-events; (**c**) fcr and micro-events concentration.

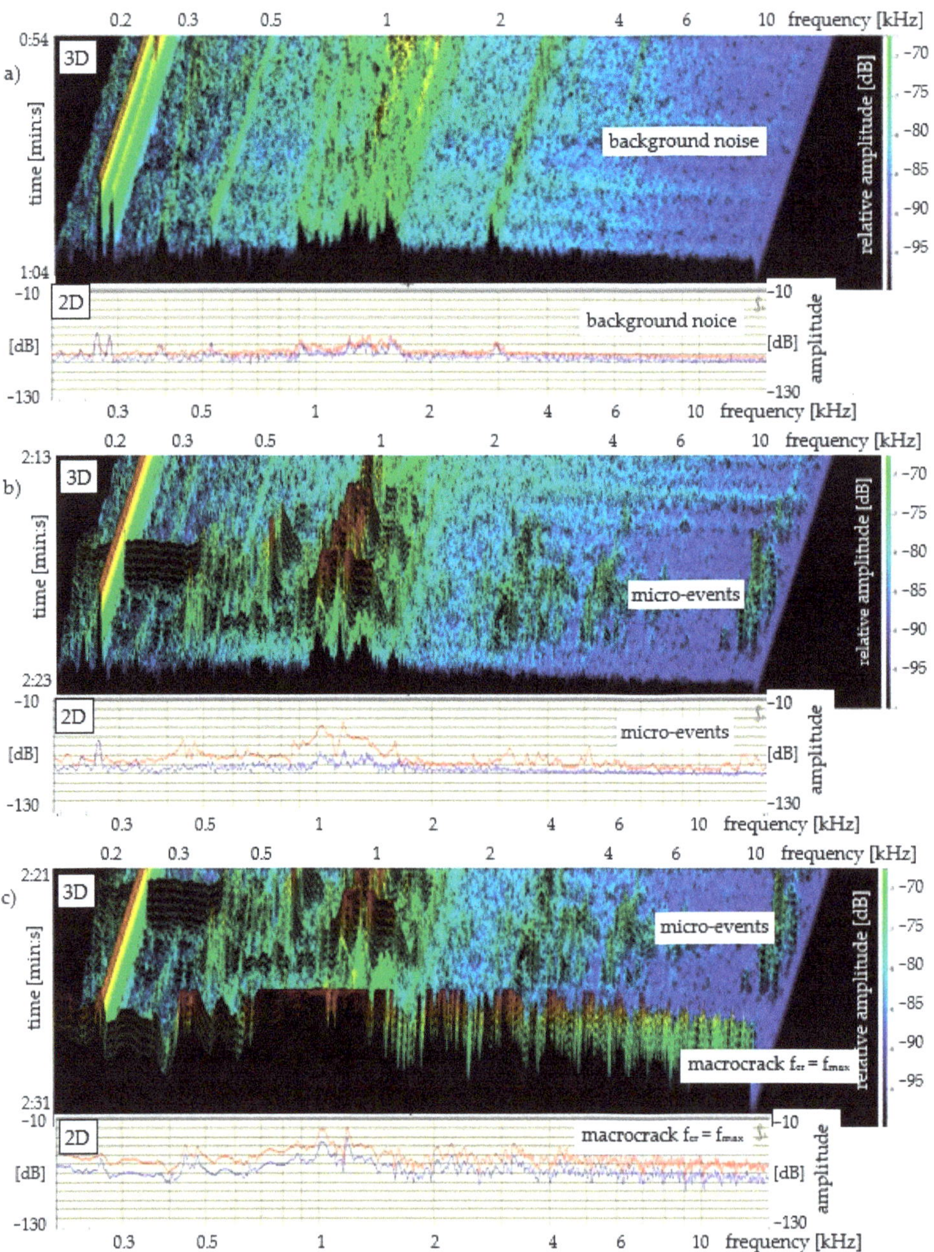

Figure 6. Concrete with traditional continuous reinforcement during the bending test and the AE, with 3D and 2D sound spectra: (**a**) background noise; (**b**) micro-events; (**c**) micro-events and macrocrack $f_{cr} = f_{max}$.

Figure 4 presents a 3D and 2D sound spectrum of a mortar specimen without reinforcement (AE measurement during the four-point bending test). The background noise spectrum is presented at the top of Figure 4a. Figure 4b shows micro-events (increase in relative amplitudes before f_{cr}). The spectrum corresponding to the critical crack (macrocrack; $f_{cr} = f_{max}$) is shown in Figure 4c. The brittle composite has been destroyed by a macrocrack after reaching the critical point $f_{cr} = f_{max}$. Based on the amplitudes and the frequency range of 0.2–20 kHz of the 2D spectrum, a detailed description can be given of each spectrum corresponding to the elastic range, strengthening control and deflection control. The load-deflection test and the AE measurement (mortar) did not begin at the same time. The duration of the load–deflection test was circa 40 s (Figure 3). The sound spectrum time range was shorter and equalled 24.4 s; this time range preceded the occurrence of f_{cr}. The sound spectrum time range was marked on the left-side part of Figure 4c, 3D. Figure 4a presents the 3D sound spectrum in the range from 0 to 11.1 s, which corresponds to the background noise. As is shown in Figure 4b,c, the micro-events begin to concentrate circa 9 s before $f_{cr} = f_{max}$ and then they increase up to $f_{cr} = f_{max}$; this 9-s period of time is within the range between 2/3 of f_{cr} and $f_{cr} = f_{max}$ (load), see Figure 3.

Figure 5 presents the 2D and 3D sound spectra for a beam made of cement paste with dispersed reinforcement $V_f = 6\%$. As is shown by the load–deflection correlation in Figure 3, the specimen is a typical ESD composite. The load–deflection test and the AE measurement (cement paste) began at the same time and lasted 138 s (Figure 3). Figure 5a presents the 3D sound spectrum of the background noise whose time range equals 10 s. Micro-events start to appear circa 27 s before f_{cr}, then they concentrate circa 8 s before f_{cr} and increase up to $f_{cr} = f_{max}$, see Figure 5b,c; this 27-s period of time (from the moment the micro-events start to appear to f_{cr}) is within the range of 2/3 of f_{cr} and $f_{cr} = f_{max}$ (load), see Figure 3.

Figure 6 presents the AE results for a concrete specimen reinforced with traditional continuous reinforcement; this specimen is characterised by the largest elastic range. The background noise spectrum is presented in Figure 6a. Figure 6b shows the spectrum of micro-events preceding $f_{cr} = f_{max}$, and Figure 6c presents the spectrum of the macrocrack $f_{cr} = f_{max}$ and the spectrum of micro-events preceding f_{cr}. The load–deflection test and the AE measurement (concrete) did not begin at the same time. The duration of the load–deflection test was circa 290 s (Figure 3). The sound spectrum time range was shorter and equalled 91 s; this time range preceded the occurrence of f_{cr}. The time was marked on the left-side part of Figure 6c, 3D. Figure 6a presents the 3D sound spectrum of the background noise whose time range equals 10 s. As is shown by Figure 6c, the micro-events begin to concentrate circa 9–10 s before the $f_{cr} = f_{max}$ point; this 9–10-s period of time is within the range of 2/3 of f_{cr} and $f_{cr} = f_{max}$ (load), see Figure 3.

4. Discussion

It has been confirmed that the load–deflection curve enables the identification of the proportionality, strengthening, deflection control and crack propagation areas, see Figure 3. The rapid decreases in the ability to carry stress that were recorded on the load–deflection curve indicate the appearance of macrocrack and fibre breaking, see Figures 3–6.

The recognition of the range of micro-events (using sound spectrum) preceding the critical crack at point $f_{cr} = f_{max}$ is the main focus of research in this paper. In cement composites (with or without reinforcement), which are characterised by a significant decrease in stress after exceeding f_{cr}, the recording of micro-events is relatively easy to recognise. The recording of micro-events begins at the point corresponding to circa 2/3 of $f_{cr} = f_{max}$ of the stress in the elastic range, see Figure 4. The concentration of micro-events and the corresponding amplitude range increase as point $f_{cr} = f_{max}$ is being approached. The above observations are illustrated in the form of a diagram in Figure 7. The analysis regarding the range of micro-events occurrence was carried out based on Figures 4–6 and the presented data.

In the case of ESD composites without the decrease in stress after exceeding f_{cr} (e.g., owing to multicracking in the cement paste with $V_f = 6\%$, see Figures 3 and 5), micro-events in the elastic range (the same as in composites $f_{cr} = f_{max}$) can be recorded. However, it should be noted that the recognition of the sound spectrum corresponding to f_{cr} as well as the spectra of micro-events preceding f_{cr} is more difficult. This results from the fact that microcracks at point f_{cr} generate smaller acoustic effects than at point $f_{cr} = f_{max}$

(composites characterised by a decrease in stress at the critical point). At the same time, the concentration of events and increase in amplitudes before f_{max} in the tested composite were also recorded. The diversified range of the strengthening control area resulting from the diversified destruction range makes it difficult to estimate the range of occurrence of micro-events before f_{max}.

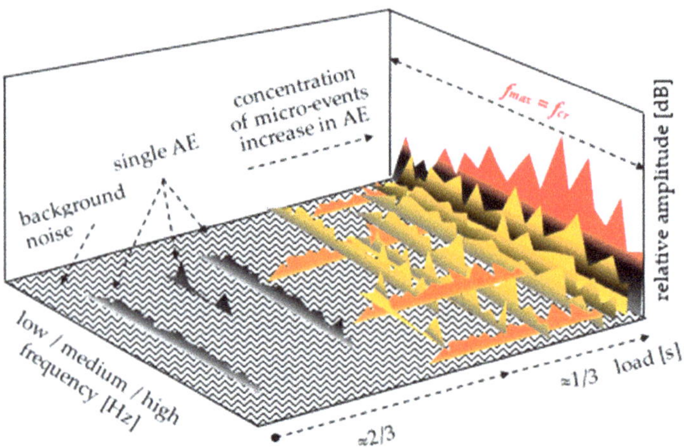

Figure 7. The range of micro-events in cement composites ($f_{cr} = f_{max}$) identified by means of a space sound spectrum.

An analysis of the time range of micro-event occurrence recorded in other papers with respect to f_{cr} and f_{max} is similar [46,47]. Micro-events were observed in all other tested specimens, small- and large-size ones, with and without cuts, and with various types of reinforcement. It seems that the conclusions drawn in this respect could be extended to the majority of cement composites.

It should be noted that the speed of crosshead displacement in the process of recognising micro-events in future tests should be significantly reduced. In addition, heads used to record AE should be located in places of applied forces and supports. It has also been noted that the frequency range should be expanded by adding low and high frequency and infrasound spectra, which will probably provide additional information, facilitating the identification of the destruction process.

Micro-events relate to changes at the microstructure level; this destruction process is not visible and requires identification in the future by means of scanning or a microscope. The occurrence of micro-events may be attributed to microcracks connected with the redistribution of stress. They may concentrate in areas subjected to tension or in places where the load is applied. The concentration of stress also occurs at the point of contact between a specimen and the supports on which the beams were placed. The confirmation or refutation of the above suggestions requires additional tests that are planned to be carried out in the near future.

The correct identification of micro-events with the corresponding range of stresses in the elastic range may reduce the level of stresses in the calculations (especially with the cyclical application of dynamic load) in order to increase the durability of structural elements, which so far has not been the focus of interest.

There are lots of variables (location of the sensor, type and geometry of the sensor and sample and background noise) that make it difficult to compare this study with other similar experiments. The identification of micro-events should be expanded for a larger number of specimens with varied geometry, which will serve as the basis for the statistical processing of the results. An attempt to use the presented method conducted under a load–deflection test to monitor building structures based on cement composites is not practical but an effort to record and identify the micro-events (before the occurrence of f_{cr}) correlated with the stress level seems to be worth making.

5. Conclusions

It has been confirmed that there is a possibility of predicting the occurrence of micro-events in cement composites, which occur at the end of the Hooke's law range $f_{cr} = f_{max}$.

It has been found that, in cement composites, for which $f_{cr} = f_{max}$, micro-events occur after exceeding 2/3 of the stress elastic range, and the concentration and amplitude of the sound spectrum increase before the critical crack occurs.

Recording micro-events in quasi-brittle cement composites (in which there is no rapid decrease in stress after exceeding critical points f_{cr} and f_{max}) is also possible, but the range of occurrence of micro-events and the correlated spectrum amplitudes are smaller, which make it more difficult to locate and recognize them.

The micro-events process is not visible and requires identification in the future by means of scanning or a microscope. This phenomenon may be connected with micro-cracks or regrouping of stresses in the structure of the material or with effects occurring at the place of the applied load, which should be confirmed in future tests.

The correct identification of micro-events with the corresponding range of stresses in the Hooke's law area may reduce the level of stresses adopted for the calculation (especially with the cyclical dynamic load), in order to increase the durability of the structural elements.

Author Contributions: Conceptualization, methodology-writing, D.L.; review and analyzed the test results J.J.; supervision and editing Z.K.; review and editing P.F. All authors have read and agreed to the published version of the manuscript.

Funding: The authors from the Brno University of Technology thanks the financial support of the Czech Science Foundation under project No. 19-09491S (MUFRAS). This research received no external funding.

Institutional Review Board Statement: Not applicable.

Informed Consent Statement: Not applicable.

Data Availability Statement: No new data were created or analyzed in this study. Data sharing is not applicable to this article.

Acknowledgments: Not applicable.

Conflicts of Interest: The author declares no conflict of interest.

References

1. Ohtsu, M. Determination of crack orientation by acoustic emission. *Mater. Eval.* **1987**, *45*, 1070–1075.
2. Ohtsu, M. The history and development of acoustic emission in concrete engineering. *Mag. Concr. Res.* **1996**, *48*, 321–330. [CrossRef]
3. Ouyang, C.S.; Landis, E.; Shh, S.P. Damage assessment in concrete using quantitative acoustic emission. *J. Eng. Mech.* **1991**, *117*, 2681–2698. [CrossRef]
4. Ranachowski, Z. Method of measurement and analysis of acoustic emission signal. *IPPT Pan Pap.* **1997**, *1*, 113. (In Polish)
5. Kucharska, L.; Brandt, A.M. Pitch-based carbon fibre reinforced cement composites. In *Materials Engineering Conference ASCE. Materials for the New Millenium*; Chong, K.P., Ed.; American Society of Civil: Washington, DC, USA; American Society of Civil Engineers (book): New York, NY, USA, 1996; Volume 1, pp. 1271–1280.
6. Brandt, A.M. Fibre reinforced cement-based (FRC) composites after over 40 years of development in building and civil engineering. *Compos. Struct.* **2008**, *86*, 3–9. [CrossRef]
7. Dzayea, E.D.; Schutterb, G.D.; Aggelisa, D.G. Monitoring early-age acoustic emission of cement paste and fly ash paste. *Cem. Concr. Res.* **2020**, *129*, 105964. [CrossRef]
8. Assi, L.; Soltangharaei, V.; Anay, R.; Ziehl, P.; Matta, F. Unsupervised and supervised pattern recognition of acoustic emission signals during early hydration of Portland cement paste. *Cem. Concr. Res.* **2018**, *103*, 216–225. [CrossRef]
9. Elaqra, H.; Godin, N.; Peix, G.; R'Mili, M.; Fantozzi, G. Damage evolution analysis in mortar, during compressive loading using acoustic emission and X-ray tomography: Effects of the sand/cement ratio. *Cem. Concr. Res.* **2007**, *37*, 703–713. [CrossRef]

10. Šimonová, H.; Topolář, L.; Schmid, P.; Keršner, Z.; Rovnaník, P. Effect of carbon nanotubes in metakaolin-based geopolymer mortars on fracture toughness parameters and acoustic emission signals. In Proceedings of the BMC 11 International Symposium on Brittle Matrix Composites, Warsaw, Poland., 28–30 September 2015; pp. 261–288.
11. Chen, B.; Liu, J. Experimental study on AE characteristics of free-point-bending concrete beams. *Cem. Concr. Res.* **2004**, *34*, 391–397. [CrossRef]
12. Ranachowski, Z.; Schabowicz, K. The contribution of fibre reinforcement system to the overall toughness of cellulose fibre concrete panels. *Constr. Build. Mater.* **2017**, *156*, 1028–1034. [CrossRef]
13. Schabowicz, K.; Gorzelańczyk, T.; Szymków, M. Identification of the degree of degradation of fibre-cement boards exposed to fire by means of the acoustic emission method and artificial neural networks. *Materials* **2019**, *12*, 656. [CrossRef] [PubMed]
14. Ranachowski, Z.; Jóźwiak–Niedźwiedzka, D.; Brandt, A.M.; Dębowski, T. Application of acoustic emission method to determine critical stress in fibre reinforced mortar beams. *Arch. Acoust.* **2012**, *37*, 261–268. [CrossRef]
15. Brandt, A.M.; Ranchowski, Z.; Zieliński, M.; Dąbrowski, M.; Sobczak, M. *Report from Tests of Cracking Resistance of Bent Cement Composite Samples*; Polish Academy of Science Institute of Fundamental Technology Problems: Warsaw, Poland, 2010.
16. Aggelis, D.G.; Mpalaskas, A.C.; Matikas, T.E. Investigation of different modes in cement-based materials by acoustic emission. *Cem. Concr. Res.* **2013**, *48*, 1–8. [CrossRef]
17. Kim, B.; Weiss, W.J. Using acoustic emission to quantify damage in restrained fibre-reinforced cement mortars. *Cem. Concr. Res.* **2003**, *33*, 207–214. [CrossRef]
18. Landis, E.; Ballion, L. Experiments to relate acoustic energy to fracture energy of concrete. *J. Eng. Mech.* **2002**, *128*, 698–702. [CrossRef]
19. Paul, S.C.; Pirskawetz, S.; Zijl, G.P.A.G.; Schmidt, W. Acoustic emission for characterising the crack propagation in strain-hardening cement-based composites (SHCC). *Cem. Concr. Res.* **2015**, *69*, 19–24. [CrossRef]
20. Reinhardt, H.W.; Weiler, B.; Grosse, C. Nondestructive testing of steel fibre reinforced concrete. *Brittle Matrix Compos.* **2000**, *6*, 17–32.
21. Shahidan, S.; Rhys Pulin, R.; Bunnori, N.M.; Holford, K.M. Damage classification in reinforced concrete beam by acoustic emission signal analysis. *Constr. Build. Mater.* **2013**, *45*, 78–86. [CrossRef]
22. Shiotani, T.; Li, Z.; Yuyama, S.; Ohtsu, M. Application of the AE improved b-value to quantitative evaluation of fracture process in Concrete Materials. *J. Acoust. Emiss.* **2004**, *19*, 118–133.
23. Soulioti, D.; Barkoula, N.M.; Paipetis, A.; Matikas, T.E.; Shiotani, T.; Aggelis, D.G. Acoustic emission behavior of steel fibre reinforced concrete under bending. *Constr. Build. Mater.* **2009**, *23*, 3532–3536. [CrossRef]
24. Watanab, K.; Niwa, J.; Iwanami, M.; Yokota, H. Localized failure of concrete in compression identified by AE method. *Constr. Build. Mater.* **2004**, *18*, 189–196. [CrossRef]
25. Yuyama, S.; Ohtsu, M. Acoustic Emission evaluation in concrete. In *Acoustic Emission-Beyond the Millennium*; Kishi, T., Ohtsu, M., Yuyama, S., Eds.; Elsevier: Amsterdam, The Netherlands, 2000; pp. 187–213.
26. Ohno, K.; Ohtsu, M. Crack classification in concrete based on acoustic emission. *Constr. Build. Mater.* **2010**, *24*, 2339–2346. [CrossRef]
27. Granger, S.; Pijaudier, G.; Loukili, A.; Marlot, D.; Lenain, J.C. Monitoring of cracking and healing in an ultra high performance cementitious material using the time reversal technique. *Cem. Concr. Res.* **2009**, *39*, 296–302. [CrossRef]
28. Ohtsu, M. Elastic wave methods for NDE in concrete based on generalized theory of acoustic emission. *Constr. Build. Mater.* **2016**, *122*, 845–855. [CrossRef]
29. Ono, K.; Gołaski, L.; Gębski, P. Diagnostic of reinforced concrete bridges by acoustic emission. *J. Acoust. Emiss.* **2002**, *20*, 83–98.
30. Parmar, D. *Non-Destructive Bridge Testing and Monitoring with Acoustic Emission (AE) Sensor Technology*; Final Report; Hampton University: Hampton, VA, USA, 2011.
31. Swit, G. Acoustic Emission Method for Locating and Identifying Active Destructive Processes in Operating Facilities. *Appl. Sci.* **2018**, *8*, 1295. [CrossRef]
32. Anay, R.; Soltangharaei, V.; Assi, L.; DeVol, T.; Ziehl, P. Identification of damage mechanisms in cement paste based on acoustic emission. *Constr. Build. Mater.* **2018**, *164*, 286–296. [CrossRef]
33. Ai, Q.; Liu, C.C.; Chen, X.R.; He, P.; Wang, Y. Acoustic emission of fatigue crack in pressure pipe under cyclic pressure. *Nucl. Eng. Des.* **2010**, *240*, 3616–3620. [CrossRef]
34. Seitl, S.; Miarka, P.; Šimonová, H.; Frantík, P.; Keršner, Z.; Domski, J.; Katzer, J. Change of fatigue and mechanical fracture properties of a cement composite due to partial replacement of aggregate by red ceramic waste. *Period. Polytech.-Civ. Eng.* **2019**, *1*, 152–159. [CrossRef]
35. Šimonová, H.; Kumpová, I.; Rozsypalová, I.; Bayer, P.; Frantík, P.; Rovnaníková, P.; Keršner, Z. Fracture parameters of alkali-activated aluminosilicate composites with ceramic precursor. In *26th Concrete Days. Solid State Phenomena*; Trans Tech Publications Ltd.: Bäch, Switzerland, 2020; pp. 73–79. ISBN 978-3-0357-1668-9. ISSN 1662-9779.
36. Juraszek, J. Influence of the spatial structure of carbon fibres on the strength properties of a carbon composite. *Fibres Text. East. Eur.* **2019**, *27*, 111–115. [CrossRef]
37. Juraszek, J. Fiber bragg sensors on strain analysis of power transmission lines. *Materials* **2020**, *13*, 1559. [CrossRef]
38. Juraszek, J. Strain and force measurement in wire guide. *Arch. Min. Sci.* **2018**, *63*, 321–334.

39. Kowalik, T.; Logoń, D.; Maj, M.; Rybak, J.; Ubysz, A.; Wojtowicz, A. Chemical hazards in construction industry. In Proceedings of the XXII International Scientific Conference on Construction the Formation of Living Environment (FORM-2019), Tashkent, Uzbekistan, 18–21 April 2019; Volkov, A., Pustovgar, A., Sultanov, T., Adamtsevich, A., Eds.; E3S Web of Conf. EDP Sciences, 2019; Volume 97, p. 03032, ISSN 2267-1242.
40. Tsangouri, E.; Michels, L.; El Kadi, M.; Tysmans, T.; Aggelis, D.G. A fundamental investigation of textile reinforced cementitious composites tensile response by Acoustic Emission. *Cem. Concr. Res.* **2019**, *123*, 105776. [CrossRef]
41. Van Steen, C.; Verstrynge, E.; Wevers, M.; Vandewalle, L. Assessing the bond behaviour of corroded smooth and ribbed rebars with acoustic emission monitoring. *Cem. Concr. Res.* **2019**, *120*, 176–186. [CrossRef]
42. Kumar Das, A.; Suthar, D.; Leung, C.K.Y. Machine learning based crack mode classification from unlabelled acoustic emission waveform features. *Cem. Concr. Res.* **2019**, *121*, 42–57.
43. Logoń, D. FSD cement composites as a substitute for continuous reinforcement. In Proceedings of the Eleventh International Symposium on Brittle Matrix Composites BMC-11, Warsaw, Poland, 28–30 September 2015; Brandt, A.M., Ed.; Institute of Fundamental Technological Research: Warsaw, Poland, 2015; pp. 251–260.
44. Logoń, D. Identification of the destruction process in quasi brittle concrete with dispersed fibres based on acoustic emission and sound spectrum. *Materials* **2019**, *12*, 2266. [CrossRef] [PubMed]
45. Logoń, D. Monitoring of microcracking effect and crack propagation in cement composites (HPFRC) using the acoustic emission (AE). In Proceedings of the 7th Youth Symposium on Experimental Solid Mechanics, YSESM '08, Wojcieszyce, Poland, 14–17 May 2008.
46. Logoń, D. The application of acoustic emission to diagnose the destruction process in FSD cement composites. In Proceedings of the International Symposium on Brittle Matrix Composites BMC-11, Warsaw, Poland, 28–30 September 2015; Brandt, A.M., Ed.; Institute of Fundamental Technological Research: Warsaw, Poland, 2015; pp. 299–308.
47. Logoń, D.; Schabowicz, K. The recognition of the micro-events in cement composites and the identification of the destruction process using acoustic emission and sound spectrum. *Materials* **2020**, *13*, 2988. [CrossRef] [PubMed]
48. Logoń, D.; Schabowicz, K. The increase in the elastic range and strengthening control of quasi brittle cement composites by low-module dispersed reinforcement—The assessment of reinforcement effects. *Materials* **2021**, *14*, 341. [CrossRef]
49. ASTM 1018. *Standard Test Method for Flexural Toughness and First Crack Strength of Fibre-Reinforced Concrete*; ASTM: Philadelphia, PA, USA, 1992; Volume 04.02.

 *materials*

Article

Reduction of Load Capacity of Fiber Cement Board Facade Cladding under the Influence of Fire

Krzysztof Schabowicz [1], Paweł Sulik [2] and Łukasz Zawiślak [1,*]

1   Faculty of Civil Engineering, Wrocław University of Science and Technology, Wybrzeże Wyspiańskiego 27, 50-370 Wrocław, Poland; krzysztof.schabowicz@pwr.edu.pl
2   Building Research Institute (Instytut Techniki Budowlanej), Filtrowa 1, 00-611 Warszawa, Poland; p.sulik@itb.pl
*   Correspondence: lukasz.zawislak@pwr.edu.pl

Abstract: The paper analyzes the issue of the reduction of load capacity in fiber cement board during a fire. Fiber cement boards were put under the influence of fire by using a large-scale facade model. Such a model is a reliable source of knowledge about the behavior of facade cladding and the way fire spreads. One technical solution for external walls—a ventilated facade—is gaining popularity and is used more and more often. However, the problem of the destruction during a fire of a range of different materials used in external facade cladding is insufficiently recognized. For this study, the authors used fiber cement boards as the facade cladding. Fiber cement boards are fiber-reinforced composite materials, mainly used for facade cladding, but also used as roof cladding, drywall, drywall ceiling and floorboards. This paper analyzes the effect of fire temperatures on facade cladding using a large-scale facade model. Samples were taken from external facade cladding materials that were mounted on the model at specific locations above the combustion chamber. Subsequently, three-point bending flexural tests were performed and the effects of temperature and the integrals of temperature and time functions on the samples were evaluated. The three-point bending flexural test was chosen because it is a universal method for assessing fiber cement boards, cited in Standard EN 12467. It also allows easy reference to results in other literature.

Keywords: ventilated facades; fire safety; fiber cement board; flexural strength; cladding

Citation: Schabowicz, K.; Sulik, P.; Zawiślak, Ł. Reduction of Load Capacity of Fiber Cement Board Facade Cladding under the Influence of Fire. *Materials* **2021**, *14*, 1769. https://doi.org/10.3390/ma14071769

Academic Editor: Hyeong-Ki Kim

Received: 2 March 2021
Accepted: 31 March 2021
Published: 3 April 2021

Publisher's Note: MDPI stays neutral with regard to jurisdictional claims in published maps and institutional affiliations.

Copyright: © 2021 by the authors. Licensee MDPI, Basel, Switzerland. This article is an open access article distributed under the terms and conditions of the Creative Commons Attribution (CC BY) license (https://creativecommons.org/licenses/by/4.0/).

1. Introduction

A ventilated facade is a modern technical solution for the exterior part of a multilayer wall. It consists of an external facade cladding that is mechanically or adhesively attached to a subframe. The subframe is mechanically attached to the exterior structural wall of the building. External facade cladding can be made of a variety of materials, e.g., fiber cement boards, concrete slabs, steel elements, ceramic and other composite elements. Facade cladding is usually installed in accordance with the individual technical design of the facade and the requirements set out by the product manufacturer. They are non-load-bearing elements, bearing only their own weight and environmental impacts such as snow, wind and temperature. External facade cladding does not ensure the airtightness of the building, but only to a certain extent ensures the protection of the external surface of the supporting wall to which the facade is fixed. A ventilated facade is a complete set of individual components that make up a system solution. The standard that sets requirements for a complete ventilated facade system is ETAG 034-1 [1], and the individual components of the whole system must additionally meet national requirements.

The most important element of a ventilated facade is the air gap between the external cladding facade and the insulation layer (mineral wool or stone wool), or the supporting wall if no insulation layer is used. The air gap, also called the ventilation air space, should be at least 20 mm according to ETAG 034-1 [1]; the literature also provides information that the ventilation air space should be in the range of 20 mm to 50 mm [2,3]. It may be reduced locally to 5–10 mm, depending on the cladding and substructure, provided that the performance function of the complete system is not affected. The most important parameter, independent of the dimension of the ventilation air space, is an appropriate possibility of airflow through the air gap. This is ensured not only by the dimension of the ventilation space but also by an appropriate number of ventilation gaps, allowing air to enter this space. Ventilation slots supply air to the ventilation air space, and should be at least 50 cm^2/1 m facade [1], assuming they are at least at the base point and at the edge of the roof.

Ventilated facades allow the facade cladding to be made with different materials, structures, textures, or colors. Due to good aesthetics and durability, a ventilated facade is increasingly used as a technical solution for the external multilayer walls of newly constructed buildings, but it also performs well in the case of buildings undergoing renovation. External facade claddings can be made of very large elements, e.g., the standard size for fiber cement boards is 1.25 × 3.10 mm^2 [2], and for HPL (high-pressure laminate) boards, 1.85 × 4.10 mm^2 [2].

This paper analyzes a ventilated facade with external facade cladding made of fiber cement boards, which are classified as fiber-reinforced composites. These composites are characterized by two phases [4,5]. The first phase is a cement matrix, based on Portland cement. The second phase of these composites is the dispersed phase, which is characterized by the distribution of fibers in a discontinuous and randomly oriented manner. In the case of fiber-reinforced composites, they offer many of the benefits of using fibers under normal conditions [6,7]. In fiber cement board mainly cellulose fibers, polyvinyl alcohol (PVA) synthetic fibers and polypropylene (PP) fibers are used. Unfortunately, in the case of fire conditions, there are few such studies, which is due to the complexity of the tests and their costly nature. The individual fibers have the following melting points: PVA (polyvinyl alcohol) synthetic fibers, about 200–220 °C [8,9]; PP (polypropylene) fibers, about 160–175 °C [8,10]; cellulose fibers, 260–270 °C [11].

Fiber cement boards have not been extensively studied in terms of fire temperatures and their behavior on the facade in case of fire. Szymkow's research [12] showed that the fibers in fiber cement boards are destroyed at 230 °C after about 3 h of exposure. In ref. [13], a decrease of about 10% in the compressive strength of concrete and fiber concrete was seen at temperatures up to 300 °C. In contrast, for fiber-reinforced cement composites, the flexural strength increases with increasing temperatures, up to about 300 °C [14]. In the case of compressive strength in fiber concrete and fiber cement, this temperature rise does not reduce the compressive strength; on the contrary, it increases the strength by evaporating water from the pores (this is confirmed by tests [8] in which high temperatures acted upon the fiber concrete sample for about 100 min). Temperatures in the range of about 300 °C, as shown in the above tests, are dangerous only for the fibers because their melting point is exceeded. Temperatures higher than 400 °C, as presented in the research by Szymkow [12], only strengthen the cement matrix over a short period of time (usually in the range of 2.5–7.5 min, depending on the sample). During this time, water evaporates from the pores, which increases the bending capacity, but after this period the fibers begin to melt and then the strength drops drastically.

Unfortunately, the temperatures of a fire affecting external facade claddings may locally reach values even exceeding 800 °C. The external curve reflecting a developed facade fire, as presented in the standard [15], represents values that limit the temperature to about 660 °C. Nevertheless, higher temperatures may be expected in the vicinity of the window lintel. Unfortunately, experimental studies are not available for fiber cement boards. Looking at the above analogies of other fiber materials, interesting conclusions are reached in ref. [13], where tests were performed for concrete and fiber-reinforced concrete. The compressive strength of concrete and fiber-reinforced concrete of class C30/37 in temperatures of 800 °C reduces by more than 90%. At 500 °C and 600 °C, the samples without fiber addition were destroyed during their annealing, while those with polypropylene fiber additions retained residual flexural strength [14]. In ref. [16]

the positive effect of using fibers to increase the flexural strength of beams subjected to a normative fire temperature curve is demonstrated. Research shows that, despite the fact that the fibers are subject to destruction at fire temperatures, the voids thus formed allow the material to withstand higher temperatures afterward. The emergency situation of a fire is shown in Figure 1.

Figure 1. View of flames escaping from a room involved in a fully developed standard fire, during a field test. (Author: E. Kotwica).

Ventilated facades, compared to ETICS (External Thermal Insulation Composite System) facade, in terms of the problems of falling facade elements during a fire, show much worse parameters [17–19]. The emergency units, which are responsible for evacuating people inside the building in the event of a fire, are particularly at risk. This problem is well known in the scientific community, and the work of the European Commission is also based on solving this problem [20,21].

There are several methods for testing the response of fiber cement boards to high temperatures. One of them was presented by Szymkow in ref. [12]—by annealing the samples in a furnace, specially prepared for this test. A similar way of affecting fiber cement board samples was adopted in ref. [22]. It is also possible to prepare a large-scale facade model. As far as this form of testing is concerned, there are many standards in the world for testing such models [20,23–25]. In most cases, they assume a fire spreads from an opening towards the facade, and simulate the window openings of a room. A hearth (fire source), defined by a temperature action standard curve, is located in a recess. Flames escape from the opening, affecting the facade and other wall elements. The standards differ in details, i.e., the type of hearth (wood cribs [20,23,24] or gas [25]), the opening dimensions, the test time, and the large-scale facade model and its dimensions. A comparison of different standards for testing life-sized facade models for fire safety is summarized in ref. [26].

2. Materials and Methods of Ventilated Facades

By analyzing the literature, deficiencies were noted in the response and destruction of fiber cement boards when exposed to high temperatures through fire. The authors decided to verify how the high temperatures from the action of fire affect the reduction of flexural strength of fiber cement board on the facade.

To perform the study, the authors prepared a model of the facade. For this purpose, they used a large-scale facade model. The facade to be analyzed was attached to a test platform made of autoclaved cellular concrete blocks, traditional masonry and poured reinforced concrete lintels.

The facade was made of 8 mm-thick fiber cement boards of natural color—not pigmented. The cladding was fastened using mechanical connections to a galvanized steel substructure. The substructure was mechanically attached to the platform through consoles. The division of the facade into facade cladding is shown in Figure 2.

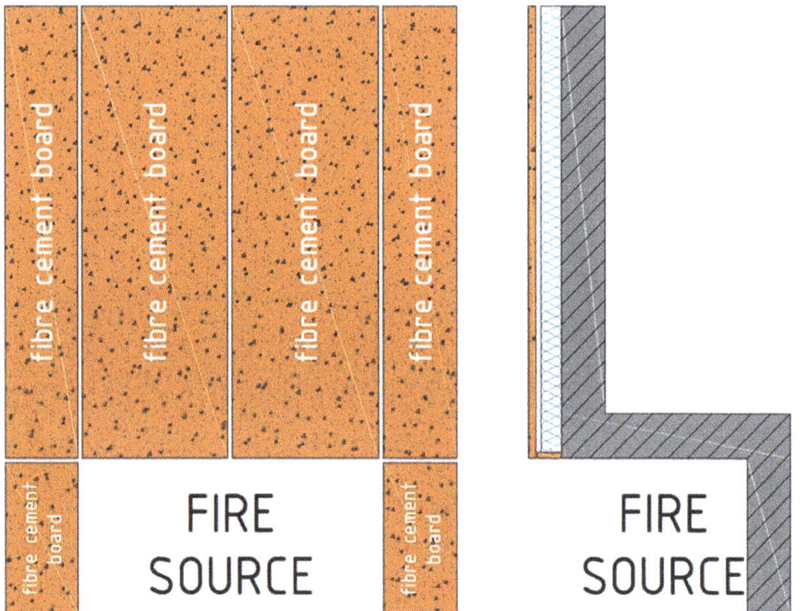

Figure 2. Division of boards on the facade—front and side view.

The full view of the large-scale facade model, as it was conducted, is shown in Figure 3a. In addition, the sand burner and the combustion chamber can be seen, through which the fire scenario is implemented. The facade cladding in the left and right part is fixed in a different way, which causes the left part to protrude beyond the face of the plane—this can be seen in Figure 3b. The technical solution for filling and maintaining the stone wool and ventilation space is shown in Figure 3c.

The impact of high temperatures on the facade was tested by means of fire and fire gases escaping from the combustion chamber—the scenario of fire in the room and fire escaping through the window opening onto the facade was carried out. The combustion chamber on the back wall contained a blower, which allowed us to reproduce the real conditions of a fire. The fire source was a sand burner that used gas as fuel—appropriately metered through a mass flow meter. The sand burner used liquid propane of 95% purity (PGNiG Polska) as fuel, with a release rate scaled in such a way that it corresponded to the fire development in accordance with the external curve, which is characteristic for facade fires as cited in EN 13501-2 [27]. International fire curves show the temperature course over time at the source location. The international fire curves, including the external curve, are shown in Figure 4. Laminar air inflow from the side of the room was provided by honeycomb.

The large-scale facade model was equipped with a set of thermocouples (named TE1-TE9) to verify and identify the damage to the exterior cladding. They were placed in accordance with the expected development and impact of flames and hot fire gases. Thermocouples were placed on the surface of the board cladding—in 3 rows at heights of 800 mm, 1600 mm and 2500 mm above the combustion chamber. These places were selected due to the expected shape of the flame coming out of the combustion chamber. The exact arrangement of thermocouples and their names are shown in Figure 5.

Figure 3. Selected details of the large-scale facade model: (**a**) actual elevation model view; (**b**) eaves problem and solution; (**c**) insulation.

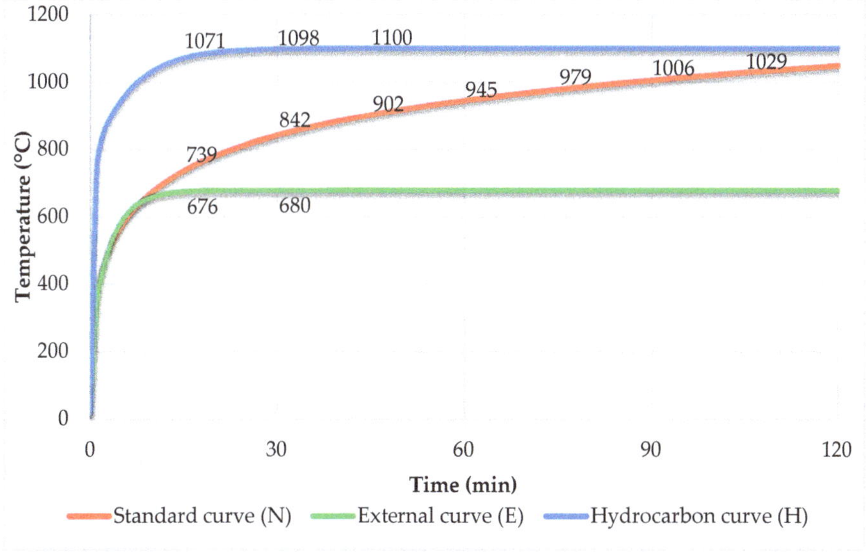

Figure 4. Summary of fire dependencies.

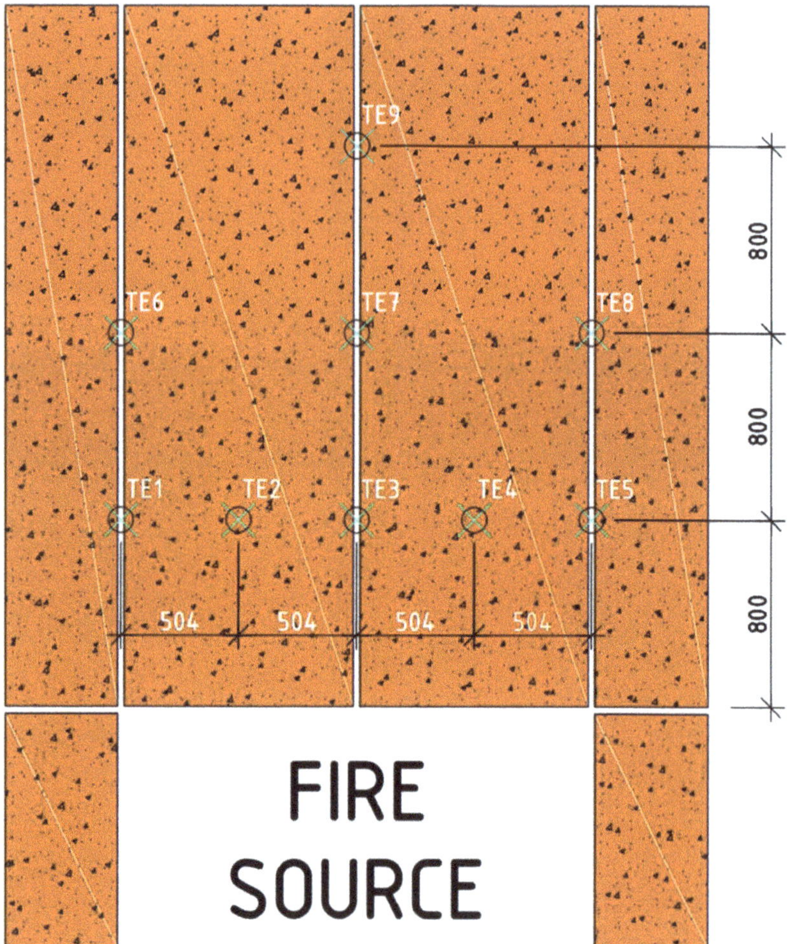

Figure 5. Location of thermocouples on the large-scale facade model—front view.

3. The Test and Results

The test scenario of heating with the sand burner corresponded to the intensity of a fire equivalent to that which would be experienced if a real fire occurred in the room located directly behind the facade (inside the building), with flames escaping from the window and affecting the facade.

The test was conducted at an ambient temperature of 22.3 °C and a relative humidity of 52%. The test began by setting up the burner and properly calibrating the supplied gas. The beginning of the test is shown in Figure 6a—uniform distribution of flames leaning against the splay is noticeable. Such a view is characteristic of the beginning of the test, where the model is not exposed to external winds and the facade cladding is not yet degraded in the initial phase. The first 6-min period of the fire brought smoke/charring of the external facade cladding, and cracks appeared on the upper splays, which are most exposed to high temperatures. Facade cladding mounted on the splays fell off first, after a time range of 9 min to 14 min. During this period, the destruction of the external facade cladding progressed, as shown in Figure 6b, where the continuing degree of destruction is shown on the already targeted facade cladding, within 600 mm of the bottom. In Figure 6c,d, the progression of facade cladding deterioration under the effect of fire temperatures can be seen; in the minutes that followed, the external facade cladding elements fell off. The side splay boards fell off at about minute 17.

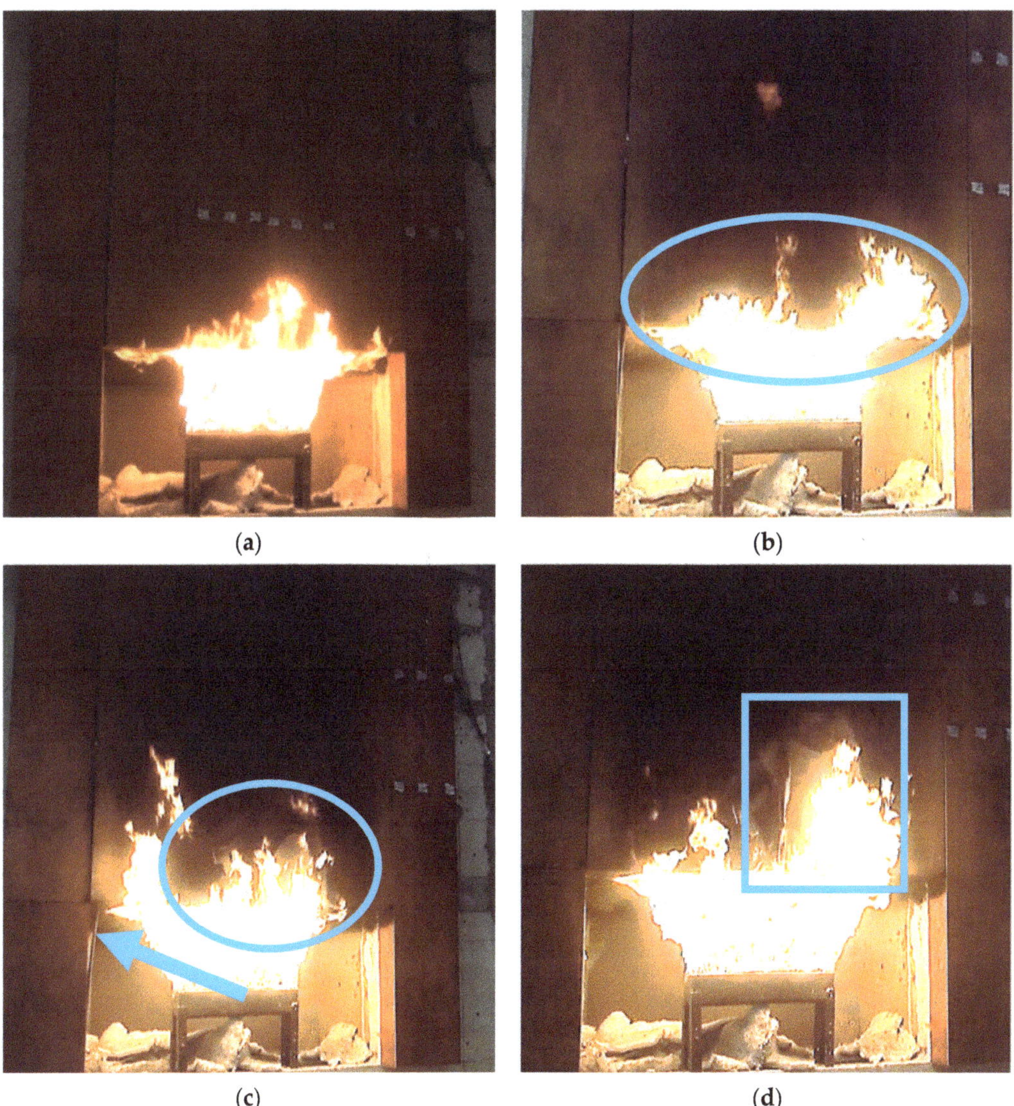

Figure 6. View of the large-scale facade model in successive minutes of the test: (**a**) start of the test; (**b**) approximately 11′00″—visible degradations; (**c**) approximately 17′00″—splay piece falling off; (**d**) approximately 20′30″—progress of destruction.

At the end of the test, the load-bearing capacity of the fiber cement boards on the higher elements was progressively exhausted as more and more elements fell off—this is shown in Figure 7a,b. Figure 7c shows the appearance of the cladding after the 60-min test. Significant deterioration of the cladding on the right side of the test model is evident.

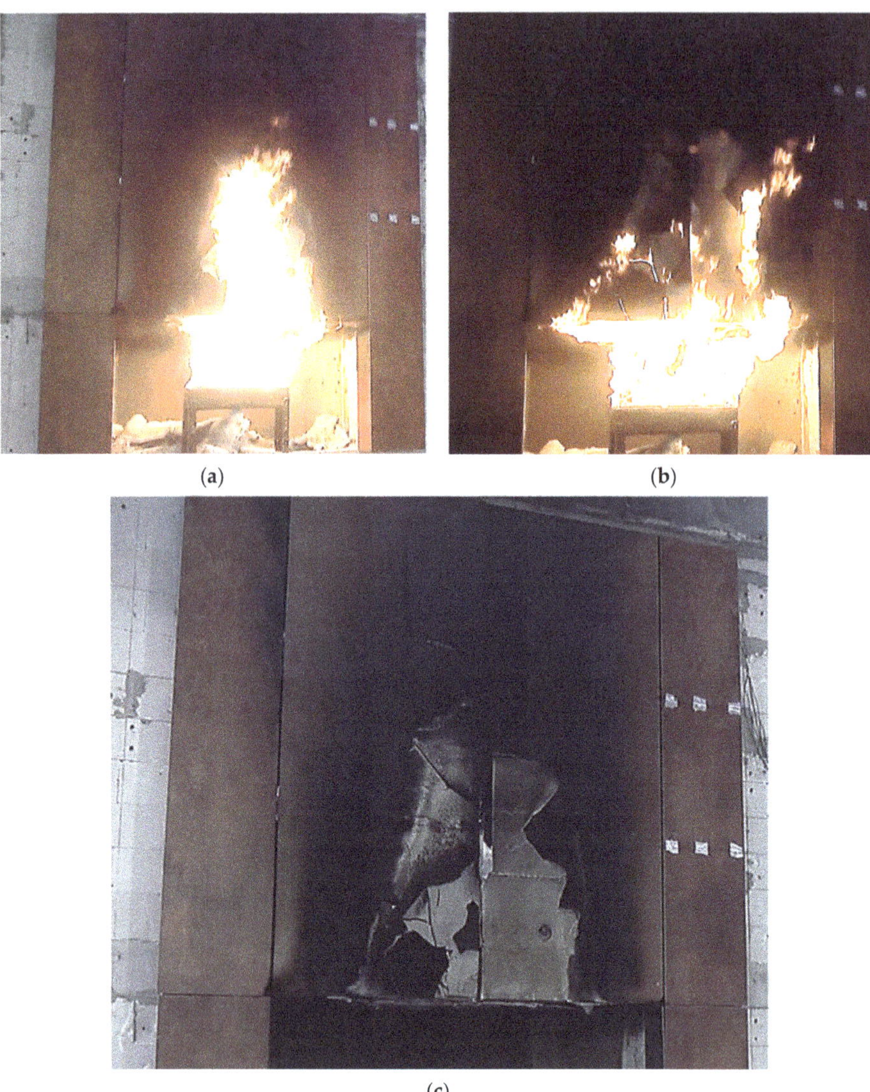

Figure 7. View of the large-scale facade model in the subsequent minutes of the test: (**a**) approximately 40′00″—high cladding degradation; (**b**) approximately 55′00″—more elements are falling off; (**c**) view of facade cladding after testing.

As shown in Figure 5, the thermocouples recorded the temperature continuously during the test. The results of these tests are shown in Figure 8. It can be observed that the temperatures for TE1 and TE5 thermocouples, and TE6 and TE8 thermocouples, are similar, operating in the low-temperature range—below 100 °C. Such temperatures have no destructive effect on fibers in fiber cement boards. The whole material did not show major signs of wear at these temperatures, either. In the case of TE2, TE3, and TE4 thermocouples a disproportion is visible, i.e., much higher temperatures prevail on the right side of the large-scale facade model—this is due to a smaller protrusion of the cladding beyond the face. Although this difference is minimal (20 mm), it directs all the flames to the right side where the TE4 thermocouple is located. It is also noticeable that the flame source has a much greater effect on thermocouples located in a non-central position, such as TE2 and TE4, than on thermocouples located centrally but higher up, such as TE7.

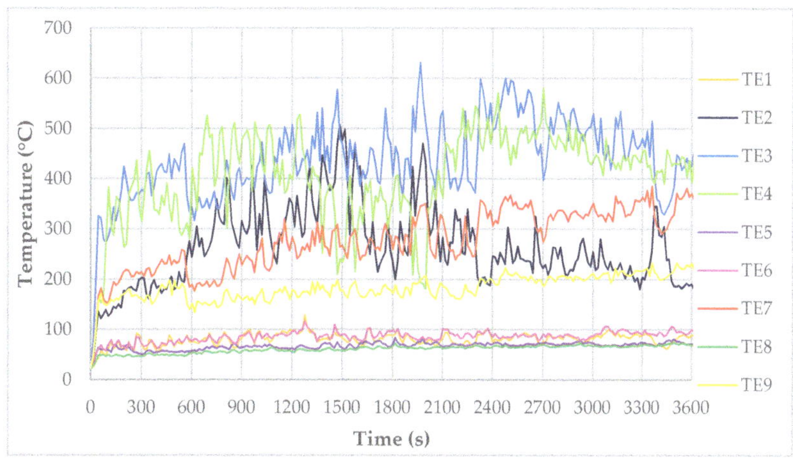

Figure 8. Temperature measurement results for thermocouples on the large-scale facade model.

4. Materials and Methods

In order to investigate the scientific issue, which was to verify the reduction in flexural strength of the fiber cement boards during fire impact, samples were taken from the model. Samples were taken directly from the large-scale facade model or from cladding elements that fell off. Samples were taken from 3 locations, shown in Figure 9, and compared to reference samples.

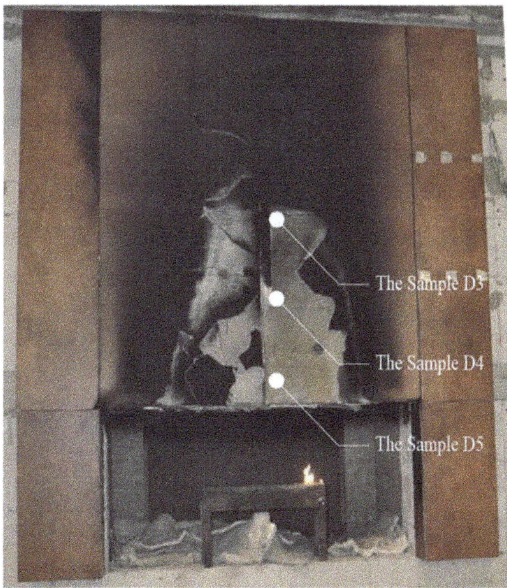

Figure 9. Sampling location.

The method in which the samples were taken and how long they were exposed to the fire is presented below:

Sample D5—fell off in approx. 13'30" of the test;
Sample D4—fell off in approx. 17'15" of the test;
Sample D3—fell off in approx. 34'00" of the test.

Modulus of rupture (*MOR*) assessment was performed according to PN-EN 12467 [28]. The dimensions of the reference samples were taken as 250×250 mm^2. For the other tested samples, the dimensions recommended by the standard [28] could not be achieved due to the recovery of the samples from the elevation model after the test and the extensive damage to these elements. The three-point bending flexural test stand is shown in Figure 10.

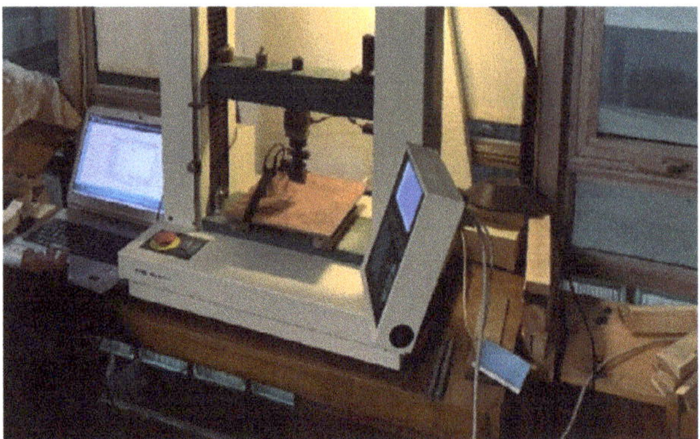

Figure 10. Three-point bending flexural testing machine.

The results of *MOR* for the reference samples are shown in Figure 11—the significant difference between the courses of the individual graphs for different samples is evident. The direction in which the samples were bent in the test (whether they were cut parallel or perpendicular to the pressing direction) was important for the test results. In the case of bending in the direction perpendicular to the pressing direction of the boards (the samples "001-D.ref" and "002-D.ref" show a much higher flexural strength), the destructive force is higher. In the case of bending in the direction parallel to the pressing direction of the boards, samples "003-D.ref" and "004-D.ref" have lower flexural strength.

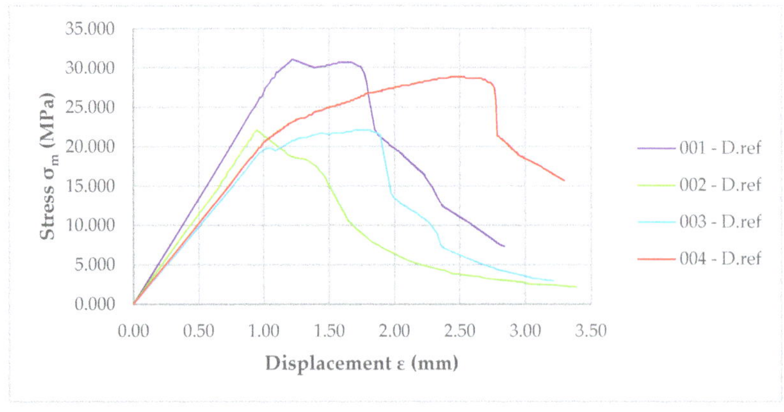

Figure 11. Graph showing the modulus of rupture (*MOR*) of the reference samples.

A total of 5 samples were taken from the large-scale facade model, and all the results of the strength test course are shown in Figure 12. All the samples are characterized by a significant reduction in strength under the three-point flexural test. In addition, they are characterized by a very rapid and sharp reduction in flexural strength after passing through the point representing the destructive forces.

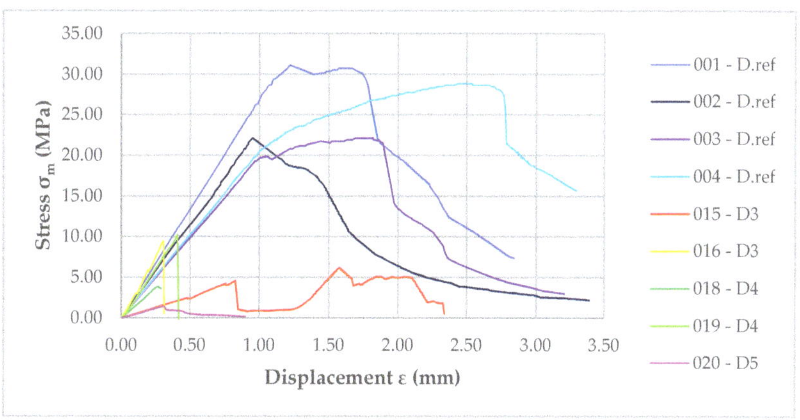

Figure 12. Graph showing *MOR* of front facade samples compared to reference samples.

The flexural strength was calculated—*MOR* (f_{max})—according to the formula in the standard [28]:

$$MOR = \frac{3Fl_s}{2b\,e^2} \tag{1}$$

where:

MOR—modulus of rupture (MPa);
F—the load (force) (N);
l_s—the length of the support span (mm);
b—sample width (mm);
e—sample thickness (mm).

The modulus of elasticity was determined based on the load of proportionality (LOP), which is the limit of applicability of Hooke's law, determined using the graphical method presented in ref. [22]. The bending modulus of elasticity was determined from the Equation:

$$E_D = \frac{Fl_s^3}{4fbe^3} \tag{2}$$

where:

E_D—Young's modulus (GPa);
F—the load (force) (N);
l_s—the length of the support span (mm);
f—flexion (mm);
b—sample width (mm);
e—sample thickness (mm).

Table 1 shows the previously calculated strength parameters, and their mean values were calculated.

Table 1. *MOR* and modulus of elasticity for individual samples.

Sample Identification	*MOR* (MPa)	Modulus of Elasticity (GPa)	Mean Value *MOR* (MPa)	Mean Value Modulus (GPa)
D.ref.	31.13	22.49		
D.ref.	22.13	19.39	26.06	18.91
D.ref.	22.13	16.61		
D.ref.	28.88	17.17		
D5	1.45	0.52	1.45	0.52
D4	10.26	2.70	3.53	2.16
D4	3.86	1.61		
D3	9,45	7.88	3.91	6.18
D3	6,20	4.48 *		

* Secondary the flexural strength when the sample is supported at all edges.

5. Discussion

D4 samples taken from a height of approximately 700 mm above the top splay above the combustion chamber show better flexural strength than the previous sample. The destruction occurs very rapidly in the testing machine, as shown in Figure 12. Unfortunately, as with the above sample, these materials do not resemble the non-degraded material. For the two samples taken, there is a significant difference between their strengths, being 10.26 MPa and 3.86 MPa, respectively. For the elasticity moduli, these values are similar. These samples were taken from a height of about 700 mm above the top splay, which roughly corresponds to the TE3 thermocouple and the temperatures shown on it (see Figure 7).

Temperatures as high as 600 °C result in the complete destruction of the sample. Cladding components in this area fall off after about 17′15″, which corresponds to an integral of the time–temperature function of about 3.8×10^5 (s·°C). The moment at which the D4 samples fall off along with the integral is shown in Figure 13.

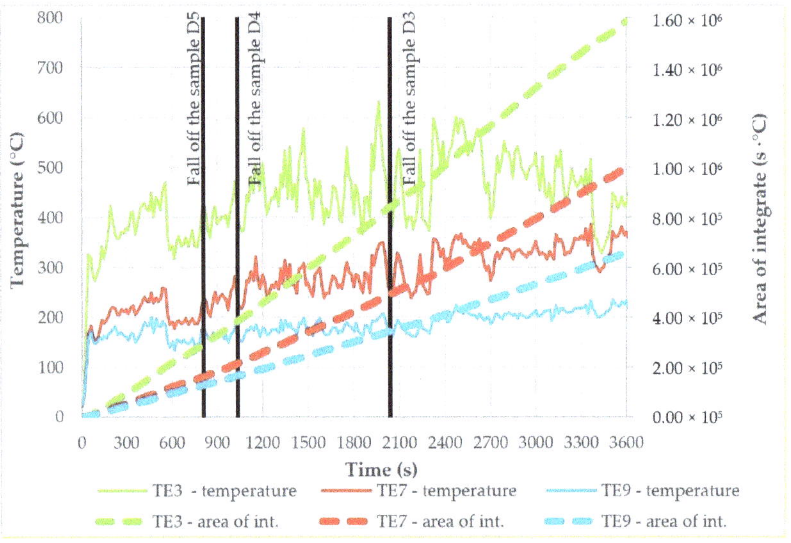

Figure 13. Temperature measurement results for TE3, TE7, and TE9 thermocouples, and the increasing integral for the temperature–time function, along with the determination of the falling-off time of individual elements.

D3 samples are from an area approximately 1200 mm above the top splay above the combustion chamber. Samples were also taken from elements that fell off the large-scale facade model. These persisted for approximately 34′ until they fell off. D3 samples show greater strength stability than those taken at locations D4 and D5. The results are shown graphically in Figure 14—despite the increase in modulus of elasticity and *MOR*, they still show reduced load capacities compared to the reference samples. In order to evaluate which integral of temperature and time function acted on it, the integral of function and time was interpolated between TE3 and TE7 thermocouples—it corresponds to the value of approx. 6.7×10^5 (s·°C). The results are presented in Table 2.

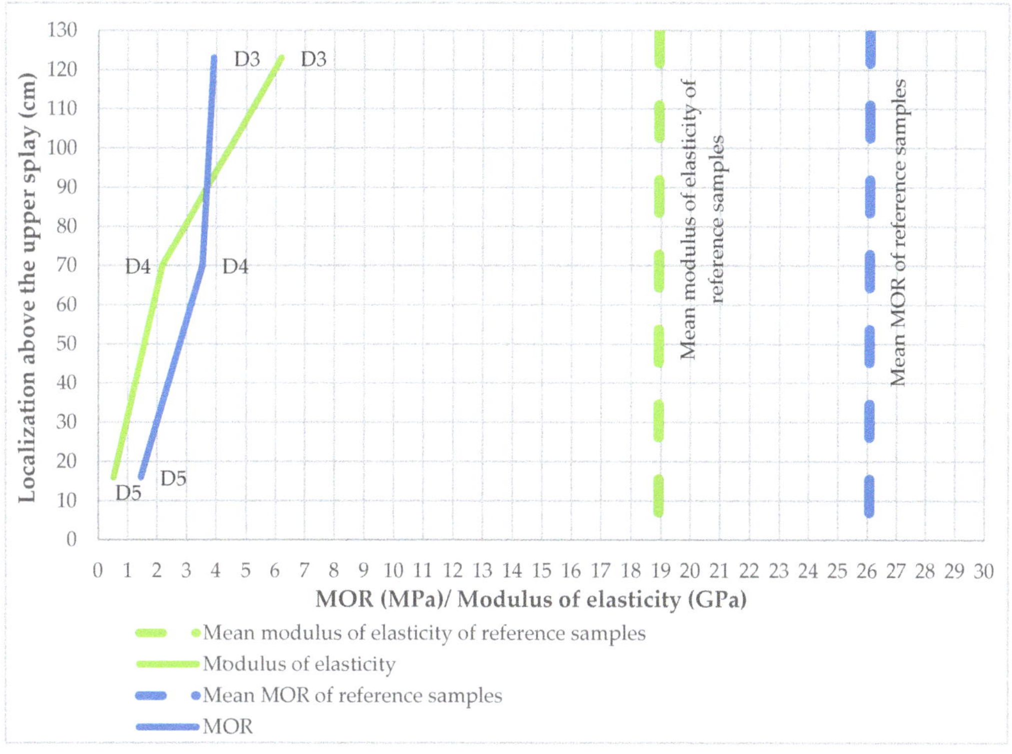

Figure 14. Reduction in MOR of the fiber cement boards depending on the localization above the upper splay.

Table 2. Characteristic data of falling samples.

Sample Identification	Time of Fall of the Sample (s)	Area of Integrate (s·°C)	Temperature′ Thermocouples in Time of Fall of the Sample (°C)
D5	810	$>2.9 \times 10^5$	>437
D4 D4	1035	3.8×10^5	403
D3 D3	2040	6.7×10^5	328

Temperature graphs for TE3, TE7 and TE9 thermocouples with their corresponding time and temperature function integrals (s·°C) are shown in Figure 13; the timeline also shows how long it takes for the individual facade elements sampled to fall off.

The results presented in Table 1 were carefully analyzed. As shown in Figure 14, samples subjected to fire, even 1200 mm above the upper splay, do not show sufficient strength, and their modulus of elasticity shows reduced values. A tendency can be noticed that the integral of the temperature and time functions, despite a significantly greater impact on D3 samples, does not show a lower elasticity modulus, and only the flexural strength is reduced. It can therefore be assumed that with such a large impact on this function, the samples remain on the facade only because of their internal predispositions, e.g., the proximity of mechanical fasteners (rivets). The samples taken at a height of approx. 200 mm or approx. 1200 mm above the splay show low flexural strength parameters.

6. Conclusions

The large-scale facade model is a great source of knowledge regarding the behavior of facade cladding during a fire and the way the fire spreads. The problem of facade cladding destruction during a fire in terms of different materials is insufficiently recognized. The authors used fiber cement boards, i.e., fiber-reinforced composite materials, as facade cladding.

In the analysis of the reduction in the load capacity of fiber cement boards, two key elements must be distinguished: the time for the element to fall off the facade cladding, and the degree of degradation. In the case of fire-induced detachment of degraded elements, temperature and exposure time are crucial, with the result that lower portions fall off first—even though the fiber melting point is exceeded over a larger area of the facade. In addition to temperature, the time at which the elements fall off is also influenced by the extent to which the cladding protrudes beyond the face of the entire facade. There is a noticeable tendency for the fire to spread in the direction where the cladding is closer to the supporting wall structure. The first few minutes consist of charring and destruction of the cladding in the model. The first pieces of significant size fall off the facade at minute 17.

The elements that fell off and that were taken for testing show insufficient flexural strength for their further use. It is also noticeable that the fragments of facade cladding that fall off from higher parts are less degraded, even though the integral responsible for temperature and time function is significantly larger. A higher temperature (in the fire range) has a more destructive effect than a temperature of about 200 °C or lower, over a much longer time period, where the integral of the time–temperature function is about 44% greater.

In terms of flexural strength, the effects of higher fire temperatures are also more significant than duration. Samples from lower parts of the model had lower values of modulus of elasticity and flexural strength.

All samples taken up to a height of about 1300 mm above the level of the combustion chamber are not suitable for reuse. Their further use in such a state would pose a significant risk to people moving beneath such an elevation.

The tests carried out showed the distribution of reduced flexural strength of fiber cement board cladding, fixed in different parts of the tested model, under the influence of fire.

The authors plan further research using large-scale facade models. Potential possible studies include an analysis of the time it takes for the fibers to degrade inside the cement matrix, and further, more global research on the behavior and depletion of load capacity.

Author Contributions: K.S. conceived and designed the experimental work; P.S. prepared the specimens and completed the experiments; Ł.Z. analyzed the test results and performed study editing. All authors discussed the results, commented on the manuscript, wrote the paper, and did the review editing. All authors have read and agreed to the published version of the manuscript.

Funding: This research received no external funding.

Institutional Review Board Statement: Not applicable.

Informed Consent Statement: Not applicable.

Data Availability Statement: The data presented in this study are available on request from the corresponding author.

Conflicts of Interest: The authors declare no conflict of interest.

References

1. *EOTA ETAG 034 Part 1: Ventilated Cladding Kits Comprising Cladding Components and Associated Fixings*; European Organisation for Technical Approvals: Brussels, Belgium, 2012.
2. Schabowicz, K. *Elewacje Wentylowane. Technologia Produkcji I Metody Badania Płyt Włóknisto-Cementowych*; Oficyna Wydawnicza Politechniki Wrocławskiej: Wrocław, Poland, 2018.
3. *EOTA ETAG 034 Part 2: Cladding Kits Comprising Cladding Components, Associated Fixings, Subframe and Possible Insulation Layer*; European Organisation for Technical Approvals: Brussels, Belgium, 2012.
4. Callister, W.D.; Tethwisch, D.G. *Materials Science and Engineering: An Introduction*, 10th ed.; Wiley: Hoboken, NJ, USA, 2005.
5. Maline, M.N.; Rżysko, J. *Mechanika Materiałów*; PWN: Warszawa, Poland, 1981.
6. Shakor, P.N.; Pimplikar, S.S. Glass Fibre Reinforced Concrete Use in Construction. *Int. J. Technol. Eng. Syst.* **2011**, *2*, 632–634.
7. Bezerra, E.M.; Joaquim, A.P.; Savastano, H. Some properties of fiber-cement composites. In Proceedings of the Conferencia Brasileira de Materiais e Tecnologias Não Convencionais: Habitações e Infra-Estrutura de Interesse Social Brasil-Nocmat, Pirassununga, Brasil, 29 October–3 November 2004.
8. Sanchayan, S.; Foster, S.J. High temperature behaviour of hybrid steel–PVA fibre reinforced reactive powder concrete. *Mater. Struct.* **2016**, *49*, 769–782. [CrossRef]
9. Abdullah Shukry, N.; Ahmad Sekak, K.; Ahmad, M.; Bustami Effendi, T. Characteristics of Electrospun PVA-Aloe vera Nanofibres Produced via Electrospinning. In *Proceedings of the International Colloquium in Textile Engineering, Fashion, Apparel and Design 2014 (ICTEFAD 2014)*; Springer: Singapore, 2014; pp. 7–12.
10. Kalifaa, P.; Chene, G.; Galle, C. High-temperature behaviour of HPC with polypropylene fibres from spalling to microstructure. *Cem. Concr. Res.* **2001**, *31*, 1487–1499. [CrossRef]
11. Schroeter, J.; Felix, F. Melting cellulose. *Cellulose* **2005**, *12*, 159–165. [CrossRef]
12. Szymków, M. Identifikacja Stopnia Destrukcji płyt Włóknisto-Cementowych pod Wpływem Oddziaływania Wysokiej Temperatury. Ph.D. Thesis, Wydział Budownictwa Lądowego i Wodnego, Politechnika Wrocławska, Poland, 2018. Raport Serii PRE nr 9/2018.
13. Bednarek, Z.; Drzymała, T. Wpływ temperatur występujących podczas pożaru na wytrzymałość na ściskanie fibrobetonu. *Zesz. Nauk. SGSP* **2008**, *36*, 61–84.
14. Drzymała, T.; Ogrodnik, P.; Zegardło, B. Wpływ oddziaływania wysokiej temperatury na zmianę wytrzymałości na zginanie kompozytów cementowych z dodatkiem włókien polipropulenowych. *Tech. Transp. Szyn.* **2016**, *23*, 82–86.
15. *EN 1363:1-2012. Fire Resistance Tests-Part 1: General Requirements*; CEN: Brussels, Belgium, 2012.
16. Al-Attar, A.; Abdulrahman, M.; Hussein, H.; Tayeh, B. Investigating the behaviour of hybrid fibre-reinforced reactive powder concrete beams after exposure to elevated temperatures. *J. Mater. Res. Technol.* **2019**, *9*, 1966–1977.
17. Sędłak, B.; Kinowski, J.; Sulik, P.; Kimbar, G. The risks associated with falling parts of glazed facades. *Open Eng.* **2018**, *8*, 147–155. [CrossRef]
18. Weghorst, R.; Hauze, B.; Guillaume, E. Determination of fire performance of ventilated facade systems on combustible insulation using LEPIR2. In Proceedings of the 14th International Fire and Engineering Conference Interflam, Windsor, UK, 4–6 July 2016.
19. Sędłak, B.; Kinowski, J.; Sulik, P. Falling parts of external walls claddings in case of fire-test method–results comparison. In Proceedings of the 2nd International Seminar for Fire Safety of Facades, MATEC Web of Conferences 46, Lund, Sweden, 11–13 May 2016.
20. *EOTA No 761/PP/GRO/IMA/19/1133/11140*; European Commision: Brussels, Belgium, 2019.
21. *Development of a European Approach to Assess the Fire Performance of Facades*; European Commision: Brussels, Belgium, 2018.
22. Veliseicik, T.; Zurauskiene, R.; Valentukeviciene, M. Determining the Impact of High Temperature Fire Conditions on Fibre Cement Boards Using Thermogravimetric Analysis. *Symmetry* **2020**, *12*, 1717. [CrossRef]
23. *BS 8414-1:2015+A1:2017. Fire Performance of External Cladding Systems. Test Method for Non-Loadbearing External Cladding Systems Applied to the Masonry Face of a Building*; British Standards Institution: London, UK, 2017.
24. *PN-90/B-02867:2013-06. Fire Protection of Buildings-The Method of Testing and Classification Principles of Fire Propagation Degree by External Walls from External Side*; Polish Standardisation Committee: Warsaw, Poland, 2013.
25. *ISO 13785-2:2002. Reaction-to-Fire Tests for Facades-Part 2: Large-Scale Test*; ISO: Geneva, Switzerland, 2002.
26. Smolka, M.; Anselmi, E.; Crimi, T.; Le Madec, B.; Moder, I.F.; Park, K.W.; Rupp, R.; Yoo, Y.-H.; Yoshioka, H. Semi-natural test methods to evaluate fire safety of wall claddings. In Proceedings of the 2nd International Seminar for Fire Safety of Facades, MATEC Web of Conferences 46, Lund, Sweden, 11–13 May 2016.
27. *EN 13501-2:2016-07. Fire Classification of Construction Products and Building Elements—Part 2: Classification Using Data from Fire Resistance Tests, Excluding Ventilation Services*; Comite Europeen de Normalisation: Brussels, Belgium, 2016.
28. *EN 12467:2012+A2:2018. Fibre-Cement Flat Sheets—Product Specification and Test Methods*; Comite Europeen de Normalisation: Brussels, Belgium, 2018.

Article

Bayesian Regularized Artificial Neural Network Model to Predict Strength Characteristics of Fly-Ash and Bottom-Ash Based Geopolymer Concrete

Sakshi Aneja [1,*], Ashutosh Sharma [1], Rishi Gupta [1] and Doo-Yeol Yoo [2]

[1] Department of Civil Engineering, University of Victoria, Victoria, BC V8W 2Y2, Canada; r.sharmaashutosh@gmail.com (A.S.); guptar@uvic.ca (R.G.)

[2] Department of Architectural Engineering, Hanyang University, Seoul 04763, Korea; dyyoo@hanyang.ac.kr

* Correspondence: sakshi.aneja141@gmail.com

Abstract: Geopolymer concrete (GPC) offers a potential solution for sustainable construction by utilizing waste materials. However, the production and testing procedures for GPC are quite cumbersome and expensive, which can slow down the development of mix design and the implementation of GPC. The basic characteristics of GPC depend on numerous factors such as type of precursor material, type of alkali activators and their concentration, and liquid to solid (precursor material) ratio. To optimize time and cost, Artificial Neural Network (ANN) can be a lucrative technique for exploring and predicting GPC characteristics. In this study, the compressive strength of fly-ash based GPC with bottom ash as a replacement of fine aggregates, as well as fly ash, is predicted using a machine learning-based ANN model. The data inputs are taken from the literature as well as in-house lab scale testing of GPC. The specifications of GPC specimens act as input features of the ANN model to predict compressive strength as the output, while minimizing error. Fourteen ANN models are designed which differ in backpropagation training algorithm, number of hidden layers, and neurons in each layer. The performance analysis and comparison of these models in terms of mean squared error (MSE) and coefficient of correlation (R) resulted in a Bayesian regularized ANN (BRANN) model for effective prediction of compressive strength of fly-ash and bottom-ash based geopolymer concrete.

Keywords: geopolymer concrete; fly-ash; bottom-ash; neural network; sustainability; industrial waste management

Citation: Aneja, S.; Sharma, A.; Gupta, R.; Yoo, D.-Y. Bayesian Regularized Artificial Neural Network Model to Predict Strength Characteristics of Fly-Ash and Bottom-Ash Based Geopolymer Concrete. *Materials* **2021**, *14*, 1729. https://doi.org/10.3390/ma14071729

Academic Editor: Krzysztof Schabowicz

Received: 5 March 2021
Accepted: 29 March 2021
Published: 1 April 2021

Publisher's Note: MDPI stays neutral with regard to jurisdictional claims in published maps and institutional affiliations.

Copyright: © 2021 by the authors. Licensee MDPI, Basel, Switzerland. This article is an open access article distributed under the terms and conditions of the Creative Commons Attribution (CC BY) license (https://creativecommons.org/licenses/by/4.0/).

1. Introduction

With a focus on decarbonization, different ways of reducing greenhouse gas emissions are being constantly explored [1]. The construction industry typically requires a huge amount of energy for its products and services and, therefore, is tagged as a carbon-intensive sector. Hence, it significantly challenges sustainable growth. In the entire spectrum of the construction industry, the production of cement alone produces the largest amount of carbon dioxide and is the second largest source of CO_2 emission worldwide. In this regard, geopolymer concrete offers a potential solution to completely overtake the role of cement in the construction industry.

The term 'geopolymer' was first used in Davidovits' work relating to the formation of polymeric Si-O-Al bonds from a chemical reaction of alkali silicates with aluminosilicate precursors. As per Duxson's model [1], the process of geo-polymerization involves three steps: (1) the dissolution of aluminosilicate materials and the release of silicate and aluminate monomers $(Si(OH)_4)$- and $(Al(OH)_4)$; (2) initial gels (mono cross-linked systems) produced by co-sharing of oxygen atoms from the reactive silicate and aluminate monomers, a process known as condensation; (3) the initial gels are converted into geopolymer gels in the last stage, a process known as polycondensation. Just like ordinary concrete,

85

geopolymer concrete can be developed by adding aggregates, and prepared with waste materials such as fly ash, glass granulated blast slag, rice husk ash to form geopolymers. By using different industrial waste materials, two problems, viz, (1) high demand for cement, (2) industrial waste management, can easily be solved.

There is a major difference between the hydration process of cement and the polymerization process in geopolymers. This is primarily due to the usage of different precursor materials. Different industrial waste materials such as glass granulated blast furnace slag (GGBFS) [2–8], Fly Ash (FA) [9–13], and metakaolin (MT) [14–23] have been used as source materials for developing geopolymer concrete, as reported by the researcher community. Geopolymer concretes are usually less workable, so much so that a nominal 90 mm slump is considered as necessary [24]. With the addition of slag in geopolymer concrete, the workability of the mix gets reduced [25]. Further, it accelerates the geo-polymerization, significantly reducing the initial and final setting times. This could be due to the formation of additional C-S-H gel during geo-polymerization [3]. The mechanical properties of GGBFS based geopolymer concrete cured under ambient conditions is greater than that of normal concrete [6]. Metakaolin based geopolymer greatly accelerates the geo-polymerization due to its high reactivity and, hence, reduced initial and final setting times are achieved [14,17]. The fine particle size of metakaolin fills pores in the matrix and significantly reduces porosity, resulting in a densified microstructure [17–19]. On the other hand, surface cracks are developed at elevated temperatures due to the movement of water from the matrix to the surface, resulting in increased water absorption [23]. The characteristics of FA based geopolymer primarily depend on the purity of raw materials and the concentration of alkali solutions, the physiochemical properties of fly ash, alkali activators, and curing conditions [10,13]. Different gels can be formed by varying the Si/Al ratio and alkali solutions, influencing the final geopolymer structure and controlling the ionic transport. Further, the hydrolyzation of fly ash depends on the alkali solution and hence the porosity of the geopolymer structure. This further impacts the movement of moisture and alkali from the geopolymer into ion solution, enhancing its mechanical strength and durability. FA based geopolymer exhibits promising resistance to chloride, sulphate and acid solutions [12]. It exhibits good efflorescence and freeze-thaw resistance as well [11].

The mechanical characteristics of any geopolymer concrete depend on multiple variables such as precursor ingredients, type of silicates, concentration, type of material as cement replacement and its quantity, amount of superplasticizers, type of curing conditions, time of curing, etc. Multi-variability of inputs complicates the process of optimization in determining proper mixture proportion while synthesizing geopolymers. Therefore, the expected results can only be obtained by properly choosing the combination of materials and correctly selecting the mix proportions. This is normally a cumbersome process involving large-scale laboratory-based experiments, a large number of materials, time, labour, and high cost [26]. This is reflected in some of the published works [27–30] wherein numerous mixes were made to find a suitable proportion for GPC of the desired characteristics, while others developed a multi-step methodology [31] to achieve high 28-day compressive strength.

Compressive strength is one of the main design parameters as mentioned in design codes and standards that indicates the ability of concrete to withstand loads. Hence, numerous empirical relationships have been reported and published for predicting the compressive strength of different types of geopolymer concrete. Traditional statistical models are ineffective in considering the actual scenarios of concrete with different constituents and the results cease to be accurate when new data differing from the original data is used. This is primarily because conventional statistical models are built with fixed equations based on limited inputs. Recently, artificial neural networks (ANN) have gained popularity in various civil engineering problems such as drying shrinkage, concrete durability, and workability of different concretes [32–37]. The ability to draw relevant inferences makes ANN a very effective prediction method. Many researchers have used ANN to predict the compressive strength of different types of geopolymer concrete with significant success [32,38–40]. However, the use of ANN for GPC and the influence of bottom ash (BA) as a replacement for cement and sand in fly ash-based geopolymer concrete has rarely been reported.

This study aims to investigate the influence of industrial waste materials such as BA on the characteristics of alkali-activated geopolymer concrete. The numerical prediction modelling for compressive strength of geopolymer concrete has been implemented with ANN. For that purpose, three algorithms, namely Levenberg Marquardt backpropagation, Bayesian Regularisation backpropagation and Scaled conjugate gradient backpropagation, have been utilized. The efficacy of each algorithm for prediction analysis has also been evaluated.

2. Artificial Neural Network Architecture

ANN is a machine learning prediction model that can predict the expected output when trained with a data-set of inputs and output. An artificial neuron is the computational unit in ANN and therefore is also known as a "computational neuron". A schematic of a computational neuron in an ANN model with three inputs and one output is shown in Figure 1. The basic operation in an ANN model involves the multiplication of input features i_1, i_2, $\ldots$, i_n with weights w_1, w_2, $\ldots$, w_n to calculate a sum of weighted inputs $i_1\,w_1 + i_2\,w_2 + \ldots + i_n x_n$. The sum of weighted inputs is compared with a certain threshold value also known as 'bias' and an output o is generated. If the weighted sum of inputs is greater than or equal to the bias, an output signal is transmitted further in the network, otherwise not.

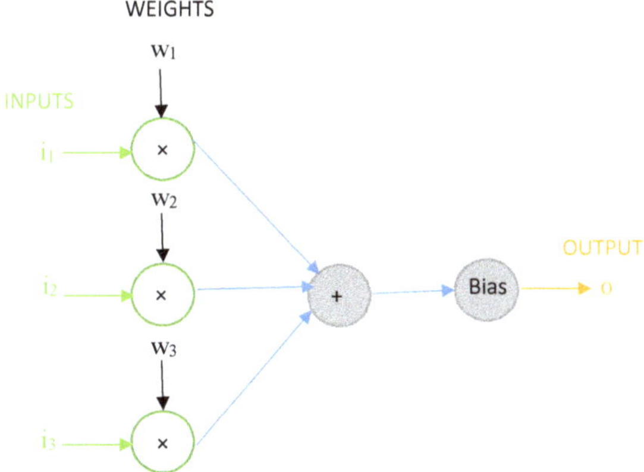

Figure 1. A computational neuron in the artificial neural network (ANN) model.

A feedforward network is an ANN with a forward flow of information from an input layer to an output layer through one or more computational layers known as hidden layers. Figure 2 shows the system model of a multilayer feedforward neural network. The input layer comprises nodes that represent the input parameters or features in the data fed into a network model. The hidden layer has numerous neurons that operate on the weighted inputs using an activation function. The output layer comprises one or more output nodes that utilize an activation function to give the estimated output y. The neurons in consecutive layers are connected. The term multilayer signifies the number of layers with an activation function. A feedforward network is said to be a single-layer network if its input layer is directly connected to the output layer. A feedforward network is said to be a two- or three-layer network if its input layer is connected to the output layer through one or two hidden layers, respectively. The input features in data are represented by i_1, i_2, $\ldots$, i_n where n is the number of input features. Hidden layers are represented by h_1, h_2, $\ldots$, h_l where l is the number of hidden layers. Each hidden layer can have multiple neurons as their elementary unit, as represented by shaded circles. The output layer is represented by a single functional unit that estimates the actual output or target t in experimental data.

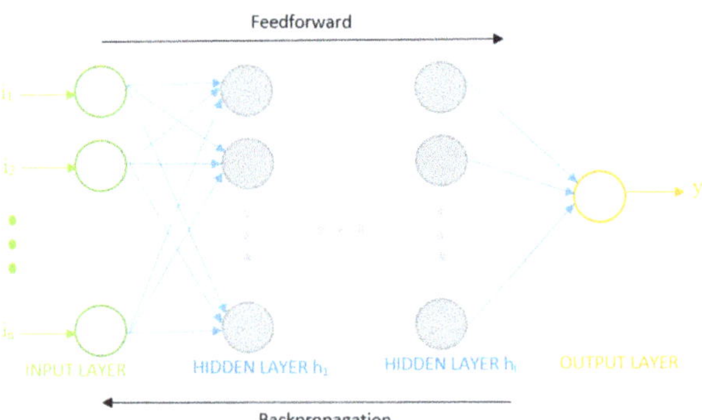

Figure 2. System model of multilayer feedforward neural network.

A backpropagation algorithm can be effectively used to train feedforward neural networks to predict an expected output that closely matches the target. The network training using backpropagation is an iterative process where each forward flow of information is followed by a backward pass that adjusts weights and biases. In each forward pass of information, the cost function which is a function of error between output and target is calculated. Gradients are obtained by differentiating the cost function for independent weights. In each iteration, gradients are calculated as a result of a chain rule and adaptive weights and biases are fed to the network to be used by the next forward flow of information. The backpropagation algorithm is aimed at reducing this cost function by finding a local minimum. This process is continued until error is minimized for efficient training and hence a better prediction model [41].

In this study, a multilayer feedforward neural network is designed and trained with Levenberg-Marquardt (LM), Bayesian Regularization (BR), and scaled conjugate gradient (SCG) backpropagation algorithms separately to identify an efficient model that can predict the compressive strength of geopolymer concrete. The multilayer feedforward neural network in this research uses the sigmoid activation function in the hidden layers and linear activation function in the output layer. Network training with LM and BR backpropagation algorithms involves Jacobian calculations while training with SCG backpropagation algorithm involves gradient calculations. LM backpropagation is the least time-consuming algorithm for training moderate-sized neural networks but consumes maximum memory. Training is stopped when the network's performance is not improving, or the network is not generalizing well. BR backpropagation consumes most of the time but can be applied to small or noisy datasets. Training is continued to the point when optimum weights are found. SCG backpropagation consumes the least memory and can be applied to any network. Training is stopped when the network's generalization is not improving further.

The BR backpropagation algorithm avoids both overfitting and overtraining as the network trains on effective network parameters or weights and does not consider the irrelevant parameters. Equation (1) provides the training objective function $F(\omega)$ used by BR, where S_ω is the sum of squared network weights and S_e is the sum of network errors. A combination of squared errors and weights is minimized until the optimum combination is achieved for which the network generalizes well. At that point, training is stopped [42].

$$F(\omega) = \alpha S_\omega + \beta S_e \tag{1}$$

The input and output parameters of geopolymer concrete specimens used to design a multilayer feedforward neural network-based prediction model in Matrix Laboratory (MATLAB) are given in Table 1. The specifications of geopolymer concrete represented by $i_1, i_2, \ldots, i_{11}$ are considered as eleven input features in the input layer of the network. Even though the features are quite commonly used in the context of geopolymer concrete, further details about what these features mean is reported in the literature [43,44]. This is not described here in order to maintain brevity. The output layer comprises a single neuron that predicts the expected compressive strength y, known as output, that is mapped to the actual compressive strength of geopolymer concrete known as target t. Table 1 also shows the minimum and maximum values of input and output data considered in this study.

Table 1. Input and output parameters used for prediction of geopolymer concrete (GPC) compressive strength.

Variables	Representation	Range
Super plasticizer (wt.%)	i_1	0–1.5
Alkaline activator/fly-ash ratio	i_2	0.4–0.6
NaOH concentration (M)	i_3	10–16
Na_2SiO_3/NaOH	i_4	1–3
NaOH (wt. in gms)	i_5	41–127
Na_2SiO_3 (wt. in gms)	i_6	93–241
Fly Ash (wt. in gms)	i_7	0–450
Bottom Ash (wt. in gms)	i_8	0–400
Coarse Aggregates (wt. in gms)	i_9	0–1400
Fine Aggregates (wt. in gms)	i_{10}	0–700
Curing Time (in Days)	i_{11}	0–28
Compressive Strength (in MPa)	t	9–65

3. Data Preparation

To develop an effective ANN model, the data was collected as input parameters as well as output parameters from previous research works published on fly ash and bottom ash-based geopolymer [9,45–57]. Further, some mixes were also developed in the laboratory to collect data. It is to be noted that compressive strengths of 7 days, 14 days, and 28 days were considered for developing the ANN model.

3.1. Data from Literature Sources

A total of 46 sets of experimental data from 15 research papers (details provided below) was collected as input to develop the ANN strength model. For this work, a total of 11 input parameters were included. These include coarse aggregates (CA), fine aggregates (FA), fly ash (FAH), bottom ash (BAH), sodium silicates (SS), sodium hydroxides (SH), sodium silicate and sodium hydroxide ratio (SS/SH), precursor powder and liquid ratio (L/S), curing time (CT) and amount of superplasticizer (S). The basic premise behind selecting these parameters is due to to their direct influence on the matrix, and consequently mechanical properties, of GPC (mainly its compressive strength). Table 2 gives the sources of data used in the study.

Table 2. Details of data and their sources.

No.	FAH	BAH	CA	FA	SH	SS	SS/SH Ratio	SS Concen-tration	L/S Ratio	S	CT	CS	Source
1	310	0	1204	649	68	102	2.5	10	0.225	6.2	0–28	43	[9]
2	404–408	0	1190–1202	640–647	41	103	2.5	14–16	0.35	6	0–28	42–45	[45]
1	417	0	927	698	92	241	2.5	15	0.4	5	0–28	43	[46]
1	400	0	1293	554	45	113	2.5	14	0.4	4	0–28	44	[47]
3	408	0	1201	647	62–68	93–103	1.5	14	0.4	4	0–28	32–48	[48]
5	408	0	1168	660	68	103	1.5	10–16	0.35	4	0–28	32–49	[49]
20	298–450	0	1100–1377	500–659	29.4–108	96–162	1.5–2.5	8–14	0.5	2–4	0–28	25.6–41	[50]
1	450	0	1150	500	108	162	1.5	12	0.6	2	0–28	35.2	[51]
2	400	0	1209–1218	651–655.9	40–45.7	100–114.3	2.5	10–14	.35	4	0–28	25.6–32.5	[52]
1	310	0	1204	649	66	108	2.5	10	0.35	4	0–28	41	[53]
1	409	0	1256	591	41	102	2.5	8	0.35	6	0–28	39	[54]
1	410	0	1100	590	40	100	2.5	14	0.55	6	0–28	38	[55]
6	414	0	1091	588	60–80	104–138	1–2	10–20	0.5	-	0–28	39–46	[56]
1	0	400	1216.1	540	66.7	133.3	2	8	0.5	8	0–28	49.3	[57]

(FAH: Fly ash, BAH: Bottom Ash, CA: Coarse aggregates, FA: Fine aggregates, SS: sodium silicate, SH: sodium hydroxide, SS/SH: Sodium Silicate and Sodium hydroxide ratio, L/S: Precursor powder and Liquid ratio, CT: Curing time, S: Superplasticizer, CS: Compressive Strength).

3.2. Data from Experiments

From the literature review, it can be seen that not many data sources are available for modelling the FA and BA based GPC. Furthermore, it can be seen that there is a wide variation in the values reported in the literature. Using such information for further mix optimization would not be straightforward. Considering experimental data as the key for validation of numerical models, extensive experimentation was carried out in the laboratory. The experiments involved three different kinds of mixes: (1) fly ash-based geopolymer mix; (2) fly ash-based geopolymer mix with bottom ash fine aggregates; (3) fly ash based geopolymer with bottom ash as a replacement for fly ash itself.

For the production of GPC, class F fly ash as a pozzolanic material obtained from Bathinda coal power plant in India was used. The physical properties of fly ash were: specific gravity: 2.4, bulk density (kg/m^3): 700, surface area (kg/m^2):19,000. The chemical composition of fly ash is given in Table 3. Bottom-ash used in this study was also obtained from the above noted thermal power plant. The specific gravity and water absorption of bottom ash was 1.39 and 31.48%. The chemical properties of bottom ash are also given in Table 3.

Table 3. Chemical composition of by-products (source: Bathinda coal power plant).

Compounds	SiO$_2$	Al$_2$O$_3$	Fe$_2$O$_3$	CaO	MgO	SO$_3$	Na$_2$O	K$_2$O	TiO$_2$	P$_2$O$_5$	Mn$_2$O$_3$
Fly Ash (%)	58.11	27.21	5.23	2.14	0.72	NA	0.5	0.5	N/A	N/A	N/A
Bottom Ash (%)	56.44	29.24	8.44	0.75	0.40	0.10	0.09	1.29	2.89	0.2	0.14

NaOH (SH) and Na$_2$SiO$_3$ (SS) were used for activation of the precursor material. Anhydrous SH powder was dissolved in water to produce SH solutions with varying molarities (10 M, 12 M, 14 M). The solution was prepared 24 h before its usage. Later, SS solutions were mixed with SH solutions at different mass ratios (1.5, 2, 2,5).

Fine aggregates and coarse aggregates obtained from local sources in Jalandhar (Punjab, India) had a relative dry density of 2.671 and 2.713, respectively, and a water absorption ratio of 0.79% and 0.69%, respectively. The coarse aggregates with a maximum size of 12.5 mm were used for preparing GPC and ordinary Portland cement concrete (OPC). The fine aggregates used in both OPC and GPC were medium-coarse sand which was suitable for multipurpose use including concrete mixtures.

The solid constituents of GPC, i.e., the aggregates, fly-ash, and bottom-ash, were first mixed in the dry condition in a rotary drum mixer for about 1 min. Next, the alkali solution was added to the solids and mixed for about 3 min, followed by 3 min rest period, then followed by 2 min of final mixing. The mixture was placed into moulds of size 150 × 150 × 150 mm and vibrated using a table vibrator for 30 s to discharge air bubbles to the surface. Then, the moulds were covered with a plastic sheet in a lab environment (approximate relative humidity range of 45–70% and approximate temperature range of 5 °C to 15 °C) and demoulded after 24 h A total of 55 mixes were prepared and the compressive strength after 7, 14, and 28 days was evaluated. The various mix designs and the experimentally evaluated compressive strength values are given below in Table 4. It should be noted that other experimental results will be reported by authors in upcoming manuscripts. The experimental results indicate that the increased concentration of sodium hydroxide (SH) from 12 to 16 improved the compressive strength for all the mix design samples [58,59]. Similarly, the increasing ratio of the sodium silicate to sodium hydroxide exhibited increased compressive strength of specimens of all mix designs. It should be noted that the ratio of alkaline to that of fly-ash was fixed at 0.4 for the entire study as it exhibited the best results in the pilot studies [60]. From the mix design, it can be seen that bottom ash was replaced with cement for GPC-1, GPC-2, and GPC-3 by 0%, 20%, and 40%. The results indicate that GPC-1 with precursor as fly ash alone exhibits the highest compressive strength which further increases with SS/SH ratio and the SH concentration. The replacement of fly-ash as a precursor with 20% and 40% reduced the compressive strength by 25% and 35% [60]. This reduction in compressive strength may be attributed to lesser polymerization of bottom ash particles in comparison to fly ash particles [61,62]. However, the replacement of fine aggregates with bottom ash for mix designs GPC-4, GPC-5, and GPC-6 by 20%, 40%, and 50% exhibited better compressive strength values than before. All the samples with 20% replacement of bottom ash with fine aggregates exhibited higher compressive strength values than the GPC-1. The results from GPC-4 samples with SH concentration of 16 exhibited 23% higher values of compressive strength than the GPC-1. All other samples from GPC-5 and GPC-6 exhibited far lesser values of compressive strength. This may be attributed to the large and porous structure of bottom ash particles inducing internal voids and cracking under loading.

Table 4. Mix proportions for Fly ash-based geopolymer concrete with bottom ash as replacement of fly ash & fine aggregates.

Mix Number	S	SS+SH/FAH	SH(M/L)	SS/SH	SH	SS	FAH	BAH	CA	FA	CS (MPa)
GPC-1 (100% FAH)	0.5	0.4	12	1.5, 2, 2.5	85.16	125.74	388	0	1170	630	35.5, 37.1, 39.9
	0.5	0.4	14	1.5, 2, 2.5	66.2	141.1	388	0	1170	630	36.2, 39.2, 41.4
	0.5	0.4	16	1.5, 2, 2.5	55.4	150.3	388	0	1170	630	36.2, 38.3, 47.1
GPC-2 (80% FAH + 20% BAH)	0.7	0.4	12	1.5, 2, 2.5	85.16	125.74	310	78	1170	630	27.7, 27.9, 29.1
	0.7	0.4	14	1.5, 2, 2.5	66.2	141.1	310	78	1170	630	31, 31.1, 31.1
	0.7	0.4	16	1.5, 2, 2.5	55.4	150.3	310	78	1170	630	33, 30, 29
GPC-3 (60% FAH + 40% BAH)	0.9	0.4	12	1.5, 2, 2.5	85.16	125.74	232	156	1170	630	23.5, 25.8, 25.2
	0.9	0.4	14	1.5, 2, 2.5	66.2	141.1	232	156	1170	630	28.7, 26.9, 25.3
	0.9	0.4	16	1.5, 2, 2.5	55.4	150.3	232	156	1170	630	28.4, 29.3, 29.6
GPC-4 (100% FAH + 20% BAH)	0.7	0.4	12	1.5, 2, 2.5	85.16	125.74	388	126	1170	504	36, 35.8, 33.2
	0.7	0.4	14	1.5, 2, 2.5	66.2	141.1	388	126	1170	504	42.3, 47.2, 49.8
	0.7	0.4	16	1.5, 2, 2.5	55.4	150.3	388	126	1170	504	45.4, 49.6,55.4
GPC-5 (100% FAH + 40% BAH)	0.7	0.4	12	1.5, 2, 2.5	85.16	125.74	388	230	1170	378	34.1, 35.4, 34.3
	0.7	0.4	14	1.5, 2, 2.5	66.2	141.1	388	230	1170	378	36.2, 36.5, 37
	0.7	0.4	16	1.5, 2, 2.5	55.4	150.3	388	230	1170	378	31, 31,37.2
GPC-6 (100% FAH + 50% BAH)	1	0.4	12	1.5, 2, 2.5	85.16	125.74	388	315	1170	315	26.2, 28.1, 29.1
	1	0.4	14	1.5, 2, 2.5	66.2	141.1	388	315	1170	315	25.5, 28, 25.1
	1	0.4	16	1.5, 2, 2.5	55.4	150.3	388	315	1170	315	25.5,29.6, 31.2

(FAH: Fly ash, BAH: Bottom Ash, CA: Coarse aggregates, FA: Fine aggregates, SS: sodium silicate, SH: sodium hydroxide, SS/SH: Sodium Silicate and Sodium hydroxide ratio, L/S: Precursor powder and Liquid ratio, CT: Curing time, S: Superplasticizer).

4. Results and Discussions

The research methodology for identifying a suitable ANN model to predict the compressive strength of geopolymer concrete is shown in Figure 3. The data of geopolymer concrete specimens, as reported in Table 1, is normalized and sampled as 70% for training, 15% for validation, and 15% for testing. The training data is presented to the network in order to predict output compressive strength closer to target compressive strength, validation data measured network generalization to keep a check on training, and testing data measured network's performance during and after training. The network optimization is aimed at obtaining a hypothesis function that predicts the compressive strength of geopolymer concrete with a minimum difference between output and target. This involves various trials and rigorous network training by varying the number of hidden layers between the input and output layer and neurons in each hidden layer.

The network's performance is analysed by training with different backpropagation algorithms such as LM backpropagation, BR backpropagation, and SCG backpropagation. The model is trained by reducing mean squared error (MSE) and as a result, increasing coefficient of correlation (R). MSE is calculated by averaging the squares of the difference between output and target. An MSE of zero signifies no error i.e., perfect condition. R represents regression values and measures the relationship between output and target. A close relationship has R = 1 in a perfect scenario. An extensive search was carried out in our study to find the optimum hidden layers, hidden neurons, and backpropagation algorithm in an endeavour to build a reliable ANN model for the prediction of compressive strength.

Different models of ANN are presented in the following discussion which is applied to geopolymer concrete data to determine an optimum model that predicts compressive strength with the lowest MSE and highest correlation between output and target.

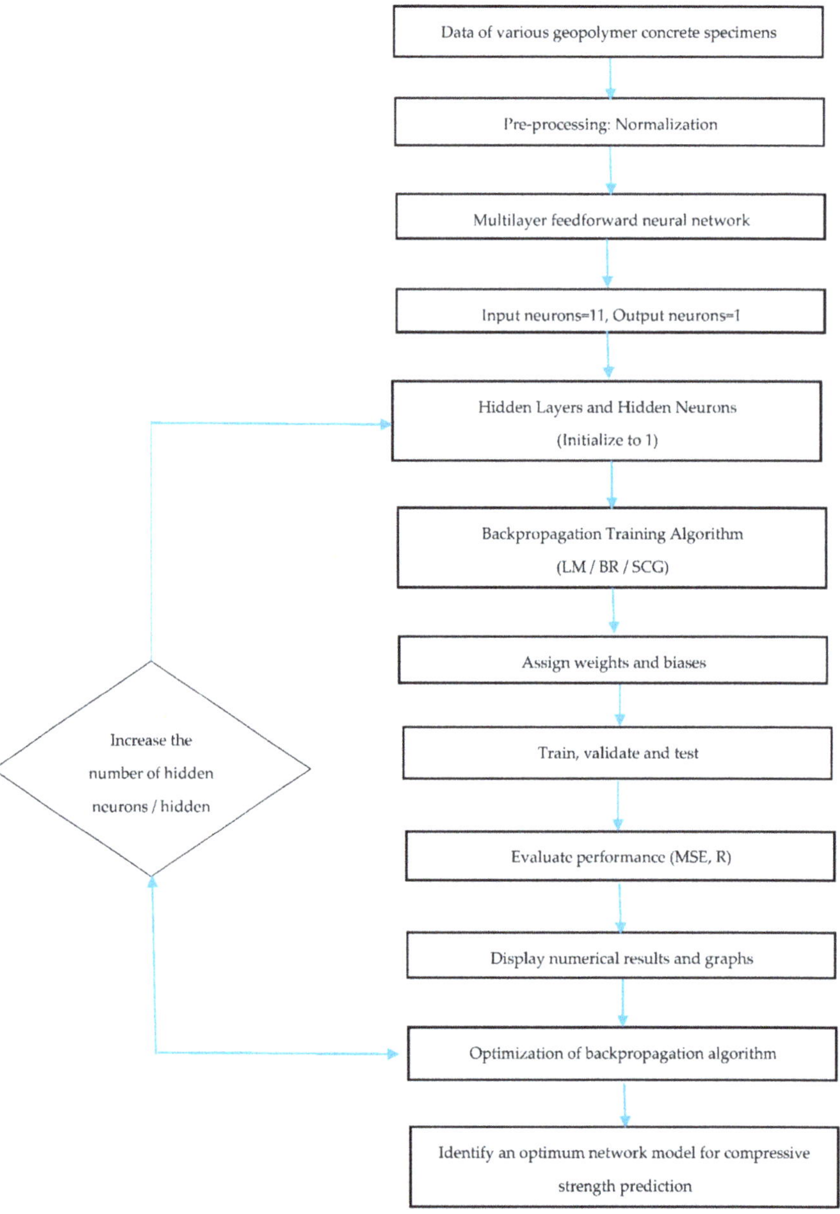

Figure 3. Step-by-step procedure to predict the compressive strength of geopolymer concrete.

4.1. Prediction Evaluation of Compressive Strength

Firstly, a two-layer feedforward neural network comprising an input layer, a hidden layer, and an output layer and trained with BR backpropagation algorithm is programmed. The network's performance is analysed by checking its ability to predict the compressive strength of geopolymer concrete with low MSE. Figure 4 shows the effect of hidden neurons on the performance of the BR trained network. It was observed that MSE decreases with an increase in the number of hidden neurons till 10, after which there is no significant decrease. This implies that the estimated compressive strength predicted by the network model can be improved by optimizing the hidden neurons. For instance, the compressive strength of geopolymer concrete can be predicted with an MSE of 1.8809 considering 5 hidden neurons, and with an MSE of 1.2098 considering 10 hidden neurons.

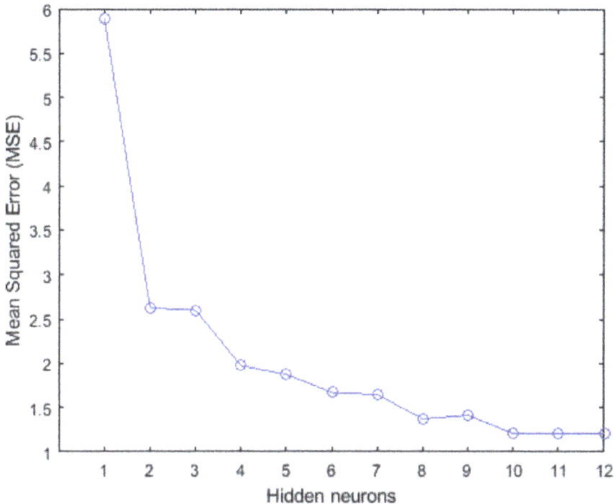

Figure 4. Effect of increasing hidden neurons on the network's performance to predict the compressive strength.

To achieve a better prediction of compressive strength, various ANN models are designed in this study by considering different backpropagation algorithms and varying the number of hidden layers and hidden neurons. Table 5 enlists ANN-I to ANN-XIV models based on two, three- and four-layer feedforward neural networks trained with BR, LM, and SCG backpropagation algorithms. ANN-I to ANN-VI are two-layer feedforward network models with a single hidden layer between the input and output layer. ANN-VII to ANN-X are three-layer feedforward network models with two hidden layers between the input and output layer. ANN-XI to ANN-XIV are four-layer feedforward network models with three hidden layers between the input and output layer.

The compressive strength prediction ability of the ANN-I to ANN-XIV network models is evaluated by analysing both MSE and R. Figures 5 and 6 show the MSE and R performance of these network models. Considering two-layer feedforward network models ANN-I to ANN-VI, it is observed that $MSE_{BR} < MSE_{LM} < MSE_{SCG}$ and $R_{BR} > R_{LM} > R_{SCG}$. This implies that network models trained with BR backpropagation are effective in predicting compressive strength. The network models trained with SCG backpropagation failed to provide a good prediction. Considering three-layer feedforward network models ANN-VII to ANN-X, it is observed that MSE decreases significantly on increasing the hidden layers, resulting in a better prediction. Further, the network models trained with BR and LM algorithms show a similar coefficient of correlation i.e., $R_{BR} \approx R_{LM}$, but BR trained networks have a better prediction of compressive strength than LM trained networks by reducing MSE, i.e., $MSE_{BR} < MSE_{LM}$. Considering four-layer feedforward network models ANN-XI to ANN-XIV, a slight reduction in MSE of LM trained networks is observed due to the addition of another hidden layer but there is no improvement in MSE of BR trained networks indicating no need for increasing hidden layers beyond 3. Again, BR trained networks outperform LM trained networks by predicting compressive strength with lesser MSE, i.e., $MSE_{BR} < MSE_{LM}$, and a comparable coefficient of correlation, i.e., $R_{BR} \approx R_{LM}$.

Table 5. Multilayer feedforward neural network models for the prediction of Compressive Strength of geopolymer concrete.

Designation	Algorithm	Number of Hidden Layers	Neurons in the Hidden Layer
ANN-I	Bayesian Regularization (BR)	1	5
ANN-II	BR	1	10
ANN-III	Levenberg-Marquardt (LM)	1	5
ANN-IV	LM	1	10
ANN-V	Scaled Conjugate Gradient (SCG)	1	5
ANN-VI	SCG	1	10
ANN-VII	BR	2	$h_1 = 10, h_2 = 5$
ANN-VIII	BR	2	$h_1 = 10, h_2 = 10$
ANN-IX	LM	2	$h_1 = 10, h_2 = 5$
ANN-X	LM	2	$h_1 = 10, h_2 = 10$
ANN-XI	BR	3	$h_1 = 10, h_2 = 10, h_3 = 5$
ANN-XII	BR	3	$h_1 = 10, h_2 = 10, h_3 = 10$
ANN-XIII	LM	3	$h_1 = 10, h_2 = 10, h_3 = 5$
ANN-XIV	LM	3	$h_1 = 10, h_2 = 10, h_3 = 10$

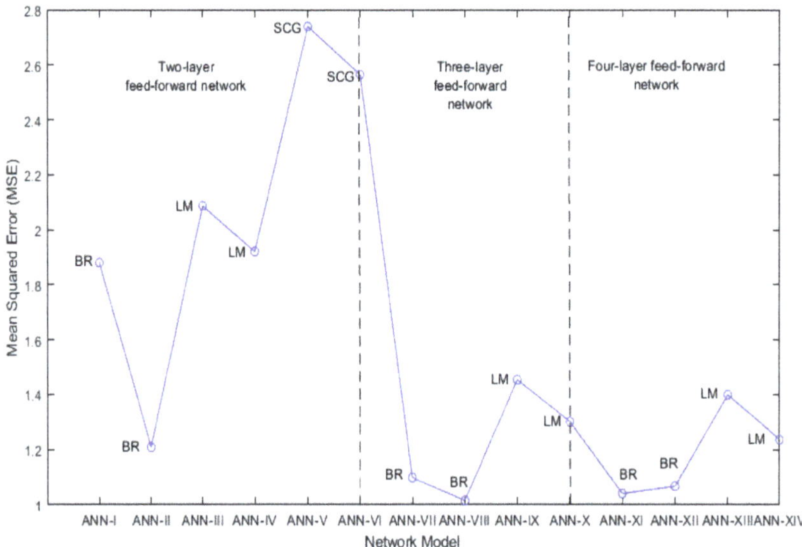

Figure 5. Mean squared error (MSE) performance comparison of two, three- and four-layer feedforward network models.

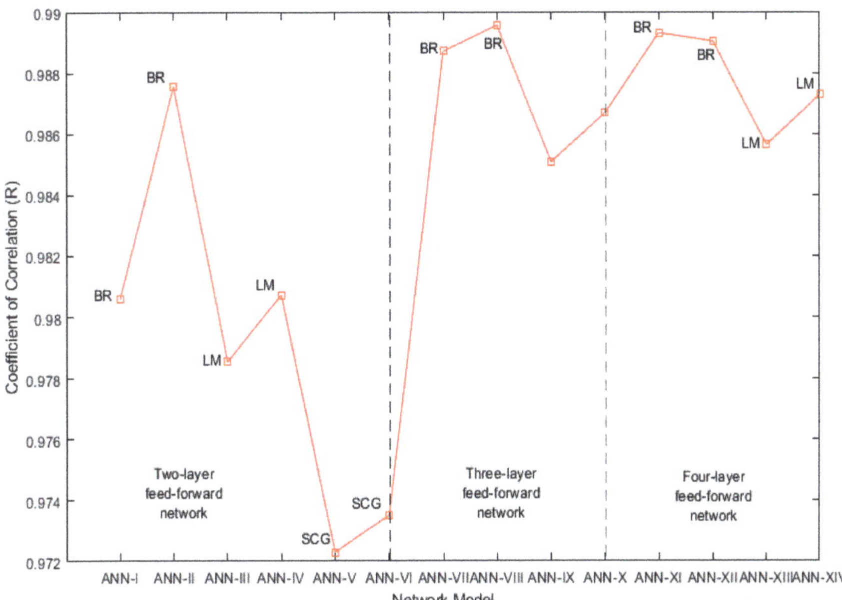

Figure 6. Correlation coefficient comparison of two, three- and four-layer feedforward network models.

During network training, LM and SCG backpropagation algorithms experience repetitive validation failures and training is stopped after six validation failures. On the contrary, a network trained with BR backpropagation can perform well on the validation data set, indicating the flexibility of prediction for unknown data. A comparison of all network models establishes the conclusion that the ANN-VIII model which is a three-layer Bayesian regularized artificial neural network (BRANN) with 10 neurons in each hidden layer is effective in predicting compressive strength of geopolymer concrete with the least MSE (1.017) and the highest R of 0.99. However, it should be noted that this performance exhibited by BRANN is at the cost of more epochs (a measure of the number of times the algorithm uses training vectors to give a hypothesis for prediction), which should not be a problem in the current era of robust and efficient hardware. This implies that the BR backpropagation algorithm improves the training of feedforward neural networks when the number of hidden layers and neurons in each hidden layer is optimized.

4.2. Performance Analysis of BRANN Prediction

The performance of three-layer BRANN to predict the compressive strength of geopolymer concrete in the ANN-VIII model is verified by checking the balance between training and non-training testing patterns. It should be noted here that non-training validation data is not recommended for performance analysis as it gives a biased estimate of prediction by stopping the network's training when performance starts to deteriorate. The intention is to get an unbiased estimate of the prediction ability of a network model built for training data by applying the same model to testing data of geopolymer concrete specimens. A network model that can predict well on testing data can predict the unknown compressive strength for any other input specifications of geopolymer concrete. However, the reliability of predicted compressive strength is dependent on the type of experimental data used for network training.

Figure 7 shows the error histogram obtained after training and testing the network model. The error on the x-axis specifies how predicted compressive strength (output) differs from the actual compressive strength of geopolymer concrete (target). Instances on the y-axis specify the number of geopolymer concrete specimens in the training or testing dataset with a specific error. Most of the errors after training and testing with three-layer BRANN lie in the range of -1.081 to 1.247. Further, three-layer BRANN can predict compressive strength for the majority of geopolymer concrete specimens with an error between -0.3051 to 0.4712, which is closer to the zero error line.

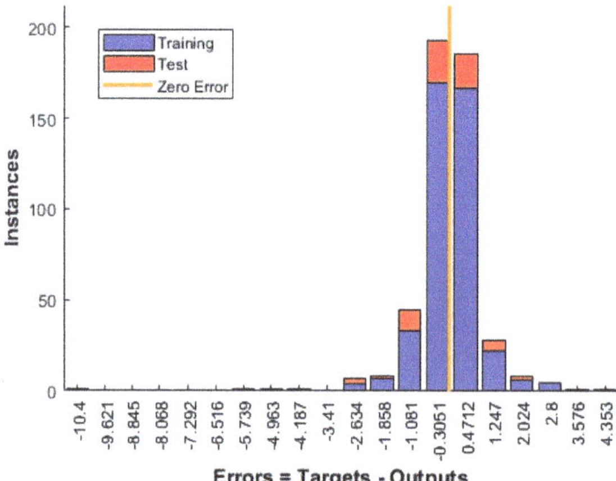

Figure 7. Error Histogram of three-layer Bayesian regularized ANN (BRANN) for GPC compressive strength prediction.

Figure 8 shows the pattern of the MSE performance of three-layer BRANN for epochs during the training and testing phase. The results indicate that, as the epochs are increased, BRANN can predict GPC compressive strength with a very low MSE due to efficient training. The prediction ability of BRANN improves to 250 epochs and remains constant afterward. The best training performance in terms of lowest MSE is highlighted with a circle corresponding to prediction with MSE of 0.92263 at epoch 330. BRANN is also able to predict the compressive strength on the testing dataset with a comparable MSE, verifying the effectiveness of GPC in the network model.

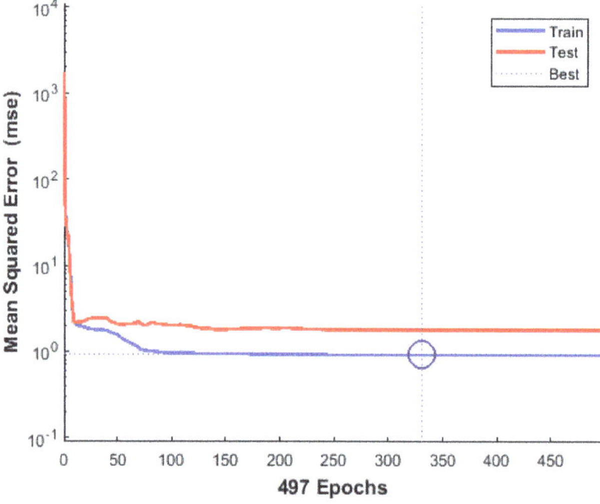

Figure 8. Mean squared error performance of three-layer BRANN for GPC compressive strength prediction.

Figure 9 displays the correlation curves obtained after applying three-layer BRANN on training, testing, and complete data of GPC specimens. A perfect fitting in an ideal scenario is represented by a dashed line at an angle of 45 degrees where output compressive strength matches the target compressive strength, i.e., coefficient of correlation R = 1. The blue, green, and red lines represent the fitting for training, testing, and entire data of GPC specimens respectively. A close relationship between output and target compressive strength with the coefficient of correlation R = 0.992, 0.979, and 0.99 for training, testing, and entire data respectively is obtained, which indicates good data fitting. This indicates the efficacy of three-layer BRANN in predicting compressive strength for any other specifications of geopolymer concrete. BRANN acts as a black-box that generates output compressive strength from input GPC specifications without defining the relationship.

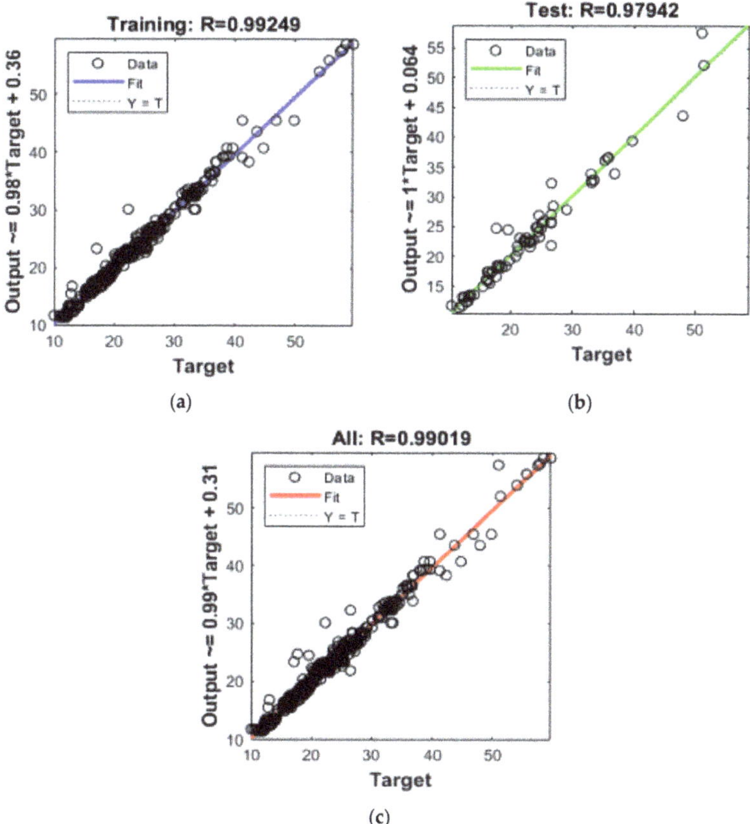

Figure 9. Correlation between predicted and actual GPC compressive strength for three-layer BRANN. (a) Training Data; (b) Testing Data; (c) All Data.

The trained BRANN model was further used with various input parameters. The eleven input parameters were set to minima, maxima, and median values of their respective ranges. The combination of input parameters resulted in more than seven hundred possible mixes. Based on the data produced, the mix predicting the maximum compressive strength was found as given in Table 6. High molarity of SH (16) leads to high compressive strength. This is a well-established relationship. However, it should be noted that high compressive strength can also be obtained by incorporating bottom ash. This indicates the efficacy of the produced BRANN model as optimized mixes can be identified by considering the interdependence of all 11 input features. The various mix designs with their predicted compressive strength lay the perfect foundation for experimental work.

Table 6. Predicted GPC compressive strength for a new set of input parameters.

CT	S	L/S	SH	SS/SH	SH Concentration	SS Concentration	FAH	BA	CA	FA	Predicted CS
28	0.5	0.4	16	3	51.73	155.2	388	124	1170	504	61.4347889

5. Conclusions

In this work, different models of multilayer feedforward neural network trained with Levenberg-Marquardt, Bayesian Regularization, and Scaled Conjugate Gradient backpropagation algorithms are used for predicting the compressive strength of geopolymer concrete with fly-ash and bottom-ash. These models are trained by optimizing mean squared error and coefficient of correlation. The proposed BRANN model is based on the experimental data collected from the literature and laboratory experiments. From the study, the following conclusions can be drawn:

1. Artificial neural network-based machine learning models are capable of predicting the strength characteristics of geopolymer concrete with fly-ash and bottom-ash.
2. MSE decreases as the number of neurons in the hidden layer increases in the feedforward neural network for estimation of compressive strength. MSE can be further reduced by increasing the number of hidden layers between the input and output layers.
3. The performance analysis for a two-layer feedforward neural network shows that $MSE_{BR} < MSE_{LM} < MSE_{SCG}$ and $R_{BR} > R_{LM} > R_{SCG}$. In this case, it should be noted that BR backpropagation outperforms LM and SCG backpropagation algorithms for the prediction of GPC compressive strength.
4. The performance analysis for a three-layer feedforward neural network indicates that both BR and LM backpropagation algorithms show a similar coefficient of correlation, i.e., $R_{BR} \approx R_{LM}$, but the BR algorithm shows better performance than the LM algorithm by reducing MSE i.e., $MSE_{BR} < MSE_{LM}$, leading to better prediction of GPC compressive strength.
5. The performance analysis for a four-layer feedforward neural network implies a slight reduction in MSE of LM trained networks due to the addition of another hidden layer, but there is no improvement in MSE of BR trained networks. Again, BR trained networks outperform LM trained networks by predicting compressive strength with lesser MSE and greater coefficient of correlation i.e., $MSE_{BR} < MSE_{LM}$ and $R_{BR} > R_{LM}$.
6. The study suggests that the three-layer BRANN model with 10 neurons in each layer is the suitable model for predicting GPC compressive strength with MSE of 1.017 and $R = 0.99$.

This work is limited to the investigation and analysis of artificial neural networks trained with backpropagation algorithms for the prediction of GPC compressive strength. The effect of temperature during curing is not considered while training the ANN model. Therefore, a possible direction for future research could be the investigation of other machine learning techniques followed by a comparative analysis of prediction performance. Furthermore, the effect of temperature on GPC compressive strength can be studied in conjunction with other input parameters.

Author Contributions: Conceptualization, S.A. and A.S.; methodology, S.A.; software, S.A.; validation, S.A., A.S., R.G. and D.-Y.Y.; formal analysis, S.A.; investigation, S.A.; resources, S.A. and A.S.; data curation, S.A. and A.S.; Writing—Original draft preparation, S.A. and A.S.; Writing—Review and editing, S.A., A.S., R.G. and D.-Y.Y.; visualization, S.A.; supervision, R.G. All authors have read and agreed to the published version of the manuscript.

Funding: This research received no external funding.

Institutional Review Board Statement: Not applicable.

Informed Consent Statement: Not applicable.

Data Availability Statement: Not available.

Acknowledgments: The authors are thankful to the University of Victoria for the consistent support.

Conflicts of Interest: The authors declare no conflict of interest. The funders had no role in the design of the study; in the collection, analyses, or interpretation of data; in the writing of the manuscript, or in the decision to publish the results.

References

1. Duxson, P.; Fernández-Jiménez, A.; Provis, J.L.; Lukey, G.C.; Palomo, A.; van Deventer, J.S.J. Geopolymer Technology: The Current State of the Art. *J. Mater. Sci.* **2007**, *42*, 2917–2933. [CrossRef]
2. Mehta, A.; Siddique, R. Sustainable Geopolymer Concrete Using Ground Granulated Blast Furnace Slag and Rice Husk Ash: Strength and Permeability Properties. *J. Clean. Prod.* **2018**, *205*, 49–57. [CrossRef]
3. Xie, J.; Wang, J.; Rao, R.; Wang, C.; Fang, C. Effects of Combined Usage of GGBS and Fly Ash on Workability and Mechanical Properties of Alkali Activated Geopolymer Concrete with Recycled Aggregate. *Compos. Part B Eng.* **2019**, *164*, 179–190. [CrossRef]
4. Yang, T.; Yao, X.; Zhang, Z.; Wang, H. Mechanical Property and Structure of Alkali-Activated Fly Ash and Slag Blends. *J. Sustain. Cem.-Based Mater.* **2012**, *1*, 167–178. [CrossRef]
5. Reddy, M.S.; Dinakar, P.; Rao, B.H. Mix Design Development of Fly Ash and Ground Granulated Blast Furnace Slag Based Geopolymer Concrete. *J. Build. Eng.* **2018**, *20*, 712–722. [CrossRef]
6. Chen, Z.; Li, J.-S.; Zhan, B.-J.; Sharma, U.; Poon, C.S. Compressive Strength and Microstructural Properties of Dry-Mixed Geopolymer Pastes Synthesized from GGBS and Sewage Sludge Ash. *Constr. Build. Mater.* **2018**, *182*, 597–607. [CrossRef]
7. Lee, W.-H.; Wang, J.-H.; Ding, Y.-C.; Cheng, T.-W. A Study on the Characteristics and Microstructures of GGBS/FA Based Geopolymer Paste and Concrete. *Constr. Build. Mater.* **2019**, *211*, 807–813. [CrossRef]
8. Nath, P.; Sarker, P.K. Effect of GGBFS on Setting, Workability and Early Strength Properties of Fly Ash Geopolymer Concrete Cured in Ambient Condition. *Constr. Build. Mater.* **2014**, *66*, 163–171. [CrossRef]
9. Mehta, A.; Siddique, R. Sulfuric Acid Resistance of Fly Ash Based Geopolymer Concrete. *Constr. Build. Mater.* **2017**, *146*, 136–143. [CrossRef]
10. Wongkeo, W.; Seekaew, S.; Kaewrahan, O. Properties of High Calcium Fly Ash Geopolymer Lightweight Concrete. *Mater. Today Proc.* **2019**, *17*, 1423–1430. [CrossRef]
11. Zhao, R.; Yuan, Y.; Cheng, Z.; Wen, T.; Li, J.; Li, F.; Ma, Z.J. Freeze-Thaw Resistance of Class F Fly Ash-Based Geopolymer Concrete. *Constr. Build. Mater.* **2019**, *222*, 474–483. [CrossRef]
12. Al-Azzawi, M.; Yu, T.; Hadi, M.N.S. Factors Affecting the Bond Strength Between the Fly Ash-Based Geopolymer Concrete and Steel Reinforcement. *Structures* **2018**, *14*, 262–272. [CrossRef]
13. Top, S.; Vapur, H.; Altiner, M.; Kaya, D.; Ekicibil, A. Properties of Fly Ash-Based Lightweight Geopolymer Concrete Prepared Using Pumice and Expanded Perlite as Aggregates. *J. Mol. Struct.* **2020**, *1202*, 127236. [CrossRef]
14. Chen, X.; Zhou, M.; Shen, W.; Zhu, G.; Ge, X. Mechanical Properties and Microstructure of Metakaolin-Based Geopolymer Compound-Modified by Polyacrylic Emulsion and Polypropylene Fibers. *Constr. Build. Mater.* **2018**, *190*, 680–690. [CrossRef]
15. Hasnaoui, A.; Ghorbel, E.; Wardeh, G. Optimization Approach of Granulated Blast Furnace Slag and Metakaolin Based Geopolymer Mortars. *Constr. Build. Mater.* **2019**, *198*, 10–26. [CrossRef]
16. Li, X.; Rao, F.; Song, S.; Corona-Arroyo, M.A.; Ortiz-Lara, N.; Aguilar-Reyes, E.A. Effects of Aggregates on the Mechanical Properties and Microstructure of Geothermal Metakaolin-Based Geopolymers. *Results Phys.* **2018**, *11*, 267–273. [CrossRef]
17. Chen, S.; Wu, C.; Yan, D. Binder-Scale Creep Behavior of Metakaolin-Based Geopolymer. *Cem. Concr. Res.* **2019**, *124*, 105810. [CrossRef]
18. Guo, L.; Wu, Y.; Xu, F.; Song, X.; Ye, J.; Duan, P.; Zhang, Z. Sulfate Resistance of Hybrid Fiber Reinforced Metakaolin Geopolymer Composites. *Compos. Part B Eng.* **2020**, *183*, 107689. [CrossRef]
19. Li, X.; Rao, F.; Song, S.; Ma, Q. Deterioration in the Microstructure of Metakaolin-Based Geopolymers in Marine Environment. *J. Mater. Res. Technol.* **2019**, *8*, 2747–2752. [CrossRef]
20. Zanotti, C.; Borges, P.H.R.; Bhutta, A.; Banthia, N. Bond Strength between Concrete Substrate and Metakaolin Geopolymer Repair Mortar: Effect of Curing Regime and PVA Fiber Reinforcement. *Cem. Concr. Compos.* **2017**, *80*, 307–316. [CrossRef]
21. Pouhet, R.; Cyr, M. Formulation and Performance of Flash Metakaolin Geopolymer Concretes. *Constr. Build. Mater.* **2016**, *120*, 150–160. [CrossRef]
22. Nuaklong, P.; Sata, V.; Chindaprasirt, P. Properties of Metakaolin-High Calcium Fly Ash Geopolymer Concrete Containing Recycled Aggregate from Crushed Concrete Specimens. *Constr. Build. Mater.* **2018**, *161*, 365–373. [CrossRef]
23. Duan, P.; Yan, C.; Zhou, W. Influence of Partial Replacement of Fly Ash by Metakaolin on Mechanical Properties and Microstructure of Fly Ash Geopolymer Paste Exposed to Sulfate Attack. *Ceram. Int.* **2016**, *42*, 3504–3517. [CrossRef]
24. Fang, G.; Ho, W.K.; Tu, W.; Zhang, M. Workability and Mechanical Properties of Alkali-Activated Fly Ash-Slag Concrete Cured at Ambient Temperature. *Constr. Build. Mater.* **2018**, *172*, 476–487. [CrossRef]

25. Singhal, D.; Jindal, B.B.; Garg, A. Mechanical Properties of Ground Granulated Blast Furnace Slag Based Geopolymer Concrete Incorporating Alccofine with Different Concentration and Curing Temperature. Available online: https://www.ingentaconnect.com/contentone/asp/asem/2017/00000009/00000011/art00006 (accessed on 3 March 2020).

26. Nguyen, K.T.; Nguyen, Q.D.; Le, T.A.; Shin, J.; Lee, K. Analyzing the Compressive Strength of Green Fly Ash Based Geopolymer Concrete Using Experiment and Machine Learning Approaches. *Constr. Build. Mater.* **2020**, *247*, 118581. [CrossRef]

27. Panda, B.; Tan, M.J. Experimental Study on Mix Proportion and Fresh Properties of Fly Ash Based Geopolymer for 3D Concrete Printing. *Ceram. Int.* **2018**, *44*, 10258–10265. [CrossRef]

28. Albitar, M.; Mohamed Ali, M.S.; Visintin, P.; Drechsler, M. Durability Evaluation of Geopolymer and Conventional Concretes. *Constr. Build. Mater.* **2017**, *136*, 374–385. [CrossRef]

29. Ferdous, M.W.; Kayali, O.; Khennane, A. A detailed procedure of mix design for fly ash based geopolymer concrete. In Proceedings of the 4th Asia-Pacific Conference on FRP in Structures (APFIS 2013), Melbourne, Australia, 11–13 December 2013; pp. 11–13.

30. Somna, K.; Jaturapitakkul, C.; Kajitvichyanukul, P.; Chindaprasirt, P. NaOH-Activated Ground Fly Ash Geopolymer Cured at Ambient Temperature. *Fuel* **2011**, *90*, 2118–2124. [CrossRef]

31. Naghizadeh, A.; Ekolu, S.O. Method for Comprehensive Mix Design of Fly Ash Geopolymer Mortars. *Constr. Build. Mater.* **2019**, *202*, 704–717. [CrossRef]

32. Dao, D.V.; Ly, H.-B.; Trinh, S.H.; Le, T.-T.; Pham, B.T. Artificial Intelligence Approaches for Prediction of Compressive Strength of Geopolymer Concrete. *Materials* **2019**, *12*, 983. [CrossRef]

33. Ukrainczyk, N.; Ukrainczyk, V. A Neural Network Method for Analysing Concrete Durability. *Mag. Concr. Res.* **2008**, *60*, 475–486. [CrossRef]

34. Bal, L.; Buyle-Bodin, F. Artificial Neural Network for Predicting Drying Shrinkage of Concrete. *Constr. Build. Mater.* **2013**, *38*, 248–254. [CrossRef]

35. Jain, A.; Jha, S.K.; Misra, S. Modeling and Analysis of Concrete Slump Using Artificial Neural Networks. *J. Mater. Civ. Eng.* **2008**, *20*, 628–633. [CrossRef]

36. Anysz, H.; Narloch, P. Designing the Composition of Cement Stabilized Rammed Earth Using Artificial Neural Networks. *Materials* **2019**, *12*, 1396. [CrossRef]

37. Kurpinska, M.; Kułak, L. Predicting Performance of Lightweight Concrete with Granulated Expanded Glass and Ash Aggregate by Means of Using Artificial Neural Networks. *Materials* **2019**, *12*, 2002. [CrossRef]

38. Mozumder, R.A.; Laskar, A.I. Prediction of Unconfined Compressive Strength of Geopolymer Stabilized Clayey Soil Using Artificial Neural Network. *Comput. Geotech.* **2015**, *69*, 291–300. [CrossRef]

39. Yadollahi, M.M.; Benli, A.; Demirboğa, R. Prediction of Compressive Strength of Geopolymer Composites Using an Artificial Neural Network. *Energy Mater.* **2015**, *10*, 453–458. [CrossRef]

40. Nazari, A.; Pacheco Torgal, F. Predicting Compressive Strength of Different Geopolymers by Artificial Neural Networks. *Ceram. Int.* **2013**, *39*, 2247–2257. [CrossRef]

41. Braspenning, P.J.; Thuijsman, F.; Weijters, A.J.M.M. *Artificial Neural Networks: An Introduction to ANN Theory and Practice*; Springer: Berlin/Heidelberg, Germany, 1995; ISBN 978-3-540-59488-8.

42. Yue, Z.; Songzheng, Z.; Tianshi, L. Regularization BP Neural Network Model for Predicting Oil-Gas Drilling Cost. In Proceedings of the 2011 International Conference on Business Management and Electronic Information, Guangzhou, China, 13–15 May 2011; Volume 2, pp. 483–487.

43. Gupta, R.; Rathod, H.M. Current State of K-Based Geopolymer Cements Cured at Ambient Temperature. *Emerg. Mater. Res.* **2015**, *4*, 125–129. [CrossRef]

44. Belforti, F.; Azarsa, P.; Gupta, R.; Dave, U. Effect of Freeze-Thaw on K-Based Geopolymer Concrete (GPC) and Portland Cement Concrete (PCC). In Proceedings of the Technology Drivers: Engine for Growth: Proceedings of the 6th Nirma University International Conference on Engineering (NUiCONE 2017), Ahmedabad, India, 23–25 November 2017; CRC Press: Boca Raton, FL, USA, 2018; p. 65.

45. Hardjito, D.; Wallah, S.E.; Sumajouw, D.M.J.; Rangan, B.V. Fly Ash-Based Geopolymer Concrete. *Aust. J. Struct. Eng.* **2005**, *6*, 77–86. [CrossRef]

46. Gunasekara, C.; Law, D.; Bhuiyan, S.; Setunge, S.; Ward, L. Chloride Induced Corrosion in Different Fly Ash Based Geopolymer Concretes. *Constr. Build. Mater.* **2019**, *200*, 502–513. [CrossRef]

47. Okoye, F.N.; Durgaprasad, J.; Singh, N.B. Effect of Silica Fume on the Mechanical Properties of Fly Ash Based-Geopolymer Concrete. *Ceram. Int.* **2016**, *42*, 3000–3006. [CrossRef]

48. Sarker, P.K.; Haque, R.; Ramgolam, K.V. Fracture Behaviour of Heat Cured Fly Ash Based Geopolymer Concrete. *Mater. Des.* **2013**, *44*, 580–586. [CrossRef]

49. Shaikh, F.U.A.; Vimonsatit, V. Compressive Strength of Fly-Ash-Based Geopolymer Concrete at Elevated Temperatures. *Fire Mater.* **2015**, *39*, 174–188. [CrossRef]

50. Chithambaram, S.J.; Kumar, S.; Prasad, M.M.; Adak, D. Effect of Parameters on the Compressive Strength of Fly Ash Based Geopolymer Concrete. *Struct. Concr.* **2018**, *19*, 1202–1209. [CrossRef]

51. Adak, D.; Sarkar, M.; Mandal, S. Structural Performance of Nano-Silica Modified Fly-Ash Based Geopolymer Concrete. *Constr. Build. Mater.* **2017**, *135*, 430–439. [CrossRef]

52. Nath, P.; Sarker, P.K. Flexural Strength and Elastic Modulus of Ambient-Cured Blended Low-Calcium Fly Ash Geopolymer Concrete. *Constr. Build. Mater.* **2017**, *130*, 22–31. [CrossRef]
53. Mehta, A.; Siddique, R. Properties of Low-Calcium Fly Ash Based Geopolymer Concrete Incorporating OPC as Partial Replacement of Fly Ash. *Constr. Build. Mater.* **2017**, *150*, 792–807. [CrossRef]
54. Pasupathy, K.; Berndt, M.; Sanjayan, J.; Rajeev, P.; Cheema, D.S. Durability of Low-calcium Fly Ash Based Geopolymer Concrete Culvert in a Saline Environment. *Cem. Concr. Res.* **2017**, *100*, 297–310. [CrossRef]
55. Jena, S.; Panigrahi, R.; Sahu, P. Mechanical and Durability Properties of Fly Ash Geopolymer Concrete with Silica Fume. *J. Inst. Eng. India Ser. A* **2019**, *100*, 697–705. [CrossRef]
56. Topark-Ngarm, P.; Chindaprasirt, P.; Sata, V. Setting Time, Strength, and Bond of High-Calcium Fly Ash Geopolymer Concrete. *J. Mater. Civ. Eng.* **2015**, *27*, 04014198. [CrossRef]
57. Saravanakumar, R.; Revathi, V. Some Durability Aspects of Ambient Cured Bottom Ash Geopolymer Concrete. *Arch. Civ. Eng.* **2017**, *63*, 99–114. [CrossRef]
58. Singh, B.; Ishwarya, G.; Gupta, M.; Bhattacharyya, S.K. Geopolymer Concrete: A Review of Some Recent Developments. *Constr. Build. Mater.* **2015**, *85*, 78–90. [CrossRef]
59. Ma, C.-K.; Awang, A.Z.; Omar, W. Structural and Material Performance of Geopolymer Concrete: A Review. *Constr. Build. Mater.* **2018**, *186*, 90–102. [CrossRef]
60. Xie, T.; Ozbakkaloglu, T. Behavior of Low-Calcium Fly and Bottom Ash-Based Geopolymer Concrete Cured at Ambient Temperature. *Ceram. Int.* **2015**, *41*, 5945–5958. [CrossRef]
61. Chindaprasirt, P.; Jaturapitakkul, C.; Chalee, W.; Rattanasak, U. Comparative Study on the Characteristics of Fly Ash and Bottom Ash Geopolymers. *Waste Manag.* **2009**, *29*, 539–543. [CrossRef]
62. ul Haq, E.; Kunjalukkal Padmanabhan, S.; Licciulli, A. Synthesis and Characteristics of Fly Ash and Bottom Ash Based Geopolymers–A Comparative Study. *Ceram. Int.* **2014**, *40*, 2965–2971. [CrossRef]

 *materials*

Article

FBG Strain Monitoring of a Road Structure Reinforced with a Geosynthetic Mattress in Cases of Subsoil Deformation in Mining Activity Areas

Janusz Juraszek [1,*], Monika Gwóźdź-Lasoń [1] and Dominik Logoń [2]

[1] Faculty of Materials, Civil and Environmental Engineering, University of Bielsko-Biala, ul. Willowa 2, 43-309 Bielsko-Biala, Poland; mgwozdz@ath.bielsko.pl
[2] Faculty of Civil Engineering, Wrocław University of Science and Technology, ul. Wybrzeże Wyspiańskiego 27, 50-370 Wrocław, Poland; dominik.logon@pwr.edu.pl
* Correspondence: jjuraszek@ath.bielsko.pl

Abstract: This paper presents implementation of purpose-designed optical fibre Bragg grating (FBG) sensors intended for the monitoring of real values of strain in reinforced road structures in areas of mining activity. Two field test stations are described. The first enables analysis of the geogrid on concrete and ground subgrades. The second models the situation of subsoil deformation due to mining activity at different external loads. The paper presents a system of optical fibre sensors of strain and temperature dedicated for the investigated mattress. Laboratory tests were performed to determine the strain characteristic of the FBG sensor-geogrid system with respect to standard load. As a result, it was possible to establish the dependence of the geogrid strain on the forces occurring in it. This may be the basis for the analysis of the mining activity effect on right-of-way structures during precise strain measurements of a geogrid using FBG sensors embedded in it. The analysis of the results of measurements in the aspect of forecasted and actual static and dynamic effects of mining on the stability of a reinforced road structure is of key importance for detailed management of the road investment and for appropriate repair and modernization management of the road structure.

Keywords: monitoring fibre Bragg grating; mining areas; strain/stress distribution

Citation: Juraszek, J.; Gwóźdź-Lasoń, M.; Logoń, D. FBG Strain Monitoring of a Road Structure Reinforced with a Geosynthetic Mattress in Cases of Subsoil Deformation in Mining Activity Areas. *Materials* **2021**, *14*, 1709. https://doi.org/10.3390/ma14071709

Academic Editor: Evangelos J. Sapountzakis

Received: 30 January 2021
Accepted: 20 March 2021
Published: 30 March 2021

Publisher's Note: MDPI stays neutral with regard to jurisdictional claims in published maps and institutional affiliations.

Copyright: © 2021 by the authors. Licensee MDPI, Basel, Switzerland. This article is an open access article distributed under the terms and conditions of the Creative Commons Attribution (CC BY) license (https://creativecommons.org/licenses/by/4.0/).

1. Introduction

The analysis of the use of systems for long-term monitoring of the state of a road structure attracts a lot of attention, especially if the construction investment is located in an area of III geotechnical category with impacts on mining activity. By providing precise real-time information on the structure and its surroundings, it is possible to evaluate the state of the structure using diagnostic and prognostic tools based on the available data. The separation of intrinsic features from comprehensive monitoring data would be the key task in the assessment of the structure state and, consequently, an effective verification of the assumed unfavourable effects of mining activity compared to real impacts. The reasons for the use of components with optical fibres as sensor elements in the monitoring of the state of structures in use are well known.

The ever-growing demand for new linear investments usually involves improvement to land with difficult geotechnical conditions. Such land is often described as low-bearing soils, including the herein analysed soils in areas affected by mining impacts [1,2]. The basic mechanisms determining the engineering methods of computational analysis of subsidence are connected with phenomena such as compaction, consolidation, changes in water conditions, or disturbances to the balance of soil massif. The road structure monitoring investigated in the paper is essential during the use of a construction investment to reduce factors generating investment risk.

Structures located in mining areas are subject to additional impacts due to surface deformation or subgrade/subsoil vibration. These impacts are transferred from the subsoil

103

to the structure, which results in the occurrence of strains and internal forces in the structure. The monitoring of the strains provides current insight into the structure operation and creates the possibility of current checks on individual limit states of the structure.

In the case of an earthen structure, the strain-related impact of mining affects a significant portion of its area. In the zone of tensile strains, the subsoil and the embankment are loosened, and large limit-state areas are created. This leads to uneven settlement of the embankment crown, local subsidence, and horizontal displacement of the road pavement, as well as to local slides of slopes and pavement damage posing a hazard to safety. The presented monitoring system for a selected road structure scheme offers completely new possibilities for the use and protection, as well as renovation and modernization of road structures. In areas with mining impacts, the road structure subgrade is reinforced to increase the load-bearing capacity, reduce the subsidence of structures, prevent the loss of stability in the form of slips or landslides, and prevent the subgrade from liquefaction. Stabilization of the subgrade structure and mitigation of the effects of mining-related deformation were observed during numerous tests.

While choosing the method of subgrade reinforcement, the land development features should also be taken into account because the application of some methods, e.g., dynamic consolidation, may have a negative impact on existing building structures (resulting in damage to buildings due to transmission of vibrations from the subsoil or changes in hydrogeological conditions)—this is another thing that confirms the necessity and usefulness of monitoring in this field.

The data collected from the conducted analyses of selected models of the reinforcement of subgrades affected by the impact of mining activity additionally create the approach to the use of anthropogenic ground as a material for the proposed geosynthetic mattresses. This type of ground has already been tested in terms of its physical and strength parameters, as well as in the environmental aspect–utilization of materials from postmining waste dumps [3]. A database is being created for the environmental analysis of the issue.

Monitoring is an interdisciplinary issue. Due to its usefulness, it has a wide range of applications in the analysis of direct and indirect variables affecting variables related to the indicated aspects: securing the interests of the investor and the contractor of a construction investment, investment risk, especially in areas with the impact of mining activity, optimal assurance of the safety of investment implementation and use, and current assessment of the technical condition of a linear structure during its operation, as well as control and the possibility of getting an early warning of deformation of engineering structures in areas with an adverse mining impact. The presented empirical approaches to the monitoring of the interaction between the subsoil and the road structure provide a great deal of data that can be properly managed in a modern approach to the structure repair and modernization management. The monitoring data can generate appropriate models of the support of the decision-making process with methods suitable for the above-specified description of differentiating factors from the perspective of the researcher and analyst. The monitoring type and methodology and the range of the obtained data continuity for the time variable generate whole sections of statistics which can be used to find the answer to a given design-, strength-, or usability-related question, as well as economic, operating, or environmental problems. In this publication, the obtained data are analysed in the aspect of the road structure protection against the mining impact using a geomattress [4,5]. For the set of data divided into main structural, protection, and reinforcement elements and monitoring measurement locations, a data-mining analysis was conducted to explore large data resources in search of systematic correlations between variables and then to evaluate the results by applying the detected patterns to new subsets of data.

The final goal of the data mining operations on the data coming from the created monitoring system is to predict the behaviour of the road structure and provide timely feedback to prevent the possibility of increasing the probability of the road structure damage. Predictive data mining gives direct economic benefits in the management of road use and renovation. The data from individual monitoring models can be used to build a model for specific patterns, together with evaluation and verification to obtain predicted values or classifications.

Many various monitoring technologies have been developed, including the sensor based on the optical fibre Bragg grating (FBG). This solution is now being investigated in terms of its possible application for road structures, herein for example in the area affected by mining activity. In 1966, the use of optical fibres for digital transmission of data was proposed by Kao, who later won the Nobel Prize for Physics in 2009.

This paper presents the results of testing the use of FBG sensors for the monitoring of the reinforced surface of a road structure affected by unfavourable static and dynamic impacts due to mining activity. The aim of the investigated monitoring system is to develop guidelines for the analysed variable that affect the comprehensive manner of utilization, operation, and management of a road structure in areas affected by tremors and subsidence due to mining.

This type of monitoring was recently investigated in [6–11] and [12]; proposals for the use of the results of already tested monitoring systems to analyse the mining activity impact on residential and commercial buildings and road structures are signalled in [13,14] or [15]. The strain measurement method based on the change in the wavelength on the Bragg grating is described in [2,8,16–19].

As it is known, the Bragg grating is defined as periodic disturbance of the effective absorption coefficient and/or the optical fibre refractive index. Based on that, quasi-distributed systems are created for the detection of structural hazards. Herein, it concerns a road structure. The operation of mines, both past and present as well as future, generates imbalance of the rock mass. On the surface of the area, this results in direct effects, both continuous and discontinuous, indirect effects, and dynamic effects. All these effects have an unfavourable impact on buildings and road structures.

The main aim of the work is to use innovative FBG sensors to determine real values of strains and forces occurring in the geomattress depending on different service loads.

2. Presentation of the Problem for the Monitoring System

The aim of the testing and analyses presented herein is to demonstrate an original approach to the monitoring of changes in the stress-and-strain state of the mining subsoil for the reinforcement subsoil layer directly interacting with the construction of the road in areas with mining impact. The development of the civil engineering sector is directly linked to the development of the technology generating new solutions for the phases of design, realization, and operation of a construction investment.

The scope of innovations for the monitoring system creates new data and new schemes of analysis, e.g., for the investment risk calculation in the management of a road structure in an area affected by mining damage. Due to a continuous measurement of specific physical quantities, it is possible to control the behaviour of a civil engineering structure [4,5,15,20,21]. The authors point to the capabilities of a system that:

(i)   identifies the actual behaviour of the subsoil under designed constant and variable loads as well as under the forecasted and unpredictable impact of mining activity;

(ii) provides data for online description of the subsoil–reinforcement element–road structure interaction for the analysed period of utilization and operation of a civil engineering structure;

(iii) is an important variable in the optimization of schedules of the structure repair and modernization and in the checking of the investment risk value depending on the variable related to the mining activity effect.

The guidelines for road design and construction binding in the EU require that earthen structures as well as surfaces should be designed and constructed so that any potential impacts and influences occurring during construction and use, including the effects of mining activity, should be carried appropriately. In these specific conditions, the structures should display adequate durability, taking account of the predicted service life, and should not succumb to destruction to an extent disproportionate to the cause. Meeting these requirements is equivalent to the need to ensure conditions in which load or usability limits are not exceeded not only in each individual element but also in the entire earthen structure together with the road surface.

The nature of damage to the road infrastructure in mining areas and of the mobilization of limit states of load capacity is of a completely different origin compared to other areas [1,4,22–24] and is the effect mainly of the susceptibility of earthen structures and surfaces to horizontal unit strain with a loosening character (ε (mm/m)). This problem is illustrated comprehensively in the diagram in Table 1 and Figures 1–3.

Table 1. Classification of mining areas in the area of continuous deformation indicators of land deformation.

Mining Area Category	Slope T (mm/m)	Curvature Radius R (km)	Horizontal Strain ε (mm/m)
0	$T \leq 0.5$	$|R| \geq 40$	$|\varepsilon| \leq 0.3$
I	$0.5 < T \leq 2.5$	$40 > |R| \geq 20$	$0.3 < |\varepsilon| \leq 1.5$
II	$2.5 < T \leq 5.0$	$20 > |R| \geq 12$	$1.5 < |\varepsilon| \leq 3.0$
III	$5.0 < T \leq 10$	$12 > |R| \geq 6$	$3.0 < |\varepsilon| \leq 6.0$
IV	$10 < T \leq 15$	$6 > |R| \geq 4$	$6.0 < |\varepsilon| \leq 9.0$
V	$15 < T$	$4 > |R|$	$9.0 < |\varepsilon|$

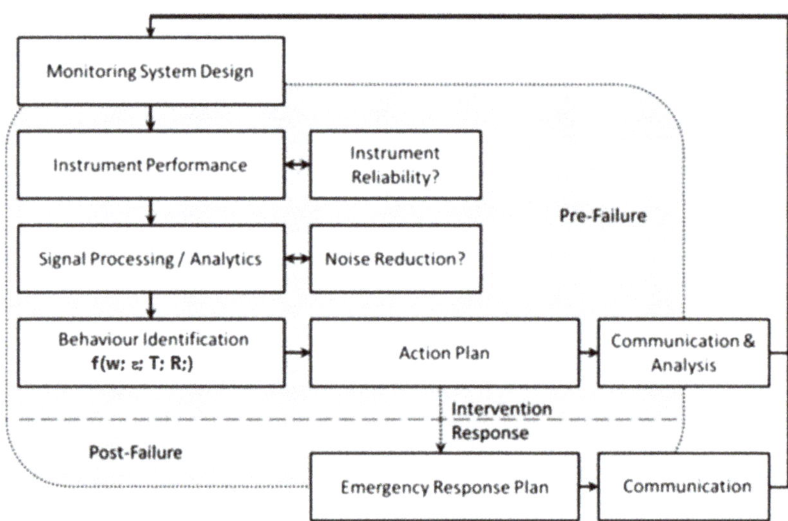

Figure 1. Scheme for the road structure monitoring process for the implementation of the safe operation process with the assumed investment risk.

At present, the majority of right-of-way structures in the road infrastructure still do not demonstrate sufficient structural resistance to such destructive impacts, which is the main cause of their damage. As stipulated by the regulations now in force, areas affected by mining activity should be protected according to the category of the mining area. The current classification of mining areas is presented, taking account of continuous and discontinuous deformation, paraseismic impacts and postmining areas. The measure of the hazard posed in a mining area by dynamic impacts both to newly erected and already existing structures is the assignment of a specific seismic zone to the area [25,26] and [27,28]. The seismic zone is described by parameters of the maximum ground vibrations that can occur in the area: acceleration and speed, as well as the subsoil design acceleration, all of which have to be taken into account in the design of civil structures in the area.

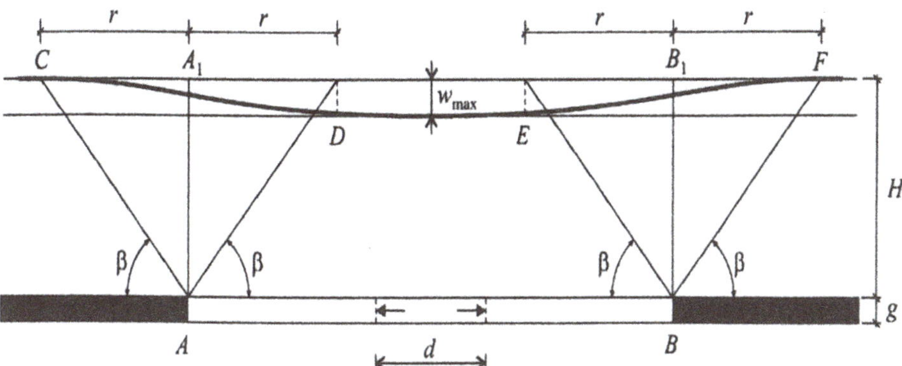

Figure 2. Deformation of the surface of the analysed land properties by factors affecting the characteristic of the range of mining impacts. Fixed mining trough in a plane deformation state: β—value of the main influence range angle; w_{max}—maximum lowering of the mining trough (m); g—thickness of the selected layer (m); a—service factor for $w_{max} = a \cdot g$

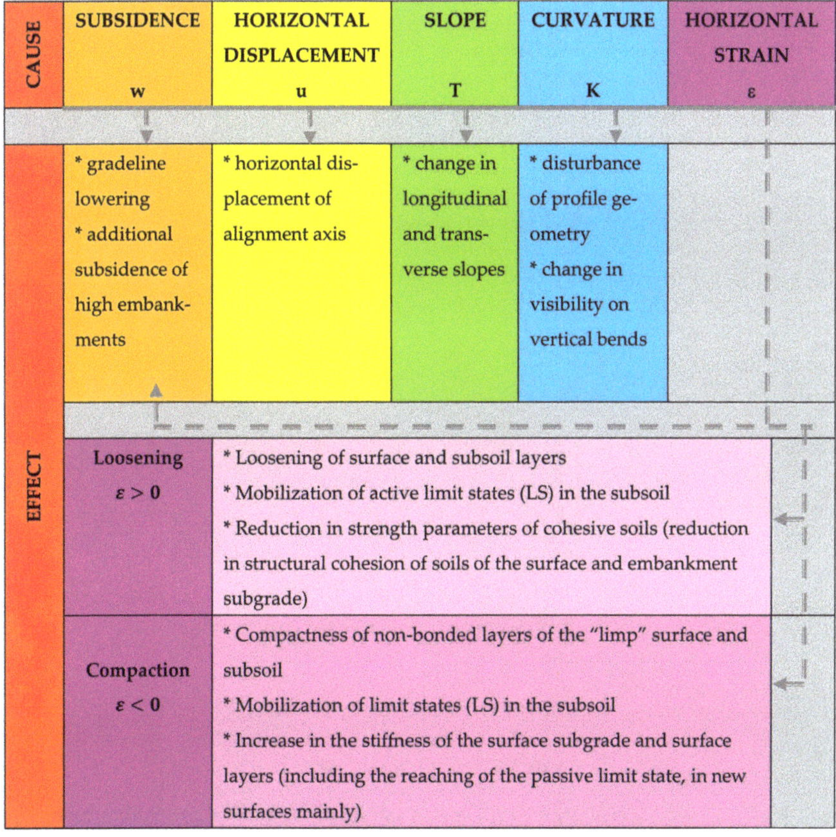

CAUSE	SUBSIDENCE	HORIZONTAL DISPLACEMENT	SLOPE	CURVATURE	HORIZONTAL STRAIN
	w	u	T	K	ε
EFFECT	* gradeline lowering * additional subsidence of high embankments	* horizontal displacement of alignment axis	* change in longitudinal and transverse slopes	* disturbance of profile geometry * change in visibility on vertical bends	
	Loosening ε > 0	* Loosening of surface and subsoil layers * Mobilization of active limit states (LS) in the subsoil * Reduction in strength parameters of cohesive soils (reduction in structural cohesion of soils of the surface and embankment subgrade)			
	Compaction ε < 0	* Compactness of non-bonded layers of the "limp" surface and subsoil * Mobilization of limit states (LS) in the subsoil * Increase in the stiffness of the surface subgrade and surface layers (including the reaching of the passive limit state, in new surfaces mainly)			

Figure 3. Cause-and-effect relationship describing the impact of indicators of the mining area continuous deformation on functional and strength parameters of the road infrastructure.

The indirect impacts affecting structures and related to the drainage of tertiary layers in the form of a large-size drainage basin are described using a single parameter: the mining area subsidence due to drainage. In areas where the impacts are not related to ground subsidence caused by mining, they do not result in mining damage to the development of the land or mining damage of a hydrogeological nature that is attributed in particular to changes in the stress-and-strain state of the ground. The definition of the mining area is directly connected to the range of the road structure interaction with the subsoil subjected to strain. According to ref. [29], this is the mass rock layer close to the surface, which is usually built of different soils, where the impact of the structure on the stresses and strains arising in the layer are considered as essential. The analysis of the results presented herein is based on the method of the Budryk-Knothe geometric-integral theory used in the majority of geomechanics in Silesia for prognostic analyses. It is the basic method of identification of kinematic limit states of investigated structures.

The analysed issues concern mining areas with subsoils reinforced with a geomattress to protect the right-of-way structure against deformation due to mining activity. The measure of the hazard posed to a mining area by impacts causing deformation is the mining area category. The category is described by indications of deformation. It is a variable that substantially affects the risk factor of the safe use of the structure. The basic indicators are subsidence—w, horizontal strain—ε (mm/m), land slope—T (mm/m), and curvature radius—R (km).

3. Experimental Setup

3.1. Test Stand Modelling

Two test stands were designed and made to achieve the research goals. The stands are equipped with optical FBG sensors of strain and displacement enabling a continuous measurement of strain in selected points of the geogrid. Test stand 1 is composed of the following layers: a 15 cm sand bed and the load-bearing mattress made of a geosynthetic grid filled with 16–32 mm aggregate with the thickness of 50 cm, compacted in layers every 25 cm. It is based partially on subsoil and partially on a reinforced concrete slab, which allows a comparison of strains in the two parts. Each subgrade has a separate FBG strain sensor embedded in the geogrid. The diagram of the test stand is shown in Figure 4.

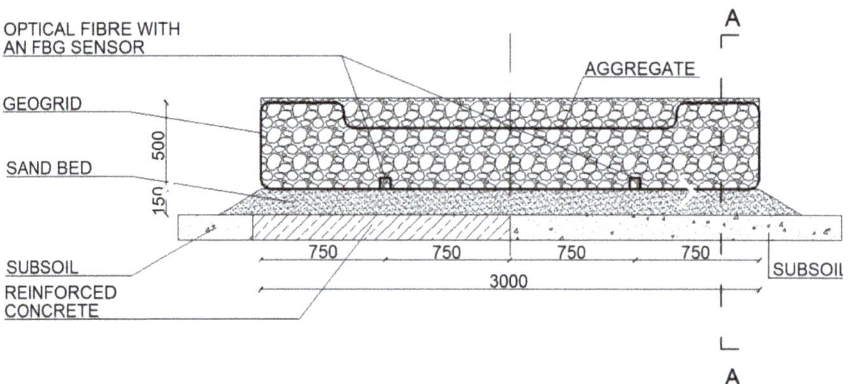

Figure 4. Diagram of test stand 1 (cm).

Test stand 2 enables an analysis of the case when there is no subgrade under the geogrid in two strips. It is composed of the following layers: geotextile, 800 × 1200 wooden pallets, and a load-bearing mattress filled with 31.5–63 mm aggregate with FBG sensors and 31.5–63 mm aggregate. The diagram of the test stand is shown in Figure 5.

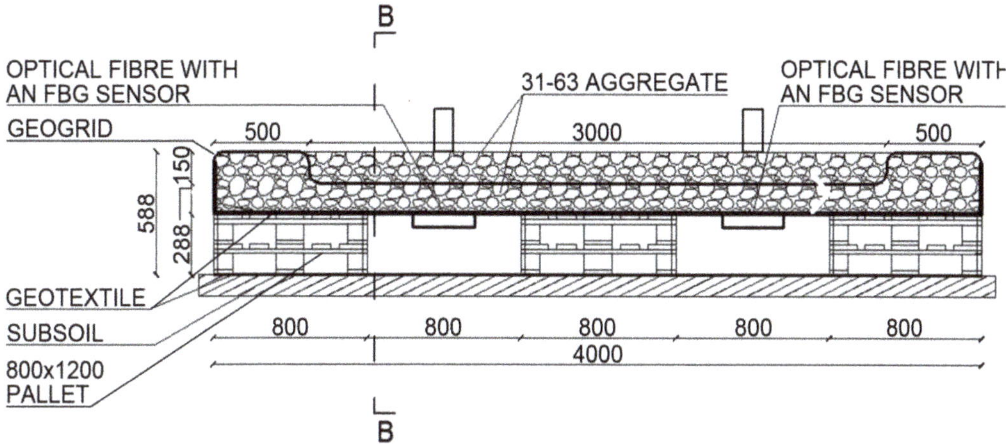

Figure 5. Diagram of test stand 2 (cm).

Two types of FBG sensors are applied: the strain sensor and the displacement sensor. Due to high values of the mattress deflection angles, the strain sensor proved useless and was no longer used in further stages of the research. On a single measuring line, the strain sensor may have 10 sensors with the wavelength difference of 5 nm with temperature compensation. A special technology of gluing the sensor into the geogrid in the middle of its span was developed. The sensor and its parameters are presented in Figure 6.

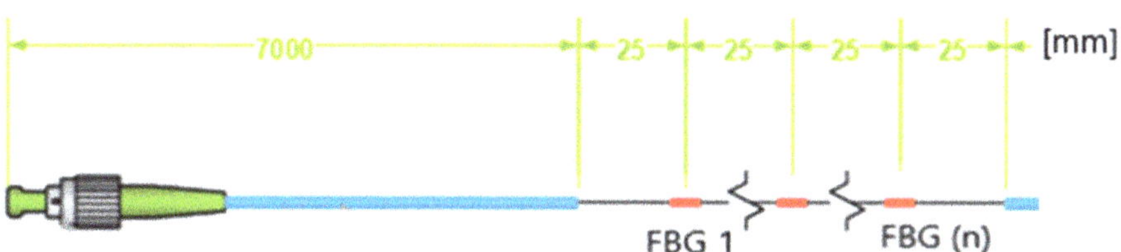

Figure 6. Scheme of fibre Bragg grating (FBG) sensor (mm).

The strain of the optical fibre sensor is the function of the wavelength (1) measured by the optical FBG-800 interrogator enabling dynamic measurements. The sampling frequency is 2000 Hz,

$$\Delta\varepsilon = \frac{\frac{\lambda_{act,strain} - \lambda_{0,\ inst,strain}}{\lambda_{0,\ inst,strain}} - B \times (T_{act} - T_{0,\ inst})}{A} \tag{1}$$

where A and B are constant Table 2.

Table 2. The strain of the optical fibre sensor is the function by the optical FBG-800 interrogator enabling dynamic measurements A and B.

Measurand	Description	A ($\mu\varepsilon^{-1}$)	B ($°C^{-1}$)
$\Delta\varepsilon$ ($\mu\varepsilon$)	Strain shift		
$\lambda_{0,inst,strain}$ (nm)	Initial strain wavelength		
$T_{0.inst}$ ($°C$)	Initial temperature	7.7584×10^{-07}	5.8929×10^{-06}
T_{act} ($°C$)	Actual temperature		
$\lambda_{act,strain}$ (nm)	Actual strain wavelength		

3.2. Plan of Investigated Experiments

The testing was divided into two main parts: the laboratory stage and in situ testing. The laboratory stage included the process of calibration of the geogrid strains with the forces occurring in the grid. The in situ testing was carried out both during the construction of the test stand and during the load tests. The test stand preparation was divided into individual stages, which had a considerable impact on the geogrid strains Table 3.

Table 3. Stages of the preparation of the test stands for the monitoring of the assumed road structure with a reinforcement element in the form of a geosynthetic mattress.

Stage	Scope of Works Leading to the Construction of the Test Stand
I	laying out the geogrid—no load
II	commencement of backfilling
III	mattress formation—first aggregate layer
IV	mattress formation—wrapping up the geogrid
V	mattress formation—second aggregate layer
VI	completion of the test stand construction
VII	tampering

The results of the geogrid strain-state testing during the construction of test stand 2 are presented in Figure 7. The first stage of laying out the geogrid produced strains at the level of about 135 µstrain. The jump in the graph at 14:04:47 h to the value of 600 µstrain proves that the geogrid backfilling process began. Backfilling was performed by hand using shovels. It lasted about 15 min, which can be seen in the graph to 14:19:00 h. The next stage was the geogrid forming and wrapping up to shape the geomattress. During that time, the geogrid strains were constant and totalled about 750 µstrain. The final stage of the test stand preparation was filling the geogrid mattress with aggregate, which can be seen in the graph between 14:32 and 14:43 h. The works were finished at 14:44 h, which can be observed in the graph in the form of a stop to the increase in strains. Strain values then stay at the level of 1050 µstrain. This is the reference level for further load tests. In the initial stage an analysis was also conducted of strains arising during the aggregate compaction with a compactor. The aggregate was compacted separately over each type of subgrade (subsoil, concrete slab).

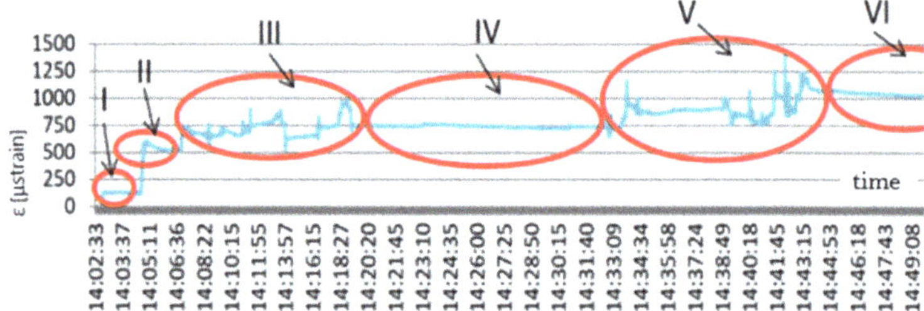

Figure 7. Scheme geogrid strains—stage zero of the monitoring: x (s); y (µstrain).

Distinct jumps in strains can be observed in Figure 8 when the compactor was directly over the sensor. After the compactor moved beyond the sensor, the strain value decreased at once. The graph also indicates that the geogrid mean strain rose cyclically. Characteristic steps (red line) appeared due to subsequent stages of the stand compaction, from 12 to 21 µstrain. Compaction of the aggregate over the part located on the natural soil resulted in different values of the geogrid strains. Distinct jumps in strains can be seen in Figure 9 when the compactor was directly over the sensor.

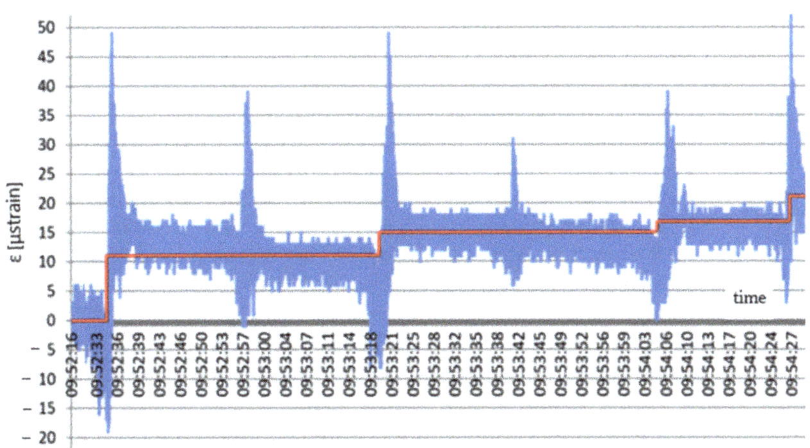

Figure 8. Scheme strains in the geogrid based on a concrete slab: x (s); y (µstrain).

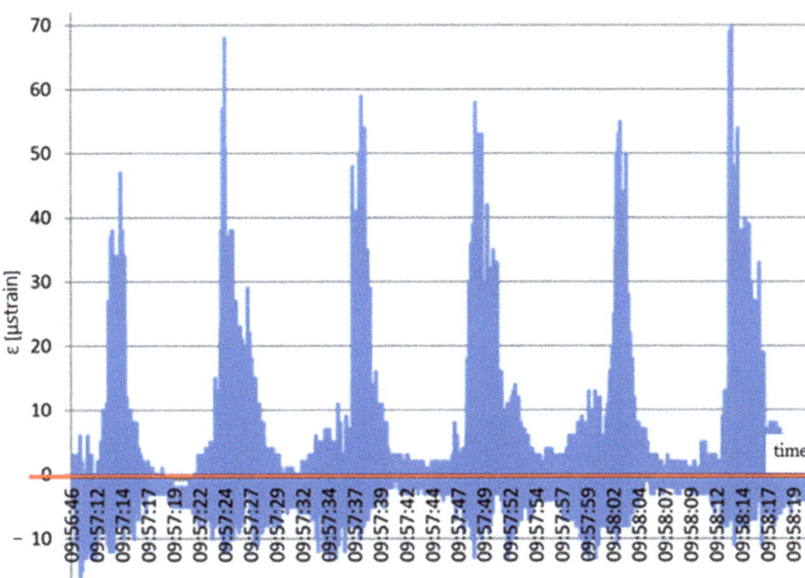

Figure 9. Scheme geogrid strains arising due to the interaction between the subsoil and the aggregate compaction process: x (s); y (µstrain).

After the compactor moved to other places, the strain value in the tested location decreased considerably. After each compaction cycle, the geogrid suffered no plastic strain, which is illustrated by the red line. The natural soil thus plays the role of an elastic subgrade.

4. Results

The following results were obtained for individual stages of the testing for the designed monitoring technology. The technology was suited to the initial data of the investigated issue concerning the analysis of un-forecasted mining activity impacts on right-of-way road structures—roads with appropriate reinforcement of the subgrade using a geosynthetic mattress.

4.1. In Situ Testing Results

In situ tests were performed including: (a) static loading of the area under the FBG monitoring, (b) dynamic loading of the tested structure, (c) loading due to the use of construction equipment–88kN ascent of an excavator, and (d) putting the excavator supports on the area of the FBG monitoring. The second experiment performed on the test stand was a static test consisting in loading the structure with concrete slabs. Standard $50 \times 50 \times 7$ cm paving slabs, each producing a 350 N load, were used for this purpose. The load was applied in layers with four slabs each. The weight of each of the layers produced a 1.4 kN load. The load was applied in places where the FBG sensor was installed. Moreover, in places under which there was a layer of pallets, an additional load was applied to additionally ballast the pallets to make the forces from the proper load to be taken over by the geogrid mainly. This additional load was not included in the weight of the layers producing the proper load. The experiment was carried out in stages by laying subsequent layers of slabs onto the stand. In the final stage of the testing, there were five layers of slabs on the stand. The total weight of the load was then 7.0 kN. The testing results are presented in Figure 10.

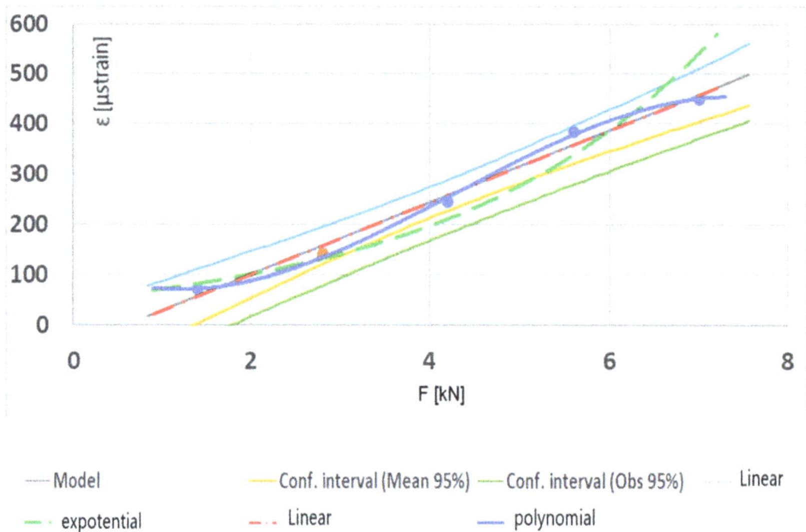

Figure 10. Geogrid strain depending on static load.

For one layer of slabs—the weight equal to 1.4 kN—the strain values totalled about 70 μstrain. After the second layer was laid, when the weight rose to 2.8 kN, the strain level in the geogrid increased to 130 μstrain. The next layer of slabs caused a further rise in strain to the value of 240 μstrain. After the fourth layer of slabs was laid, the load totalled 5.6 kN, and the strain level rose to 380 μstrain. Lastly, the fifth layer of slabs resulted in another rise in strain to the value of 450 μstrain. The next experiment carried out on the field test stand was the free-fall test of concrete slabs simulating a dynamic load of the type of the Heaviside step function. Next, $50 \times 50 \times 7$ cm concrete slabs were dropped from different heights of 30, 60, 90, and 120 cm, respectively. The idea of the testing is presented in Figure 11, and the testing results are shown in Figure 12.

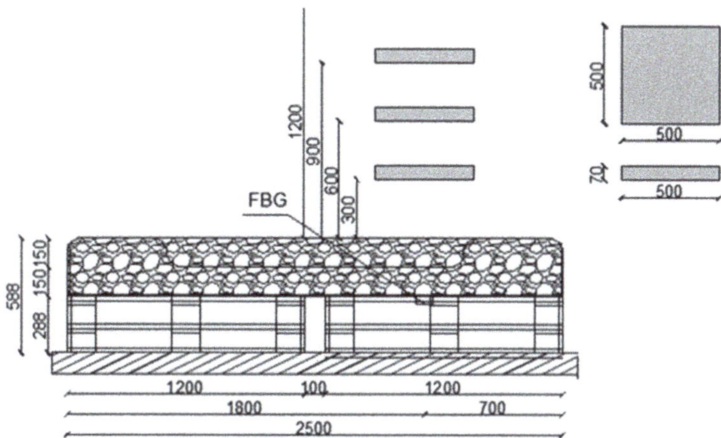

Figure 11. Idea of the testing—Heaviside step function (cm).

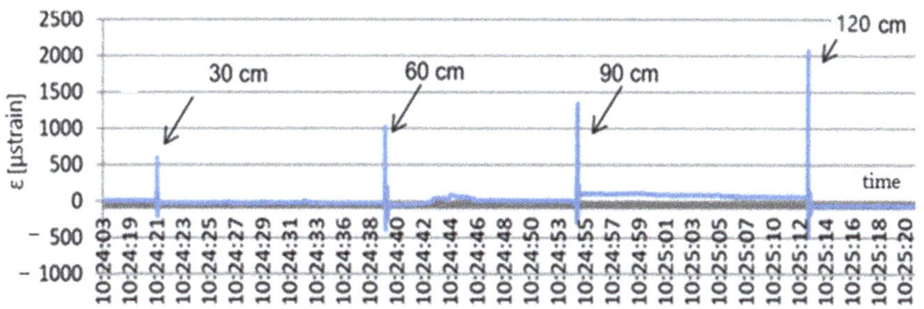

Figure 12. Dynamic test results: x (s); y (µstrain).

The free fall of the slabs from different heights caused jump changes in the strain value. At a free fall from the height of 30 and 60 cm, the slab caused a strain jump to the value of 600 and 1000 µstrain, respectively. The slab free fall from 90 and 120 cm caused a jump in the geogrid strain value to 1400 and 2050 µstrain, respectively. This means that the impact of falling slabs caused a jump change in the values of the geogrid strains. The geogrid then returned to the initial state. In the graph it can also be noticed that after the slab free fall from 90 cm the geogrid plastic strain occurred and the strain level being the result of the next impact totalled 100 µstrain. In addition, in this experiment with impulse loads, the FBG sensors performed very well. They made it possible to analyse the geomattress strain state and provided an answer to the question whether or not the mattress had been damaged and whether or not the loads had produced plastic strain values. The next experiment performed on the field test stand for the geogrid monitoring was the run of the CASE Super R backhoe loader. The machine mass totals 8850 kg. The backhoe loader ran onto and left the stand three times. The loader run was controlled to ensure that its rear wheel was directly over the sensor. The loader ran onto the stand to cause maximum strain of the geogrid. Figure 13 shows the geogrid strain testing results during the loader run onto the test stand.

During the first run of the loader the maximum strain value totalled 1200 µstrain. It then dropped to the level of −200 µstrain to reach during the next run the maximum value of about 1380 µstrain. During the break between the loader second and last run onto the test stand, strains fell to the value of −300 µstrain. The last run produced a jump in strain to the value of 4300 µstrain. After the loader left the stand, the strain values dropped again to the level of −370 µstrain. This is the response of the geomattress system to its considerable strain.

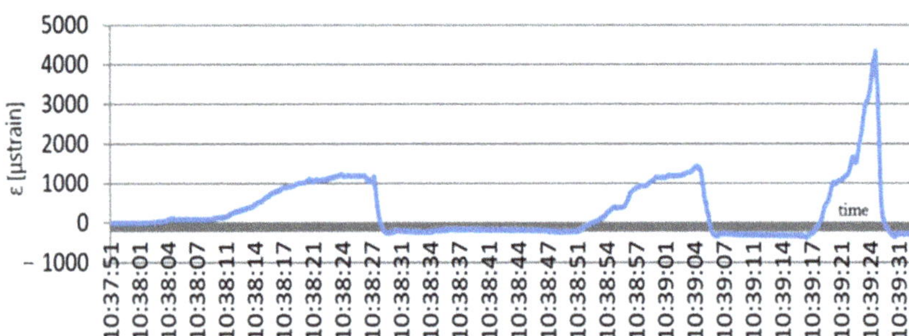

Figure 13. Geogrid strains during the backhoe loader run; x (s); y (µstrain).

The last experiment on the test stand was also carried out using the CASE Super R backhoe loader. This time the excavator supports were spread out and raised the machine. The supports were placed over the sensor to produce the highest loads of the area covered by the FBG monitoring and obtain the highest strains of the geogrid. The strains arising when the construction machine was placed on the geogrid are presented in Figure 14.

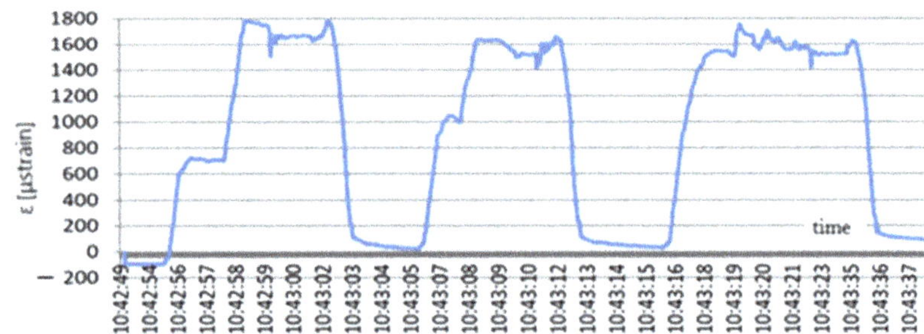

Figure 14. Change in the geogrid strain values due to the impact of the loader weight; x (s); y (µstrain).

Raising the loader on the supports caused an increase in the geogrid strain. After the first raising the strain level totalled 1780 µstrain. When the loader left the stand, the strain values decreased almost to zero to reach the level of 1650 µstrain at the next raising. After the loader left the stand, the strain level dropped again to the value of 50 µstrain. In the last test of raising the loader, the obtained strain value totalled 1770 µstrain. The drop in strain below zero from 10:43:42 h proves that the loader left the stand.

4.2. Laboratory Testing Results

The aim of the laboratory testing was experimental determination of relations between strains and loads (axial forces) in a unidirectional grid. For this purpose, a series of tensile tests of specimens from the geosynthetic grid were performed. The reference to the standard force was the dynamometer of a strength-testing machine (HOUNSFIELSD TEST EQUIPMENT 0133 model H50KS, Shakopee, MN, USA with Accuracy Class 0.5 (Figure 15). The geogrid was the same as the one used in the field stand structure. The rupture test was performed according to standard EN ISO 10319:2008 geosynthetics–wide-width tensile test. Thirty (30) specimens 200 mm wide and 200 mm long were prepared for the testing to ensure the nominal length of 100 mm according to 6.3.3 PN-EN ISO 10319:2008. As the geogrid is reinforced unidirectionally, the specimen was ruptured in the longitudinal direction only. Special holders—elastomeric inserts—were designed to hold the strength-testing machine, which ensured that the loading process was carried out correctly. Subsequent reference testing was carried out in the same manner, except that the geogrid was equipped with an optical FBG sensor permanently glued into the grid using special technology.

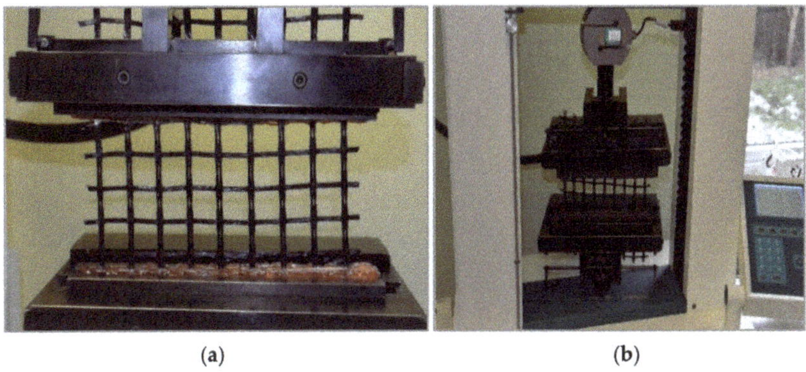

(a)   (b)

Figure 15. Geogrid tensile test: (**a**) fastening of the geogrid, (**b**) force sensor.

Such embedding of the sensor into the geogrid worked very well in the case of field testing. The FBG sensor was glued into a 620-mm-long geogrid specimen to ensure the nominal length of 500 mm. Rubber inserts were fixed at the ends of the specimen to ensure the best assembly. The inserts were joined to the geogrid using a cyanoacrylate-based adhesive. The results of calibration tests are presented in Figure 16.

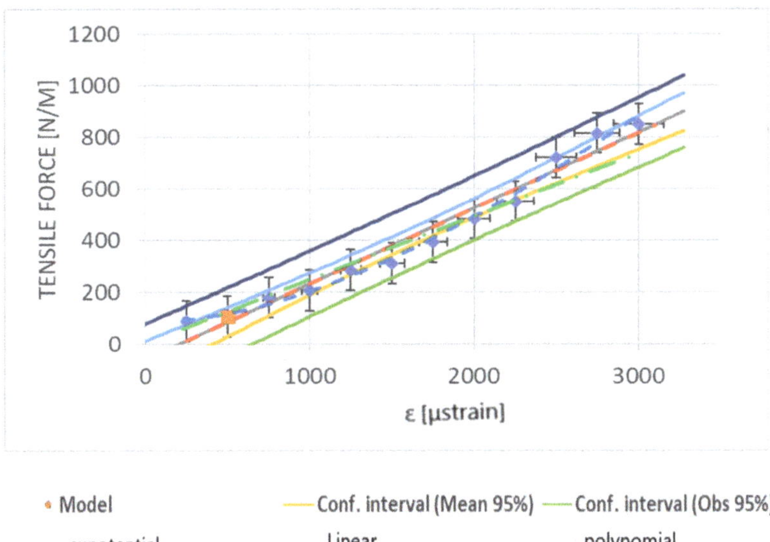

Figure 16. Calibration testing of the geogrid with an optical FBG sensor.

The obtained calibration chart makes it possible to determine the value of the load occurring in the geogrid based on the measurement of the grid strain using an FBG sensor.

5. Discussion

The paper presents an innovative method of measuring strains of the geosynthetic grid using optical FBG sensors. The strain measurement, together with determination of the forces acting in the geogrid, is a very important element of the monitoring of the condition of a geotechnical structure and areas affected by mining damage. The values of strains and loads occurring in the geogrid were determined experimentally. The stand construction and the backfilling with aggregate caused strains at the level of 1050 μstrain corresponding to the load of 220 N/m. This strain state was adopted as the initial one. The highest strain values were caused by dynamic loads in the form of the Heaviside impulse of 2050 μstrain comparable to the excavator ascent onto the mattress. The lowest strain values of 70 μstrain were recorded during aggregate compaction on natural soil. The optical FBG sensors belong to the family of the fibre-optic sensors (FOS), and they perform very well in difficult terrain and environmental conditions. Individual FBG sensors can be connected easily to obtain a comprehensive measurement system of the entire structure. The sensors can be used to measure the structure strains on the one hand and the temperature of the structure or the moisture level on the other. However, their most important task is the strain measurement as strains are the factor that has the most essential impact on the safety of the use of a structure.

A very important value of the work is determination of real values of the geogrid strains and real values of forces acting in the grid. This is important as it creates the possibility of the geogrid stress-and-strain state evaluation. The works published so far do not refer to the values of forces in the geogrid and do not enable the grid stress-and-strain state evaluation. Most of them concentrate on monitoring the condition of a few key geotechnical structures, including soil nail systems, slopes, and piles, without any reference to the issue of the stress-and-strain state in the analysed elements. Ref. [30] points to the possibility of using the results of the grid strain measurement as input or boundary data for a numerical experiment using the finite difference or the finite element method. This will be the subject of the authors' further work. Many works concentrate on the FBG sensor mounting method. Two solutions can be distinguished. The first consists in applying an optical fibre with Bragg gratings over an intermediate and properly secured element. This works in structures with low deformability and is not proposed here because it would generate too large a measurement error. The second solution is to embed the optical fibre sensor in the tested structure using an adhesive. A special gluing technique was used in the analysed geogrid to embed in it not only the FBG strain sensor but, significantly enough, also a telecommunication optical fibre to ensure data transmission and the sensor supply with light. Indicating directions of further development, a proposal is made in [29] to construct 3D sensors. The research results point to the possibility of constructing orthogonal strain-measuring systems, 90- and 120-degree strain rosettes, as well as 3D systems. Another issue raised in to-date publications is the cost of the FBG monitoring system [30]. The experimentally determined strain values [19,31,32] are comparable to the values occurring in the geogrid. Undoubtedly, monitoring systems based on optical FBG sensors are relatively expensive. Low-cost solutions are being sought. The authors tested a new generation of 1 Hz optical interrogators with FBG sensors operating at the wavelength of 800 nm. Such a solution is equivalent to sensors working at the wavelength of 1550 nm and, most importantly, is twice as cheap as classical solutions. Attention should also be drawn to the unprecedented possibility of integrating different fibre-optic sensors to perform parallel measurements of strains, the forces occurring in the geogrid and the current temperature and moisture content. Another advantage of measurements based on optical fibre sensors is their fatigue strength, long service life, and the multiplexing capacity.

6. Conclusions

Appropriately processed, the data obtained from the geogrid monitoring can become the database for the analysis of risk presented by the operation of a structure in cases of un-forecasted mining activity impacts. Known forces of the interaction along the mining-activity–reinforcement-system–road-structure scheme in the monitored time interval should considerably improve the management of road structures in areas affected by unfavourable impacts of mining activity. The probability defining the risk factor related to the use of a road structure is usually expressed as frequency, i.e., the number of expected events in a unit of time. For structures provided with appropriate monitoring and covered by permanent analysis of variable values qualifying the structure for specific mining area categories, the frequencies of unfavourable and random events may vary by many orders of magnitude for the analysed logarithmic scale. The categories of the consequences of on-line monitoring of the work of a road structure can also be expressed in many matrices of risk related to the safety of the structure. The assumed schemes of data analysis give theses with very good results. Subsequent planned testing of buildings and structures in areas with forecasted mining impacts can become the basis for the verification of both the adopted monitoring methodology and the system of analysis of obtained data in the aspect of risk related to the use of an investment and optimization of schedules with the scope, type, and frequency of the performed repair and modernization management of the structure. Moreover, data are collected all the time for cost analysis of such monitoring of the structure. For the preliminarily developed risk matrices for selected types of mining damage repair work taking account of probability and consequence classes, the initial results are very promising. The combination of optical FBG sensors with geosynthetic materials and their application in road structures make them undoubtedly one of the innovative methods of strain measurement, and the presented calibration methodology enables the assessment of loads occurring in the geogrid.

Summarizing the initial assumptions, the adopted models, and the obtained final results, together with a discussion of current research trends, the following main points can be mentioned:

a.   The applied system of FBG sensors enables precise and continuous monitoring of strains and loads of the road structure reinforcement in the form of a geomattress which takes over the impacts from the mining activity areas.

b.   Compared to the existing systems, the presented monitoring system is characterized by exceptional resistance to difficult environmental conditions.

c.   In the future, the monitoring information package can become the basis for a system warning against mining hazards presented to existing engineering structures. Such information is now indispensable for statistical analyses of the ground behaviour in the interaction with an engineering structure.

The growing set of monitoring data should be analysed in all interdisciplinary issues that are connected with the discussed monitoring system intended for road structures in areas affected by the impact of mining activity. Datasets are being created for a detailed environmental, economic, and performance analysis of the investigated investment in the form of a monitoring system intended for a road structure reinforced with a geosynthetic material.

Author Contributions: Conceptualization, J.J.; methodology, J.J.; software, M.G.-L.; validation, J.J. investigation, M.G.-L. and J.J.; resources, J.J.; writing—original draft preparation, M.G.-L. and D.L.; writing—review and editing, J.J. and M.G.-L.; visualization, M.G.-L. and D.L.; supervision, J.J. All authors have read and agreed to the published version of the manuscript.

Funding: This research received no external funding.

Institutional Review Board Statement: Not applicable.

Informed Consent Statement: Not applicable.

Data Availability Statement: No new data were created or analyzed in this study. Data sharing is not applicable to this article.

Conflicts of Interest: The authors declare no conflict of interest.

References

1. Gwozdz-Lason, M. Multiplication analysis of the cause, form and extent of damage to buildings in areas with mining impact. *IOP Conf. Ser. Mater. Sci. Eng.* **2019**, *603*, 042093. [CrossRef]
2. Qin, J.-Q.; Yin, J.-H.; Zhu, Z.-H.; Tan, D.-Y. Development and application of new FBG mini tension link transducers for monitoring dynamic response of a flexible barrier under impact loads. *Measurement* **2020**, *153*, 107409. [CrossRef]
3. Gwozdz-Lason, M. Odpady kopalniane jako wytrzymały grunt antropogeniczny wykorzystywany w nowoczesnych konstrukcjach geotechnicznych—The anthropogenic ground come from mining wastes used in modern geotechnical structures). *Saf. Eng. Anthropog. Objects* **2018**, 3–9.
4. Gwozdz-Lason, M. Effect of active mining impact on properties with engineering structures—Forecast and final result discrepancies. *IOP Conf. Ser. Earth Environ. Sci.* **2019**, *221*, 012103. [CrossRef]
5. Francken, L.; Clauwaert, C. Characterization and structural assessment of bound materials for flexible road structures. In Proceedings of the 6-th International Conference on Asphalt Pavements, Ann Arbor, MI, USA, 13–17 July 1987; pp. 130–140.
6. Zhou, Z.; Liu, W.Q.; Huang, Y.; Wang, H.P.; He, J.P.; Huang, M.H.; Ou, J.P. Optical fiber Bragg grating sensor assembly for 3D strain monitoring and its case study in highway pavement. *Mech. Syst. Signal Process.* **2012**, *28*, 36–49. [CrossRef]
7. Liu, Z.; Zhang, Y.A.; Shen, L.Y.; Qian, J.W. Detection of pipeline curvature with FBG sensors twisted. *J. Shanghai Univ.* **2006**, *12*, 450–456.
8. Wang, H.P.; Liu, W.Q.; Zhou, Z.; Wang, S.H.; Li, Y.; Guo, Z.W. The behavior of a novel raw material-encapsulated FBG sensor for pavement monitoring. In Proceedings of the 2011 International Conference on Optical Instruments and Technology: Optical Sensors and Applications, Beijing, China, 6–9 November 2011.
9. Juraszek, J. Residual magnetic field for identification of damage in steel wire rope. *Arch. Min. Sci.* **2019**, *64*, 79–92.
10. Juraszek, J. Residual magnetic field non-destructive testing of gantry cranes. *Materials* **2019**, *12*, 564. [CrossRef] [PubMed]
11. Juraszek, J. Strain and force measurement in wire guide. *Arch. Min. Sci.* **2018**, *63*, 321–334.
12. Kadela, M. Systemy monitorowania obiektów liniowych na terenach górniczych (Line objects monitoring systems in mining areas). In Proceedings of the Materiały konferencyjne XII Dni Miernictwa Górniczego i Ochrony Terenów Górniczych, Gliwice, Poland, 12–14 June 2013; pp. 163–172.
13. Nosenzo, G.; Whelan, B.E.; Brunton, M.; Kay, D.; Buys, H. Continuous monitoring of mining induced strain in a road pavement using fiber Bragg grating sensors. *Photonic Sens.* **2012**, *3*, 144–158. [CrossRef]
14. Buddery, P.; Morton, C.; Scott, D.; Owen, N. A continuous roof and floor monitoring systems for tailgate roadways. In Proceedings of the 18th Coal Operators' Conference, Mining Engineering, Wollongong, Australia, 7–9 February 2018; pp. 72–81.
15. Mokrosz, R. *Obiekty Budowlane na Terenach Górniczych*; Inżynier Budownictwa: Katowice, Poland, 2015.
16. Loizos, A.; Plati, C.; Papavasilio, V. Fiber optic sensors for assessing strains in cold in-place recycled pavements. *Int. J. Pavement Eng.* **2013**, *14*, 125–133. [CrossRef]
17. Juraszek, J. Fiber Bragg sensors on strain analysis of power transmission lines. *Materials* **2020**, *13*, 1559. [CrossRef]
18. Juraszek, J. Hoisting machine brake linkage strain. *Arch. Min. Sci.* **2018**, *63*, 583–597.
19. Hong, C.-Y.; Zhang, Y.-F.; Zhang, M.-X.; Leung, L.M.G.; Liu, L.-Q. Application of FBG sensors for geotechnical health monitoring, a review of sensor design, implementation methods and packaging techniques. *Sens. Actuators A Phys.* **2016**, *244*, 184–197. [CrossRef]
20. Chudek, M.; Strzałkowska, P.; Flisowski, A. Weryfikacja metody prognozowania deformacji górotworu wykorzystują funkcje wpływów całkowalną przez kwadraturę. *Zeszyt Naukowy Politechniki Śląskiej* **2007**, *276*.
21. Kłosek, K. Nawierzchnia I budowle ziemne autostrat na terenach górniczych. *Miesiecznik WUG* **2006**, *146*, 19–28.
22. Kadela, M. Model of multiple-layer pavement structure-subsoil system. *Bull. Pol. Acad. Sci. Tech. Sci.* **2016**, *64*, 751–762. [CrossRef]
23. *Wytyczne Wzmacniania Podłoża Gruntowego w Budownictwie Drogowym (Guidelines for Subsoil Reinforcement in Road Construction)*; GDDP-IBDiM: Warsaw, Poland, 2002.
24. *Metody Stosowania Geosyntetyków do Budowy i Wzmocnienia Nawierzchni i Ziemnych Budowli Drogowych (Geo-Synthetics Application Methods for the Construction and Reinforcement of Surfaces and Earthen Road Structures)*; IBDiM: Warsaw, Poland, 2003.
25. Myrczek, J.; Juraszek, J.; Tworek, P. Risk management analysis in construction enterprises in selected regions in Poland. *Tech. Trans.* **2020**, *117*, 1–13. [CrossRef]
26. Gwozdz-Lason, M. Analysis by the residual method for estimate market value of land on the areas with mining exploitation in subsoil under future new building. *IOP Conf. Ser. Earth Environ. Sci.* **2017**, *95*, 042064. [CrossRef]
27. Standards Australia. *Risk Management—Principles and Guidelines*. AS/NZ ISO 31000. 2009. Available online: https://www.iso.org/standard/43170.html (accessed on 20 February 2021).
28. Gwozdz-Lason, M. The cost-effective and geotechnic safely buildings on the areas with mine exploitation. In Proceedings of the 17th International Multidisciplinary Scientific GeoConference SGEM 2017, Albena, Bulgaria, 29 June–5 July 2017; pp. 877–884, ISBN 978-619-7105-00-1. [CrossRef]
29. ITB 2006 Instrukcje, Wytyczne, Poradniki, 416/2006. In *Projektowanie Budynków na Terenach Górniczych*; Wydawnictwo Instytutu Techniki Budowlanej: Warsaw, Poland, 2006.
30. Zhu, H.-H.; Shi, B.; Zhang, C.-C. FBG-Based Monitoring of Geohazards: Current Status and Trends. *Sensors* **2017**, *17*, 452. [CrossRef]

31. Ren, Y.-W.; Yuan, Q.; Chai, J.; Liu, Y.-L.; Zhang, D.-D.; Liu, X.-W.; Liu, Y.-X. Study on the clay weakening characteristics in deep unconsolidated layer using the multi-point monitoring system of FBG sensor arrays. *Opt. Fiber Technol.* **2021**, *61*, 102432. [CrossRef]
32. Briançon, L.; Villard, P. Design of geosynthetic-rein-forced platforms spanning localized sinkholes. *Geotext. Geomembr.* **2008**, *26*, 416–428. [CrossRef]

 *materials*

Article

Elements of Pathway for Quick and Reliable Health Monitoring of Concrete Behavior in Cable Post-Tensioned Concrete Girders

Lukasz Bednarz [1,*], Dariusz Bajno [2], Zygmunt Matkowski [1], Izabela Skrzypczak [3] and Agnieszka Leśniak [4]

[1] Faculty of Civil Engineering, Wroclaw University of Science and Technology, Wybrzeże Wyspiańskiego 27, 50-370 Wrocław, Poland; zygmunt.matkowski@pwr.edu.pl

[2] Faculty of Civil and Environmental Engineering and Architecture, UTP University of Science and Technology, Al. prof. S. Kaliskiego 7, 85-796 Bydgoszcz, Poland; dariusz.bajno@utp.edu.pl

[3] Faculty of Civil and Environmental Engineering and Architecture, Rzeszow University of Technology, Powstanców Warszawy 12, 35-082 Rzeszow, Poland; izas@prz.edu.pl

[4] Faculty of Civil Engineering, Cracow University of Technology, Warszawska 24, 31-155 Kraków, Poland; alesniak@l7.pk.edu.pl

* Correspondence: lukasz.bednarz@pwr.edu.pl

Abstract: The paper discusses the problems connected with long-term exploitation of reinforced concrete post-tensioned girders. The scale of problems in the world related to the number of cable post-tensioned concrete girders built in the 1950s and still in operation is very large and possibly has very serious consequences. The paper presents an analysis and evaluation of the results of measurements of the deflection and strength and homogeneity of concrete in cable–concrete roof girders of selected industrial halls located in Poland, exploited for over 50 years. On the basis of the results of displacement monitoring in the years 2009–2020, the maximum increments of deflection of the analyzed girders were determined. Non-destructive, destructive, and indirect evaluation methods were used to determine the compressive strength of concrete. Within the framework of the indirect method recommended in standard PN-EN 13791, a procedure was proposed by the authors to modify the so-called base curve for determining compressive strength. Due to the age of the analyzed structural elements, a correction factor for the age of concrete was taken into account in the strength assessment. The typical value of the characteristic compressive strength is within the range 20.3–28.4 MPa. As a result of the conducted tests, the concrete class assumed in the design was not confirmed, and its classification depended on the applied test method. The analyzed girders, in spite of their long-term exploitation, can be still used for years on the condition that regular periodical inspections of their technical condition are carried out. The authors emphasize the necessity for a permanent and cyclic diagnostic process and monitoring of the geometry of girders, as they are expected to operate much longer than was assumed by their designers.

Keywords: post-tension; cable; concrete; girder; diagnostic; destructive test; non-destructive test; structural health monitoring; safety

Citation: Bednarz, L.; Bajno, D.; Matkowski, Z.; Skrzypczak, I.; Leśniak, A. Elements of Pathway for Quick and Reliable Health Monitoring of Concrete Behavior in Cable Post-Tensioned Concrete Girders. *Materials* **2021**, *14*, 1503. https://doi.org/10.3390/ma14061503

Academic Editor: Theodore Matikas

Received: 27 January 2021
Accepted: 15 March 2021
Published: 18 March 2021

Publisher's Note: MDPI stays neutral with regard to jurisdictional claims in published maps and institutional affiliations.

Copyright: © 2021 by the authors. Licensee MDPI, Basel, Switzerland. This article is an open access article distributed under the terms and conditions of the Creative Commons Attribution (CC BY) license (https://creativecommons.org/licenses/by/4.0/).

1. Introduction

In many cases in construction practice, it may not be possible to carry out a precise assessment of the technical condition of buildings and their components using traditional, commonly used methods. Such situations will concern buildings and structures with a large number of structures very sensitive to environmental conditions, for which the assumed technical life has already significantly exceeded their current service life. It will not be practicable to replace these elements with new ones due to the large number of buildings constructed with such and similar technologies. An example of such facilities are industrial halls built using prefabricated elements made in prestressed concrete technology.

Post-tensioning techniques have been widely applied in construction to design bridge piers and decks, building slabs, and long-span girders [1]. Post-Tensioned structural

121

elements are very often used in bridges i roof construction industrial halls due to their ability to economically span long widths while providing an aesthetically structure [2,3]. Due to these advantages and effectiveness of the structural behavior post-tensioned pre-stressed concrete girders have been constructed for the last half century, [4]. Cable post-tensioned concrete girders are reinforced concrete constructions which are preloaded with prestressing forces either after the concrete has partially or completely hardened in the molds, or directly on site with the use of cables, anchored at the ends of the elements [5,6]. The cables (prestressing tendons) are inserted into the channels mainly located inside, so that the strength of the element can be used rationally in each section.

Tension presses (jacks) are used to prestress the structure. The space between the tendons and the walls of the cable tunnel is filled with an injection material (e.g., cement–water grout) of sufficient strength to protect the steel of the tendons against corrosion and to transfer the prestressing forces to the structure.

With a few exceptions, these structures have a segmental structure, joined directly on site during assembly by means of welded joints, filled with mortar. In spite of the impression made by their seemingly massive external appearance, these elements are quite "delicate" and have a long service life (approx. 50–75 years), hence the serious problem of their safe use; i.e., maintenance in an appropriate technical condition has already become a topic of discussion [2,7–9].

The beginnings of tests on prestressed reinforced concrete structures date to the second half of the 20th century [2]. Research was conducted both in Europe and in the United States. The introduction of this new type of reinforced concrete structure was intended to eliminate the disadvantages of concrete and reinforced concrete, which was characterized by very low tensile strength and high susceptibility to cracking. The lack of materials with the required properties meant that these structures had not been used before. The development of high-strength steel for prestressing tendons and the possibility to repair war damage more quickly (after World War II) brought about a change, mainly in industry, for the piercing of long-span halls. In 1952, the International Association of Prestressed Structures was founded, which in 1988 became part of the International Concrete Association. The first industrial-scale use of prestressed concrete began in the 1950s, mainly for the manufacture of prefabricated girders, roof and floor slabs, and railway sleepers [10].

There are not many studies in the literature of specific standard guidelines for the testing and maintenance of these girders apart from a few technical manuals or handbooks, e.g., [11–15]. Current knowledge on the serviceability of cable post-tensioned concrete girders is based on the assessment of their technical condition. The perspective of maintaining building objects in an adequate technical condition is one of the most important problems during their service life [16]. Mostly visual inspection, non-destructive testing, measurements, calculations, and when necessary, in-depth material tests are carried out for the diagnosis of technical conditions. The methods of material testing of structures are quite well known and are described in the literature [17–21]. Depending on the degree of their invasiveness, these can be divided into destructive, semi-destructive, and non-destructive methods [22]. Unfortunately, some of the material destructive tests (DT) may not be feasible due to the possibility of damaging the tested cable post-tensioned concrete structures [23,24]. Aging of materials, particularly of concrete, as well as the influence of concrete aging mechanisms such as ASR (alkali–silica reaction), DEF (delayed ettringite formation) or ACR (alkali–carbonate reaction) are separate issues.

Hence, the need has arisen to propose an appropriate and reliable methodology for testing materials via non-destructive testing (NDT) [25–31]. Specialized NDT tests can only be performed by a limited number of main research centres (laboratories and accredited institutes), and the scope of the need is enormous, because the period 1960–1980 saw growth in this type of design.

Monitoring the long-term behaviour of civil structures is of great importance for damage prevention and design improvement [32]. Technical condition assessments carried out on the basis of continuous monitoring of a structure make it possible to control the ultimate and serviceability limit states and thus the risk of possible overloading. Structural monitoring may be defined as all techniques and methods aimed at measuring, among other things, the deformation of structural elements as a function of time. Geometric monitoring techniques usually involve measurement of deformation and displacement using various methods, including lasers [33].

The application of an appropriate approach to the monitoring of structures, as proposed, e.g., in [34,35] and in particular, the monitoring of cable post-tensioned concrete girders [33,36] as well as NDT material testing and repair of such elements [37] as a long-term solution, may lead to benefits that cannot be achieved otherwise, saving time and avoiding costs. Attention may also be drawn to the increasing spread of new, innovative materials for the protection and strengthening of concrete structures, such as nanomaterials [38,39] or composite materials [40–42]. In addition, an appropriate approach to repair may help avoid negative environmental impacts [43,44], service interruptions, overloading of the nearby infrastructure, etc. An additional feature guaranteeing the usefulness of structure monitoring is non-invasive, constantly updating information about its condition and the possibility to react in advance, which increases its safety and durability and thus extends its service life.

This paper presents selected problems of the evaluation of the technical condition of cable post-tensioned concrete girders. In particular, the path of their quick monitoring is presented. The results of deflection measurements and tests of compression strength and uniformity of concrete were evaluated. The subject of consideration was selected roof girders of two industrial halls located in southern Poland.

2. Materials and Methods

2.1. Description of Seleceted Cable Post-Tensioned Concrete Girders

The structures proposed for the case study are divided into two groups: typified and non-typified (individually designed) cable post-tensioned concrete girders (type: KBO—monolithic one-piece girders and KBOS and KBS—girders assembled on the building site from prefabricated elements) with spans ranging from 15 to 27 m. Their shape is an arched upper chord, "broken" into ~3 m long sections, and a straight lower chord (tie beam), which are connected by vertical posts (Figures 1 and 2). The top chord of these girders has a parabolic shape, while the bottom chord has a rectilinear shape. The prestressing cables 12 Ø5, the number of which depends on the type of girder, are routed in the inner channels of the bottom chord.

The subject of the study are the girders built-in in two halls located in southern Poland: Hall A—17 girders with a span of 27 m (girder type KBS/27—Figures 1 and 2c), Hall B—23 girders with a span of 24 m (girder type KBOS/24—Figures 1 and 2a). The post-tensioned concrete girders used in Halls A and B are located in an environment with a weak (low) degree of aggressiveness and class "B" susceptibility to corrosion hazards; therefore, they are subject to obligatory periodical technical inspections, which should be carried out at intervals of no longer than 18 months. In [45], it is stated that if any irregularities should occur in the external appearance of these structures, a change in their condition of use or in the type and amount of loads, technical inspections should be carried out once or at intervals of no longer than five years. Under the regulations applicable in Poland, building structures with a surface area of more than 1000 m^2 or a roof area of more than 2000 m^2 require inspection twice a year.

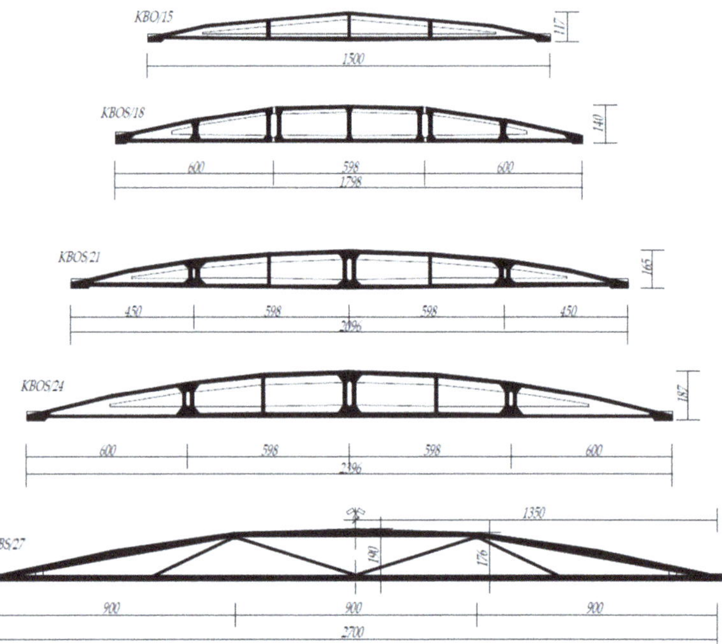

Figure 1. Form and dimensions of example cable post-tensioned concrete girders.

Prefabricated ribbed roof slabs are usually supported on the top chords of the girders for hall buildings. The original roof covering, laid directly on the roof plates, consisted of at least three layers of roofing felt (it had been repaired many times), two layers of insulating fiberboards with a total thickness of approx. 2.5 cm, and a levelling layer of cement mortar with a thickness of approx. 1.5–2.0 cm. Subsequent thermo-modernization was carried out on the existing roof covering, which also increased the permanent load on the roof. This modernization consisted in laying hard mineral wool up to 20 cm thick and covering it with two layers of weldable roofing paper. Installations (lighting, ventilation, heating) were suspended from the lower girder strips, and roof windows (skylights) with a steel structure and single reinforced glass filling (Figure 2c) were rebuilt over the central strip of Hall A.

2.2. Diagnostic Methods Description

Diagnostic tests allow evaluation of the condition of reinforced concrete prestressed girder structures, and the degree of deterioration of individual elements, and determination of the characteristics of concrete and reinforcement, and the causes of damage, and prediction of the durability of the structure [46]. In the diagnostic testing of building structures or their elements, the actual condition and the actual conditions of their operation should be taken into account, and current standards should be used in the assessment. Former standards that were the basis for the design of these structures should be treated as technical knowledge [45]. As far as the types of diagnostics are concerned, we can mention: periodic diagnostics (required by the operation of the object); ad hoc diagnostics (after finding significant irregularities in the work of elements); and full diagnostics (in connection with the planned modernization of the object). General principles of diagnostics of existing prestressed girder roof structures list the following activities [11]: visual inspection of girders and damage inventory, examination of the degree of filling of cable channels, examination of the distribution and quality of reinforcement, geodetic examination of girder deflections, macroscopic examination of concrete, imparting of concrete strength and uniformity, corrosion examination of concrete, and evaluation of cable corrosion hazard. Testing should include [11]: visual observation and assessment, "in situ" testing of structures or their components, and laboratory testing.

Figure 2. Factory buildings with a large number of cable post-tensioned concrete girders: (**a**) production and assembly hall; (**b**,**c**) production hall; (**d**) storage hall for powdered materials.

In this paper, the authors will present selected measures, including: visual inspection of girders (visual method), geodetic investigation of girder deflections, and selected concrete tests. These are elements of the developed simple and effective monitoring of the state of the girders, which the authors have been using for over ten years. It is a method of cyclic, precise updating of the measured deformations combined with macroscopic examination of all elements. In particular, the examination of the quality and class of concrete connections of prefabricated elements and the extent of corrosion according to the pathway are presented in Figure 3. Material and static deformation analysis is carried out at the same time of the year, under similar recorded climatic conditions taking into account the level of protection of structures susceptible to thermal movement (e.g., is there insulation, and, if so, what kind?).

The use of a proven and reliable fast-track procedure (including material analysis) is necessary when a large number of girders are involved in a single structure or assembly (there can be up to 150 girders in a single object). The tests performed should be supported by an appropriate computational algorithm. From a financial perspective, this may not be cost-effective unless it is done automatically or when changes are significant. This is especially the case if you use an already known algorithm of calculation.

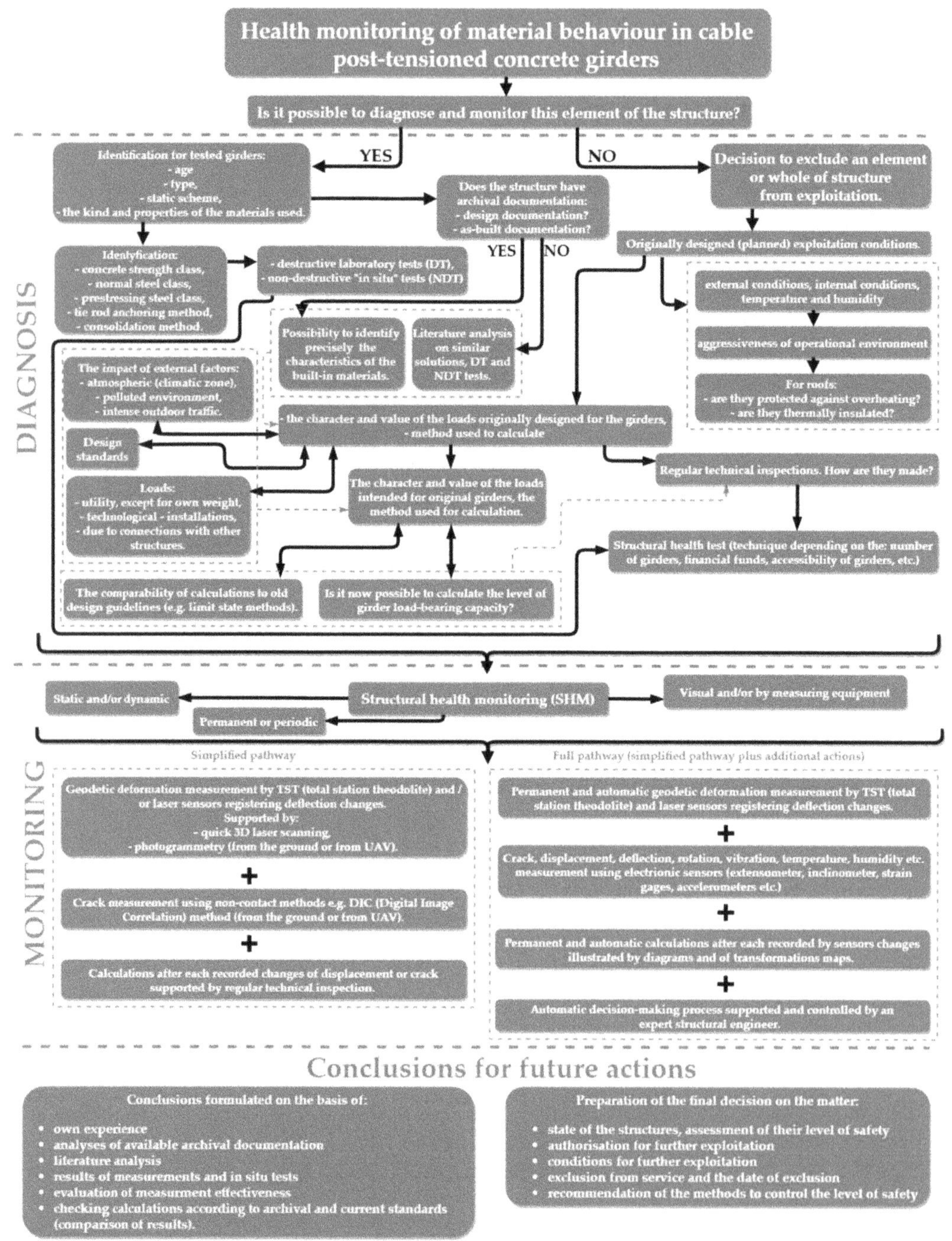

Figure 3. Flowchart for quick and reliable material health monitoring and structure behaviour.

The ultimate limit state is the design and check for the safety of a structure and its users by limiting the stress that materials experience. It would be difficult to determine these values in a simple (measured) way without performing verifiable strength calculations, analyzing the actual static schemes of the structure as well as its verticality, and the flatness and deformation of the girders (taking into account imperfections) and their current properties and parameters, in this case, mainly concrete. The assessment of the serviceability limit state of buildings in intensive use and their structures consists, among others, in comparing the actual values of deformation of their geometry with the permissible values in relation to previous testing seasons. For this purpose, generally available measuring instruments or devices are used.

Technical inspections should be carried out immediately any changes in the appearance of even a single girder are noticed. Additional load on the roof (addition of approx. 0.4 kN/m^2 on the whole surface of the roof) as part of thermomodernization work of hall B and in the case of hall A replacement of the carpentry of the roof windows (subtraction of approx. 0.5 kN/m on the whole surface of the skylights) already represent a requirement to check the capacity of the girder.

The proposed pathway for quick and reliable health monitoring of material behavior in cable post-tensioned concrete girders (Figure 3) is very helpful. These types of structures are resilient, yet sensitive to environmental conditions and aging processes, especially if they have already exceeded their service life. The authors of this paper think that further abandonment of cable post-tensioned concrete girders in their current location will be possible only if the procedures described in the above-mentioned algorithm are followed, which will make it possible to ensure their appropriate level of reliability. An essential element of this process will be to successively carry out reliable material tests, the results of which will decide the fate of these structures and that of the entire facility.

2.3. Visual Method

A visual inspection of the girders includes a preliminary assessment of the structural condition. During a visual inspection carried out as part of the periodic technical inspection, particular attention should be paid to:

- cracking and deformation of prestressed structures in monolithic sections and assembly joints, as well as the state of supports,
- defects in the concrete in the prestressed structures themselves and in the elements joined to them as well as the corrosion of the connecting elements,
- scratches and deformation of roof slabs, especially on their undersides, near the ends of girders, as well as roof leaks, with particular attention paid to the location of cable anchorage zones,
- the accumulation of dirt (dust) on the elements and the extent and weight of suspended installations (including: the change of their location and weight, replacement, or extension), information boards and other loads,
- cracks and defects in the glazing of skylights or replacement of their constituent elements with heavier ones,
- use of the building in accordance with its original purpose,
- thermal and humidity parameters of the building envelope and thermal and humidity conditions of the internal environment,
- possible influence of dynamic loads on the whole structure of the building, e.g., connected with the production technology of the plant.

In the case of regular inspection by visual methods, e.g., by UAV, it is possible to use the very accurate and fast DIC (Digital Image Correlation) optical method for the analysis of the state of preservation of the material (discussed in more detail, e.g., in [47–50]).

2.4. Deflection Measurements

Systematic measurement of girder deflections, using precision levelling, makes it possible to assess the correctness of girder behavior and can signal possible failure risks. One should always strive to make deflection measurements under comparable conditions. Maximum allowable deflection and displacement values should be taken according to the recommendations of investors, supported by standard requirements, e.g., [51]. In general, for reinforced and prestressed concrete structures, these should not exceed 1/250 of their theoretical span, assuming that the permanent loads on flat roofs vary from 30 to 50% (thus, the variable loads are in the range of 50 to 70%). Under the assumptions given above, it should be noted that in extreme cases, the maximum allowable deflection should not exceed 75 mm for a 27 m span girder (67 mm for a 24 m span member).

Based on the permissible vertical deflection values, it should be considered with what accuracy they can be reliably determined in reality. It seems that an accuracy of 10% of the measured deflections is sufficient from the point of view of the user and the small cost of the measuring equipment. Therefore, for a 10 mm deflection, the accuracy of the measuring equipment should not be more than ± 1.0 mm.

It is important to bear in mind that the strain measuring system should be able to withstand external environmental factors such as strong gusts of wind, rolling or dropping loads, etc. during the measurement. In addition, periodic measurements of deformation should be carried out in similar outdoor as well as indoor climatic conditions. The influence of temperature on the operation of the structure must not be neglected and must be taken into account in the final measurement results.

Commonly known displacement measurement methods have their limitations, of which one should be aware before the type of monitoring is chosen. The method of measuring vertical displacement relative to the floor or stop crane platform should be immune to measurement interference, and resistant to measurement interference, which can be difficult in buildings with heavy traffic or production. Measurements would then have to be taken during periods of production stoppage. This method of horizontal measurement under the roof does not affect the use of the building. Detailed possibilities for such a measurement methodology are presented in [52].

Measurements of the amount of deflection can indicate those that show excessive disturbing deformation. In the given questionable cable post-tensioned concrete girders, the state of preservation of the material and its parameters should be thoroughly checked (using the NDT method). The complete measurements and tests are used for the overall assessment of the elements and to assign them to a specific class:

- I—good condition,
- II—acceptable condition,
- III—poor condition,
- IV—very bad condition.

Classification of the technical condition of the girder/girders to classes III and IV necessitates conducting extensive technical and material tests and carrying out urgent repair recommendations. For the technical condition of an object/element classified to classes I and II, on-going maintenance work and subsequent inspections in the current cycle (according to the established schedule) are required, as well as repair and removal of defects.

2.5. Concrete Testing Methods

The compressive strength of concrete is a basic parameter which proves its quality and durability. Achieving the desired quality of concrete involves appropriate design of concrete mixes [53,54], proper manufacturing [55], development of innovative research methods that aid concrete design aimed at obtaining appropriate properties and durability [56], and development of methods for analyzing assessment results both during production and in existing constructions [57,58].

The evaluation of concrete strength in existing building structures can be carried out in accordance with standard PN-EN-13791 [59], which organizes the rules for evaluating the compressive strength of concrete structures and precast concrete products. The method of determining the characteristic value of the strength of concrete in structures based on core boring can be determined in accordance with PN-EN 13791 [59], which is based on the recommendations of PN-EN 206 [60], thus determining the properties of concrete based on compatibility criteria.

The standard recommendations for strength assessment in structures according to PN-EN-13791 [59] concern three cases:

- Demonstration of compliance of the compressive strength of concrete in a structure, that is, concrete already embedded in the structural element (e.g., precast concrete), but without the use of "standard" samples. The indirect methods are used here, that is, those that do not "destroy" the structure, and at the same time are cheaper than traditional sampling.
- Evaluation of the quality of concrete in the case of non-compliance of the compression strength conditions, which was carried out with the use of standard samples or when execution errors are found during the execution of work.
- Evaluation of the technical condition of existing structures when they are to be modernized or redesigned.

The PN-EN 13791 standard [59] provides formulas for the analysis of test results, while the performance of the tests themselves is described in the four-tool standards of the PN-EN 12504 series [61–64]. They concern core drilling and sclerometric pull-out and ultrasonic testing. The compressive strength of concrete in structures is evaluated by non-destructive and/or destructive methods.

2.5.1. Non-Destructive Methods—Sclerometric and Ultrasonic Tests

The development of non-destructive testing (NDT) of materials in civil engineering is mainly concerned with the detection of flaws and defects in concrete elements and structures [22]. In indirect methods, values other than strength are measured. To obtain reliable information about the compressive strength of concrete, it is necessary to calibrate the adopted test method proposed in PN-EN 13791 [59]. The standard provides for the use of indirect methods (NDT), in which the following quantities are measured:

- rebound number (sclerometric method)—PN-EN 12504-2 [62],
- pull-out force ("pull-out" method)—PN-EN 12504-3 [63],
- ultrasonic wave speed (ultrasonic method)—PN-EN 12504-4 [64].

In practice, many different methods using different measuring instruments are used. Characteristics of NDT methods of concrete are presented in Table 1.

In the case study presented in this paper, sclerometric and ultrasonic tests were used. These methods are used often and successfully to evaluate the quality of concrete in existing structures [65,66]. The accuracy of estimating the strength of concrete with sclerometric and ultrasonic methods is about 15–20%, while with the drilling method accuracy it is about 10%. When using all these methods in one element, the accuracy increases and amounts to 5–10% [67].

2.5.2. Destructive Methods—Concrete Cores Test

In the case of the direct method (destructive method/DT), tests are carried out on samples cut out of the structure through core drilling. A description of the testing is given in the tooling standard PN-EN 12504 series [61] concerning core drilling. These boreholes should have a diameter from 50 to 150 mm, while those with a diameter from 100 to 150 mm will allow the use of the direct correlation relation obtained in the strength test in relation to the tests carried out on standard specimens, while diameters less than 100 mm require the use of correction factors. Correlation relationships relating to specimen dimensions, quantity, appearance, and the influence of these factors on the interpretation of the results obtained are given in PN-EN 13791 [59].

Specimens should be drilled in the structure, from those places where section weakening is least likely to affect the load-bearing capacity of the structure, then tested in the laboratory. Nondestructive testing should cover from 2% (for good homogeneity) to 100% (for poor homogeneity) of the girders in a given structure [45]. Examples of the application of the "direct" method can be found in the literature [68,69].

Table 1. Classification of non-destructive (NDT) concrete testing methods.

Group of Methods	Type of Method	Test Instruments	Tested Parameter	Specified Concrete Properties	Comments	Advantages	Disadvantages
Sclerometric methods	Impact methods	Poldi hammer, HPS hammer	imprint depth and diameter	hardness, compressive strength	historical methods, rarely used today	fixed interaction energy	necessity of measuring the imprint diameter increases the testing time
	Rebound methods	Sclerometers (Schmidt hammer L, N, M, PT type)	number of rebounds of the impact mass acting on the component under test with constant energy	hardness, compressive strength	method often used today	fixed interaction energy, measurement speed	influence of concrete age on reflection number
Acoustic methods	Ultrasonic method	ultrasonic pulse velocity testers, material samplers	speed of the ultrasonic wave that spreads in the test material	concrete compressive strength, concrete homogeneity, defect detection (defectoscopy)	two ultrasonic heads: transmitting and receiving	totally non-destructive method	influence of reinforcement on ultrasonic wave velocity measurement, discrepancies in results
	Ultrasonic tomography method	ultrasonic tomographs	measurement of the propagation of elastic waves (ultrasonic) induced by a multi-head antenna	detection of various types of material imperfections, with one-sided access	multiple transmitter/ receiver heads	totally non-destructive method, high accuracy	high work intensity of the tests, discrepancies in results
	AE Acoustic Emission	AE instruments	energy and amplitude of the acoustic wave generated in a component as a result of its loading or deformation	defectoscopy	the method is mainly used for steel elements, less frequently for concrete elements	totally non-destructive method, high accuracy	high work intensity of the tests, discrepancies in results
	Echo method	Specialised measuring equipment	transition time of the pulse reflected from the opposite surface	defectoscopy	one transmitter/ receiver head	totally non-destructive method, high accuracy	high work intensity of the tests, discrepancies in results
	Hammer method	Specialised measuring equipment	velocity of the ultrasonic wave excited in the component under test	defectoscopy	acoustic impulse is generated by hitting the surface of the test piece using a hammer	totally non-destructive method, high accuracy	high work intensity of the tests, discrepancies in results
Quasi-destructive metod	Methods of pulling out an anchor or bolt anchored in concrete	Equipment: - pull off, - pull out, - lock test	pull-out force	compressive strength	there is local damage of approx. 5 cm depth from the surveyed surface	possibility of fixing anchors in concrete as well as in hardened concrete	high work intensity of fixing the anchors in hardened concrete
	Borehole method	Drill rig with core bit, testing machine	destructive force of specimens recovered from structures by means of boreholes	compressive strength	need to collect concrete samples	high accuracy	high work intensity of the sampling

3. Results

3.1. Visual Method Results

All analyzed girders were visually inspected for structural and material damage using a visual method (using an elevator or more easily and quickly using a 4K camera with a resolution of 4000 × 3000 pixels, mounted on a DJI Inspire 2 unmanned aerial vehicle (UAV) with a gimbal (Figure 4).

Figure 4. Visual inspection using an unmanned aerial vehicle (UAV).

Examples of cable post-tensioned concrete girder damage in Hall A identified as a result of the inspection shown in Figure 4.

During the operation of Hall A, damage occurred to the lower flange of the girder when the oversized structure was moved. The workers either did not notice this or ignored it. This coincided with the significant addition of loads during the replacement of a window skylight. The deflections of this girder were observed in the measurement results. At present, it is not known at what exact time (within one year) the damage shown in Figure 5 occurred. As can be observed, the damage is significant, the crack size is on the centimeter level (Figure 5a,d,e), the girder is twisted (Figure 5b,f), and there is significant damage of the concrete (Figure 5c). No cracks were observed on the other elements that did not meet the conditions of the standard [51], i.e., at exposure class X0 greater than 0.2 mm. Considering the non-aggressive exposure class, these values were not considered dangerous.

Figure 5. Examples of cable post-tensioned concrete girder damage in the Hall A inventoried from: (**a**) a basket lift; (**b**–**f**) using UAV.

Was this a direct result of the exceptional load applied or did it propagate over time? Such situations may occur during the continuous operation of a production plant. During the following periodic inspection of the hall, the danger presented by a damaged bottom chord became apparent. Cracks, delamination, and deformation of the lower prestressed chord had completely changed its static scheme. Instead of compressing the concrete section, the tension of the cable caused it to be compressed by a large eccentric force that extended beyond the core of the chord section. It was only by chance that the damage to one girder, whose lower flange had lost its load-bearing capacity, did not lead to a widespread building disaster. After this event, two adjacent girders were carefully observed, for which deformation diagrams are shown in Figures 6–8.

3.2. Deflection Measurements

In the analyzed Halls A and B, measurements of girder deflections have been carried out since 2009. The measurements were performed with the trigonometric levelling method, using an electronic total station (Leica TCRP 1203 R100). The height differences between three points on each cable post-tensioned concrete girder were measured. The points were marked with markers on the lower flanges of the concrete girders, two at the ends and one in the middle of the span. Measurements were taken with comparable accuracy ±1 mm. The reference level for measurements was the altitude level of the hall floor, the fixed benchmark of the national PL-KRON86-NH network with the value of + 105.50 m above sea level for Hall A, and +114.10 m above sea level for Hall B. Girder maximum deflection increments Δf in the middle of the span, monitored from 2009 to 2020 in Hall A was 25 mm, and in Hall B 32 mm. The results are summarized in Figures 6–8 for Hall A and Figures 9–11 for Hall B.

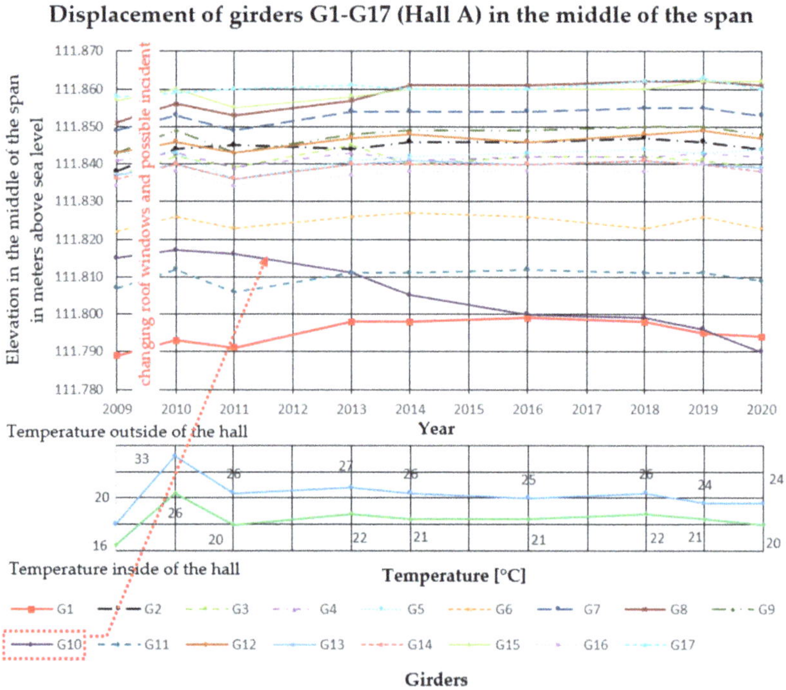

Figure 6. Displacement of girders in the middle of the span in Hall A. Visible significant change in the deflections of G10 girder detail shown in Figure 5.

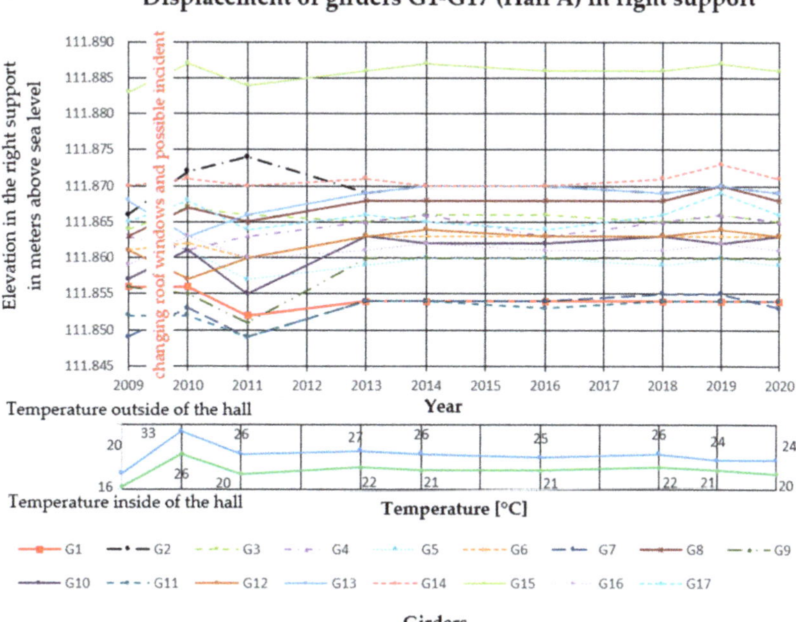

Figure 7. Displacement of the right side of the girders in Hall A.

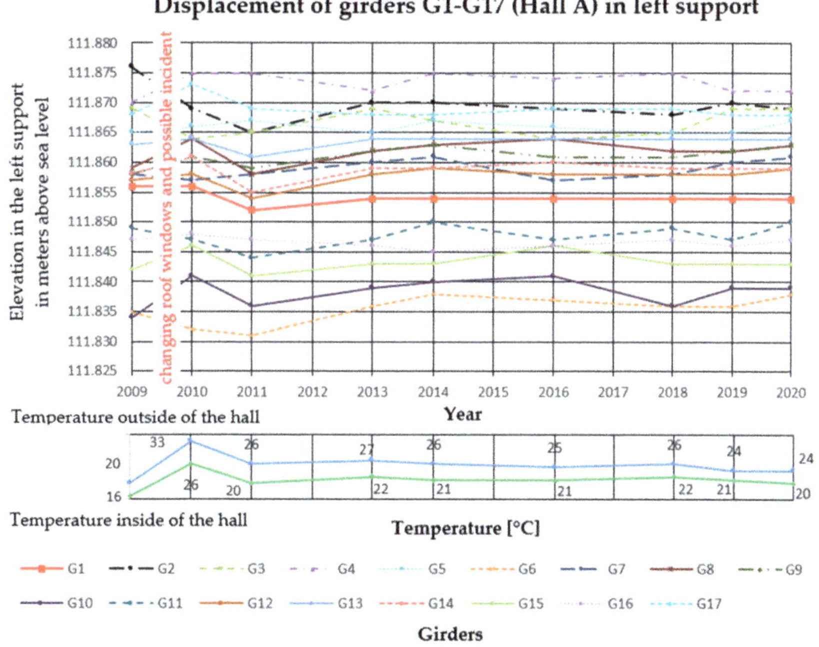

Figure 8. Displacement of the left site of the girders in Hall A.

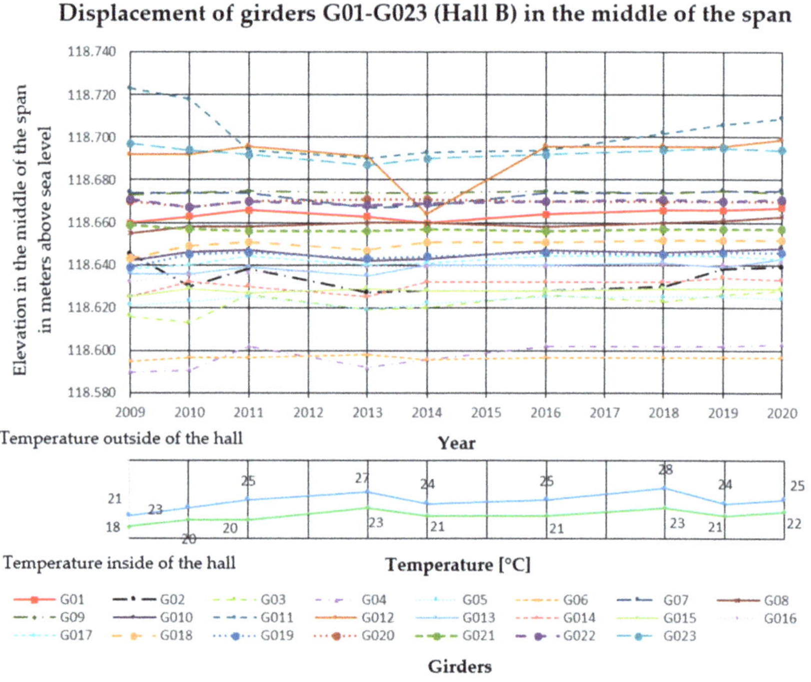

Figure 9. Displacement of the girders in the middle of the span in Hall B.

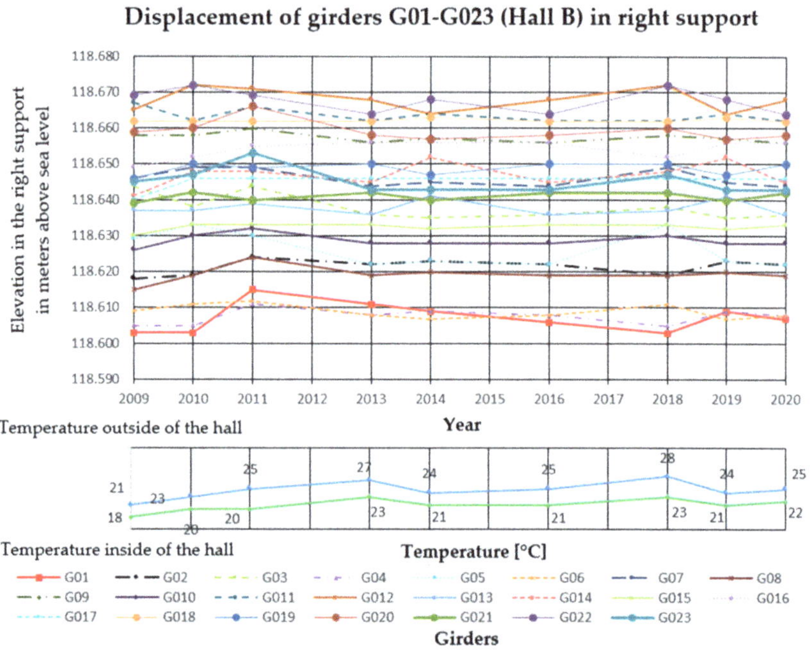

Figure 10. Displacement of the right side of the girders in Hall B.

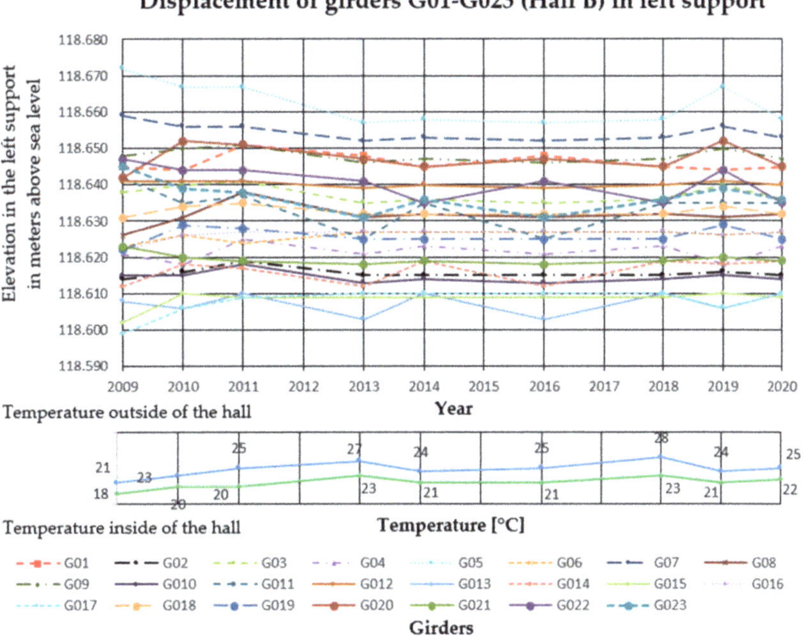

Figure 11. Displacement of the left side of the girders in Hall B.

3.3. Sclerometric Test Results

Current procedures for testing reinforced and prestressed concrete structures using the sclerometric method are included in PN-EN 12504 [62]. According to the introductory note in the standard [62], the reflection number determined by the sclerometric method can be used to assess the homogeneity of concrete in a structure and to determine areas and sections of the structure where the concrete is of poor quality or where its quality has deteriorated. However, it is made clear that the test method is not considered an alternative to the determination of the compressive strength of concrete [70], but with the use of proper correlation, it can allow the estimation of the strength of a structure [59]. Thus, the base curve provided in EN 13791 [59] cannot be used in the assessment of concrete strength without appropriate scaling on boreholes taken from the structure.

For this reason, the recommendations of B-06250 [71], which allows non-destructive methods to evaluate the strength and quality of concrete, were used in the evaluation of both compressive strength and quality.

Testing of four roof girders was carried out with a Schmidt N-type sclerometer. In each structural element, measurements were made on lateral (vertical) surfaces at 15 measuring points. The scaling curve was determined by selecting a hypothetical scaling curve on the basis of reflection number tests and compressive strength (determined by the failure method). As a criterion for the selection of the hypothetical scaling curve, the condition was adopted that the value of the mean relative square deviation was $v_k < 12\%$. The hypothetical scaling curve was assumed in the form of Equation (1):

$$f_c = 0.0418 \times L^2 - 0.932 \times L + 7.5 \text{ (MPa)} \tag{1}$$

where:

f_c—compressive strength of concrete,
L—number of rebound.

For this equation, the value $v_k = 23.6\% > 12\%$ was too high. Therefore, correction factor c was introduced, calculated as the ratio of the average strength of the samples taken in the press ($f_{c\ mean}$) to the average strength of the specimens determined from the hypothetical scaling curve ($f_{c,h\ mean}$). This ratio was $c = f_{c\ mean} : f_{c,h\ mean} = 1.15$. The corrected scaling curve was therefore of the form of Equation (2).

$$f_c = 1.15 \times (0.0418 \times L^2 - 0.932 \times L + 7.5)\ (\text{MPa}) \tag{2}$$

For this curve, $v_k = 10.5\% < 12\%$. This curve could be used to determine the compressive strength of the concrete based on the reflection number L. The log of sclerometric measurements for one prestressed element (G11 from Hall A) is shown in Table 2.

Table 2. Sclerometric test results (girder G11, Hall A).

No.	Testing Angle α	Mean Rebound Value L_{mv}	Equivalent Mean Rebound Value $L_{mv(\alpha-0)}$	Strength from the Curve f_{ci}	Corrected Strength $f_{ci}' = 0.6 \cdot f_{ci}$
1	0	46.6	46.6	63.1	37.9
2	0	45.4	45.4	59.1	35.6
3	0	45.2	45.2	58.4	35.1
4	0	45.8	45.8	60.4	36.3
5	0	45.4	45.4	59.1	35.5
6	0	46.8	46.8	63.8	38.3
7	0	44.8	44.8	57.1	34.3
8	0	45.4	45.4	59.1	35.5
9	0	45.8	45.8	60.4	36.3
10	+90°	44.4	45.8	60.4	36.3
11	+90°	44.6	48.0	67.9	40.8
12	+90°	40.6	43.8	53.9	32.4
13	0	45.2	45.2	58.4	35.1
14	0	45.2	45.2	58.4	35.1
15	0	45.4	45.4	59.1	35.5

Average compressive strength	$f_{c\ mean} = 36.0$ MPa
Standard deviation	$s = 1.92$ MPa
Relative standard deviation	$v_k = 5.3\%$

Tested component: cable post-tensioned concrete roof girder. Age of concrete: more than 1000 days, correction factor 0.6. Measuring device: N-type Schmidt sclerometer. Test angle α: 0° (horizontal test), +90° (test from bottom site).

The characteristic compressive strength can be determined with formulas defined in [71]:

1st criterion: $f_{i,min} \geq f_{ck\ cube}$,
2nd criterion: $f_{cm} \geq 1{,}2\,f_{ck\ cube}$,
3rd criterion: $f_{cm} \geq f_{ck\ cube} + 1.64 \cdot s$

Thus, according to [71], the minimum value from the values should be taken as the concrete class identified with the characteristic compressive strength of concrete:

$f_{ck} \leq 32.4$ MPa
$f_{ck} \leq 36.0$ MPa$/1.2 = 30.0$ MPa
$f_{ck} \leq 36.0$ MPa$—1.64 \cdot 1.92$ MPa $= 32.8$ MPa.

Based on sclerometric tests, the characteristic compressive strength value is 30.0 MPa. Using the recommendations from [71,72], the homogeneity coefficient of concrete can be calculated using Equation (3):

$$k = 1 - 1.645 \times v_k \tag{3}$$

Based on the obtained value of homogeneity, the coefficient is $k = 0.91$ and relative standard deviation $v_k = 5.1\%$. According to [72], the concrete quality for the selected element, girder G11 in Hall A, is very good.

The homogeneity of hardened concrete from sclerometer tests for the four girders was determined and the results are given in Table 3.

Table 3. The homogeneity of hardened concrete from sclerometer test (girders G10 & G11—Hall A, girders G05 & G018—Hall B).

Girder	Homogeneity of Hardened Concrete According to [72]
G10—Hall A	$k = 0.90 > 0.8$; Very good
G11—Hall A	$k = 0.91 > 0.8$; Very good
G05—Hall B	$k = 0.89 > 0.8$; Very good
G018—Hall B	$k = 0.92 > 0.8$; Very good

Both the estimated characteristic compressive strength of 30.0 MPa for girder G11 in Hall A and the assessed concrete quality based on homogeneity factor $k = 0.91$ was very good and the quality of other considered beams (Table 3) seem to be overestimated values; therefore, other methods of determining the compressive strength of concrete, such as ultrasonic or borehole methods, should be used.

3.4. Ultrasonic Tests Results

Ultrasonic tests were used as an alternative non-destructive method to estimate the characteristic compressive strength and to evaluate the homogeneity of concrete, which is allowed by PN-B 06250 [71].

Ultrasonic tests were carried out according to the principles given in PN-EN12504-4 [64]. A CT1 m with heads at a frequency of 100 kHz was used in the ultrasonic tests. The velocity was measured using the so-called crossed and semi-direct methods (Figure 12); the number of measurement sites per element was 15.

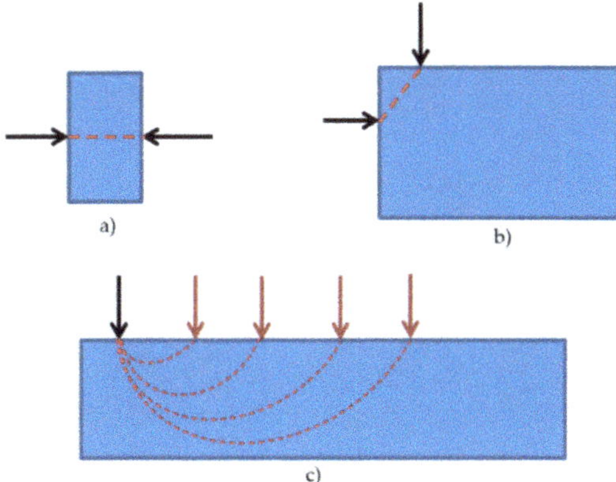

Figure 12. Wave velocity measurement—ultrasonic testing schema: (**a**) crosswise method; (**b**) semi-direct method; (**c**) surface method.

The layout of the measurement bases is shown in Figure 13.

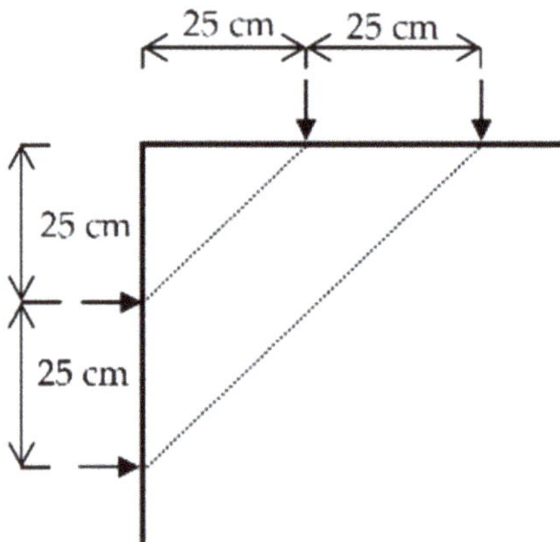

Figure 13. Scheme of transducer positioning in ultrasonic testing.

The scaling curve was determined by selecting a hypothetical scaling curve analogous to the ultrasonic method. The final corrected scaling curve was in the form of Equation (4).

$$f_c = 4.125 \times v_L{}^2 - 12.18 \times v_L + 7.2 \ (MPa) \tag{4}$$

where:

v_L—ultrasound wave speed (km/s).

A detailed example log of the ultrasonic tests carried out for one of the girders (G11 from Hall A) is summarized in Table 4.

Table 4. Ultrasonic test results (girder G11, Hall A).

No.	Length of the Measurement Path s (mm)	Pulse Transition Time t (µs)	Ultrasound Wave Speed v_L (km/s)	Concrete Compressive Strength f_c (Mpa)
1	210	48.1	4.4	32.6
2	210	47.2	4.4	34.7
3	205	46.9	4.4	32.8
4	200	46.6	4.3	30.9
5	205	46.7	4.4	33.2
6	205	46.1	4.4	34.6
7	210	46.1	4.6	37.3
8	210	46.3	4.5	36.8
9	205	46.4	4.4	33.9
10	210	46.3	4.5	36.8
11	205	46.2	4.4	34.4
12	210	46.5	4.5	36.3
13	200	46.9	4.3	30.3
14	205	46.7	4.4	33.2
15	210	46.9	4.5	35.4
Average compressive strength				$f_{c\ mean} = 42.0$ Mpa
Standard deviation				$s = 2.11$ MPa
Relative standard deviation				$v_k = 6.2\%$

Based on the results of ultrasonic tests, the characteristic compressive strength of concrete can be estimated using the recommendations of standard [71]. According to [71], the minimum value from the values should be taken as the characteristic compressive strength of concrete:

$f_{ck} \leq 30.2$ MPa
$f_{ck} \leq 34.2$ MPa$/1.2 = 28.5$ MPa
$f_{ck} \leq 34.2$ MPa$—1.64 \cdot 2.11$ MPa $= 30.7$ MPa.

Based on ultrasonic tests, the characteristic compressive strength is 28.5 MPa.

Using the recommendations of [73], the homogeneity coefficient can be calculated using Equation (5):

$$k = 1 - 1.645 \times v_k \tag{5}$$

Based on the obtained value of homogeneity coefficient $k = 0.90$ and relative standard deviation $v_k = 6.2\%$, according to [73] the concrete quality for the selected element: girder G11 in Hall A, is very good.

The homogeneity of hardened concrete from ultrasonic tests for the four girders was determined from the recommendations in [73,74]. The results are given in Table 5.

Table 5. The homogeneity of hardened concrete from ultrasonic test (girders G10 & G11—Hall A, girders G05 & G018—Hall B).

Girder	Homogeneity of Hardened Concrete According	
	to [73]	to [74]
G10—Hall A	$k = 0.89 > 0.8$; Very good	$v = 4.0$ m/s $\in$ (3.5–4.5) m/s; Good
G11—Hall A	$k = 0.90 > 0.8$; Very good	$v = 4.4$ m/s $\in$ (3.5–4.5) m/s; Good
G05—Hall B	$k = 0.87 > 0.8$; Very good	$v = 3.8$ m/s $\in$ (3.5–4.5) m/s; Good
G018—Hall B	$k = 0.92 > 0.8$; Very good	$v = 4.4$ m/s $\in$ (3.5–4.5) m/s; Good

Both the estimated characteristic compressive strength of 28.5 MPa for girder G11 in Hall A and the evaluated concrete quality based on uniformity coefficient $k = 0.90$ were very good and the quality of other considered beams (Table 5) seem to be values divergent from the values obtained in the sclerometer test; therefore, the other method recommended in [59] using correlation relationships of destructive and nondestructive methods and the test based method recommended in [75] should be used.

3.5. Concrete Core Tests Results

Strength tests of concrete cores, cut directly from existing building structures, are considered to be the most reliable. This situation changed significantly since the publication of the standard [59], which clarified the rules for assessing the compressive strength of concrete in structures and precast concrete products. The standard PN-EN 12504-1 [61] organized the procedures related to the collection and preparation of samples cut from core borings for testing. In our case, a Hilti rig with Ø100 mm and Ø150 mm core drills was used.

United Kingdom recommendations [76] suggest that one should aim to take cores with the smallest possible ratio of h/d = 1.0–1.2 due to the low cost of drilling, structural repairs associated with testing, and variability of concrete properties along the elevation and the influence of "scale effect". In the case in question, h/d = 1.0 and the value of the scale factor was assumed to be 1.0 [76].

An important issue in destructive testing is the determination of the minimum number of specimens to be tested. PN-EN 13791 [59] treats this issue very generally and emphasizes that from the statistical point of view and with respect to safety requirements, it is recommended to use as many boreholes as practicable to evaluate the compressive strength of concrete in the structure. In the case under study, two samples each were taken (in the support zone of the beams) for four selected beams, for a total of eight core samples.

The concrete samples taken by means of cores were used to determine the actual compressive strength of concrete in a destructive manner. The samples were prepared and then tested in a destructive method in a strength press. The compressive strengths of the concrete are presented in Table 6.

Table 6. Determining the strength of structural concrete (girders G10 & G11—Hall A, girders G05 & G018—Hall B).

Girder/Sample no.	Destructive Force F (kN)	Concrete Compressive Strength f_c (MPa)
G10—Hall A/001	254.0	33.1
G10—Hall A/002	248.0	32.7
G11—Hall A/003	267.0	34.2
G11—Hall A/004	260.0	33.1
G05—Hall B/005	279.0	36.3
G05—Hall B/006	264.0	33.6
G018—Hall B/007	283.0	36.0
G018—Hall B/008	286.6	36.4
Average compressive strength of concrete		f_{cm} = 34.4 MPa
Standard deviation		s = 1.6 MPa
Relative standard deviation		v_k = 4.5%

The method recommended in [59] was used to determine the compressive strength based on a correlation relationship determined using a limited number of core borehole test results (number of samples $n < 18$) and sclerometric test results.

To determine the correlation relationship between the compressive strength of concrete in the structure and the indirect test result, the correlation relationship of the strength obtained for cores drilled from the structure and the test results performed for nondestructive testing must be determined, thus obtaining pairs of test results. Determination of the correlation relationship involves fitting a straight line or curve via regression analysis of pairs of results obtained from the test program. The indirect method measurement result is considered a variable value and the determined compressive strength of concrete in the structure is a function of the variable. The standard [59] emphasizes that the correlation relationship is determined assuming the possibility of a ten per cent underestimation of strength. The standard notes further state that the correlation relationship used to determine the strength of concrete provides the required standard level of safety, where 90% of the strength values are expected to be higher than the value determined from the relationship. The base curves provided in the standard are defined as the lower envelopes of the relationship between the non-destructive test results (in this case reflection number L) and the concrete strength $f_{c,cyl}$. It should be noted that the standard base curves were determined for $f_{c,cube}$ values determined on 150 mm cube specimens tested after 28 days of curing under laboratory conditions.

Since the study obtained average L_{mv} reflection numbers in a range specified as 43.8–48.0, a standard base curve [59] from Equation (6) was adopted for scaling:

$$f_L = 1.37 \times L - 34.5 \tag{6}$$

The evaluation of concrete strength consists in shifting the basic base curve (Equation (5)) to an appropriate level, determined by core drilling and non-destructive testing. The value of the shift of the basic Δf of the base curve depends on the mean value of the differences $\delta f_{m(n)}$ and the factor k_1 associated with the number of measurements and is calculated from Equation (7).

$$\Delta f = \delta f_{m(n)} - k_1 \times s \tag{7}$$

where:

$\delta f_{m(n)}$—the mean value of the difference of the compressive strength determined on the core bore samples and the strength determined from the base curve (Equation (6)),
s—standard deviation of strength differences $\delta f_{m(n)}$,
k_1—coefficient taken from the standard table, for $n \geq 9$.

From Equation (6), the compressive strength of concrete (corresponding to the strength of concrete on 150 mm cubes) can be calculated anywhere in the structure, then the characteristic compressive strength of concrete dependent on the number of measurements can be determined. See Equation (8):

$$f_{is,R} = f_R + \Delta f \tag{8}$$

The classification of concrete into a given strength class [59] is based on calculation of the strength from the scaled base curve described by Equation (8) and divided by a factor $\beta_{cc}(t)$ that takes into account the age of the concrete after 28 days of curing [51] (Equations (9) and (10)):

$$f_{cm}(t) = \beta_{cc}(t) \times f_{cm} \tag{9}$$

and:

$$\beta_{cc} = exp\left[s_c \times \left(1 - \sqrt{\frac{28}{t}}\right)\right] \tag{10}$$

where:

t—age of structure (days),
s_c—cement class.

In the standard, the method of determining the shift parameter Δf depending on the standard deviation s of the differences $\delta f_{m(n)}$ and the parameter (statistic) k_1. remains a matter of debate. In this procedure, the values of the parameter k_1. are assumed to correspond to the coefficients used in consistency criteria based on the operational characteristic function method, analogous to [60], which should not be used for correlation analyses. In place of the proposed baseline curve shift parameter, it would be more reasonable to use generally accepted principles of mathematical statistics and a shift parameter that depends on the statistics of the assumed distribution. Current recommendations assume a large number of samples $n \geq 9$ as the basis for scaling the f_R-R. relationship. For approximate scaling, the number of pairs of results $n \geq 3$ may already be sufficient, similar to national (Polish) regulations applied to the sclerometric and ultrasound methods [72,73]. In the analyzed case, the number of results from core samples is $n = 8$, so on the basis of national experience [71–73] the authors proposed a modification of the standard method using as k_1 as the value of statistics recommended in [75]. In the calculation, the confidence level $\gamma = 0.75$ and $n = 8$, and the statistic value read from [75] is $k_1 = 3.27$. The calculation of the base curves is shown in Table 7.

Table 7. Calculations for the base curve offset parameter.

Girder/Sample no.	Compressive Strength		$\delta f = f_{is} - f_L$ (MPa)
	Core with Correction Value 0.85 f_{is} (MPa)	Basic Curve—Equation (5) f_L (MPa)	
G10—Hall A/001	38.9	22.8	6.1
G10—Hall A/002	38.5	22.1	6.4
G11—Hall A/003	40.2	24.7	5.1
G11—Hall A/004	38.9	22.8	6.1
G05—Hall B/005	42.7	28.3	3.3
G05—Hall B/006	39.5	23.6	5.6
G018—Hall B/007	42.4	27.8	3.6
G018—Hall B/008	42.8	22.1	3.2
Average value			$\delta f_{m(8)} = 4.9$ MPa
Standard deviation is			s = 1.34 MPa < 3.0 MPa
and for calculation according to [44]			
standard deviation can be use			s = 3.0 MPa

According to Equation (7), the value of the parameter Δf of the base curve is $\Delta f = 4.9 - 3.27 \cdot 3.0 = -4.9$ MPa, and the corrected equation of the base curve determined by Equation (5) will finally take the form (11):

$$f_{is,L} = f_L + \Delta f = 1.73\,L - 34.5 - 4.9 = 1.73\,L - 39.4 \tag{11}$$

Equation (11) is valid only in the interval of rebound numbers $44.8 - 2 = 42.8 < L < 46.8 - 2 = 44.8$, because a modified base curve has been defined for this interval.

The purpose of the analyses was to determine the characteristic strength of concrete in the structure using the indirect method based on the modified base curve (Figure 14), so the calculated values of concrete compressive strength for the base curve, modified base curve, and core specimen methods are shown in Table 8.

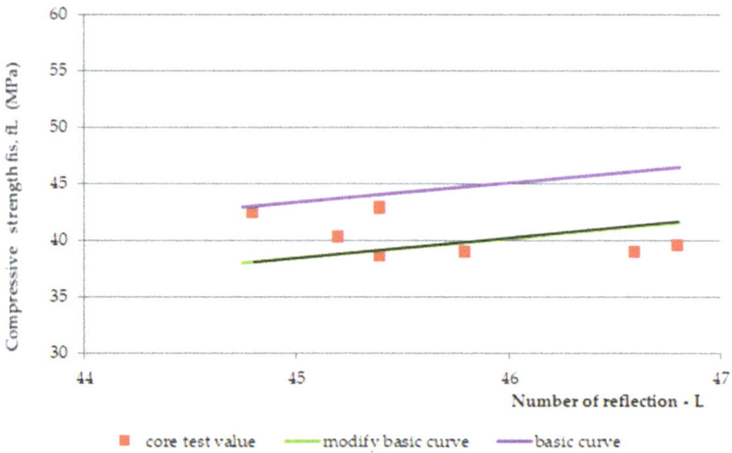

Figure 14. Test results according to the indirect method taking into account the results of destructive and non-destructive tests.

Table 8. Compressive strength obtained from destructive and non-destructive testing.

Girder/Sample No.	Core Vale f_{is} (MPa)	Compressive Strength Determined from the Curve	
		Basic Curve Equation (6)	Modified Equation (11)
G10—Hall A/001	38.9	46.1	41.2
G10—Hall A/002	38.5	44.0	39.1
G11—Hall A/003	40.2	43.7	38.8
G11—Hall A/004	38.9	44.7	39.8
G05—Hall B/005	42.7	44.0	39.1
G05—Hall B/006	39.5	46.5	41.6
G018—Hall B/007	42.4	43.0	38.1
G018—Hall B/008	42.8	44.0	39.1
Average value—$f_{c(8)\,is,\,cube}$ (MPa)	40.5	44.5	39.6
Lowest value—$f_{clowest,cube}$ (MPa)	38.5	43.0	38.1
Standard deviation s (MPa)	1.8	1.2	1.2
3.0 MPa, and for calculation according to [44] standard deviation is $s = 3.0$ MPa			

The characteristic strength of concrete according to [59] was evaluated using a modified base curve. The criterion for the compliance of the strength of the inspected concrete with the required characteristic strength with the number of specimens $n < 15$ is presented as follows:

1st criterion: $f_{ck,is} = f_{cm\ (8),is,cube} - 4\text{MPa} = 39.6 - 4\ \text{MPa} = 35.6\ \text{MPa}$

2nd criterion: $f_{ck,is} = f_{c\ lowest,cube} + 4\text{MPa} = 38.1 + 4\ \text{MPa} = 42.1\ \text{MPa}$

The characteristic strength of the concrete is 35.6 MPa. After assuming that class N cement was used to make the concrete and the age of the concrete was more than 50 years, the value of the coefficient considering the age of the concrete $\beta_{cc}(t) = 1.438$ was determined from Equations (9) and (10); hence, the strength of the concrete was 35.6 MPa/1.438 = 24.8 MPa.

When using the modification of the method, the formal conditions of the core boreholes should be checked. The coefficient of variation of the reflection number at the locations of the drilled wells is $v_L = 2.0\% < 15\%$ (Table 7), and the quotient of the average compressive strength obtained from the curve and the wells is equal to $(39.6/40.5\ \text{L}) - 100\% = 2\% < 15\%$. The wells can be considered authoritative and fully representative. The obtained value of relative standard deviation was $v_k = 2.2\%$, so the developed modified base curve can be considered the basic scaling curve for the case in question and the small sample size is $n = 8$.

4. Discussion

4.1. Deflection Measurement Discussion

In many cases in construction practice, it may not be possible to carry out a precise assessment of the technical condition of buildings and their components using traditional, commonly used methods.

The authors have been observing cable post-tensioned concrete girders for several years, cyclically evaluating the material and geometric changes occurring in them. In the analyzed cases, the girder maximum deflection increment Δf in the middle of the span, monitored from 2009 to 2020 in Hall A was 25 mm (about 23% of the maximum deflection) and in hall B, 32 mm (about 33% of the maximum deflection). These values are high for this type of structure because of the lack of concerning amount of girder deformation since the beginning of the hall exploitation. The first measured deflections occurred in 2009 and these values were taken as a reference for later measurements and analyses. According to the authors, because of the age of the elements and incomplete measurement data, the constant measured deflection of the girders in the middle of the span should not exceed 35% of the permissible value. One of the aims of the research was to monitor the changes of displacement magnitude in the same points with changing sign, in similar conditions of the internal and external climate. The change of the ordinates of the measuring points exceeding the permissible values will indicate the overloading of the girder structure, the settling of supports, or the change of tension force in cables. It will become the reason for the detailed examination of this particular element (not all of them) after its temporary protection.

Of course, such tests should not only concern roof structures, but also all other elements which are responsible for their proper support and ensure the spatial stiffness of the objects and their individual structures, and thus affect their deflections.

The hidden defects of built materials as well as the ageing processes will determine their current technical condition, which will constantly change over time to their disadvantage. Periodic examinations, carried out on an ongoing basis, allow for the continued periodic use of the facilities, but at some point they should indicate the approaching end of their service life. Assessment of the parameters of the materials which make up these structures on its own will not be sufficient here, as it will be difficult or even impossible to reach all their locations. It will also not be possible here to assess the tension of cables due to their inaccessibility or to reliably assess the degree and extent of corrosion of steel cable strands (apart from single random tests) due to their number, as there are dozens or even hundreds of them in such halls. In the absence of developed testing methods for this type of structures, which are estimated to be even more than 60 years old, it is necessary to use those that monitor the maintenance of their geometric shapes not only in comparison with the permissible deformations but also through the analysis of these changes over time.

4.2. Concrete Test Discussion

The obtained values of the characteristic compressive strength were compared with the values of the characteristic strength calculated on the basis of the guidelines contained in Annex D of PN-EN 1990 [75] concerning research-assisted design.

According to Annex D, the procedure according to D.7.2 and D.7.3 can be directly applied to estimate the characteristic values. It is recommended that the estimation of test results should be carried out on the basis of statistical methods using existing (statistical) information about the type of distribution used and the associated parameters. Equation (12) can be used to estimate the characteristic value according to EN 1990 [75]:

$$f_{ck} = f_{cm} \left(1 - k_n \times v_k\right) \tag{12}$$

where:

v_k—coefficient of variation of the sample; $v_k = s/f_{cm}$
k_n—a factor that depends on the sample size and confidence level (at the minimum confidence level $\gamma = 0.75$ and $n = 8 \rightarrow k_n = 5.07$).

The resulting characteristic compressive strength of concrete according to Equation (12) will be $f_{ck} = 35.9$ MPa. The characteristic value of the compressive strength of concrete can also be determined according to EN 1990 [75] in Annex D and the theoretical value of the standard deviation $\sigma = 4.86$ MPa, calculated from the relationship $f_{cm} = f_{ck} + 8$ according to [51].

With an assumed a priori standard deviation of 4.86 MPa, the standard deviation estimator s can be obtained using the Mellin function (Equation (13)):

$$s = \sigma \times \left(\sqrt{\frac{2}{n}} \times \frac{\Gamma\left(\frac{n-1}{2}\right)}{\Gamma\left(\frac{n}{2}\right)} \right) \tag{13}$$

where:

$\Gamma(\ldots)$—gamma function value,
n—sample size,
σ—population standard deviation.

To determine the characteristic compressive strength of concrete according to [75], we use Equation (14):

$$f_{ck} = f_{cm} - k_n \times \sigma \times \left(\sqrt{\frac{2}{n}} \times \frac{\Gamma\left(\frac{n-1}{2}\right)}{\Gamma\left(\frac{n}{2}\right)} \right) \tag{14}$$

Under the assumption regarding the knowledge of standard deviation, the value of coefficient will be $k_n = 3.27$, while the characteristic value of compressive strength according to Equation (14) for core samples will be $f_{ck} = 33.0$ MPa.

After considering the age parameter of concrete, the characteristic compressive strength for the core test is, 35.9/1.438 = 29.4 MPa and 33.0/1.438 = 27.0 MPa respectively.

To determine the differences in the characteristic compressive strength values obtained by different methods, the values from each destructive and non-destructive method are summarized in a graph (Figure 15). In the characteristic compressive strength values obtained by non-destructive methods, sclerometric and ultrasonic, a coefficient related to the age of concrete is also included (Equations (9) and (10)).

The typical value (of the compressive strength characteristic values, understood as $(f_{ck(5)} - s; f_{ck(5)} - s)$, is in the range of 20.3–28.4 MPa, and the relative variation of the compressive strength characteristic value measured by the coefficient of variation is 17% > 15%, which indicates the average variation of the compressive strength characteristic results obtained using different methods.

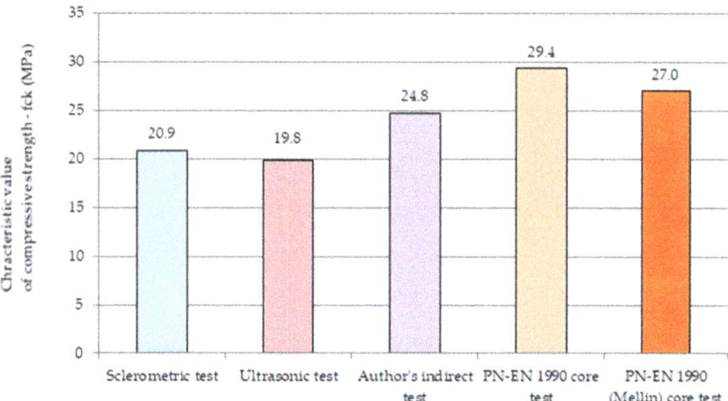

Figure 15. Characteristic values of compressive strength obtained using different methods.

On the basis of Figure 15, it can be stated that, based on the recommendations [71], the characteristic values of compressive strength obtained using non-destructive methods, i.e., sclerometric and ultrasonic, are correct and smaller than those obtained using the destructive method or the author's mean method. It should be stressed that in non-destructive methods these are characteristic values of compressive strength, i.e., concrete class determined according to the standard used during the construction of cable concrete girders under consideration, taking into account the age of concrete. The values obtained indicate the correctness of the standard compliance criteria recommended at that time and the possibility their use in the evaluation of concrete in the structure using non-destructive methods.

Based on the results of the tests (Figure 15) of the characteristic compressive strength of concrete of the analyzed structural members using certain methods, i.e., sclerometric, ultrasonic, and indirect with the use of the author's proposal to modify the base curve using the procedures recommended in [75], the obtained concrete class is C15/20. In the case of applying the destructive method (core drilling), the concrete in the analyzed girders can be classified to class C20/25. Moreover, according to the author's proposed method, the concrete is on the border between classes C15/20 and C20/25. The design class of concrete for the analyzed girders is C25/30, so the requirement for the design class of concrete is not fulfilled at present. Evaluating the concrete according to the standard recommendations [59], the concrete in the analyzed structure is one (destructive method) or two classes (non-destructive methods) lower than the class assumed for the design concrete.

The choice of test method should be determined by the reliability of the strength estimation. In the presented situation for cable concrete girders, it is very difficult to decide where to take the samples, because they are obtained from where access is the easiest, there are no collisions with the existing reinforcement (in this case tendons and their anchorage,) and also soft reinforcement, or from where the impediments to the use of the object are the least, hence the rationale for using primarily non-destructive methods. Such situations are often the result of a lack of prior visual inspection of the structure and assessment of potential obstructions. In the case under analysis, a total of $n = 8$ samples were taken; however, $n = 2$ samples were taken for one element. This number is less than the $n = 3$ recommended in the standard, which would allow the determination of the characteristic compressive strength of concrete and the subsequent qualification of concrete to a given class. According to [75], non-destructive methods are not an alternative to destructive testing, but only a supplement in the case of a limited number of boreholes. However, in the case of prestressed or post-tensioned concrete structures, it is difficult to take core samples, so only non-destructive methods and the application of appropriate compliance criteria should be considered for the proper classification of concrete.

Sampling of $n = 3$ is allowed by the standard and is justified when the tests concern a single element or a small fragment of the structure. When using the direct method, a minimum of 4–6 core holes should be made in single elements, and when the assessment concerns several elements (substrings, beams, and slabs or columns) then a minimum of 9–18 holes should be made and a combination of indirect methods of testing should be used. Analyzing the obtained values of coefficients of variation for compressive strength, it can be stated that they are less than 15%, which indicates a very good homogeneity of compressive strength of hardened concrete. Very good homogeneity in compression is also evidenced by the obtained values of the homogeneity coefficient for non-destructive methods: sclerometric and ultrasonic.

The calculation procedures recommended in the standards proposed in this paper can enable the proper classification of concrete. It should be noted that using the criteria given in the standard [60], which apply to manufactured concrete, one can obtain dangerously inflated concrete strengths in a structure. When a small sample size of $n < 15$ is available, the recommendations of the standard [59] should be used. Using the compatibility criteria given in [59], the cores taken from the structure allow the characteristic strength of the concrete to be determined, which must be taken into account when determining the safety of the structure. In this case, the characteristic strength of the concrete in the structure must be "reduced" to the concrete strength on the 28th day of concrete curing in laboratory conditions by dividing it by a factor of 0.85 and taking into account the age of the concrete.

Implementation of European standards in the field of testing procedures (PN-EN 12504 series [61–64]) and concrete qualifications [59] has significantly clarified the situation in the field of the diagnostics of existing reinforced concrete structures. The gap in national (Polish) regulations concerning the methodology of direct destructive testing has been filled and at the same time new procedures for performing indirect non-destructive tests have been given. Both the standard for manufactured concrete [60] and the standard for testing concrete in a structure [59] as well as other subject standards (for non-destructive testing) clearly emphasize that non-destructive methods cannot be used without appropriate scaling on core samples taken from the structure. This highlights the problem of concrete evaluation in post-tensioned or prestressed concrete structures, where it is very difficult to take a sample of a statistically significant number and to use only non-destructive methods. When determining the number of samples, one should not be guided by economic criteria but by the needs of the tests performed and the reliability of the results obtained. Usually, limiting the number of samples to the absolute minimum ($n = 3$) results in the necessity of additional tests. To exclude doubts arising at the stage of drawing conclusions, in situ tests should be attended by representatives of the interested parties, whose task it is to both confirm the location of the test points and visually evaluate the samples and, if possible, observe the test procedures.

5. Conclusions

A reliable condition assessment of cable post-tensioned concrete girder structures is very difficult to perform in their "natural" operating environment. Nevertheless, it is necessary to maintain such structures in a safe technical condition. A serious problem is the age of these structures, sometimes exceeding 60 years. The danger associated with this may appear suddenly and uncontrollably, leading to the collapse of entire facilities without any warning signals. A big problem is the assessment of the technical condition of covered and inaccessible elements of girders, mainly steel tendons and their anchorages. Therefore, people carrying out such investigations are required not only to have advanced theoretical knowledge, but also practical experience, expertise, and above all common sense. Summarizing the content of the paper, it can be concluded that:

- monitoring of existing building structures containing prestressed structures is essential for economic reasons, safety of use, durability, and determination of the moment until which they can be used;

- the methods proposed in this paper are useful or even necessary to determine and monitor the current technical condition of such objects "in situ";
- the method of testing deformation and displacements of prestressed elements is the basic method, which makes it possible to assess the technical condition of prestressed elements, whereas non-destructive strength testing may be an auxiliary and supplementary method.

Those persons responsible for the exploitation of such objects should take special care to maintain the internal microclimate at a stable level and prevent any adverse phenomena (e.g., leaks in roof coverings) which could accelerate corrosion, mainly of invisible steel structures "covered" in the girder body.

Author Contributions: Conceptualization, L.B., D.B., and Z.M.; methodology, L.B.; software, D.B.; validation, L.B., D.B., Z.M., A.L., and I.S.; formal analysis, L.B. and D.B.; investigation, L.B. and D.B.; resources, L.B., D.B., and Z.M.; data curation, L.B. and Z.M.; writing—original draft preparation, L.B., D.B., and Z.M.; writing—review and editing, L.B, D.B., Z.M., A.L., and I.S.; visualization, L.B.; supervision, D.B. and Z.M.; project administration, L.B.; funding acquisition, D.B. All authors have read and agreed to the published version of the manuscript.

Funding: This research received no external funding.

Institutional Review Board Statement: Not applicable.

Informed Consent Statement: Not applicable.

Data Availability Statement: Data sharing is not applicable.

Conflicts of Interest: The authors declare no conflict of interest.

References

1. Yun, Y.M. Evaluation of Ultimate Strength of Post-Tensioned Anchorage Zones. *J. Adv. Concr. Technol.* **2005**, *3*, 149–159. [CrossRef]
2. Im, S.B.; Hurlebaus, S. Non-destructive testing methods to identify voids in external post-tensioned tendons. *KSCE J. Civ. Eng.* **2012**, *16*, 388–397. [CrossRef]
3. Mao, W.; Gou, H.; He, Y.; Pu, Q. Local Stress Behavior of Post-Tensioned Prestressed Anchorage Zones in Continuous Rigid Frame Arch Railway Bridge. *Appl. Sci.* **2018**, *8*, 1833. [CrossRef]
4. Shin, K.-J.; Lee, S.-C.; Kim, Y.Y.; Kim, J.-M.; Park, S.; Lee, H. Construction Condition and Damage Monitoring of Post-Tensioned PSC Girders Using Embedded Sensors. *Sensors* **2017**, *17*, 1843. [CrossRef] [PubMed]
5. Oh, B.H.; Kim, K.S. Shear Behavior of Full-Scale Post-Tensioned Prestressed Concrete Bridge Girders. *ACI Struct. J.* **2004**, *101*, 176–182. [CrossRef]
6. Gunter, B.; Galo, G.; Mercado, E.; Baibas, D. Post-Tensioned Concrete Girders. In *Bridge Design Practice*, 4th ed.; Bustamente, Y., Ed.; State of California Department of Transportation, Caltrans: Sacramento, CA, USA, 2015; Chapter 7.
7. Calvi, G.M.; Moratti, M.; O'Reilly, G.J.; Scattarreggia, N.; Monteiro, R.; Malomo, D.; Calvi, P.M.; Pinho, R. Once upon a Time in Italy: The Tale of the Morandi Bridge. *Struct. Eng. Int.* **2019**, *29*, 198–217. [CrossRef]
8. Rania, N.; Coppola, I.; Martorana, F.; Migliorini, L. The Collapse of the Morandi Bridge in Genoa on 14 August 2018: A Collective Traumatic Event and Its Emotional Impact Linked to the Place and Loss of a Symbol. *Sustain. J. Rec.* **2019**, *11*, 6822. [CrossRef]
9. Invernizzi, S.; Montagnoli, F.; Carpinteri, A. Fatigue assessment of the collapsed XXth Century cable-stayed Polcevera Bridge in Genoa. *Procedia Struct. Integr.* **2019**, *18*, 237–244. [CrossRef]
10. Nawy, E.G. *Prestressed Concrete*; Pearson Education: Sydney, Australia, 2011.
11. Runkiewicz, L. *Technical Condition Assessment of a Post-Tensioned Reinforced Concrete Girders*; Instructions, Guidelines, Handbooks No. 353/2018; Building Research Institute: Warsaw, Poland, 2018.
12. Breen, J.E. *Anchorage Zone Reinforcement for Post-Tensioned Concrete Girders*; American Association of State Highway and Transportation Officials, United States, Federal Highway Administration, National Research Council (U.S.), Transportation Research Board: Washington, DC, USA, 1994; Volume 356.
13. *Highway Structures and Bridges. Inspection and Assessment*; CS 465 Management of Post-Tensioned Concrete Bridges (Formerly BD 54/15); Highways England: London, UK, 2020.
14. Japan International Cooperation Agency. *Quality Control Manual for Prestressed Concrete Girder*; Ministry of Construction, the Republic of the Union of Myanmar, Japan International Cooperation Agency: Tokyo, Japan, 2019.
15. Salas, R.M.; West, J.S.; Breen, J.E.; Kreger, M.E. *Conclusions, Recommendations and Design Guidelines for Corrosion Protection of Post-Tensioned Bridges*; Report No. FHWA/TX-04/0-1405-9; Texas Department of Transportation: Austin, TX, USA, 2004.
16. Nowogońska, B.; Korentz, J. Value of Technical Wear and Costs of Restoring Performance Characteristics to Residential Buildings. *Buildings* **2020**, *10*, 9. [CrossRef]

17. Bungey, J.H.; Grantham, M.G. *Testing of Concrete in Structures*; CRC Press: Boca Raton, FL, USA, 2006.
18. Malhotra, V.; Carino, N.J. *Handbook on Nondestructive Testing of Concrete*; CRC Press: Boca Raton, FL, USA, 2003.
19. Gorzelańczyk, T.; Schabowicz, K.; Szymków, M. Tests of Fiber Cement Materials Containing Recycled Cellulose Fibers. *Materials* **2020**, *13*, 2758. [CrossRef] [PubMed]
20. Szymanowski, J. Evaluation of the Adhesion between Overlays and Substrates in Concrete Floors: Literature Survey, Recent Non-Destructive and Semi-Destructive Testing Methods, and Research Gaps. *Buildings* **2019**, *9*, 203. [CrossRef]
21. Hooton, R.D. Future directions for design, specification, testing, and construction of durable concrete structures. *Cem. Concr. Res.* **2019**, *124*, 105827. [CrossRef]
22. Schabowicz, K. Non-Destructive Testing of Materials in Civil Engineering. *Materials* **2019**, *12*, 3237. [CrossRef] [PubMed]
23. Hoła, J.; Schabowicz, K. State-of-the-art non-destructive methods for diagnostic testing of building structures – anticipated development trends. *Arch. Civ. Mech. Eng.* **2010**, *10*, 5–18. [CrossRef]
24. Malek, J.; Kaouther, M. Destructive and non-destructive testing of concrete structures. *Jordan J. Civ. Eng.* **2014**, *159*, 1–10.
25. McCann, D.; Forde, M. Review of NDT methods in the assessment of concrete and masonry structures. *NDT E Int.* **2001**, *34*, 71–84. [CrossRef]
26. Sbartaï, Z.-M.; Breysse, D.; Larget, M.; Balayssac, J.-P. Combining NDT techniques for improved evaluation of concrete properties. *Cem. Concr. Compos.* **2012**, *34*, 725–733. [CrossRef]
27. Bassil, A.; Chapeleau, X.; LeDuc, D.; Abraham, O. Concrete Crack Monitoring Using a Novel Strain Transfer Model for Distributed Fiber Optics Sensors. *Sensors* **2020**, *20*, 2220. [CrossRef]
28. Runkiewicz, L. Application of non-destructive testing methods to assess properties of construction materials in building diagnostics. *Archit. Civ. Eng. Environ.* **2009**, *2*, 79–86.
29. Hoła, A.; Sadowski, Ł. A method of the neural identification of the moisture content in brick walls of historic buildings on the basis of non-destructive tests. *Autom. Constr.* **2019**, *106*, 102850. [CrossRef]
30. Gerwick, B.C., Jr. *Construction of Prestressed Concrete Structures*; John Wiley & Sons: Hoboken, NJ, USA, 1997.
31. Naaman, A.E. *Prestressed Concrete Analysis and Design: Fundamentals*; Techno Press 3000: Sarasota, FL, USA, 2012.
32. Guo, T.; Chen, Z.; Lu, S.; Yao, R. Monitoring and analysis of long-term prestress losses in post-tensioned concrete beams. *Measurement* **2018**, *122*, 573–581. [CrossRef]
33. Han, K.; DeGol, J.; Golparvar-Fard, M. Geometry- and Appearance-Based Reasoning of Construction Progress Monitoring. *J. Constr. Eng. Manag.* **2018**, *144*, 04017110. [CrossRef]
34. Bednarz, Ł.J.; Jasieńko, J.; Nowak, T.P. Test monitoring of the Centennial Hall's dome, Wroclaw (Poland). In *Optics for Arts, Architecture, and Archaeology V*; International Society for Optics and Photonics: Bellingham, WA, USA, 2015; Volume 9527, p. 95270C. [CrossRef]
35. Bednarz, Ł.J.; Jasieńko, J.; Rutkowski, M.; Nowak, T.P. Strengthening and long-term monitoring of the structure of an historical church presbytery. *Eng. Struct.* **2014**, *81*, 62–75. [CrossRef]
36. Coronelli, D.; Castel, A.; Vu, N.A.; François, R. Corroded post-tensioned beams with bonded tendons and wire failure. *Eng. Struct.* **2009**, *31*, 1687–1697. [CrossRef]
37. Terzioglu, T.; Karthik, M.M.; Hurlebaus, S.; Hueste, M.B.D.; Maack, S.; Woestmann, J.; Wiggenhauser, H.; Krause, M.; Miller, P.K.; Olson, L.D. Nondestructive evaluation of grout defects in internal tendons of post-tensioned girders. *NDT E Int.* **2018**, *99*, 23–35. [CrossRef]
38. Ghaffary, A.; Moustafa, M.A. Synthesis of Repair Materials and Methods for Reinforced Concrete and Prestressed Bridge Girders. *Materials* **2020**, *13*, 4079. [CrossRef] [PubMed]
39. Ghahari, S.; Assi, L.N.; Alsalman, A.; Alyamaç, K.E. Fracture Properties Evaluation of Cellulose Nanocrystals Cement Paste. *Materials* **2020**, *13*, 2507. [CrossRef] [PubMed]
40. Al-Hamrani, A.; Kucukvar, M.; Alnahhal, W.; Mahdi, E.; Onat, N.C. Green Concrete for a Circular Economy: A Review on Sustainability, Durability, and Structural Properties. *Materials* **2021**, *14*, 351. [CrossRef] [PubMed]
41. Pino, V.; Nanni, A.; Arboleda, D.; Roberts-Wollmann, C.; Cousins, T. Repair of Damaged Prestressed Concrete Girders with FRP and FRCM Composites. *J. Compos. Constr.* **2017**, *21*, 04016111. [CrossRef]
42. Fang, H.; Bai, Y.; Liu, W.; Qi, Y.; Wang, J. Connections and structural applications of fibre reinforced polymer composites for civil infrastructure in aggressive environments. *Compos. Part B Eng.* **2019**, *164*, 129–143. [CrossRef]
43. De Domenico, D.; Urso, S.; Borsellino, C.; Spinella, N.; Recupero, A. Bond behavior and ultimate capacity of notched concrete beams with externally-bonded FRP and PBO-FRCM systems under different environmental conditions. *Constr. Build. Mater.* **2020**, *265*, 121208. [CrossRef]
44. Wałach, D. Analysis of Factors Affecting the Environmental Impact of Concrete Structures. *Sustain. J. Rec.* **2021**, *13*, 204. [CrossRef]
45. Runkiewicz, L.; Sieczkowski, J. Research and assessment of reinforced concrete roof tensile girders in exploited facilities. *Przegląd Bud. Buildging Rev.* **2018**, *89*, 41–46.
46. Hoła, J.; Runkiewicz, L. Methods and diagnostic techniques used to analyse the technical state of reinforced concrete structures. *Struct. Environ.* **2018**, *4*, 309–336. [CrossRef]
47. Dang, J.; Haruta, D.; Shrestha, A.; Endo, H.; Matsunaga, S.; Kasai, A.; Wang, X. Arial patrol for bridge routine and post-earthquake emergency inspection using small aerial photography UAV. In Proceedings of the 5th International Symposium on Advances in Civil and Environmental Engineering Practises for Sustainable Development, Galle, Sri Lanka, 16 March 2017; Volume 2018, p. 21.

48. Mrówczyńska, M.; Grzelak, B.; Sztubecki, J. Unmanned Aerial Vehicles as a Supporting Tool of Classic Land Surveying in Hard-to-Reach Areas. In *International Scientific Siberian Transport Forum*; Springer: Cham, Switzerland, 2019; pp. 717–729.

49. Lacidogna, G.; Piana, G.; Accornero, F.; Carpinteri, A. Multi-technique damage monitoring of concrete beams: Acoustic Emission, Digital Image Correlation, Dynamic Identification. *Constr. Build. Mater.* **2020**, *242*, 118114. [CrossRef]

50. Chen, G.; Liang, Q.; Zhong, W.; Gao, X.; Cui, F. Homography-based measurement of bridge vibration using UAV and DIC method. *Measurement* **2021**, *170*, 108683. [CrossRef]

51. *PN-EN 1992-1-1: 2008 Eurocode 2: Design of Concrete Structures*; PKN: Warszawa, Poland, 2008.

52. Bednarz, Ł. Flat roof structures monitoring—Requirements. *Inżynier Budownictwa* **2020**, *2*, 63–66.

53. Gorzelańczyk, T.; Pachnicz, M.; Różański, A.; Schabowicz, K. Identification of microstructural anisotropy of cellulose cement boards by means of nanoindentation. *Constr. Build. Mater.* **2020**, *257*, 119515. [CrossRef]

54. Skrzypczak, I.; Kokoszka, W.; Buda-Ożóg, L.; Kogut, J.; Słowik, M. Environmental Aspects and Renewable Energy Sources in the Production of Construction Aggregate. In Proceedings of the International Conference on Advances in Energy Systems and Environmental Engineering (ASEE17), Wrocław, Poland, 2–5 July 2017; EDP Sciences: Ulis, France, 2017; p. 00160.

55. Al Ajmani, H.; Suleiman, F.; Abuzayed, I.; Tamimi, A. Evaluation of Concrete Strength Made with Recycled Aggregate. *Buildings* **2019**, *9*, 56. [CrossRef]

56. Wang, D.; Liu, G.; Li, K.; Wang, T.; Shrestha, A.; Martek, I.; Tao, X. Layout Optimization Model for the Production Planning of Precast Concrete Building Components. *Sustain. J. Rec.* **2018**, *10*, 1807. [CrossRef]

57. Domagała, L. Size Effect in Compressive Strength Tests of Cored Specimens of Lightweight Aggregate Concrete. *Materials* **2020**, *13*, 1187. [CrossRef] [PubMed]

58. Skrzypczak, I.; Kokoszka, W.; Zięba, J.; Leśniak, A.; Bajno, D.; Bednarz, L. A Proposal of a Method for Ready-Mixed Concrete Quality Assessment Based on Statistical-Fuzzy Approach. *Materials* **2020**, *13*, 5674. [CrossRef] [PubMed]

59. PKN. *PN-EN 13791: 2019-12 Assessment of In-Situ Compressive Strength in Structures and Precast Concrete Components*; PKN: Warszawa, Poland, 2019.

60. PKN. *PN-EN 206: 2013+A1: 2016 Concrete—Specification, Performance, Production and Conformity*; PKN: Warszawa, Poland, 2017.

61. PKN. *PN-EN 12504-1: 2019-08 Testing Concrete in Structures—Part 1: Cored Specimens—Taking, Examining and Testing in Compression*; PKN: Warszawa, Poland, 2019.

62. PKN. *PN-EN 12504-2: 2012 Testing Concrete in Structures—Part 2: Non-Destructive Testing—Determination of Rebound Number*; PKN: Warszawa, Poland, 2012.

63. PKN. *PN-EN 12504-3: 2006 Testing Concrete in Structures—Part 3: Determination of Pull-Out Force*; PKN: Warszawa, Poland, 2006.

64. PKN. *PN-EN 12504-4: 2005 Testing Concrete in Structures—Part 4: Determination of Ultrasonic Pulse Velocity*; PKN: Warszawa, Poland, 2005.

65. Bergamo, O.; Campione, G. Experimental Investigation for Degradation Analysis of an RC Italian Viaduct and Retrofitting Design. *Pr. Period. Struct. Des. Constr.* **2020**, *25*, 05020010. [CrossRef]

66. Stawiski, B.; Kania, T. Tests of Concrete Strength across the Thickness of Industrial Floor Using the Ultrasonic Method with Exponential Spot Heads. *Materials* **2020**, *13*, 2118. [CrossRef] [PubMed]

67. Schickert, G. Critical reflections on non-destructive testing of concrete. *Mater. Struct.* **1984**, *17*, 217–223. [CrossRef]

68. Panahandeh, M.; Hashemolhosseini, H.; Eftekhar, M.R.; Hashemolhosseini, A.H.; Baghbanan, A. Obtaining the strength parameters of concrete using drilling data. *J. Build. Eng.* **2021**, *38*, 102181. [CrossRef]

69. Sathish, S.; Ramkumar, S.; Geetha, M. Drilling performances and wear characteristics of coated drill bits during drilling reinforced concrete. *Int. J. Appl. Ceram. Technol.* **2019**, *16*, 357–366. [CrossRef]

70. PKN. *PN-EN 12390-3: 2019-07 Testing Hardened Concrete—Part 3: Compressive Strength of Test Specimens*; PKN: Warszawa, Poland, 2019.

71. PKNMiJ. *PN-B-06250: 1988 Ordinary Concrete*; PKNMiJ: Warszawa, Poland, 1988.

72. ITB. *Instructions for Using Schmidt Hammers for Non-Destructive Quality Control of Concrete*; No. 510; ITB: Warszawa, Poland, 1977.

73. ITB. *Instructions for Using Ultrasonic Method for Non-Destructive Quality Control of Concrete in Structure*; No. 209; ITB: Warszawa, Poland, 1977.

74. Jaggerwal, H.; Bajpai, Y. Estimating the Quality of Concrete Bridge Girder Using Ultrasonic Pulse Velocity Test. *Int. J. Comput. Eng. Res. (IJCER)* **2014**, *4*, 2250–3005.

75. PKN. *PN-EN 1990. Eurocode: Basis of Structural Design*; PKN: Warszawa, Poland, 2005.

76. Concrete Society. *Concrete Core Testing for Strength*; Technical Report No 11; Concrete Society: London, UK, 1987.

 *materials*

Article

Testing Crack Resistance of Non-Load-Bearing Ceramic Walls with Door Openings

Tomasz Kania [1,*]**, Valery Derkach** [2] **and Rafał Nowak** [3]

[1] Faculty of Civil Engineering, Wrocław University of Science and Technology, Wybrzeże Wyspiańskiego 27, 50-370 Wrocław, Poland
[2] Research Enterprise for Construction "Institute BelNIIS", 15 "B", F. Skoriny str., 220076 Minsk, Belarus; institute@belniis.by
[3] Faculty of Civil and Environmental Engineering, West Pomeranian University of Technology, 70-311 Szczecin, Poland; rnowak@zut.edu.pl
*   Correspondence: Tomasz.Kania@pwr.edu.pl; Tel.: +48-71-352-84-52

Abstract: Cracking in non-load-bearing internal partition walls is a serious problem that frequently occurs in new buildings within the short term after putting them into service or even before completion of construction. Sometimes, it is so considerable that it cannot be accepted by the occupiers. The article presents tests of cracking in ceramic walls with a door opening connected in a rigid and flexible way along vertical edges. The first analyzes were conducted using the finite element method (FEM), and afterward, the measurements of deformations and stresses in walls on deflecting floors were performed on a full scale in the actual building structure. The measurements enabled to determine floor deformations leading to cracking of walls and to establish a dependency between the values of tensile stresses within the area of the door opening corners and their location along the length of walls and type of vertical connection with the structure.

Keywords: partition walls; brick walls; bending strength; cracking

Citation: Kania, T.; Derkach, V.; Nowak, R. Testing Crack Resistance of Non-Load-Bearing Ceramic Walls with Door Openings. *Materials* **2021**, *14*, 1379. https://doi.org/10.3390/ma14061379

Academic Editor: Dolores Eliche Quesada

Received: 12 February 2021
Accepted: 8 March 2021
Published: 12 March 2021

Publisher's Note: MDPI stays neutral with regard to jurisdictional claims in published maps and institutional affiliations.

Copyright: © 2021 by the authors. Licensee MDPI, Basel, Switzerland. This article is an open access article distributed under the terms and conditions of the Creative Commons Attribution (CC BY) license (https://creativecommons.org/licenses/by/4.0/).

1. Introduction

Cracking of partition walls in buildings is a frequent phenomenon. According to the statistical data available in the literature [1–3], the displacement of their supporting elements is responsible for 60–70% of damage to walls in Central and Eastern Europe. It results partially from the lack of possibility to analyze accurately spatial relocations of building structures. Application of oversized elements of load-bearing structures in order to limit the cracking of partition walls is an action unjustified economically. Small cracks of filling elements in the building generally are not the causes for concern; however, the commonness of this type of defects, their scale, and often considerable width of cracks are the reasons for expedited repair work, which is often completed still before the structures are handed over for use. The scale of this phenomenon leads to the question about the possibility and methods of counteracting it, the validity of research conducted within this scope, and consistency of requirements both with respect to acceptable deformations of structures supporting partition walls and methods of completing properly their circumferential connections and elements having an effect on cooperation with the building structure. In the countries of Central and Eastern Europe, in the residential buildings, non-load-bearing walls are being made mainly with the use of masonry elements. The percentage of individual building materials for the construction of this type of wall in Poland is shown in Figure 1 [3].

Despite the growing popularity of the gypsum plasterboard lightweight walls in the countries of Central and Eastern Europe, this technology is currently used primarily to make partitions in office and service buildings. In residential buildings constructed in 2020 in Poland, over 95% of non-load-bearing walls were made in masonry technology.

151

Moreover, 27.5% of walls were made of ceramic elements, which indicates the essence of the problem of their cracking undertaken in the presented research.

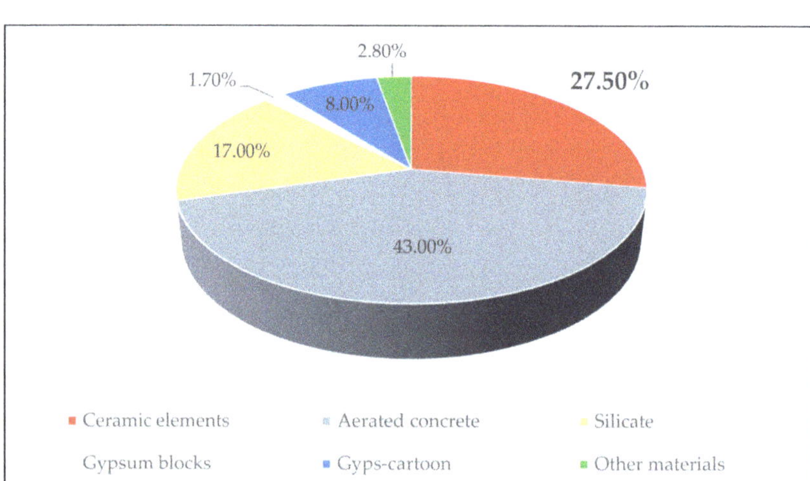

Figure 1. Distribution of the use of wall materials for the construction of non-load-bearing walls in residential buildings in Poland in 2020.

There have been studies concerning the influence of foundation deformations on the cracking of ceramic walls. It can be concluded from the first research analyses concerning the operation of masonry walls on flexible supports that in order to protect them from cracking, the ratio of deflection of the supporting structure to its span should not exceed 1/2000, and bending tensile strength of the wall should not be lower than 0.21 MPa [4]. In his study [5], Beranek recommends that the bending tensile strength of the walls should be a minimum of 0.1–0.3 MPa. The study [6] presents results of tests concerning the strength of walls with and without window and door openings. The models were made at a scale of 1:3 versus actual dimensions of walls, using appropriately reduced ceramic wall elements. The models were laid on double-span reinforced concrete beams, and they were tested in two stages. In the first stage, a uniformly distributed load was applied using a spreader beam to the top edge of the wall. Then, relocation of the reinforced concrete beam that was a support of the wall in the middle of its span was forced. In the model without openings, the first crack appeared when the ratio of the beam deflection to its span was 1/1000. In the models with openings, cracks were observed slightly later when the ratio of deflection to span was 1/947. Openings in the wall had a significant effect on the layout of cracks. The study [7] presents results of tests and analyses of walls made of full ceramic bricks. It was concluded that admissible deflection of the supporting structure depends on whether it is made with openings or without them. In the case of walls without openings, it was recommended that deflection of the supporting structure should not exceed 1/500 of the span, and in walls with openings, 1/1000 of the span. In the study [8], tests of the wall made of ceramic brick 3.11 m long, 0.98 m high, and 0.1 m wide were published. The wall was built on an I-section steel beam and the test consisted of loading the wall from the top using the force simulating the load from the ceiling and forcing the relocation of the steel supporting beam. The first crack appeared at the deflection of the order of 2.5 mm that is approximately 1/1200 of the wall span. The study [9] quotes tests of full walls and walls with openings made of masonry bricks on a full scale. The walls were constructed on a steel beam deflecting with the increase in loads applied through the reinforced concrete tie beam to the top edge of the wall. In order to prevent cracking of the brick wall supported on the floor or beam element, it was recommended that the requirement of not exceeding its limit deformation characterized by the shape deformation angle of the wall Θ should be fulfilled. The studies [10–12] indicate results of tests on ceramic partition walls on a full scale. It was concluded that they were characterized by lower floor bending strength than the previously described scaled partitions made of ceramic elements. All described results are presented in Table 1.

Table 1. Critical deflections and stresses in the state of cracking of brick partition walls.

Ordinal Number	Author and Date	Wall Type and Description	Bending Tensile Strength	Ratio of Deflection of the Supporting Structure to Its Span at Cracking
1	Meyerhof G., 1953 [4]	External brick walls without openings; different dimensions; analytical approach	0.21	1/2000
2	Beranek 1983 [5]	External brick walls with openings; different dimensions; analytical, and experimental approach	0.1–0.3	1/2000
3	Rolanda et al., 2003 [6]	Brick wall without opening in a scale of 1:3, experimental approach	-	1/1000
4	Pfeffermann et al. 1981 [7];	Brick wall with an opening in a scale of 1:3, experimental approach	-	1/946
5	Loots et al.l 2004 [8]	Brick wall without opening in a scale of 1:2.5, experimental approach	-	1/1200
6	Piekarczyk 2019 [9]	Brick walls with (a) and without opening (b) asymmetrically loaded, experimental laboratory approach	-	1/1700 (a)1/2800 (b)

A commonly used method of limiting cracking of infilling masonry walls is their expansion from the upper ceiling, for example, by filling the gap with polyurethane foam. The thickness of the expansion joint depends on the calculated value of the ceiling deflection. The most common vertical connection is steel anchoring, which is strengthening the partition in the direction perpendicular to the wall surface [13]. In the studies carried out until now, few tests of destroying non-load-bearing walls in real conditions, i.e., building structures undergoing shape deformations on a full scale, were shown. Therefore, the authors of the article undertook this task.

The appearance of cracks in partition walls with deflection of supporting floors is considered in standards as exceeding the limit condition of their usability [13–16]. Regulations concerning the limits of deflections in structures constituting support under masonry walls are often available in standards concerning the design of reinforced concrete structures, and more rarely, in standards concerning masonry structures. Selected requirements for permissible floor deformations are presented in Table 2.

The American standard ACI 318-14 [17] and its earlier editions, in the situation in which non-structural elements are designed on the floor (e.g., masonry partition walls), allows deflection values not exceeding 1/480 of the effective span of the floor. On the other hand, the American standard ACI 530-08/ASCE 5-08/TMS 402-08 [18] requires that deflections of beams and lintels from constant and variable loads should be limited to 1/600 of the span, or 7.6 mm, where the lower value is decisive. In the British standard BS 5628-2:2005 [19], it is assumed that deflection of the wall supporting structure should not exceed 1/500 of the span, or 20 mm. The German standard DIN 1045-1 [20] intended for designing reinforced concrete structures limits deflections of the floor with partition walls based on it to 1/500 of the span and imposes minimum floor thicknesses. In the Belgium standard NBN B 03-003 [21], the following recommendations concerning limit deflection of the wall supporting structure are assumed: unreinforced walls with openings—l/1000, unreinforced walls without openings or reinforced walls with openings—l/500, reinforced walls—l/350, movable walls—l/250. In the EU standard EC 6 [25], it was stipulated that usability of masonry structural elements cannot be deteriorated by the behavior of other structural elements, such as deflections of floors or walls. On the other hand, the standard does not define any rules for checking these deflections or does not indicate any limit deflections of the structures on which the walls would be constructed.

Table 2. Comparison of permissible deflection values of the supporting structure under partition walls.

Ordinal Number	Standard No	Country/Region	Maximal Ratio of Deflection of the Supporting Structure to Its Span or Maximal Deflection	Additional Information
			[-] or [mm]	
1	ACI 318-08 [17]	USA	1/480	-
2	ACI-530-08/ASCE 5-08/TMS 402-08 [18]	USA	1/600 or 7.6 mm	the lower value is decisive
3	BS 5628-2 [19]	UK	1/500 or 20 mm	the lower value is decisive
4	DIN 1045-1 [20]	Germany	1/500	-
5	NBN B 03-003 [21]	Belgium	1/1000 or 1/500 *	*—valid for unreinforced walls with openings
6	EN 13747: 2005 [22]	European Union	1/350 or1/500 *	*—valid for brick walls
7	EN 1992-1-1: 2008 [23]	European Union	1/250 or 1/500 *	*—value valid for the deflections affecting the non-load-bearing walls
8	PN-B-03264: 2002 [24]	Poland		

The presented literature review shows differences in the results of crack resistance of masonry walls due to deflection of supports. The requirements included in the standards [17–24] regarding the impact of building structure on non-load-bearing walls provide different permissible values for ceiling slab deflections. The reference documents do not fully describe the impact of the location of door openings in relation to the length of the walls, the conditions of the slab support, and the conditions of contact between the wall and the surrounding structures on their cracking resistance [10–16,26,27]. Therefore, research on the behavior of partition walls on flexible supports is still necessary. The impact of these factors on the stress state and cracking of self-supporting partition walls with door openings, made of ceramic bricks, supported on a concrete slab is the subject of this article.

2. Materials and Methods

In order to carry out the undertaken research study, three types of tests and analyses were performed. At first, tests of the wall samples in laboratory conditions were conducted in order to obtain material data needed for conducting numerical analyses using the finite element method (FEM). After completion of the numerical analyses, tests of full-scale walls in a building structure were carried out. The order and scope of the conducted research study are pointed below.

1. Laboratory tests: Their purpose was to characterize the used materials and obtain the strength parameters of the walls in accordance with the protocol provided in the standards [28–32]. The scope of the research included, among others, the following mechanical parameters of masonry materials and masonry samples:

- Test code LAB 1—mechanical parameters of the ceramic bricks and mortar;
- LAB 2—compressive strength of the tested wall samples in the horizontal and vertical directions;
- LAB 3—tensile strength of the tested wall samples in the horizontal and vertical directions;
- LAB 4—shear strength of the tested wall and its internal friction coefficient.

2. Numerical analyses performed using the Finite Element Method (FEM) in the ANSYS program environment: Mechanical parameters obtained on the basis of the above-mentioned laboratory tests were used for modeling the walls. The full mechanical characteristics of the examined walls used in modeling were the reason for the application of the homogeneous isotropic material model. Numerical analyses were carried out for walls 3.07 m high, 5.75 m long, and 12 cm thick with a door opening 2.1 m high and 0.99 m wide. The following two wall models were analyzed due to the types of their connection with the building structure along their vertical edges:

- FEM TEST 1—free connection in the horizontal and vertical direction of the wall surface;
- FEM TEST 2—rigid (model of toothing connection with perpendicular load-bearing walls).

3. Tests of the full-scale partitions conducted on walls constructed on reinforced concrete floors in a real building: Partitions have been prepared with the use of the same materials and protocols as in laboratory measurements and FEM analysis. Two partitions with door openings in the middle of their span and the same dimensions and types of vertical connections as in FEM analysis has been tested, which are as follows:

- BUILD 1—vertical connection of the wall with flexible steel anchors restraining relocation only in the direction transversal to its surface;
- BUILD 2—rigid toothing connection.

2.1. Materials

The walls for all of the tests presented in this research: laboratory (BelNIIS laboratory, Brest, Belarus), numerical, and full scale were built using hollow (18%) ceramic brick, class M12.5 (FCP, Brest, Belarus). This is the most common ceramic brick element used for the interior partitions within the apartment in the Belarus (with a wall thickness of 12 cm). Dimensions and layout of holes for the ceramic brick are shown in Figure 2.

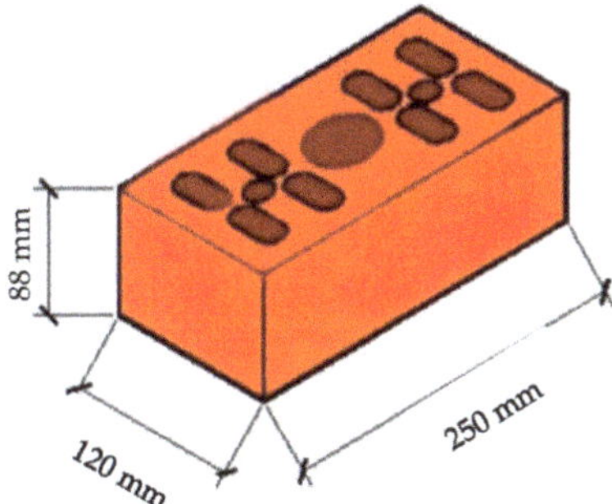

Figure 2. Dimensions and layout of holes for ceramic brick class M12.5 (FCP, Brest, Belarus) used in the tests.

Tested walls have been carried out on standard concrete mortar with compressive strength of 10.9 MPa, with mortar joints 10–15 mm thick. For the preparation of masonry mortars, a factory-made dry mixture was used (FCP, Brest, Belarus).

First, the masonry elements and mortar were tested (LAB 1). The compressive strength of the mortar at the time of testing was established in accordance with the requirements of EN 1015-11 [25].

Wall samples have been prepared in the laboratory and tested on the stand of own production, with the use of a 3000 kN hydraulic press (Pneumat P3000, Minsk, Belarus). Dial gauges (Baker V2, Pune, India, 0.01 mm) were used to measure the displacements of the tested samples. The production of the masonry samples, their curing, testing, and processing of test results were carried out in accordance with the standard EN 1052-1 [26–28]. The specimens were tested with an axial compressive load acting perpendicularly and parallel to the direction of horizontal mortar joints (LAB 2). In Figure 3, the diagram of the test stand of the compressive strength in axial load is directed perpendicularly to the horizontal mortar joints of the wall.

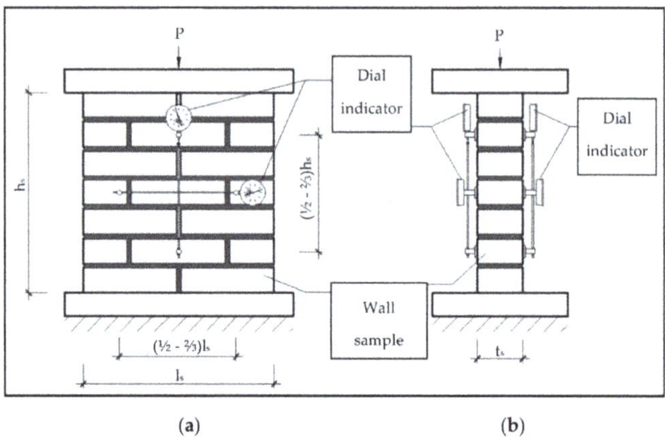

Figure 3. Compressive strength test in the direction perpendicular to the horizontal mortar joints of the wall samples: (**a**) front view and (**b**) side view.

Wall samples were also tested for tensile strength in the direction along and across to the horizontal mortar joints (LAB 3). In the case when the destruction occurs due to the rupture of the masonry element (Figure 4), the tensile strength of the wall sample is determined by Equation (1) as follows:

$$f_{t1} = 0.5 \cdot f_{bt} \cdot 1/(1 + t_m/h_u),\tag{1}$$

where f_{bt} is the tensile strength of the masonry element, t_m is the thickness of mortar joints, and h_u is the height of the masonry element.

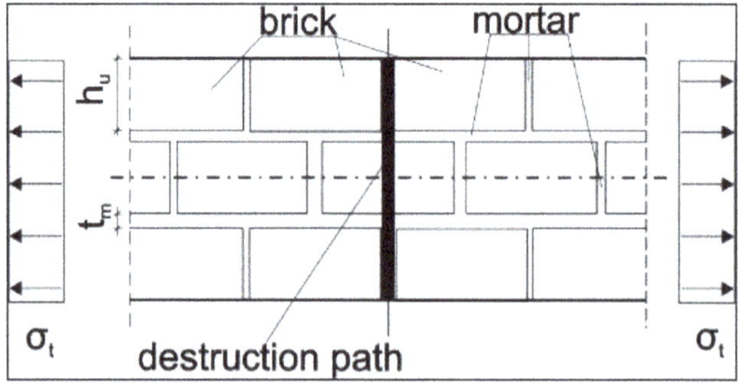

Figure 4. Destruction of the masonry wall sample exposed to axial tension with cracking along the joints and masonry elements.

If the destruction passes along the mortar joints only, the tensile strength of the wall sample is set according to Equation (2) as follows:

$$f_{t2} = f_{v0} \cdot u_j / (h_u + t_m),$$ (2)

where f_{v0} is the initial shear strength of the masonry in the plane of the horizontal mortar joints (tangential bond), which is determined in the laboratory conditions and u_j is the distance between the crack in joints and in masonry elements, according to Figure 5.

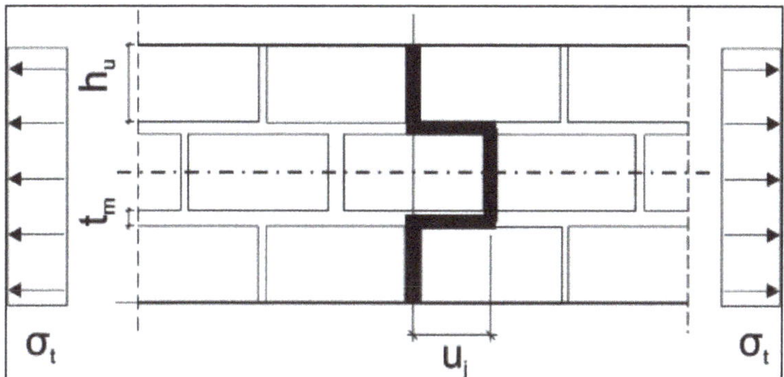

Figure 5. Destruction of the masonry wall sample exposed to axial tension with cracking along only the joints.

The minimal value obtained from the tests calculations is taken as a characteristic strength of the masonry under axial tension.

Initial shear strength of the masonry (tangential adhesion f_{v0}) and the internal friction coefficient (*tg α*) have been established on the basis of the thawing test of masonry samples exposed to the simultaneous action of compressing and shear stresses, according to EN 1052-3 [29] (LAB 4), which is illustrated on Figure 6.

Figure 6. Testing of the initial shear strength and the internal friction coefficient of the masonry wall.

Since the magnitude of the compressive stresses during the tests was variable, it was possible to test the dependency between shear forces $f_{v,i}$ and compressive forces $f_{c,i}$. The strength of clean-cut (tangential adhesion) f_{v0} was established by extrapolating the graph to the ordinate $f_{c,i} = 0$.

2.2. Numerical Calculations

Numerical calculations were performed with the assumption of fully completed expansion joints from the floor on top edges of the walls, for loads from self-weight of partition walls. Calculations were conducted using the finite element method (FEM) in the computational environment of ANSYS program (AES, Canonsburg, PA, USA). The decision of using ANSYS environment for numerical analysis results from its extensive computing environment with the structural analysis module, wide library of finite elements, and experience of the authors. For the purposes of the analyses presented in this study, numerical tests of models were carried out until the serviceability limit state was exceeded. Due to the size of the analyzed problem and the comparative analyses for structures in natural conditions carried out in the next phase of the research, global FEM analyses were performed. The ceramic wall was modeled as a homogenous, isotropic material. It was possible to use macromodeling because of the wide range of wall sample tests that allowed for the implementation of mechanical properties. Macromodel results had been confirmed by the results of experimental tests. Choosing the linear flexible model was accepted for the description of material properties. The accepted criterion for crack formation is that the main tensile stresses exceed the tensile strength of the masonry in the corresponding direction.

The finite elements (FE) were implemented from the available ANSYS program library. The choice of FE was caused by different properties of the modeled elements of the structure (frame and filling wall). Pillars and beams of the frame were modeled using dual-node BEAM3 elements, with three degrees of freedom in each node (in the directions of the horizontal and vertical axis and rotation). The walls subject to the analyses were modeled using quadruple-node PLANE182 elements with two degrees of freedom in each node (translations in horizontal and vertical nodal directions) predefined for the surface elements as walls. The elements were defined with respect to coordinates, wall thickness, and elastic properties of the material. The dimensions of the wall FE are 50 mm × 50 mm. In the corners of the door opening and in the zone of contact with the adjacent construction, FE has been thickened up to the dimension of 10 mm × 10 mm. Division of the meshes has been made with FE densification within the zone of expected cracking of the material.

During the calculations, the problem of non-linear contact between the wall and elements of the building structure frame was solved. The non-linearity of contact between elements results from variable adhesion and friction of contacting surfaces in the deformation conditions. In the accepted computational models, the contact between the walls and the adjacent structures was simulated using the surface-to-surface contact finite elements (CFE). Contacting structures were considered as deformable bodies, the contacting surfaces of which form a contact pair. The target surface (frame elements) was modeled by FE TARGE169, and the contact surface (filling) was modeled by FE CONTA171. The calculations of friction used the basic Coulomb–Mohr model. In this model, two contacting surfaces can have shear stresses of a certain magnitude due to their interaction before the slip phase. This condition is known as sticking. The Coulomb–Mohr friction model determines the equivalent shear stress τ, in which the sliding on the surface first represents a part of the contact pressure (3) as follows:

$$\tau = \mu \cdot p + c, \tag{3}$$

where μ is the coefficient of friction, p is contact pressure, and c is tangential adhesion. The μ and c values are a property of the contact surface material. As soon as the limit shear stress is exceeded, the two contact surfaces slide against one another. This state is known as sliding. The grip–slip calculation determines when a point changes from grip to slip and vice versa.

2.3. Testing of Full-Scale Walls

Experimental in situ tests were conducted on full-scale non-load-bearing partition walls in a framed structure residential building under construction. Two masonry walls were tested (height H = 3.07 m, length L = 5.75 m, and thickness 12 cm) with a door opening (2.1 m high and b = 0.99 m wide) in the middle of their span. The strips of walls between the opening and ceiling were reinforced at the bottom using three steel bars (diameter 12 mm) anchored in horizontal joints along the lengths of 250 mm. This type of reinforcing the wall strip above the door opening is a typical solution in the region of Minsk (Belarus) and some also appear in Poland. An expansion joint (30 mm thick) was made between the ceiling and the top edge of the wall. This type of horizontal connection between the infilling wall and the ceiling allows us to dilate these elements. It is the most commonly used connection solution in multi-story buildings in Central and Eastern Europe, studied in [10,13]. The walls were built on the floor made of prefabricated reinforced concrete multi-channel slabs (FCP, type: $240 \times 90 \times 24$ cm^3, Brest, Belarus) supported on reinforced concrete spandrel beams of the building frame. One of the walls was connected with cross walls using flexible steel anchors that restrained relocation only in the direction transversal to its surface (BUILD 1). The second wall was rigidly connected with the brick load-bearing cross walls by constructing toothings with it (BUILD 2). The scheme of the measurement station is shown in Figure 7.

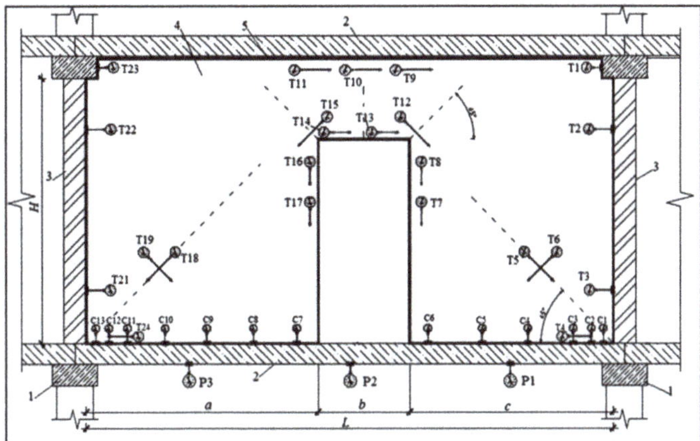

Figure 7. General scheme of the tested partition walls with positioned units for measurements of deformations and relocations: 1—reinforced concrete spandrel beam, 2—multi-channel ceiling slabs made of reinforced concrete, 3—masonry cross walls, 4—partition wall under test, and 5—horizontal expansion joint (T—dial deformation gauges, accuracy to 0.001 mm, C—dial relocation gauges, accuracy to 0.01 mm, P—floor deflection gauges, accuracy to 0.01 mm).

Figure 8 presents photos of the testing stand after positioning the measuring tools. During testing of the partition wall, deflection of the floor was forced by loading it in the middle of the span using hydraulic actuators, lifting capacity 100 kN (Pneumat P100, Minsk, Belarus). In order to transfer the load to the floor under the wall, steel threaded spacers were placed between the actuators and ceilings (Figure 8b).

(a)

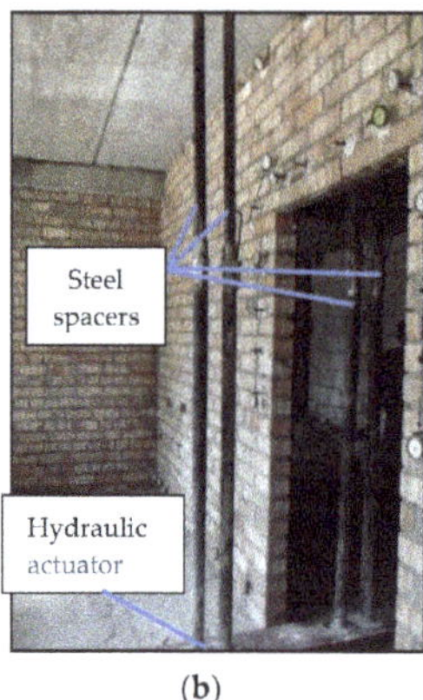

(b)

Figure 8. View of the partition walls under study (**a**) with fitted measuring tools and (**b**) steel spacers for floor loading.

The floor was loaded gradually with an increase of deflection at each stage by 1 mm up to the value at which cracking of the partition wall occurred. At each loading stage, the floor deflections, vertical relocations of the bottom edge of the partition walls, and thickness of the gap appearing between them were measured. The thickness of the gap between the wall and the floor was measured using the set of steel feeler gauges (Kafer M2/20 T, Villingen-Schwenningen, Germany) with an accuracy of 0.1 mm. This allowed the determination of the length of contact between the wall and the floor. In addition, using the mechanical dial gauges (Kafer FM1000/5 T, Villingen-Schwenningen, Germany), deformation of the wall was measured with the accuracy of 0.001 mm in the direction of the trajectory of the main tensile stresses as the most dangerous as far as cracks of the wall are concerned. The location of the gauges was established based on numerical calculations of the walls under test (Figures 7 and 8). The size of the measurement base depended on the direction of deformation measurement and included at least a wall element with masonry joints. Measurement readings were taken immediately after the achievement of a specific floor deflection level and after exposure of the load, within 15–20 min.

3. Results and Discussion

3.1. Results of Laboratory Measurements

Results of laboratory tests (LAB 1) on ceramic wall elements, which are the objects of further analyses are shown in Table 3.

Table 3. Average values of strength characteristics of ceramic bricks with a voidness V = 18%.

Average Normalized Strength according to EN 772-1 [1]	Compressive Strength in the Direction Horizontal to the Stretcher Surface	Flexural Tensile Strength	Shear Strength	Axial Tensile Strength
$f_{by,mv}$ (MPa)	$f_{bx,mv}$ (MPa)	$f_{btb,mv}$ (MPa)	$f_{bv,mv}$ (MPa)	$f_{bt,mv}$ (MPa)
18.37	7.50	2.97	2.81	0.99

[1] In the direction perpendicular to the stretcher surface.

The compressive strength of the mortar at the time of testing was established in accordance with the requirements of EN 1015-11, reaching the value of 3.1 MPa. Results of determining the strength and deformation characteristics of masonry walls (LAB 2–LAB 4) under compression along and across horizontal mortar joints have been presented in Table 4.

Table 4. Characteristic values of strength and deformation properties of the tested wall samples under compression along horizontal (x) and vertical (y) directions.

Compressive Strength of Masonry Walls (MPa)		$\dfrac{f_{cy,mv}}{f_{cx,mv}}$	Short-Term Modulus of Elasticity E (MPa)		$\dfrac{E_{y,mv}}{E_{x,mv}}$	Lateral Deformation Coefficient	
$f_{cy,mv}$	$f_{cx,mv}$		$E_{y,mv}$	$E_{x,mv}$		$v_{xy,mv}$	$v_{yx,mv}$
5.2	3.5	1.49	5400	5642	0.96	0.26	0.27

Results of laboratory tests concerning tensile and friction properties of the analyzed brick walls are shown in Table 5.

Table 5. Characteristic values of mechanical properties for the tested partition walls.

Initial Shear Strength	Internal Friction Coefficient	Tensile Strength across Supporting Joints	Tensile Strength along Supporting Joints
$f_{v0,obs}$ (MPa)	$Tg\ \alpha$ (-)	$f_{w,obs}$ (MPa)	$f_{t,cal}$ (MPa)
0.18	0.63	0.16	0.22

Presented values of mechanical properties for the wall samples were used for numerical calculations shown in Section 3.2.

3.2. Results of Numerical Calculations

During the first stage of the analyses (FEM TEST 1), the calculations were performed for the walls, the vertical edges of which were not connected with vertical load-bearing structures of the building, with a door opening in the middle of their span. In Central and Eastern Europe, the most popular method of connecting non-load-bearing masonry walls along vertical edges is the use of flexible steel anchors, restraining relocation only in the direction transversal to its surface. The tests were performed in a flat stress state. For this reason, in the FEM TEST 1 calculations for this type of connection, the model free of connection in the horizontal and vertical directions in the plane of the wall surface has been used. It was determined that during deflection of the floor, the main maximum tensile stresses were concentrated in top corners of the openings at an angle of approximately 45° to the supporting joint (Figure 9).

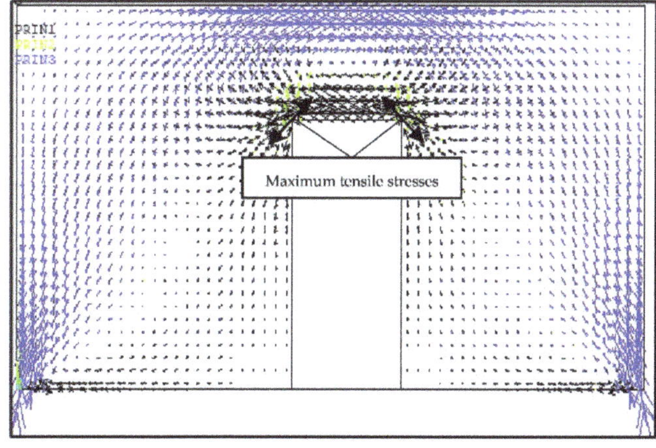

Figure 9. Trajectories of the main stresses in the non-load-bearing partition wall with door opening during deflection of the floor.

After exceeding of admissible limit stress, shape deformations of the wall may lead both to diagonal and horizontal cracks of the wall in this zone. According to the conducted analysis, it was found as expected, that morphology of these cracks depends, among other things, on the following factors:

- the distance of door opening from the vertical edge of the wall;
- the ratio of its length to its height;
- wall strength and deformability;
- method of connecting the wall with vertical load-bearing structures of the building.

The presented factors are applicable to all technologies of masonry infill walls.

In the bottom zones of the wall, tensile stresses also appear, which have much lower values than in corners of the opening, however. The presence of the opening leads to the reduction of the main tensile stresses σ_1, occurring in the zones of contact with the floor compared to the wall without opening. Values of maximum tensile stresses σ_1 in the zone of contact between the wall and the floor constitute approximately 5–6% of the maximum value of contact compressive stresses σ_c in this zone.

As expected, on the basis of the conducted numerical tests, it was determined also that the level of wall stresses caused by deflection of the floor to a considerable extent depends on the elasticity properties of the wall. This results from the cooperation of the walls with the floor as a statically undefined structure system in which materials of the walls and floors have different deformability features, both with respect to temporary and long-term loads. The lower the flexural modulus E, the lower the values of tensile stresses in corners of the opening and in the zone of contact between the wall and the floor. The achieved dependency between the stresses in the door opening zone and flexural modulus E of the wall is presented in Figure 10.

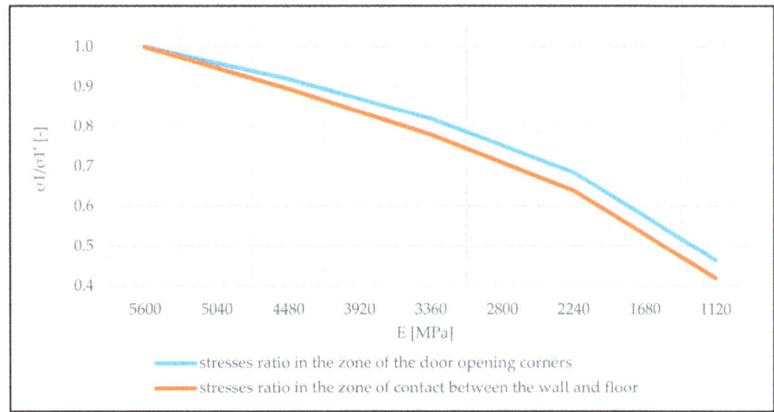

Figure 10. Graph of the dependency between $\sigma_1/\sigma_{1'}$ ratio and wall modulus of elasticity E [MPa] (σ_1—main tensile stress with a lowered modulus of elasticity and $\sigma_{1'}$ —stress for modulus of elasticity E = 5600 MPa).

From these regularities, it can be concluded that the stress level in non-load-bearing walls can be decreased by using the masonry elements with lower flexural modulus. It can be also stated that in walls with joints of regular thickness 10–12 mm, reduction of stresses is possible by using mortar characterized by high flexibility and deformability.

In the wall in which the opening is moved to one of the vertical edges, the maximum values of tensile stresses σ_1 are located near the corner adjacent to the longer section of the wall (Figure 11).

The closer the opening is to the wall edge, the higher the values of the main tensile stresses in the zone of the corner on the longer side of the wall. Figure 12 shows the curve of maximum values of σ_1 in the corner versus the location of the door opening.

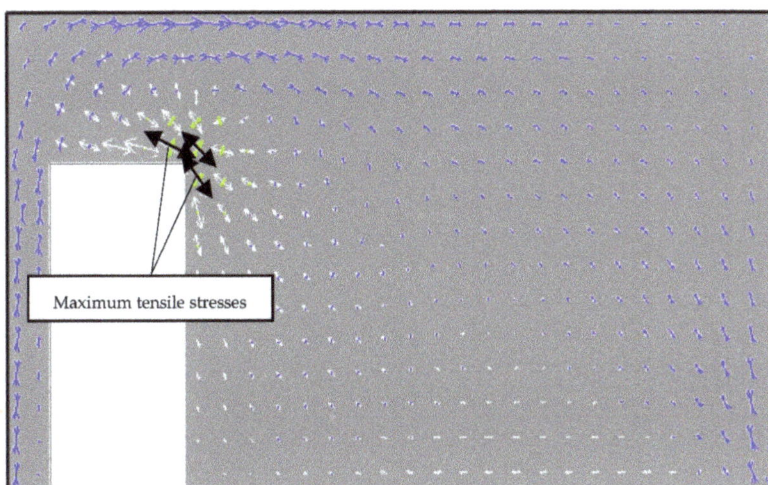

Figure 11. Trajectories of the main stresses in the wall with door opening moved to the vertical edge of the wall.

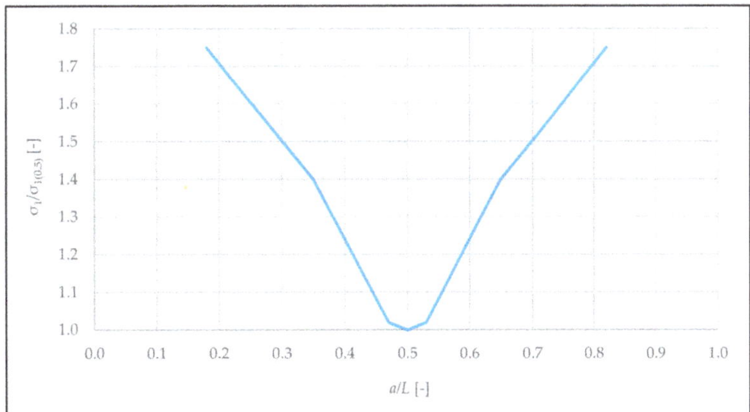

Figure 12. Graph of the dependency between $\sigma_1/\sigma_{1\,(0.5)}$ ratio and door opening localization on the length of the wall. $\sigma_{1\,(0.5)}$—main tensile stresses in the zone of the door opening corners, situated in the middle part of the wall ($a/L = 0.5$), a—distance from the vertical edge of the wall to the opening center, and L—length of the wall.

The level of the main tensile stresses σ_1 in the corners of the opening in walls not connected along horizontal edges with the load-bearing structure of the building is considerably affected by the value of the coefficient of friction ($tg\,\alpha$) between the masonry wall and floor. Figure 13 presents the dependence $\sigma_1/\sigma_{1(0)}$—$tg\,\alpha$, where $\sigma_{1(0)}$ are the main tensile stresses in the zone of door corners at $tg\,\alpha = 0$.

With the increase of friction coefficient $tg\,\alpha$ from 0 (no friction) to 1 (rigid connection with the floor), stresses σ_1 in the zone of the door opening corners decrease more than five times. On the other hand, tangential stresses operating in the zone of contact between the wall and floor increase. Exceeding $f_{v0,obs}$ (shear strength stresses of the wall along supporting joints) may lead to horizontal shearing of the wall in the zone of contact with the floor. Prevention of this is possible by transferring spreading forces T directly to load-bearing structural elements of the building along vertical joints of the wall.

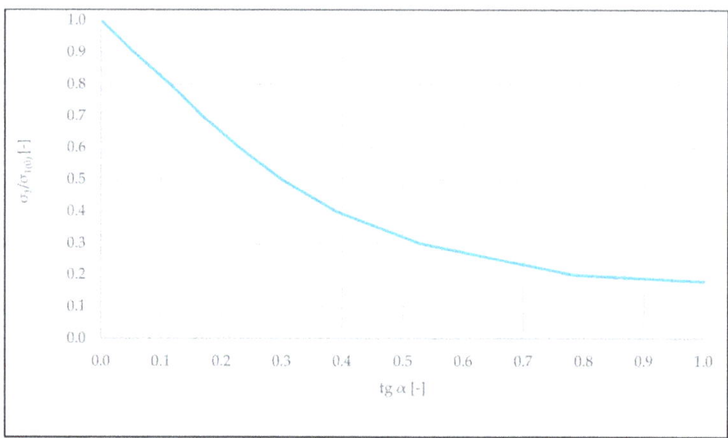

Figure 13. Graph of $\sigma_1/\sigma_{1\,(0)}$—$tg\,\alpha$ dependency ($\sigma_{1(0)}$—the main tensile stresses in the zone of door corner at the coefficient of friction $tg\,\alpha$ equal to 0).

According to the completed numerical tests, in the case of the rigid connection of the wall with vertical load-bearing structures of the building (FEM TEST 2), with the increase of the floor deflection, the main tensile stresses increase only in the zone of contact between the wall and floor. In the door opening corners, the values of σ_1 are insignificant and practically do not depend on the floor deflection level. In addition, a comparative analysis of the main stresses in walls with the door opening and full walls was conducted. It was concluded that with equal values of floor deflection, the main tensile stresses in the zone of contact with the floor in the wall with door opening were approximately 25% lower than in the wall without opening. This results from higher flexural rigidity of the wall without opening and thus its lower ability to adapt to deflection of the floor.

3.3. Experimental Tests in the Facility

Figure 14 presents the graphs of deflection for reinforced concrete floor and vertical relocations of the bottom edge of the partition wall not connected with cross walls at the first and last stage of loading (BUILD 1).

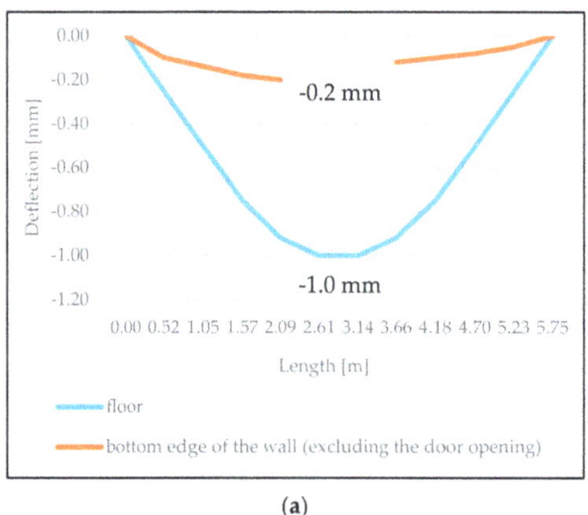

(a)

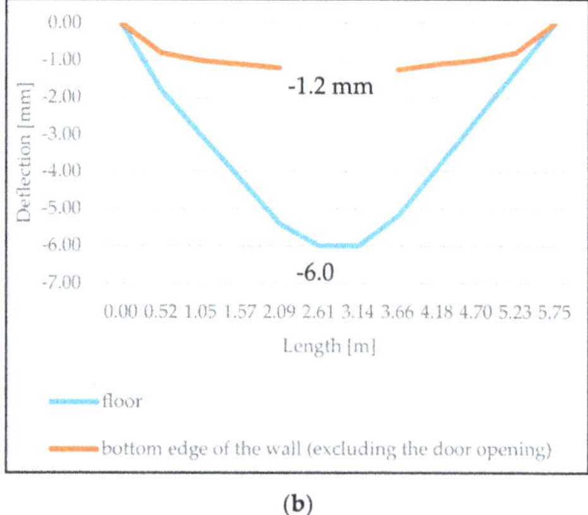

(b)

Figure 14. Graphs of deflection of the floor and bottom edge of the partition wall not connected with cross walls. (a) Wall at the first stage of loading and (b) wall at the last stage of loading.

As it can be concluded from the graphs, as early as the first stage of the floor loading, a gap appears between the floor and bottom edge of the walls, and the thickness of this gap increases with the increase of the floor deflection. This is illustrated by dependencies between maximum vertical relocations of the bottom edge of the walls u_a (at the end of the edge of the vertical door openings) and deflection of the floor u in the middle of its span shown in Figure 15. During the final stage of loading, the maximum deflection of the floor exceeded maximum vertical relocations of the bottom edge of the walls by approximately five times. Vertical relocations of the bottom edge of the partition wall stiffened using cross walls (BUILD 2) were 1.5–2 times lower than for the wall not connected with cross walls. The comparison of experimental and theoretical vertical relocations of bottom edges of the walls indicates that with deflection of the floor in the middle of its span $u = 1$–5 mm, the discrepancy in the values of these relocations did not exceed 15%.

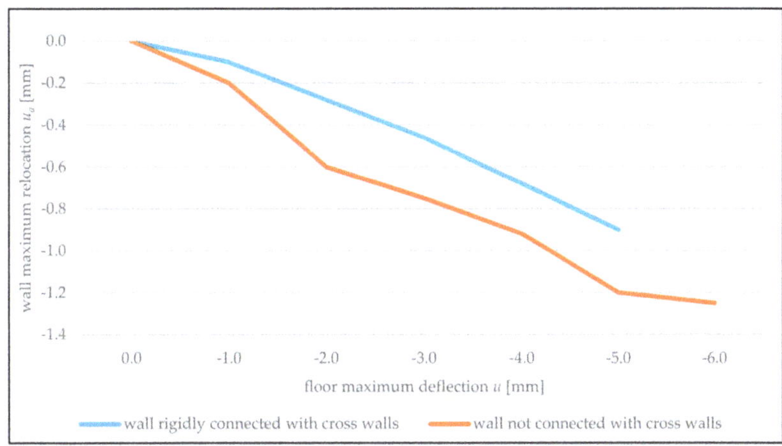

Figure 15. Graphs of maximum vertical relocations of the bottom edge of the wall (u_a), depending on maximum deflection of the floor (u).

Distortion of the contact between the floor and bottom edge of the wall starts at the very beginning of the floor bending. As a result of the loosening of the wall from the floor, with the increase of its deflections, the length of contact between the wall and floor decreases. For example, with the floor deflection in the middle of its span of $u = 1.0$ mm, the length of its contact with the wall not connected with cross walls was 28 cm, and with the deflection of $u = 6.1$ mm, the length of contact decreased down to the value of 9.3 cm. In the case of the wall rigidly connected with cross walls with the floor deflection of $u = 1.2$ mm, the length of contact between the floor and wall was approximately 10 cm, and with the deflection of $u = 5.3$ mm, this length was 8 cm. Graphs of the relative length of the zone of contact between the wall and floor (l_{cont}/L) depending on relative deflection u/L of the floor are shown in Figure 16.

Any change of the length of contact between the partition wall and floor causes redistribution of contact stresses. Their concentration occurs in the zones with minimum floor deflections that are in supporting zones. In these zones, the wall is loaded with contact stresses causing its local pressing perpendicularly to supporting joints. This can lead to local crushing of the wall or the occurrence of diagonal cracks in corner areas of the partition walls. One positive effect of this redistribution is the reduction of bending moments in the floor from loading caused by partition walls. The local nature of load transfer from partition walls to floors with distortion of contact between them is not always included in standard provisions. For instance, according to standard [33], 60% of the self-weight of the partition wall with door opening are transferred to the floor as a uniform linear load along the wall length. On the other hand, the remaining 40% are transferred to the floor in the form of concentrated forces applied on the section at 1/3 length of the wall from floor supports to the door opening. In European standards, the load from partition walls is usually assumed to be linear or surface. These recommendations are contradictory to the presented results from the completed tests.

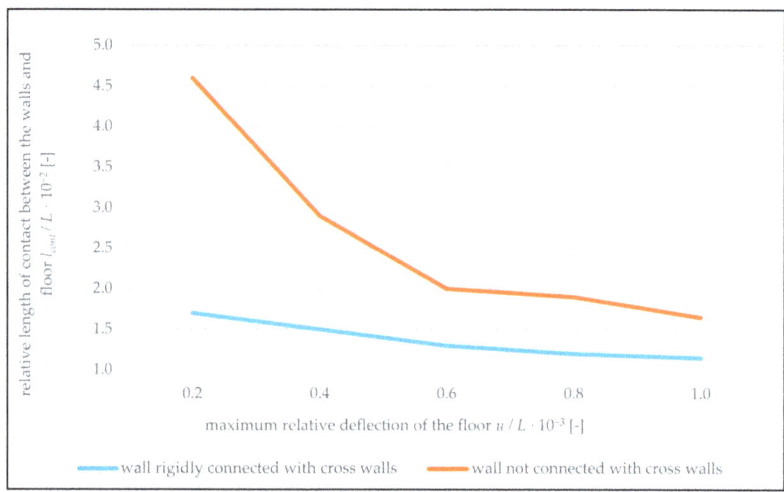

Figure 16. Graphs of the relative length of contact between the walls and floor l_{cont}/L, depending on maximum relative deflection of the floor u/L.

Figure 17 illustrates the dependencies of normal stresses operating in the direction parallel to the supporting joints on the bottom and top edge of the strip of wall above the opening on the floor deflection value u. These stresses are defined as a product of Young's modulus E and wall deformations ε obtained based on measurements using gauges T13 and T10 (Figure 6).

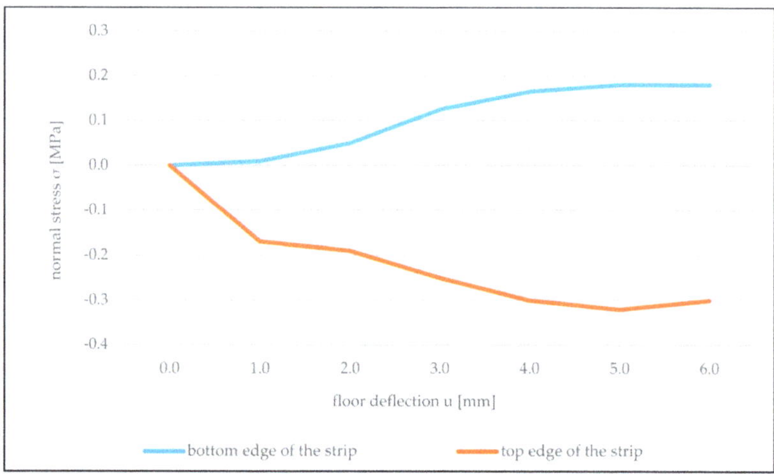

Figure 17. Graphs of changes in normal stresses on edges of the wall strip above the opening in the wall not connected with cross walls, depending on the floor deflection u.

Compression of the top edge of the strip and expansion of the bottom one indicates the eccentric load on the strip in the direction of its length, which was caused by balanced spreading forces T occurring on the contact of the wall with the floor. Values of tensile stresses on the bottom edge of the strip at maximum floor deflections are similar to the tensile strength of the wall along supporting joints $f_{t,cal}$ = 0.22 MPa (Table 3).

In the case of rigid connection of the partition wall with cross walls, the wall strip above the opening also acted as eccentrically compressed (Figure 18). The entire cross section of the strip was compressed, and the value of maximum compressing stresses was approximately three times lower than in the case of the partition wall not connected with cross walls (Figure 17). It is connected with the reduction of horizontal deformations of the wall by vertical structural elements of the building.

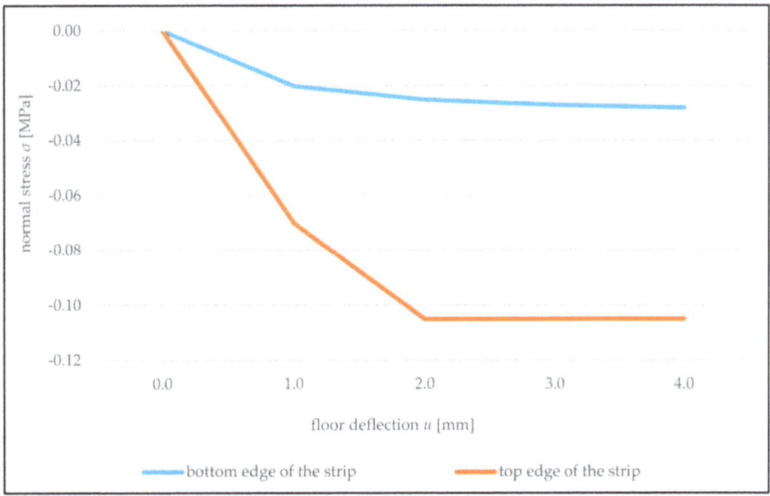

Figure 18. Graphs of changes in regular stresses on edges of the wall strip above the opening in the wall rigidly connected with cross walls, depending on the floor deflection u.

It was established also that near the contact with the floor, expansion of the wall along supporting joints occurred. In the wall rigidly connected along vertical edges, the values of expansion deformations were 1.5–2 times lower than in the wall connected freely. In both cases, stresses caused by these deformations did not exceed the tensile strength of the wall along supporting joints ($f_{t,cal}$ = 0.22 MPa, according to Table 1). In the wall not connected with cross walls, using gauges T7, T8, T16, and T17 (Figure 6), expansion deformations of the wall were recorded along vertical edges of the opening in the direction perpendicular to supporting joints. Values of these deformations were increasing when the door opening corners were approached. The most strained areas of the wall are corners of the opening where maximum tensile stresses operate at an angle of 45° to vertical supporting joints. Dependencies of relative wall deformations ε in this direction recorded using gauges T12 and T15 (Figure 6) on the wall deflections u are shown in Figure 19. In the case of the wall connected with cross walls, the wall contributed to compression both along vertical edges of the door opening and at an angle of 45° to supporting joints in the zone of its corners. This results from stiffening of the partition wall with load-bearing cross walls because in top areas of the partition wall, according to measurements using gauges T1, T2, T22, and T23 (Figure 6), the wall contributed to expansion along supporting joints, while in bottom areas, according to gauges T3 and T21, it contributed to compression.

Damage of the partition wall not connected with cross walls took place as a result of cracking of the wall in the zone of the door opening corners. The wall cracks occurred in the form and sequence shown in Figure 20.

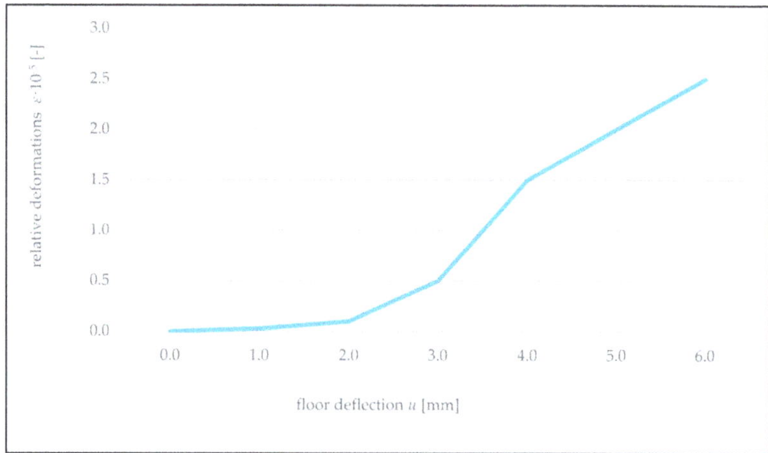

Figure 19. The dependency of floor deflection u on the relative deformations ε of the wall at an angle of 45° in corners of the door opening for the wall not connected with cross walls.

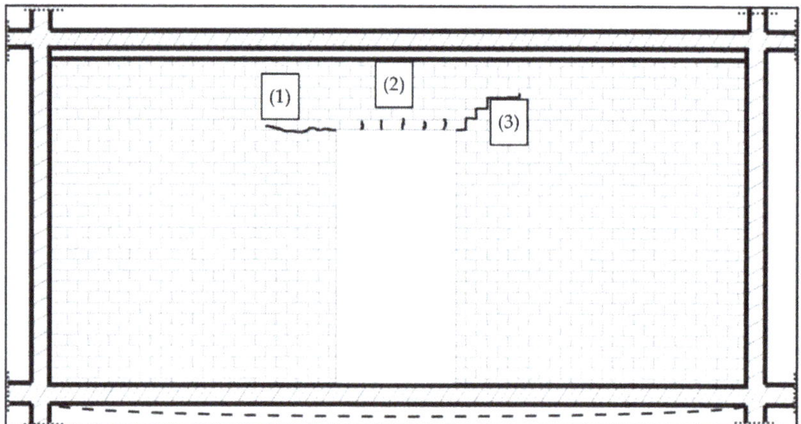

Figure 20. Character and sequence of wall cracks on the wall not connected with cross walls; (1)—horizontal crack; (2)—vertical crack; and (3)—oblique crack.

Horizontal crack (1) appeared in the supporting join, vertical cracks (2) appeared in vertical joints, and oblique crack (3) had a shape of stairs matching vertical and supporting joints. The cracks appeared suddenly, with deflection of the floor in the middle of its span at the value above 6 mm, that is, relative deflection of $u/L = 1/958$. Crack (2) appeared with tensile stresses on the bottom edge of the wall strip above the opening $\sigma = 0.18$ MPa, close to the tensile strength of the wall along supporting joints $f_{t,cal} = 0.22$ MPa (Table 3). Crack (3) appeared at maximum values of the wall deformation of $\varepsilon = 2.5 \times 10^{-5}$, that is, stress $\sigma = 0.14$ MPa comparable with the tensile strength of the wall across supporting joints $f_{w,obs} = 0.16$ MPa (Table 5).

As mentioned already, in the wall rigidly connected with cross walls (BUILD 2), compressive stresses occurred in the area of the opening corners and in the wall strip above the opening (Figure 16) at all stages of the floor loading. It had a significant effect on the crack resistance of the wall. At the maximum achieved deflection of the floor u = 12 mm, that is, at the relative deflection of $u/L = 12/5750 = 1/479$, no cracks or damage was noticed in the examined wall.

3.4. Research Findings and Their Application

On the basis of the FEM TEST 1 numerical analyses, the relationships between the wall flexural modulus and tensile stresses in the area of the door opening were derived. It has been confirmed that the use of masonry elements or a mortar with a lower elasticity coefficient leads to a reduction in the tensile stress value in the corner zone of the door opening. The derived dependencies allow for the calculation of theoretical stresses for the given elasticity of the designed partition walls.

The relationship between friction coefficient $tg\ \alpha$ (between the tested wall and foundation) and the maximum tensile stresses in the area of the door opening has been obtained. It has been calculated that with the increase of friction coefficient $tg\ \alpha$ from 0 (no friction) to 1 (rigid connection with the floor), stresses in the zone of the door opening corners decrease more than five times.

The relationship between the location of the door opening along the wall length and the value of the maximum tensile stresses in the area above the door opening was derived. It has been shown that the location of the door opening close to the edge of the wall leads to an increase in the tensile stress value by 175% compared to the location of the door opening in the middle of the wall length.

The conducted analyses may have practical application for the calculation of tensile stresses in non-load-bearing walls made of brick elements with a door opening and in the design of this type of structure.

Numerical analysis of the wall with rigid connection with vertical structure (FEM TEST 2) indicated that with the increase of the floor deflection, the main tensile stresses increase only in the zone of contact between the wall and floor. In the door opening corners, the values of σ_1 are insignificant and practically do not depend on the floor deflection level.

On the basis of the experimental tests, it was found that the value of the relative deflection of the foundation u/L, which led to the cracking of the wall with the door opening in the middle of the span, freely connected along the vertical edges (BUILD 1), was 1/958. The BUILD 1 experiment (carried out on a natural scale in a real building) showed that the deformation of the foundation that is causing cracking of the non-load-bearing masonry wall was comparable with the results of wall tests on the 1:3 scale carried out by Pfeffermann [7]. The rigid connection of the vertical edges of the wall with the building structure performed in the BUILD 2 experiment allowed us to avoid cracking in the extent of the relative foundation deflection u/L of 1/479. This value allows the avoidance of cracking of walls erected on foundations with a maximum deflection in the range provided by all standards referred to in Table 2 [17–24]. Relying on these results, the authors would suggest to project and perform the masonry partitions with rigid connection along their vertical edges.

In the future, it is planned to carry out measurements and experiments on the walls with different masonry elements, different static schemes of the foundation, and varied patterns of openings in the walls.

Results of the presented research are limited to the analyzed type of materials and connections of the walls with the structure of the building.

4. Conclusions

The performed tests and analyses allowed for the conclusions presented below.

- The most strained area of masonry partition walls supported on reinforced concrete floors are corners of door openings together with the strip of wall above them;
- Cracking of the wall in this area occurs mainly as a result of tensile stresses, which appeared, in the case of the lack of rigid connection, between vertical edges of the wall with adjacent vertical load-bearing structures;
- The appearance of a gap between the bottom edge of the wall and the floor results in redistribution of load favorable for the floor from the weight of the partition wall, which concentrates in supporting zones of the floor;

- In the case of the completed experimental tests, walls made of ceramic brick with the span of 6 m and vertical edges connected freely with the building structure cracked at the relative deflection of the floor under the wall, with the value equal to 1/958 of the floor's length;
- Deformation of the foundation equal to 1/958 that is causing cracking of the tested wall BUILD 1 was comparable with the results of wall tests on the 1: 3 scale carried out by Pfeffermann [7];
- Application of rigid connection along vertical edges of the walls, while maintaining appropriate expansion joints along both horizontal edges, allowed the avoidance of cracking of the wall across the entire measuring range. The wall was not damaged at the floor deflection equal to 1/479 of its length;
- Floor deflection of 1/479 does not exceed the limit deflection values provided in the standards;
- The measured tensile strength of the tested walls in the horizontal direction was 0.22 MPa;
- Strength value obtained in laboratory tests of the wall samples corresponded to destructive values measured for the full-scale wall built in the construction site;
- The numerical analyses allowed the determination of dependency between the change in location of the door opening along the length of the tested walls and tensile stresses in the zone of its corners;
- It was concluded that the value of the main tensile stresses in the corners of the opening in walls not connected along vertical edges with the load-bearing structure of the building depends on the friction coefficient (tg α) between the masonry wall and the deflecting floor. With the increase of friction coefficient tgα from value 0 (free connection) to 1 (rigid connection in the horizontal direction), the values of tensile stresses in the zone of the door opening corner decrease by five times;
- The conducted analyses may have practical application for the calculation of tensile stresses in non-load-bearing walls made of brick elements with a door opening and in the design of this type of structure.

Author Contributions: Conceptualization, T.K., V.D., and R.N.; methodology, V.D. and T.K.; software, V.D. and R.N.; validation, T.K., V.D., and R.N.; formal analysis, T.K and V.D.; investigation, T.K. and V.D.; writing—original draft preparation, T.K., V.D., and R.N.; writing—review and editing, T.K.; visualization, T.K. All authors have read and agreed to the published version of the manuscript.

Funding: This research received no external funding.

Institutional Review Board Statement: Not applicable.

Informed Consent Statement: Not applicable.

Conflicts of Interest: The authors declare no conflict of interest.

References

1. Drobiec, Ł. Repair of the cracks and reinforcement of the brick walls. In Proceedings of the 30th Conference Structural Designer Workshop, Szczyrk, Poland, 25–28 March 2015; pp. 324–398. (In Polish).
2. Małyszko, L.; Orłowicz, R. Selected methods of repairing scratched masonry walls. *Build. Rev.* **2008**, *12*, 40–46.
3. Production of major construction products in December 2020. European Statistical System, Statistics Poland. Available online: https://stat.gov.pl/en/topics/industry-construction-fixed-assets/construction/ (accessed on 20 February 2021).
4. Meyerhof, G. Some Recent Foundation Research and its Application to Design. *Struct. Eng.* **1953**, *31*, 151–167.
5. Beranek, W. The Prediction of Damage to Masonry Buildings Caused by Subsoil Settlements. *Heron* **1987**, *32*, 55–93.
6. Pfeffermann, D. Deformations admissibles dans le batiment. *Notes d'Information Technol.* **1981**, *132*, 29.
7. Rolanda, O.G., Jr.; Ramal, H.M.A.; Correa, M.R.S. Experimental and numerical analysis of masonry load-bearing walls subjected to differential settlements. In Proceedings of the 9th North American Masonry Conference, Clemson, SC, USA, 1–4 June 2003; pp. 134–145.
8. Loots, J.J.; van Zijl, G.P.A.G. Experimental verification of settlement induced damage to masonry walls. Proceedings of 13th International Brick and Block Masonry Conference, Amsterdam, The Netherlands, 4–7 July 2004.

9.	Piekarczyk, A. Cracking and Failure Mechanism of Masonry Walls Loaded Vertically and Supported by Deflecting Structural Member. Proceedings of World Multidisciplinary Civil Engineering-Architecture-Urban Planning Symposium 2018, Prague, Czech Republic, 18–22 June 2018.

10.	Kania, T.; Stawiski, B. Research on crack formation in gypsum partitions with doorway by means of FEM and fracture mechanics. Proceedings of World Multidisciplinary Civil Engineering-Architecture-Urban Planning Symposium–WMCAUS, Prague, Czech Republic, 12–16 June 2017.

11.	Derkach, V.N. Investigations of the stress-strain state of stone partitions during floor deflection. *Ind. Civ. Constr.* **2013**, *6*, 62–66.

12.	Brameshuber, W.; Beer, I.; Kang, B.-G. Untersuchungen zur Vermeidung von Rißschäden bei nicht tragenden Trennwänden. *Mauerwerk* **2007**, *11*, 54–62. [CrossRef]

13.	Stawiski, B. *Masonry Structures. Repairs and Reinforcements*; Polcen: Warsaw, Poland, 2014. (In Polish)

14.	Dmochowski, G.; Szolomicki, J. Technical and Structural Problems Related to the Interaction between a Deep Excavation and Adjacent Existing Buildings. *Appl. Sci.* **2021**, *11*, 481. [CrossRef]

15.	Jasinski, R. Identification of Stress States in Compressed Masonry Walls Using a Non-Destructive Technique (NDT). *Materials* **2020**, *13*, 2852. [CrossRef] [PubMed]

16.	Jasiński, R.; Drobiec, Ł.; Mazur, W. Validation of Selected Non-Destructive Methods for Determining the Compressive Strength of Masonry Units Made of Autoclaved Aerated Concrete. *Materials* **2019**, *12*, 389. [CrossRef]

17.	American Concrete Institute. *ACI 318-08: Building Code. Requirements for Structural Concrete and Commentary*; ACI Committee 318; American Concrete Institute: Farmington Hills, MI, USA, 2008.

18.	Masonry Standards Joint Committee. *ACI-530-08/ASCE 5-08/TMS 402-08: Building Code. Requirements for Masonry Structures*; Masonry Standards Joint Committee: Farmington Hills, MI, USA, 2004.

19.	UK National Standards Body. BS 5628-2: Code of practice for the use of masonry. In *Part 2: Structural Use of Reinforced and Prestressed Masonry*; UK National Standards Body: London, UK, 2005.

20.	Deutsches Institut für Normung. DIN 1045-1: Tragwerke aus Beton, Stallbeton und Spannbeton. In *Teil 1: Bemessung und Konstruktion*; Deutsches Institut für Normung: Berlin, Germany, 2008.

21.	Bureau de Normalisation. NBN B 03-003: Deformations des structures. In *Valeurs limites de deformation—Batiments*; Bureau de Normalisation: Brussel, Belgium, 2003.

22.	European Committee for Standardization. *EN 13747: 2010. Precast Concrete Products—Floor Plates for Floor Systems*; European Committee for Standardization: Brussels, Belgium, 2010.

23.	European Committee for Standardization. *EN 1992-1-1: 2008. Eurocode 2: Design of concrete structures - Part 1-1: General rules and rules for buildings*; European Committee for Standardization: Brussels, Belgium, 2008.

24.	Polish Committee for Standardization. *PN-B-03264: 2002: Reinforced and prestressed concrete structures - Static calculations and design*; Polish Committee for Standardization: Warsaw, Poland, 2002.

25.	European Committee for Standardization. *EN 1996-1-1 EN 1996-1-1: Eurocode 6: Design of Masonry Structures - Part 1-1: General Rules for Reinforced and Unreinforced Masonry Structures*; European Committee for Standardization: Brussels, Belgium, 2005.

26.	Raj, A.; Borsaikia, A.C.; Dixit, U.S. Evaluation of Mechanical Properties of Autoclaved Aerated Concrete (AAC) Block and its Masonry. *J. Inst. Eng. India Ser. A* **2020**, 315–325. [CrossRef]

27.	Shabbara, R.; Nedwella, P.; Zhangjian, W. Influence of silica fume content on the properties of aerated concrete. In Proceedings of the 36th Cement and Concrete Science Conference At: Royal welsh collage of Music and Drama, Cardiff, Welsh, 5–6 October 2016.

28.	European Committee for Standardization. *EN 1015-11 Methods of Test for Mortar for Masonry-Part 11: Determination of Flexural and Compressive Strength of Hardened Mortar*; European Committee for Standardization: Brussels, Belgium, 2019.

29.	European Committee for Standardization. *EN 1052-1 Methods of Test for Masonry-Part 1: Determination of Compressive Strength*; European Committee for Standardization: Brussels, Belgium, 1998.

30.	Drobiec, Ł.; Jasiński, R.; Mazur, W.; Jonkiel, R. The effect of the strengthening of AAC masonry walls using FRCM system. *Cem. Wapno Beton* **2020**, *25*, 376–389. [CrossRef]

31.	Drobiec, Ł.; Jasiński, R.; Mazur, W.; Rybraczyk, T. Numerical Verification of Interaction between Masonry with Precast Reinforced Lintel Made of AAC and Reinforced Concrete Confining Elements. *Appl. Sci.* **2020**, *10*, 5446. [CrossRef]

32.	European Committee for Standardization. *EN 1052-3 Methods of Test for Masonry-Part 3: Determination of Initial Shear Strength*; European Committee for Standardization: Brussels, Belgium, 2002.

33.	Stroyizdat. *SNiP 2.08.01-85 Manual for the Design of Residential Buildings. Issue 3: Constructions of Residential Buildings, Dwellings of the State Committee for Architecture and Construction*; Stroyizdat: Moscow, Russia, 1989. (In Russian)

 *materials*

Article

The Impact of Water and Road Salt with Anti-Caking Agent on the Stiffness of Select Mixes Used for the Road Surface

Piotr Mackiewicz * and Eryk Mączka

Faculty of Civil Engineering, Wrocław University of Science and Technology, 50-370 Wrocław, Poland; eryk.maczka@pwr.edu.pl
* Correspondence: piotr.mackiewicz@pwr.edu.pl; Tel.: +48-71-320-45-57

Abstract: An original experimental method was used to investigate the influence of water and road salt with anti-caking agent on the material used in pavement construction layers. This method allowed for monitoring material changes resulting from the influence of water and road salt with anti-caking agent over time. The experiment used five different mineral road mixes, which were soaked separately in water and brine for two time intervals (2 days and 21 days). Then, each sample of the mix was subjected to tests of the complex module using the four-point bending (4PB-PR) method. The increase in mass of the soaked samples and the change in value of the stiffness modulus were analyzed. Exemplary tomographic (X-ray) imaging was performed to confirm the reaction of the road salt and anti-caking agent (lead agent) with the material. Based on measurements of the stiffness modulus and absorption, the correlations of the mass change and the value of the stiffness modulus were determined, which may be useful in estimating the sensitivity of mixes to the use of winter maintenance agents—e.g., road salt with anti-caking agent (sodium chloride). It was found that the greatest changes occur for mixes intended for base course layers (mineral cement mix with foamed asphalt (MCAS) and mineral-cement-emulsion mixes (MCE)) and that the smallest changes occur for mixes containing highly modified asphalt (HIMA).

Keywords: brine; sodium chloride; stiffness modulus; X-ray

Citation: Mackiewicz, P.; Mączka, E. The Impact of Water and Road Salt with Anti-Caking Agent on the Stiffness of Select Mixes Used for the Road Surface. *Materials* **2021**, *14*, 1345. https://doi.org/10.3390/ma14061345

Academic Editor: Francesco Canestrari

Received: 30 January 2021
Accepted: 8 March 2021
Published: 10 March 2021

Publisher's Note: MDPI stays neutral with regard to jurisdictional claims in published maps and institutional affiliations.

Copyright: © 2021 by the authors. Licensee MDPI, Basel, Switzerland. This article is an open access article distributed under the terms and conditions of the Creative Commons Attribution (CC BY) license (https://creativecommons.org/licenses/by/4.0/).

1. Introduction

In winter, the pavement layers are exposed to various environmental factors: heavy rainfall, snowfall, and rapid temperature change (changing the state of rainwater). Unfavorable conditions mobilize road services to ensure the safety of traffic on the road by the use of various chemical agents, in particular, road salt with anti-caking agent (sodium chloride (NaCl)). Road salt with anti-caking agent is an agent commonly used to remove snow and ice from roads [1]. The first successful use of salt for de-icing the surface course took place in the USA on New Hampshire roads in the winter of 1938. Since then, sodium chloride has been widely used in Northeast Europe and North America [2]. Salt use has increased significantly since its first use, e.g., to around 20 million tons of salt per year on roads in North America [3–5].

Brine for winter road maintenance should have a concentration of 20–25%, while the moistened pavement salt should contain 30% brine (NaCl or $CaCl_2$ solution) with a concentration of 20–25% and 70% dry NaCl salt [6]. This form is characterized by rapid and effective melting of ice and snow to a temperature of -9 °C [7]. In many European countries, including Poland, road salt with anti-caking agent is commonly used, which is a mix of sodium chloride (96% NaCl), calcium chloride (2.5% $CaCl_2$), and 0.2% anti-caking agent (potassium ferrocyanide ($C_6FeK_4N_6$)). The additive prevents the formation of lumps of salt due to improper storage conditions or excessive exposure to moisture and low hygroscopic properties of the salt.

The influence of salt, although ensuring safety conditions on the pavement, significantly affects the strength properties of pavement layers. This applies not only to the

173

upper surface of the surface course but also to the deeper layers. Rainwater as well as brine solution, which is mixed with snow or water, reaches the surface and within the road in various ways:

1. Through surface damage, e.g., fatigue cracks resulting from operation (mainly top-down)—the mix moves by gravity and is pressed under pressure from vehicle wheels to the inside of the crack.
2. During reconstruction of the pavement in winter, when vehicle traffic is allowed on a milled surface course. The solution penetrates directly into the upper part of the bonding layers or the substructure as a result of pressure and friction from vehicle wheels.
3. As a result of the inter-layer leakage and poor shape conditions of the ground shoulder.
4. As a result of aerosol dispersion of saline water droplets in the air by passing vehicles at high speed and settling on other areas of the road, including those that were not covered with salt.

It is worth mentioning that the action of water and salt, influencing the change in material strength and material properties, also leads to crucial changes in fatigue life. Salt is hygroscopic and attracts water. In winter maintenance, the surface is filled with water saturated with salt. Salt asphalt concrete can hold up to 10 percent extra water, which will expand upon subsequent freezing, causing additional destruction. In addition, it may have a negative impact on the environment itself (effects of soil acidification in the vicinity of the road lane) and cause corrosion of bridges, vehicles, road infrastructure elements, and roadside green. Despite the fact that the impacts are seasonal, it is assumed that excess and accumulation of chloride ions in a cyclical manner accelerates degradation of the material through maintenance.

Freezing and thawing cycles admittedly are a key factor when it comes to damage to the asphalt pavement in winter. Temperature fluctuations cause the moisture to constantly freeze and melt. When temperatures are above freezing, melting snow or rainwater seep into small cracks in the pavement. When re-freezing, it expands and cracks the pavement. The impact of road salt (mechanisms and chemical reaction between salt and material) and its influence on the pavement are probably the second key factors; therefore, it cannot be skipped. It is worth highlighting the sensitivity of the impact: the greater the impact, the worse the condition of the pavement (various other damages) and the traffic intensity.

2. State of Knowledge Review

The first studies related to the applicability of these maintenance measures for road surfaces date back to the late 1980s [8]. Susan et al., in their work, tried to develop a mathematical model describing defrosting of an ice-covered surface course depending on the location of the agent used, temperature, and its amount. On the basis of the analyses, the authors indicated that sodium chloride is a cheap and effective means of defrosting large areas in a short time in relation to the comparable calcium chloride. For several years since 1987, the simultaneous influence of water and road salt (sodium chloride) on road mixes has not been dealt with. The widespread availability and action of salt and the failure to register its negative effects resulted in little interest in this subject. It was only as a result of severe winters that the attention was refocused on measures supporting the improvement of road traffic safety in periods of heavy snowfall and frost. In 2002, Hassan et al. [9] published an article in which, based on the indirect tensile test, various freeze–thaw cycles, and the share of de-icing agents, they indicated that greater damage to bituminous materials (both in aggregate and binder) is a result of the action of various road salts and other agents compared to the action of water alone.

In 2004, Kosior et al. [10] analyzed concrete and concrete with the addition of an asphalt binder used in road pavements in terms of material resistance to the applied freeze–thaw cycles with the use of water and, separately, a road salt solution with a concentration of 3%. The research showed that concrete with the addition of an asphalt binder better withstands the destructive activity of salt expressed by mass loss from 16 to 18 times depending on the cement used.

In the work of [11] in 2008, the authors analyzed various concrete mixes that were distinguished by different road salt content. It was suggested that, on the basis of the imaging performed, it can be presumed that chlorides affect the properties of the concrete mix. A prototype technique was used to detect the presence of chlorides and water in the layers of the pavement structure, based on wave measurements.

In 2013, Serin et al. [12] published interesting work on the resistance of bitumen mixes to water and frost without the presence of salt in very cold regions. Attempts have been made to reflect the in situ conditions by using a different number of freeze–thaw cycles. Using ultrasound methods, scientists found a significant impact of the change in the state of water concentration on the tested mineral–asphalt mixes, for which it has been estimated that the damage to the material can reach up to 35%.

In 2014, Amini et al. analyzed the mass change of fine-grained bitumen mixtures, taking into account freeze–thaw cycles for an aqueous solution and the presence of various chemicals (including road salt) [13]. Based on mass changes and Marshall's research, it was indicated that the tested bituminous mixtures are very susceptible to water and salt—they lose from 17 to 52% of their original strength and up to 22% of their original mass. In the same year, Liu et al. [14] published a paper that analyzed the effect of the brine of various concentrations (e.g., 3%, 5%, 10%, and 15%) on the stiffness determined in a static three-point bending test. It has been shown that the use of sodium chloride lowers the freezing point of water and that chlorides have an unfavorable effect on asphalt concrete, reducing its strength by up to 50%.

In [15], the authors analyzed the solubility of salt depending on the temperature. It was found that the solubility stabilizes at 0 °C at the concentration level of 35.8%.

In 2016, Hossain et al. [16] measured the thermal condition of various pavements in winter depending on the type of de-icing agent used (including salt) and different amounts of snow. Measurements have shown that the salt's effectiveness is greater on asphalt concretes than on ordinary concretes. Tsang et al. [17] presented research on porous asphalt concrete, in which they analyzed various combinations of freeze–thaw cycles with the participation of water and the salts melted in it: calcium chloride II, magnesium II chloride, and sodium chloride I—road salt. It has been proven that sodium chloride has the greatest impact on mass change, and its loss after freeze–thaw cycles can reach up to 41%. It was indicated that the effects of de-icing agents were not checked in the time perspective, which is the desired issue.

The results of the experimental method studying the effect of salt (calcium chloride and its hydrates) on unpaved roads—breakstone and soil surfaces—were developed in 2019 by Almasi et al. [18]. They showed that the use of calcium chloride reduces soil moisture while reducing the value of the plasticity index by up to 25%. The effect increased caking of the ground layers and lowered load capacity.

Saltanovsa et al. [19] analyzed the effect of road salt on the efficiency of the method analyzing changes in the electromagnetic field, which allows for determining the condition of bituminous pavements (damage monitoring). The experiment in laboratory and in situ conditions showed that the presence of road salt significantly disrupts the efficiency of the presented method and decreases its effectiveness. For 1% brine concentration, the measurement error is 20%, which makes the developed method unsuitable for use.

In 2020, a key work by Gandara et al. [20] was published, in which the authors focused on the influence of road salt on the properties and parameters of 3 types of bituminous mix combinations (PA 16—porous asphalt on 50/70 ordinary asphalt and AC16 asphalt concrete): immersed in a salt solution, as an additive to aggregate and aggregate additive and soaking in salt. Various salt concentrations of 3.5%, 5%, and 10% were taken into account, the soaking time was 3 days, and the test temperature was 20 °C. The authors showed that mixes treated with salt as a result of soaking reduce their initial stiffness modulus (about 6%), but the changes are not as significant as when salted aggregate was applied to the composition of the mixture and then soaking was implemented (decrease by 70%). The fatigue life was also assessed, which for samples made of salted aggregate and soaked in brine decreased by an average of 50%.

A similar topic was analyzed by Zamanillo et al. [21]. The effect of brine with a percentage concentration of 5% at a temperature of 20 °C was investigated during the 2-day test on the bitumen AC 16 subjected to soaking and freeze–thaw cycles. The Indirect Tension Strength (ITS) method was used to test the stiffness, and the four-point bending (4PB-PR) test was used to determine the fatigue life. Moreover, the water and frost resistance test procedure (ITSR) was used. The research shows that the ITSR index for samples soaked in salt with samples soaked in water decreased by 1%, that the stiffness modulus decreased by 5%, and that the fatigue life decreased by about 30%.

The literature review shows that the topic of the influence of salt on the mixtures used in pavement construction layers is and should be constantly analyzed, if only due to various new road materials and complex conditions of impact. However, current studies mostly focus on the study of water and frost activities (freeze–thaw cycles) and their mutual correlation. There is a lack of water and de-icing agents' analysis, which refers to, e.g., rain influence, or humid or chemical aggression. The vast majority of items mentioned define the results of tests carried out on typical asphalt mixtures. The conditions for the impact of de-icing agents and various levels of brine concentrations (especially those that may occur in the winter after intensive salting of the pavement, e.g., 20%) were not analyzed. No correlation was sought for their influence on the mechanical parameters of the mixtures— the stiffness modulus determined, e.g., by the 4PB-PR method (which reflects well the behavior of pavement layers under load). Different combinations of temperature soaking and solution flow simulation were also not used.

The aim of this article is to show, on the basis of experimental studies, the sensitivity of the action of water and road salt to the stiffness of various asphalt mixtures in the long term. Environmental variables simulating in situ conditions were used. In the evaluation of material changes in mixes, the change in sample mass and the change in the stiffness modulus were analyzed. Using the X-ray tomographic method, the extent of material destruction is shown.

3. Materials and Methods

Three specific type of mixes for top asphalt pavement layers and two mixes for base coarse using recycle asphalt milling and, additionally, an innovative hydraulic binder were prepared for the experiment. This binder consisted of 40% CEM I 32.5R cement, 20% hydrated lime ($Ca(OH)_2$), and mainly 40% dusty cement byproducts. Mixes containing this binder have been optimized for composition and material properties in previous research work [22–25]. Table 1 presents the detailed compositions of the analyzed mixes.

Mixes formed for prismatic beams with dimensions of 60 mm × 50 mm × 380 mm were placed in a closed container with a maximum capacity of 130 L. Inside the container, on its walls, was placed four independent pumps resistant to road salt with an anti-caking agent with an average pumping capacity of 7500 L/h, forcing steady circulation of the solution. The introduction of these devices was aimed at two aspects:

1. The brine concentration must be constant and well-distributed everywhere in the container (no local concentration points).
2. Forced circulation causes cyclical pressure of the solution on the material—an approximation of in situ conditions, in which the vehicle wheel "forces" the mixture under pressure into the road material.

Water- and salt-resistant sensors for temperature measurement were installed at five locations in the container near the bottom, which allowed us to track the changes in temperature during the test and when a fixed amount of road salt with an anti-caking agent was added to the water. The entire set was placed in the thermal chamber. The container was filled with water in such a way to know its quantity, density at the test temperature, and that the soaked samples were immersed to a depth of about 18 cm. The scheme of the test stand at the stage of soaking the samples is shown in Figures 1 and 2.

Table 1. Road mixes used in experiment.

Mix	Bond Material
AC11S	Asphalt D50/70: 5.90% Amfibolit 8/11: 32.40% Amfibolit 5/8: 10.50% Sjenit 2/5: 13.80% Sjenit 0/2: 31.20% 0/1 milled stone extender: 6.20% Porosity: 3.70%
AC22WMS (Pmb)	Asphalt 25/55-60: 4.98% 16/22 gabbro grit: 28.50% 11/16 gabbro grit: 9.50% 8/11 gabbro grit: 8.55% 4/8 gabbro grit: 8.55% 2/5 gabbro grit: 15.20% 0/2 gabbro crushed sand: 17.10% 0/1 milled stone extender: 7.60% Porosity: 2.38%
AC22WMS (HIMA)	Asphalt 25/55-80 HIMA: 4.98% 16/22 gabbro grit: 28.50% 11/16 gabbro grit: 9.50% 8/11 gabbro grit: 8.55% 4/8 gabbro grit: 8.55% 2/5 gabbro grit: 15.20% 0/2 gabbro crushed sand: 17.10% 0/1 milled stone extender: 7.60% Porosity: 2.97%
MCAS	Asphalt destruct 0/10: 37.6% Natural ginning aggregate 0/2: 9.4% Crushed ginning aggregate 0/31.5: 47% Foamed asphalt 70/100: 3% Hydraulic binder: 3% (40% CEM cement I 32,5R, 20% hydrated lime (Ca(OH)2, 40% dusty cement byproducts) Porosity: 13.21%
MCE	Asphalt destruct 0/10: 34.4% Natural ginning aggregate 0/2: 8.6% Crushed ginning aggregate 0/31.5: 43% Asphalt emulsion 60/40: 5% Hydraulic binder: 3% (40% CEM cement I 32,5R, 20% hydrated lime (Ca(OH)2, 40% dusty cement byproducts) Water: 6% Porosity: 14.02%

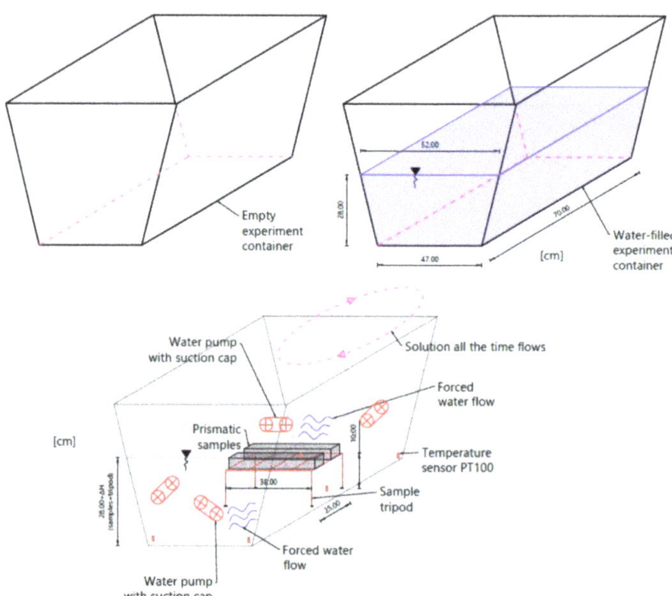

Figure 1. Experiment idea diagram.

Figure 2. Fully armed container prepared to experiment.

Of the nine specimens prepared for this study (for each type of mix), four were assigned to soaking in water and five were assigned to soaking in brine (target brine percentage concentration of the saline solution was 20%). The temperature at which the samples were soaked was 10.0 ± 0.2 °C. Samples of all mixes were accurately weighed prior to soaking. Then, each sample of a given mixture was subjected to tests of the complex module using the 4PB-PR method according to [26] at a temperature of 10.0 °C. The mixes were then soaked at two-time intervals (2 days and 21 days). The changes in mass and the stiffness modulus were analyzed. The mass measurements of the samples were performed according to the procedure described in [27]. Figure 3 shows the test stand for testing the stiffness modulus.

Figure 3. Testing machine and first sample series.

The principles adopted in the research method were established on the basis of phenomenon occurring in the winter and basic technological practices used in Poland. The equivalent temperature of 10 °C was adopted for the design of pavement layers using catalogue methods, e.g., [28,29], and was used in the stiffness test of most mixes [26]. The applied percentage of 20% brine was a compromise between saturated and unsaturated solutions. In fact, in winter, there is never a situation where we get saturation or supersaturation of the solution. This would be the case with a thin layer of snow/ice and too frequent road salt with an anti-caking agent. The task of water pumps with a flow of 7500 L/h was to simulate the interaction of variable snow/ice–salt environmental conditions related to the interaction of vehicle wheels.

The last part of the research was to visualize the behavior of the material after soaking in brine on one selected mix and to show the salt concentration on the surface of the material. For this purpose, observations were made in computed tomography using X-ray radiation.

4. Results

4.1. Preliminary Research

In the field of preliminary examination (taking into account the variability of the concentration of the salt solution on the pavement), the change in water temperature with the change in salt concentration by 10% and 20% was analyzed.

Sample masses (m_s, g) were calculated on the basis of the solution concentration (C_p, %) and the mass of water (m_w, g) according to dependency (1):

$$m_s = \frac{C_p \cdot m_w}{100\% - C_p} \ [\text{g}] \tag{1}$$

The calculations and measured values are shown in Table 2, while the graph of monitored values is shown in Figure 4.

Table 2. Brine data and measured temperature variability.

		Brine			Temperature Shock	
Step	Temperature (°C)	Total Water Mass, m_w (g)	Percentage Concentration (Target Value), C_p (%)	Total Mass of Gritting Salt with Anti-caking Agent, m_s (g)	Time (s)	Temperature Noticed (°C)
1	10.0	96,534.9	10.0%	10,726.1	0	10.0
					60	9.5
					120	9.3
					180	9.2
					240	9.2
2	10.0	96,534.9	20.0%	24,133.7	480	9.3
					560	8.8
					640	8.6
					720	8.5
					800	8.5
					1040	8.5
					1200	8.6
					2400	8.6
					4800	8.9
					5400	8.9
					7200	9.3
					9000	9.5
					10,800	9.7
					12,600	10.0

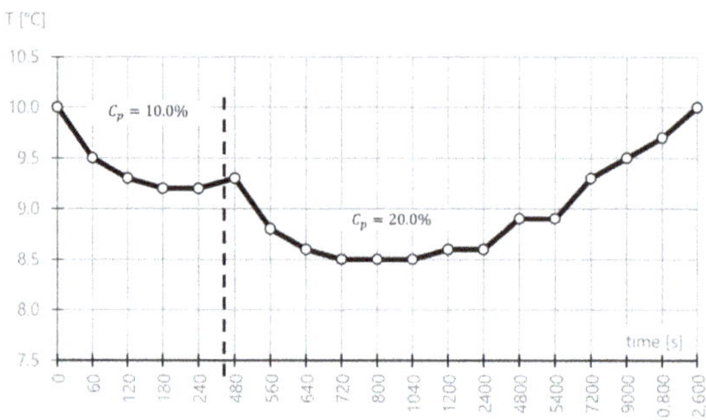

Figure 4. Temperature variability via adding road salt.

The measurements show that, at the peak moment of saturation (for Cp = 10%), the temperature dropped by about 1.5 °C. For an increased concentration of 20%, temperature stability took place for approximately 12,600 s. Thanks to these preliminary examinations, the period needed to soak the samples in brine in order to obtain the thermal state of equilibrium and to maintain the consistency of the results of all tested mixes.

It should be mentioned that each of the prepared mixtures was soaked first in water only and then in the brine solution. This procedure was to simulate the initial impact of wet snow/ice, and then the conditions after the salt was spilt on the pavement.

4.2. Results for AC11S

First, the AC11S mix containing 50/70 plain soft asphalt was analyzed. Before soaking, all samples were weighed and their initial stiffness was determined, and then the selected samples were soaked in water and brine. The results of the mass measurement related to absorption of the solution and the relative changes are presented in Table 3. Figure 5 shows the comprehensive results of the mass change for AC11S.

Table 3. Mass variability of AC11S via soaking in water and brine.

Specimen Index	Mass					Mass Change			
	Dry Specimen	Water Soaked		Brine Soaked		Water Soaked		Brine Soaked	
	Initial State	2 Days	21 Days	2 Days	21 Days	2 Days	21 Days	2 Days	21 Days
	(g)	(g)	(g)	(g)	(g)	(g)	(g)	(g)	(g)
1	3000.1	3006.1	3010.1			6.0	10.0		
2	2886.2	2892.3	2896.1			6.1	9.9		
3	2954.4	2960.5	2964.6			6.1	10.2		
4	2926.7	2934.9	2938.5			8.2	11.8		
5	2963.0			2974.3	2981.2			11.3	18.2
6	2855.8			2864.6	2871.1			8.8	15.3
7	2944.2			2953.2	2960.5			9.0	16.3
8	2879.3			2889.1	2896.1			9.8	16.8
9	2997.4			3007.7	3013.8			10.3	16.4
Average	2934.1	2948.5	2952.3	2937.8	2944.5	6.60	10.48	9.84	16.60
Standard deviation	48.5	41.2	41.4	53.3	53.2	0.9	0.8	0.9	0.9
The coefficient of variation (%)	1.65%	1.40%	1.40%	1.81%	1.81%	14.01%	7.38%	9.24%	5.66%

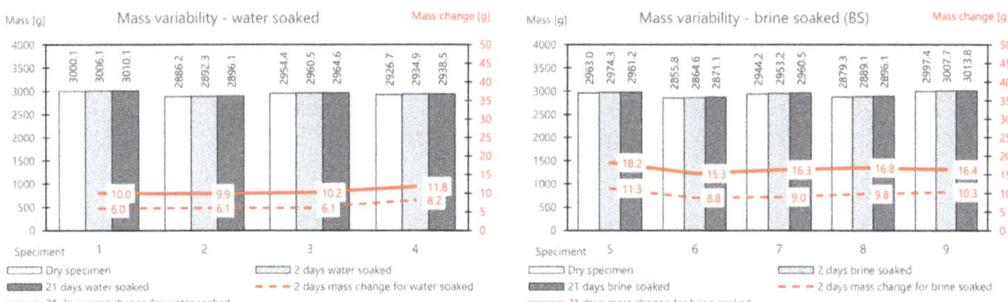

Figure 5. AC11S mix mass variability.

On the basis of the performed measurements, it should be concluded that the tested samples of the mixture are homogeneous—the dry mass variation index is less than 2%. Water absorption in the samples after two days of soaking is on average 6.6 g, while in brine, it is 9.84 g. The difference between the mass soaked in water and the mass soaked in brine (after 21 days) is 6.12 g. This is the average amount of salt deposited on the surface and in deeper areas of the samples. It should be noted that the salt enters the pores and reacts with the material, affecting its stiffness modulus. In a further part of this work, the change in the stiffness modulus for samples in a dry state after soaking in water and brine at two-time intervals (2 and 21 days) was analyzed. The results are presented in Table 4 and Figure 6.

Table 4. AC11S mix stiffness variation via water and brine soaking.

Specimen Index	Stiffness					Stiffness Change			
	Dry Specimen	Water Soaked		Brine Soaked		Water Soaked		Brine Soaked	
	Initial State	2 Days	21 Days	2 Days	21 Days	2 Days	21 Days	2 Days	21 Days
	(MPa)	(MPa)	(MPa)	(MPa)	(MPa)	(MPa)	(MPa)	(MPa)	(MPa)
1	11,270	10,987		10,604		283	666		
2	10,659	10,478		10,241		181	418		
3	10,962	10,682		10,365		280	597		
4	10,798	10,556		10,122		242	676		
5	11,120		10,113		8369			1007	2751
6	11,003		10,267		8255			736	2748
7	10,781		9807		7831			974	2950
8	10,638		9635		7837			1003	2801
9	11,001		10,148		8325			853	2676
Average	10,914.7	10,675.8	9994.0	10,333.0	8123.4	246.5	589.3	914.6	2785.2
Standard deviation	199.8	193.9	235.1	178.5	239.1	41.1	103.4	105.4	91.5
The coefficient of variation (%)	1.83%	1.82%	2.35%	1.73%	2.94%	16.68%	17.56%	11.53%	3.29%

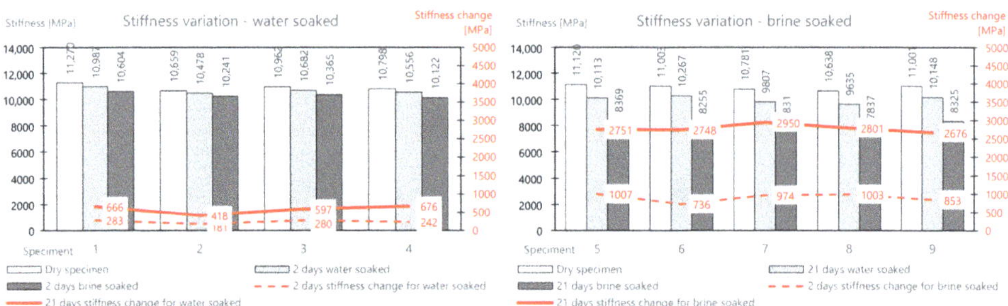

Figure 6. AC11S mix stiffness variation via water and brine soaking.

The tests of the dry batch showed that the samples had similar values of the stiffness modulus, on average 10,915 MPa. After soaking the samples in water, a slight decrease in the modulus was observed both after 2 days and 21 days. Ultimately, the stiffness modulus changed by an average of about 590 MPa after the full test period (which is about 5.4% of the initial value). The brine soaking compared to the water treatment showed a clear and significant change in the initial stiffness modulus. Its value dropped by 915 MPa (about 8.4%) after 2 days, and after 21 days it dropped by as much as 2785 MPa (about 26%) compared to the initial value.

4.3. Results for AC WMS 22 (Pmb 25/55-60)

Another tested mix was the high modulus AC 22 WMS mix made on the common modified asphalt Pmb 25/55-60. The material is commonly used for the bonding layer and for the framework subjected to heavy and very heavy movement.

As part of the experiment, nine prismatic specimens were also prepared. Then, the research material was selected—a part (four bars) was intended for soaking in water, and a part (five bars) was intended for brine at a fixed concentration. The developed results related to mass changes as a result of the interaction between water and salt are presented in Table 5 and presented in Figure 7.

Table 5. Mass variability of AC 22 WMS(Pmb) via soaking in water and brine.

Specimen Index	Mass					Mass Change			
	Dry Specimen	Water Soaked		Brine Soaked		Water Soaked		Brine Soaked	
	Initial State	2 Days	21 Days	2 Days	21 Days	2 Days	21 Days	2 Days	21 Days
	(g)	(g)	(g)	(g)	(g)	(g)	(g)	(g)	(g)
1	2911.2	2915.8	2919.5			4.6	8.3		
2	2910.2	2916.0	2920.3			5.8	10.1		
3	2907.6	2911.8	2914.8			4.2	7.2		
4	2983.6	2989.0	2993.4			5.4	9.8		
5	2919.4			2929.0	2936.9			9.6	17.5
6	2898.0			2907.8	2917.1			9.8	19.1
7	2928.6			2939.2	2948.9			10.6	20.3
8	2908.8			2919.6	2929.7			10.8	20.9
9	2969.4			2979.7	2988.7			10.3	19.3
Average	2926.3	2933.2	2937.0	2935.1	2944.3	5.00	8.85	10.22	19.42
Standard deviation	28.2	32.3	32.6	24.6	24.5	0.6	1.2	0.5	1.2
The coefficient of variation (%)	0.96%	1.10%	1.11%	0.84%	0.83%	12.65%	13.24%	4.48%	5.99%

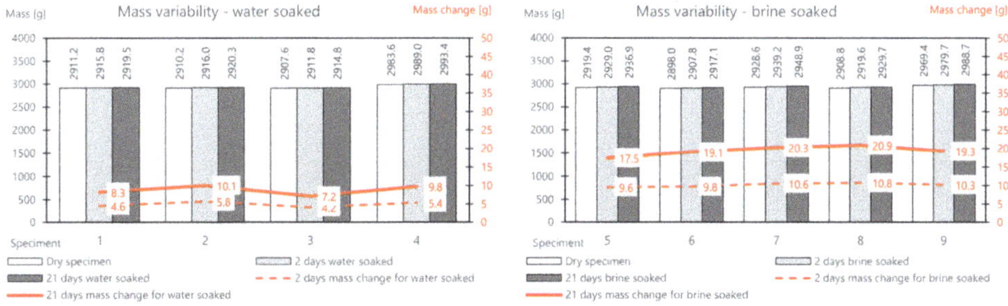

Figure 7. ACWMS22(Pmb) mix mass variability.

The mass change as a result of water absorption after two days of soaking is on average 5.0 g and that for brine is 10.22 g. The mass of the deposited salt after 2 days is 5.22 g, and after 21 days, it is 10.57 g. The change in the stiffness modulus is presented in Table 6 and Figure 8.

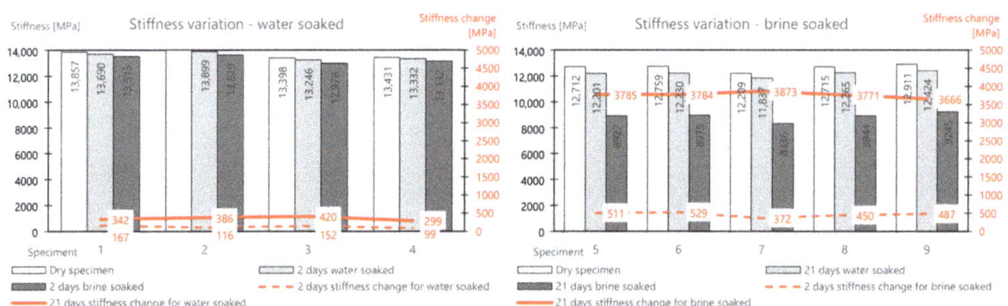

Figure 8. ACWMS22(Pmb) mix stiffness variation via water and brine soaking.

Table 6. AC22WMS(Pmb) mix stiffness variation via water and brine soaking.

Specimen Index	Stiffness					Stiffness Change			
	Dry Specimen	Water Soaked		Brine Soaked		Water Soaked		Brine Soaked	
	Initial State	2 Days	21 Days	2 Days	21 Days	2 Days	21 Days	2 Days	21 Days
	(MPa)	(MPa)	(MPa)	(MPa)	(MPa)	(MPa)	(MPa)	(MPa)	(MPa)
1	13,857	13,690		13,515		167	342		
2	14,015	13,899		13,629		116	386		
3	13,398	13,246		12,978		152	420		
4	13,431	13,332		13,132		99	299		
5	12,712		12,201		8927			511	3785
6	12,759		12,230		8975			529	3784
7	12,209		11,837		8336			372	3873
8	12,715		12,265		8944			450	3771
9	12,911		12,424		9245			487	3666
Average	13,111.9	13,541.8	12,191.4	13,313.5	8885.4	133.5	361.8	469.8	3775.8
Standard deviation	562.9	265.1	193.2	267.2	298.1	27.2	45.6	55.6	65.8
The coefficient of variation (%)	4.29%	1.96%	1.59%	2.01%	3.36%	20.38%	12.60%	11.83%	1.74%

As for the previous mix, after soaking in water only, a slight decrease in the stiffness modulus was observed both after 2 days (134 MPa) and 21 days (362 MPa). The mix, therefore, appears to be water-resistant. The brine-soaked treatment compared to the water treatment showed a clear and significant change in the stiffness modulus. Its value decreased by 470 MPa (about 3.7%) after 2 days, and after 21 days, it decreased by as much as 3776 MPa (about 30%) compared to the initial value.

4.4. Results for ACWMS 22 (Pmb 25/55-80 HIMA)

The AC 22 WMS mix was made of highly modified asphalt Pmb 25/55-80 highly modified asphalt (HIMA), which is characterized by both high fatigue strength and resistance to permanent deformation. The composition and raw material of the ACWMS 22 (HIMA) mineral mix are identical compared to the previously discussed ACWMS 22 (Pmb) mix. The only difference is the type of asphalt binder used. The developed results related to absorption of the solution after 2 and 21 days are presented in Table 7 and Figure 9.

Table 7. Mass variability of AC 22 WMS(HIMA) via soaking in water and brine.

Specimen Index	Mass					Mass Change			
	Dry Specimen	Water Soaked		Brine Soaked		Water Soaked		Brine Soaked	
	Initial State	2 Days	21 Days	2 Days	21 Days	2 Days	21 Days	2 Days	21 Days
	(g)	(g)	(g)	(g)	(g)	(g)	(g)	(g)	(g)
1	3068.8	3079.8	3088.7			11.0	19.9		
2	3040.8	3049.8	3056.9			9.0	16.1		
3	3044.2	3052.2	3058.4			8.0	14.2		
4	3013.0	3021.8	3028.8			8.8	15.8		
5	3031.4			3049.0	3063.3			17.6	31.9
6	3023.0			3037.2	3048.7			14.2	25.7
7	2998.8			3013.4	3025.2			14.6	26.4
8	3033.0			3049.9	3063.6			16.9	30.6
9	2998.4			3013.6	3025.3			15.2	26.9
Average	3027.9	3050.9	3058.2	3032.6	3045.2	9.20	16.53	15.70	28.30
Standard deviation	21.4	20.5	21.2	16.2	17.2	1.1	2.1	1.3	2.5
The coefficient of variation (%)	0.71%	0.67%	0.69%	0.54%	0.56%	12.01%	12.59%	8.43%	8.67%

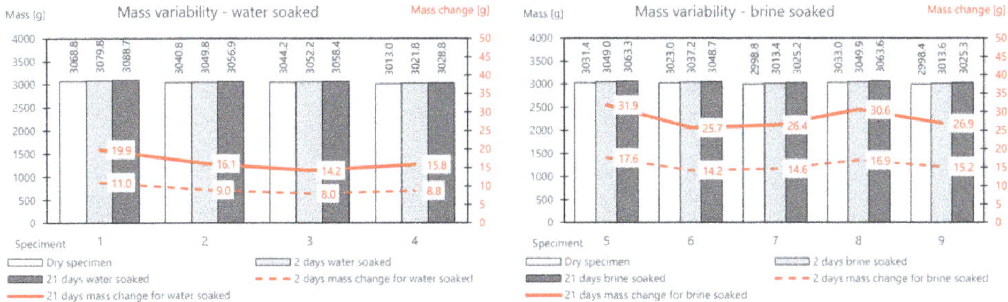

Figure 9. ACWMS22(HIMA) mix mass variability.

The mass change as a result of water absorption after two days of soaking is on average 9.2 g, and that for brine is 15.7 g. The mass of the deposited salt after 2 days is 6.5 g, and after 21 days, it is 11.8 g. This is a greater value than for the AC11S and ACWMS 22 mix (Pmb). The change in the stiffness modulus is presented in Table 8 and Figure 10.

Table 8. AC22WMS(HIMA) mix stiffness variation via water and brine soaking.

Specimen Index	Stiffness					Stiffness Change			
	Dry Specimen	Water Soaked		Brine Soaked		Water Soaked		Brine Soaked	
	Initial State	2 Days	21 Days	2 Days	21 Days	2 Days	21 Days	2 Days	21 Days
	(MPa)	(MPa)	(MPa)	(MPa)	(MPa)	(MPa)	(MPa)	(MPa)	(MPa)
1	10,875	10,782		10,612		93	263		
2	12,047	11,961		11,689		86	358		
3	11,885	11,821		11,521		64	364		
4	10,834	10,759		10,593		75	241		
5	11,365		10,897		9864			468	1501
6	11,340		10,985		9711			355	1629
7	11,508		11,051		9963			457	1545
8	11,253		10,851		9526			402	1727
9	12,233		11,817		10,654			416	1579
Average	11,482.2	11,330.8	11,120.2	11,103.8	9943.6	79.5	306.5	419.6	1596.2
Standard deviation	462.0	562.5	355.2	504.8	384.7	11.0	55.1	40.6	77.7
The coefficient of variation (%)	4.02%	4.96%	3.19%	4.55%	3.87%	13.85%	17.97%	9.67%	4.87%

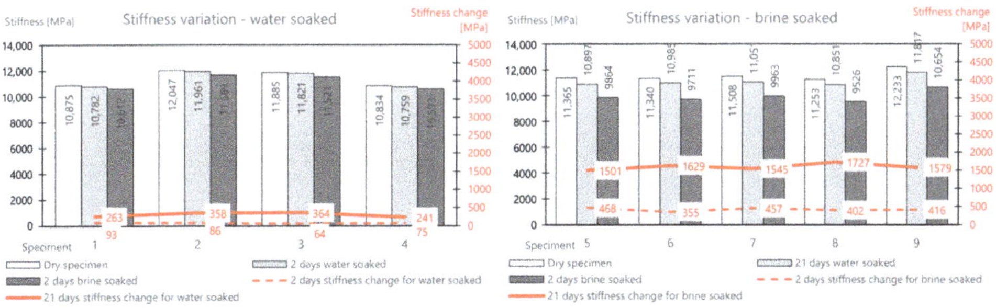

Figure 10. ACWMS22(HIMA) mix stiffness variation via water and brine soaking.

After soaking in water only, a slight decrease in stiffness modulus was also observed after both 2 days (80 MPa) and 21 days (307 MPa). Brine treatment compared to water treatment again showed a clear and significant change in the stiffness modulus. Its value dropped by 420 MPa (about 3.6%) after 2 days, and after 21 days, it dropped by 1596 MPa (about 14%) compared to the initial value.

4.5. Results for MCAS

The mineral cement mix with foamed asphalt (MCAS) mixture, intended mainly for base course layers, showed large changes in the mass of samples soaked in water and brine (Table 9 and Figure 11).

Table 9. Mass variability of MCAS via soaking in water and brine.

Specimen Index	Mass					Mass Change			
	Dry Specimen	Water Soaked		Brine Soaked		Water Soaked		Brine Soaked	
	Initial State	2 Days	21 Days	2 Days	21 Days	2 Days	21 Days	2 Days	21 Days
	(g)	(g)	(g)	(g)	(g)	(g)	(g)	(g)	(g)
1	2366.2	2415.4	2455.3			49.2	89.1		
2	2458.6	2506.0	2543.4			47.4	84.8		
3	2353.4	2402.0	2439.9			48.6	86.5		
4	2435.2	2478.0	2512.2			42.8	77.0		
5	2427.2			2587.0	2716.4			159.8	289.2
6	2546.8			2671.2	2772.0			124.4	225.2
7	2470.4			2608.0	2719.5			137.6	249.1
8	2543.8			2666.4	2765.7			122.6	221.9
9	2399.1			2534.2	2638.2			135.1	239.1
Average	2444.5	2450.4	2487.7	2613.4	2722.4	47.00	84.36	135.90	244.90
Standard deviation	65.0	43.1	42.0	51.3	47.9	2.5	4.5	13.3	24.2
The coefficient of variation (%)	2.66%	1.76%	1.69%	1.96%	1.76%	5.34%	5.32%	9.78%	9.89%

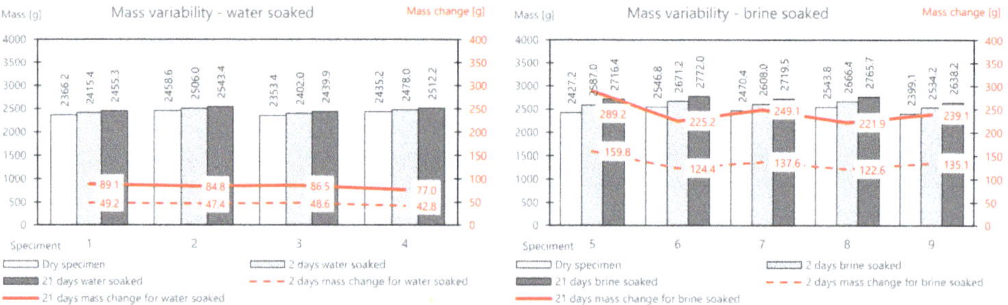

Figure 11. MCAS mix mass variability.

The mass change as a result of water absorption after two days of soaking is on average 47 g, and that for brine is 135.9 g. The mass of the deposited salt after 2 days is 88.9 g, and after 21 days, it is 160.54 g. There is a significant influence from the different structures of this mix related to greater porosity. In the following part, the change in the stiffness modulus is analyzed. The results are presented in Table 10 and Figure 12.

Table 10. MCAS mix stiffness variation via water and brine soaking.

Specimen Index	Stiffness					Stiffness Change			
	Dry Specimen	Water Soaked		Brine Soaked		Water Soaked		Brine Soaked	
	Initial State	2 Days	21 Days	2 Days	21 Days	2 Days	21 Days	2 Days	21 Days
	(MPa)	(MPa)	(MPa)	(MPa)	(MPa)	(MPa)	(MPa)	(MPa)	(MPa)
1	5312	5098		4469		214	843		
2	5417	5236		4782		181	635		
3	4999	4754		4275		245	724		
4	5799	5568		4976		231	823		
5	4860		4321		2250			539	2610
6	5054		4596		2139			458	2915
7	5048		4612		2085			436	2963
8	5801		5245		2977			556	2824
9	5103		4659		2514			444	2589
Average	5265.9	5164.0	4686.6	4625.5	2393.0	217.8	756.3	486.6	2780.2
Standard deviation	325.0	291.9	303.3	271.4	327.3	23.9	83.3	50.5	154.3
The coefficient of variation (%)	6.17%	5.65%	6.47%	5.87%	13.68%	10.97%	11.01%	10.38%	5.55%

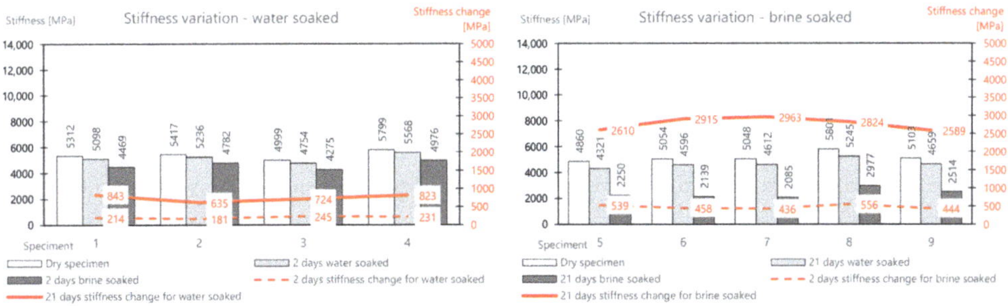

Figure 12. MCAS mix stiffness variation via water and brine soaking.

After soaking in water alone, a greater decrease (than for the previous mixes) of the modulus is also observed after both 2 days (218 MPa) and 21 days (756 MPa). The brine treatment compared to the water treatment again showed a clear and significant change in the stiffness modulus. Its value dropped by 487 MPa (about 9.4%) after 2 days, and after 21 days, it dropped by 2780 MPa (about 54%) compared to the initial value.

4.6. Results for MCE

The mineral-cement-emulsion (MCE) mix is also intended for base course layers and shows large changes in the mass of the samples soaked in water and brine (Table 11 and Figure 13).

The mass change due to the absorption of water after two days of soaking is on average 20 g, and that for brine is 26.72 g (about half as much as for MCAS). The mass of the deposited salt after 2 days is 6.92 g, and after 21 days, it is 12.6 g. A slightly smaller influence of the less porous structure of the mix is visible.

The change in the stiffness modulus was further analyzed. The results are presented in Table 12 and Figure 14.

Table 11. Mass variability of MCE via soaking in water and brine.

Specimen Index	Mass					Mass Change			
	Dry Specimen	Water Soaked		Brine Soaked		Water Soaked		Brine Soaked	
	Initial State	2 Days	21 Days	2 Days	21 Days	2 Days	21 Days	2 Days	21 Days
	(g)	(g)	(g)	(g)	(g)	(g)	(g)	(g)	(g)
1	2486.4	2504.0	2518.3			17.6	31.9		
2	2450.6	2469.2	2483.9			18.6	33.3		
3	2621.0	2643.2	2660.5			22.2	39.5		
4	2511.6	2532.4	2549.0			20.8	37.4		
5	2528.4			2556.8	2579.8			28.4	51.4
6	2497.4			2523.8	2545.2			26.4	47.8
7	2620.2			2643.4	2662.2			23.2	42.0
8	2637.6			2664.4	2686.1			26.8	48.5
9	2498.8			2527.6	2549.8			28.8	51.0
Average	2539.1	2537.2	2552.9	2583.2	2604.6	19.80	35.53	26.72	48.13
Standard deviation	64.8	65.2	66.3	59.2	58.5	1.8	3.1	2.0	3.4
The coefficient of variation (%)	2.55%	2.57%	2.60%	2.29%	2.25%	9.12%	8.68%	7.42%	7.00%

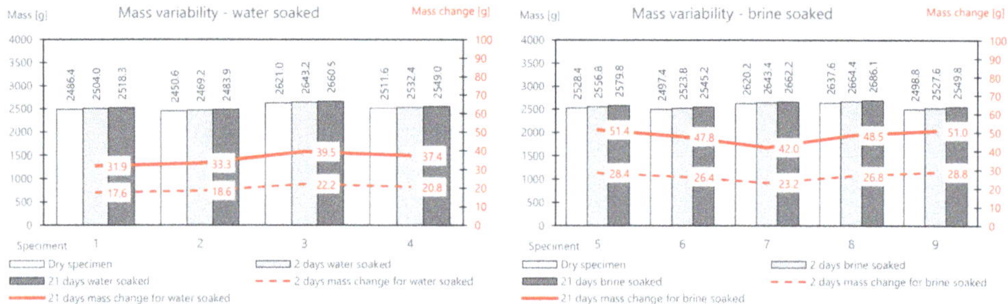

Figure 13. MCE mix mass variability.

Table 12. MCE mix stiffness variation via water and brine soaking.

Specimen Index	Stiffness					Stiffness Change			
	Dry Specimen	Water Soaked		Brine Soaked		Water Soaked		Brine Soaked	
	Initial State	2 Days	21 Days	2 Days	21 Days	2 Days	21 Days	2 Days	21 Days
	(MPa)	(MPa)	(MPa)	(MPa)	(MPa)	(MPa)	(MPa)	(MPa)	(MPa)
1	4301	4150		3899		151	402		
2	4018	3869		3587		149	431		
3	4258	4132		3854		126	404		
4	4328	4171		3836		157	492		
5	4125		3774		2597			351	1529
6	3986		3620		2509			366	1477
7	4424		3901		2594			523	1831
8	3899		3348		2171			551	1729
9	4222		3814		2694			408	1528
Average	4173.4	4080.5	3691.4	3794.0	2512.7	145.8	432.3	439.8	1618.5
Standard deviation	166.5	122.9	194.3	121.7	180.9	11.8	36.3	82.0	136.6
The coefficient of variation (%)	3.99%	3.01%	5.26%	3.21%	7.20%	8.08%	8.41%	18.65%	8.44%

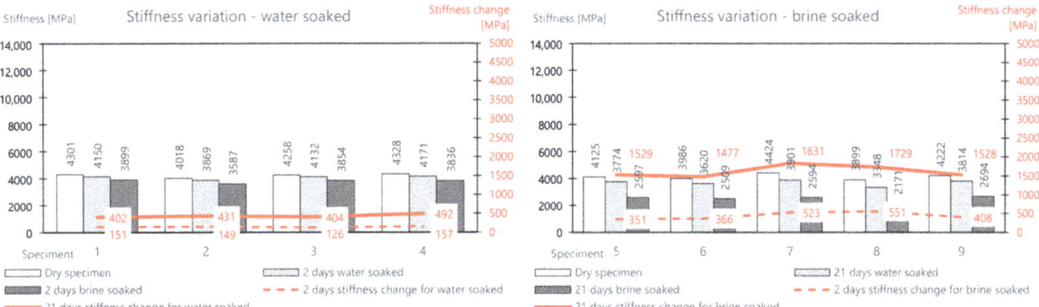

Figure 14. MCE mix stiffness variation via water and brine soaking.

After soaking in water alone, about half the decrease (compared to MCAS) in the module was observed after both 2 days (146 MPa) and 21 days (432 MPa). Brine soaked maintenance, compared to water, showed a change in the stiffness modulus by 440 MPa (about 11%) after 2 days, and after 21 days, it decreased by 1619 MPa (about 39%) compared to the initial value.

It can therefore be concluded that road salt with an anti-caking agent significantly weakened the tested mixes by lowering the stiffness modulus. This observation proves that the action of water and brine leads to uniform destruction of the material. It depends on the composition of the mixtures and their structure. Later in this article, the results for all tested mixtures were collected and compared.

4.7. Tomography Imaging Salt in a Selected Mix

The results of the research, and in particular the mass changes, confirm that salt has a destructive effect and penetrates the structure of the mixes. As part of a further part of the analysis, imaging using computed tomography was performed to analyze the depth at which salt is deposited and how it reacts with it while reducing its strength parameters.

Computed tomography scanning was performed on a sample of the AC11S mix that was soaked in saline for 21 days. After data preparation and data acquisition, two key shots of the x-rayed material were taken—shown in 3 planes at the very edge of the material and shown in 3 planes passing through the center of the material. Analyzing the results, 4 color ranges were selected, which directly define the variable structure of the x-rayed material. Orange is the color of the mineral mix, white is the asphalt binder, and purple is the material degraded by road salt with anti-caking agent. Black means air voids. The test machine used together with the sample is presented in Figure 15, while the views discussed are presented in Figures 16 and 17, which show, respectively, the side, front, and top sections of the exposed sample.

The main salt clusters are on the surface of the X-rayed element at a maximum depth of 2.7 mm. The volumetric analysis with the use of tomographic imaging showed that, for the AC11S mix, about 5.51% of the material was degraded by the effect of salt. It is worth recalling that, for this mix, after 21 days of soaking in brine, about 6.12 g of salt deposited on the surface and in deeper areas of the sample. At a later stage of this work, the authors will conduct research for other mixtures in order to determine the correlation between mass changes and the volume of material destruction.

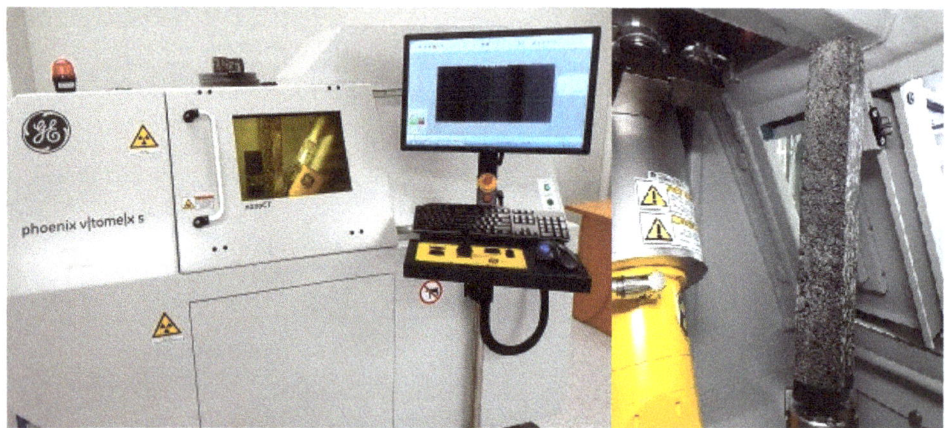

Figure 15. Computed tomography and mounted specimen.

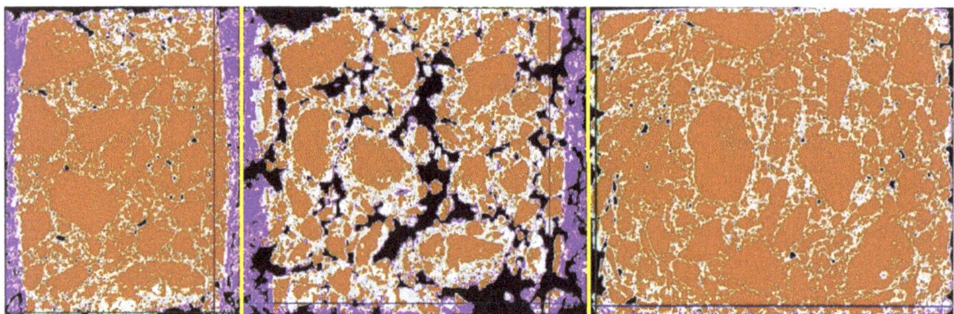

Figure 16. Edge specimen view.

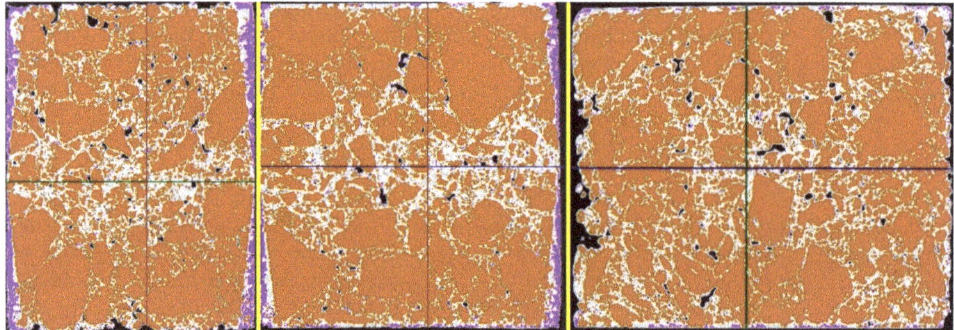

Figure 17. Middle specimen view.

5. Discussion

In terms of the impact of changes in the interaction of road salt with anti-caking agent and water, five different road mixes used in the pavement were analyzed.

High Modulus Asphalt Concrete (HMAC) mixes were used in the article contain polymer modified asphalt—Pmb 25/55-60 and 25/55-80 (HIMA). Due to modification in the asphalt matrix, which contains different amounts of SBS polymer, the required stiffness modulus for typical HMAC mixes (with common asphalt like 20/30) used in the binder course—14,000 MPa—or base course—11,000 MPa—is difficult to reach or to be attained. It is caused by the internal properties of the binder that lead the material to being more flexible. Nowadays, it is a general problem for producers, distributors, and scientists because the admission of Pmb and HIMA binder's usage into HMAC mixes creates a problem related to the required value of stiffness modulus in some regulations [30]. In this article, the stiffness modulus value for AC22WMS with common Pmb 25/55-60 samples reaches or is close to 14,000 MPa and that for AC22WMS with HIMA asphalt (25/55-80) is close to 12,000 MPa.

In this order (before the mixtures were tested), it was examined how road salt affected the temperature change. Water as a liquid is a set of water molecules that are mutually arranged in relation to each other and are connected with each other by hydrogen bonds. A chemical reaction takes place when salt is added to the water. The hydrogen bonds are broken, and the water molecule is arranged with appropriate charges against the dissolved sodium Na+ cation and the Cl− chloride anion. The observations from the experiment confirm (Figure 4) that, after adding road salt (e.g., sprinkling an iced surface with salt), the system needs external energy for the entire dissolution process to take place—an endothermic reaction takes place (the reaction enthalpy changes). When heat is "taken" from the existing state—the temperature drops (for road materials, this situation introduces a risky shrinkage that activates the cracking process occurring as a result of energy changes in the system). The effect of lowering the temperature (using heat for the reaction) is to lower the freezing point of water. However, it should be remembered that the entire process is temporary—after dissolving the salt, the temperature of the system stabilizes.

High reproducibility of the results was found, obtaining similar results for samples of the same material. Figure 18 shows the percentage change in the value of the stiffness modulus for the tested mixtures under various conditions (soaking in water and salt after 2 and 21 days). It was found that the greatest changes occur for the base course mixtures: MCAS and MCE, which after 21 days reach 54% and 39%, respectively. The mixture of the high stiffness modulus with the highly modified asphalt AC22WMS (HIMA) showed the lowest sensitivity to salt. It should be noted that the effect of water itself also has a destructive effect on the mixtures and reduces the modulus value from 2% to 14%.

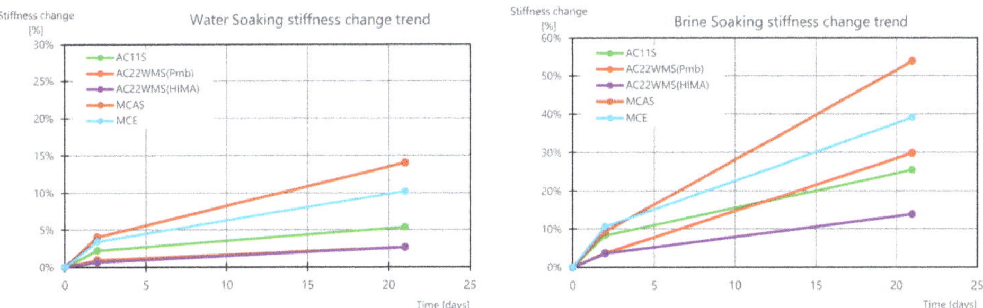

Figure 18. Results of the percentage changes in the stiffness modulus for different mixes over time.

Figure 19 shows the absolute stiffness modulus value changes. The largest change in the value of the module was found for the 21st day of brine soaking. It is about three to four times greater than for two days of soaking in water. The greatest change was achieved for the AC22WMS (Pmb), AC11S, and MCAS mixes.

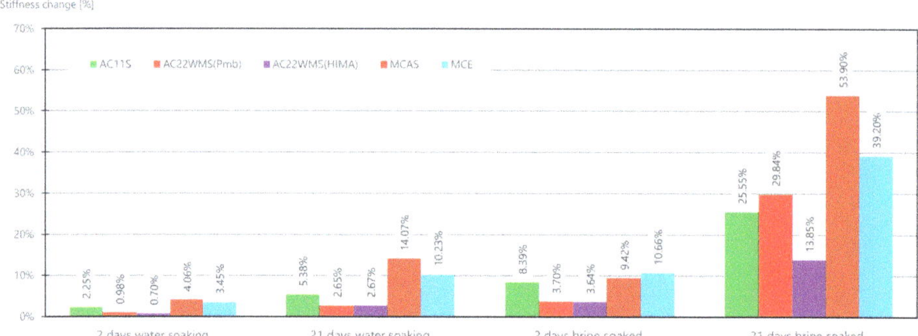

Figure 19. Stiffness modulus change results for various mixes.

By analyzing the absorption of water and brine for various mixes, it can be concluded that the mixes of MCAS and MCE show high absorption of water and salt solutions. Their mass changes are 9.9% and 1.9%, respectively. Although they are intended for the base course layers, in the absence of full adhesion of the layers and significant surface damage to the upper layers (cracks, potholes), they will not be resistant to salt treatments in winter maintenance.

The other mixes are also characterized by a lower degree of absorption (up to 1%), but they constitute the upper layers of the pavement and the degradation process for them will be equally intense and dangerous.

The effect of water and salt on mixtures is reflected in the reduction in the value of the stiffness modulus; therefore, for soaked mixtures, correlations between modulus change and mass change as a result of soaking phenomenon were developed (Figure 20).

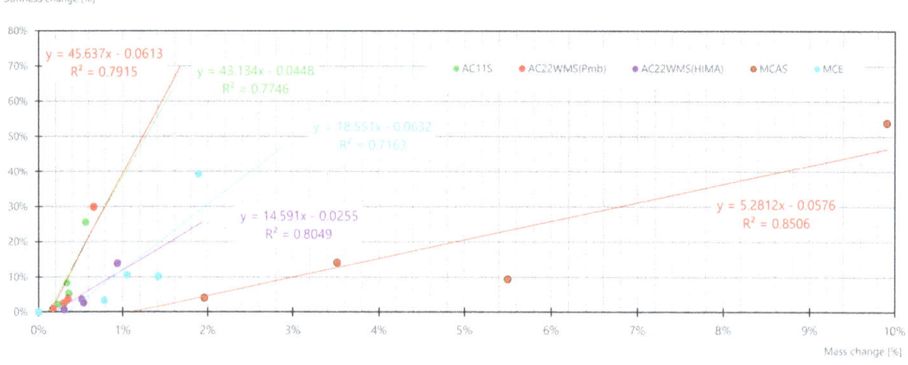

Figure 20. Correlations between modulus change and mass change as a result of the soaking phenomenon for various mixes.

On the basis of the obtained dependencies, it is possible to estimate the change in the value of the stiffness modulus on the basis of the mass change associated with the absorption of the salt solution. It should be noted that mixes AC11S and AC22WMS (Pmb) have similar dependencies.

6. Conclusions

Our analyses showed that brine has a fundamental impact on changes in the mechanical properties of road materials. It causes an increase in mass and a change in the stiffness modulus. We found that a chemical reaction that occurs when road salt is added to water causes a temporary temperature fall. Probably the shrinkage during this reaction could accelerate material degradation (including aging) or could modify material rheological properties. However, it should be confirmed by other separate research.

The image recorded in tomography allows for indicating the extent of the impact of the salt solution and for determining the area of material degradation. The X-ray method can be effective in assessing the quality and, consequently, the durability of materials exposed to the effects of road salts. Changes in the structure are directly related to changes in the stiffness modulus. Salt depositing on the surface and reacting with the asphalt weakens the original material, which can cause reduced fatigue life.

We found that the greatest changes occurred for mixes intended for the base layers: MCAS and MCE. Modules changed after 21 days to 54% and 39%, respectively. Among the five tested mixes, the mix characterized by high stiffness modulus with highly modified asphalt AC22WMS (HIMA) showed the lowest sensitivity to salt. Therefore, it is necessary to recognize the properties of highly modified asphalt, thanks to which the WMS mix is able to withstand the salt solution well.

The determined correlations between mass change and changes in the value of the stiffness modulus may be useful in estimating the sensitivity of mixes to the application of winter maintenance treatments with road salt. Water with salt will penetrate the structure of mixtures intended for upper layers and substructures as well as in damaged and strongly cracked wear layers. It can cause a change in the value of modules and deterioration of durability. It is therefore important to pay attention to the tightness and state of damage of the pavement layers before using large amounts of salt in winter maintenance of road surfaces. For salt concentrations of 10–20%, significant changes in asphalt mixes and, consequently, in the pavement durability should be expected. Fatigue analyses will be the subject of further work by the authors.

Author Contributions: Conceptualization, E.M.; methodology, E.M.; software, E.M. and P.M.; validation, P.M. and E.M.; formal analysis, P.M. and E.M.; investigation, E.M.; resources, E.M.; data curation, P.M; writing—original draft preparation, E.M. and P.M.; writing—review and editing, E.M. and P.M; visualization, P.M. and E.M.; supervision, P.M.; project administration, E.M.; funding acquisition, P.M. All authors have read and agreed to the published version of the manuscript.

Funding: This work was supported by the National Center for Research and Development and the General Directorate for National Roads and Motorways in Poland (the grant Use of materials derived from recycling—Development of Road Innovation (RID)).

Institutional Review Board Statement: Not applicable.

Informed Consent Statement: Not applicable.

Data Availability Statement: Not applicable.

Conflicts of Interest: The authors declare no conflict of interest.

References

1. Godwin, K.S.; Hafner, S.D.; Buff, M.F. Long-term trends in sodium and chloride in the Mohawk River, New York: The effect of fifty years of road-salt application. *Environ. Pollut.* **2003**, *124*, 273–281. [CrossRef]
2. Kelly, V.R.; Lovett, G.M.; Weathers, K.C.; Findlay, S.E.G.; Strayer, D.L.; Burns, D.I.; Likens, G.E. Long-term sodium chloride retention in a rural watershed: Legacy effects of road salt on streamwater concentration. *Environ. Sci. Technol.* **2008**, *42*, 410–415. [CrossRef] [PubMed]
3. Government of Canada. *Priority Substances List Assessment Report—Road Salts*; Environment Canada: Ottawa, AN, Canada, 2001; ISBN 9780662310181.
4. *Guidelines for the Selection of Snow and Ice Control Materials to Mitigate Environmental Impacts*; Transportation Research Board: Washington, DC, USA, 2007; ISBN 978-0-309-42233-8.
5. Kelly, V.R.; Findaly, S.E.G.; Schlesinger, W.H.; Menking, K.; Morrill Chatrchyan, A. Road Salt—Moving Toward the Solution. 2010. Available online: https://www.caryinstitute.org/sites/default/files/public/reprints/report_road_salt_2010.pdf (accessed on 14 January 2021).
6. Industrial experimental department of road and bridge engineering Sp. z o.o. General Technical Specification D—10.10.01c Combating Winter Slippy on the Road (Polish), Warsaw. 2004. Available online: https://www.bip.powiattorunski.pl/plik,43153,ost-pdf.pdf (accessed on 28 January 2021).

7. Rodrigues, P.M.; Rodrigues, R.M.; Costa, B.H.; Martins, A.A.T.; Da Esteves Silva, J.C. Multivariate analysis of the water quality variation in the Serra da Estrela (Portugal) Natural Park as a consequence of road deicing with salt. *Chemom. Intell. Lab. Syst.* **2010**, *102*, 130–135. [CrossRef]

8. Trost, S.E.; Heng, F.J.; Cussler, E.L. Chemistry of Deicing Roads: Breaking the Bond between Ice and Road. *J. Transp. Eng.* **1987**, *113*, 15–26. [CrossRef]

9. Hassan, Y.; Abd El Halim, A.O.; Razaqpur, A.G.; Bekheet, W.; Farha, M.H. Effects of Runway Deicers on Pavement Materials and Mixes: Comparison with Road Salt. *J. Transp. Eng.* **2002**, *128*, 385–391. [CrossRef]

10. Kosior-Kazberuk, M.; Jezierski, W. Surface scaling resistance of concrete modified with bituminous addition. *J. Civ. Eng. Manag.* **2004**, *10*, 25–30. [CrossRef]

11. Hugenschmidt, J.; Loser, R. Detection of chlorides and moisture in concrete structures with ground penetrating radar. *Mater. Struct.* **2008**, *41*, 785–792. [CrossRef]

12. Özgan, E.; Serin, S. Investigation of certain engineering characteristics of asphalt concrete exposed to freeze–thaw cycles. *Cold Reg. Sci. Technol.* **2013**, *85*, 131–136. [CrossRef]

13. Amini, B.; Tehrani, S.S. Simultaneous effects of salted water and water flow on asphalt concrete pavement deterioration under freeze–thaw cycles. *Int. J. Pavement Eng.* **2014**, *15*, 383–391. [CrossRef]

14. Liu, Y.H.; Zhang, H.; Wang, X.L.; Li, L. The Weakening Effect of the Snow-Melting Agent on the Performance of Municipal Asphalt Pavement in the Severe Cold Region. *AMR* **2014**, *953–954*, 1604–1608. [CrossRef]

15. Pavuluri, S. *Kinetic Approach for Modeling Salt Precipitation in Porous-Media*; Independent Study; Universitat Stuttgart: Stuttgart, Germany, 2014.

16. Hossain, S.K.; Fu, L.; Hosseini, F.; Muresan, M.; Donnelly, T.; Kabir, S. Optimum winter road maintenance: Effect of pavement types on snow melting performance of road salts. *Can. J. Civ. Eng.* **2016**, *43*, 802–811. [CrossRef]

17. Tsang, C.; Shehata, M.H.; Lotfy, A. Optimizing a Test Method to Evaluate Resistance of Pervious Concrete to Cycles of Freezing and Thawing in the Presence of Different Deicing Salts. *Materials* **2016**, *9*, 878. [CrossRef] [PubMed]

18. Almasi, S.A.; Khabir, M.M. Experimental evaluation of calcium chloride powder effect on the reduction of the pavement surface layer performance. *CEJ* **2019**, *28*, 61–72. [CrossRef]

19. Saltanovs, R.; Rubenis, A.; Krainyukov, A. Influence of Constructive Materials of Road Cover on Magnetic Field Dispersion of Wireless Power Transmission Systems. In *Reliability and Statistics in Transportation and Communication*; Kabashkin, I., Yatskiv, I., Prentkovskis, O., Eds.; Springer International Publishing: Cham, Switzerland, 2019; pp. 214–223. ISBN 978-3-030-12449-6.

20. Juli-Gándara, L.; Vega-Zamanillo, Á.; Calzada-Pérez, M.Á.; Teijón-López-Zuazo, E. Effect of Sodium Chloride on the Modulus and Fatigue Life of Bituminous Mixtures. *Materials* **2020**, *13*, 2126. [CrossRef] [PubMed]

21. Vega-Zamanillo, Á.; Juli-Gándara, L.; Calzada-Pérez, M.Á.; Teijón-López-Zuazo, E. Impact of Temperature Changes and Freeze—Thaw Cycles on the Behaviour of Asphalt Concrete Submerged in Water with Sodium Chloride. *Appl. Sci.* **2020**, *10*, 1241. [CrossRef]

22. Skotnicki, L.; Kuzniewski, J.; Szydlo, A. Stiffness Identification of Foamed Asphalt Mixtures with Cement, Evaluated in Laboratory and In Situ in Road Pavements. *Materials* **2020**, *13*, 1128. [CrossRef] [PubMed]

23. OWSIAK, Z.; CZAPIK, P.; Zapała-Sławeta, J. Testing the cement, hydrated lime and cement by-pass dust mixtures hydration. *Roads Bridges—Drog. Mosty* **2020**, 135–147. [CrossRef]

24. OWSIAK, Z.; CZAPIK, P.; Zapała-Sławeta, J. Properties of a Three-Component Mineral Road Binder for Deep-Cold Recycling Technology. *Materials* **2020**, *13*, 3585. [CrossRef]

25. Skotnicki, Ł.; Kuźniewski, J.; Szydło, A. Research on the Properties of Mineral–Cement Emulsion Mixtures Using Recycled Road Pavement Materials. *Materials* **2021**, *14*, 563. [CrossRef] [PubMed]

26. *PKN/KT 212. Bituminous Mixtures—Test Methods—Part 26: Stiffness: PN-EN 12697-26:2018-08*; PKN: Warsaw, Poland, 2018.

27. *PKN/KT 212. Bituminous Mixtures—Test Methods—Part 6: Determination of Bulk Density of Bituminous Specimens: PN-EN 12697-6:2020-07*; PKN: Warsaw, Poland, 2020.

28. Judycki, J.; Jaskula, P.; Alenowicz, J.; Dołżycki, B.; Jaczewski, M.; Ryś, D.; Stienss, M. *Catalog of Typical Pavement Constructions for Flexible and Semi-Rigid Pavement*; GDDKiA: Warsaw, Poland, 2014. (In Polish)

29. Szydło, A.; Mackiewicz, P.; Wardęga, R.; Krawczyk, B. *Catalog of Typical Rigid Pavement Structures*; GDDKiA: Warsaw, Poland, 2014. (In Polish)

30. Błażejowski, K.; Ostrowski, P.; Wójcik-Wiśniewska, M.; Baranowska, W. Mixes and Pavements with Orbiton HIMA Asphalts (Polish), Płock, 2020. Available online: https://www.orlen-asfalt.pl/PL/InformacjeTechniczne/Strony/Nasze-Publikacje.aspx (accessed on 1 May 2020).

 *materials* 

Article

Analysis of Acoustic Emission Signals Recorded during Freeze-Thaw Cycling of Concrete

Libor Topolář *, Dalibor Kocáb, Luboš Pazdera and Tomáš Vymazal

Faculty of Civil Engineering, Brno University of Technology, Veveří 95, 602 00 Brno, Czech Republic;
Dalibor.Kocab@vutbr.cz (D.K.); pazdera.l@fce.vutbr.cz (L.P.); Tomas.Vymazal@vutbr.cz (T.V.)
* Correspondence: Libor.Topolar@vutbr.cz; Tel.: +420-541-147-664

Abstract: This manuscript deals with a complex analysis of acoustic emission signals that were recorded during freeze-thaw cycles in test specimens produced from air-entrained concrete. An assessment of the resistance of concrete to the effects of freezing and thawing was conducted on the basis of a signal analysis. Since the experiment simulated testing of concrete in a structure, a concrete block with the height of 2.4 m and width of 1.8 m was produced to represent a real structure. When the age of the concrete was two months, samples were obtained from the block by core drilling and were subsequently used to produce test specimens. Testing of freeze-thaw resistance of concrete employed both destructive and non-destructive methods including the measurement of acoustic emission, which took place directly during the freeze-thaw cycles. The recorded acoustic emission signals were then meticulously analysed. The aim of the conducted experiments was to verify whether measurement using the acoustic emission method during Freeze-thaw (F-T) cycles are more sensitive to the degree of damage of concrete than the more commonly employed construction testing methods. The results clearly demonstrate that the acoustic emission method can reveal changes (e.g., minor cracks) in the internal structure of concrete, unlike other commonly used methods. The analysis of the acoustic emission signals using a fast Fourier transform revealed a significant shift of the dominant frequency towards lower values when the concrete was subjected to freeze-thaw cycling.

Keywords: acoustic emission method; freeze-thaw cycles; concrete; signal analysis; short-time Fourier transform; fast Fourier transform

Citation: Topolář, L.; Kocáb, D.; Pazdera, L.; Vymazal, T. Analysis of Acoustic Emission Signals Recorded during Freeze-Thaw Cycling of Concrete. *Materials* **2021**, *14*, 1230. https://doi.org/10.3390/ma14051230

Academic Editor: Krzysztof Schabowicz

Received: 29 January 2021
Accepted: 3 March 2021
Published: 5 March 2021

Publisher's Note: MDPI stays neutral with regard to jurisdictional claims in published maps and institutional affiliations.

Copyright: © 2021 by the authors. Licensee MDPI, Basel, Switzerland. This article is an open access article distributed under the terms and conditions of the Creative Commons Attribution (CC BY) license (https://creativecommons.org/licenses/by/4.0/).

1. Introduction

Concrete has been used in construction practically since the discovery of cement. During their service life, concrete elements are exposed to various degradation factors, such as mechanical and chemical influences, rapid temperature changes, etc. The knowledge of the influence of temperature changes on the quality of concrete during its service life is crucial both from the scientific point of view and in particular from the point of view of practical application in construction. In addition to the degrading impacts of very high temperatures, the alternation of positive and negative temperatures is one of the most destructive operating factors for a number of concrete products [1,2]. Freeze-thaw (F-T) cycles may adversely, as well as very quickly affect the durability of concrete structures. From a research point of view, it is appropriate to monitor the behaviour of concrete already during the actual exposure to freeze-thaw cycles [3–5].

In the case of concrete, which is in contact with water (examples of such structures include railway sleepers, road panels, water tanks, etc.), capillarity causes water to enter the pore structure of the concrete. At negative ambient temperatures, water begins to turn into crystalline ice and then to ice, which has a larger volume than water by about 9%. This presence of water inside the concrete structure exerts pressures that can significantly and irreversibly damage the concrete [6].

The assessment of the resistance of concrete to freeze-thaw cycles is typically based only on monitoring of certain mechanical properties, such as strengths, moduli of elasticity, etc. These properties are obtained after a certain number of F-T cycles using destructive or non-destructive methods. Subsequently, the observed changes in the properties (often a decrease) are evaluated with regard to the values obtained before freezing [7]. In the case of destructive tests, the sample is therefore not monitored continuously, but the behaviour of concrete is assessed on the basis of the results of individual test specimens. Despite the fact that the test specimens comprise one set of identical specimens produced from the same concrete, this can lead to inaccurate conclusions. This is due to the fact that the estimation of concrete behaviour is based on a statistical interpretation of the results of similar, but not exactly identical, specimens. When using non-destructive testing methods [8–11], monitoring of the condition of a concrete sample is conducted on the same specimens—the change of the monitored property is observed on individual specimens during the test at discrete time intervals (always after a certain number of completed F-T cycles). This process exhibits a lower degree of error than destructive testing methods. However, it is ideal to monitor the test specimens continuously even during freeze-thaw cycles, i.e., throughout the entire test. The acoustic emission method allows such an approach. The analysis of the recorded acoustic emission signals can be successfully used for a more detailed evaluation of the behaviour of the tested materials [12,13].

Acoustic emission (AE) is a real-time non-destructive testing method which can be employed to monitor the formation of cracks in concrete [14]. AE signals correspond to sound waves that emerge during the formation of cracks in a material. When a crack forms in concrete, energy is released and part of this energy is scattered in the form of an acoustic wave [15,16]. Conversely, acoustic waves and the corresponding energy are released when material damage occurs due to freeze-thaw cycles [4,17,18]. The AE signal parameters (e.g., number of counts, amplitude, frequency, etc.) can provide an effective tool to determine the degree of material damage during F-T cycles [19]. In general, the power of AE signals and their parameters depend on the amount of the released energy, the source, the distance and orientation of the source in relation to the location of the AE sensors [20]. Testing using AE relies on AE sensors continuously recording AE signals that are generated by the formation of damage in a material during its loading (e.g., F-T cycles). In cement-based composite materials, the source of AE activities can be found either in the cement binder or in the interfacial transition zone (ITZ) [20].

The aim of this manuscript is to compare the behaviour of test specimens of different sizes (produced from core samples drilled from a concrete block) during freezing and thawing. The interpretation of the frost resistance of concrete will be performed using the outputs of the acoustic emission method, resp. by analysing AE signals generated during 100 F-T cycles. For comparison, results of traditional non-destructive and destructive tests performed after every twenty-fifth F-T cycle will also be presented.

Another main goal of the described experiment was to determine a way to record the first changes to the quality of concrete, or the first signs of damage to the internal structure of concrete during exposure to F-T cycles. For this reason, air-entrained concrete was selected for the experiment since it was expected to exhibit a lesser degree of damage when exposed to F-T cycles and, above all, very gradual progression of the damage.

2. Experiment Description and Setup

The composition of the employed concrete is given in Table 1. The water/cement coefficient of the concrete was 0.46. The basic properties of fresh concrete were determined: density according to [21] was 2290 kg/m^3, flow according to [22] was 460 mm, slump according to [23] was 180 mm, air content according to [24] was 5.0% and the temperature of the fresh concrete was 28 °C.

Table 1. Theoretical composition of the concrete.

Component	kg/m^3 of Fresh Concrete
Cement CEM I 42.5 R (HeidelbergCement Group, Mokrá, Czech Republic)	390
Sand 0–4 mm (HeidelbergCement Group, Tovačov, Czech Republic)	810
Natural crushed aggregate 4–8 mm (HeidelbergCement Group, Luleč, Czech Republic)	160
Natural crushed aggregate 8–16 mm (HeidelbergCement Group, Olbramovice, Czech Republic)	760
Water	185
Superplasticising admixture (Sika CZ, Brno, Czech Republic)	1.0
Air-entraining admixture (Sika CZ, Brno, Czech Republic)	0.6
Workability enhancing admixture (Sika CZ, Brno, Czech Republic)	1.6

A 2.4 m high concrete block with floor plan dimensions of 1.8 m × 0.45 m was produced from the air-entrained concrete in an exterior in an open space. Concreting was conducted vertically into the wall formwork and the concrete was compacted using an immersion vibrator. After concreting, the block was covered with a damp cloth and then a PE foil. During the first two days after concreting, the cloth under the PE foil was regularly moistened. The concrete block was left in the formwork for one week, the formwork was then removed, and the concrete of the block was not treated any further.

Approximately two months after concreting, core samples for the production of the test specimens were drilled from the concrete block and had a diameter of 100 mm or 150 mm and the same length of 450 mm (block width). Cylinder and prism test specimens were then cut from the core samples using a diamond circular saw with constant water cooling. The described experiment included 4 sets of test specimens—cylinders with a diameter and length of 100 mm (marked CS100), cylinders with a diameter and length of 150 mm (marked CS150), cylinders with a diameter of 100 mm and length of 200 mm (marked C200) and prisms with dimensions 95 mm × 95 mm × 380 mm (marked P95, cut from the core sample with a diameter of 150 mm). Each set comprised 9 test specimens, which were divided into three groups of three. The first group were reference specimens, which were not exposed to freezing and thawing. The second group of specimens was subjected to 50 F-T cycles, and the last third group of specimens was subjected to 100 F-T cycles. Different properties and their development in relation to the number of conducted F-T cycles were monitored on the individual sets of the test specimens. Tensile splitting strength, according to [25], was determined on the CS100 and CS150 sets and these specimens were also subjected to the measurement using the AE method during F-T cycles. In the case of the C200 and P95 sets, the relative dynamic modulus of elasticity (RDM) was determined using the ultrasonic pulse velocity method and the resonance method, according to [26,27]. The P95 set was also subjected to the determination of flexural strength according to [28]. At least 3 test specimens were used for all destructive tests (even more in the case of non-destructive tests). Only continuous measurement of acoustic emission involved two specimens for each specimen size due to a limited number of AE sensors. Despite that, the results from both sensors on one specimen size did not differ statistically significantly.

The frost resistance test of concrete was conducted according to the standard [29]. The selected procedure is less time demanding than, for example, the procedure according to [26], the temperature range is larger than, for example, in the procedure according to [27] (which theoretically accelerates the concrete degradation process) and the freezing and thawing process can be very easily repeated since the times of the individual cycle parts are strictly defined. Equipment KD 20 (manufactured by EKOFROST s.r.o., Olomouc, Czech Republic, see [30]) was used for the test since it allows setting of the required freezing and thawing intervals and the test is performed automatically. One F-T cycle consists of freezing in air at −15 °C to −20 °C (the negative temperature we selected was always −18 °C) and thawing in a water bath at +20 °C. Freezing to the desired temperature takes 0.5 h and the negative temperature is then maintained for 3.5 h. Heating in a water bath takes 2 h. The total time of one F-T cycle is therefore 6 h + approximately 15 min (filling and draining water into the vessel of the KD 20 equipment where the test specimens are placed). The test was interrupted after every 25 F-T cycles, which lasts approximately one week, and respective test specimens were subjected to measurement of RDM or tensile splitting strength and flexural strength (after 50 and 100 FT cycles), the measured data were saved, and recording of the measurement using acoustic emission was re-initiated.

The measurement using the AE method was conducted continuously during cycling. One AE sensor was glued to each upper face of two CS100 test specimens and two CS150 test specimens, see Figure 1. The glue bond of the sensors was checked after every 25 F-T cycles.

The acoustic emission activity was generated by material damage during F-T cycles. Monitoring the AE activity was done by a multi-channel unit DAKEL XEDO (ZD Rpety-Dakel, Rpety, Czech Republic) [31] with the following input parameters: threshold value for counts was 200 mV, for individual AE hits then 400 mV, sampling of AE hits was set to 4 MHz, frequency range from 10–2000 kHz. The utilised sensors are hermetically sealed and have an IP68 degree of protection with increased frost resistance. The sensors are equipped with an integrated preamplifier and the total gain was 65 dB. The use of these extremely wideband sensors is intentional, in particular to filter out unwanted signals during postprocessing. The sensors do not have a model number because they were custom made in a limited number by the company DAKEL. The test also employed two monitoring sensors, which were placed on materials whose structure is not influenced by F-T cycles. These monitoring sensors were used to filter out false signals from the environment during post-processing. This approach provided pure signals from the individual test specimens. The AE sensors were attached to the specimens with an ethyl-based adhesive—it is a rubber-filled, resilient product with increased flexibility and peel resistance, with moisture resistance and a temperature range for use from −40 °C to +100 °C. The AE sensors, including fine-tuning of their mounting, were thoroughly tested in 2013, well before the first freezing experiments were conducted, see [4]. The first experiments also employed mechanical mounting, which we subsequently abandoned as it was not necessary. The AE sensors are tested and inspected every year and their characteristics remain unchanged. One problem occurred during the experiment and related to the computer's internal memory. Although it seemed that everything was progressing correctly and the record was being saved, the actual situation was different. Unfortunately, it was not possible to retrieve data from the interval between the 25th and 50th F-T cycle—the data were unusable.

Figure 1. Arrangement of the test specimens with AE sensors in the KD 20 equipment during F-T cycles.

As has already been mentioned, after every 25th F-T cycles, RDM was determined according to the relation:

$$RDM = \frac{X_N^2}{X_0^2} \cdot 100\%,\qquad(1)$$

where RDM is the relative modulus of elasticity of the concrete in %, X_N is the respective dynamic quantity after N performed F-T cycles and X_0 is the same quantity on the same test specimen before the start of the frost resistance test [26]. The dynamic quantity in this case is either the ultrasonic pulse velocity (UPV) in km/s, or the natural frequency of longitudinal oscillation f_L in kHz. UPV was measured using a Pundit PL-200 instrument manufactured by Proceq SA (Proceq AG, Schwerzenbach, Switzerland) [32] with 150 kHz probes. Each test specimen was measured in three longitudinal lines and UPV was determined as the average of these three measurements. The determination of f_L was done using a Handyscope HS4 oscilloscope manufactured by TiePie engineering (Sneek, Netherlands) [33].

Strength was determined on 3 test specimens from each set—either tensile splitting strength (CS100 and CS150) or flexural strength (P95) before the start of freezing and thawing, after the 50th and then after the 100th F-T cycle. The results were used to calculate the relative strength (RS) in a manner analogous to RDM,

$$RS = \frac{f_N}{f_0} \cdot 100\%\qquad(2)$$

where RS is the relative strength in %, f_N is the average strength (tensile splitting or flexural) after N performed F-T cycles, and f_0 is the same average strength found on 3 specimens not exposed to F-T cycles. The relative flexural strength $RS(F)$ is the main criterion for the evaluation of frost resistance of concrete to F-T cycles in the standard [29]. The strength tests were conducted in a DELTA 6-300 testing machine manufactured by FORM+TEST Seidner and Co. GmbH (Riedlingen, Germany) [34].

AE signals are evaluated, for example, on the basis of counts, amplitude height, frequency, etc. The least complicated approach is to add up the counts over a threshold level. This threshold can be set, which allows control of the minimum threshold, and if exceeded, i.e., one count, this forms one pulse, which is counted by the counter. One AE hit can create several counts and their number depends on the set threshold level [35]. When counting an AE hit, it is necessary to rectify and filter the high frequency pulse of the hit. The number of counts is a so-called cumulative parameter, from which cumulative curves are obtained. The frequency and amplitude band of acoustic emission is wide, from units in Hz to high ultrasonic frequencies in MHz [36]. The pulse shape and the decrease in amplitude depend on the geometry of the test specimen and on its material properties, or on the degree of material damage. To determine the cause of an AE hit, it is necessary to perform a frequency analysis of the spectrum of the recorded AE signal. AE sensors are designed to receive surface waves, which are then converted to electrical signals. These signals are amplified, filtered and saved. The measuring process of the AE system begins at the moment when the value of the amplified and filtered analogue signal exceeds the set threshold level.

3. Results

The results of tensile splitting and flexural strengths are shown in Tables 2 and 3 then presents the relative expression of not only the strengths (RS) but also the moduli of elasticity (RDM) in relation to the number of performed F-T cycles. Mechanical properties that are obtained in a destructive way (i.e., on different test specimens) only estimate concrete behaviour—they certainly do not reflect the behaviour of individual test specimens during freezing and thawing. This may lead to inaccurate test results. Table 3 indicates that the F-T cycles had practically no impact on the tensile splitting strength of the concrete. Flexural strength was influenced by freezing and thawing to a greater degree since after 100 F-T cycles, the strength dropped to almost 85% of its original value. However, this significant decrease, when compared to the other monitored properties of the concrete, is partly caused by one specimen that achieved much lower strength than the other two specimens in the set. This may, thus, constitute the already mentioned inaccuracy in the evaluation of concrete.

On the other hand, relative dynamic moduli of elasticity, as an example of non-destructive testing, characterize the behaviour of test specimens during an entire frost resistance test of concrete. However, they do so only at discrete time points (at the moment of measurement after N F-T cycles), not continuously. With regard to the dynamic modulus of elasticity, the results in Table 3 show that the F-T cycles did not practically affect the concrete at all. The maximum recorded decrease amounted to less than 3 percentage points.

Table 2. Average tensile splitting f_{ct} and flexural f_{cf} strengths of the test specimens after 0, 50 and 100 F-T cycles.

Strength-Specimen Set	Number of F-T Cycles		
	0 (REF)	50	100
f_{ct}-CS100	3.55	3.45	3.60
f_{ct}-CS150	3.20	3.25	3.25
f_{cf}-P95	5.1	4.6	4.3

Table 3. Relative development of the monitored strengths and moduli of elasticity of concrete in relation to the number of F-T cycles in percent.

RS/RDM-Specimen Set	Number of F-T Cycles			
	25	50	75	100
RS(TS)-CS100	-	97.2	-	101.4
RS(TS)-CS150	-	101.6	-	101.6
RS(F)-P95	-	91.7	-	85.3
RDM(U)-P95	99.5	100.3	100.3	101.8
RDM(U)-C200	99.3	100.9	101.8	102.2
RDM(FL)-P95	98.1	97.1	97.8	98.4
RDM(FL)-C200	98.2	98.8	99.3	99.9

In contrast to the abovementioned tests, the AE method allows monitoring of the given state of each test specimen throughout the entire test. The AE method therefore indirectly describes the behaviour of test specimens during F-T cycles. To evaluate the acoustic emission records, the number of counts (over the set threshold level) in its cumulative form (AE_{cum}) during F-T cycles was initially selected. The course of the cumulative number of counts over time shows AE activity, with higher AE activity corresponding to higher damage/failure, as is shown for example in [37]. The graph in Figure 2 presents a very similar character of AE_{cum} courses during the first 25 F-T cycles for both sizes of the test specimens. The cumulative curves for individual specimens do not differ significantly even in the later course of F-T cycles. This was one of the reasons why the experimentally obtained data of cumulative counts were interpolated with straight lines. The slope of the lines for both sizes of the specimens did not differ much during the first 25 F-T (Table 4). A change in the slopes of the interpolated lines occurs only during 50–75 F-T cycles, when the increase for larger specimens CS150 is significantly higher. Bearing in mind that the mechanical properties have not been affected (Table 3), this indicates a formation of small failures in large numbers. Nevertheless, in the last stages of the freezing cycles (from 75 to 100 F-T cycles), the smaller specimens CS100 exhibit a higher growth of AE activity, as can be seen in Table 4.

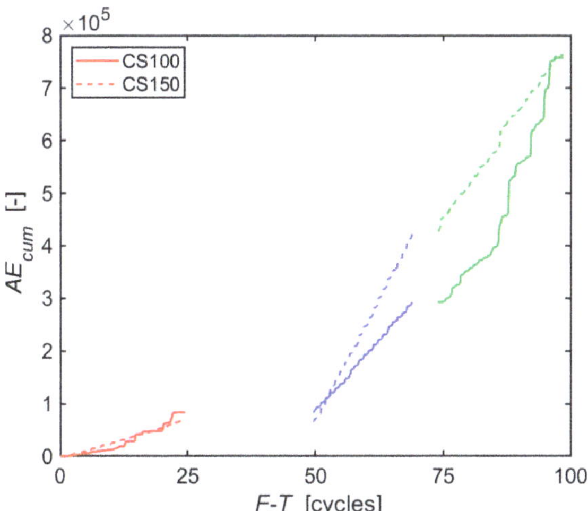

Figure 2. Cumulative number of AE counts during F-T cycles.

In the case of CS100 specimens, it can be further concluded from Figure 2 that the character of the slope from the third series of F-T cycles is maintained even at the beginning of the fourth series of F-T cycles, i.e., after the 75th F-T cycle. Approximately at the 85th F-T cycle, the slope begins to break to a higher value than shown in Table 4 (that is why that moment shows the lowest value of the coefficient of determination).

Table 4. Slope of the line interpolated with cumulative AE counts during F-T cycles (the number after the slash represents the coefficient of determination R^2).

Set of Specimens	F-T Cycles			
	0–25	25–50	50–75	75–100
CS100	540/0.922	-	1622/0.998	3086/0.947
CS150	433/0.997	-	2768/0.999	2070/0.996

It can be stated that the most commonly used AE parameter in general, which AE counts undoubtedly is, indicated in this case only the breaking points of the ongoing damage to the material. Therefore, it was appropriate to conduct a more detailed analysis of the individual AE hits. Focus was directed at the height of the amplitude, the change of the position of the dominant frequency, and the attenuation of the spectral density of the individual AE hits.

The amplitude of the recorded AE hits indicates the degree of the emerging material damage. More significant damage to the material structure generates a higher signal amplitude, as shown for example in [38]. The height of the amplitude (see Figure 3) now shows more significant differences between the individual sizes of the tested specimens. While the amplitudes remain the same, within the measurement error, during the first 25 F-T cycles, a higher amplitude was recorded for the CS100 specimens in the subsequent cycles. This trend may indicate a greater degree of damage to these smaller specimens. This in fact corresponds to the changes of the slope in Table 4, respectively in the graph in Figure 2.

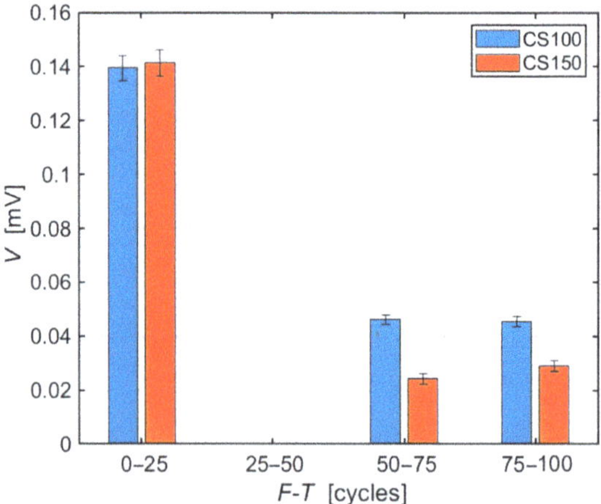

Figure 3. Average amplitude height of AE hits during F-T cycles, error bars represent sample standard deviations.

Another parameter that can be analysed from AE hits is the dominant frequency, which is obtained using the fast Fourier transform (FFT) from the time spectrum, which is also utilised in [39]. The graph in Figure 4 demonstrates how the dominant frequency for each recorded AE hit shifts towards low values during freezing and thawing. This shift is caused by the mechanical wave passing through an increasingly more damaged environment [40,41]. It can be observed that during the first 25 F-T cycles, the dominant frequencies reach values around 200 kHz and as the number of F-T cycles increases, the dominant frequency decreases. In the case of CS100 specimens, these average values are gradually 194, 142 and 61 kHz, and in the case of CS150 specimens then 199, 162 and 72 kHz.

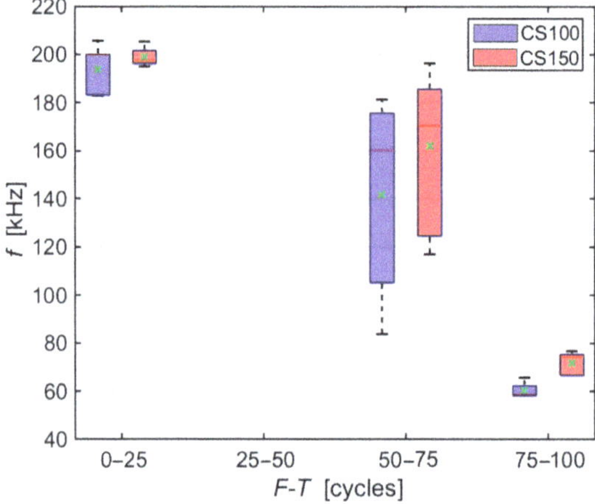

Figure 4. Box plot of development of dominant frequency values of AE hits during F-T cycles.

The following series of evaluations, include typical AE hits recorded in individual stages of the F-T cycles for the individual sizes of the test specimens. The time-frequency spectrum of the power spectral density [42] (Figures 5c, 6c, 7c, 8c, 9c and 10c) was calculated using a Short Time Fourier Transform with a Kaiser window. On the horizontal axis, the graph shows the time shift of the frequency, which is projected on the vertical axis. The size of the power spectral density is determined by the colour tones. The decibel scale is basically a linearization of the logarithm, i.e., [43],

$$x_{dB} = 20 \cdot log\left(\frac{x_V}{x_k}\right), \tag{3}$$

where x_{dB} is the value in decibels, x_V is the value (in this case) in Volts and x_k is the reference value also in Volts.

The Short Time Fourier Transform (STFT) spectrum is defined as [43],

$$S(\tau, f) = \int\limits_{-\infty}^{\infty} x(t)w(t - \tau)e^{-i2\pi ft}dt \tag{4}$$

where $w(\tau)$ is the window function (Kaiser in this case), $x(t)$ is the observed function, $S(\tau,f)$ is the resulting complex spectrum, f is the frequency and τ is the time shift (location).

Evaluation uses the absolute spectrum value [43],

$$|S(\tau, f)| = S(\tau, f) \cdot S(\tau, f)^* \tag{5}$$

where $S(\tau,f)^*$ is the complex combined function.

Selected typical AE signals recorded in CS100 specimens are presented in graphs in Figures 5–7 (all presented graphs are provided only for illustration and better understanding for the reader).

The series of graphs in Figure 5, relating to CS100 specimens during the first 25 F-T cycles, shows that the maximum spectrum values at the beginning of the signal (frequency band 120–230 kHz) are relatively quickly attenuated. With regard to the most significant frequency of 201 kHz, attenuation of spectral density reaches 62 dB/(Hz·ms). The longest occurring frequency in the signal is 70 kHz, for approximately 1 ms.

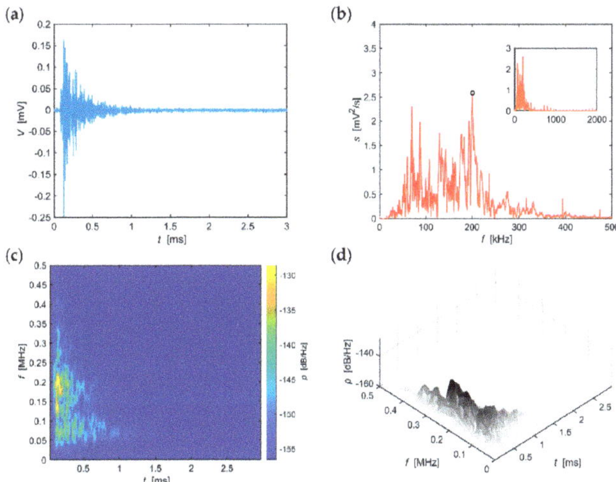

Figure 5. Visualisation of the recorded AE signal during 0–25 F-T cycles from specimens CS100: (**a**) time course of the amplitude, (**b**) frequency spectrum, (**c**) spectrogram STFT-2D image, (**d**) spectrogram STFT-3D image.

The series of graphs in Figure 6, which apply to CS100 specimens from the 50th to the 75th F-T cycle, includes significant values of the frequency spectrum in the frequency range 50–350 kHz, however, the range 50–220 kHz is the most important region. At these frequencies, the signal is attenuated very slowly. Attenuation of spectral density reaches 11 dB/(Hz·ms) at the most significant frequency of 184 kHz. The longest sounding signals were found in the frequency range 170–210 kHz (for approximately 3 ms) and then in the frequency region around 120 kHz (for approximately 2.5 ms).

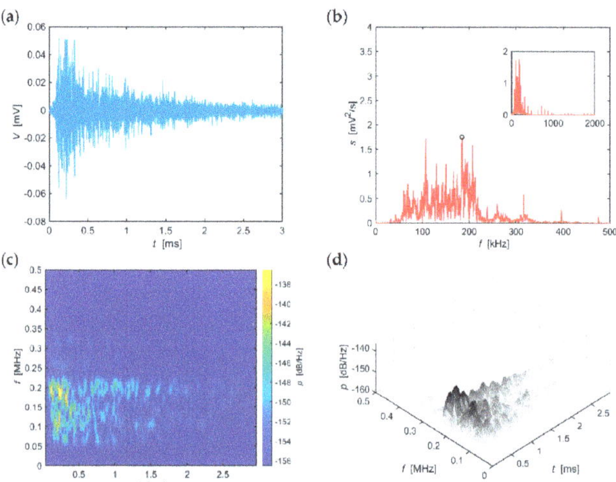

Figure 6. Visualisation of the recorded AE signal during 50–75 F-T cycles from specimens CS100: (**a**) time course of the amplitude, (**b**) frequency spectrum, (**c**) spectrogram STFT-2D image, (**d**) spectrogram STFT-3D image.

The time-frequency spectrum, see the series of graphs in Figure 7, reveals two significant frequency bands 50–100 kHz (attenuated after more than 3 ms) and 170–220 kHz (attenuated slightly below 3 ms). Attenuation of spectral density of the most significant frequency of 66 kHz was 4 dB/(Hz·ms). These data apply to CS100 specimens during the last 25 F-T cycles.

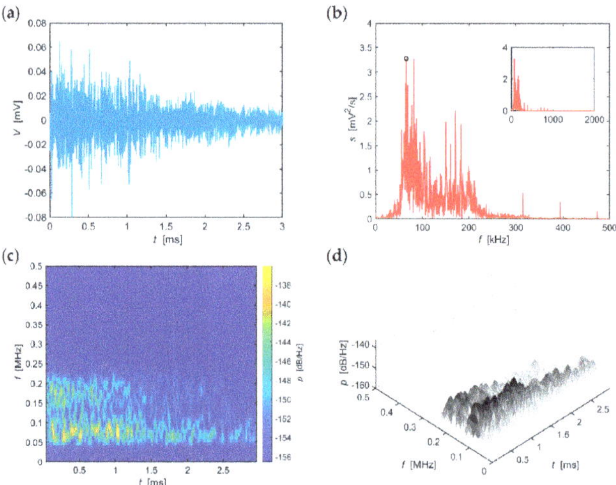

Figure 7. Visualisation of the recorded AE signal during 75-10 F-T cycles from specimens CS100: (**a**) time course of the amplitude, (**b**) frequency spectrum, (**c**) spectrogram STFT-2D image, (**d**) spectrogram STFT-3D image.

Selected typical AE signals recorded from larger specimens CS150 are shown in the graphs in Figures 8–10 (all presented graphs are provided only for illustration and better understanding for the reader).

The series of graphs in Figure 8, describing CS100 specimens during the first 25 F-T cycles, indicates that the significant frequency band is 50–300 kHz, of which the region between 160–250 kHz is extreme with regard to amplitudes. It is evident that the attenuation at higher frequencies is greater than at lower frequencies. Attenuation of spectral density at the most significant frequency of 185 kHz is 43 dB/(Hz·ms). The signal length is relatively short, approximately up to 1.5 ms.

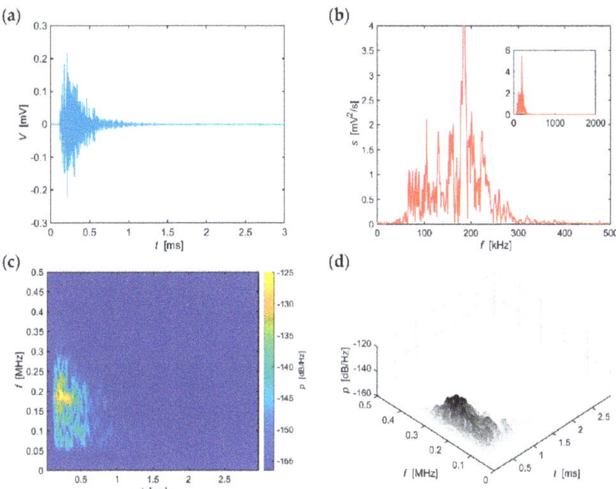

Figure 8. Visualisation of the recorded AE signal during 0–25 F-T cycles from specimens CS150: (**a**) time course of the amplitude, (**b**) frequency spectrum, (**c**) spectrogram STFT-2D image, (**d**) spectrogram STFT-3D image.

The significant frequency spectrum, see the series of graphs in Figure 9, is in a very narrow band 140–230 kHz but exhibits relatively long duration—more than 3 ms. At the most significant frequency of 171 kHz, attenuation of spectral density is 6 dB/(Hz·ms), however, the spectrum shows apparent rapid attenuation—approximately 1 ms. These data apply to CS150 specimens between the 50th and 75th F-T cycle.

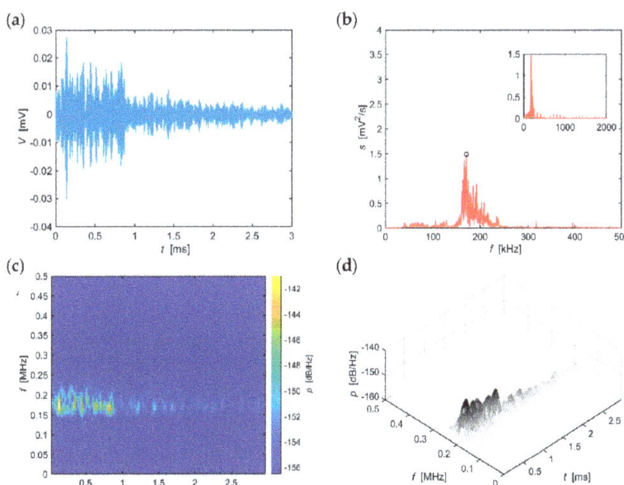

Figure 9. Visualisation of the recorded AE signal during 50–75 F-T cycles from specimens CS150: (**a**) time course of the amplitude, (**b**) frequency spectrum, (**c**) spectrogram STFT-2D image, (**d**) spectrogram STFT-3D image.

Two frequency packets are apparent in the series of graphs in Figure 10, describing the behaviour of CS150 test specimens during the last 25 F-T cycles, the more significant one at 50–120 kHz and the less significant one at 120–240 kHz. The maximum amplitude value occurs as late as 0.7 ms from the beginning of the signal. From that point in time, attenuation at the most significant frequency (75 kHz) can be determined—attenuation of spectral density is 6 dB/(Hz·ms).

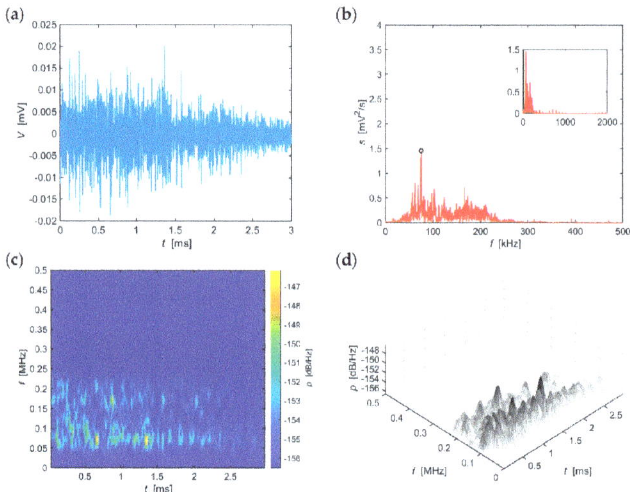

Figure 10. Visualisation of the recorded AE signal during 75–100 F-T cycles from specimens CS150: (**a**) time course of the amplitude, (**b**) frequency spectrum, (**c**) spectrogram STFT-2D image, (**d**) spectrogram STFT-3D image.

The typical AE signals of both test specimen sizes during the entire 100 F-T cycles (Figure 5 to Figure 10) confirm the conclusions inferred from the graphs in Figures 3 and 4. That is, the amplitude decreases with the increasing number of F-T cycles and the dominant frequency shifts towards lower values.

The time-frequency power spectra also indicate a gradual decrease of the spectral density attenuation values, for CS100 specimens it is gradually 62, 11 and 4 dB/(Hz·ms), for CS150 specimens then 43, 6 and 6 dB/(Hz·ms). This decrease in attenuation is probably caused by the progressing damage to the internal structure of the concrete, or to its cementitious matrix, where the compromised matrix bonds that have been oscillated cannot be easily and quickly attenuated. Since they are loose, they oscillate slightly longer than at the beginning of the frost resistance test. In addition, there is a decrease in the frequency and amplitude of the generated AE signals. All these phenomena then result in an accelerating decrease of the spectral density attenuation values.

4. Discussion

There have not been many papers published recently that deal with the continuous monitoring of the behaviour of concrete specimens during F-T cycles. The papers that describe continuous measurement using the AE method, and at the same time, concern cement-based materials deal only with standard AE parameters, see, e.g., [44–46]. Other studies deal with different materials, such as water-saturated ceramics ([47], this study utilizes equipment from the same manufacturer as in the manuscript submitted by us) or materials for asphalt roads, see [48]. However, the acoustic emission method is typically used to compare the behaviour of materials after F-T cycles during mechanical loading, which is described, for example, in [49–52]. No other known study showed a frequency analysis of individual AE hits that recorded directly during F-T cycles.

5. Conclusions

The abovementioned results from the typical test procedures indicate that no significant changes in the quality of the concrete were recorded after 100 F-T cycles. The dynamic modulus of elasticity exhibited a maximum decrease of less than 3 percentage points (for RDM(FL) of the P95 specimen set after 50 F-T cycles) and mere 1.6 percentage points after 100 F-T cycles (also for RDM(FL) of the P95 specimen set). There was no decrease in tensile splitting strength due to freezing and thawing. The only standard parameter that exhibited some decrease was flexural strength. In this case, the decrease to 85.3% of the original value can be partly contributed to one test specimen of three test, which achieved lower strength. This confirms, inter alia, the problem with evaluation of concrete on the basis of a comparison of results obtained on different test specimens.

Based on both RS and RDM, the test specimens therefore appear to be almost intact by frost. Even the commonly presented cumulative number of AE counts does not indicate any significant difference between smaller (CS100) and larger (CS150) test specimens. However, a detailed analysis of the recorded AE signals that were generated directly during the F-T cycles in the test specimens proves the situation is different:

- There is a decrease in the average amplitude during freezing cycles in the case of both specimen sizes;
- There is an apparent difference between larger and smaller specimens in the height of the amplitude after between 75 and 100 F-T cycles;
- There is an evident shift in the values of the dominant frequency to lower values in the case of both specimen sizes;
- There is a noticeable difference between larger and smaller specimens in the shift of the dominant frequency from the 75th to the 100th F-T cycle;
- There is a decrease in the spectral density attenuation values during F-T cycles, with a higher decrease observed in the case of smaller CS100 test specimens.

It has been demonstrated that while conventional test procedures do not reveal significant changes in the internal structure of the tested concrete even after 100 F-T cycles, changes still occur in the material and can subsequently lead to irreversible damage of the material. It is, thus, appropriate to employ not only traditional test procedures, but also less traditional ones, such as the AE method, which can safely detect the formation of changes (damage) in the structure of concrete already at the beginning.

The innovative contribution of the manuscript lies in the application of the methods of frequency analysis of recorded AE hits during F-T cycles. It was discovered that the typical AE parameters (such as the cumulative number of counts) and commonly used construction testing methods are not sensitive enough to determine the degree of damage of concrete specimens during F-T cycles. In particular, in the case of air-entrained concrete, which exhibits very low degree of damage, is it necessary to employ modern tools for the analysis of AE hit signals. The submitted manuscript provides one of the first insights into the issue of structural health monitoring. It is possible that some testing procedures and methods will have to be revised in the future since the developments in building materials are still advancing and some current diagnostic tools will no longer be sufficient for the detection of developing damage. This manuscript could be one of the pioneers in this context.

Author Contributions: D.K. and L.T. designed and performed the experiments; D.K., L.T. and L.P. analyzed the data; D.K. and L.T. wrote the manuscript; L.P. and T.V. revised the manuscript; T.V. project administration. All authors have read and agreed to the published version of the manuscript.

Funding: This research was funded by Czech Science Foundation, GA 19-22708S "New approaches to predicting air-entrained concrete durability by means of determination of pore size distribution".

Institutional Review Board Statement: Not applicable.

Informed Consent Statement: Not applicable.

Data Availability Statement: The data presented in this study are available on request from the corresponding author. The data are not publicly available due to ongoing research.

Conflicts of Interest: The authors declare no conflict of interest.

References

1. Mastori, H.; Piluso, P.; Haquet, J.-F.; Denoyel, R.; Antoni, M. Limestone-siliceous and siliceous concretes thermal damaging at high temperature. *Constr. Build. Mater.* **2019**, *228*, 116671. [CrossRef]
2. Marcantonio, V.; Monarca, D.; Colantoni, A.; Cecchini, M. Ultrasonic waves for materials evaluation in fatigue, thermal and corrosion damage: A review. *Mech. Syst. Signal. Process.* **2019**, *120*, 32–42. [CrossRef]
3. Zou, Y.; Shen, X.; Zuo, X.; Xue, H.; Li, G. Experimental study on microstructure evolution of aeolian sand concrete under the coupling freeze–thaw cycles and carbonation. *Eur. J. Environ. Civ. Eng.* **2020**, 1–16. [CrossRef]
4. Pazdera, L.; Topolar, L. Application acoustic emission method during concrete frost resistance. *Russ. J. Nondestruct. Test.* **2014**, *50*, 127–131. [CrossRef]
5. Farnam, Y.; Bentz, D.; Sakulich, A.; Flynn, D.; Weiss, J. Measuring Freeze and Thaw Damage in Mortars Containing Deicing Salt Using a Low-Temperature Longitudinal Guarded Comparative Calorimeter and Acoustic Emission. *Adv. Civ. Eng. Mater.* **2014**, *3*, 20130095. [CrossRef]
6. Shang, H.-S.; Yi, T.-H. Freeze-Thaw Durability of Air-Entrained Concrete. *Sci. World J.* **2013**, *2013*, 1–6. [CrossRef] [PubMed]
7. Setzer, M.J.; Janssen, G.M.T. Final Report of RILEM TC 176-IDC: Internal Damage of Concrete due to frost action. *Mater. Struct.* **2004**, *37*, 740–742.
8. Korenska, M.; Pazdera, L.; Ritickova, L. Resonant inspection—Interesting non-destructive testing tools for determine quality of tested specimens. In *Previous Experience and Current Innovations in Non-Destructive Testing, Proceedings of the 6th In-ternational Conference of the Slovenian-Society-for-Non-Destructive-Testing, Portoroz, Slovenia, 13–15 September 2001*; Slovenian Society for Non-destructive Testing: Portoroz, Slovenia, 2001; pp. 45–48, ISBN 9789619061015.
9. Plšková, I.; Chobola, Z.; Matysik, M. Assessment of ceramic tile frost resistance by means of the frequency inspection method. *Ceram. Silikaty* **2011**, *55*, 176–182.
10. Smutny, J.; Nohal, V. Vibration Analysis in the Gravel Ballast by Measuring Stone Method. *Akustika* **2016**, *25*, 22–28.
11. Petraskova, V. Correctness Testing and Equality Testing in Evaluation of Acoustic Laboratory Measurements. *Akustika* **2016**, *26*, 20–28.
12. Zhou, Z.; Rui, Y.; Zhou, J.; Dong, L.; Chen, L.; Cai, X.; Cheng, R. A New Closed-Form Solution for Acoustic Emission Source Location in the Presence of Outliers. *Appl. Sci.* **2018**, *8*, 949. [CrossRef]
13. Topolář, L.; Pazdera, L.; Kucharczyková, B.; Smutný, J.; Mikulášek, K. Using Acoustic Emission Methods to Monitor Cement Composites during Setting and Hardening. *Appl. Sci.* **2017**, *7*, 451. [CrossRef]
14. Farhidzadeh, A.; Dehghan-Niri, E.; Salamone, S.; Luna, B.; Whittaker, A. Monitoring Crack Propagation in Reinforced Concrete Shear Walls by Acoustic Emission. *J. Struct. Eng.* **2013**, *139*, 04013010. [CrossRef]
15. Qin, L.; Ren, H.-W.; Dong, B.-Q.; Xing, F. Acoustic Emission Behavior of Early Age Concrete Monitored by Embedded Sensors. *Materials* **2014**, *7*, 6908–6918. [CrossRef]
16. Yuyama, S.; Li, Z.-W.; Ito, Y.; Arazoe, M. Quantitative analysis of fracture process in RC column foundation by moment tensor analysis of acoustic emission. *Constr. Build. Mater.* **1999**, *13*, 87–97. [CrossRef]
17. Ranz, J.; Aparicio, S.; Romero, H.; Casati, M.J.; Molero, M.; González, M. Monitoring of Freeze-Thaw Cycles in Concrete Using Embedded Sensors and Ultrasonic Imaging. *Sensors* **2014**, *14*, 2280–2304. [CrossRef] [PubMed]
18. De Kock, T.; Boone, M.A.; De Schryver, T.; Van Stappen, J.; Derluyn, H.; Masschaele, B.; De Schutter, G.; Cnudde, V. A Pore-Scale Study of Fracture Dynamics in Rock Using X-ray Micro-CT Under Ambient Freeze–Thaw Cycling. *Environ. Sci. Technol.* **2015**, *49*, 2867–2874. [CrossRef]
19. Landis, E.N.; Baillon, L. Experiments to Relate Acoustic Emission Energy to Fracture Energy of Concrete. *J. Eng. Mech.* **2002**, *128*, 698–702. [CrossRef]
20. Farnam, Y.; Geiker, M.R.; Bentz, D.; Weiss, J. Acoustic emission waveform characterization of crack origin and mode in fractured and ASR damaged concrete. *Cem. Concr. Compos.* **2015**, *60*, 135–145. [CrossRef]
21. Comité Européen de Normalisation. *EN 12350-6: Testing Fresh Concrete—Part 6: Density*; CEN: Brussels, Belgium, 2009.
22. Comité Européen de Normalisation. *EN 12350-5: Testing Fresh Concrete—Part 5: Flow Table Test*; CEN: Brussels, Belgium, 2009.
23. Comité Européen de Normalisation. *EN 12350-2: Testing Fresh Concrete—Part 2: Slump-Test*, 1st ed.; CEN: Brussels, Belgium, 2009.
24. Comité Européen de Normalisation. *EN 12350-7: Testing Fresh Concrete—Part 7: Air Content—Pressure Methods*; CEN: Brussels, Belgium, 2009.
25. Comité Européen de Normalisation. *EN 12390-6: Testing Hardened Concrete—Part 6: Tensile Splitting Strength of Test Specimens*; CEN: Brussels, Belgium, 2009.
26. Comité Européen de Normalisation. *CEN/TR 15177: Testing the Freeze-Thaw Resistance of Concrete—Internal Structural Damage*; CEN: Brussels, Belgium, 2006.
27. American Society for Testing and Materials. *ASTM C666/C666M—15: Standard Test Method for Resistance of Concrete to Rapid Freezing and Thawing*; ASTM International: West Conshohocken, PA, USA, 2015.

28. Comité Européen de Normalisation. *EN 12390-5: Testing Hardened Concrete—Part 5: Flexural Strength of Test Specimens*; CEN: Brussels, Belgium, 2009.

29. Czech Office for Standards. *Metrology and Testing. ČSN 73 1322: Determination of Frost Resistance of Concrete*; UNMZ: Prague, Czech Republic, 1968. (In Czech)

30. Ekofrost.cz. Available online: https://www.ekofrost.cz/ (accessed on 13 November 2020).

31. Dakel.cz. Available online: http://dakel.cz/index.php?pg=prod/dev/xedo_en (accessed on 10 January 2019).

32. Proceq.com. Available online: https://www.proceq.com/uploads/tx_proceqproductcms/import_data/files/Pundit%20PL-2_Sales%20Flyer_English_high.pdf (accessed on 6 May 2020).

33. Tiepie.com. Available online: https://www.tiepie.com/en/usb-oscilloscope/handyscope-hs4 (accessed on 23 August 2020).

34. Formtest.de. Available online: https://www.formtest.de/en/Products/Machines/DELTA-6-300.php (accessed on 3 December 2019).

35. Nazarchuk, Z.; Skalskyi, V.; Serhiyenko, O. *Acoustic Emission*; Springer International Publishing: Berlin/Heidelberg, Germany, 2017; pp. 1–294.

36. Deutsches Institut für Normung. *EN 13554:2011-04: Non-Destructive Testing—Acoustic Emission Testing—General Principles*; DIN: Berlin, Germany, 2011.

37. Kahirdeh, A.; Sauerbrunn, C.; Modarres, M. Acoustic emission entropy as a measure of damage in materials. In Proceedings of the Technologies and Materials for Renewable Energy, Environment and Sustainability: TMREES, Beirut, Lebanon, 15–18 April 2016.

38. Nair, A.; Cai, C. Acoustic emission monitoring of bridges: Review and case studies. *Eng. Struct.* **2010**, *32*, 1704–1714. [CrossRef]

39. Ahmad, A.; Schlindwein, F.S.; Ng, G.A. Comparison of computation time for estimation of dominant frequency of atrial electrograms: Fast fourier transform, blackman tukey, autoregressive and multiple signal classification. *J. Biomed. Sci. Eng.* **2010**, *3*, 843–847. [CrossRef]

40. Balayssac, J.-P.; Garnier, V. *Non-Destructive Testing and Evaluation of Civil Engineering Structures*, 1st ed.; Elsevier: Amsterdam, The Netherlands, 2017; pp. 1–376, ISBN 9780081023051.

41. Matysik, M.; Plšková, I.; Chobola, Z. Estimation of Impact-Echo Method for the Assessment of Long-Term Frost Resistance of Ceramic Tiles. *Adv. Mater. Res.* **2014**, *1000*, 285–288. [CrossRef]

42. Bugno, A.A.-; Swit, G.; Krampikowska, A. Assessment of Destruction Processes in Fibre-Cement Composites Using the Acoustic Emission Method and Wavelet Analysis. *IOP Conf. Ser. Mater. Sci. Eng.* **2019**, *471*, 032042. [CrossRef]

43. Boashash, B. Time-Frequency synthesis and filtering. In *Time-Frequency Signal Analysis and Processing*; Elsevier: Amsterdam, The Netherlands, 2016; pp. 637–691.

44. Qian, Y.; Farnam, Y.; Weiss, J. Using acoustic emission to quantify freeze–thaw damage of mortar saturated with NaCl solutions. In Proceedings of the 5th International Conference on the Durability of Concrete Structures, Lafayette, India, 30 June–1 July 2014.

45. Todak, H.N.; Tsui, M.; Ley, M.T.; Weiss, W.J. Evaluating freeze-thaw damage in concrete with acoustic emissions and ultrasonics. In *Springer Proceedings in Physics*; Springer International Publishing: Berlin/Heidelberg, Germany, 2017; Volume 179, pp. 175–189.

46. Farnam, Y.; Bentz, D.; Hampton, A.; Weiss, W.J. Acoustic Emission and Low-Temperature Calorimetry Study of Freeze and Thaw Behavior in Cementitious Materials Exposed to Sodium Chloride Salt. *Transp. Res. Rec. J. Transp. Res. Board* **2014**, *2441*, 81–90. [CrossRef]

47. Húlan, T.; Knapek, M.; Csáki, Š.; Kušnír, J.; Šmilauerová, J.; Dobroň, P.; Chmelík, F.; Kaljuvee, T.; Uibu, M. The Formation of Microcracks in Water-Saturated Porous Ceramics During Freeze–Thaw Cycles Followed by Acoustic Emission. *J. Nondestruct. Eval.* **2021**, *40*, 1–11. [CrossRef]

48. Behnia, B.; Buttlar, W.; Reis, H. Evaluation of Low-Temperature Cracking Performance of Asphalt Pavements Using Acoustic Emission: A Review. *Appl. Sci.* **2018**, *8*, 306. [CrossRef]

49. Shahidan, S.; Pullin, R.; Bunnori, N.M.; Zuki, S.S.M. Active crack evaluation in concrete beams using statistical analysis of acoustic emission data. *Insight Non-Destr. Test. Cond. Monit.* **2017**, *59*, 24–31. [CrossRef]

50. Gorzelańczyk, T.; Schabowicz, K. Effect of Freeze-Thaw Cycling on the Failure of Fibre-Cement Boards, Assessed Using Acoustic Emission Method and Artificial Neural Network. *Materials* **2019**, *12*, 2181. [CrossRef] [PubMed]

51. Wu, Y.; Li, S.; Wang, D. Characteristic analysis of acoustic emission signals of masonry specimens under uniaxial compression test. *Constr. Build. Mater.* **2019**, *196*, 637–648. [CrossRef]

52. Yang, Z.; Weiss, W.J.; Olek, J. Water Transport in Concrete Damaged by Tensile Loading and Freeze–Thaw Cycling. *J. Mater. Civ. Eng.* **2006**, *18*, 424–434. [CrossRef]

 *materials*

Article

Fibre Optic FBG Sensors for Monitoring of the Temperature of the Building Envelope

Janusz Juraszek * and Patrycja Antonik-Popiołek

Faculty of Materials, Civil and Environmental Engineering, University of Bielsko-Biala, 43-309 Bielsko-Biala, Poland; pantonik@ath.bielsko.pl
* Correspondence: jjuraszek@ath.bielsko.pl; Tel.: +48-33-8279191

Abstract: Standard sensors for the measurement and monitoring of temperature in civil structures are liable to mechanical damage and electromagnetic interference. A system of purpose-designed fibre optic FBG sensors offers a more suitable and reliable solution—the sensors can be directly integrated with the load-bearing structure during construction, it is possible to create a network of fibre optic sensors to ensure not only temperature measurements but also measurements of strain and of the moisture content in the building envelope. The paper describes the results of temperature measurements of a building 2-layer wall using optical fibre Bragg grating (FBG) sensors and of a three-layer wall using equivalent classical temperature sensors. The testing results can be transmitted remotely. In the first stage, the sensors were tested in a climatic test chamber to determine their characteristics. The paper describes test results of temperature measurements carried out in the winter season for two multilayer external walls of a building in relation to the environmental conditions recorded at that time, i.e., outdoor temperature, relative humidity, and wind speed. Cases are considered with the biggest difference in the level of the relative humidity of air recorded in the observation period. It is found that there is greater convergence between the theoretical and the real temperature distribution in the wall for high levels (~84%) of the outdoor air relative humidity, whereas at the humidity level of ~49%, the difference between theoretical and real temperature histories is substantial and totals up to 20%. A correction factor is proposed for the theoretical temperature distribution.

Keywords: monitoring; wall temperature; fibre bragg grating sensors

Citation: Juraszek, J.; Antonik-Popiołek, P. Fibre Optic FBG Sensors for Monitoring of the Temperature of the Building Envelope. *Materials* **2021**, *14*, 1207. https://doi.org/10.3390/ma14051207

Academic Editor: David N. McIlroy

Received: 29 January 2021
Accepted: 25 February 2021
Published: 4 March 2021

Publisher's Note: MDPI stays neutral with regard to jurisdictional claims in published maps and institutional affiliations.

Copyright: © 2021 by the authors. Licensee MDPI, Basel, Switzerland. This article is an open access article distributed under the terms and conditions of the Creative Commons Attribution (CC BY) license (https://creativecommons.org/licenses/by/4.0/).

1. Introduction

Practical temperature and humidity calculations are based on the values of outdoor and indoor temperatures and characteristics of the building envelope materials. Apart from temperature, however, they usually fail to take account of the variability of more weather, environmental and other factors.

The classical method of analysing the temperature distribution in the building envelope consists of embedding temperature sensors (typically electrical resistance sensors) in it during the building construction or renovation. This solution was applied during the renovation of a three-layer building envelope by placing sensors in four layers [1] The testing focused on the temperature distribution and the possibility of the occurrence of water vapour condensation. The applied three-layer system was composed of a brickwork wall, polyurethane insulation with an aluminium foil layer, an air gap, and a brickwork wall, which worked well. Another way is to scan the building envelope surface using a thermographic camera. The methodology of the testing is discussed in detail in [2]. The study also presents the method of a theoretical calculation of temperature on the outer surface of a multilayer building envelope exposed to 24-h changes in ambient temperature caused by solar radiation and measured using a thermographic camera.

The FBG sensors are the latest solution to use for temperature measurements. The use of fibre optic sensors to measure temperatures in tight walls to monitor potential

211

underground leakages of fluids is discussed in [3–6]. The test results made it possible to state that the temperature of the fibre optic sensors varied with a change in the seepage rate. The results indicate unequivocally that the use of optic fibre sensors in the building envelope is one of the best solutions that will enable the most accurate analysis of the temperature distribution and of the way the distribution is affected by other external factors.

An interesting series of thermal testing of glued laminated timber beams used in the construction industry performed with a measuring system based on the fibre Bragg grating (FBG) principle was carried out in a thermal chamber [7]. The chamber can generate specific cycles of temperature and humidity to simulate the structural behaviour in real environmental conditions. The tested FBG system is suitable for load-carrying timber (glued laminated timber beams) and concrete structures [8]. The second part of the paper presents the testing of mechanical loads of a glued and laminated timber beam with embedded FBG sensors.

The fibre Bragg grating is an element in a glass or polymer fibre with periodic modulation of the refractive index along with the fibre core. The fundamentals of the FBG-based method of strain and temperature measurements are presented in [9,10].

Fibre optic sensors have many advantages, such as high accuracy, ease of installation, small size, flexibility, cost-effectiveness, resistance to electromagnetic interference and corrosion, multiplexing capability, and great potential in long-term continuous measurements [11–16].

An intelligent system of monitoring a concrete paving slab subjected to the impact of the outdoor temperature is presented in [17]. Truly distributed fibre optic temperature and strain sensors and an innovative interferometer-based fibre optic inclinometer were installed in the paving slab to monitor and evaluate the thermal curing process.

The results confirm that the paving slab top surface responds to heating/cooling very fast. The temperature of the bottom surface lags behind the temperature of the top, which results in a rapid rise and then a slow reduction in the temperature difference between the top and the bottom.

A large number of quasi-distributed FBG sensors were presented in [18–20]. The aim of the testing was to develop and demonstrate an integrated comprehensive system for the monitoring of thermal shrinkage of the roadway surfacing. The technology has been used for measuring the temperature, strain, and deflection of surfacing since 2003.

It should be noted that the fibre Bragg grating is a promising measuring technology for future applications of sensor systems. The FBG sensor housing should be compatible with the housing of the electrical sensor so that it can be installed easily in systems of instrumentation [21].

Other implementations of FBG sensors made by the author included their application for strain analysis of power transmission line towers, strain analysis of the hoisting machine brake lever, and strain monitoring of a residential building made in the polystyrene concrete technology [22,23]. Interesting study on the temperature distribution of the earth's surface was presented in the article [24].

Summing up the results of the literature survey, it should be emphasised that conventional temperature sensors have certain limitations, such as poor operational life, low resolution, sensitivity to environmental factors-strong alkalinity, and water permeability (building walls), in particular, low zero stability and high measurement noise. FBG sensors are an alternative method of monitoring building structures due to a number of significant advantages, e.g., high accuracy and sensitivity, flexibility, resistance to electromagnetic interference, and harsh environmental conditions; they are also multiplexable (a large number of sensors can be connected to one telecommunication fibre optic cable, which means simpler cabling systems). A very interesting feature is the possibility of measuring different physical quantities, for example, temperature and strain, using one optical interrogator. The FBG technique is more expensive compared to classical methods of temperature measurement, but considering the technological progress in the field of interrogators and sensors, especially with the wavelength of 800 nm, it will become comparable in terms of price offering much greater research possibilities at the same time.

The approach based on special optical FBG temperature sensors to monitor the temperature distribution in a building wall has never been proposed before. Another novelty is the application of optical FBG sensors to measure not only the wall temperature but also the wall strain. The proposed FBG sensors have a very small diameter, which is of special significance because they can be installed in existing walls of already operated buildings. Temperature measurements in external walls of buildings are necessary because based on them, it is possible to select the optimal heating system depending on the wall structure and specific climatic conditions of the building location. Moreover, the knowledge of real delays of the temperature distribution in the wall due to changes in the outdoor temperature makes it possible to take adequate action in the HVAC system in advance. Furthermore, the knowledge of the real temperature distribution in the wall enables verification of the theoretical (standard) temperature distribution, the introduction of a correction factor, better selection of the heating system parameters, and more precise identification of places in the wall where water vapour condensation occurs. The rationale for presenting walls with traditional and optical FBG temperature sensors in the paper was the need to draw attention to a rather high failure frequency of the former. What is also important, previous studies do not take up the issues of the life and failure frequency of classical sensors intended for measurements of temperature. These measurements are carried out on a long-term basis (over a period of many years). A failure of a temperature sensor and the time drift create serious problems in the analysis of results. The operational analysis of classical temperature sensors embedded in the building walls conducted by the authors over the period of four years showed their 19% failure rate. In a building that is already operated their replacement is practically impossible, which is a significant problem. The number of different sensors installed in the analysed facility was about 3000. The only feasible option in an already operated building is to introduce optical FBG sensors into the wall in place of the damaged classical ones. It should also be noted that in the same period of observation no failure occurred of any of the optical FBG sensors. FBG sensors can be installed in place of damaged traditional sensors due to their small dimensions, and that was conducted also in the case of the analysed three-layer wall, which may not have been described precisely enough. FBG sensors are equivalent to classical sensors and they also demonstrate a number of advantages that the latter are practically unable to achieve. The system of temperature measurements in the building wall presented in the paper and based on optical FBG sensors ensures unprecedented measuring stability, operational durability, and the possibility of creating networks. Moreover, it enables strain and humidity measurements. The significant development of the fibre optic technology, and especially of the design of optical interrogators and fibre optic sensors, will make it possible to introduce into the building structure a "nervous system" that comprehensively and simultaneously monitors the temperature, strain, stress, or vibration of the building walls. Considering the latest technological progress in the field of optical interrogators and optical FBG sensors, their cost is comparable to classical methods of temperature measurement in the building wall, whereas the computational cost is acceptable for large facilities. The outdoor temperature in both cases of the walls under analysis was obtained from a weather station, while the indoor temperature was measured with a sensor type Testo 605i. The theoretical distribution of temperature is determined in the paper-based analytical formulas and is included in the EN ISO 13788:2013-05 standard.

Striving to reduce energy consumption is increasingly associated with the introduction of "smart" technologies into civil engineering structures and the application of systems monitoring and controlling the processes occurring in buildings. The paper presents a proposal for purpose-designed FBG sensors intended for the measurement of temperature in the building envelope. The analysis concerns a three-layer building envelope.

A thermal analysis was conducted of individual layers of the building envelope depending on varying outdoor weather conditions. From the point of view of the building envelope heating and cooling, it is essential what external factors affect the temperature in it and how long it takes the envelope to respond to changes in the outdoor temperature. FBG sensors were used, which—as previously stated—are a very promising measuring technology for future applications in systems of sensors to measure temperatures, strain, displacement, and even moisture content. Experimentally, through-bore sets of FBG sensors were made to measure temperature. The sensors were fixed in a special housing.

Strain measurement of the building structural elements by means of FBG sensors makes it possible to detect the formation of cracks in the building structure, strains arising during the foundation settlement, analysis of volumetric shrinkage of the concrete floor and reinforced concrete columns, or analysis of strains due to standard loads or load tests. Because research in this field falls into a different category, the results will be presented in a separate paper. The paper includes example measurement results of the wall strain recorded using FBG sensors.

2. Materials and Methods and Preliminary Testing

The tests were performed on two selected walls—an external cavity wall (three-layer wall) and an external insulated solid wall (two-layer wall) of a university building. The cross section of the walls is presented in Figures 1 and 2, respectively (λ-coefficient of thermal conductivity and thickness of the layers).

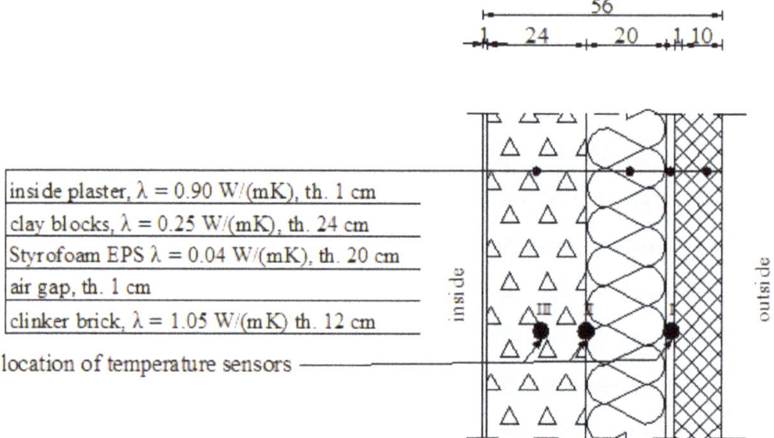

Figure 1. Three-layer wall cross section.

The first wall consists of hollow clay blocks, a thermal insulation layer, and clinker bricks. The second—of hollow clay blocks and thermal insulation finished with plaster. These wall types are the most common solutions for the two- and three-layer systems. The first wall was analysed using classical temperature sensors, whereas the second was tested with FBG sensors. The environmental conditions in the terms of air temperature, wind speed, and air humidity were monitored by a classical meteorological station located on the university campus. A temperature monitoring system was constructed inside the building envelope walls. The system is based on fibre Bragg grating (FBG) temperature sensors. The essence of the solution is to introduce FBG sensors at the appropriate points of the cross section of the wall of the structure. In the case of the building envelope in the form of a wall, the appropriate places are the borders of individual layers and points in the middle of a given layer. The places were pre-determined using numerical simulations.

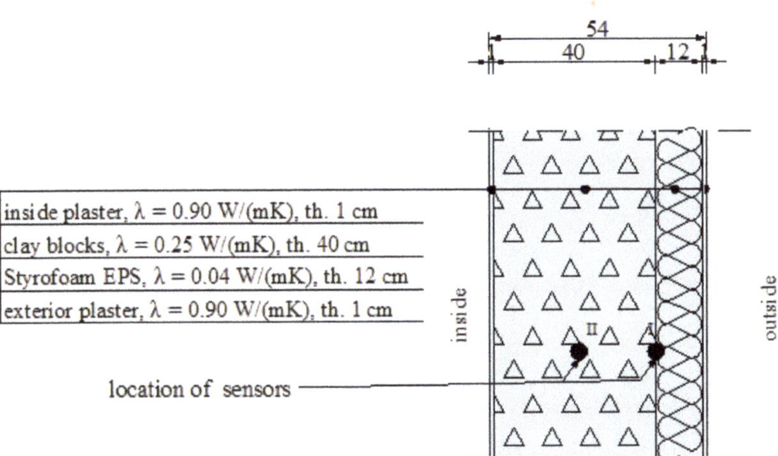

Figure 2. Two-layer wall cross section.

The FBG-based set of temperature sensors is presented in Figure 3. Based on preliminary testing, the total measuring base was adopted as L = 450 mm. The system of FBG sensors was fixed in the building envelope using special holders.

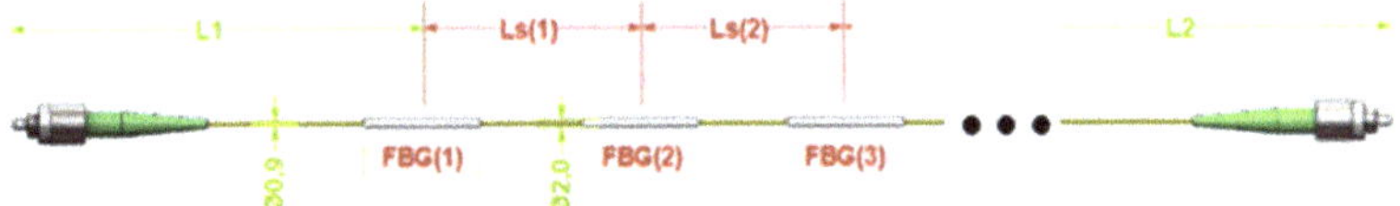

Figure 3. Fibre Bragg grating (FBG) temperature sensors.

The measuring system calibration is presented in Table 1.

Table 1. Measuring system calibration.

Temperature Accuracy (Guaranteed)	Temperature Accuracy (Typical)	Temperature Precision (Guaranteed)	Temperature Precision (Typical)	Ingress Protection Rating
1 °C	<1 °C	±0.3 °C	±0.2 °C	IP 67

The measuring part of each sensor is a Bragg grating characterised by a specific wavelength and embedded in an optical fibre. Due to the sensor identifiability by the optical interrogator, the difference of at least 5 nm has to be kept between the wavelengths of each Bragg grating. A temperature change in a selected point of the building envelope is closely related to the change in the Bragg grating wavelength. This relationship is described by the following Equation (1):

$$T = T_{S1}\left(\frac{\lambda_{T,act} - \lambda_{T,ref}}{\lambda_{T,ref}}\right)^2 + T_{S2}\left(\frac{\lambda_{T,act} - \lambda_{T,ref}}{\lambda_{T,ref}}\right) + T_{S3} \tag{1}$$

The description of constant calibration parameters is presented in Table 2, whereas the values of the calibration constants for individual sensors are listed in Table 3.

Table 2. Marking of constant calibration parameters.

Measurand	Description
T (°C)	Temperature
$\lambda_{T,act}$ (nm)	Actual temp. wavelength
$\lambda_{T,ref}$ (nm)	Reference temp. wavelength
T_{S1} (°C)	Temperature coefficient 1
T_{S2} (°C)	Temperature coefficient 2
T_{S3} (°C)	Temperature coefficient 3

Table 3. Calibration coefficients.

T_{s1} (°C)	T_{s2} (°C)	T_{s3} (°C)	$\lambda_{T,ref}$ (nm)
−2,477,681.366	52,768.19076	22.50476755	1,548,514.494
−2,569,888.019	53,819.29229	22.50525754	1,548,477.574
−2,311,1030.106	53,588.17046	22.5027883	1,548,633.785

Apart from dedicated fibre optic sensors, the system includes a 2 kHz FBG-800 optical interrogator, a recorder, special software, a multiplier and telecommunication fibres. An option is also possible with a wireless transfer of measurement results from the interrogator. The sampling frequency in the case of the temperature measurement in the building envelope was set at the level of 1 Hz, which gives 172,000 measurements per day. A special program was developed to sample the results every 1 min.

The constructed fibre optic system intended for the measurement of temperature inside the external wall was first tested in laboratory conditions. The tests consisted of setting known values of temperature in the climatic chamber containing the set of sensors and presented in Figure 4. The tests were carried out in the Laboratory of Geosynthetics of the University of Bielsko-Biala, Poland. The laboratory is equipped with a climatic chamber with a certified temperature measurement system.

Figure 4. Climatic chamber.

The temperature inside the chamber was additionally monitored using a temperature sensor with an accuracy of 0.1 °C. The essence of the calibration calculations was to record first the heating and then the cooling process in the range of temperatures from $-20\,^\circ$C to $40\,^\circ$C. The process was repeated to determine the measurement uncertainty and hysteresis. The testing resulted in characteristics illustrating wavelength variations for every sensor. The characteristics are presented in Figure 5. The curves illustrate changes in the FBG sensor wavelength depending on temperature changes in the climatic chamber.

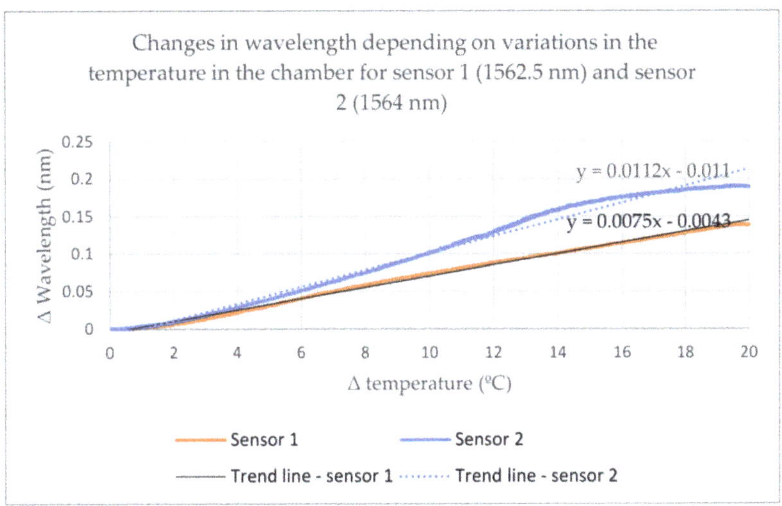

Figure 5. Changes in wavelength depending on temperature changes—sensor 1 and sensor 2.

The obtained results confirm the findings of other studies, where a change by 1 degree caused a change in the wavelength by 0.007–0.01 nm [7].

3. Results

During the testing, a high-resolution thermographic camera (type: Testo 872, Testo Sp. Z o.o., Warsaw, Poland) was also used to validate the temperature inside and outside the building. The differences in temperature between the surface of the building envelope and ambient air are determined with a thermographic camera from a single thermogram presenting the image of the envelope surface and of the object receiving the air temperature (cf. Figure 6). The object receiving the air temperature should be characterised by low thermal capacity (the change in its temperature will follow the variations in the air temperature) and a matt surface with a high emissivity value. The object should be placed about 20–30 cm away from the surface of the imaged wall. A folded sheet of matt paper can be used for example. The temperature differences specified thermographically should be determined as differences in mean temperature values in a certain field and not in a measuring point. This approach improves accuracy considerably. Because the value of the temperature difference is derived from the thermogram, quantitative thermographic measurements of buildings should be carried out using high-accuracy cameras. It is also significant that the method of thermal imaging is absolutely non-invasive (it does not disturb the investigated temperature field and has no destructive impact on the object whatsoever), and it is possible to perform remote measurements, which is of special importance in the case of testing historical building or buildings where direct access to the surface is difficult. The biggest advantages of thermography are the possibility of making fast measurements and the visual form of the result—the thermogram. The air temperature and humidity were monitored inside the room using the Testo 605i probe. The results of the external wall temperature measurements performed using a thermographic camera are presented in Figure 6. The lowest recorded value was $-7.1\,^\circ$C.

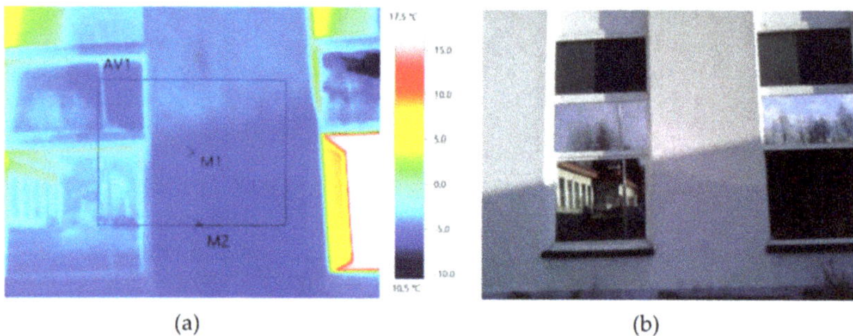

(a)   (b)

Figure 6. Temperature measurement of the wall from the outside using a thermographic camera (**a**) and the outside view of the wall (**b**).

Table 4 presents the values of temperature of the external wall of the analysed building envelope—points M1 and M2.

Table 4. Temperature values on the surface of the external wall, obtained using the thermographic camera.

Date: 12 January 2021	Time: 8:58:37	Outdoor Temperature: $-3.1\,^{\circ}$C
Measured quantity		Temperature ($^{\circ}$C)
Measuring point 1		-6.5
Measuring point 2		-7.1
Mean value of the surface measurements		-5.9

The results of the measurements of the wall temperature from the inside (the laboratory room) are presented in Figure 7. To determine the air temperature in the room adjoining the analysed building envelope, an additional measurement was performed using the camera with a white sheet of paper placed 40 cm away from the wall in the bottom left-hand corner. The measured temperature of the surface of the paper sheet, i.e., of the room air in this point totalled 18.4 °C, whereas the wall temperature was included in the range of 17.5–18.4 °C (cf. Figure 7).

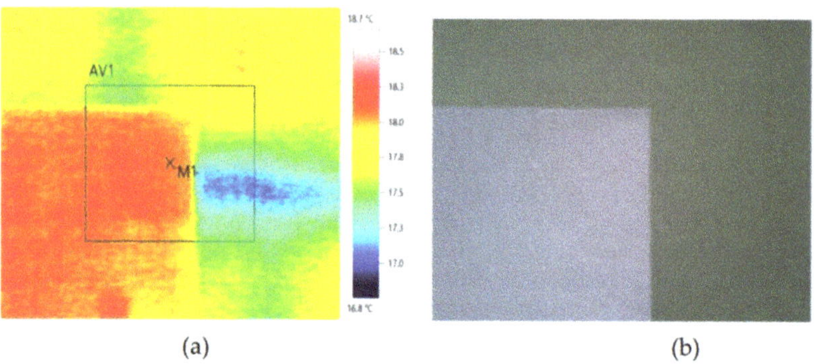

(a)   (b)

Figure 7. Temperature measurement of the wall from the inside using a thermographic camera (**a**) and the view of the wall and the white sheet of paper (**b**).

3.1. Two-Layer Wall

The testing of the temperature distribution using the FBG sensors was started in the two-layer wall. Two fibre optic sensors were placed on the border of the layers. Depending on the needs, the sensors could be moved along the building envelope thickness. This technological solution makes it possible to reduce the number of FBG sensors to the minimum. Moreover, the temperature sensors can be used many times. The space between the inside of the building envelope and the guide of the unit with the FBG sensors was filled with a special gel. This substantially reduces the costs of the FBG temperature monitoring due to the possibility of the multiple uses of the sensor unit. FBG sensors are equivalent to classical temperature measurement sensors used in the three-layer wall.

Considering the large number of data obtained for the two-layer wall, a 24-h measurement was performed. The outdoor temperature in the analysed periods varied in the range from $-1\ °C$ to $4\ °C$. Figure 8 illustrates the temperature distribution determined based on previously performed calibration tests in the calibration chamber.

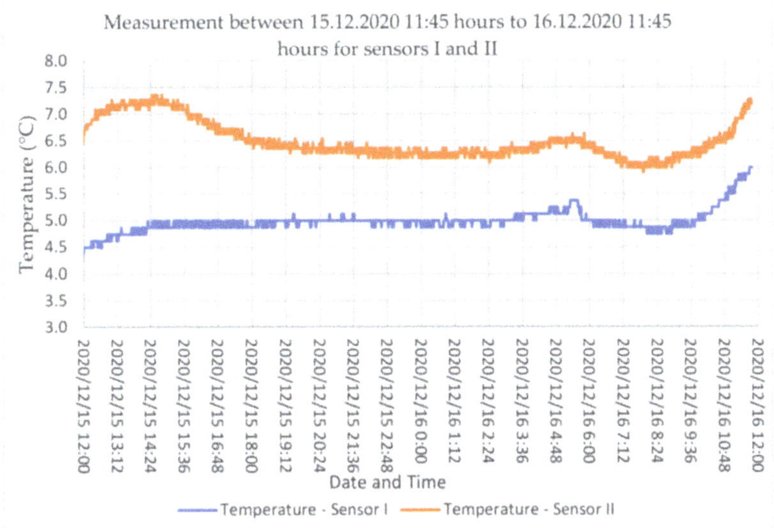

Figure 8. The 24-h temperature distribution for FBG sensors placed in the building envelope.

The temperature distribution determined experimentally inside the building envelope based on the measurements performed using fibre optic sensors and a thermographic camera is presented in Figure 9.

3.2. Three-Layer Wall

The temperature distribution was also analysed for the three-layer wall. At the stage of the building construction, classical temperature sensors were introduced with relevant infrastructure. During the building operation, sensor II was damaged and replaced with an equivalent FBG sensor. The environmental data related to the air outdoor temperature and relative humidity were obtained from a weather station. The process of temperature changes in the building envelope was analysed for extreme temperature conditions. Compared to positive values of temperature, the temperature in the building envelope was much more stable in the case of negative temperatures. Analysing the distributions determined experimentally, it can be noticed that the temperature inside the envelope does not fall below $-8\ °C$ even if the outdoor temperature keeps at the level of $-14\ °C$ for more than 5 h. It is only after the outdoor temperature rose to the temperature level in the building envelope that a slight increase in temperature could be observed between the external layer (1) and the thermal insulation. From the moment when the outdoor temperature rises at 8:15, 3 h pass before the temperature inside the building envelope starts to increase (cf. Figures 10 and 11).

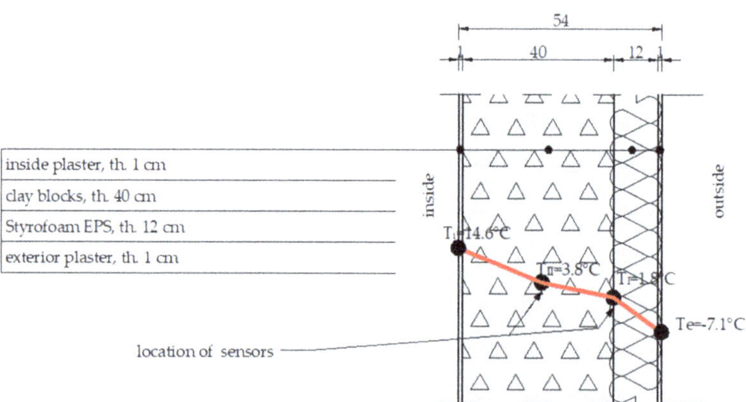

Figure 9. Temperature distribution in the building envelope during the thermographic camera measurement.

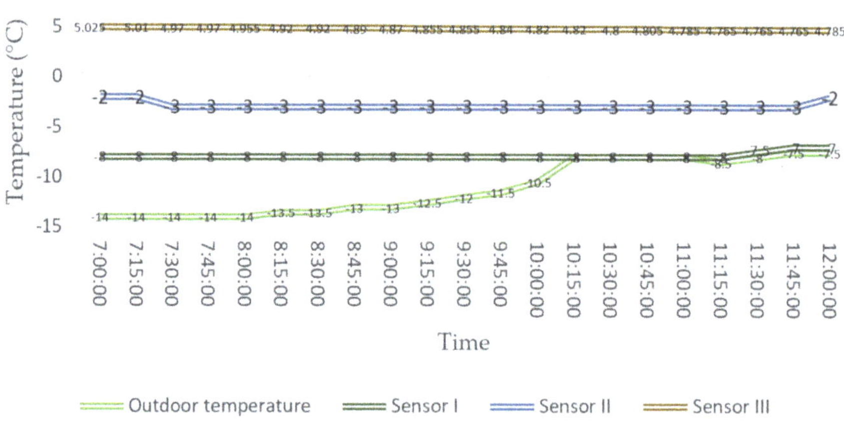

Figure 10. The 24-h distribution of outdoor temperature and temperature inside the building envelope.

In the case of positive values of the outdoor air temperature, the temperature in the building envelope is less stable and rises due to heat accumulation to levels higher than the outdoor temperature (cf. Figure 12). At the maximum outdoor temperature of 35 °C, the temperature in the air gap reaches over 38 °C. The delay in the temperature change in layer 2 due to the temperature drop from 16:45 h onwards totals 3 h. The delay totals about 3 h both for negative and positive values of outdoor temperatures.

The performed analyses also drew attention to differences between the theoretical and the real temperature distribution. They become visible especially if the air relative humidity is included in the range of 40–50% (cf. Figure 13). The theoretical and real temperature distributions in the building envelope obtained for the ambient air relative humidity of 46% and outdoor temperature of 6 °C are presented in Figure 13. Figure 14 presents the distributions for a relative humidity of 91% and outdoor temperature of 9 °C. For high values of relative humidity (90–100%), a good agreement can be observed between the theoretical and experimental temperature distribution (agreement up to 1 °C). The difference between the values determined experimentally (14.57 °C) and theoretically (18.06 °C) for the case with relative humidity at the level of 46% and the air outdoor temperature of 6 °C totals 3.49 °C. It is an essential difference, which reappeared in subsequent measurements.

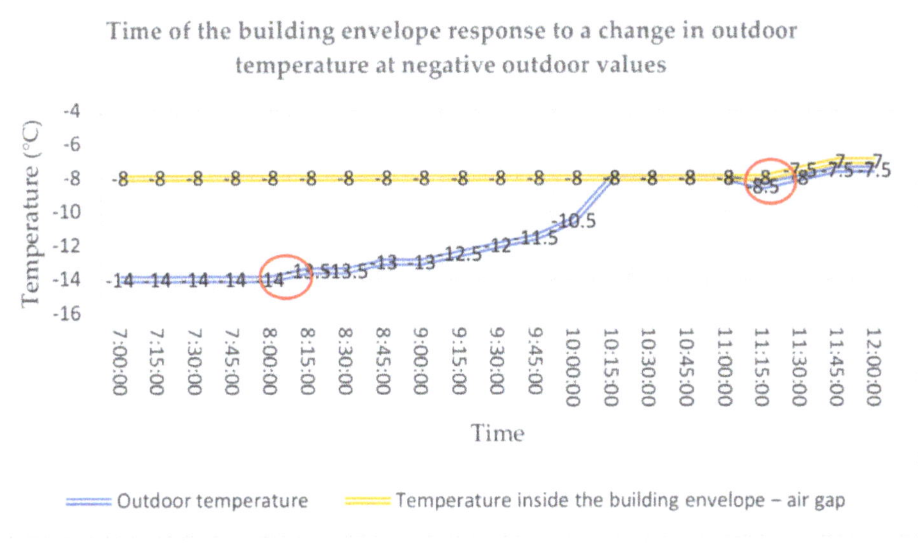

Figure 11. Time of the building envelope response to a change in outdoor temperature at negative outdoor values.

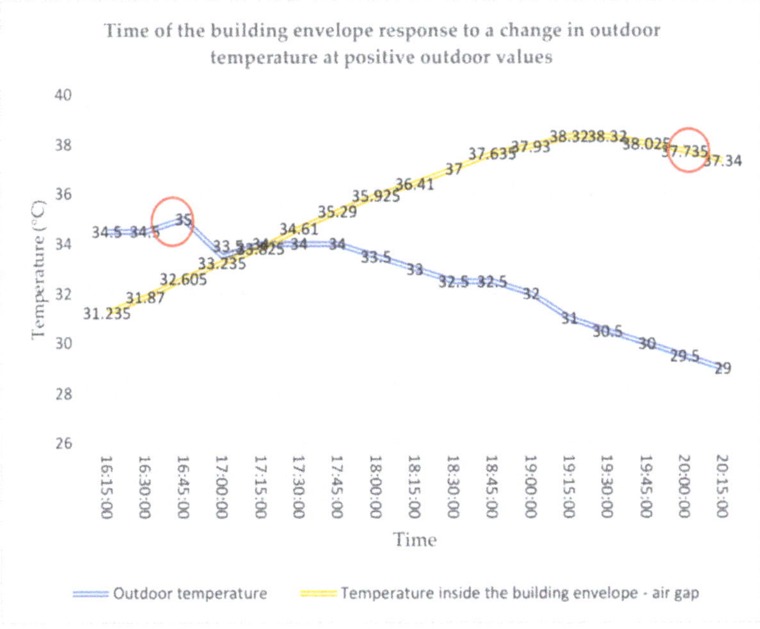

Figure 12. Time of the building envelope response to a change in outdoor temperature at positive outdoor values.

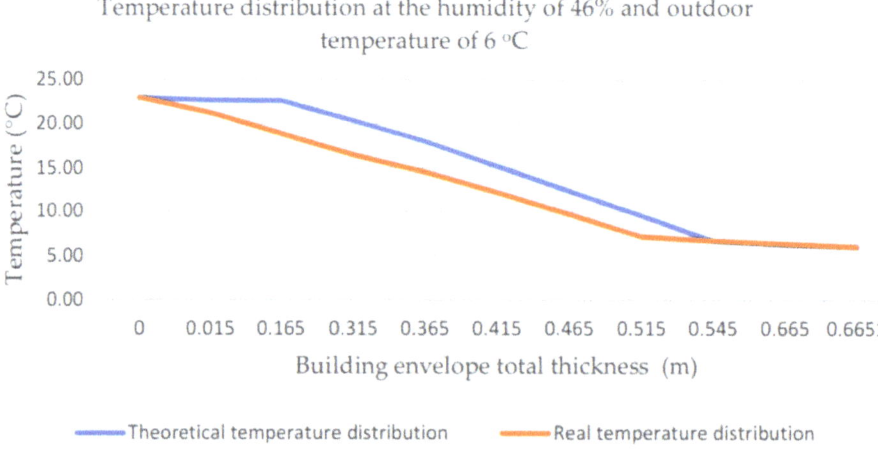

Figure 13. Temperature distribution at the humidity of 46% and outdoor temperature of 6 °C.

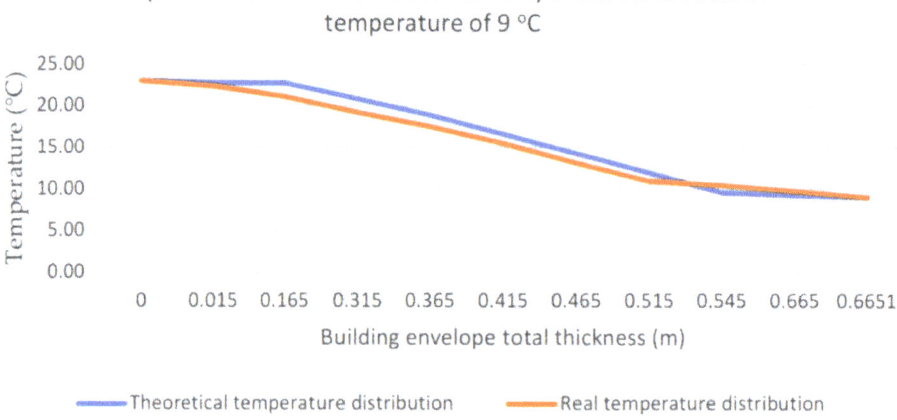

Figure 14. Temperature distribution at the humidity of 91% and outdoor temperature of 9 °C.

As already mentioned, optical FBG sensors also make it possible to determine strain values. Example strain measurement results obtained for a wall under a 5 kN load from the floor slab placed above it are presented in Figure 15. A relatively small strain of about 7 µstrain (a millionth) can be observed together with n measurements performed by the FBG-800 optical interrogator with a measuring frequency of 100 Hz. This type of sensor enables structural monitoring of building structures.

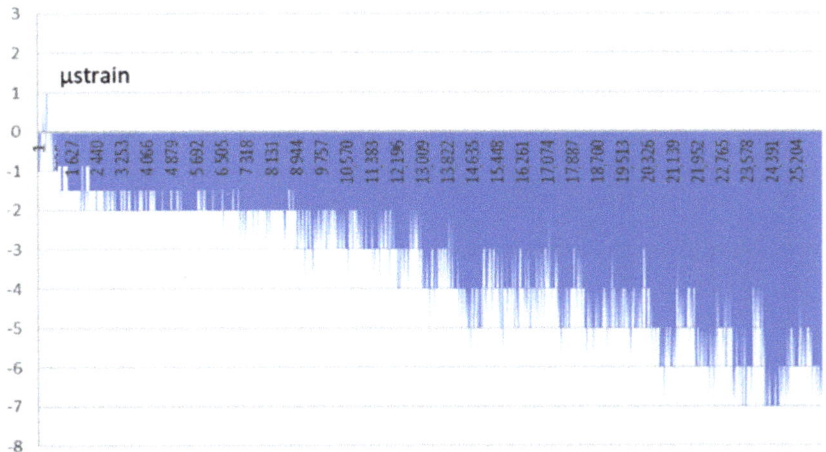

Figure 15. Strain in the building envelope.

4. Discussion

Correlation analysis (linear correlation) was conducted to establish the relationship between the air outdoor temperature and the temperature in the three-layer envelope of the building. The correlation is very strong (>75%) and the results are listed in Table 5 with respect to individual sensors embedded in the building envelope. The correlation strength decreases with the analysis of subsequent sensors located closer to the inside of the room.

Table 5. Linear correlation between the temperature inside the building envelope and the outdoor temperature.

Temperature Sensor in the Envelope	Linear Correlation ($p < 0.05000$ N = 2689) in the Winter Period
	Outdoor Temperature
Sensor I	0.937465
Sensor II	0.843755
Sensor III	0.760560

Based on the temperature differences inside the analysed building envelope between the values determined experimentally and theoretically, a correction function was developed for temperature distributions with the air relative humidity higher than 60% in the form of a fifth-degree polynomial. The polynomial is expressed as follows:

$$\Delta t = y = 83.772x^4 - 48.362x^3 - 50.793x^2 + 29.504x + 0.4736 \tag{2}$$

where x is the total thickness of the building envelope from the internal wall to the analysed point of temperature determination (usually the border between the layers).

To correct the theoretical temperature in a given point of the building envelope, a correction factor has to be added as follows:

$$t_x = t_t + \Delta t_L \tag{3}$$

where

t_x—corrected temperature;

t_t—theoretical temperature;

Δt_L—correction factor for the theoretical temperature distribution at a low humidity value.

Figure 16 presents the implementation of the polynomial approximation of the temperature differences of the envelope.

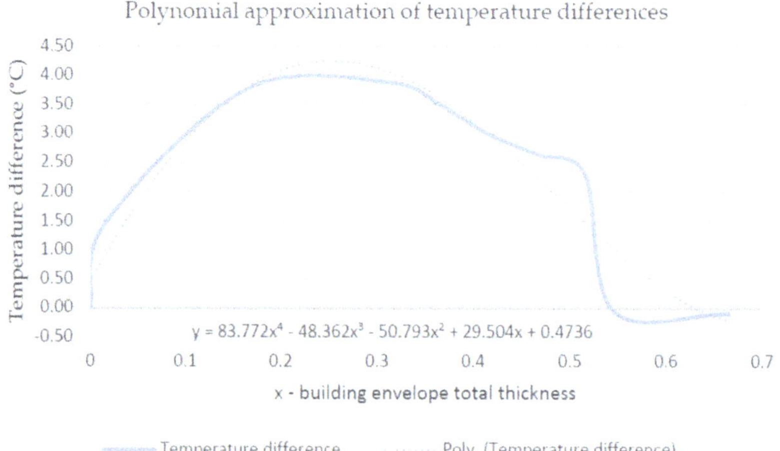

Figure 16. Polynomial approximation of temperature differences.

Table 6 presents polynomial approximation for an example of the distribution of the theoretical temperature in the analysed envelope.

Table 6. Polynomial approximation of the temperature distribution.

Theoretical Temperature on the Interface of Layers	Real Temperature	Temperature Difference	Polynomial Approximation of Differences (Correction)	Post-Correction Temperature on the Interface of Layers
23.00	23.00	0.00	0.47	22.53
22.72	21.27	1.45	0.90	21.82
22.68	18.91	3.77	3.80	18.88
20.37	16.50	3.88	4.04	16.33
18.06	14.57	3.50	3.61	14.45
15.22	12.23	2.99	3.00	12.23
12.39	9.74	2.65	2.26	10.12
9.55	7.25	2.30	1.48	8.07
6.71	6.76	−0.05	1.03	5.68
6.33	6.42	−0.09	−0.21	6.54
6.09	6.09	0.00	−0.21	6.29

The developed approximation function applied to other temperature distributions makes it possible to correct the theoretical distribution and bring it closer to the real one. Figures 17 and 18 present the theoretical and the real temperature distribution together with the distribution corrected using the developed correction factor for two selected temperature measurements at the air humidity of 46% and 24% and the outdoor temperature of 9 °C and 8 °C, respectively.

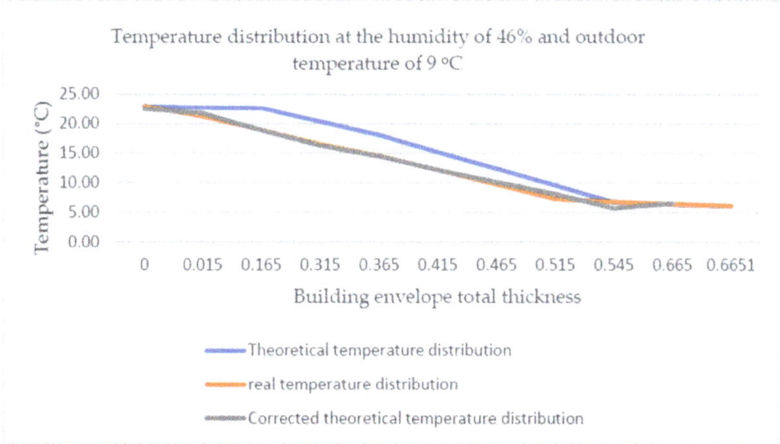

Figure 17. Temperature distribution at the humidity of 46% and outdoor temperature of 9 °C together with the corrected distribution.

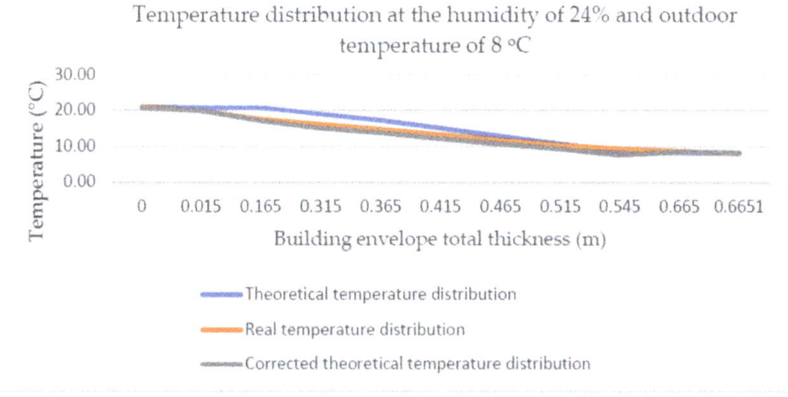

Figure 18. Temperature distribution at the humidity of 24% and outdoor temperature of 8 °C together with the corrected distribution.

5. Conclusions

The paper presents an innovative method of monitoring the temperature of the building envelope using FBG temperature sensors. The following conclusions can be drawn:

1. It is demonstrated that the fibre optic technology based on multiplexed FBG sensors enables effective measurement of temperature in the building envelope. The thermal boundary conditions on the outside of the analysed envelope can be obtained by means of measurements performed using a thermographic camera. FBG sensors can be used instead of classical temperature sensors, especially if the latter types are damaged. The hybrid method of the measurement of temperature distributions consists of implementing various measuring methods, which enhances the effectiveness of performed measurements and may contribute to a reduction in the costs of conducting the experiment;

2. The calibration tests demonstrate a linear dependence between temperature and the wavelength measured using fibre optic sensors;

3. Differences are observed between the theoretical and the real temperature distribution in the building envelope for the ambient air relative humidity level of 40–50%;
4. A correction factor is proposed to correct the theoretical temperature distribution in the building envelope. The factor makes it possible to eliminate the difference between the theoretical and the real temperature distribution;
5. Considering the latest technological progress in the field of optical interrogators, the cost is comparable to classical methods of temperature measurements in the building wall. In fact, it can even be included in the low-cost group, while the computational cost is acceptable for large facilities;
6. In the future, the system presented in the paper will enable comprehensive thermal and structural monitoring of building structures. The determined real thermal delays occurring in the wall will make it possible to optimise the building heating/cooling strategy. The knowledge of temperature distributions in the building walls depending on climatic conditions will enable the selection of a heating system suitable for a given wall configuration, which may reduce energy consumption.

Author Contributions: Conceptualization, J.J.; methodology, J.J.; software, P.A.-P.; validation, J.J. investigation, P.A.-P. and J.J.; resources, J.J.; writing—original draft preparation, P.A.-P.; writing—review and editing, J.J. and P.A.-P.; visualization, P.A.-P.; supervision, J.J. All authors have read and agreed to the published version of the manuscript.

Funding: This research received no external funding.

Institutional Review Board Statement: Not applicable.

Informed Consent Statement: Not applicable.

Data Availability Statement: The data presented in this study are available on request from the corresponding author.

Conflicts of Interest: The authors declare no conflict of interest.

References

1. Ahola, S.; Lahdensivu, J. Long term monitoring of repaired external wall assembly. In Proceedings of the 11th Nordic Symposium on Building Physics (NSB2017), Trondheim, Norway, 11–14 June 2017.
2. Marino, B.M.; Muñoz, N.; Thomas, L.P. Calculation of the external surface temperature of a multi-layer wall considering solar radiation effects. *Energy Build.* **2018**, *174*, 452–463. [CrossRef]
3. Liu, T.; Sun, W.; Kou, H.; Yang, Z.; Meng, O.; Zheng, Y.; Wang, H.; Yang, X. Experimental Study of Leakage Monitoring of Diaphragm Walls Based on Distributed Optical Fiber Temperature Measurement Technology. *Sensors* **2019**, *19*, 2269. [CrossRef]
4. Juarez, J.C.; Maier, E.W.; Choi, K.N.; Taylor, H.F. Distributed fiber-optic intrusion sensor system. *J. Lightwave Technol.* **2005**, *23*, 2081–2087. [CrossRef]
5. Minardo, A.; Bernini, R.; Zeni, L. Distributed Temperature Sensing in Polymer Optical Fiber by BOFDA. *IEEE Photonics Technol. Lett.* **2014**, *26*, 387–390. [CrossRef]
6. Bao, X.Y.; Liang, C. Recent progress in distributed fiber optic sensors. *Sensors* **2012**, *12*, 8601–8639. [CrossRef]
7. Čápováa, K.; Velebila, L.; Včeláka, J.; Dvořákb, M.; Šašekb, L. Environmental Testing of a FBG Sensor System for Structural Health Monitoring of Building and Transport Structures. In Proceedings of the ICSI 2019 The 3rd International Conference on Structural Integrity, Procedia Structural Integrity, Funchal, Madeira, Portugal, 2–5 September 2019.
8. Velebil, L.; Čápová, K.; Včelák, J.; Kuklík, P.; Demuth, J.; Dvořák, M. Mechanical Stress Monitoring of Timber and Concrete Structures by Fibre Optic Sensors. In Proceedings of the WCTE 2018, Seoul, Korea, 20–23 August 2018.
9. Kersey, A.D. A Review of Recent Developments in Fiber Optic Sensor Technology. *Opt. Fiber Technol.* **1996**, *2*, 291–317. [CrossRef]
10. Othonos, A.; Kyriacos, K. *Fiber Bragg Gratings: Fundamentals and Applications in Telecomunications and Sensing*; Artech House Print on Demand: Norwood, MA, USA, 1999; ISBN 0-89006-344-3.
11. Chong, K.; Carino, N.; Washer, G. Health monitoring of civil infrastructures. *Smart Mater. Struct.* **2003**, *12*, 483. [CrossRef]
12. Ansari, F. Fiber optic health monitoring of civil structures using long gage and acoustic sensors. *Smart Mater. Struct.* **2005**, *14*, S1–S7. [CrossRef]
13. Cosentino, P.; Eckroth, W.; Grossman, B. Analysis of fiber optic traffic sensors in flexible pavements. *J. Transp. Eng.* **2003**, *129*, 549–557. [CrossRef]
14. Liu, W.; Wang, H.; Zhou, Z.; Li, S.; Ni, Y.; Wang, G. Optical fiber based sensing system design for the health monitoring of multi-layered pavement structure. In Proceedings of the SPIE 8199, 2011 International Conference on Optical Instruments and Technology: Optical Sensors and Applications, Beijing, China, 6–9 November 2011.

15. Zhong, R.; Guo, R.; Deng, W. Optical-fiber-based smart concrete thermal integrity profiling: An example of concrete shaft. *Adv. Mater. Sci. Eng.* **2018**, *2018*, 9290306. [CrossRef]
16. Sun, X.; Du, Y.; Liao, W.; Ma, H.; Huang, J. Measuring the heterogeneity of cement paste by truly distributed optical fiber sensors. *Constr. Build. Mater.* **2019**, *225*, 765–771. [CrossRef]
17. Liao, W.; Zhuang, Y.; Zeng, C.; Deng, W.; Huang, J.; Ma, H. Fiber Optic Sensors Enabled Monitoring of Thermal Curling of Concrete Pavement Slab: Temperature, Strain and Inclination. *Measurement* **2020**, *165*, 108203. [CrossRef]
18. Weng, X.; Zhu, H.-H.; Chen, J.; Liang, D.; Shi, B.; Zhang, C.-C. Experimental investigation of pavement behavior after embankment widening using a fiber optic sensor network. *Struct. Health Monit. Int. J.* **2014**, *14*, 46–56. [CrossRef]
19. Loizos, A.; Plati, C.; Papavasiliou, V. Fiber optic sensors for assessing strains in cold in-place recycled pavements. *Int. J. Pavement Eng.* **2013**, *14*, 125–133. [CrossRef]
20. Wang, J.; Tang, J.; Chang, H. Fiber Bragg grating sensors for use in pavement structural strain-temperature monitoring. In Proceedings of the SPIE 6174, Smart Structures and Materials 2006: Sensors and Smart Structures Technologies for Civil, Mechanical, and Aerospace Systems, San Diego, CA, USA, 27–28 February 2006.
21. Hirayama, N.; Sano, Y. Fiber Bragg grating temperature sensor for practical use. *ISA Trans.* **2000**, *39*, 169–173. [CrossRef]
22. Juraszek, J. Fiber Bragg Sensors on Strain Analysis of Power Transmission Lines. *Materials* **2020**, *13*, 1559. [CrossRef] [PubMed]
23. Juraszek, J. Hoisting machine brake linkage strain. *Arch. Min. Sci.* **2018**, *63*, 583–597.
24. Zhang, B.; Zhang, M.; Hong, D. Land surface temperature retrieval from Landsat 8 OLI/TIRS images based on back-propagation neural network. *Indoor Built Environ.* **2019**, *30*, 1–17. [CrossRef]

Article

Comparative Analysis of Slip Resistance Test Methods for Granite Floors

Ewa Sudoł [1,*], Ewa Szewczak [2] and Marcin Małek [3]

[1]   Construction Materials Engineering Department, Instytut Techniki Budowlanej, 00-611 Warszawa, Poland
[2]   Group of Testing Laboratories, Instytut Techniki Budowlanej, 00-611 Warszawa, Poland; e.szewczak@itb.pl
[3]   Faculty of Civil Engineering and Geodesy, Military University of Technology in Warsaw, 01-476 Warsaw, Poland; marcin.malek@wat.edu.pl
*   Correspondence: e.sudol@itb.pl; Tel.: +48-22-56-64-286

Abstract: This paper attempts to compare three methods of testing floor slip resistance and the resulting classifications. Polished, flamed, brushed, and grained granite slabs were tested. The acceptance angle values (α_{ob}) obtained through the shod ramp test, slip resistance value (SRV), and sliding friction coefficient (μ) were compared in terms of the correlation between the series, the precision of each method, and the classification results assigned to each of the three obtained indices. It was found that the evaluation of a product for slip resistance was strongly related to the test method used and the resulting classification method. This influence was particularly pronounced for low roughness slabs. This would result in risks associated with inadequate assessments, which could affect the safe use of buildings facilities.

Keywords: slip resistance; granite floor; slip resistance value; ramp test; acceptance angle; sliding friction coefficient; comparability of test methods

Citation: Sudoł, E.; Szewczak, E.; Małek, M. Comparative Analysis of Slip Resistance Test Methods for Granite Floors. *Materials* **2021**, *14*, 1108. https://doi.org/10.3390/ma14051108

Academic Editor: Neven Ukrainczyk

Received: 31 January 2021
Accepted: 23 February 2021
Published: 27 February 2021

Publisher's Note: MDPI stays neutral with regard to jurisdictional claims in published maps and institutional affiliations.

Copyright: © 2021 by the authors. Licensee MDPI, Basel, Switzerland. This article is an open access article distributed under the terms and conditions of the Creative Commons Attribution (CC BY) license (https://creativecommons.org/licenses/by/4.0/).

1. Introduction

Slip resistance of granite floors is a performance that determines the fulfilment of basic requirement no. 4 (safety and accessibility in use), which, according to Annex I to Regulation (EU) No 305/2011 of the European Parliament and of the Council (CPR) [1], is one of the seven basic requirements to be met by construction works as a whole and by their separate parts. The construction works must be designed and built in such a way that they do not present unacceptable risks of accidents or damage in service or in operation, such as slipping.

Individual European countries have defined more or less specific requirements in this respect. In the UK, the criterion used is the pendulum test value (PTV) of at least 36 units [2]. In Germany, a classification based on the acceptable angle value has been developed, expressed in classes from R9 to R13 [3]. In Italy, the value of the dynamic coefficient of friction for which an acceptable threshold of more than 0.4 has been established is considered for evaluation [4]. In Poland, the issue of slip resistance of granite floors in rooms intended for permanent human occupation is regulated by technical conditions that should be met by buildings and their location [4]. They indicate that the surface of entrances to buildings, external and internal stairs and ramps, passageways in the building and floors in rooms intended for human occupancy, and garage floors should be made of materials that do not pose a slipping hazard. Detailed evaluation criteria are further provided in the Ministry of Investment and Development's guide, indicating a PTV of at least 36 units [5].

The inadequate slip resistance of a floor carries a risk of slipping. According to the Polish Central Statistical Office, it is, together with trips and falls, one of the leading causes of injuries [6]. These data correspond with the results of analyses conducted by Kemmlert and Lundholm [7] on behalf of the Swedish Council for Occupational Safety and

229

Health, which indicated that slips, trips, and falls account for 17–35% of accidents. Their consequences can be very serious, and treatment can be lengthy and expensive. The most common result is a sprain or fracture of a limb, but more severe cases, such as concussion, have also been reported. It is estimated that one in five slip and fall accidents result in injuries that cause at least one month period of incapacity for work. The slip resistance of floors can be considered a socioeconomically relevant problem.

Slip resistance of floors is determined by several factors, including the properties of the material from which the floor is made, the conditions of its use, the psychophysical state of the user, and the properties of footwear [8,9]. Statistical data indicates that accidents due to slipping most often occur in the autumn–winter period, mainly in public buildings [6] where, as indicated by the analysis of contemporary architectural trends, large-format stone or ceramic tiles with high gloss and smooth surface dominate (Figure 1).

(**a**) (**b**)

Figure 1. Contemporary floor in public facilities: (**a**) a commercial function, (**b**) a service function.

One of the most popular flooring solutions used in public facilities, granite slabs, was used for this study. Slabs of medium-grained Strzegom granite, light gray in color and obtained from the Strzegom–Sobótka massif deposits located in the Sudetes Foreland block, were used. Strzegom granite slabs have been used for flooring in buildings such as the Peace Palace in The Hague, the Palace of Culture and Science in Warsaw, the 10th-Anniversary Stadium in Warsaw, the Congress Centre in Berlin, underground stations in Warsaw, Berlin, and Vienna, numerous office and commercial buildings, railroad stations, underground passages, as well as boulevards, bridges, markets, and squares [10].

Evaluation of slip resistance is conducted using a variety of test methods. Consequently, it is expressed in different parameters that form the basis for independent classifications. Among the most popular is the classification based on slip resistance value (SRV; Figure 2), the acceptance angle value determined by the ramp test (Figure 3), and the dynamic friction coefficient (μ) (Figure 4).

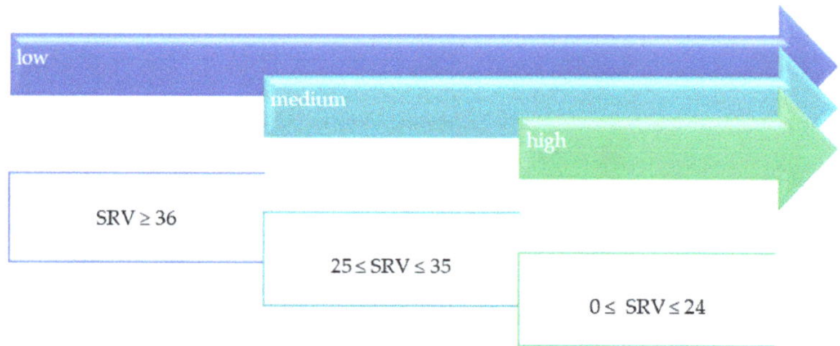

Figure 2. Slip risk in relation to slip resistance value (SRV).

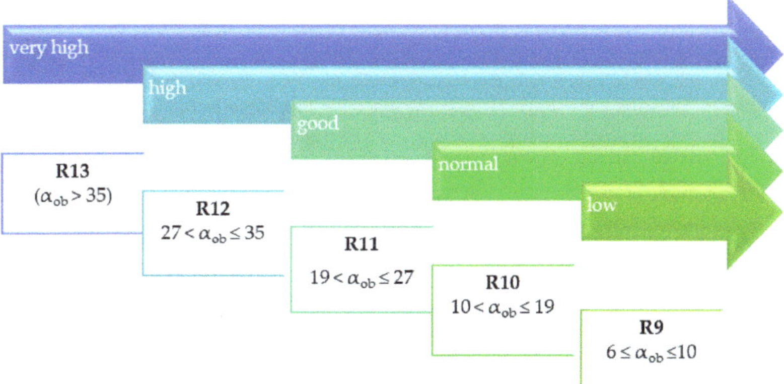

Figure 3. Slip resistance depending on the antislip class and the corresponding values of the acceptance angle (α_{ob}) determined by the shoe foot method.

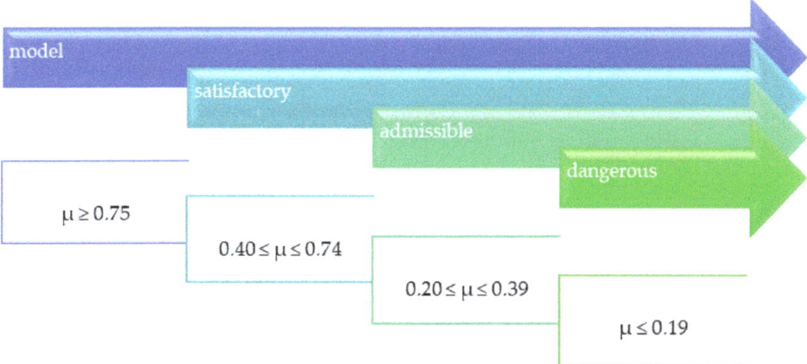

Figure 4. Slip resistance of floors depending on the value of the dynamic coefficient of friction μ.

Classification based on SRV, also referred to as PTV, was developed by the UK Slip Resistance Group [2] and introduced into the Health and Safety Executive guidelines [11]. It shows the risk of slipping depending on the value of SRV (Figure 2). It is assumed that the probability of slipping on a floor with an SRV $\geq$ 36 is 1:1,000,000, while it increases up to 1:20 with an SRV < 24 [2].

The values of the acceptance angle (α_{ob}) determined by the shod ramp test are the basis for determining the slip resistance class. There are five classes in accordance with DIN 51130 [3]. Solutions classified as R9 have the lowest slip resistance, while those corresponding to class R13 have the highest (Figure 3).

A separate classification was developed based on the value of the dynamic coefficient of friction (μ) (Figure 4). It is assumed that the floors for which the value of μ is higher than 0.75 can be regarded as antislip [12], while μ above 0.4 is regarded as an acceptable value [4].

The classification scales associated with the various slip resistance test methods are not compatible [13–15]. The feelings of people walking on the floors are not always reflected by the values of the coefficients obtained by methods using only test equipment, as shown Choi et al.'s work [14]. An additional source of confusion is the fact that subjective evaluations are expressed on a nominal scale, while measurement results (e.g., sliding friction coefficient (μ)) are expressed on ratio or interval scales.

Attempts are being made to create new classification methods. For example, Çoşkun [16] proposed a new sliding risk scale on natural stones surfaces. Cluster analysis was performed by the author using the *k*-means method. This resulted in a better resolution classification (results were assigned to more classes than in the case of standard methods used). While it would seem appropriate to create a new scale with better resolution that is universal, habits of a particular type of classification and requirements in higher-level documents expressed in the classification scales used in a particular country may prevail in different countries. In view of this, the best way for unification might be to create a matrix of result dependencies and classifications in the existing set, rather than expanding the set.

Following this pattern, this study compares the results attributed to three different classifications. One of the research methods used is a subjective evaluation, but the study results' obstacle of nominal scale is removed. In the ramp test, the result is expressed in a physical unit of measure (tilt angle expressed in degrees), although its evaluation is typically subjective. The acceptance angle values obtained through the ramp test, slip resistance value, and sliding friction coefficient (μ) are compared in terms of the correlation between the series, the precision of each method, and the classification results assigned to each of the three obtained indices. A full comparison of classifications and the creation of possible conversion factors will be possible after results are obtained for a large group of different materials.

2. Materials and Methods

2.1. Materials

Granite slabs with different processing textures were used for study, namely polished (PO), flamed (PL), brushed (SZ), and grained (GR). Their characteristics, including roughness parameters, are shown in Table 1.

Table 1. Characteristics of granite slabs.

Sample's Mark	Processing Texture	Roughness Parameters, µm				Surface Characteristics
		S_a	S_z	R_a	R_z	
PO	polishing	4.05	213.87	1.85	25.96	high degree of smoothness, shine
PL	flaming	37.28	149.19	25.93	130.47	appearance close to natural fracture, with clear changes in the surface of the quartz grains caused by temperature and flame
SZ	brushed	38.20	556.38	27.11	294.39	clear roughness with abrasive scratches
GR	graining	97.11	839.01	83.75	418.39	even but rough surface with characteristic regular concaves and convexities

An Olympus OLS4100 laser scanning digital noncontact microscope (Olympus, Tokyo, Japan) was used to measure roughness parameters. *Sa* and *Ra* values, expressing roughness parameters, were determined according to ISO 3274 [17] and ISO 4288 [18]. The use of the noncontact method (laser beam) in the procedure of measuring the parameters of the geometric structure of the surface, especially the roughness profile, significantly improves the accuracy of the measurement by eliminating the effect of rounding the measuring tip used in the contact method. A $5\times$ objective lens was utilized at $864\times$ total magnification in mixed observation mode. Measurement resolution (laser measurement) was 200 nm. The observed area was 2560–320 µm. The raw *Ra* and *Sa* values were utilized to obtain bulk surface roughness information. *Ra* was measured in 10 different regions, with 10 profiles chosen from each region approximately equidistant from one another. Five equidistant measurements were taken along the sample's length. The sample was then rotated roughly 180°, and a subsequent five additional measurements were taken. The total number of *Ra* and *Sa* measurements were arithmetically averaged to obtain the final value of *Ra* and *Sa*. To verify the robustness of the method, reproducibility tests were conducted on five samples.

2.2. Slip Resistance Value

The SRV test was conducted using an instrument (WESSEX, Aldershot, UK) referred to as a British pendulum (Figure 5). The test technique used was in compliance with EN 14231 [19], which is in accordance with CEN/TS 16165 [20] Annex C. The test was to determine the energy loss of the slider due to friction against the test surface. A Type 57 (CEN) slider made of 55–61 International Rubber Hardness Degrees (IRHD) rubber (WESSEX, Aldershot, UK) was used, with a width of 76.2 mm and a slid length of 126 mm. The frictional force between the slider and the test surface was determined by measuring the pendulum deflection while the slider was moving using the C scale. Before testing, the instrument was calibrated using reference substrates, namely glass, a reference plate, and polishing paper. Measurements were conducted under dry and wet conditions (after wetting both the sample and the slider with distilled water). The test was conducted on two samples of a given solution, with 10 measurements in dry conditions and 10 measurements in wet conditions taken in each series.

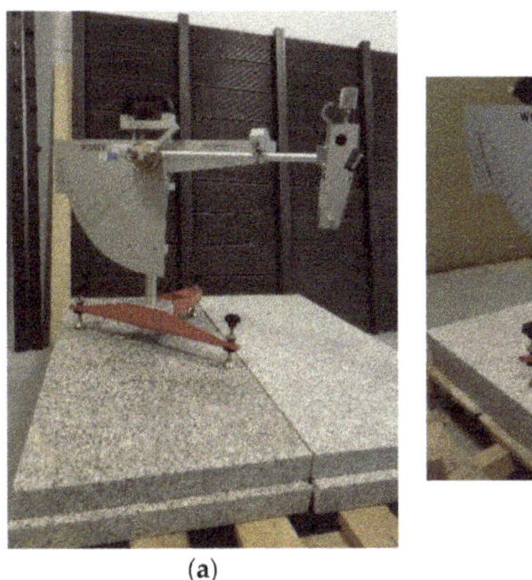

(a)   (b)

Figure 5. SRV test: (**a**) polished granite slabs (PO) in wet condition, (**b**) brushed granite slabs (SZ) in dry condition.

2.3. Ramp Test

Ramp test was performed using the shoe foot method according to CEN/TS 16165 [20] Annex B, corresponding to DIN 51130 [3]. The test was to determine the acceptance angle (α_{ob}), which is the maximum angle of the sample in relation to the level at which a person walking on the floor begins to slip. The researchers walked on ramp (Gabrielli SRL, Florence, Italy) in an upright posture, forward and backward. At the same time, the angle of the sample was changed from a horizontal position (Figure 6a) to an angle at which the researcher no longer felt confident and could not continue walking (Figure 6b). The tests were conducted independently by two researchers. The subjectivity of their experiences was reduced using calibration liners, and the resulting corrections were incorporated into the acceptance angle value. Three standard liners were used for the calibration process. The liners' acceptance angles were 8.7, 17.3, and 27.3°, respectively. Each person walked on each standard liner three times, and the mean calibration acceptance angle values were determined. Each individual correction value $\Delta\alpha$ was calculated as a difference between the liners' acceptance angle and the calibration acceptance angles. Each of the individual correction value $\Delta\alpha$ was less than the critical differences ($\leq 3.0°$). If one of the absolute values was greater, the test person in question would be excluded from the test. Correction value (*Dj*) was calculated from the values obtained from the calibration liners' values. The calculation of *Dj* was carried out as follows:

$$D_j = \Delta\alpha - \frac{1}{\sqrt{2}}$$

(1)

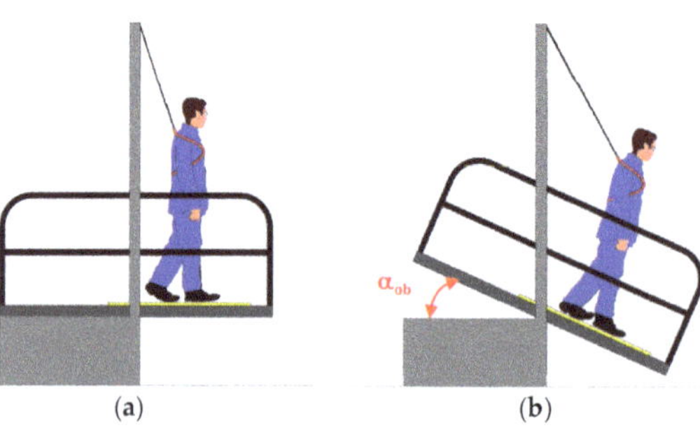

Figure 6. Ramp test: (**a**) starting position, (**b**) position at maximum acceptable angle α_{ob}.

The test samples were covered with engine oil during testing, and the researchers wore standardized footwear with properly profiled rubber sole. The contact pressure values in the ramp test hovered around the level (1.8–2.0) N/cm^2 when in static state.

2.4. Sliding Friction Coefficient

Sliding friction coefficient (μ) test was performed according to CEN/TS 16165 [20] Annex D. A tribometer (GTE Industrieelektronik GmbH, Viersen, Germany) equipped with sliders imitating shoe heels, exercising a total contact pressure of 9 ± 1 N/cm^2 when in static state, was used (Figure 7). Moving along two intersecting paths with a constant speed of 0.2 m/s, the device recorded the frictional force between the slider and the sample. The dynamic coefficient of friction (μ) was calculated as the quotient of the frictional force and the contact force of the slider on the sample. Two different sets of sliders were used in the study. The first set consisted of three sliders with styrene–butadiene rubber (SBR) with a density of 1.23 g/cm^3 and a Shore D hardness of 50. In the second set, the rear slider was made of SBR and the front sliders were made of tanned leather with a density of 1.0 g/cm^3 and a Shore D hardness of 60. Prior to testing, the instrument was calibrated with reference substrates, namely glass, high-pressure laminate (HPL), and ceramic tile. Measurements were carried out in dry and wet conditions (after wetting the sample with demineralized water). The test was carried out on two samples of a given type, with 10 measurements in dry conditions and 10 measurements in wet conditions taken for each sample.

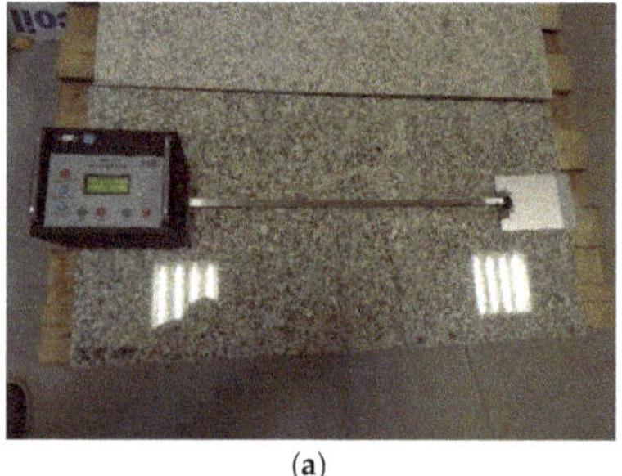 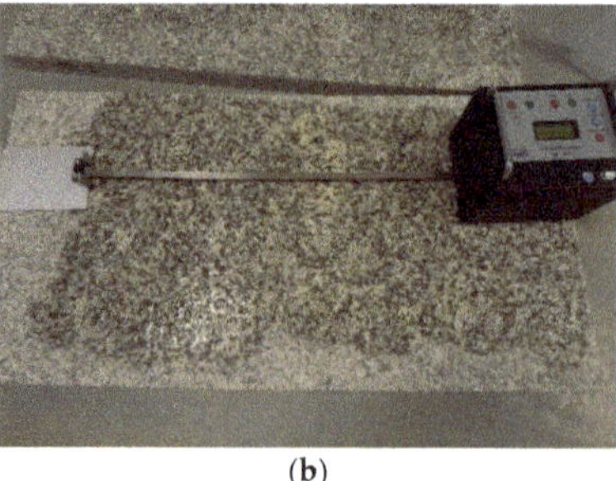

Figure 7. Dynamic coefficient of friction measurement: (**a**) PO series in dry condition, (**b**) flamed (PL) series in wet condition.

3. Results

3.1. Slip Resistance Value

The SRV obtained in this work is summarized in Figure 8 against the arithmetic mean of the profile deviation from the mean line (*Ra*) expressing the surface roughness. Comparing these results to the criteria developed by the UK Slip Resistance Group [2], it can be concluded that an SRV $\geq$ 36, which is taken as an indicator of low slip risk (Figure 2), was achieved in dry conditions by all the tested solutions. The SRV depended on the type of treatment texture [21,22]. The lowest SRV value was obtained for PO with *Ra* equal to 1.85 µm (60 units), followed by PL with *Ra* equal to 25.9 µm (77 units) and SZ with *Ra* equal to 27.1 µm (93 units). The highest SRV value was for GR with *Ra* equal to 83.7 µm (98 units). The increase in *Ra* was generally accompanied by a nearly proportional increase in SRV for polished, flamed, and brushed slabs. For the grained slabs, the increase in *Ra* did not fully translate into SRV values. The above may be due to the specific surface profile of the grained slabs. Out of the tested range, only this type of slab has concavities and convexities with a circular shape, which may cause a different adhesion of the slider in the PTV test. As in other works [13,23], a significant reduction in SRV was observed under wet conditions. However, it remained satisfactorily above 36 units for the flamed, brushed, and grained slabs. There was a decrease to 22 units for polished tiles, indicating that the floor poses a very high risk of slipping under these conditions. The SRV results obtained under dry conditions corresponded with the study of Karaca et al. [24], who obtained SRV values ranging from 42 to 74 units for granite slabs with slightly lower roughness than those tested in this study. In contrast, the slip resistance value determined in this study under wet conditions was significantly more favorable. In the aforementioned work, they ranged from 9 to 12 units.

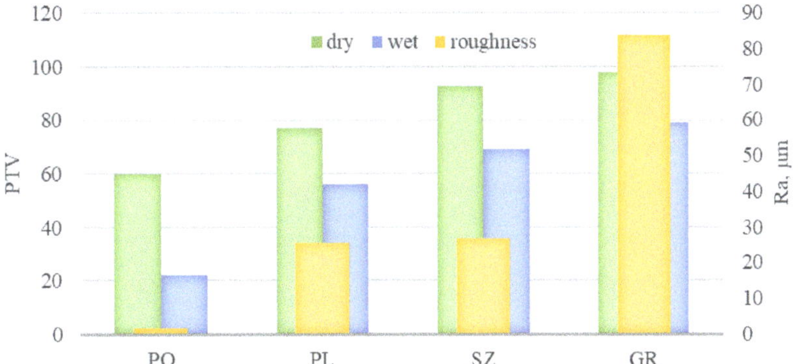

Figure 8. Results of pendulum test value (PTV) slip resistance tests for PO, PL, SZ, and grained (GR) granite slabs under dry and wet conditions against surface roughness expressed as *Ra*.

3.2. Ramp Test

The above gradation of the slip resistance of granite slabs in terms of SRV values was quite well reflected in the acceptance angle values determined by the shod ramp method (Figure 9). It was found that the least resistant to slipping in terms of α_{ob} values were PO slabs, which according to [3] should be regarded as out-of-class ($\alpha_{ob} \leq 6°$), followed by PL of class R10 ($10° < \alpha_{ob} \leq 19°$) and GR series of class R12 ($27° < \alpha_{ob} \leq 35°$). The highest antislip properties were exhibited by SZ slabs with α_{ob} equal to 34.7°; however, this also places them in class R12. The above classification leads to the conclusion that the resistance of PL series slabs can be considered normal, while SZ and GR series slabs can be considered high (Figure 3), which generally corresponds to the classification according to SRV in dry conditions. However, polished slabs with very low roughness (*Ra* equal to 1.85 µm) are noteworthy. In the SRV test in dry conditions, they obtained a result of $\geq$ 36, which translates into a high score of slip resistance according to the criteria of the UK Slip Resistance Group [2]. The result of an $\alpha_{ob} \leq 6°$, on the other hand, should be considered as disqualifying the solution in the context of floor application.

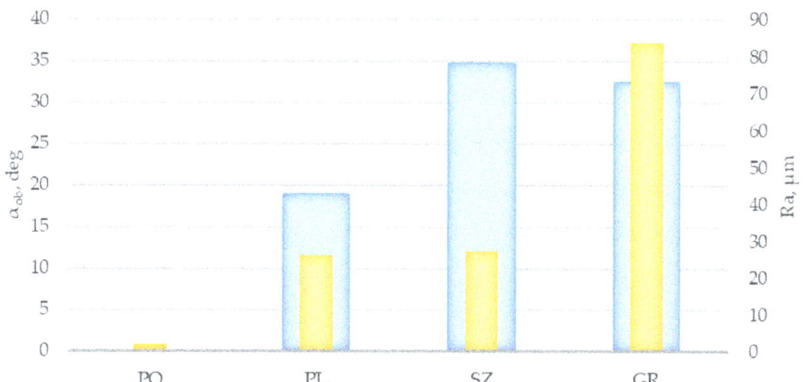

Figure 9. Test results of the acceptance angle (α_{ob}) with the shoe foot method for PO, PL, SZ, and GR granite slabs against surface roughness expressed by *Ra*.

3.3. Sliding Friction Coefficient

The third method used in this study to verify slip resistance was to measure the dynamic friction coefficient, which was conducted under conditions analogous to those used in the slip resistance test. The results obtained with the rubber slider set (μ_R) are shown in Figure 10 and those with the leather slider set (μ_L) are shown in Figure 11. In both cases, the dynamic friction coefficient values were close to the value of slip resistance. The lowest μ_R and μ_L values under both dry and wet conditions were obtained for PO slabs at 0.57 and 0.39 and 0.46 and 0.42, respectively, followed by PL slabs at 0.58 and 0.54 and 0.58 and 0.47, respectively. The μ_R and μ_L results obtained for brushed and grained slabs were nearly the same level under dry conditions at 0.68 and 0.73 and 0.67 and 0.72, respectively, while the values under wet conditions were 0.66 and 0.60 and 0.67 and 0.59, respectively. Analyzing the obtained μ_R and μ_L results in the context of the criteria [25] all tested solutions except for the PO series in the test with rubber sliders can be assigned to the level of $0.40 \leq \mu \leq 0.74$, which means that the slabs showed a satisfactory slip resistance in both in dry and wet conditions (Figure 4).

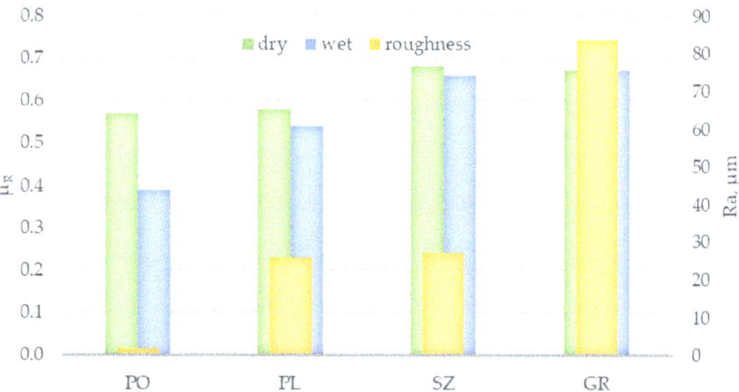

Figure 10. Results of dynamic friction coefficient (μ_R) of PO, PL, SZ, and GR granite slabs, determined in dry and wet conditions using a set of rubber sliders against the background of surface roughness expressed by *Ra*.

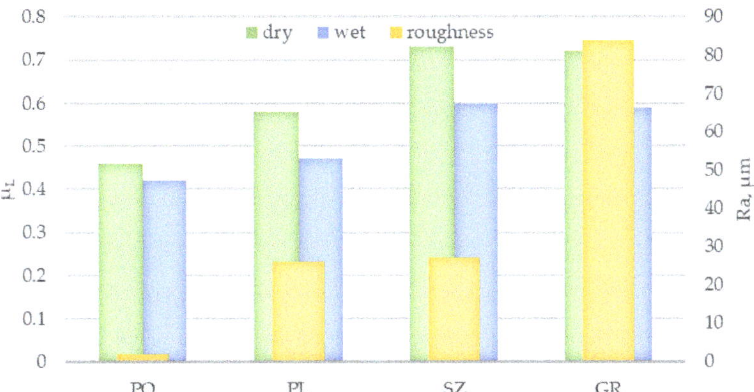

Figure 11. Results of dynamic friction coefficient (μ_L) of PO, PL, SZ, and GR granite slabs, determined in dry and wet conditions using a set of leather sliders against the background of surface roughness expressed by *Ra*.

4. Discussion

The variety of methods for evaluating slip resistance creates the need to address the relationship between test results obtained by different methods and establish links between classification criteria related to these methods. This issue should be based on testing as many flooring materials as possible.

For measurement methods for which the results are metrologically comparable (i.e., relate to the same physical quantity and are metrologically traceable to the same reference), the issue is relatively straightforward: a method compatibility assessment can be applied to determine whether the differences between results of measurements obtained by different methods are insignificant [23,26].In the present case of slip resistance evaluation, the test methods were determined by the test process's convention, and their results do not refer to the same physical quantity. The compatibility assessment of the methods in such a case is not justified, and other indirect methods should be used to assess the differences in trueness and precision.

A comparison of the acceptance angle values (α_{ob} using ramp test), SRV, μ_R (sliding friction coefficient using rubber slider), and μ_L (sliding friction coefficient using leather slider) should also take into account the fact that α_{ob} tests were carried out on a surface covered with engine oil, while SRV and μ tests were carried out on dry surface and on surface wetted with water. Dry surface and wet surface test results differed but were subject to the same classification. Due to the fact that the ramp test was conducted on four surfaces (PO, PL, SZ, and GR) and the other tests were conducted on the same surfaces but with their number doubled using wet surface and dry surface, we used a method of comparison that did not treat the wet and dry surface as separate surfaces but rather considered the tests for wet surface and dry surface as separate test methods.

Pearson's correlation coefficient within each pair of results was used to compare the results obtained by the seven methods thus defined. The results are presented in Table 2. These were determined for 80 results in each method. All values were greater than the critical value at the confidence level $\alpha = 0.05$ [27].

Table 2. Pearson's correlation coefficient between test methods.

Test Mark	SRV (dry)	SRV (wet)	μ_R (dry)	μ_R (wet)	μ_L (dry)	μ_L (wet)
Ramp test	0.977	0.970	0.837 *	0.988	0.979	0.955
SRV (dry)	-	0.984	0.863 *	0.990	0.978	0.963
SRV (wet)	-	-	0.783 *	0.981	0.949	0.914 *
μ_R (dry)	-	-	-	0.852 *	0.879 *	0.904 *
μ_R (wet)	-	-	-	-	0.978	0.953
μ_L (dry)	-	-	-	-	-	0.974

If we consider that the results were obtained for four levels (PO, PL, SZ, and GR), the critical value of Pearson's coefficient for $n = 4$ is much higher and values marked * are smaller than the critical value, but in nonstatistical evaluation such values are usually considered as good correlation.

The interpretation of the Pearson's coefficient depends on the purpose and context. For advanced and precise measurement methods, the values shown in the table could be considered insufficient confirmation of the correlation between the results. Each of the methods used in this experiment is characterized by specific arrangements and many noncontrollable factors, namely factors that were not precisely determined in the test model. These factors may cause variabilities taking the form of differences between the results under repeatability and reproducibility conditions.

To compare tests in terms of the dispersion of the results obtained, repeatability standard deviation (s_r) and reproducibility standard deviation (s_R) according to ISO 5725-2 [28] are used. However, when the results obtained by different methods are not comparable, standardization of the results would have to be used to obtain information about the differences in precision of the methods. However, classical standardization unifies the dispersion of test results to a standard deviation value of 1. Thus, in this case, quotient transformations where the normalizing values are maximal values for each test method were used. This kind of transformation retains the differences in means and standard deviations [27].

The resulting repeatability standard deviations (s_r) and reproducibility standard deviation (s_R) calculated from the results undergoing quotient transformations are shown in Figure 12.

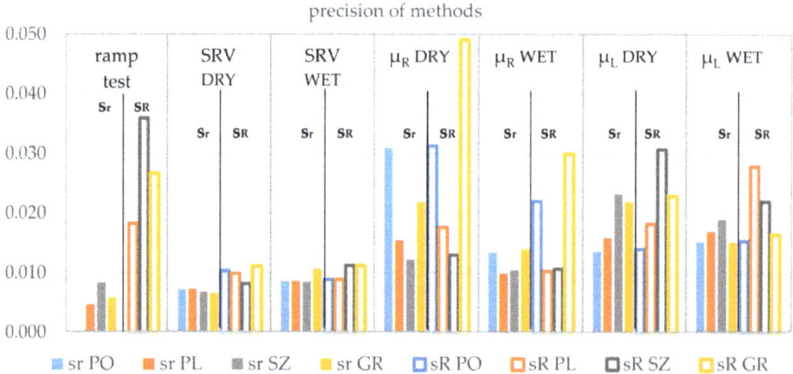

Figure 12. Precision of methods: repeatability standard deviation (s_r) and reproducibility standard deviation (s_R) for ramp test, test of SRV for dry and wet surfaces, sliding friction coefficient using rubber slider (μ_R), and sliding friction coefficient using leather slider (μ_L) for dry and wet surfaces. The values of s_r and s_R were obtained from the results subjected to quotient transformations (maximal values for each test method were used as normalizing values).

Analysis of the graph shown in Figure 12 indicates that the ramp test method has fairly good repeatability compared to other methods. It defines the angle at which the examiner researcher stopped feeling confident, so the judgment given by the same researcher (expressed by the value of the angle, α_{ob}) was not significantly different for the same substrate surface. When another person conducted the research test, their feelings about safety might have varied somewhat, despite their initial mental and physical state. This was confirmed by statistical analysis of the difference between the variance of reproducibility (s_R) and repeatability (s_r), which is the component of variance derived from intergroup (interlaboratory) differences:

$$s_L{}^2 = s_R{}^2 - s_r{}^2 \tag{2}$$

The SRV test had a low standard deviation value of repeatability and reproducibility, and the precision of the method depended very little on the surface on which the test was performed. On the other hand, methods based on the determination of the μ factor showed clear differences in precision depending on the surface tested. Table 3 shows the values of the s_L deviation.

Table 3. Values of the intergroup deviation (s_L). The values of s_L for which $s_L{}^2$ had not met the condition described by Equation (2) are marked with gray shading.

Test Mark	s_L (PO)	s_L (PL)	s_L (SZ)	s_L (GR)
Ramp test	0.000	0.018	0.035	0.026
SRV (dry)	0.007	0.007	0.004	0.009
SRV (wet)	0.002	0.002	0.007	0.003
μ_R (dry)	0.005	0.008	0.004	0.044
μ_R (wet)	0.017	0.003	0.001	0.027
μ_L (dry)	0.003	0.009	0.020	0.006
μ_L (wet)	0.002	0.022	0.011	0.006

To determine the statistical significance of differences in the precision of individual methods (ramp test, slip resistance value (for dry and wet surfaces), friction coefficients (μ_R for dry and wet surfaces and μ_L for dry and wet surfaces), and differences in the precision of testing of individual surfaces (PO, PL, SZ, and GR)), chi-squared test (χ^2) was performed for $s_L{}^2$ variances.

The following criterion was used:

$$\chi^2 = \frac{n s_L^2}{\sigma_0^2} < \chi_{\alpha,n}^2 \tag{3}$$

where n is the number of degrees of freedom, $n = N - 1$ ($N = 20$ for each value of $s_L{}^2$). σ_0^2 is the variance of the population. As this value is unknown, it was taken as the mean of all $s_L{}^2$ scores. $\chi_{\alpha,n}^2$ is the critical value for the significance level $\alpha = 0.05$ and the number of degrees of freedom n.

Thus, the χ^2 test was used to confirm the hypothesis that the differences of the interlaboratorty component of variance $s_L{}^2$ of the results obtained for different types of surfaces and different types of tests were not statistically significant.

In Table 3, the values of s_L for which $s_L{}^2$ had not met the condition described by Equation (3) are marked with gray shading. The greatest number of deviating values of interlaboratory variance was found for two surfaces: SZ and GR. This may mean difficulty in ensuring the reproducibility of the results of tests carried out on such surface. As previously established, the SRV method was characterized by the best (as a set for all surfaces) reproducibility.

The precision of the test method, particularly the interlaboratory variance, is extremely important because poor reproducibility of the method increases the risk of incorrect evaluation. From this point of view, on the basis of the presented results, the SRV method should be considered the best (the lowest risk of different assessment by different laboratories).

On the other hand, to get a complete picture of the slip resistance test method, the classification method assigned to it should also be considered, including the classification resolution.

Different scales of safety risk classifications (Figures 2–4) were assigned to the ramp test, slip resistance value, and sliding friction test methods. Figure 13 shows the test results against the classifications assigned to the methods.

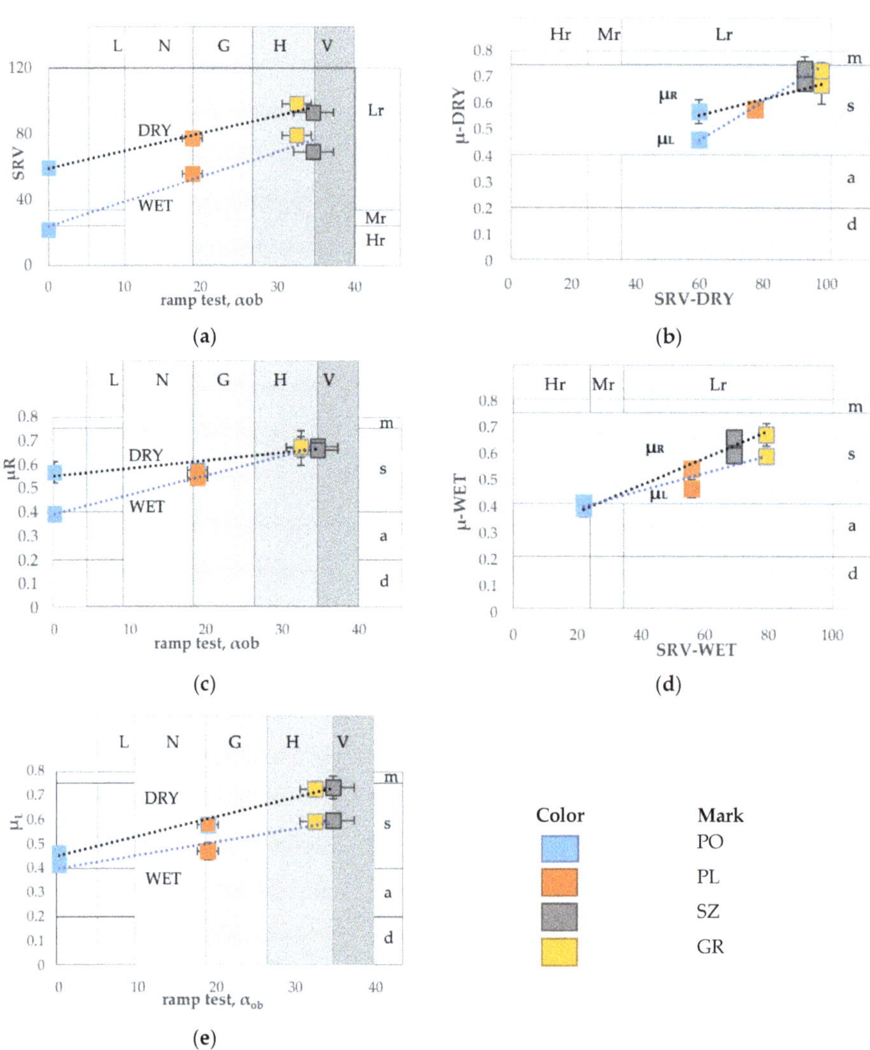

Figure 13. Slip evaluation for four surfaces (PO, PL, SZ, and GR) depending on the test method and associated classification criteria. (**a**) SRV for dry and wet surfaces versus α_{ob} results of the ramp test. (**b**) Sliding friction coefficient using rubber slider (μ_R) for dry and wet surfaces versus results of the ramp test. (**c**) Sliding friction coefficient using leather slider (μ_L) for dry and wet surfaces versus results of the ramp test. (**d**) Dry surface μ_R and μ_L versus SRV. (**e**) Wet surface μ_R and μ_L versus SRV.

The ramp test classification based on acceptance angle values (α_{ob}) had the highest resolution. The three surfaces PL, SZ, and GR were assigned to different classes, with the results for PL and SZ surfaces being on the edge of different classes. The PO surface results were below the lower limit of the acceptance angle, indicating an unacceptable risk of slipping.

In the scales assigned to the SRV and sliding friction coefficient tests, PL, SZ, and GR in both wet and dry states as well as the PO floor in dry state were in the same class, i.e., low risk according to the SRV value, while the μ_R and μ_L values were in the satisfactory class. Only the wet PO floor in the SRV test results reached the high slip risk class. The μ_R and μ_L values were on the borderline between admissible and satisfactory classes.

The designations for individual classifications were as follows: for ramp test (floor classes): L: low (R9), N: normal (R10), G: good (R11), H: high (R12), and V: very high (R13); for SRV (risk of the slip): Lr: low risk, Mr: medium risk, and Hr: high risk; for sliding friction coefficient (floor classes): d: dangerous, a: admissible, s: satisfactory, and m: model.

The above results indicate that the test method and classification method strongly influenced the slip risk assessment. This supports the thesis regarding the need for uniformity of testing and evaluation rules. Various paths may lead to consistency in assessments, all requiring thorough research. This includes assigning a specific test and classification method to the floor usage conditions; using only one method for all slip resistance assessments; and creating a new research method and a new, consistent classification. A transitional phase could be to provide the possibility of converting results and classifications obtained with different methods to harmonized indices, which could allow comparisons of products in terms of slip risk in situations where slip resistance results are obtained using different methods. This paper is an outline of the problem and an embryo of that phase, showing the relationship between the results and classifications for one type of material with differently prepared surfaces. Creating a matrix to compare slip resistance results would have to involve a large number of test results obtained for different flooring materials. The authors intend to continue to work in this direction.

5. Conclusions

This study analyzed the slip resistance of polished, flamed, brushed, and grained granite slabs in terms of slip resistance value, acceptance angle in shod ramp test, and sliding friction coefficients and found that the evaluation of the products in terms of slip resistance was strongly related to the applied testing and classification method.

The test method's influence was particularly visible with regard to products with very low roughness, such as polished slabs. These solutions were classified as low slip risk for slip resistance value in dry conditions and satisfactory for dynamic friction coefficient. In the ramp test, they obtained an acceptable angle value at a level that prevented them from being classified in the lowest slip resistance class.

The test methods were compared in terms of accuracy and resolution of classification. Neither method had been well assessed on both criteria. The SRV method showed the highest precision, resulting in the lowest risk of different classification by different laboratories, but the resolution of the classification assigned to this method was too low. The ramp test method had a resolving classification, but the test method reproducibility appeared to be the worst.

The choice of test and classification method is often dictated by national regulations related to specific applications. However, a question still remains as to which of the assessment methods provides the lowest risk of slippage. According to the authors of this paper, it is not guaranteed by even the most precise method that classifies all surfaces as providing a low risk of slippage. Thus, the safest of the discussed methods of testing and classification, both from the point of view of producer and user risk, seems to be the ramp test.

The testing and classification results indicate an important need to standardize testing methods and classification methods or to create a matrix that allows comparison of results obtained by different methods.

The authors will continue research on other types of flooring materials.

Author Contributions: Conceptualization, E.S. (Ewa Sudoł); methodology, E.S. (Ewa Sudoł) and E.S. (Ewa Szewczak); formal analysis, E.S. (Ewa Sudoł) and E.S. (Ewa Szewczak); investigation, E.S. (Ewa Sudoł) and E.S. (Ewa Szewczak); resources, E.S. (Ewa Sudoł), E.S. (Ewa Szewczak), and M.M.; writing—original draft preparation, E.S. (Ewa Sudoł), E.S. (Ewa Szewczak), and M.M.; writing—review and editing, E.S. (Ewa Sudoł) and E.S. (Ewa Szewczak); visualization, E.S. (Ewa Sudoł). and E.S. (Ewa Szewczak). All authors have read and agreed to the published version of the manuscript.

Funding: This research was funded by Ministerstwo Nauki i Szkolnictwa Wyższego (NZM-059/2020).

Data Availability Statement: The data presented in this study are available on request from the corresponding author.

Acknowledgments: Special thanks to Cezary Strąk and Andrzej Nowacki for technical support.

Conflicts of Interest: The authors declare no conflict of interest.

References

1. Regulation (EU) No 305/2011 of the European Parliament and of the Council. Available online: https://eur-lex.europa.eu/legal-content/EN/TXT/?uri=uriserv:OJ.L_.2011.088.01.0005.01.ENG&toc=OJ:L:2011:088:TOC (accessed on 15 January 2021).
2. The UK Slip Resistance Group Guidelines. The Assesment of Floor Slip Resistance. Available online: www.ukslipresistance.org.uk (accessed on 15 January 2021).
3. Deutsches Institut für Normung. *DIN 51130:2014 Testing of Floor Coverings-Determination of the Anti-Slip Property-Workrooms and Fields of Activities with Slip Danger-Walking Method-Ramp Test*; Deutsches Institut für Normung: Berlin, Germany, 2014.
4. Bellopede, R.; Marini, P.; Karaca, Z.; Gokce, M.V. Relationship between slipperiness and other characteristics of stones used as flooring slabs. *J. Mater. Civ. Eng.* **2016**, *28*. [CrossRef]
5. Ministerstwa Inwestycji i Rozwoju Standardy Projektowania Budynków dla Osób Niepełnosprawnych. Available online: https://www.gov.pl/web/rozwoj-praca-technologia/standardy-dostepnosci-budynkow-dla-osob-z-niepelnosprawnosciami (accessed on 15 January 2021).
6. Państwowa Inspekcja Pracy: Praktyczne Sposoby Zapobiegania Potknięciom I Poślizgnięciom. Available online: www.pip.gov.pl (accessed on 15 January 2021).
7. Kemmlert, K.; Lundholm, L. Slips, trips and falls in different work groups—With reference to age and from a preventive perspective. *Appl. Ergon.* **2001**, *32*, 149–153. [CrossRef]
8. Kim, I.-J. Investigation of floor surface finishes for optimal slip resistance performance. *Saf. Health Work* **2018**, *9*, 17–24. [CrossRef] [PubMed]
9. Sariisik, A. Safety analysis of slipping barefoot on marble covered wet areas. *Saf. Sci.* **2009**, *47*, 1417–1428. [CrossRef]
10. Granit-Strzegom. Available online: http://granit-strzegom.com.pl/realizacje (accessed on 15 January 2021).
11. Health and Safety Executive L24 Workplace Health, Safety and Welfare. Workplace (Health, Safety and Welfare) Regulations Approved Code of Practice and Guidance. Available online: https://www.hse.gov.uk (accessed on 15 January 2021).
12. Nemire, K.; Johnson, D.A.; Vidal, K. The science behind codes and standards for safe walkways: Changes in level, stairways, stair handrails and slip resistance. *Appl. Ergon.* **2016**, *52*, 309–316. [CrossRef] [PubMed]
13. Terjék, A.; Dudás, A. Ceramic Floor Slipperiness Classification—A new approach for assessing slip resistance of ceramic tiles. *Constr. Build. Mater.* **2018**, *164*, 809–819. [CrossRef]
14. Choi, S.-K.; Kudoh, R.; Koga, J.; Mikami, T.; Yokoyama, Y.; Takahashi, H.; Ono, H. A comparative evaluation of floor slip resistance test methods. *Constr. Build. Mater.* **2015**, *94*, 737–745. [CrossRef]
15. Li, K.W.; Chang, W.R.; Leamon, T.B.; Chen, C.J. Floor slipperiness measurement: Friction coefficient, roughness of floors, and subjective perception under spillage conditions. *Saf. Sci.* **2004**, *42*. [CrossRef]
16. Çoşkun, G. A new slip safety risk scale of natural stones with statistical K-means clustering analysis. *Arab. J. Geosci.* **2018**, *11*, 799. [CrossRef]
17. International Organization for Standardization ISO. *ISO 3274:1996 Geometrical Product Specifications (GPS)—Surface Texture: Profile Method—Nominal Characteristics of Contact (Stylus) Instruments*; International Organization for Standardization ISO: Geneva, Switzerland, 1996.
18. International Organization for Standardization ISO. *ISO 4288:1996 Geometrical Product Specifications (GPS)—Surface Texture: Profile Method—Rules and Procedures for The Assessment of Surface Texture*; International Organization for Standardization ISO: Geneva, Switzerland, 1996.
19. European Committee for Standardization CEN. *EN 14231:2003 Natural Stone Test Methods-Determination of the Slip Resistance By Means of the Pendulum Tester*; European Committee for Standardization CEN: Brussels, Belgium, 2003.
20. European Committee for Standardization CEN. *CEN/TS 16165:2012 Determination of Slip Resistance of Pedestrian Surfaces; Methods of Evaluation*; European Committee for Standardization CEN: Brussels, Belgium, 2012.
21. Chang, W.-R.; Kim, I.-J.; Manning, D.P.; Bunterngchit, Y. The role of surface roughness in the measurement of slipperiness. *Ergonomics* **2001**, *44*, 1200–1216. [CrossRef] [PubMed]

22. Liu, J.; Guan, B.; Chen, H.; Liu, K.; Xiong, R.; Xie, C. Dynamic model of polished stone value attenuation in coarse aggregate. *Materials* **2020**, *13*, 1875. [CrossRef] [PubMed]
23. Chen, C.C.; Chen, Z.X.; Chang, C.L.; Lin, F.L. The slip-resistance effect evaluation of floor roughness under different liquid viscosity. *Procedia Manuf.* **2015**, *3*, 5007–5013. [CrossRef]
24. Karaca, Z.; Gürcan, S.; Gökçe, M.V.; Sivrikaya, O. Assessment of the results of the pendulum friction tester (EN 14231) for natural building stones used as floor-coverings. *Constr. Build. Mater.* **2013**, *47*, 1182–1187. [CrossRef]
25. Deutsches Institut für Normung. *DIN 51131:2014 Testing of Floor Coverings-Determination of the Anti-slip Property-Method for Measurement of the Sliding Friction Coefficient*; Deutsches Institut für Normung: Berlin, Germany, 2014.
26. Szewczak, E.; Winkler-Skalna, A.; Czarnecki, L. Sustainable test methods for construction materials and elements. *Materials* **2020**, *13*, 606. [CrossRef] [PubMed]
27. Statistics Solutions. Available online: https://www.statisticssolutions.com/table-of-critical-values-pearson-correlation (accessed on 15 January 2021).
28. International Organization for Standardization ISO. *ISO 5725-2:1994 Accuracy (Trueness and Precision) of Measurement Methods and Results: Part 2: Basic Method for the Determination of Repeatability and Reproducibility of a Standard Measurement Method*; International Organization for Standardization ISO: Geneva, Switzerland, 1994.

 *materials*

Review

Non-Destructive Corrosion Inspection of Reinforced Concrete Using Ground-Penetrating Radar: A Review

Ksenija Tešić, Ana Baričević * and Marijana Serdar

Department of Materials, Faculty of Civil Engineering, University of Zagreb, 10000 Zagreb, Croatia; ksenija.tesic@grad.unizg.hr (K.T.); marijana.serdar@grad.unizg.hr (M.S.)
* Correspondence: ana.baricevic@grad.unizg.hr

Abstract: Reduced maintenance costs of concrete structures can be ensured by efficient and comprehensive condition assessment. Ground-penetrating radar (GPR) has been widely used in the condition assessment of reinforced concrete structures and it provides completely non-destructive results in real-time. It is mainly used for locating reinforcement and determining concrete cover thickness. More recently, research has focused on the possibility of using GPR for reinforcement corrosion assessment. In this paper, an overview of the application of GPR in corrosion assessment of concrete is presented. A literature search and study selection methodology were used to identify the relevant studies. First, the laboratory studies are shown. After that, the studies for the application on real structures are presented. The results have shown that the laboratory studies have not fully illuminated the influence of the corrosion process on the GPR signal. Also, no clear relationship was reported between the results of the laboratory studies and the on-site inspection. Although the GPR has a long history in the condition assessment of structures, it needs more laboratory investigations to clarify the influence of the corrosion process on the GPR signal.

Keywords: ground-penetrating radar (GPR); non-destructive techniques (NDT); corrosion of reinforcement

Citation: Tešić, K.; Baričević, A.; Serdar, M. Non-Destructive Corrosion Inspection of Reinforced Concrete Using Ground-Penetrating Radar: A Review. *Materials* **2021**, *14*, 975. https://doi.org/10.3390/ma14040975

Academic Editor:
Krzysztof Schabowicz
Received: 7 January 2021
Accepted: 15 February 2021
Published: 19 February 2021

Publisher's Note: MDPI stays neutral with regard to jurisdictional claims in published maps and institutional affiliations.

Copyright: © 2021 by the authors. Licensee MDPI, Basel, Switzerland. This article is an open access article distributed under the terms and conditions of the Creative Commons Attribution (CC BY) license (https://creativecommons.org/licenses/by/4.0/).

1. Introduction

Every structure, depending on its intended purpose, must be designed and constructed so that during its lifecycle it fulfils the basic requirements for structures and other requirements, namely, the conditions prescribed by the Building Act [1]. Unfortunately, experience has shown that a large number of concrete structures show significant signs of degradation after only 20 to 30 years due to the joint action of mechanical and environmental effects [2]. The causes of degradation are mainly the consequence of corrosion, which on a global scale increases the annual maintenance costs to more than 3% of the world's Gross Domestic Product (GDP) [3]. The maintenance and strengthening of bridges in Europe alone require £215 million, not including the costs of redirection and organization of traffic [4]. The unsystematic approach to maintenance, especially of infrastructure, contributes to its premature deterioration and has a negative impact on safety and reliability. Particularly worrying is the fact that today, the resources invested in maintenance and repair are higher than the cost of construction [5]. The question therefore arises: how to stop or delay the degradation of global infrastructure?

The concern resulting from the problems outlined led to the development of strategies to mitigate the consequences of the corrosion process. At the design level, strategies are mainly aimed at improving the durability properties of the concrete cover in terms of its thickness and quality [6]. Other strategies aim at the preventive use of corrosion inhibitors, corrosion-resistant steel or other surface treatments [7–9]. However, these have limited ability to solve the corrosion problem of existing structures.

One of the most promising approaches to delay the degradation of existing structures is the extensive use of non-destructive techniques (NDT). Increased inspection frequency

245

and coverage of larger inspection areas could lead to timely detection of deterioration and, in sum, better decisions in the maintenance of the structure. In this regard, the progress in the development of NDT methods towards visualization of results leads to the increased use of advanced NDT methods in the future [10]. Many NDT methods are currently available; however, this paper focuses on the application of ground-penetrating radar (GPR) for corrosion inspection of reinforced concrete structures.

Originally, the radar was designed for military use [11]. Today, its application has expanded to various disciplines, such as civil engineering, hydrogeology, archaeology, etc. [12,13]. When combined with other non-destructive methods, it is feasible for evaluating the condition assessment of the concrete structures [14,15] with increased effectiveness and speed of inspection. The laboratory investigations have shown that the corrosion process could be monitored based on observing the changes in the GPR signal [16–20]. Moreover, GPR has a history in the assessment of corrosion and corrosion-related pathologies for on-site inspection [21–25]. The main difference between these two approaches is that the laboratory investigation was mainly focused on discrete corrosion characterization. The on-site investigation is based on the observation of several simultaneous effects, namely the variation of moisture, chlorides, and the formation of corrosion products and cracks.

Previous studies have focused on reviewing the general application of GPR in civil engineering [26,27] or have focused on on-site inspection for a specific type of construction [28]. To date, there is no comprehensive critical study that evaluates the use of GPR for corrosion assessment of reinforced concrete. The main objective of this review paper is to identify all relevant publications on corrosion assessment of reinforced concrete using ground-penetrating radar. The authors have attempted to gain more understanding of the relationship between laboratory testing and the application of GPR on-site. This prompted the authors to organize the paper into the following sections. Section 2 presents the details of the literature search in terms of the databases used, the search terms and the rationale for the publication screening. Section 3 deals with the main topic. It is introduced with the corrosion process and the main principles of GPR, presenting the characteristics for corrosion monitoring. Furthermore, it is divided into sections dealing with laboratory and on-site inspections, with each section ending with conclusions. Finally, Section 4 summarizes the work with recommendations for future studies.

2. Methodology

As a first step, a systematic literature search was conducted. Relevant studies were searched in the databases of Web of Science [29] and Scopus [30] over a period between 1 January 2000 and 30 October 2020. Initially, the authors began the search with the terms [(ground penetrating radar OR GPR) AND (corrosion) AND (concrete)]. The authors found that a number of studies for on-site assessment of concrete structures using GPR were excluded. They suggest that this is because some of the studies looked at the causes and consequences of corrosion (e.g., delamination), and it appears that the term corrosion was not appropriate in this case. For this reason, the term deterioration was included in the database search. The authors found that the terms [(ground penetrating radar OR GPR) AND (corrosion OR deterioration) AND (concrete)] expanded the number of studies so that a better overview of GPR application could be created. Duplicates were then removed, and the authors briefly reviewed titles and abstracts and excluded studies that did not meet the following criteria: (a) the study is published in English, (b) the full version of the study is available to the authors and (c) GPR was used to evaluate reinforcement corrosion and corrosion consequences (e.g., studies in which GPR was used only to determine cover thickness were excluded). Full-text articles were obtained, and further selection excluded studies that were not relevant or were beyond the scope. The final selection included 69 studies. Figure 1 shows the steps described.

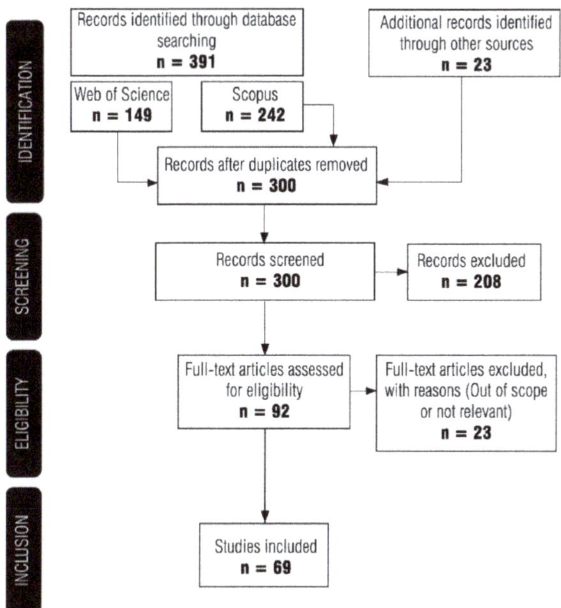

Figure 1. Steps in the review process for the selection of articles.

3. Corrosion Monitoring Using Ground-Penetrating Radar

As mentioned earlier, the main causes of degradation are mainly the result of corrosion of reinforcement [31]. The corrosion of steel in concrete is a balanced electrochemical mechanism [32] between anodic and cathodic reactions that occur on the surface of the reinforcing steel. The anodic reaction, the oxidation of iron, occurs in an environment where the protective passive film of steel is not stable. The instability of the protective layer is related to the changes in the surrounding concrete and the main cause of these changes are processes such as chloride penetration or carbonation [32,33]. The time required for the breakdown of the passive film is called the initiation period in Tuutti's corrosion model [34]. The further development of corrosion is called the propagation period and involves crack initiation, as a result of expansive stresses around the bar induced by rust formation. The progressive corrosion leads to spalling of the concrete and reduction of the cross-section of the reinforcement, which may compromise the load-bearing capacity of the structure [35]. Most corrosion assessment techniques are electrochemical-based [36,37]. In the field assessment of corrosion probability, the half-cell potential (HCP) and electrical resistivity (ER) are most used.

The description of the half-cell method and interpretation of the results are given in ASTM C876 [38] and RILEM recommendations [37]. The Wenner probe is commonly used to determine the electrical resistivity [39,40]. The resistivity values can be used to estimate the corrosion risk [41]. Although these methods have long been used successfully in the condition assessment of concrete structures, they have some drawbacks. The half-cell potential is a semi-destructive technique, so it requires a connection to the reinforcement. This is a limitation when a large area is to be inspected. Measuring electrical resistivity does not provide information about the reinforcement, only about the corrosive environment. Also, large areas require a lot of time for inspection. These issues can be overcome by using GPR.

Ground-penetrating radar is a non-destructive technique that emits electromagnetic waves into the material, with the main objective of locating the buried objects underneath the surface [12]. Nowadays, its scope broadens to a wide range of materials, and among others is concrete [26]. The emitted electromagnetic wave propagates through the host material, as far as it encounters an interface between different materials, whereupon it is reflected back. The predominant types of GPR antennas used for civil engineering investigations are air-coupled and ground-coupled. The second type implies contact of the antenna with the ground and has a better penetration depth. The reflected wave is recorded with the receiving antenna and the recording is called an A-scan (Figure 2, right). When a wave is transmitted, the receiver first records a direct wave propagating through the air from the transmitter to the receiver. Then, a portion of the electromagnetic wave is reflected off the surface of the material. In a ground-coupled system, these two components superimpose to form the wave, called direct coupling, Figure 2. The rest of the wave energy passes through the material until it reaches the material with different dielectric properties. The electromagnetic wave is then reflected, and the receiver records it as a reflected wave. Therefore, the attributes of the A-scan that provide information about the target are the amplitude of the reflected wave and the travel time from the transmitter to the receiver. In addition, the most common representation of the results obtained with GPR is a B-scan, the two-dimensional slice that represents the area under investigation along the line.

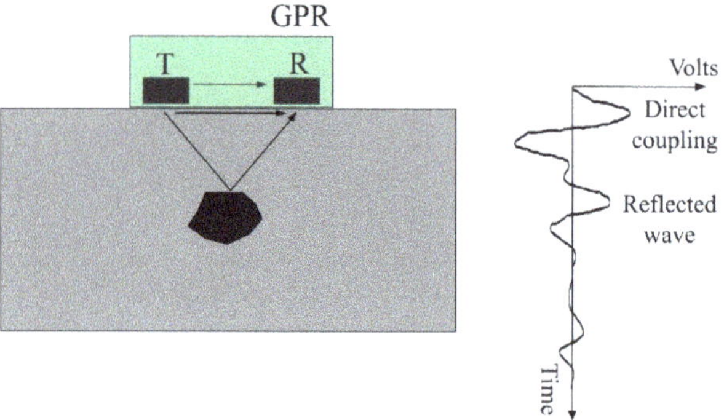

Figure 2. Recorded signals for ground-coupled antenna.

The strength of the reflected wave depends on the properties of the host material. The properties that determine the behavior of electromagnetic waves in the material are its dielectric properties—dielectric permittivity (ε) and electrical conductivity (σ) [42]. Signal losses are mainly due to electrical conduction and dielectric relaxation [43]. Electrical conduction arises from the motion of free charges, while dielectric relaxation arises from the rotation of polar molecules. At the microscopic scale, friction occurs between particles due to these motions, resulting in energy dissipation. In summary, the propagation behavior of electromagnetic waves strongly depends on the composition of the pore solution. Changes in dielectric properties can be expected in the presence of moisture and/or chlorides in the concrete. The presence of water molecules and chlorides in pores results in an overall loss of energy and signal [43–45]. However, this attenuation is primarily caused by the presence of chlorides in the pore solution [46] as a result of the increased electrical conductivity of concrete. Changes in the concrete microstructure caused by carbonation can also affect the GPR response. The most noticeable ones are due to a reduction in porosity and ion exchange in the pore solution. It has been reported that carbonation causes a decrease in the dielectric permittivity, resulting in reduced attenuation [47].

3.1. Laboratory Simulated Corrosion Inspection

Corrosion assessment using ground-penetrating radar is still a novel approach, therefore only a limited number of studies have been conducted under laboratory conditions (Table 1). There are many challenges to ensure a suitable experimental setup for such a study, starting with the criteria for corrosion initiation, corrosion monitoring and corrosion probability assessment. In order to simulate natural corrosion under laboratory conditions, various techniques are often used to accelerate the process, such as impressed current technique, artificial climatic environment, accelerated migration tests, etc. [48]. The most commonly used method for corrosion acceleration is the impressed current technique, which is based on exposing the embedded reinforcement to the electric current provided by an external power supply. The current density and exposure time are controlled to achieve different degrees of corrosion [49,50]. Besides, corrosion can be enhanced by creating favorable conditions such as high temperature, high humidity and cycles of wetting–drying [51]. Even if these methods tend to simulate corrosion well, it is inevitable that the artificial conditions for corrosion to occur will differ from natural conditions. Recognition of these limitations is important to ensure adequate correlation between accelerated corrosion investigations and on-site corrosion assessments using GPR. Therefore, Table 1 summarizes the corrosion probability studies conducted to date that consider both corrosion acceleration methods and GPR signal attributes analysis. Two experimental setups were found: (a) GPR attributes acquired before and after the corrosion process, and (b) GPR attributes monitored during the corrosion acceleration process. The second setup is more significant as it ensures information about the different stages of corrosion, starting from the depassivation of the steel to the appearance of cracks.

Table 1. Previous laboratory investigations on the influence of corrosion on the ground-penetrating radar (GPR) attributes.

Study	Year	Technique for Accelerated Corrosion Test	Method of Acquiring GPR Attributes	Current Density, i ($\mu A/cm^2$)	Dimension of Specimens (m)	GPR (GHz) [1]
Hubbard et al. [16]	2003			-	$1.25 \times 1 \times 0.25$	1.2
Raju et al. [53]	2018		Before and after corrosion acceleration	-	$0.76 \times 0.38 \times 0.203$	2.6
Zaki et al. [54]	2018			-	$1 \times 0.5 \times 0.2$	2
Lai et al. [56]	2010			-	$1.5 \times 0.5 \times 0.5$	1.5 and 2.6
Zhan et al. [57]	2011	Impressed current technique		165,000	$0.45 \times 0.14 \times 0.135$	1
Lai et al. [58]	2011		Monitoring during corrosion acceleration	340	-	1.5 and 2.6
Lai et al. [17]	2013			260 and 760	$1.5 \times 0.5 \times 0.5$	1.5 and 2.6
Hong et al. [18]	2014			424	$1.5 \times 1.5 \times 0.3$	2.6
Hong et al. [19]	2015			125	$0.8 \times 0.8 \times 0.24$	2.6
Wong et al. [20]	2019			650 [2]	$0.548 \times 0.4 \times 0.15$	2
Hasan et al. [55]	2016	Corroded rebars immersed in emulsion	Before and after corrosion acceleration	-	Water oil emulsions	2.6
Sossa et al. [52]	2019	Corroded rebars cast in concrete		-	$0.3 \times 0.08 \times 0.08$	1.6
		Curing chamber		-	$0.3 \times 0.2 \times 0.07$	

[1] All of the antennas are ground-coupled. [2] Level of current density was lowered in the latter stage of experiment.

One of the first studies to have acknowledged the GPR potential for corrosion detection was published in 2003 by Hubbard et al. [16]. The rebar was subjected to an accelerated corrosion process for 10 days. The results showed that corrosion causes a reduction in amplitude, which they attributed to the scattering and attenuation of waves due to the roughness at the corroded interface between concrete and rebar. This was also confirmed in Reference [52], where the amplitude of the signal was also reduced in specimens where the previously corroded bar was cast in concrete. To extend these observations, additional specimens were subjected to an accelerated corrosion process in an environmental chamber at different corrosion levels. It was found that corrosion causes a decrease in amplitude for each corrosion level. This is explained by the signal scattering in the concrete cover zone caused by the presence of cracks, corrosion products and the roughness of the corroded bar. In Reference [53], the influence of the diameter of the anode bar on the signal reflection was also observed. The increase in amplitude was associated with corrosion development but was also increased with the increase in diameter. In contrast, Reference [54] attributed lower amplitudes to the presence of corrosion products, but also indicated that the decrease could be influenced by the accumulation of chlorides in the concrete cover zone.

In Reference [55], concrete properties were simulated by oil emulsions with different dielectric permittivity, where corroded bars were immersed in the emulsions. However, in such an experimental setup, all results are based on the theories and therefore cannot faithfully represent real structures.

3.1.1. Long-Term Corrosion Monitoring

Long-term monitoring of corrosion may ensure a better understanding of its effect on GPR signal attributes. Several studies [17–20,56–58] have been conducted to distinguish and correlate significant attributes of corrosion development and GPR signal. The conclusions from these studies are divided into: (1) initiation phase, (2) formation of cracks and (3) spalling of concrete cover.

Initiation Phase

The application of electrical current in the accelerated corrosion process induces the faster motion of chloride ions due to the electrical potential gradient, and this mechanism is called migration [32]. As this process thrives, a decrease in GPR amplitude is reported by Lai et al. [17]. According to the authors' assertation, the accumulation of chloride ions around the anode absorbs the energy of the electromagnetic wave, which also causes the delay of the wave.

Formation of Cracks

After the initiation phase, the researchers had noticed a steady trend of change in the signal's attributes until the wide crack is visible on the concrete surface. This phase is characterized by the formation of corrosion products that migrate into the surrounding concrete. The increased amplitude of the reflected wave was reported in Reference [57]. The rebars were subjected to external power supply until a longitudinal crack was visible on the concrete surface. The authors claimed that the migration of corrosion products into the shallower concrete cover zone enlarges the intersection points of the signal with different interfaces—concrete, microcracks, corrosion products—and leads to the increased amplitude. This was also confirmed in References [17,56,58]. In Reference [18], the experiment was set up to exclude the effects of moisture and chlorides on the GPR signal. The specimens were stored for two months to achieve stable moisture and chloride content before accelerated corrosion. Here, the increased amplitude of the GPR signal was then attributed to the effect of corrosion only. This was also outlined in Reference [20].

Spalling

In addition to the effect of corrosion development on the GPR amplitude, the effect of crack propagation and the occurrence of wide cracks on the amplitude of the GPR signal was also noted by Lai et al. [56]. A decrease in amplitude was observed after the occurrence of a wide longitudinal crack. This was explained by the scattering of signal energy caused by additional irregularities when a wide crack propagates through the concrete cover. The extension of the experimental setup was presented in Reference [20] to obtain a better representation of the crack influence on the signal amplitude.

3.1.2. Conclusions from Laboratory Simulated Corrosion Inspection

The corrosion process can be divided into the initiation and propagation phases, with each phase having specific effects on concrete microstructures. From the long-term corrosion monitoring experiments, the influence of corrosion promotion on GPR attributes was summarized, as in Figure 3.

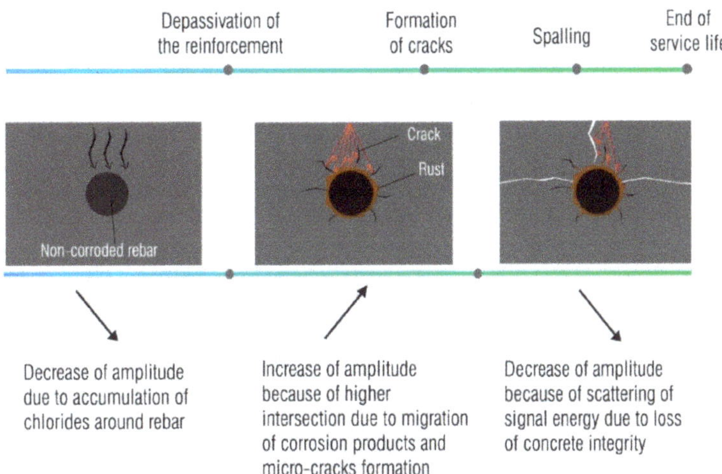

Figure 3. Changes in the GPR signal during corrosion process.

The effects shown in Figure 3 are determined by the accelerated corrosion processes, in particular the formation of corrosion products on the surface of the reinforcing bar, their diffusion into the concrete cover and thus, crack propagation. These effects modify the amplitude in terms of different reflection coefficient and different dielectric properties of the concrete cover. The ability of corrosion products to migrate depends on the moisture content in the concrete cover since their movement is favored in the presence of moisture [59]. Similarly, the ability of their migration depends on the duration of the acceleration process. When accelerated corrosion with high current density is established in a short timeframe, it leads to increased crack width due to the sudden accumulation of corrosion products and increased pressure around the reinforcement [50,60]. Therefore, an appropriate current density should be selected to ensure the best possible simulation of natural conditions still within a reasonable timeframe [50,61].

The results of these studies also indicate that laboratory simulated corrosion studies mentioned above do not provide relevant results unless the level of corrosion is properly described. In these studies, results were recorded before and after corrosion acceleration and conflicting results were reported. Some authors reported higher amplitudes at the end of the experiments, while others reported lower amplitudes. Since these studies differ in terms of experiments setup, corrosion level and induced damage, it is possible that the observed changes in the GPR signal were recorded during different stages of the corrosion processes.

3.2. On-Site Corrosion Inspection

Most of the published research focuses on the application of GPR to the assessment of bridge decks, while other structures are represented to a lesser extent (tunnels, buildings, wharves, etc.). In terms of geographic location, most studies using GPR are conducted in the United States of America. The authors are sure that this is also a consequence of the existence of relevant standards [62].

The use of ground-penetrating radar data for condition assessment of concrete bridges dates back to the early 1980s [63]. The amplitude of the ground-penetrating radar signal during an inspection is affected by the presence of structural elements, variations in cover depth, moisture, chlorides [22,64,65] and other variables. Therefore, a simple approach to evaluating changes in the GPR signal is not practical. Coexisting influences with other phenomena, such as variations in moisture and chlorides, are unavoidable and prevent the development of methods for direct location of corroded rebar. Therefore, indirect methods are used in which the localization of corroded areas is correlated with the areas of high signal attenuation. Attenuation has been found to be strongly related to the increased conductivity around the rebar [22] caused by the accumulated chloride ions and corrosion products [25].

The inspection of concrete piers and wharves is very similar to the inspection of bridge decks where moisture and chlorides are the main causes of deterioration. The use of ground-penetrating radar has also been reported in the inspection of tunnels. It was noted that the complicated design and compound deterioration mechanisms of these structures made a simple corrosion assessment impossible. Instead, the data was used to assess the overall condition.

Quantification of attenuation can be determined by numerical analysis of the signal or by visual analysis of B-scans. These are discussed in more detail in the following sections.

3.2.1. Numerical Analysis of GPR Attributes

The ASTM standard [62] for the evaluation of concrete bridge decks using ground-penetrating radar proposes two procedures for numerical analysis using GPR data. The first procedure is based on considering the reflection amplitude from the bridge deck bottom and the bridge deck surface. The second procedure considers the reflection amplitudes from the top reinforcement layer. In most cases, the reflection amplitudes from the top reinforcement are considered for corrosion evaluation. The amplitude is derived from the A-scan.

Numerical analysis is usually performed by normalizing the amplitude, which represents the deterioration rate, and is calculated as follows [66]:

$$\text{Normalized amplitude[dB]} = -20 \log \frac{\text{signal amplitude}}{\text{reference signal amplitude}} \tag{1}$$

The signal amplitudes are compared to the reference signal amplitude which is usually the amplitude with the lowest degree of attenuation and represents sound concrete [66]. The GSSI [67] suggests 32,767 for 16-bit data and 2,147,483,648 for 32-bit data as the reference signal amplitude. This approach may be inconvenient when a concrete structure is in an advanced stage of deterioration and high attenuation is primarily detected. The differences between amplitudes are then smaller and the deterioration could be underestimated [68,69]. On the other hand, if the structure is in a relatively good condition, the attenuation may be misinterpreted. In this case, the attenuation could come from a different source, namely the variation of the concrete cover thickness. In such cases, it is recommended to consider the whole amplitude and not only the attenuation zones [70]. Other approaches have also been used, with Dinh et al. [25] using the average direct coupling wave as the reference amplitude. According to Pashoutani et al. [71], the use of a constant value of a reference amplitude does not take into account the contribution of concrete surface quality to the signal amplitude, even to the normalized amplitude. Therefore, a normalization procedure was proposed in which each signal amplitude is normalized to its own direct coupling amplitude. In order to eliminate the influence of the cover depth variation on the signal amplitude, Barnes et al. [21] demonstrated an amplitude correction method. It was shown that subtracting the depth-dependent amplitude loss gives a better correlation of ground-penetrating radar amplitude maps with ground truth results than maps without correction.

The method is based on determining the linear-dependent function of signal loss from the two-way travel time (TWTT) for the 90th percentile value of normalized amplitude. The 90th percentile value of the normalized amplitude is supposed to represent the sound concrete where the attenuation is mainly caused by the propagation of the signal through the dielectric material, i.e., the dielectric loss [25]. After correction of the amplitude, the attenuation should represent the signal loss due to chloride and moisture, i.e., the conductive loss. The method was improved after it was found that the conductive loss was also depth-dependent, so that an additional correction was necessary [25]. Two automated methods for depth correction have also been proposed [72]. In these studies, the correction was performed at the two-way travel time level. A more accurate correction could be performed if the linear function is determined using the real reinforcement depth instead of the two-way travel time [71]. This procedure requires the determination of the real velocities of the signal.

Obviously, it is of interest to establish a threshold for attenuation that is suitable for identifying the area of deterioration. However, a universally applicable threshold has not been established. It is usually based on the experience of the analyst and is related to a specific case [21]. There have been several attempts to relate the attenuation, mostly in comparison with thresholds of other methods. In References [73–76], ground-penetrating radar data were correlated with half-cell potential data, with the aim of determining the threshold value of attenuation. In Reference [75], the ROC (Receiver Operating Characteristic) curve was used, and in Reference [76], the author used statistical parameters to obtain the threshold value. In the second paper, the relationship between the percentage of corroded area, based on the results obtained with the half-cell potential at several bridges, and the product of the mean and skewness of the amplitude of the ground-penetrating radar, was established. The relationship can be used to predict the corroded area based on the analysis of the statistical parameters of the amplitudes. In some studies [77,78], the k-means clustering method was used to determine thresholds values.

Numerical analysis is generally used to obtain the deterioration map, which in most cases is the spatial distribution of normalized amplitudes. The main steps of the numerical approach in the condition assessment of concrete bridge decks are shown in Figure 4. The ability of the numerical approach to provide autonomous assessment of concrete structures using GPR is one of the reasons for its predominant use, while the algorithms for automatic reinforcement selection can be found in References [79,80].

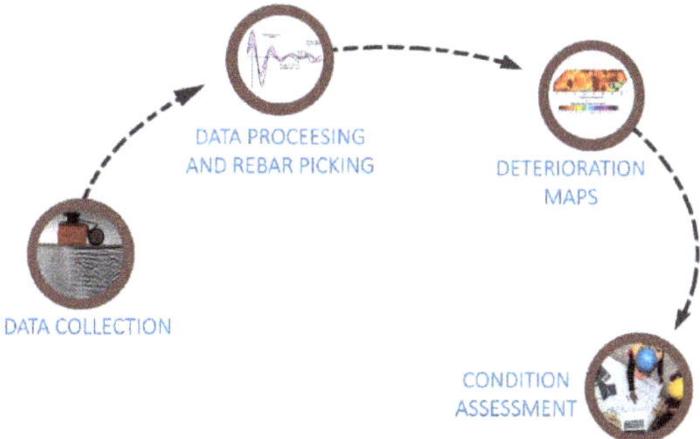

Figure 4. Numerical approach in condition assessment of concrete bridge decks.

The results obtained by periodical inspections can be collected in databases, so that the correlation of successive data allows continuous monitoring of the progress of deterioration. Dinh et al. [65] also proposed a method based on comparing the complete waveform (amplitudes and shapes of the electromagnetic wave) at a point with baseline data. The advantage of this method is that it excludes non-corrosion attenuation causes. However, baseline data is required for proper detection of deterioration, which is often not available.

The method was improved in Reference [81] and the waveform was compared with the simulated waveform. The simulated waveform has the original direct wave, but it has no reflected wave, so it simulates full attenuation. The higher similarity with the simulated wave correlates with a higher degree of deterioration. Hong et al. [82] proposed a method to monitor the corrosion process by comparing different GPR data using the image registration technique.

Despite the deterioration maps, the statistical distribution of amplitudes had shown the relationship with the condition of the structure. In Reference [83], where several bridges with different environmental conditions were observed, it was found that after correcting the amplitude depth, the distribution of the amplitude of the sound deck was symmetrical with high kurtosis. In contrast, severely damaged concrete exhibited higher dispersion of amplitude distribution with lower value of kurtosis. This statistical dependence has been previously confirmed [74]. An automated crack tracking method based on the analysis of the processed amplitude of the ground-penetrating radar was presented in Reference [84]. The model considers the amplitude compared with the threshold value. The final result of the model is a three-dimensional (3D) visualization of the cracks, which provides the possibility to evaluate their geometry. However, the reliability of the model depends on the threshold value, which is difficult to determine accurately.

3.2.2. Visual-Based or Combined Analysis of GPR Attributes

In addition to the numerical approach, a visual or combined visual and numerical approach has been supported by a number of authors [22,77,85]. The visual approach implies the visual analysis of B-scans. This method is highly dependent on the expertise of the analyst, especially in the case of severely damaged structures [86], so the final conclusion is prone to error. As noted by some authors [22], numerical analysis of amplitudes misinterprets most anomalies that alter the signal and are not causes of deterioration (surface anomalies, reinforcement spacing, reinforcement depth, structural variations). Due to of these drawbacks, a method is proposed in which an analyst reviews the ground-penetrating radar profiles (B-scans), considers the reflections of the reinforcement and concrete surfaces and marks the boundaries of deteriorated areas. The profiles are processed, and the final output is the corrosion map. The detailed procedure is described in Reference [87]. This method was improved to overcome the subjective opinion of analysts in visual-based interpretation [85]. A set of if/then rules was created to locate anomalies that alter the signal but do not indicate deterioration. Dinh et al. [77] used visual analysis of ground-penetrating radar profiles as a tool to determine the number of condition categories as input to the k-means clustering method. It is a combined method: after determining the number of condition categories, the amplitudes of the signal are grouped and thresholds between the groups are determined. The corrosion map obtained in this way was used for the deterioration modelling of concrete bridge decks presented in Reference [88]. Dawood et al. [89] presented an improved visual-based analysis of ground-penetrating radar data for the detection of air and water voids in tunnels. Moreover, an evaluation flowchart based on inspection of pier structure considering B-scans and GPR signal energy was proposed in Reference [90].

3.2.3. Condition Assessment by Combination of Multiple NDT

Ground-penetrating radar has a number of advantages over other non-destructive techniques (NDT), and it is not surprising that it has shown much interest in replacing other techniques. It is completely non-destructive, and it is rational to give it precedence over other techniques that make surveying slow and less efficient. In the next sections, a brief overview is shown on current research results obtained by comparing GPR data with other test methods, such as electrical resistivity (ER), half-cell potential (HCP), chain drag (CD), hammer sounding (HS), infrared thermography (IRT), acoustic emission (AE) and impact-echo (IE). These studies are summarized in Table 2.

Electrical resistivity and half-cell potential are fundamental tools for determining the probability of corrosion in the condition assessment of concrete structures. A good correlation has been found in the analysis of electrical resistivity and GPR data [81,91–93]. However, such behavior is to be expected as both techniques are affected by the conductivity of the concrete [91].

The comparison between HCP and GPR data can be found in References [21,24,69,73,74,81,92,93]. All observations were obtained by superimposing the signal attenuation and potential maps. In most studies, a good correlation was found since the attenuation is indicative of a corrosive environment and coincides with the areas of extremely negative half-cell potentials [92]. However, when the degree of deterioration is low, the ground-penetrating radar could overestimate corroded areas [69].

Other techniques can also serve for condition assessment and correctly predict potential deterioration due to corrosion propagation. These techniques include chain drag (CD), hammer sounding (HS), infrared thermography (IRT), acoustic emission (AE) and impact-echo (IE). Compared to the chain-drag method, the ground-penetrating radar is effective while the deterioration level ranged between 10% and 50% [69]. However, the divergence between the result of the ground-penetrating radar and the acoustic scanning system was observed in Reference [94], where the authors investigated the suitability of these techniques for delamination detection. The area of high attenuation was larger than the delaminated area detected by the acoustic system because the ground-penetrating radar generally detects the deterioration earlier than the acoustic system. The GPR can detect deterioration before delamination occurs. Also, the comparative feasibility study on delamination detection using ground-penetrating radar (GPR) and infrared thermography (IRT) based on ROC (Receiver Operating Characteristic) analysis showed that IRT is more reliable than GPR in detecting delamination [95]. However, the contribution of IRT is limited to a shallow cover depth, while GPR can provide a deeper insight. Also, the usefulness of GPR in predicting repair quantities was presented in Reference [96], where the results of ground-penetrating radar matched the depth of removal measured by LiDAR (Light Detection and Ranging) method after hydro-demolition.

Table 2. Review of studies that combined GPR with other techniques.

Study	Year	Other Techniques	GPR (GHz) Air-Coupled	GPR (GHz) Ground-Coupled	Main Findings
		Comparison with other NDT			
Barnes et al. [24]	2000	HCP, CD	1	-	Agreement on spatial distribution of deteriorated areas; 65.1% and 66.2% correctly predicted deteriorated areas compared to HCP and CD, respectively.
Scott et al. [97]	2003	IE, CD	2.4	1.5	GPR systems could not detect whole delaminated areas.
Barnes et al. [69]	2004	HCP, CD	1	-	GPR was effective in predicting damaged areas when the degree of deterioration is between 10% and 50%.
Rhazi et al. [73]	2007	HCP	-	1.5	The values for the degree of attenuation were proposed based on the correlation with HCP.

Table 2. *Cont.*

Study	Year	Other Techniques	GPR (GHz)		Main Findings
			Air-Coupled	Ground-Coupled	
Barnes et al. [21]	2008	HCP, CD	-	1.5	The correlation between GPR and HCP and CD was improved after the depth correction.
Maser et al. [74]	2012	HCP, IE, HS	1 and 2	1.5 and 2.6	The agreement between GPR and HCP was 90.2%, and between GPR and IE was 79.3%.
Simi et al. [98]	2012	IE, CD	-	2	Moisture and corrosion maps produced with commercial software showed good spatial agreement with IE and CD.
Gucunski et al. [91]	2013	ER	-	1.5	The good agreement between GPR and ER; 95% of the locations where ER $\leq$ 40 kΩcm agreed with the location where GPR amplitude was <15 dB.
Pailes et al. [93]	2015	ER, HCP, IE, CD, HS	-	1.5	The best spatial agreement compared to different NDT was between GPR and ER, and GPR and sounding techniques (CD and HS).
Dinh et al. [81]	2017	ER, HCP, IE	-	1.5	Correlation between GPR and other NDT was determined by a traditional numerical analysis and a method based on comparison with a simulated waveform; better agreement was found using ER and HCP than IE.
Sun et al. [94]	2018	AE, CD	-	1.5	GPR showed a larger deteriorated area than AE. GPR detected deteriorated areas near joints, while AE did not.
Sultan et al. [95]	2018	HS, IRT	-	1.6	Compared to the IRT, GPR was less accurate in detecting delamination.
Dinh et al. [92]	2019	ER, HCP	-	1.5	GPR maps produced by the method based on SAFT showed good correlation with HCP and ER. In one case, the correlation with ER was better than with HCP.
Combination with other NDT					
Maser [99]	2009	GPR, IRT	-	-	The combination of GPR and IRT was effective in condition assessment. The GPR assisted the IRT in detecting deeper delamination.

Table 2. *Cont.*

Study	Year	Other Techniques	GPR (GHz)		Main Findings
			Air-Coupled	Ground-Coupled	
Gucunski et al. [23]	2010	GPR, ER, HCP, IE, USW	-	1.5	This combination of NDT can characterize different levels of deterioration. GPR brought effectiveness in the speed of inspection as the fastest technology from these five.
Gucunski et al. [100]	2013	GPR, ER, IE, USW	-	2	GPR deterioration maps were effectively implemented in a robotic system for bridge deck evaluation.
Alani et al. [101]	2014	GPR, deflection and vibration system	-	2	GPR results were combined with the deflection and vibration system to create a FEM model; GPR was used to locate rebar and detect cracks and potential moisture areas.
Kim et al. [102]	2016	GPR, ER, IE	-	2	GPR results were combined with ER and IE to calculate the condition index for estimation of service life.
Abu Dabous [103]	2017	GPR, IRT	-	1.6	Maps obtained with GPR and IRT were overlapped to form areas of possible delamination; the detected area was used to determine the condition rating.
Omar et al. [104]	2018	GPR, IRT	1	1.6	A method based on the integrated results obtained with GPR and IRT was proposed.
Ahmed et al. [105]	2018	GPR, ER, HCP, IE	-	-	Data fusion model from GPR, ER, HCP and IE maps was developed; fusion was on pixel and feature level.
Solla et al. [106]	2019	GPR, IRT	-	2.3	The paper proposes a procedure for anomaly detection based on joint observation of GPR signal and IRT temperature.
Kilic et al. [107]	2020	GPR, IRT, laser distance sensor, camera	-	2	The effectiveness of the integrated techniques was demonstrated on a bridge; GPR was used to detect water leakage, large cracks and corrosion.
Rashidi et al. [108]	2020	GPR, ER, HCP, IE, USW	-	1.5	The results from NDT were used to determine condition indices calculated using divergence from the ideal distribution using the Jensen–Shannon method.

In a very detailed study, Omar et al. [109] presented the weaknesses and advantages of the most commonly used methods for condition assessment of concrete bridges. The conclusion was that none of the commonly used techniques are able to detect active corrosion, delamination and vertical cracking simultaneously, so that the most reliable condition assessment lies in a combination of multiple techniques. Such an approach ensures accurate condition assessment as deterioration can be detected from its onset to an advanced stage [23].

The simultaneously used non-destructive techniques usually consider methods such as ground-penetrating radar, electrical resistivity, half-cell potential, ultrasonic surface waves, impact-echo, etc. In References [14,100,110], an example of integration of different non-destructive testing methods in robotic systems, RABIT (Robotics Assisted Bridge Inspection Tool), was presented, which ensures real-time visualization of the concrete deck condition. Here, the evaluation is supported by a Jensen–Shannon probability method that focuses on the determination of the condition index [108]. Additional support in the interpretation of GPR data for delamination detection can be provided by infrared thermography (IRT) [99,103,104,107]. Solla et al. [106] demonstrated the technique to inspect a military base in an advanced stage of corrosion with visible signs of damage such as cracking and spalling. The results obtained with GPR were combined with the IRT technique. The corrosion assessment was based on the observation of GPR signal attenuation, changes in signal velocity and amplitude polarity. Overall, high signal attenuation was declared to indicate the presence of mineral salts and moisture, while reverse reflection polarity could be a sign of voids. The same parameters have been used in the assessment of wastewater plants [111], although the corrosion process is different in this case.

Deterioration modelling was part of the study in Reference [102], in which deterioration curves were developed based on the condition assessment of 10 bridges. Similar assessments were carried out by Alani et al. [101], where finite element models were constructed based on inputs from ground-penetrating radar and the deflection and vibration sensor system. In Reference [105], a data fusion model for bridge deck evaluation was developed based on the combination of the results from the ground-penetrating radar, half-cell potential method, electrical resistivity method and impact-echo method. In Reference [112], the ground-penetrating radar data combined with the capacitive technique and the impact-echo method were correlated with durability indicators for the overall assessment of the wharf.

3.2.4. Conclusions from the On-Site Corrosion Inspection

The previous section has shown that corrosion assessment in on-site corrosion testing is mostly based on the assessment of the attenuated areas identified by signal amplitude analysis. Most of the studies are carried out on the bridge decks. In terms of comparison with other NDT, GPR has been compared with various techniques used for the service life condition assessment of the structures, Figure 5.

The high correlation between the electrical resistivity and the attenuation maps obtained with ground-penetrating radar is to be expected, as the signal depends on the material properties, so the conductive medium generated by moisture and chlorides changes its properties. In general, the GPR has shown good agreement with the HCP. However, there are certain situations where the GPR does not agree very well with the HCP. In cases where moisture and chlorides provide a favorable environment for corrosion, but their concentration is not sufficient to start corrosion, the GPR and HCP maps may differ. The applicability of ground-penetrating radar in detecting delamination is also uncertain [113]. In many cases, it does not detect delamination directly, and the assessment is based on the localization of deteriorated areas [114]. Moreover, the visual signs of delamination are not always visible on B-scans [87]. If the delamination is too thin to be detected by the antenna, it will not show any detectable change on the scan.

In summary, additional information, such as the age of the structure or the environmental conditions, may be helpful in analyzing GPR results. Moreover, this additional information can be obtained with other NDT, so a suitable combination of NDT can be a very powerful tool for the condition assessment of concrete structures.

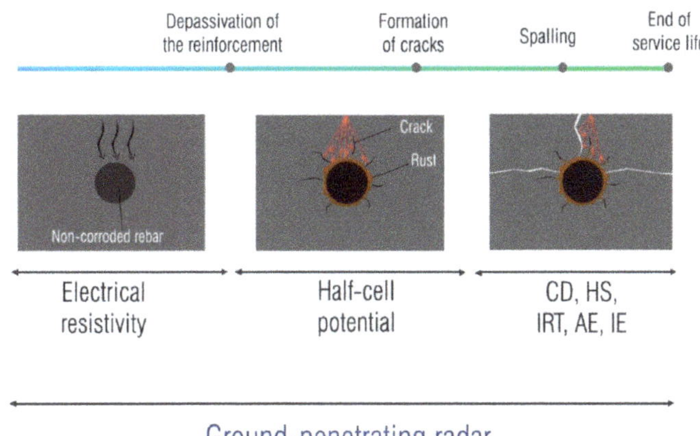

Figure 5. Multiple NDT used for the condition assessment.

4. Conclusions

This paper investigated the evaluation of corrosion probability in concrete using ground-penetrating radar. The study analyzed laboratory and on-site investigations and the results were related to the evolution of the corrosion process. Advantages and recommendations for future research are presented below.

GPR is a completely non-destructive method, which gives it an advantage over other techniques for corrosion assessment of reinforced concrete. Its ability to examine large areas in a short time, together with providing information on the depth and spacing of reinforcement, makes it a multifunctional NDT. The literature review identified certain challenges in the use of GPR for corrosion assessment, one of the main being the understanding of the influence of concrete conditions on GPR parameters. In fact, in most laboratory studies, moisture and chloride content were controlled after depassivation of the reinforcement. On-site in real conditions, variations of moisture and chloride content are inevitable, which makes the detection of corroded areas based only on the observation of amplitude potentially ambiguous. Since opposing data have been reported in the literature, further laboratory studies are needed to show the influence of the change in dielectric properties of the concrete cover on the GPR amplitude and the change in reflection coefficient due to the formation of corrosion products and their migration. Since an absolute comparison of studies is difficult due to the variance in experimental design and the degree of damage induced by the accelerated corrosion process, further studies should correlate the degree of damage with the change in GPR amplitude. Obtaining concluding results from the proposed research topics could enable the use of GPR as a stand-alone tool for detecting corroded areas, moving from its use for the detection of corrosive environment towards detection of corrosion itself.

In conclusion, as the knowledge of the effect of corrosion on the GPR signal increases, GPR will be a very valuable tool for condition assessment of reinforced concrete structures. This method will certainly be improved, leading to an upgrade of the construction management system and making the assessment more reliable with reduced maintenance costs.

Author Contributions: Conceptualization, K.T., A.B. and M.S.; methodology, K.T. and A.B.; investigation, K.T. and A.B.; data curation, K.T.; writing—original draft preparation, K.T.; writing—review and editing, A.B. and M.S.; visualization, K.T.; supervision, A.B.; project administration, M.S.; funding acquisition, M.S. All authors have read and agreed to the published version of the manuscript.

Funding: This research was funded by the European Union through the European Regional Development Fund's Competitiveness and Cohesion Operational Program, grant number KK.01.1.1.04.0041, project "Autonomous System for Assessment and Prediction of infrastructure integrity (ASAP)".

Institutional Review Board Statement: Not applicable.

Informed Consent Statement: Not applicable.

Data Availability Statement: Not applicable.

Conflicts of Interest: Authors declare no conflict of interest. The funders had no role in the design of the study; in the collection, analyses, or interpretation of data; in the writing of the manuscript, or in the decision to publish the results.

References

1. Croatian, P. *Building Act, NN 153/2013*; Official Gazette: Zagreb, Croatia, 2017.
2. Mehta, B.P.K.; Burrows, R.W. Building Durable Structures in the 21st Century. *Concr. Int.* **2001**, *23*, 57–63.
3. Bossio, A.; Monetta, T.; Bellucci, F.; Lignola, G.P.; Prota, A. Modeling of concrete cracking due to corrosion process of reinforcement bars. *Cem. Concr. Res.* **2015**, *71*, 78–92. [CrossRef]
4. Yan, L.; Chouw, N. Behavior and analytical modeling of natural flax fibre-reinforced polymer tube confined plain concrete and coir fibre-reinforced concrete. *J. Compos. Mater.* **2013**, *47*, 2133–2148. [CrossRef]
5. Vecchio, F.J.; Bucci, F. Analysis of Repaired Reinforced Concrete Structures. *J. Struct. Eng.* **1999**, *125*, 644–652. [CrossRef]
6. Navarro, I.J.; Yepes, V.; Martí, J.V.; González-Vidosa, F. Life cycle impact assessment of corrosion preventive designs applied to prestressed concrete bridge decks. *J. Clean. Prod.* **2018**, *196*, 698–713. [CrossRef]
7. Cao, Y.; Dong, S.; Zheng, D.; Wang, J.; Zhang, X.; Du, R.; Song, G.; Lin, C. Multifunctional inhibition based on layered double hydroxides to comprehensively control corrosion of carbon steel in concrete. *Corros. Sci.* **2017**, *126*, 166–179. [CrossRef]
8. Luo, H.; Su, H.; Dong, C.; Li, X. Passivation and electrochemical behavior of 316L stainless steel in chlorinated simulated concrete pore solution. *Appl. Surf. Sci.* **2017**, *400*, 38–48. [CrossRef]
9. Pan, X.; Shi, Z.; Shi, C.; Ling, T.C.; Li, N. A review on concrete surface treatment Part I: Types and mechanisms. *Constr. Build. Mater.* **2017**, *132*, 578–590. [CrossRef]
10. Ohtsu, M. Introduction. In *Innovative AE and NDT Techniques for On-Site Measurement of Concrete and Masonry Structures*; Ohtsu, M., Ed.; Springer: Dordrecht, Germany, 2016; pp. 1–4.
11. Núñez-Nieto, X.; Solla, M.; Lorenzo, H. Applications of GPR for Humanitarian Assistance and Security. In *Civil Engineering Applications of Ground Penetrating Radar*; Pajewski, L., Benedetto, A., Eds.; Springer: New Delhi, India, 2015; pp. 301–326.
12. Annan, A.P. Electromagnetic Principles of Ground Penetrating Radar. In *Ground Penetrating Radar: Theory and Applications*; Jol, M.H., Ed.; Elsevier B.V.: Amsterdam, The Netherlands, 2009; pp. 1–40.
13. Daniels, D.J. Introduction. In *Ground Penetrating Radar*, 2nd ed.; The Institution of Electrical Engineers: London, UK, 2004; pp. 1–11.
14. Gucunski, N.; Basily, B.; Kim, J.; Duong, T.; Maher, A.; Dinh, K.; Azari, H.; Ghasemi, H. Assessing Condition of Concrete Bridge Decks by Robotic Platform RABIT for Development of Deterioration and Predictive Models. In Proceedings of the 8th International Conference on Bridge Maintenance, Safety and Management (IABMAS), Foz do Iguacu, Brazil, 26–30 June 2016.
15. Reichling, K.; Raupach, M.; Wiggenhauser, H.; Stoppel, M.; Dobmann, G.; Kurz, J. BETOSCAN—Robot controlled non-destructive diagnosis of reinforced concrete decks. In Proceedings of the NDTCE'09, Non Destructive Testing in Civil Engineering, Nantes, France, 30 June–3 July 2009.
16. Hubbard, S.S.; Zhang, J.; Monteiro, P.J.M.; Peterson, J.E.; Rubin, Y. Experimental Detection of Reinforcing Bar Corrosion Using Nondestructive Geophysical Techniques. *ACI Mater. J.* **2003**, *100*, 501–510. [CrossRef]
17. Lai, W.W.L.; Kind, T.; Stoppel, M.; Wiggenhauser, H. Measurement of Accelerated Steel Corrosion in Concrete Using Ground-Penetrating Radar and a Modified Half-Cell Potential Method. *J. Infrastruct. Syst.* **2013**, *19*, 205–220. [CrossRef]
18. Hong, S.; Lai, W.W.L.; Wilsch, G.; Helmerich, R.; Helmerich, R.; Günther, T.; Wiggenhauser, H. Periodic mapping of reinforcement corrosion in intrusive chloride contaminated concrete with GPR. *Constr. Build. Mater.* **2014**, *66*, 671–684. [CrossRef]
19. Hong, S.; Lai, W.W.L.; Helmerich, R. Experimental monitoring of chloride-induced reinforcement corrosion and chloride contamination in concrete with ground-penetrating radar. *Struct. Infrastruct. Eng.* **2015**, *11*, 15–26. [CrossRef]
20. Wong, P.T.W.; Lai, W.W.L.; Sham, J.F.C.; Poon, C. Hybrid non-destructive evaluation methods for characterizing chloride-induced corrosion in concrete. *NDT E Int.* **2019**, *107*. [CrossRef]
21. Barnes, C.L.; Trottier, J.F.; Forgeron, D. Improved concrete bridge deck evaluation using GPR by accounting for signal depth-amplitude effects. *NDT E Int.* **2008**, *41*, 427–433. [CrossRef]
22. Tarussov, A.; Vandry, M.; De La Haza, A. Condition assessment of concrete structures using a new analysis method: Ground-penetrating radar computer-assisted visual interpretation. *Constr. Build. Mater.* **2013**, *38*, 1246–1254. [CrossRef]
23. Gucunski, N.; Romero, F.; Kruschwitz, S.; Feldmann, R.; Abu-Hawash, A.; Dunn, M. Multiple complementary nondestructive evaluation technologies for condition assessment of concrete bridge decks. *Transp. Res. Rec.* **2010**, 34–44. [CrossRef]

24. Barnes, C.L.; Trottier, J.F. Ground-Penetrating Radar for Network-Level Concrete Deck Repair Management. *J. Transp. Eng.* **2000**, *126*, 257–262. [CrossRef]

25. Dinh, K.; Gucunski, N.; Kim, J.; Duong, T.H. Understanding depth-amplitude effects in assessment of GPR data from concrete bridge decks. *NDT E Int.* **2016**, *83*, 48–58. [CrossRef]

26. Lai, W.W.L.; Dérobert, X.; Annan, A.P. A review of Ground Penetrating Radar application in civil engineering: A 30-year journey from Locating and Testing to Imaging and Diagnosis. *NDT E Int.* **2018**, *96*, 58–78. [CrossRef]

27. Tosti, F.; Ferrante, C. Using Ground Penetrating Radar Methods to Investigate Reinforced Concrete Structures. *Surv. Geophys.* **2020**, *41*, 485–530. [CrossRef]

28. Abu Dabous, S.; Feroz, S. Condition monitoring of bridges with non-contact testing technologies. *Autom. Constr.* **2020**, *116*. [CrossRef]

29. Clarivate Analytics Web of Science. Available online: www.webofknowledge.com (accessed on 30 October 2020).

30. Science Direct Scopus. Available online: https://www.scopus.com (accessed on 30 October 2020).

31. Beushausen, H.; Torrent, R.; Alexander, M.G. Performance-based approaches for concrete durability: State of the art and future research needs. *Cem. Concr. Res.* **2019**, *119*, 11–20. [CrossRef]

32. Bertolini, L.; Elsener, B.; Pedeferri, P.; Redaelli, E.; Polder, R. *Corrosion of Steel in Concrete: Prevention, Diagnosis, Repair*, 2nd ed.; Wiley-VCH Verlag GmbH & Co.: Weinheim, Germany, 2013.

33. Broomfield, J.P. *Corrosion of Steel in Concrete: Understanding, Investigation and Repair*, 2nd ed.; Taylor and Francis: London, UK, 2003.

34. Tuutti, K. *Corrosion of Steel in Concrete*; Swedish Cement and Concrete Research Institute: Stockholm, Sweden, 1982.

35. Alexander, M.; Beushausen, H. Durability, service life prediction, and modelling for reinforced concrete structures—Review and critique. *Cem. Concr. Res.* **2019**, *122*, 17–29. [CrossRef]

36. Andrade, C.; Alonso, C.; Gulikers, J.; Polder, R.; Cigna, R.; Vennesland, O.; Salta, M.; Raharinaivo, A.; Elsener, B. Test methods for on-site corrosion rate measurement of steel reinforcement in concrete by means of the polarization resistance method. *Mater. Struct.* **2004**, *37*, 623–643. [CrossRef]

37. Elsener, B.; Andrade, C.; Gulikers, J.; Polder, R.; Raupach, M. Half-cell potential measurements—Potential mapping on reinforced concrete structures. *Mater. Struct.* **2003**, *36*, 461–471. [CrossRef]

38. ASTM C876-91. *Standard Test Method for Half-Cell Potentials of Uncoated Reinforcing Steel in Concrete*; ASTM International: West Conshohocken, PA, USA, 1999.

39. Hornbostel, K.; Larsen, C.K.; Geiker, M.R. Relationship between concrete resistivity and corrosion rate—A literature review. *Cem. Concr. Compos.* **2013**, *39*, 60–72. [CrossRef]

40. Song, H.W.; Saraswathy, V. Corrosion monitoring of reinforced concrete structures—A review. *Int. J. Electrochem. Sci.* **2007**, *2*, 1–28. [CrossRef]

41. Polder, R.; Andrade, C.; Elsener, B.; Vennesland, O.; Gulikers, J.; Weidert, R.; Raupach, M. Test methods for on site measurement of resistivity of concrete. *Mater. Struct.* **2000**, *33*, 603–611. [CrossRef]

42. Cassidy, N.J. Electrical and Magnetic Properties of Rocks, Soils and Fluids. In *Ground Penetrating Radar: Theory and Applications*; Jol, H.M., Ed.; Elsevier B.V.: Amsterdam, The Netherlands, 2009; pp. 41–72.

43. Laurens, S.; Balayssac, J.P.; Rhazi, J.; Arliguie, G. Influence of concrete relative humidity on the amplitude of ground-penetrating radar (GPR) signal. *Mater. Struct.* **2002**, *35*, 198–203. [CrossRef]

44. Sbartaï, Z.M.; Laurens, S.; Balayssac, J.; Arliguie, G.; Ballivy, G. Ability of the direct wave of radar ground-coupled antenna for NDT of concrete structures. *NDT E Int.* **2006**, *39*, 400–407. [CrossRef]

45. Hugenschmidt, J.; Loser, R. Detection of chlorides and moisture in concrete structures with ground penetrating radar. *Mater. Struct.* **2008**, *41*, 785–792. [CrossRef]

46. Senin, S.F.; Hamid, R. Ground penetrating radar wave attenuation models for estimation of moisture and chloride content in concrete slab. *Constr. Build. Mater.* **2016**, *106*, 659–669. [CrossRef]

47. Dérobert, X.; Villain, G.; Balayssac, J.P. Influence of concrete carbonation on electromagnetic permittivity measured by GPR and capacitive techniques. *J. Environ. Eng. Geophys.* **2018**, *23*, 443–456. [CrossRef]

48. Ahmad, S. Techniques for inducing accelerated corrosion of steel in concrete. *Arab. J. Sci. Eng.* **2009**, *34*, 95–104.

49. Malumbela, G.; Moyo, P.; Alexander, M. A step towards standardising accelerated corrosion tests on laboratory reinforced concrete specimens. *J. South African Inst. Civ. Eng.* **2012**, *54*, 78–85.

50. El Maaddawy, T.A.; Soudki, K.A. Effectiveness of Impressed Current Technique to Simulate Corrosion of Steel Reinforcement in Concrete. *J. Mater. Civ. Eng.* **2003**, *15*, 41–47. [CrossRef]

51. Yuan, Y.; Ji, Y.; Shah, S.P. Comparison of Two Accelerated Corrosion Techniques for Concrete Structures. *ACI Struct. J.* **2007**, *104*, 344–347.

52. Sossa, V.; Pérez-Gracia, V.; González-Drigo, R.; Rasol, M.A. Lab non destructive test to analyze the effect of corrosion on ground penetrating radar scans. *Remote Sens.* **2019**, *11*, 2814. [CrossRef]

53. Raju, R.K.; Hasan, M.I.; Yazdani, N. Quantitative relationship involving reinforcing bar corrosion and ground-penetrating radar amplitude. *ACI Mater. J.* **2018**, *115*, 449–457. [CrossRef]

54. Zaki, A.; Johari, M.A.; Hussin, W.M.A.W.; Jusman, Y. Experimental Assessment of Rebar Corrosion in Concrete Slab Using Ground Penetrating Radar (GPR). *Int. J. Corros.* **2018**, *2018*. [CrossRef]

55. Hasan, M.I.; Yazdani, N. An experimental study for quantitative estimation of rebar corrosion in concrete using ground penetrating radar. *J. Eng.* **2016**, *2016*. [CrossRef]
56. Lai, W.W.L.; Kind, T.; Wiggenhauser, H. Detection of accelerated reinforcement corrosion in concrete by ground penetrating radar. In Proceedings of the XIII Internartional Conference on Ground Penetrating Radar, Lecce, Italy, 21–25 June 2010.
57. Zhan, B.J.; Lai, W.W.L.; Kou, S.C.; Poon, C.S.; Tsang, W.F. Correlation between accelerated steel corrosion in concrete and ground penetrating radar parameters. In Proceedings of the International RILEM Conference on Advances in Construction Materials Through Science and Engineering, Hong Kong, China, 5–7 September 2011.
58. Lai, W.W.L.; Kind, T.; Wiggenhauser, H. Using ground penetrating radar and time-frequency analysis to characterize construction materials. *NDT E Int.* **2011**, *44*, 111–120. [CrossRef]
59. Hong, S. GPR-Based Periodic Monitoring of Reinforcement Corrosion in Chloride- Contaminated Concrete. Ph.D. Thesis, TU Berlin, Berlin, Germany, 2015.
60. Said, M.E.; Hussein, A.A. Induced Corrosion Techniques for Two-Way Slabs. *J. Perform. Constr. Facil.* **2019**, *33*. [CrossRef]
61. Altoubat, S.; Maalej, M.; Shaikh, F.U.A. Laboratory Simulation of Corrosion Damage in Reinforced Concrete. *Int. J. Concr. Struct. Mater.* **2016**, *10*, 383–391. [CrossRef]
62. ASTM D6087-08. *Standard Test Method for Evaluating Asphalt-Covered Concrete Bridge Decks Using Ground Penetrating Radar*; ASTM International: West Conshohocken, PA, USA, 2008.
63. Saarenketo, T. NDT Transportation. In *Ground Penetrating Radar: Theory and Applications*; Jol, H.M., Ed.; Elsevier B.V.: Amsterdam, The Netherlands, 2009; pp. 395–444.
64. Belli, K.M.; Birken, R.A.; Vilbig, R.A.; Wadia-Fascetti, S.J. Simulated GPR investigation of deterioration in reinforced concrete bridge decks. In Proceedings of the Symposium on the Application of Geophysics to Engineering and Environmental Problems 2013, Denver, CO, USA, 17–21 March 2013.
65. Dinh, K.; Zayed, T.; Romero, F.; Tarussov, A. Method for analyzing time-series GPR data of concrete bridge decks. *J. Bridg. Eng.* **2015**, *20*. [CrossRef]
66. Diamanti, N.; Annan, A.P.; Redman, J.D. Concrete Bridge Deck Deterioration Assessment Using Ground Penetrating Radar (GPR). *J. Environ. Eng. Geophys.* **2017**, *22*. [CrossRef]
67. GSSI. *RADAN 7 Manual*; Geophysical Survey Systems, Inc.: Nashua, NH, USA, 2015.
68. Barnes, C.L.; Trottier, J.F. Phenomena and conditions in bridge decks that confound ground-penetrating radar data analysis. *Transp. Res. Rec.* **2002**, 57–61. [CrossRef]
69. Barnes, C.L.; Trottier, J.F. Effectiveness of Ground Penetrating Radar in Predicting Deck Repair Quantities. *J. Infrastruct. Syst.* **2004**, *10*, 69–76. [CrossRef]
70. Parrillo, B.; Roberts, R. Bridge deck condition assessment using ground penetrating radar. In Proceedings of the 9th European Conference on NDT (ECNDT), Berlin, Germany, 25–29 September 2006.
71. Pashoutani, S.; Zhu, J. Ground Penetrating Radar Data Processing for Concrete Bridge Deck Evaluation. *J. Bridg. Eng.* **2020**, *25*. [CrossRef]
72. Romero, F.A.; Barnes, C.L.; Azari, H.; Nazarian, S.; Rascoe, C.D. Validation of Benefits of Automated Depth Correction Method Improving Accuracy of Ground-Penetrating Radar Deck Deterioration Maps. *Transp. Res. Rec.* **2015**, 100–109. [CrossRef]
73. Rhazi, J.; Dous, O.; Laurens, S. A New Application of the GPR Technique To Reinforced Concrete. In Proceedings of the 4th Middle NDT Conference and Exhibition, Manama, Bahrain, 2–5 December 2007.
74. Maser, K.; Martino, N.; Doughty, J.; Birken, R. Understanding and detecting bridge deck deterioration with ground-penetrating radar. *Transp. Res. Rec.* **2012**, 116–123. [CrossRef]
75. Martino, N.; Maser, K.; Birken, R.; Wang, M. Determining ground penetrating radar amplitude thresholds for the corrosion state of reinforced concrete bridge decks. *J. Environ. Eng. Geophys.* **2014**, *19*, 175–181. [CrossRef]
76. Martino, N.; Maser, K.; Birken, R.; Wang, M. Quantifying Bridge Deck Corrosion Using Ground Penetrating Radar. *Res. Nondestruct. Eval.* **2016**, *27*, 112–124. [CrossRef]
77. Dinh, K.; Zayed, T.; Moufti, S.; Shami, A.; Jabri, A.; Abouhamad, M.; Dawood, T. Clustering-Based Threshold Model for Condition Assessment of Concrete Bridge Decks with Ground-Penetrating Radar. *Transp. Res. Rec.* **2015**, 81–89. [CrossRef]
78. Alsharqawi, M.; Zayed, T.; Shami, A. Ground penetrating radar-based deterioration assessment of RC bridge decks. *Constr. Innov.* **2020**, *20*, 1–17. [CrossRef]
79. Dinh, K.; Gucunski, N.; Duong, T.H. An algorithm for automatic localization and detection of rebars from GPR data of concrete bridge decks. *Autom. Constr.* **2018**, *89*, 292–298. [CrossRef]
80. Ma, X.; Liu, H.; Wang, M.L.; Birken, R. Automatic detection of steel rebar in bridge decks from ground penetrating radar data. *J. Appl. Geophys.* **2018**, *158*, 93–102. [CrossRef]
81. Dinh, K.; Gucunski, N.; Kim, J.; Duong, T.H. Method for attenuation assessment of GPR data from concrete bridge decks. *NDT E Int.* **2017**, *92*, 50–58. [CrossRef]
82. Hong, S.; Wiggenhauser, H.; Helmerich, R.; Dong, B.; Dong, P.; Xing, F. Long-term monitoring of reinforcement corrosion in concrete using ground penetrating radar. *Corros. Sci.* **2017**, *114*, 123–132. [CrossRef]
83. Rhee, J.Y.; Shim, J.; Kee, S.H.; Lee, S.Y. Different Characteristics of Radar Signal Attenuation Depending on Concrete Condition of Bare Bridge Deck. *KSCE J. Civ. Eng.* **2020**, *24*, 2049–2062. [CrossRef]
84. Benedetto, A. A three dimensional approach for tracking cracks in bridges using GPR. *J. Appl. Geophys.* **2013**, *97*, 37–44. [CrossRef]

85. Abouhamad, M.; Dawood, T.; Jabri, A.; Alsharqawi, M.; Zayed, T. Corrosiveness mapping of bridge decks using image-based analysis of GPR data. *Autom. Constr.* **2017**, *80*, 104–117. [CrossRef]

86. Martino, N.; Maser, K. Comparison of air-coupled GPR data analysis results determined by multiple analysts. In Proceedings of the SPIE Conference on Health Monitoring of Structural and Biological Systems, Las Vegas, NV, USA, 21–24 March 2016.

87. Dinh, K.; Zayed, T.; Tarussov, A. GPR image analysis for corrosion mapping in concrete slabs. In Proceedings of the Canadian Society of Civil Engineering 2013 Conference, Montreal, QC, Canada, 29 May–1 June 2013.

88. Alsharqawi, M.; Zayed, T.; Abu Dabous, S. Integrated condition rating and forecasting method for bridge decks using Visual Inspection and Ground Penetrating Radar. *Autom. Constr.* **2018**, *89*, 135–145. [CrossRef]

89. Dawood, T.; Zhu, Z.; Zayed, T. Deterioration mapping in subway infrastructure using sensory data of GPR. *Tunneling Undergr. Sp. Technol.* **2020**, *103*. [CrossRef]

90. Sham, J.F.C.; Wallace, W.L.L. Diagnosis of reinforced concrete structures by Ground Penetrating Radar survey-case study. In Proceedings of the 9th International Workshop on Advanced Ground Penetrating Radar (IWAGPR), Edinburgh, UK, 28–30 June 2017.

91. Gucunski, N.; Parvardeh, H.; Romero, F.; Pailes, B.M. Deterioration progression monitoring in concrete bridge decks using periodical NDE surveys. In Proceedings of the Second Conference on Smart Monitoring, Assessment and Rehabilitation of Civil Structures (SMAR 2013), Istanbul, Turkey, 9–11 September 2013.

92. Dinh, K.; Gucunski, N.; Zayed, T. Automated visualization of concrete bridge deck condition from GPR data. *NDT E Int.* **2019**, *102*, 120–128. [CrossRef]

93. Pailes, B.M.; Gucunski, N. Understanding Multi-modal Non-destructive Testing Data Through the Evaluation of Twelve Deteriorating Reinforced Concrete Bridge Decks. *J. Nondestruct. Eval.* **2015**, *34*, 1–14. [CrossRef]

94. Sun, H.; Pashoutani, S.; Zhu, J. Nondestructive evaluation of concrete bridge decks with automated acoustic scanning system and ground penetrating radar. *Sensors* **2018**, *18*, 1955. [CrossRef] [PubMed]

95. Sultan, A.A.; Washer, G.A. Comparison of Two Nondestructive Evaluation Technologies for the Condition Assessment of Bridge Decks. *Transp. Res. Rec.* **2018**, *2672*, 113–122. [CrossRef]

96. Varnavina, A.V.; Sneed, L.H.; Khamzin, A.K.; Torgashov, E.V.; Anderson, N.L. An attempt to describe a relationship between concrete deterioration quantities and bridge deck condition assessment techniques. *J. Appl. Geophys.* **2017**, *142*, 38–48. [CrossRef]

97. Scott, M.; Rezaizadeh, A.; Delahaza, A.; Santos, C.G.; Moore, M.; Graybeal, B.; Washer, G. A comparison of nondestructive evaluation methods for bridge deck assessment. *NDT E Int.* **2003**, *36*, 245–255. [CrossRef]

98. Simi, A.; Manacorda, G.; Benedetto, A. Bridge deck survey with high resolution Ground Penetrating Radar. In Proceedings of the 14th International Conference on Ground Penetrating Radar (GPR), Shanghai, China, 4–8 June 2012.

99. Maser, K.R. Integration of ground penetrating radar and infrared thermography for bridge deck condition testing. In Proceedings of the NDTCE'09, Non Destructive Testing in Civil Engineering, Nantes, France, 30 June–3 July 2009.

100. Gucunski, N.; Maher, A.; Ghasemi, H. Condition assessment of concrete bridge decks using a fully autonomous robotic NDE platform. *Bridg. Struct.* **2013**, *9*, 123–130. [CrossRef]

101. Alani, A.M.; Aboutalebi, M.; Kilic, G. Integrated health assessment strategy using NDT for reinforced concrete bridges. *NDT E Int.* **2014**, *61*, 80–94. [CrossRef]

102. Kim, J.; Gucunski, N.; Dinh, K. Similarities and differences in bare concrete deck deterioration curves from multi NDE technology surveys. In Proceedings of the SPIE Conference on Health Monitoring of Structural and Biological Systems, Las Vegas, NV, USA, 21–24 March 2016.

103. Abu Dabous, S.; Yaghi, S.; Alkass, S.; Moselhi, O. Concrete bridge deck condition assessment using IR Thermography and Ground Penetrating Radar technologies. *Autom. Constr.* **2017**, *81*, 340–354. [CrossRef]

104. Omar, T.; Nehdi, M.L.; Zayed, T. Rational Condition Assessment of RC Bridge Decks Subjected to Corrosion-Induced Delamination. *J. Mater. Civ. Eng.* **2018**, *30*. [CrossRef]

105. Ahmed, M.; Moselhi, O.; Bhowmick, A. Two-tier data fusion method for bridge condition assessment. *Can. J. Civ. Eng.* **2018**, *45*, 197–214. [CrossRef]

106. Solla, M.; Lagüela, S.; Fernández, N.; Garrido, I. Assessing rebar corrosion through the combination of nondestructive GPR and IRT methodologies. *Remote Sens.* **2019**, *11*, 1705. [CrossRef]

107. Kilic, G.; Caner, A. Augmented reality for bridge condition assessment using advanced non-destructive techniques. *Struct. Infrastruct. Eng.* **2020**. [CrossRef]

108. Rashidi, M.; Azari, H.; Nehme, J. Assessment of the overall condition of bridge decks using the Jensen-Shannon divergence of NDE data. *NDT E Int.* **2020**, *110*. [CrossRef]

109. Omar, T.; Nehdi, M.L.; Zayed, T. Performance of NDT Techniques in Appraising Condition of Reinforced Concrete Bridge Decks. *J. Perform. Constr. Facil.* **2017**, *31*. [CrossRef]

110. Gucunski, N.; Basily, B.; Kim, J.; Yi, J.; Duong, T.; Dinh, K.; Kee, S.H.; Maher, A. RABIT: Implementation, performance validation and integration with other robotic platforms for improved management of bridge decks. *Int. J. Intell. Robot. Appl.* **2017**, *1*, 271–286. [CrossRef]

111. Manhães, P.M.B.; Araruna Júnior, J.T.; Chen, G.; Anderson, N.L.; dos Santos, A.B. Ground penetrating radar for assessment of reinforced concrete wastewater treatment plant. *J. Civ. Struct. Heal. Monit.* **2020**. [CrossRef]

112. Villain, G.; Sbartaï, Z.M.; Derobert, X.; Garnier, V.; Balayssac, J.P. Durability diagnosis of a concrete structure in a tidal zone by combining NDT methods: Laboratory tests and case study. *Constr. Build. Mater.* **2012**, *37*, 893–903. [CrossRef]
113. Hoegh, K.; Khazanovich, L.; Worel, B.J.; Yu, H.T. Detection of subsurface joint deterioration. *Transp. Res. Rec.* **2013**, 3–12. [CrossRef]
114. Gucunski, N.; Romero, F.; Shokouhi, P.; Makresias, J. Complementary Impact Echo and Ground Penetrating Radar Evaluation of Bridge Decks on I-84 Interchange in Connecticut. In Proceedings of the Geo-Frontiers Congress 2005, Austin, TX, USA, 24–26 January 2005.

 *materials*

Article

Testing and Assessing Method for the Resistance of Wood-Plastic Composites to the Action of Destroying Fungi

Anna Wiejak and Barbara Francke *

Building Research Institute, Filtrowa 1, 00-611 Warsaw, Poland; a.wiejak@itb.pl
* Correspondence: b.francke@itb.pl

Abstract: Durability tests against fungi action for wood-plastic composites are carried out in accordance with European standard ENV 12038, but the authors of the manuscript try to prove that the assessment of the results done according to these methods is imprecise and suffers from a significant error. Fungi exposure is always accompanied by high humidity, so the result of tests made by such method is always burdened with the influence of moisture, which can lead to a wrong assessment of the negative effects of action fungus itself. The manuscript has shown a modification of such a method that separates the destructive effect of fungi from moisture accompanying the test's destructive effect. The functional properties selected to prove the proposed modification are changes in the mass and bending strength after subsequent environmental exposure. It was found that intensive action of moisture measured in the culture chamber of about $(70 \pm 5)\%$, i.e., for 16 weeks, at $(22 \pm 2)\ °C$, which was the fungi culture, which was accompanying period, led to changes in the mass of the wood-plastic composites, amounting to 50% of the final result of the fungi resistance test, and changes in the bending strength amounting to 30–46% of the final test result. As a result of the research, the correction for assessing the durability of wood-polymer composites to biological corrosion has been proposed. The laboratory tests were compared with the products' test results following three years of exposure to the natural environment.

Keywords: wood-plastic composites; methods of testing resistance to fungi; methods of assessment

Citation: Wiejak, A.; Francke, B. Testing and Assessing Method for the Resistance of Wood-Plastic Composites to the Action of Destroying Fungi. *Materials* **2021**, *14*, 697. https://doi.org/10.3390/ma14030697

Academic Editor: Ramón Nóvoa
Received: 5 January 2021
Accepted: 30 January 2021
Published: 2 February 2021

Publisher's Note: MDPI stays neutral with regard to jurisdictional claims in published maps and institutional affiliations.

Copyright: © 2021 by the authors. Licensee MDPI, Basel, Switzerland. This article is an open access article distributed under the terms and conditions of the Creative Commons Attribution (CC BY) license (https://creativecommons.org/licenses/by/4.0/).

1. Introduction

In recent years greater attention has been paid to the aspects of sustainable development in construction. It is assumed that buildings are designed, constructed, and dismantled to enable sustainable use of natural resources, ensures the building structures' long life, and allows environmentally friendly raw and secondary materials. On the one hand, there is a drive to develop materials that are more susceptible to biodegradation processes [1], but on the other hand, works are carried out on improving the durability of building materials. Although accelerated biodegradation of different types of materials improves the effectiveness of waste management, the drive to protect building materials against the impact of microorganisms results from two factors. First and foremost, the corrosion which occurs due to biodegradation may pose a hazard to the life and health of the users of these materials because of the toxic influence of fungi [2,3]. The other aspect means the need to ensure the required durability of the materials, which has a tremendous impact on their appearance. The issue applies to wood, too. Natural fibers have low densities and are biodegradable, and they are highly available and cheap, making them more attractive than traditional synthetic fibers. Still, there are some disadvantages of using these fibers, i.e., low thermal stability, susceptibility to moisture absorption, and biological degradation [4]. Wood-plastic composites [5], called WPC in short, used for terrace surfaces and wall cladding, result from the efforts to develop a product whose performance combines the advantages of wood and plastics [1]. The polymer material in WPCs is the matrix, whereas particles of plant origin are the filler. Typical polymer matrices include polyethylene, polypropylene, polyvinylchloride, and high-density polyethylene (HDPE), whereby polypropylene (PP) is most prevalent in Poland. The fillers include but are not limited to various lignocellulose-based particles, usually wood, in the form of flour, fibers, or fine chips. Other fibers of plant origin such as rice hulls [6] are also used. The process of fiber modification results in the end product being less capable of water absorption. It is crucial from the point of view of resistance to biological factors that develop on a damp substrate [7,8]. The decay can be caused by fungi, which develop favorable conditions for their growth [9]. Optimum environmental conditions mean humidity measured in the culture chamber of ca. 70%, adequate temperature, and pH ranging from 5.6 to 6.5 [10]. Wood-plastic composites are more resistant to biodeterioration than untreated wood; however, several laboratory and field studies have shown that the wood in these materials remains susceptible to decay for example Silva et al. [11], Morris and Cooper [12], Mankowski and Morrell [13], Pendleton et al. [14], Verhey et al. [15], and Schirp et al. [16]. Mentioned reports indicate that decay does occur, but destruction rates are slower than found with untreated wood of the same species. While moisture levels eventually reach the point at which biological attack is possible, the wetting rate is slow and most of the moisture is confined to a zone within 5 mm of the WPC surface [11,17]. Wang and Morrell [18] stated that the rate of moisture sorption and ultimately decay rate depends on the wood's particle size and geometry, wood/polymer ratio, and the presence of other compounds that may repel water. Ibach et al. [17], in contrast to the information mentioned above, have done clinical magnetic resonance imaging and found a significant amount of water distributed randomly all across the board cross-sections of the decayed WPC, including the core of the samples. They explained that this phenomenon was caused by both the transport of moisture by fungal mycelia to a particular area of activity and by the generation of water as decay fungi metabolize wood, releasing bound water.

The development of specific fungi species depends on the substrate's character and properties where they develop (the content of minerals, salination) [19]. The polymer material's susceptibility to biodegradation processes results, e.g., from the polymer's chemical structure and molecular weight, its physical and chemical characteristics, and the kind and intensity of the microorganisms' impact [20,21]. Some literature findings indicate that thermoplastic material fully closes the composite's wooden component, protecting it from moisture and decomposition caused by fungi, because synthetic polymers that typically demonstrate a lack or negligible susceptibility to biodegradation are the matrix of polymer composites [22]. It results, e.g., from the chemical structure and hydrophobic character of the material surface. Many papers suggest that natural fiber coating with a polymer matrix is incomplete [23,24]. It turns out that the fibers reach moisture content levels that enable fungi attack [25]. Schirp et al. [16] stated that if the wood filler's moisture content can be kept out or at least below 20%, WPC decay may be prevented. This could be achieved by complete encapsulation of wood particles by the plastic matrix, hydrophobation of the WPC surface, or by chemical the wood substrate's modification. They also stated that voids between wood and plastic represent entry points and proliferation pathways for microbes; thus, these should be eliminated or reduced. Morris and Cooper [12] were the first to prove the presence of touchwood and discoloration caused by fungi on WPC terrace decking profiles. In the conditions of use, terrace decking profiles made of wood-plastic composites are susceptible to simultaneous influence of biological corrosion, variable positive and negative temperatures in the presence of water and moisture, UV radiation, and additional mechanical functional loads [21,22]. Previous studies assumed that deterioration of the measured property by more than 50% disqualified the material for further use [26].

Of course, when evaluating the tests' results, it is necessary to consider how they were performed. Biological corrosion tests are mainly performed using two methods. In North America, the soil-block test for wood [27–29] has been adopted for fungal durability tests of WPC in which weight loss serves as an indicator of decay. In Europe, the agar-block test, according to ENV 12038 [30], is commonly used in fungal decay testing. While American Standards [27,28] aims at determining the natural decay resistance of woods,

ENV 12038 [29] is intended to determine the efficiency threshold of wood preservatives against wood-decay fungi. Schirp and Wolcott [30] compared the North American and European methods for WPC fungal durability testing. The agar-block test was modified such that no support rods were employed to accelerate moisture uptake by WPC specimens that had not been pre-treated, only steam-sterilized in an autoclave. It was determined that modified agar-and soil-block tests are equally suited for determining weight loss in WPC, but that agar-block tests can be completed in a shorter period. Basidiomycetes fungi are recommended in the method according to ENV 12038 [29]. The group Basidiomycetes fungi, mentioned above, belong among others white-rot decay *Coriolus versicolor* and brown rot decay fungi, *Coniophora puteana*. Lignin and cellulose fibers are the fungi's food. *Coriolus versicolor* fungus attack lignin primarily, whereas *Coniophora puteana* fungus decomposes cellulose [31]. The mold growth is observed on composites; the mold uses contaminants on the surface as its food and reduces the material appearance rather than damage the material [32].

The aims of this manuscript are twofold: first to prove that the assessment of tests results obtained in tests done by European method used in fungal decay testing does not allow separate the destructive effect of fungi from the destructive effect of the moisture accompanying the test, second: propose a new method of assessing the test results. The research subject of the research was the decking WPC profiles used only on the ground's surface, without contact with the soil; therefore soil-block method was not included in the tests. A fungus from the Basidiomycetes group called *Coniophora puteana* was the trigger of biological corrosion in the study. The presented tests were carried out in two stages. In the first stage, the changes in the reference material characteristics under the influence of the fungus mentioned above were determined in laboratory conditions. On the second stage, the assessment was extended to cover aging tests carried out in natural conditions for three years, i.e., including climate loads. The natural tests were carried out under climate load conditions corresponding to mid- European transitional climate Since the literature reports [4,5,20,23,33–35] suggested that the negative impact of destroying fungi might cause changes in the mechanical characteristics of the tested materials and a change in their weight, the change in the characteristics after the impact of aging factors was taken as a biodegradation measure; the research methodology intended to assess the characteristics was verified. The obtained results showed that intensive action of moisture of about $(70 \pm 5)\%$, i.e., for 16 weeks, at (22 ± 2) °C, which was accompanying the fungi culture period, led to changes in the mass of the wood-plastic composites, amounting to 50% of the final result of the fungi resistance test, and changes in the bending strength amounting to $30\% \div 46\%$ of the final test result.

2. Materials and Methods

2.1. Materials

Wood-plastic composites are primarily used in Poland for the production of decking profiles intended for balconies and terraces. That is why decking profiles were used for the tests. Since the most common solutions employed for the production of wood-plastic composites include polyethylene (PE), polypropylene (PP), polyvinyl chloride (PVC), and high-density polyethylene (HDPE), materials with such matrixes were used for the tests. Products with similar content of wood material were used in all tested cases.

The following materials were used in all tests:

➢ Sample I—decking profiles made of a composite containing wood flour (45%), polyvinyl chloride (PVC) (45%), pigments (5%), stabilizers, absorbents, fillers (5%),
➢ Sample II—decking profiles made of a composite containing wood flour (60%), HDPE plastic (30%), stabilizers and pigments (10%),
➢ Sample III—decking profiles made of composite with low content of wood flour (49%) and polypropylene (PP),
➢ Sample IV—decking profiles made of a composite containing wood flour (49%) and polyvinyl chloride (PVC) (51%).

Additionally, water absorption values of the samples I-IV are shown in Table 1. These values were determined after 14 and 28 days of immersing the samples in distilled water, at temperature (22 ± 2) °C. Mentioned characteristics are supplemented with samples' water absorption values after removing fungi, i.e., after completion of the environmental exposures.

Table 1. Water absorption of samples I–IV.

Test Sample	Water Absorption, % m/m		After Completion of Environmental Exposures (after Removal of the Fungi)
	After Immersion in Water for:		
	14 Days	28 Days	
I	5.78	6.93	16.98
II	3.47	4.90	15.40
III	5.44	4.14	16.01
IV	5.12	6.32	16.50

2.2. Environmental Exposure

To determine the influence of destroying *Basidiomycetes* fungi, i.e., *Coniophora puteana*, on the durability of wood-plastic composites, compared to other functional impacts, the samples were exposed to various aging factors. *Coniophora puteana* fungus was selected because it is typically used in tests on wood protected with preservatives and it causes relatively large losses of the wood mass. That is why it was considered useful for the assessment of wood-plastic composites. The test samples were exposed according to the diagram shown in Figure 1.

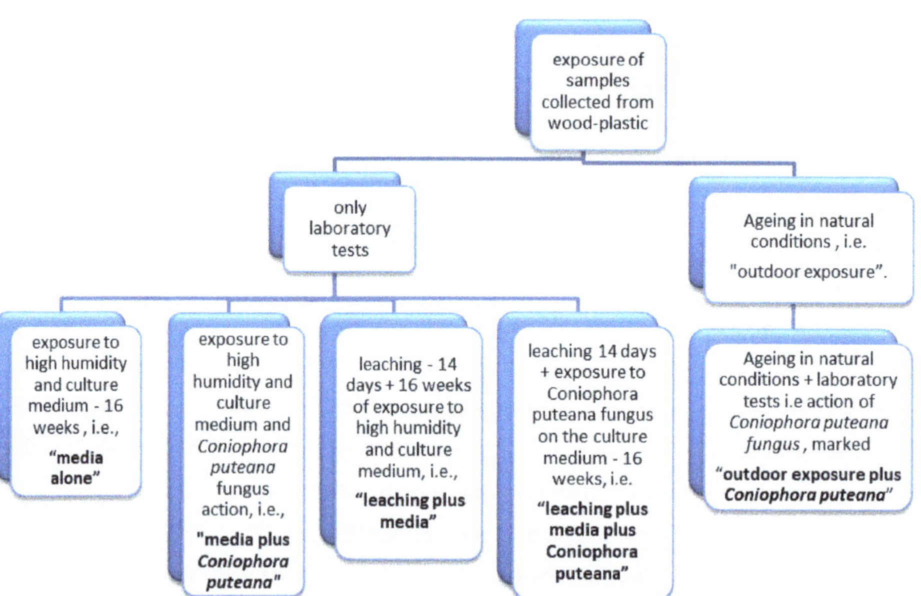

Figure 1. Schematic presentation of the samples of wood-plastic composite exposure patterns.

The exposures mentioned above involved:

- exposure to high humidity and culture medium—16 weeks (further marked as "media alone"). The samples are exposed in pairs in a Kolle flask on a culture medium, i.e., a substance composed of 40 g of malt extract, 35 g of agar, and water to 1000 mL at the test conditions at (22 ± 2) °C and (70 ± 5)% humidity, for 16 weeks. The medium (nutrient) is the food and source of moisture for the fungi during the test and simultaneously affects the test samples.
- leaching—14 days + 16 weeks of exposure to high humidity and culture medium (further marked as "leaching plus media"). The samples were immersed in water at (20 ± 2) °C for 14 days, according to the description given in EN 84 [36]. The samples were soaked in water, five volumes of water for one volume of sample, for two weeks. The water was replaced nine times during the cycle. After the end of the cycle, the samples were placed in flasks on maltose agar medium and incubated in the culture chamber at (22 ± 2) °C and (70 ± 5)% for 16 weeks. The leaching aging test according to EN 84 [36] is obligatorily used in tests of impregnated wood for exterior applications;
- exposure to high humidity and culture medium, and *Coniophora puteana* fungus action (further marked as "media plus *Coniophora puteana*")—16 weeks. The samples were placed in Kolle flasks in two, on glass plates on the culture medium completely covered with fungus *Coniophora puteana*. The flasks with the samples were placed in the culture chamber and incubated for 16 weeks at (22 ± 2) °C and (70 ± 5)% humidity'
- leaching 14 days + exposure to *Coniophora puteana* fungus on the culture medium— 16 weeks (further marked as "leaching plus media plus *Coniophora puteana*"). The samples were leaching in water for 14 days and then were placed in Kolle flasks in two, on glass plates on the culture medium completely covered with fungus *Coniophora puteana*. The flasks with the samples were placed in the culture chamber and incubated for 16 weeks at (22 ± 2) °C and (70 ± 5)% humidity;
- aging in natural conditions (further marked as "outdoor exposure"). Three mocks made of decking profiles screwed to composite ground beams were prepared in the test station in natural conditions, on the supports made of hollow bricks. After three years of exposure, the test samples were cut out from the profiles. The arrangement of the samples during the test in natural conditions is shown in Figure 2.
- aging in natural conditions and action of *Coniophora puteana fungus* (further marked as "outdoor exposure plus *Coniophora puteana*"). After exposure in natural conditions for three years, test samples were cut out from the profiles and placed in test vessels on the medium covered with *Coniophora puteana* fungus and exposed in a culture chamber for 16 weeks, as described above.

Figure 2. The view of wood-plastic decking profiles during exposure in natural conditions.

All types of composite decking profiles selected for the tests were exposed according to the first three patterns. Aging tests in natural conditions were carried out only for samples marked as sample III and sample IV.

Ten samples (10 mm × 80 mm, 5 mm thick) cut out from the face profiles' surface any type of material (I-IV) were exposed according to each of the exposures listed above.

Simultaneously with the test samples, control samples were also used to determine the fungus activity. Ten virulens control test specimens (50 mm × 25 mm × 15 mm) with Scots pine wood were placed in Kolle flasks in two, on glass plates on the culture medium completely covered with fungus *Coniophora puteana* and were incubated like research samples. The loss in mass of the samples was obtained 38% ÷ 45% with the required ≥20%.

2.3. Methods of Tests

2.3.1. Testing the Change in the Mass

Before exposing the samples to the aging process, their mass was determined. The samples cut out from the decking profiles were seasoned in a climatic chamber at (20 ± 2) °C and (65 ± 5)% humidity until obtaining a constant mass. Eight samples of each type of WPC material were dried in an oven at (103 ± 2) °C to constant mass and the initial dry matter (m_o) of the test samples was calculated. The samples were exposed according to the patterns mentioned above. Following the exposure to fungus, the surface of the samples was cleaned of the mycelium. Once the aging processes were completed, the samples were dried at (103 ± 2) °C, and the end dry mass (m) of the samples was determined. The loss of the samples' mass was the measure of the material decay caused by each exposure and was calculated based on the following Formula (1):

$$U = (m_o - m) \times 100/m_o \tag{1}$$

For all tested samples the assessment was carried out after the following exposures:

- exposure to high humidity and culture medium—16 weeks,
- exposure to high humidity and culture medium and *Coniophora puteana* fungus action—16 weeks,
- leaching—14 days + 16 weeks of exposure to high humidity and culture medium,
- leaching 14 days + exposure to *Coniophora puteana* fungus on the culture medium—16 weeks.

Additional tests were carried out for samples III and IV after aging in natural conditions and for three years, followed by exposure for 16 weeks in high humidity conditions on the medium overgrown by fungus.

2.3.2. Bending Strength Tests

The bending strength tests of the samples following their exposure to aging factors were carried out according to EN ISO 178 [37], using eight samples of each material and for each test option. The tests were carried out at (23 ± 2) °C once different samples' exposure patterns were completed. The samples' surfaces were cleaned of any surface mycelium and dried to constant mass at (20 ± 2) °C and (65 ± 5)% humidity before the tests. The bending strength of the samples not subjected to aging (σ) was additionally identified to determine the change in the bending strength after exposure to aging processes (σ_i), according to the Formula (2):

$$Z_\sigma = [(\sigma_i - \sigma)/\sigma] \times 100 \tag{2}$$

The tests were carried out on INSTRON strength tester after:

- exposure to high humidity and culture medium—16 weeks,
- leaching—14 days + 16 weeks of exposure to high humidity and culture medium,
- leaching 14 days + exposure to Coniophora puteana fungus on the culture medium—16 weeks.

Additional tests were carried out for samples III and IV after:

- aging in natural conditions,
- aging in natural conditions for three years, followed by 16 weeks of exposure to fungus on the culture medium.

3. Results

Table 2 summarizes the percent change in the mass after exposure to, sequentially: leaching, exposure to fungus, leaching and exposure to fungus, and three years of exposure in natural conditions at later exposure to the fungus.

Table 2. Effect of combination of exposure over agar in Kollar flask, 14 days of leaching, and 16 weeks of exposure to *Coniophora puteana* on mass loss of selected wood plastic composites.

Test Sample	Average Mass Loss after Exposure to % m/m/Coefficient of Variation [%]				
	Media Alone	Leaching Plus Media	Media Plus *Coniophora puteana*	Leaching Plus Media Plus *Coniophora puteana*	Outdoor Exposure Plus Media Plus *Coniophora puteana*
I	−1.29/(13.0)	−1.4/(6.4)	−2.3/(18.7)	−2.8/(17.9)	-
II	−1.34/(35.8)	−1.7/(28.2)	−2.4/(3.8)	−2.5/(19.6)	-
III	−1.58/(10.1)	−1.4/(25.7)	−2.7/(5.6)	−2.6/(5.4)	−2.98/(13.8)
IV	−1.01/(8.9)	−1.2/(2.5)	−2.6/(5.4)	−2.8/(5.4)	−2.85/(10.2)

Figure 3 shows samples of wood-plastic composites after 16 weeks of exposure to *Coniophora puteana* fungus.

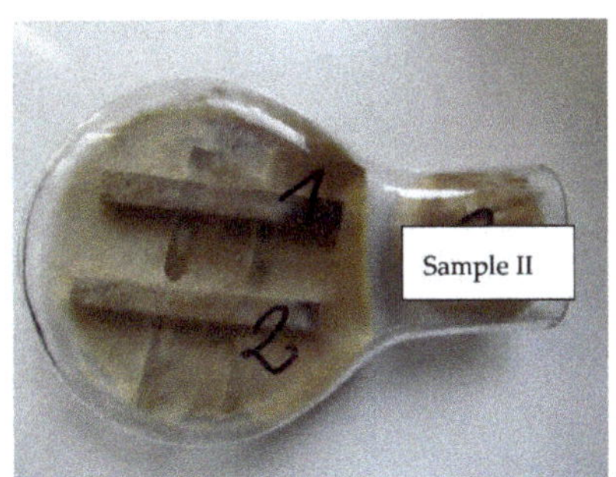

a)

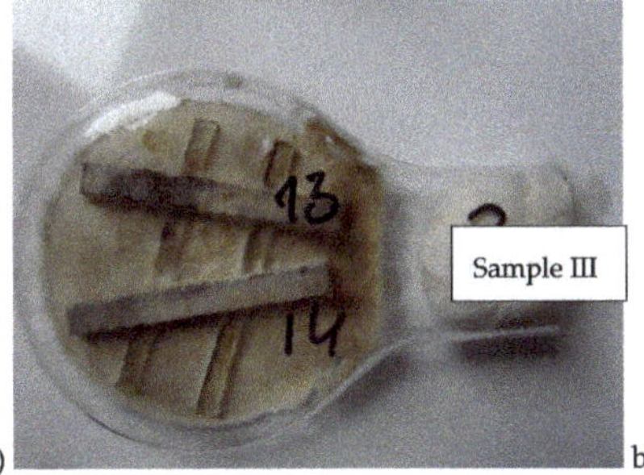

b)

Figure 3. Samples made of the wood-plastic composite after 16 weeks of exposure on Coniophora puteana fungus (**a**) sample II, (**b**) sample III.

Table 3 summarizes the bending strength test results for the exposed samples.

Table 3. Effect of combination of exposure over agar in Kollar flask, 14 days of leaching, and 16 weeks of exposure to *Coniophora puteana* on mean bending strength values of selected wood plastic composites.

Test Sample	Bending Strength after Exposure: MPa/Coefficient of Variation [%]					
	Original	Media Alone	Leaching Plus Media	Leaching Plus Media Plus *Coniophora puteana*	Outdoor Exposure	Outdoor Exposure Plus Media Plus *Coniophora puteana*
I	36.5/(8.8)	32.5/(4.6)	28.7/(6.6)	24.2/(10.7)	-	-
II	35.4/(11.0)	32.7/(5.5)	30.1/(3.7)	27.3/(4.4)	-	-
III	29.3/(2.7)	26.5/(3.4)	23.2/(4.3)	23.3/(7.3)	29.0/(6.6)	23.1/(3.4)
IV	36.4/(4.1)	33.1/(6.3)	31.7/(2.2)	28.2/(7.1)	35.8/(5.9)	27.8/(4.7)

4. Discussion

Figure 4 shows the percent changes in the samples' mass in reference to their original mass, resulting from different exposure patterns. The obtained results suggest that exposure in high humidity conditions of ca. $(70 \pm 5)\%$ at $(22 \pm 2)\,^{\circ}\mathrm{C}$ in the presence of culture medium causes a significant change in the mass (weight) of wood-plastic composite samples already during testing the material susceptibility to biological corrosion. The change in the mass of wood-plastic composites following long-term exposure to increased humidity amounts to 1.31% on average and increases slightly, within the measurement error range, if the samples are soaked in water for 14 days before the exposure. In the latter case, the mean loss of the mass is 1.35%.

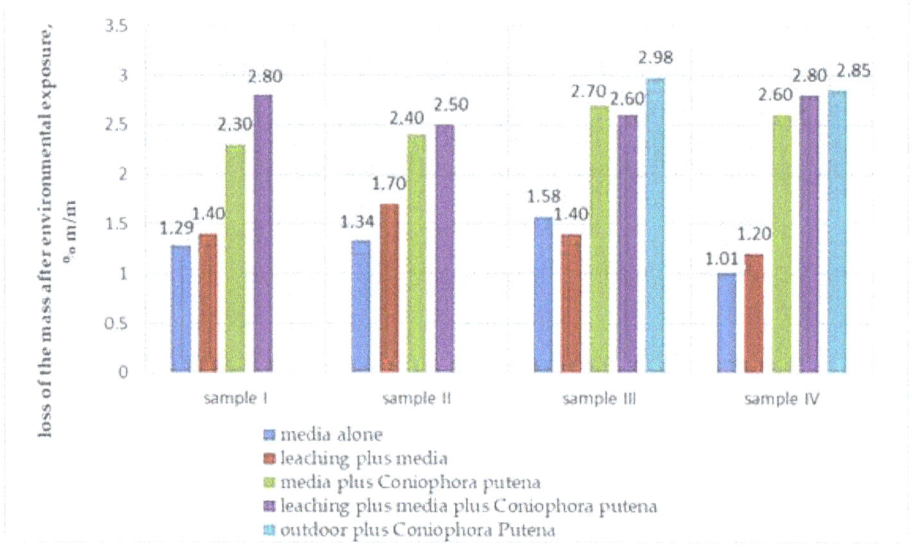

Figure 4. Influence of environmental exposure on the loss of mass in wood-plastic composites.

Further comparison of the effects of exposure to different factors leads to a conclusion that the mean loss of the mass caused by exposure to fungus, following the exposure on the culture medium, amounts to 2.5%, whereby 50% of the value, i.e., 1.31%, is the negative influence of the test conditions, namely high humidity in the presence of the medium. The presented calculation suggests that the loss of the mass caused by the fungus' actual impact amounts to 1.19%. The same is true for another factor—exposure to fungus after subjecting the samples to a long-term water impact. In this case, the mean values of the change in the mass after combined exposure to all the factors mentioned above amounts to 2.68%, which means that the fungus influence on the destructive process is 1.33% for the value of the loss in the wood-plastic composite mass.

The decay mechanism of WPC is often described in the literature [38] as the formation of a network of larger voids mainly through the connection of microvoids that are inherently present in the material. The mentioned process occurs as a result of an action of moisture alongside fungal activity or without the presence of fungi, only in moist conditions combined with elevated temperature. It was proved [39] that conditioning not only supplies the moisture needed for fungal growth but also effectively creates larger voids, assumably formed by the interconnection of smaller voids, which accelerates the decay process.

The exposure to natural use factor has shown to have no significant impact on the *Coniophora puteana* fungus growth. This has been confirmed by the observations made by Sun et al. [39], showing that the period of 3 years of storage of WPC in natural conditions is insufficient for fungal growth on the surface because of the low water absorption. They found out that in the field, an extensive time was required to initiate fungal colonization that results in wood weight loss, which may explain the deficiencies in some laboratory evaluations of WPC for fungal decay resistance [39]. Considering the above, samples after being exposed to natural conditions were exposed to Coniophora under laboratory conditions for 16 weeks. It seems that the influence of the fungus on the samples previously exposed to natural atmospheric conditions for 3 years on the increase in the loss of wood-plastic composite mass is minor. The comparison of the same samples subjected to natural aging and then exposed to the fungus on the sample medium with the samples exposed only to the fungus on the culture medium in laboratory conditions clearly shows that the increase in the loss of the mass following environmental exposure amounts to 0.27%, which means it is within the measurement error range.

The obtained values of the loss of mass after exposure to aging factors were compared with the content of wood flour in the composite. Figure 5 shows the graphic representation of the values.

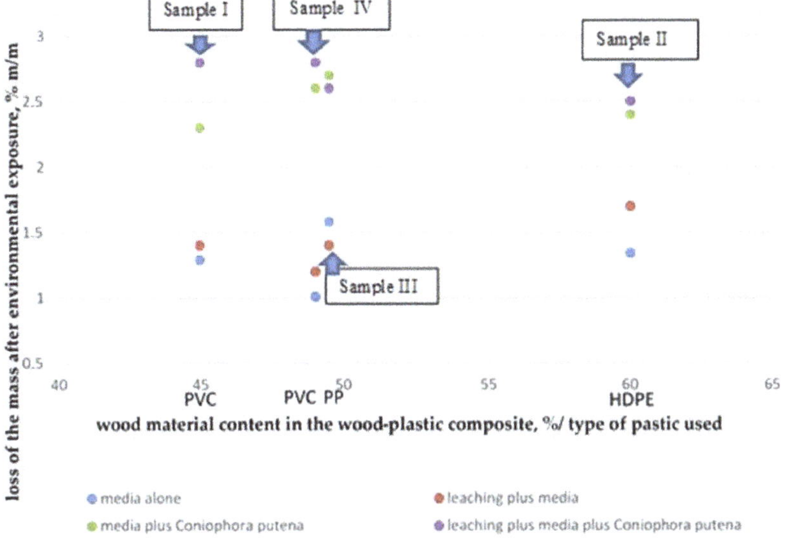

Figure 5. Comparison of the losses of mass after environmental exposure compared with the wood material content in the composite and polymer type in the matrix.

The summary reveals that the increase in the wood flour quantity has no significant influence on the composite susceptibility to the loss of mass due to biological corrosion, and the wood flour is the ingredient most sensitive to such exposure effects. Unfortunately, this conclusion does not confirm previous literature reports [13], but the obtained result may be affected by stabilizers, the composition of which is not known to the authors of the manuscript. The most significant losses of mass following the exposure to *Coniophora puteana* fungus were discovered for decking profiles with the polyvinyl chloride matrix, even though in both cases the content of the wood flour was the lowest among all tested products. The obtained results indirectly confirm the previous literature findings [19] that the thermoplastic material in a polypropylene and high-density polyethylene matrix closes the composite's wooden component and thus protects it from moisture and decay caused by fungi.

The results obtained for a polyvinyl chloride matrix (samples I and IV) are not so unequivocal. Sample I and sample IV are characterized by the highest absorbability values and they are accompanied by the loss of mass after exposure to environmental factors (Figure 6).

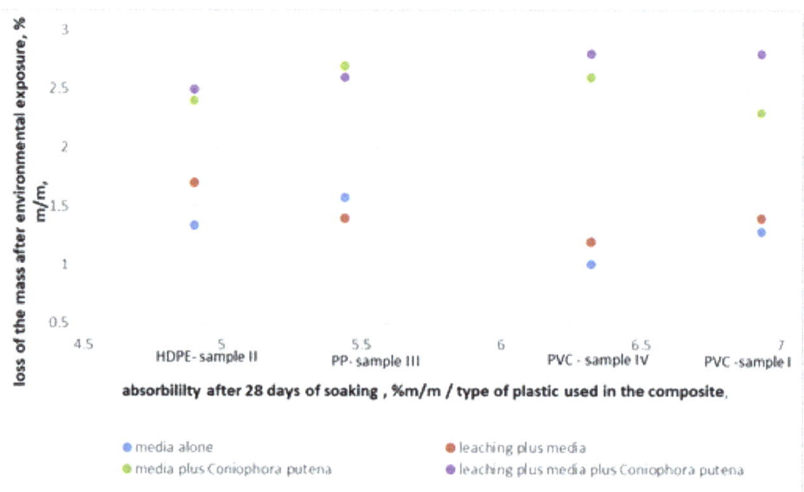

Figure 6. Comparison of the losses of mass after environmental exposure compared with the composite absorbability and polymer type in the matrix.

The observations mentioned above from laboratory studies confirm the observation related to the product's behavior during natural conditions. Continuous dampness, i.e., using the product in shaded areas, with limited air circulation, at a simultaneous water accumulation possibility, often contributed to the surfaces being overgrown with house fungi. Not only does it deteriorate the appearance but may also cause a loss of the product's performance.

The presented tests confirmed that the destructive action of *Coniophora puteana* fungus in the environments mentioned above, observed as a loss of mass, has a minor impact on the product life and should not contribute to the damage discovered during regular inspections of the building substance. Comparing the results with the limit values of the acceptable loss of mass for impregnated wood amounting to 3% reveals that the composites fall within the range, even if the influence of higher humidity exposure during the test on the final results is not eliminated.

In light of the above, the gradual loss of the tested material's strength can be even more worrying. The test composites' bending strength was tested on the following stage of the study, after their environmental exposure (Figure 7). It is evident that the negative impact of the environment with higher humidity of ca. $(70 \pm 5)\%$ in the presence of the culture medium has an equally strong influence on reducing the bending strength of wood-plastic composites and on their loss of mass. In all studied cases, the test products' bending strength decreased from 7.6% up to 11% in the environment with $(70 \pm 5)\%$ humidity, which amounts to 30–46% of the total test result. This action is intensified when the medium exposure in high humidity conditions is preceded by 14 days of immersion in water. It reduced the bending strength from 12.9% to 21.4%. In both presented cases the medium is the food for the fungi and maintains a specific moisture level. The obtained values of the bending strength reduction after the exposure mentioned above should not pose any significant problems related to the maintenance of structures made of the tested products within the evaluated range of functional loads. These findings are consistent with the results of studies and other scientists. Ibach et al. [38] also showed that the major factors degrading the mechanical properties of the tested composite under detected field exposure conditions were elevated temperature and moisture exposure, whereas UV radiation from the sun had a low impact, if any, on the flexural properties of a WPC board exposed for almost 10 years.

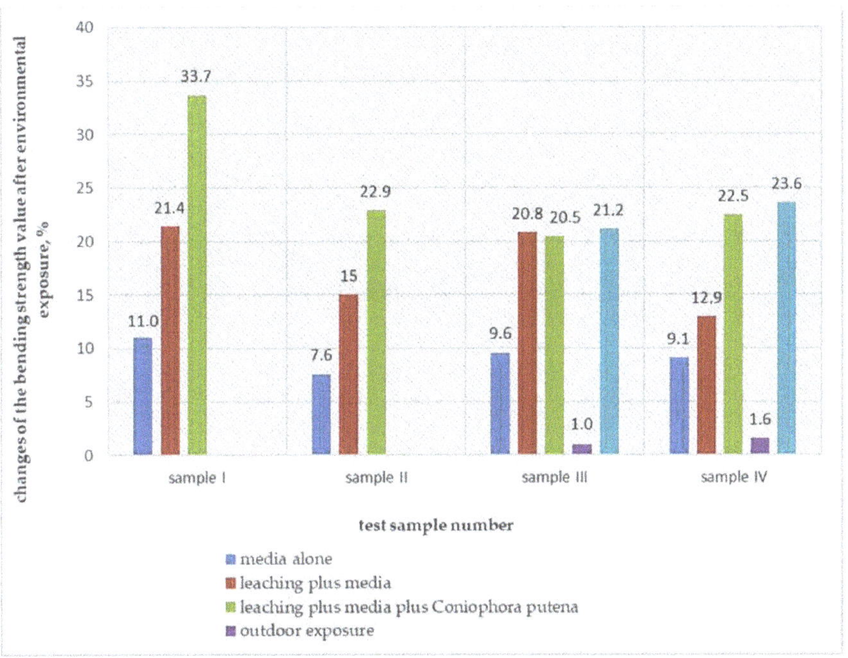

Figure 7. Reduction in the bending strength of wood-plastic composites after environmental exposure.

Thus, it seems justified to eliminate the influence of the exposure condition on the test result final assessment.

Taking the above into consideration, the authors proposed to apply Equation (3) for the evaluation of the wood-plastic composite susceptibility to the action of destroying fungi, defined by the change in the mass and bending strength after environmental exposure; it was also suggested to neglect the negative influence of the initial, 16 weeks' exposure to the high humidity of $(70 \pm 5)\%$, at $(22 \pm 2)\ °C$, in the presence of the medium.

$$Z_{mg} = (Z_x - Z_p) \tag{3}$$

where:

Z_{mg}—loss of mass/bending strength reduction as a result of exposure to house fungi, %

Z_x—loss of mass/bending strength reduction after combined exposure of all studied aging factors, %

Z_p—loss of mass/bending strength reduction after exposure to high humidity environmental conditions—$(70 \pm 5)\%$, at $22 \pm 2\ °C$, in the presence of the medium, %.

The proposal is also justified by changes in the wood-plastic composite bending strength after three years of exposure to natural environmental conditions. After the exposure mentioned above, the changes in the bending strength value did not exceed 2%. Once the samples were exposed to the fungus, and consequently the exposure on the culture medium, the obtained bending strength values were similar to those for the samples subjected to washing out in laboratory conditions for 14 days, followed by 16 weeks of exposure to the fungus. The same is true for the assessment of the change in the wood-plastic composite mass after environmental exposure. These observations conclude that for the assessment of wood-plastic composite resistance to *Coniophora puteana* fungus similar research outcomes can be obtained by substituting a long-term exposure cycle in the natural condition with a much quicker test carried out in the laboratory conditions.

5. Conclusions

The paper presents tests' results on four randomly selected commercial wood-polymer composites exposed to environmental impacts related to *Coniophora puteana* fungi's action on the reference products' durability. Based on the obtained results, the following conclusions can be drawn:

- The proposed assessment method of wood-polymer composites resistance to destroying fungi introduces bending strength test as a supplement to the weight change assessment of samples exposed to environmental impacts both in the laboratory and natural conditions. The assessment of the bending strength decrease renders more accurate results than the method involving loss of mass because the coefficient of variation is lower for such series of results,
- The manuscript presents a modification of assessment of the resistance of wood-polymer composites to destroying fungi which allow one to separate the destructive effect of fungi from the destructive effect of the moisture accompanying the test. It was proved the negative impact of high humidity for 16 weeks constitutes a significant percentage of the test result. In the case of the mass change assessment, it exceeds 50%, whereas for the bending strength change assessment it ranges from 30% to 46%.
- The method of testing wood-plastic composites' resistance to destroying fungi in laboratory conditions, presented in the paper, renders the results similar to those after three years of use in a natural environment and then infecting them with destroying fungi.

The performed tests confirm the high resistance of wood-plastic composites to destroying fungi.

Author Contributions: Conceptualization, A.W. and B.F.; methodology, A.W. and B.F.; formal analysis, A.W. and B.F.; investigation, A.W.; writing—original draft preparation, A.W. and B.F.; visualization, A.W. and B.F. All authors have read and agreed to the published version of the manuscript.

Funding: This research received no external funding.

Institutional Review Board Statement: Not applicable.

Informed Consent Statement: Not applicable.

Data Availability Statement: https://biblioteka.itb.pl.

Conflicts of Interest: The authors declare no conflict of interest. The funders had no role in the design of the study; in the collection, analyses, or interpretation of data; in the writing of the manuscript, or in the decision to publish the results.

References

1. Prochoń, M.; Witczak, M.; Biernacka, A. *Wood as a Component of Polymer Biocomposites (In Polish: Drewno Jako Składnik Biokompozytów Polimerowych)*; Technical University of Łódź: Łódź, Poland, 2017.
2. Andersson, M.A.; Nikulin, M.; Koljalg, U.; Andersson, M.C.; Rainey, F.; Reijula, K.; Hintikka, E.L.; Salkinoja-Salonen, M. Bacteria, molds, and toxins in water-damaged building materials. *Appl. Environ. Microbiol.* **1997**, *63*, 387–393. [CrossRef]
3. Fisher, M.C.; Henk, D.A.; Briggs, C.J.; Brownstein, J.S.; Madoff, L.C.; McCraw, S.L.; Gurr, S.J. Emerging fungal threats to animal, plant and ecosystem health. *Nature* **2012**, *484*, 186–194. [CrossRef]
4. Tascioglu, C.; Goodell, B.; Lopez-Anido, R.; Peterson, M.; Halteman, W.; Jellison, J. Minitoring fungal degradation of E-glass/phenolic fiber reinforced polymer (FRP) composites used in wood reinforcement. *Int. Biodeter. Biodegrad.* **2003**, *51*, 157–165. [CrossRef]
5. Eder, A.; Carus, M. Global trends in wood-plastic composites (WPC). *Bioplast. Mag.* **2013**, *8*, 16–17.
6. Penczek, S.; Pretula, J.; Lewiński, P. Polymers from renewable raw materials, biodegradable polymers (in Polish: Polimery z odnawialnych surowców, polimery biodegradowalne). *Polimery* **2013**, *58*, 833–958.

7. Orhan, Y.; Buyukgungor, H. Enhancement of biodegradability of disposable polyethylene in controlled biological soil. *Int. Biodeterior. Biodegrad.* **2000**, *45*, 49–55. [CrossRef]

8. Leja, K.; Lewandowicz, G. Polymer biodegradation and biodegradable polymers—A review. *Pol. J. Environ. Stud.* **2010**, *19*, 255–266.

9. Gautam, R.; Bassi, A.S.; Yanful, E.K. A review of biodegradation of synthetic plastic and foams. *Appl. Biochem. Biotechnol.* **2007**, *141*, 85–108. [CrossRef]

10. Wołejko, E.; Matejczyk, M. Civil and Environmental Engineering Reports. *Bud. I Inżynieria Środowiska* **2011**, *2*, 191–195.

11. Silva, A.; Gartner, B.L.; Morrell, J.J. Towards the development of accelerated Methods for Assessing the Durability of wood Plastic Composites. *J. Test. Eval.* **2007**, *35*, 1–28. [CrossRef]

12. Morris, P.I.; Cooper, P. Recycled plastic/wood composite lumber attacked by fungi. *For. Prod. J.* **1998**, *48*, 86–88.

13. Mankowski, M.; Morrell, J.J. Patterns of fungal attack in wood plastic composites following exposure in a soil block test. *Wood Fiber Sci.* **2000**, *32*, 340–345.

14. Pendleton, D.E.; Hoffard, T.A.; Addock, T.; Woodward, B.; Wolcott, M.P. Durability of an extruded HDPE/wood composite. *For. Prod. J.* **2002**, *52*, 21–27.

15. Verhey, S.; Laks, P.; Richter, D. Laboratory decay resistance of woodfiber/thermoplastic composites. *For. Prod. J.* **2001**, *51*, 44–49.

16. Schirp, A.; Ibach, R.E.; Pendelton, D.E.; Wolcott, M.P. *Biological Degradation of Wood-Plastic Composites (WPC) and Strategies for Improving the Resistance of WPC against Biological Decay*; ACS Symposium Series; American Chemical Society: Washington, DC, USA, 2008; pp. 480–507.

17. Ibach, R.; Sun, G.; Gnatowski, M.; Glaeser, J.; Leung, M.; Haight, J. Exterior decay of wood–plastic composite boards: Characterization and magnetic resonance imaging. *For. Prod. J.* **2015**, *66*, 4–17. [CrossRef]

18. Wang, W.; Morrell, J.J. Water sorption acharacteristics of two wood-plastic composites. *For. Prod. J.* **2004**, *54*, 209–212.

19. Nuhoglu, Y.; Oguz, E.; Uslu, H.; Ozbek, A.; Ipekoglu, B.; Ocak, I.; Hasenekoglu, I. The accelerating effects of the microorganisms on biodeterioration of stone monuments under air pollution and continental-cold climatic conditions in Erzurum. *Turk. Sci. Total Environ.* **2006**, *364*, 272–283. [CrossRef]

20. Coffta, G.; Borysiak, S.; Doczekalska, B.; Garbarczyk, J. Resistance of polypropylene-wood composites to fungi (in Polish: Odporność kompozytów polipropylen-drewno na rozkład powodowany przez grzyby). *Polimery* **2006**, *51*, 276–279.

21. Falk, R.; Lundin, T.; Felton, C. The Effects of Weathering on Wood-Thermoplastic Composites Intended for Outdoor Applications. In Proceedings of the 2nd Annual Conference Durability and Disaster Mitigation in Wood-Frame Housing, Madison, WI, USA, 6–8 November 2000; pp. 175–179.

22. Faruk, O.; Błędzki, A.K.; Fink, H.P.; Sain, M. Biocomposites reinforced with natural fibers: 2000–2010. *Prog. Polym. Sci.* **2012**, *37*, 1552–1596. [CrossRef]

23. Kamdem, D.P.; Jiang, H.; Cui, W.; Freed, J.; Matuana, L.M. Properties of wood plastic composites made of recycled DHPE and wood Flour from CCA- treated wood removed from service. *Composites Part A* **2004**, *35*, 347–355. [CrossRef]

24. Yildiz, U.C.; Yildiz, S.; Gezer, E.D. Mechanical properties and decay resistance of wood polymer composites prepared from fast growing species in Turkey. *Bioresour. Technol.* **2005**, *96*, 1003–1011. [CrossRef] [PubMed]

25. Clemons, C.M.; Ibach, R.E. Effects of processing method and moisture history on laboratory fungal resistance of wood-HDPE composites. *Prod. J.* **2004**, *54*, 50–57.

26. Sobków, D.; Barton, J.; Czaja, K.; Sudoł, M.; Mazoń, B. Reasearch on the resistance of materials to the effects of the natural environment (in Polish: Badania odporności materiałów na działanie czynników środowiska naturalnego). *Chemik* **2014**, *68*, 347–354.

27. ASTM D-2017. *Standard Test Method for Accelerated Laboratory Test of Natural Decay Resistance of Wood*; American Society for Testing and Materials: Philadelphia, PA, USA, 2005.

28. ASTM D-1413. *Standard Test Method for Wood Preservatives by Laboratory Soil-Block Cultures*; American Society for Testing and Materials: Philadelphia, PA, USA, 2007.

29. ENV 12038. *Durability of Wood and Wood-Based Products—Wood-Based Panels—Method of Test for Determining the Resistance against Wood-Destroying Basidiomycetes*; European Committee for Standardization (CEN): Brussels, Belgium, 2002.

30. Schirp, A.; Wolcott, M.P. Influence of fungal decay and moisture absorption on mechanical properties of extruded wood-plastic composites. *Wood Fiber Sci.* **2005**, *37*, 643–652.

31. Seefeldt, H.; Braun, U. Burning behavior of wood-plastic composite decking boards in end-use conditions; the effects of geometry, material composition and moisture. *J. Fire Sci.* **2012**, *30*, 41–54. [CrossRef]

32. Naumann, A.; Seefeldt, H.; Stephan, I.; Braun, U.; Noll, M. Material resistance of flame retarded wood-plastic composites against fire and fungal decay. *Polym. Degrad. Stab.* **2012**, *97*, 1189–1196. [CrossRef]

33. Chmielnicki, B.; Jurczyk, S. WPC composites as an alternative to wood products (in Polish: Kompozyty WPC jako alternatywa dla wytworów z drewna). *Przetwórstwo Tworzyw* **2013**, *19*, 477–484.

34. Ashori, A.; Behzad, H.M.; Tarmian, A. Effects of chemical preservative treatments on durability of wood flour/HDPE composites. *Composites* **2013**, *47*, 308–313. [CrossRef]

35. Friedrich, D.; Luible, A. Investigations on ageing of wood-plastic composites for outdoor applications: A meta-analysis using empiric data derived from diverse weathering trials. *Constr. Build. Mater.* **2016**, *124*, 1142–1152. [CrossRef]

36. EN 84. *Wood Preservatives. Accererated Ageing of Treted Wood Prior to Biological Testing. Leaching Procedure*; European Committee for Standardization (CEN): Brussels, Belgium, 1997.
37. EN ISO 178. *Plastics—Determination of Flexural Properties*; European Committee for Standardization (CEN): Brussels, Belgium, 2019.
38. Sun, G.; Ibach, R.E.; Faillace, M.; Gnatowski, M.; Glaeser, J.A.; Haight, J. Laboratory and exterior decay of wood–plastic composite boards: Voids analysis and computed tomography. *Wood Mater. Sci. Eng.* **2017**, *12*, 263–278. [CrossRef]
39. Ibach, R.; Gnatowski, M.; Sun, G.; Glaeser, J.; Leung, M.; Haight, J. Laboratory and environmental decay of wood–plastic composite boards: Flexural properties. *Wood Mater. Sci. Eng.* **2018**, *13*, 81–96. [CrossRef]

 *materials*

Article

Windblown Sand-Induced Degradation of Glass Panels in Curtain Walls

Yuxi Zhao [1], Rongcheng Liu [1,*], Fan Yan [1], Dawei Zhang [1] and Junjin Liu [2]

1   College of Civil Engineering and Architecture, Zhejiang University, Hangzhou 310058, China; yxzhao@zju.edu.cn (Y.Z.); 21612175@zju.edu.cn (F.Y.); dwzhang@zju.edu.cn (D.Z.)
2   China Academy of Building Research, Beijing 100000, China; liujunjin@cabrtech.com
*   Correspondence: rc_liu@zju.edu.cn

Abstract: The windblown sand-induced degradation of glass panels influences the serviceability and safety of these panels. In this study, the degradation of glass panels subject to windblown sand with different impact velocities and impact angles was studied based on a sandblasting test simulating a sandstorm. After the glass panels were degraded by windblown sand, the surface morphology of the damaged glass panels was observed using scanning electron microscopy, and three damage modes were found: a cutting mode, smash mode, and plastic deformation mode. The mass loss, visible light transmittance, and effective area ratio values of the glass samples were then measured to evaluate the effects of the windblown sand on the panels. The results indicate that, at high abrasive feed rates, the relative mass loss of the glass samples decreases initially and then remains steady with increases in impact time, whereas it increases first and then decreases with an increase in impact angle such as that for ductile materials. Both visible light transmittance and effective area ratio decrease with increases in the impact time and velocities. There exists a positive linear relationship between the visible light transmittance and effective area ratio.

Keywords: degradation of glass panels; effective area ratio; relative mass loss; visible light transmittance; windblown sand

Citation: Zhao, Y.; Liu, R.; Yan, F.; Zhang, D.; Liu, J. Windblown Sand-Induced Degradation of Glass Panels in Curtain Walls. *Materials* **2021**, *14*, 607. https://doi.org/ 10.3390/ma14030607

Academic Editor: Krzysztof Schabowicz
Received: 14 December 2020
Accepted: 21 January 2021
Published: 28 January 2021

Publisher's Note: MDPI stays neutral with regard to jurisdictional claims in published maps and institutional affiliations.

Copyright: © 2021 by the authors. Licensee MDPI, Basel, Switzerland. This article is an open access article distributed under the terms and conditions of the Creative Commons Attribution (CC BY) license (https:// creativecommons.org/licenses/by/ 4.0/).

1. Introduction

Glass panels are widely used, play an important role in curtain walls, and directly affect the performance of curtain wall structures. One cause of glass panel degradation is windblown sand. As they are exposed to windblown sand, glass panels are affected continuously by grit. Slight damage (such as hollows and scratches) may occur and gradually accumulate, affecting the safety and serviceability of the panels. The deterioration behaviors of glass panels are of significant interest to engineers and researchers in the context of curtain wall structures, e.g., for the reasonable maintenance and timely repair or for the replacement of glass panels and prolongation of the service time of glass curtain walls.

There have been many studies on the degradation behaviors of building materials owing to the effects of windblown sand. Li [1] measured the parameters for evaluating the windblown sand resistance of concrete using an airflow sand-carrying jet test method. Wang et al. [2,3] and Ju [4] studied the windblown sand resistance of cement paste, mortar, and concrete under different impact wind velocities and impact angles using the same test method and determined the mass loss rate and period when entering the steady erosion stage. Jiang [5] and Guo [6] studied the variation rule of the relative mass loss, impact angles, and impact velocities of cement mortar under the effect of salt soaking erosion and multiple factors of freeze–thaw and dry–wet cycles in a compound salt solution. Zhang [7] modified the epoxy resin with the addition of SiC micro-powder or both SiC micro-powder and polyurethane toughener and evaluated the mechanical properties and erosion wear properties of the modified epoxy resin. Based on the Taguchi method, Hao

279

et al. [8] found that the impact angle had the largest impact on concrete erosion, which reached 72.48%. Hao et al. [9], Li [10], and Zhang et al. [11,12] used the same method to simulate windblown sand and analyzed the surface morphology by scanning electron microscopy (SEM), thereby studying the degradation behaviors and mechanisms of the surface coatings on steel structures and nano-titanium dioxide films on glass surfaces, respectively. There are many experimental studies on the erosion wear of new materials, e.g., glass fiber/unsaturated polyester composites [13], WC-Co cemented carbide [14], glass-ceramic composite coating [15], fluoroplastic-steel composite tubes [16], carbon fiber reinforced polymer [17], and glass fiber reinforced plastics [18].

The windblown sand that degrades glass materials has also been studied. Li [19] conducted an experimental study and finite element simulations of the windblown sand erosion of toughened glass under temperature and humidity fluctuations and ultraviolet rain. Fan et al. [20] studied the erosion mechanism, surface pattern, and process performance of micro-abrasive water jets on brittle glass. Hao et al. [21] designed the experiment of toughened glass based on the Taguchi method. It was found that the degree of effect on the relative mass loss of toughened glass was mainly due to the impact angle, the impact velocity, and the abrasive feed rate. Zhao [22] and Hao et al. [23] studied the damage mechanism of tempered glass by windblown sand erosion under the combined effect of the freeze–thaw cycle and ultraviolet radiation. Bouzid and Azari [24], and Ismail et al. [25] explored mechanisms of glass degradation. They pointed out that erosion pits caused by mass loss are formed when transverse cracks of the subsurface stratum develop along the direction parallel to the surface until they intersect the surface; radial cracks are formed when cracks on the subsurface stratum develop upward along the vertical direction of the surfaces, ultimately intersecting with the surface (as shown in Figure 1). Ismail et al. [25] also found an exponential function relationship between the sizes of the damage hollows and impact velocities; the exponents were between 1.65 and 2.

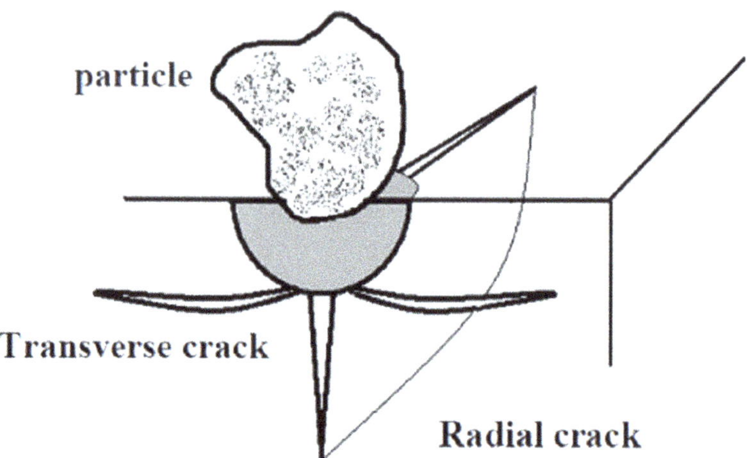

Figure 1. Erosion damage mechanisms of glass [25].

The degradation of glass panels under a gas–solid two-phase flow was explored by Hao et al. [26] using the airflow sand-carrying jet test method. The results showed that there were many damage modes in the damage process of glass panels by Scanning Electron Microscopy (SEM). Then, Hao et al. [26] pointed out that the meso-damage modes of engineering glass have two types: a combination of low-angle micro-cutting and high-angle elastic-plastic deformation or a combination of low-angle micro-cutting and damage from high-angle fatigue crack propagation.

In summary, the test methods for windblown sand damage are relatively mature and have been introduced into the research field of glass panel damage. However, since the existing studies on the windblown sand damage of glass panels have focused on the degradation mechanism, there have been few studies on the degradation behavior (s) of glass panels. Therefore, in this study, the windblown sand-induced degradation of glass panels was studied using the airflow sand-carrying jet test method. Focusing on the safety and serviceability changes in glass panels, our experiments explored the long-term performance degradation of glass panels caused by windblown sand. Based on the value obtained, the tendencies of the relative mass loss, the visible light transmittance, and effective area ratio have been obtained, which predict the state of glass panels under service conditions.

2. Experimental Program

2.1. Materials

The test material was a 6 mm thick piece of float glass; float glass is widely used in curtain walls, owing to its good smoothness and transparency. The basic information on the material is shown in Table 1, based on the initial state of the glass panels.

Table 1. Basic information of glass.

Thickness T/mm	Gravity Density γ_g/(kN/m^3)	Initial Visible Transmittance τ_{v0}/%	Design Strength f_g/(N/mm^2)	Modulus of Elasticity E_g/(N/mm^2)	Poisson Ratio V
6	25.6	92.34	28	0.72×10^5	0.20

2.2. Specimens

The specimens were 50 mm × 50 mm float glass panels with 6 mm thickness, as listed in Table 2. The specimens of the P_a group were used for studying the mass loss of the glass panels, whereas the specimens of the P_s group were used for studying the visible light transmittance and effective area ratio.

Table 2. Specimens of the glass panels.

Specimen Type	Specimen Number	Abrasive Feed Rate m (g/s)	Impact Velocities v (m/s)	Impact Angles a (°)	Impact Time t (s)
P_a	P_a-1-30	15.5	17.40	30	
	P_a-1-60	15.5	17.40	60	
	P_a-1-90	15.5	17.40	90	0, 30, 60, 90, 120
	P_a-2-30	15.0	26.34	30	
	P_a-2-60	15.0	26.34	60	
	P_a-2-90	15.0	26.34	90	
P_s	P_s-1-30	15.5	17.40	30	
	P_s-1-60	15.5	17.40	60	
	P_s-1-90	15.5	17.40	90	
	P_s-2-30	15.0	26.34	30	
	P_s-2-60	15.0	26.34	60	0, 2.5, 5, 7.5, 10
	P_s-2-90	15.0	26.34	90	
	P_s-3-30	14.8	35.28	30	
	P_s-3-60	14.8	35.28	60	
	P_s-3-90	14.8	35.28	90	

2.3. Experimental Equipment

A sandblasting machine was applied to simulate windblown sand damage. An air compressor was used as the air power source, as schematically illustrated in Figure 2.

The high-speed air provided by the air compressor flowed through the pipe connected to the sandbox, causing local low pressure. Then, sand in the sandbox was sucked into the pipeline owing to the pressure difference, forming the sand flow. The gas–solid two-phase flow mixed with the high-speed flow and the sand flow impinged on the samples.

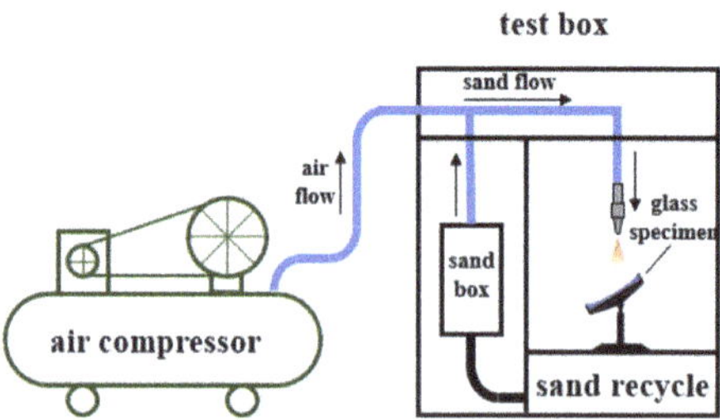

Figure 2. Schematic drawing of the sandblasting equipment.

In the testing box, the specimens were clamped at different heights and angles using a specimen bracket, as shown in Figure 3. The angles and distances of the specimens were adjustable. After the specimens were clamped, the angles (i.e., the acute angles between the upper surface of each specimen and central axis of the corresponding nozzle) were adjusted to the target impact angles listed in Table 2. The distance between the center of the upper surface of each specimen and the nozzle was adjusted to 20 cm.

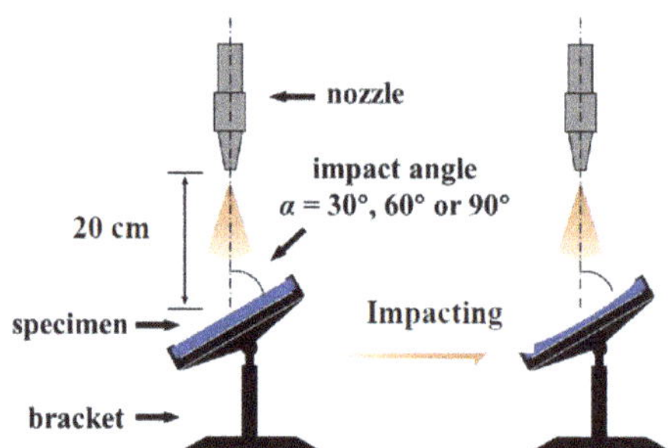

Figure 3. Location of the glass sample.

2.4. Testing Methods

The specimens were clamped on the brackets, as shown in Figure 3, and then, they were adjusted to the target heights and impact angles. To keep the air–sand phase flow stable, it was necessary to keep the sandblasting machine on for two minutes before eroding the specimens. After the impact time of each group of P_a and P_s specimens reached the target time in Table 2, the specimens were removed from the specimen bracket.

2.4.1. Mass Loss

An electronic balance was used to weigh the remaining mass of the P_a specimens. The mass was weighed after blowing off the remaining sand and debris on the surface to ensure that the weighed mass did not contain the masses of such sand and debris.

2.4.2. Visible Light Transmittance

After weighing the mass, the surfaces of the P_s specimens were cleaned with water and an acetone solution. Then, the spectral transmittance of the P_s specimens was measured by the visible light transmittance ratio and shading coefficient detector.

2.4.3. Meso-Morphology

The surface morphology of the P_s specimens was observed via TM3000 desktop SEM (Hitachi, Japan), and photographs were taken. The photographs were then processed using an image processing software (Image J, version 1.48, developed by National Institutes of Health) to extract the damaged area. The percentage of damaged areas with respect to the total area was calculated using the software.

3. Results and Discussion

3.1. Damage Mode

The mechanism of glass panel degradation is characterized by accumulation of the meso-damage of the glass panel. Hao et al. [26] conducted an experimental study on windblown sand damage of glass panels, pointing out that the meso-damage modes of engineering glass have two types: a combination of low-angle micro-cutting and high-angle elastic-plastic deformation or a combination of low-angle micro-cutting and damage from high-angle fatigue crack propagation. In our study, these damage modes can be observed on the surface of the damaged specimens using SEM, as shown in Figure 4, and are similar to those described by Hao et al. [26]. Ismail et al. [25] and Li [27] introduced these meso-damage modes and their generation mechanisms. Based on these studies, the damage modes in Figure 4 can be classified into three modes: cutting, smashing, and plastic deformation. **Cutting** is the removal of surface material caused by the tangential forces of the impact particles. **Smashing** is material breaking caused by the normal force of the impact particles, as transverse microcracks in the subsurface layer propagate to the surface. **Plastic deformation** is the plastic extrusion deformation caused by the normal force of the impact particles.

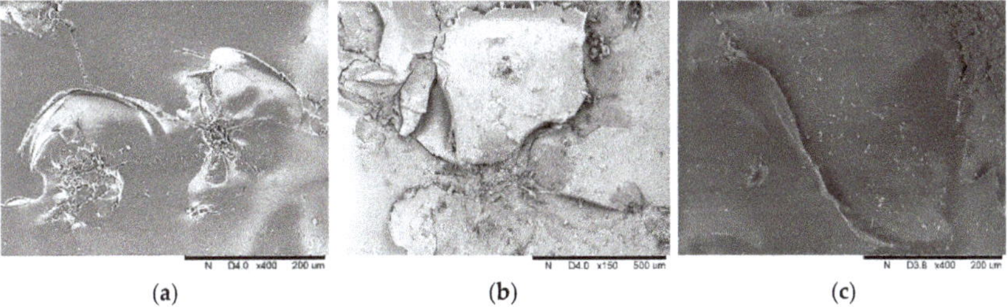

(a)   (b)   (c)

Figure 4. Damage modes of glass specimens: (**a**) cutting, Ps-10-30; (**b**) smashing, Ps-10-90; and (**c**) plastic deformation, Ps-12-30.

These three damage modes coexist in the erosion process and are affected by the impact angle and impact velocity. When the impact angles are small, the tangential force of the impact particles is dominant and the cutting mode develops well. When the impact angles are large, the normal force plays a larger role, and smashing and plastic deformation become the main failure modes. In addition, with the growth of the wind force, the impact velocity of the particles increases and the plastic deformation develops better than the smashing mode (brittle damage), owing to the increase in strain rate at high-speed impact [27].

3.2. Relative Mass Loss

The degree of erosion impact is evaluated based on the relative mass loss. The relative mass loss of the specimens is related to the impact time, impact angles, and impact velocities and is defined as follows:

$$E = \frac{\Delta M}{mT} \tag{1}$$

where E represents the relative mass loss (mg/g), ΔM is the mass loss of the specimens after impact (mg), m is the abrasive feed rate (g/s), and T is the impact time (s).

3.2.1. Effect of Impact Time on Relative Mass Loss of Glass Specimens

By subtracting the mass measured at each time from that measured at the initial time (impact time of 0 s), the mass loss ΔM of the P_a specimens at each impact time t can be obtained. Then, according to Equation (1), the relative mass loss is calculated at different impact velocities and impact angles with the increase in impact time, as shown in Figure 5.

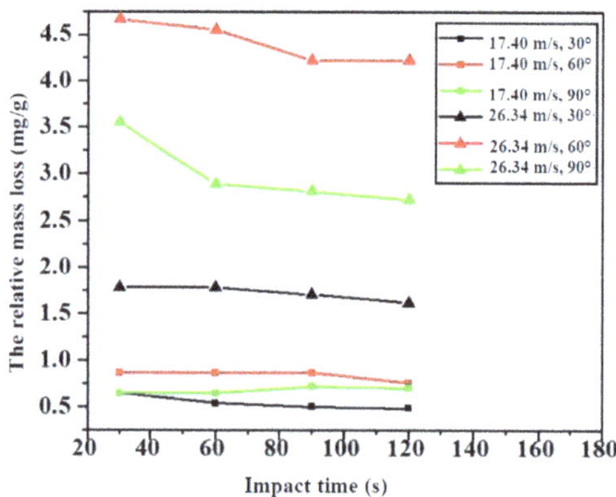

Figure 5. Relative mass loss of glass specimens with an increase in impact time.

From Figure 5, the relative mass loss of the specimens can be divided into two stages: decreasing stage and steady stage. Nearly all curves have a tendency to proceed in two stages, except for the group at 17.4 m/s and 90°. This can be explained as follows: at the beginning, the surfaces of the glass specimens are smooth, meaning that cutting (at low impact angles) and smashing/plastic deformation (at high impact angles) develop well. Therefore, the value of the relative mass loss is high. Then, with the specimens degraded, the surfaces gradually become rough. Rough surfaces are detrimental to the development of cutting, causing a decrease in the relative mass loss at low impact angles. As for the specimens at high impact angles, the rough surfaces caused by the sand of the specimens appear as many erosion pits, which cause the particles to bounce off in all directions and to collide with the incident particles, reducing the impact velocities of incident particles. Therefore, it causes a reduction in the relative mass loss of the specimens. Therefore, the relative mass loss of the specimens at high impact angles gradually decreases. After degradation for a long time, the rough surfaces remain stable, corresponding to the steady stage.

It can also be seen from Figure 5 that the relative mass loss at 26.34 m/s is higher than that at 17.4 m/s, which should be evident: degradation at a high kinetic energy is more significant than that at low kinetic energy. Thus, the kinetic energy of sand particles at 26.34 m/s is higher than that at 17.4 m/s, causing a higher relative mass loss.

3.2.2. Effect of Impact Angles on Relative Mass Loss of Glass Specimens

There are two main types of materials with respect to fracture damage modes: ductile and brittle. According to Sheldon and Finnie [28], the ductile and brittle mode curves for the erosion (relative mass loss) with an increase in the impact angle are shown in Figure 6. The relative mass loss of a ductile material includes two stages: in the initial stage, the relative mass loss increases with an increase in impact angle. In the second stage, the relative mass loss decreases with an increase in the impact angle. Therefore, the maximum relative mass loss of the ductile materials appears at an impact angle of 20° to 30°. In contrast to ductile materials, the relative mass loss of brittle materials increases with an increase in the impact angle, and the maximum relative mass loss appears at an impact angle of 90°.

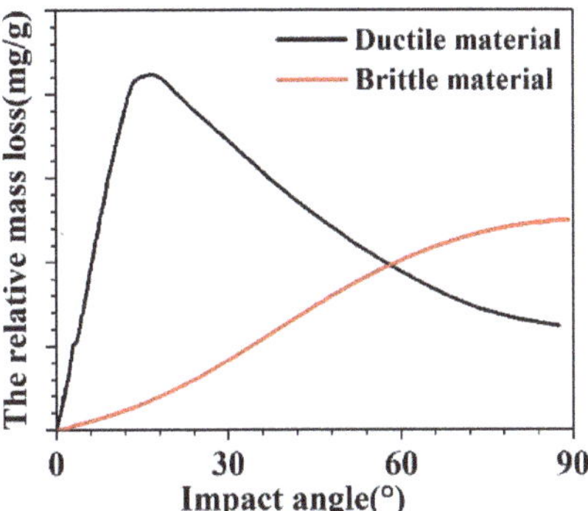

Figure 6. Relative mass loss of ductile and brittle materials with an increase in impact angle [28].

The relative mass losses at different impact velocities and impact times with the increase in impact angle are shown in Figure 7. It can be seen from Figure 7 that the relative mass loss initially increases with an increase in the impact angle and then decreases.

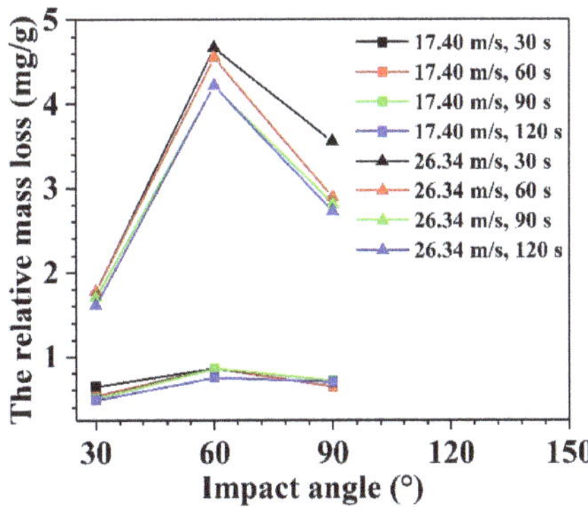

Figure 7. Relative mass loss of glass specimens with increase in impact angle.

Despite being a typical brittle material, the tendency of the float glass in Figure 7 is not in accordance with the curve of typical brittle materials (shown in Figure 6). This can be explained as follows. As compared with 110 g/min (1.83 g/s) in the reference of Hao et al. [29], the abrasive feed rates in this experiment are very high, i.e., 15.5 g/s at an impact velocity of 17.4 m/s and 15.0 g/s at 26.34 m/s. According to Section 3.2.1, the rough surfaces (caused by the sand) of the specimens exhibit many erosion pits, limiting the reversal of sand particles and reducing the mass loss of the specimens at high impact angles. The higher the impact angles, the lower the mass loss. Therefore, the relative mass loss at 90° is much lower than that at 60°.

3.3. Visible Light Transmittance

The full spectral transmittance τ (λ) of the P_s specimens was measured by a visible light transmittance ratio and shading coefficient detector, as shown in Figure 8a. The visible spectral transmittance τ (λ) (wavelength from 380 nm to 780 nm) of the P_s specimens was intercepted from the full spectral transmittance τ (λ) for analysis, as shown in Figure 8b. Based on this, the visible light transmittance τ_v for each specimen was calculated in accordance with the specification in GB/T 2680-1994 [30].

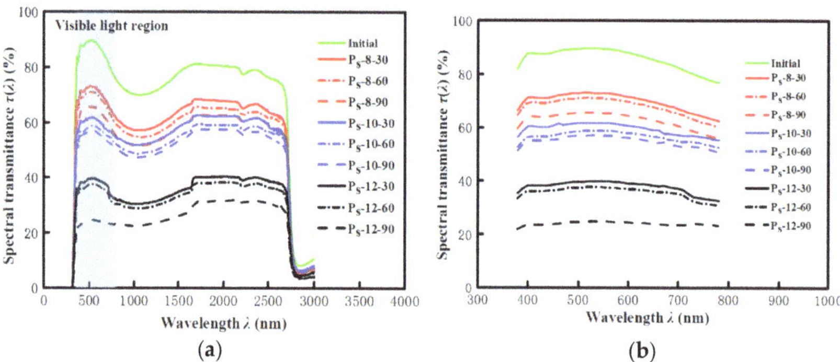

Figure 8. Spectral transmittance τ (λ) of the glass specimens without and with erosion for 5 s: (**a**) full spectral transmittance τ (λ) (300 nm $\leq \lambda \leq$ 3000 nm) of glass specimens without and with erosion for 5 s and (**b**) visible spectral transmittance τ (λ) (380 nm $\leq \lambda \leq$ 780 nm) of glass specimens without and with erosion for 5 s.

The visible light transmittance values are obtained for the P_s specimens, as shown in Figure 9. It can be seen from Figure 9 that the visible light transmittance τ_v of the specimens decreases gradually with an increase in the impact time owing to the damage accumulation. In addition, the visible transmittance τ_v decreases with an increase in the impact velocity v. This is the result of the increase in erosion damage strength. As known, a higher velocity means higher destructive power, causing the decrease in visible light transmittance.

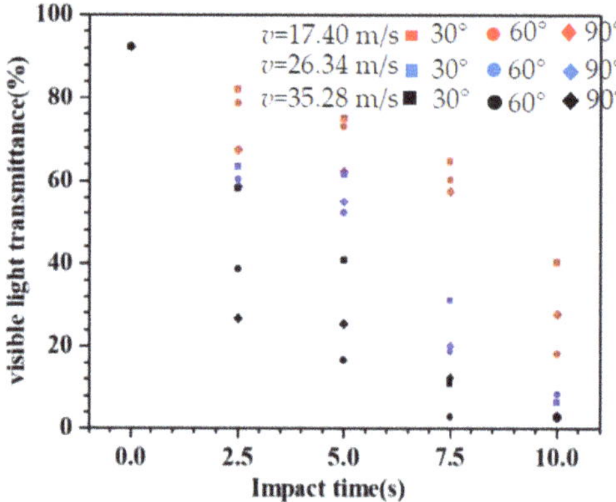

Figure 9. Visible light transmittance of the glass specimens τ_v.

3.4. Effective Area Ratio

After the damage area is extracted by the Image J software, the percentage of damaged areas to the total area is calculated using the software. The value after deducting the proportion of damaged areas from the total area (100%) is defined as the effective area ratio R_{AE} (%).

The effective area ratio was obtained with the image J software, as shown in Figure 10. From Figure 10, it can be seen that the effective area ratio R_{AE} decreases with the increase in impact time, as expected; R_{AE} also decreases with increases in the impact angle α and impact velocity v.

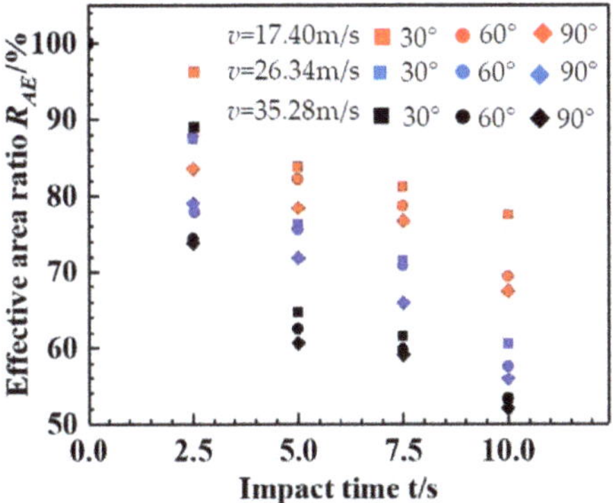

Figure 10. Relationship between the effective area ratio R_{AE} of glass specimens and impact time t.

The decrease in effective area ratio R_{AE} with the increase in impact velocity v is the result of the increase in erosion damage strength. The decrease in effective area ratio R_{AE} with the increase in impact angle α may be owing to the fact that, when the normal direction component of the impact particles is large, additional cracks will appear on the subsurface stratum, promoting further development of cracks. It may make more areas of the surface vulnerable to damage. With an increase in the impact angle α, the normal direction component of the impact particles increases; thus, smashing and plastic deformation develop better, resulting in a decrease in the effective area ratio.

Assuming that the effective area ratio R_{AE} of the panel is 100% without damage, then each R_{AE}-t curve passes through the point (0, 100). By fitting the experimental data in Figure 8 with data processing software (Origin 8.0, developed by OriginLab and 1stOPT8.0, developed by 7D-Soft High Technology Inc.), an equation for the effective area ratio and time can be obtained, as follows:

$$R_{AE} = \frac{1}{0.01 + K_{RAE}t} \tag{2}$$

where K_{RAE} is a parameter related to the impact angle α and impact velocity v and is defined as the effective area erosion coefficient. Based on the test data, the values of K_{RAE} are obtained by fitting, as follows:

$$K_{RAE} = \begin{cases} 3.04 \times 10^{-4} & v = 17.40 \text{ m/s} & \alpha = 30° \\ 4.21 \times 10^{-4} & v = 17.40 \text{ m/s} & \alpha = 60° \\ 4.91 \times 10^{-4} & v = 17.40 \text{ m/s} & \alpha = 90° \\ 5.98 \times 10^{-4} & v = 26.34 \text{ m/s} & \alpha = 30° \\ 6.87 \times 10^{-4} & v = 26.34 \text{ m/s} & \alpha = 60° \\ 7.79 \times 10^{-4} & v = 26.34 \text{ m/s} & \alpha = 90° \\ 8.63 \times 10^{-4} & v = 35.28 \text{ m/s} & \alpha = 30° \\ 1.02 \times 10^{-4} & v = 35.28 \text{ m/s} & \alpha = 60° \\ 1.07 \times 10^{-4} & v = 35.28 \text{ m/s} & \alpha = 90° \end{cases}$$

According to the above values of K_{RAE}, an equation can be obtained by data processing software (1st OPT) fitting, as follows:

$$K_{RAE} = -1.169 \times 10^{-4} + 7.616 \times 10^{-6} v l n v + 2.328 \times 10^{-4} \sin^{2.5} \alpha \tag{3}$$

where $\sin \alpha$ reflects the erosion effect of certain physical quantities in the normal direction of the specimens and "$vlnv$" reflects the relationship between the erosion effect and erosion impact velocity v.

3.5. Relationship between Visible Transmittance and Effective Area Ratio

It is reasonable that, with an increase in the damaged area, the visible transmittance τ_v will decrease. Therefore, the visible transmittance τ_v is positively correlated with the effective area ratio R_{AE}. The quantitative relationship between the visible transmittance τ_v and effective area ratio R_{AE} is discussed below.

Figure 11 plots the tested data of the visible transmittance τ_v and effective area ratio R_{AE} for each specimen. As shown, an approximately positive linear relationship exists between τ_v and R_{AE}.

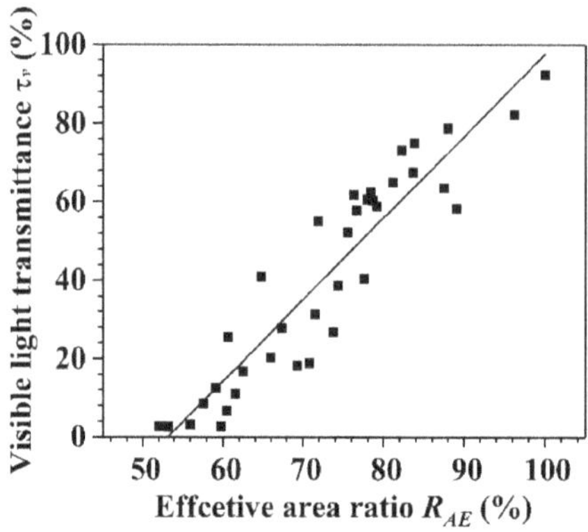

Figure 11. Relationship between visible light transmittance τ_v and effective area ratio R_{AE}.

To obtain their quantitative relationship, the analysis needs to be extended to a non-damaged condition of the glass specimen. For a non-damaged specimen, the effective area ratio is defined as 100% whereas the visible light transmittance is 92.34%, as shown in Table 1. However, if the impact time is long enough, the entire surface of the glass panel will be damaged and the effective area ratio can be regarded as 0; however, the panel still transmits light. This means that the relationship between the visible light transmittance and effective area ratio is different between low and high effective area ratios. Therefore, it is assumed that, when the effective area ratio is 0, the visible light transmittance of the specimen has a threshold. This threshold needs to be studied in the future.

Based on the above boundary conditions, a simplified equation for the visible light transmittance τ_v and effective area ratio R_{AE} is obtained using data processing software, as follows:

$$\tau_v(R_{AE}) = -110.902 + 2.085 R_{AE} \left(54.419 \leq R_{AE} \leq 97.431,\ R^2 = 0.872\right) \tag{4}$$

where the value of the effective area ratio R_{AE} ranges from 54.419 to 100. The value of 54.419 is determined based on the minimum visible light transmittance. The value of 97.431 is determined based on the maximum visible light transmittance.

According to Equation (4), for every 1% decrease in the effective area ratio R_{AE}, the visible light transmission ratio τ_v decreases by 2.085% when the effective area ratio ranges from 54.419 to 97.431.

4. Conclusions

Based on a simulation of a sandstorm using a gas–solid two-phase flow, a windblown sand experiment was conducted to explore the degradation law (s) of glass panels. The following conclusions can be drawn:

(1) There are three damage modes in the glass panels subject to windblown sand: cutting, smashing, and plastic deformation. At low impact angles, the cutting mode predominates, whereas under high impact angles, the smashing and plastic defor-

mation modes are dominant. In addition, with the growth of the wind force, the impact velocity of the particles increases and the plastic deformation develops better than the smashing mode (brittle damage), owing to the increase in strain rate at high-speed impact.

(2)   With an increase in the impact time, the relative mass loss initially decreases and then remains steady. In contrast, with an increase in the impact angle, the relative mass loss initially increases and then decreases, which exhibits the properties of ductile materials.

(3)   With increases in the time or impact velocity, the visible light transmittance decreases gradually owing to damage accumulation, as expected.

(4)   The tendency of the variation in the effective area ratio under different situations is similar to that of the visible light transmittance. There exists an approximately positive linear relationship between the visible light transmittance and effective area ratio.

In summary, the experiment simulated the degradation of glass panels subjected to windblown sand. We can get the tendencies of the relative mass loss, the visible light transmittance, and effective area ratio, which are similar to the degradation of glass panels under service condition. However, further study is needed on damage accumulation to predict the service life of glass panels.

Author Contributions: Every author has contributed substantially to the work reported. We adopted the following division of labor: conceptualization and resources, Y.Z. and D.Z.; methodology, data curation and validation, F.Y. and R.L.; writing—original draft preparation, R.L.; writing—review and editing, Y.Z.; supervision and project administration, J.L. All authors have read and agreed to the published version of the manuscript.

Funding: This research was funded by the National Key Research and Development Program of China (Nos. 2017YFC0806100): research on anti-fall and safety performance improvement technology for urban building envelope.

Data Availability Statement: Data is contained within this article. The data presented in this study are available. Anyone in need can use it or contact with me directly.

Conflicts of Interest: The authors declare no conflict of interest.

References

1.   Li, Y.J. Estimation method of sand abrasion in hydraulic structures. *J. Hydraul. Eng.* **1989**, *7*, 60–66. [CrossRef]
2.   Wang, Y.P.; Ju, C.C.; Wang, Q.C. Experimental Study on the Solid Particle Erosion of Concrete, Mortar and Cement Paste under Blown Sand Environment. *China Railw. Sci.* **2013**, *34*, 21–26. [CrossRef]
3.   Wang, Y.P. Experimental Study on Erosive Wear for Concrete and Protective Materials under Strong Wind-Sand Environment in Northwest. Master's Thesis, Zhejiang University, Hangzhou, China, 2016.
4.   Ju, C.C. Study on the Erosion Wear of Concrete and Protective Material. Master's Thesis, Lanzhou Jiaotong University, Lanzhou, China, 2014.
5.   Jiang, N. Experimental Study on Windblown Sand Erosion of Concrete under the Effect of Salt Soaking Erosion. Master's Thesis, Inner Mongolia University of Technology, Hohhot, China, 2016.
6.   Guo, J. Experimental Study on Sand Erosion of Cement Mortar Subjected to Multiple Factors of Freeze-Thaw and Dry-Wet Cycles in Compound Salt Solution. Master's Thesis, Inner Mongolia University of Technology, Hohhot, China, 2017.
7.   Zhang, Y.S. Erosion and Wear Protection Material for Concrete Structure in Strong Wind-Sand Environment. Master's Thesis, Lanzhou Jiaotong University, Lanzhou, China, 2019.
8.   Hao, Y.H.; Fan, J.C.; Xing, Y.M.; Jiang, N.; Guo, J.; Liu, Y.C.; Feng, Y.J. Study of Concrete under Salt Freeze-Thaw Cycles and Wind-Blown Sand Erosion by Using Tiguchi Method. *J. Build. Mater.* **2017**, *20*, 124–129. [CrossRef]
9.   Hao, Y.H.; Xing, Y.M.; Yang, S.T. Erosion-wear Behavior of Steel Structure Coating Subject to Sandstorm. *Tribology* **2010**, *30*, 26–31. [CrossRef]
10.   Li, J. Study on Sand Erosion Resistance of Steel Structure Coatings under Multi-Factors Influence. Master's Thesis, Inner Mongolia University of Technology, Hohhot, China, 2019.
11.   Zhang, B.; Ji, G.J.; Liu, Q.S.; Shi, Z.M. Study on Erosion Behavior of TiO_2 Nano Thin Films Coated on Glass by Sandstorm. *J. Synth. Cryst.* **2016**, *45*, 1670–1676. [CrossRef]
12.   Zhang, B. Study on the Sand Erosion Behavior of TiO_2 Thin Films. Master's Thesis, Inner Mongolia University of Technology, Hohhot, China, 2017.

13. Yang, S.C.; Li, K.; Duan, Y.X.; Yuan, C. Anti-Wind and Sand Erosion Properties of GF/Unsaturated Polyester Composite. *Acta. Mater. Compos. Sin.* **2011**, *28*, 77–82. [CrossRef]
14. Jiang, J.W.; Shi, M.; Yin, M.Y.; Yu, X.G. Influencing Factors and Mechanism of Erosion Wear of WC-Co Cemented Carbide. *China Tungsten Ind.* **2020**, *35*, 48–55. [CrossRef]
15. Ma, Z.; Li, J.; Yan, C.J.; Li, Z.C. Erosion Wear of Glass-ceramic Composite Coating from Fly Ash. *Bull. Chin. Ceram. Soc.* **2013**, *32*, 1684–1687. [CrossRef]
16. Yuan, R.F. Experimental Study on Erosion Wear Behaviors of Fluoroplastic-Steel Composite Tubes. Master's Thesis, Zhejiang University, Hangzhou, China, 2018.
17. Wang, S. High-Speed Erosion Tests on Carbon Fiber Reinforced Polymer Plates. *Sci. Technol. Innov. Her.* **2020**, *17*, 106–110. [CrossRef]
18. Liu, G. Study on the Erosion Wear Behaviors of the GFRP Impacted by Air Flow Containing Sands. Master's Thesis, Hefei University of Technology, Hefei, China, 2013.
19. Li, H. Experimental Study and Finite Element Simulation on Sand Erosion of Tempered Glass under Temperature and Humidity Fluctuations and Ultraviolet Rain. Master's Thesis, Inner Mongolia University of Technology, Hohhot, China, 2018.
20. Fan, J.M.; Fan, C.M.; Wang, J. Erosion Mechanism of Brittle Glass by Micro-Abrasive Water Jet. *Diam. Abras. Eng.* **2010**, *30*, 1–5.
21. Hao, Y.H.; Wu, R.G.; Zhao, C.G.; Guo, X.; Ya, R.H. Effect of Weathering on Erosion Resistance of Toughened Glass Based on Taguchi Method. *Bull. Chin. Ceram. Soc.* **2020**, *39*, 2980–2986. [CrossRef]
22. Zhao, C.G. Experimental Study on the Damage Mechanism of Tempered Glass by Sand Erosio under the Combined Effect of Freeze-Thaw Cycle and Ultraviolet Radiation. Master's Thesis, Inner Mongolia University of Technology, Hohhot, China, 2018.
23. Hao, Y.H.; Guo, X.; Zhao, C.G.; Wu, R.G. Mechanics Mechanism of Wind-sand Erosion Damage on Tempered Glass Surface under Combined Freeze-thaw Cycles and Ultraviolet Irradiation. *Surf. Technol.* **2020**, *49*, 188–197. [CrossRef]
24. Bouzid, S.; Azari, Z. Glass Damage by Impact. *J. Ningbo Univ. Nat. Sci. Eng. Ed.* **2003**, *16*, 354–362. Available online: https://kns.cnki.net/kcms/detail/detail.aspx?FileName=NBDZ200304005&DbName=CJFQ2003 (accessed on 16 October 2020).
25. Ismail, J.; Zaïri, F.; Naït-Abdelaziz, M.; Bouzid, S.; Azari, Z. Experimental and numerical investigations on erosion damage in glass by impact of small-sized particles. *Wear* **2011**, *271*, 817–826. [CrossRef]
26. Hao, Y.H.; Li, H.; Ya, R.H.; Zhao, C.G. Erosion Performance and Damage Mechanism of Engineering Glass under the Action of Gas-solid Two-phase Flow. *Bull. Chin. Ceram. Soc.* **2017**, 36. [CrossRef]
27. Li, Z.P. Research on Simulation and Experiment of Grinding Optical Glass with Single Diamond Girt. Master's Thesis, Harbin Institute of Technology, Harbin, China, 2013.
28. Sheldon, G.; Finnie, I. On the ductile behaviour of nominally brittle materials during erosive cutting. *J. Eng. Ind.* **1966**, *88*, 387–392. [CrossRef]
29. Hao, Y.H.; Ya, R.H.; Liu, Y.L.; Li, H. Erosion damage mechanism analysis of tempered glass in a wind-blown sand environment. In Proceedings of the 2016 2nd International Conference on Energy Equipment Science and Engineering (Advances in Energy Science and Equipment Engineering), Guangzhou, China, 12 November 2016; Volume 2.
30. GB/T 2680-1994. *Determination of Light Transmittance, Solar Direct Transmittance, Total Solar Energy Transmittance and Ultraviolet Transmittance for Glass in Building and Related Glazing Factors*; State Bureau of Technical Supervision: Beijing, China, 1994.

 *materials*

Article

Research on the Properties of Mineral–Cement Emulsion Mixtures Using Recycled Road Pavement Materials

Łukasz Skotnicki *, Jarosław Kuźniewski and Antoni Szydło

Roads and Airports Department, Faculty of Civil Engineering, Wroclaw University of Science and Technology, Wyb. Stanislawa Wyspianskiego 27, 50-370 Wroclaw, Poland; jaroslaw.kuzniewski@pwr.edu.pl (J.K.); antoni.szydlo@pwr.edu.pl (A.S.)
* Correspondence: lukasz.skotnicki@pwr.edu.pl; Tel.: +48-71-320-45-38

Abstract: The reduction in natural resources and aspects of environmental protection necessitate alternative uses of waste materials in the area of construction. Recycling is also observed in road construction where mineral–cement emulsion (MCE) mixtures are applied. The MCE mix is a conglomerate that can be used to make the base layer in road pavement structures. MCE mixes contain reclaimed asphalt from old, degraded road surfaces, aggregate improving the gradation, asphalt emulsion, and cement as a binder. The use of these ingredients, especially cement, can cause shrinkage and cracks in road layers. The article presents selected issues related to the problem of cracking in MCE mixtures. The authors of the study focused on reducing the cracking phenomenon in MCE mixes by using an innovative cement binder with recycled materials. The innovative cement binder based on dusty by-products from cement plants also contributes to the optimization of the recycling process in road surfaces. The research was carried out in the field of stiffness, fatigue life, crack resistance, and shrinkage analysis of mineral–cement emulsion mixes. It was found that it was possible to reduce the stiffness and the cracking in MCE mixes. The use of innovative binders will positively affect the durability of road pavements.

Keywords: complex modulus; shrinkage analysis; reclaimed asphalt; mineral–cement emulsion mixtures; cement dusty by-products (UCPPs)

Citation: Skotnicki, u.; Kuźniewski, J.a.; Szydło, A. Research on the Properties of Mineral–Cement Emulsion Mixtures Using Recycled Road Pavement Materials. *Materials* **2021**, *14*, 563. https://doi.org/10.3390/ma14030563

Academic Editor: Lizhi Sun
Received: 22 December 2020
Accepted: 22 January 2021
Published: 25 January 2021

Publisher's Note: MDPI stays neutral with regard to jurisdictional claims in published maps and institutional affiliations.

Copyright: © 2021 by the authors. Licensee MDPI, Basel, Switzerland. This article is an open access article distributed under the terms and conditions of the Creative Commons Attribution (CC BY) license (https://creativecommons.org/licenses/by/4.0/).

1. Introduction

The issue of environmental protection is very important, especially due to the progressive degradation and exploitation of the surrounding nature. For the construction of road pavements, not only new, unprocessed materials can be used but also those from recycling. Recycling used in road construction provides many tangible benefits: it reduces the need for mineral resources, lowers the cost of aggregate transport, and significantly reduces or even completely eliminates waste landfill from damaged road surfaces.

Recycling of road surfaces enables the reuse of road materials, which, after appropriate grading and mixing with binders such as asphalt or cement with their appropriate percentage, create full-value material products.

1.1. Protecting the Environment by Using Recycled Materials

For economic and ecological reasons, at the end of the last century in Europe, attempts were made to explain the problem of using building rubble as a building component. It was shown in [1] that the rubble from recycled masonry intended for the production of cement concrete should be precisely sorted. Despite the lower compressive strength, higher water absorption, and thus lower frost resistance of cement concrete containing aggregate derived from masonry, laboratory tests and experience in construction practice have clearly shown the suitability of recycled aggregate for the production of structural concrete also.

Aggregates recovered from the demolition of single-family houses, including foundations and walls, did not pose a threat to the environment when used in layers of road

291

surfaces not bound with binder. As a mineral component, they did not meet the requirements specified in Spanish standards (grain size, Los Angeles abrasion), however, in the mineral mix, the recycled aggregates showed better structural behavior and less degradation and higher module values than the mix of natural aggregates. On the other hand, the durability was not lower than that obtained for natural aggregates [2].

Chinese experiments [3,4] confirmed the possibility of using concrete slag and brick slag for the construction of highway embankment surfaces. However, the humidity of the embankment should be controlled, as damaged or improperly drained pavement elements and slope inclinations may rapidly increase the humidity in the areas of water infiltration, and may also lead to its uncontrolled settlement. Construction and demolition waste (CDW) can be used in road embankments according to [5–7]. The use of demolition waste materials (CDW) in unbound layers of the foundation is confirmed by studies [8–13].

In Beijing, the authors of [14] analyzed the impact of construction waste on the function of the subgrade. Based on the determination of the degree of compaction, observation of settlement, and plate loading test, it has been shown that recycled CDW aggregates with appropriate sorting and appropriate construction technologies are useful. Based on the life cycle assessment, it was concluded that the use of recycled CDW can provide better environmental quality and economic benefits compared to direct disposal.

In [15], the results emphasize the possible technical benefits of using CDW materials instead of natural raw materials in applications such as roadside or unpaved roads. The monitored parameters were density, bearing capacity, and frost resistance. Assessment of the physical state and mechanical properties of CDW as waste from landfills [16] showed the possibility of using them as a base material.

Recycled mixes containing aluminum waste obtained better mechanical values (compressive strength, California bearing ratio (CBR) parameter) than the recycled mixes without this waste content [17].

The addition of metallic waste in the form of steel fibers to cement mortars resulted in a decrease in their bulk density and an increase in porosity [18]. Moreover, these metal wastes can modify the electrical resistance and thermal conductivity of the mortars, regardless of the type and amount of metal wastes. Such materials can be adopted, e.g., in self-de-icing road pavements where controlled temperature of the road layers is necessary.

In turn, the use of recycled fibers reduces the cost of the finished product by up to 50% in relation to the use of virgin fiber [19], without deteriorating the mechanical properties.

1.2. Waste Disposal—Asphalt, Rubber, and Concrete Reclaimed Waste

Chinese researchers [20] have shown in their research that reclaimed asphalt together with the recovered stabilized cement substrate can be used as a secondary raw material of aggregate for the preparation of cement-stabilized mixes in the cold recycling technology. Although the increase in the content of recycled materials resulted in a decrease in the value of mechanical properties, these mixes were characterized by sufficient durability and good performance.

The use of used tires for mixes with modified asphalt in pavement construction, according to [21,22], results in a large energy saving and reduction of carbon dioxide emissions. Such a solution provides many economic and environmental benefits, leading to the sustainable development of pavement infrastructure. Car rubber in the form of crumbs as a modifier improves the rheological and mechanical properties of rubber–asphalt mixtures [21]. Small pieces of rubber mixed with the soil obtain a beneficial effect in geotechnical engineering applications, such as providing better elastic deformations, improved shear strength, increased permeability, and better dynamic characteristics [23].

Increasing the reclaimed asphalt pavement (RAP) content in the mix reduces the strength of the mixtures but has no significant effect on drying shrinkage, erosion susceptibility, and capillary flow characteristics [24].

In Flanders (Belgium), up to 20% of coarse natural aggregates, as a component of concrete mixes, can be replaced with high-quality materials from recycling concrete aggregates [25].

In Malaysia [26], the results of the tests of intermediate tensile strength and modulus of elasticity showed that the addition of recycled concrete aggregate in the amount of 40% of the mineral mix to asphalt mixes is optimal and recommended.

In Hong Kong, a mix of recycled fine aggregate (RFA) and recycled concrete aggregate (RCA) [27] was used for construction applications using the coupled Taguchi–RSM optimization approach. Durability results confirmed the possibility of using this type of waste in building structures.

In [28], the authors showed that, for concretes with compressive strength up to 45 MPa, the type of aggregate used did not matter. It was based on several hundred designs of mixes with recycled concrete and natural aggregate.

Roughness and load-bearing capacity tests carried out on two sections of unpaved rural road showed no significant differences [29]. The surface layer of one section was made of a layer of recycled concrete aggregate and the surface layer of the other section was made of limestone aggregate. This type of waste can be used as an alternative to natural aggregates on unpaved country roads.

A significant influence of asphalt binders on physical and mechanical parameters, such as intermediate tensile strength, creep modulus, stiffness, or free space content, of mixes recycled using the cold method was demonstrated in the works [30,31]. The type of binder had a significant impact on the compaction properties of the tested mixes, which resulted in obtaining different values of mechanical parameters. The obtained results confirmed the possibility of applying cold recycled mixes with foamed asphalt to the layers of pavement structures [32].

1.3. Waste Utilization—Dusts

In Denmark [33], the environmental impact of stored ash and ash in road structures was estimated. The ecotoxicity in water showed a slight difference in their environmental impact. Dusts from electric furnaces do not deteriorate the strength parameter of asphalt concrete and at the same time improve its stiffness [34]. The use of dust as a substitute for Portland cement in the production of concretes dramatically reduces the compressive strength but results in lower sound permeability, which can be used for acoustic insulation [35]. Dust-containing mixes provide better thermal comfort, contributing to a reduction in operating energy in buildings.

Not only does the type of waste [36,37] used have an impact on the properties of road mixes, but the order and method of mixing and the conditions for mixing and compacting these components also have an impact on the obtained values of mechanical parameters, such as intermediate tensile strength or dynamic modules [38–40].

Road managers and policy makers should bear in mind the economic and environmental effects of mining rock raw materials, the demand for which is constantly growing. Only the configuration of an optimal model describing the production processes of obtaining rock raw materials or providing material data by producers [41,42] will enable the supply of aggregates in an economical and environmentally friendly manner and eliminate legal, technical, and market barriers [27].

1.4. Effect of Waste on Shrinkage and Stiffness

According to [43], it was noticed that the cracking sensitivity of mixes with recycled concrete decreased, despite a slightly lower tensile strength, compared to that of mixes containing natural gravel and sand. Moderate content of recycled aggregate (<30%) has little effect on tensile strength values, while complete replacement of natural aggregate with recycled materials reduces tensile strength by 20%.

Improvement of the properties of recycled mixes can be achieved not only by appropriate additives [44,45] but also thanks to "carbonization," which improved mechanical properties and resistance to dry shrinkage for mixes containing recycled concrete aggregate (RCA) [46,47]. The carbonation process is relatively long but provides a more effective yield improvement.

It was observed that with the increase in the content of RFA and RCA, the contraction value increased [27,48], and with the increase of the water–cement ratio (WCR), the coefficient of the contraction value decreased. Similar observations are presented in [20]. The increase in the content of recycled aggregate resulted in an increase in optimal humidity, and its increase, in turn, increased the shrinkage value.

In [49], it was shown that cold recycled mix (CRM) achieved similar dry shrinkage values to cement stabilized macadam (CSM) mixes with only virgin aggregate. Higher cement content increases the dry shrinkage, which was also observed in [24]. As the temperature rises, the value of thermal shrinkage (TS) increases. CRM blends show worse TS resistance than CSM blends. Lower cement content is beneficial for the improvement of TS. Significant addition of coarse natural aggregate improves mechanical strength, frost resistance, resistance to dry shrinkage (DS), and CRM resistance to TS. However, the addition of fine aggregate makes CRMs more prone to shrinkage cracking, and freezing cycle failure [49]. The setting time is crucial in the initial phase; therefore, optimal humidity of the recycled layer must be maintained during this process to avoid high shrinkage on drying [24].

The properties of the mixes with the content of Portland cement were compared to mixes with lime-based low-carbon cementitious materials in [50]. It was found that not only the shrinkage value was lower in the mixes based on lime, but their strength increased after adding sodium sulfate to them. The RCA-containing blends achieved similar performance to the virgin aggregate blends through internal bonding caused by the release of moisture contained in the porous aggregate to allow further hydration of the cement [48]. Lower shrinkage was also observed in carbonized cement–slag mortars than in standard cement–slag mortars [46]. In [51], granulated blast furnace slag was replaced with fly ash and various types of fibers, and it was found that the addition of these materials did not adversely affect the setting time of the cement, slightly reduced strength, and reduced shrinkage compared to the slag mix. The mixes with fibers were characterized by increased strength and obtained the required resistance to freezing. Attempts were made to use kitchen waste oil and asphalt emulsion in mixes with recycled concrete aggregate stabilized with cement [52]. The results of the compressive and tensile strength tests showed that both additives caused an acceptable reduction in the strength of the mix in the early curing stage (7 days), but later, the treated mix showed a relatively fast increase in force to compensate for this deficiency, with a slightly higher value in the mix with the asphalt emulsion. These additives also reduced the stiffness of the mix and the amount of shrinkage. Moreover, the use of used oil and asphalt emulsion is ecologically friendly.

2. Materials and Methods

2.1. Purpose of the Research

The main objective of the research presented in this article is to assess the impact of an innovative binder on the physical, mechanical, and rheological properties of a recycled base layer made of mineral–cement emulsion (MCE) mixes. The result of this recognition is the development of a dedicated binder in the form of a hydraulic binder for the MCE substructure. The flowchart of the research approach is shown in Figure 1.

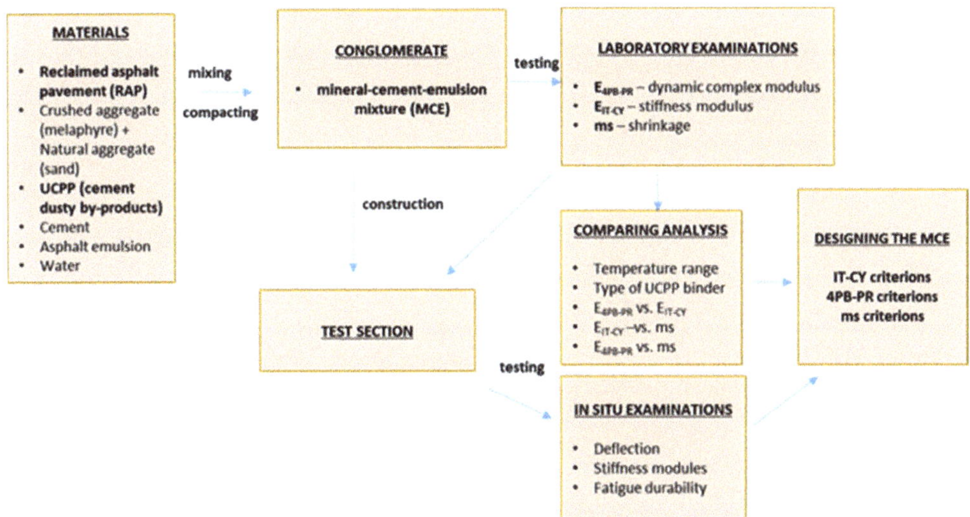

Figure 1. Research methodology flowchart.

2.2. MCE Material as a Conglomerate for the Subbase Layers

The tests were performed for a total of 14 MCE blend variants, i.e., seven combinations for the fine-grained blend (MCE-D-1V, MCE-D-2V, MCE-D-3V, MCE-D-4C, MCE-D-5C, MCE-D- 6C, MCE-D-7C) and seven for coarse-grained mix (MCE-G-1V, MCE-G-2V, MCE-G-3V, MCE-G-4C, MCE-G-5C, MCE-G-6C, MCE-G-7C). The number of combinations equal to seven results from the plan of the simplex-centroid experiment [53,54]. Each point of the experiment describes the composition of the binder formed by combining the three basic components (cement, slaked lime, and cement dusty by-products). They differed from each other in the content of three components: cement, slaked lime, and cement dusty by-products (UCPPs). Cement dusty by-products are fillers made from wastes (blast furnace slags) obtained from heating plants. The test plan also used two reference mixes (MCE-D-Ref, MCE-G-Ref) using Portland cement as a binder.

The MCE mixtures are based on three main ingredients–aggregate, cement, and asphalt emulsion. The MCE mixtures are semi-rigid materials. When in the material there is only the cement, this material has high stiffness. After adding the asphalt emulsion and replacing part of the cement with cement dusty by-products (UCPPs), the reduction in stiffness and improvement in the fatigue durability of the road construction layer will be possible.

The compositions of the mixes were determined as assumed, i.e., for two types of mixes: fine-grained (marked with the symbol D) and coarse-grained (marked with the symbol G). The fine-grained mix consisted of natural grading aggregate with continuous graining 0/31.5 mm (melaphyre), natural grading aggregate with continuous graining 0/2 mm (sand), and recycled basalt aggregate with asphalt binder (recycled asphalt) 0/10 mm. On the other hand, the coarse-grained mixture contained natural grading aggregate with continuous graining 0/31.5 mm, natural grading aggregate with continuous graining 0/2 mm, and recycled basalt aggregate with asphalt binder (recycled asphalt) 0/31.5 mm.

This binder was designed to reduce stiffness and ensure service life for the MCE mix. The compositions of individual binders are presented in Table 1 and graphically in Figure 2 [54].

Table 1. Designation and composition of the composed binders.

Symbol of Binder	Ingredient (–)		
	Cement	Lime	UCPP
1V	0.20	0.20	0.60
2V	0.20	0.60	0.20
3V	0.60	0.20	0.20
4C	0.20	0.40	0.40
5C	0.40	0.20	0.40
6C	0.40	0.40	0.20
7C	0.33	0.33	0.33
Ref	1	0	0

UCPP—cement dusty by-product.

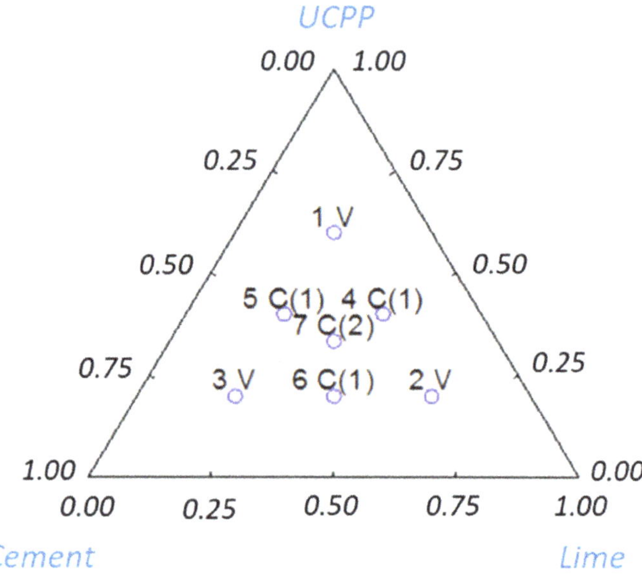

Figure 2. Plan of a simplex-centroid experiment [53,54].

The adopted research plan in the field of physical, mechanical, and rheological properties was aimed at assessing the impact of the composition of the innovative binder on the change in the properties of the recycled base, which resulted in the development of nomograms (regression models) that were used to optimize the binder composition. The optimization made it possible to determine the most favorable configuration of the binder composition, taking into account the limitation of the stiffness of the foundation and ensuring the appropriate operational durability of the recycled foundation. The binder of the selected composition was then used to perform the experimental section, which enabled the assessment of its suitability in real conditions, i.e., as a binder in the foundation layer made of MCE in the construction of the road pavement.

2.3. The Scheme of the Recycled Substructure Composition

The design of the MCE mix should be correlated with the design of the pavement structure and the organization of works, depending on the method of its execution. The following materials were used to produce mineral–cement emulsion mixes: reclaimed asphalt, grading aggregate, hydraulic binder, asphalt emulsion, and water.

2.3.1. Selection of the Optimal Grading Curve

The strength and deformation properties of MCE mixes are significantly influenced by the grain size composition of the mineral mix itself, in accordance with the requirements of the MCE mixes designing manual [55].

The conducted analyses show that the grading curve of the fine-grained and coarse-grained mineral mix meets the conditions of the MCE boundary curves and that the percentage of its individual components (natural aggregate and reclaimed aggregate) in both types of mixes is the same. In the design of mineral mixes, the percentage of asphalt waste was assumed on a level of 40%, natural aggregate 0/31.5 mm in the amount of 50%, and natural aggregate 0/2 mm in the amount of 10%—Table 2.

Table 2. Composition of mineral mixes.

Components of the Mineral Mix	Fine (MCE_D)	Coarse (MCE_G)
	(%)	(%)
0/10 reclaimed asphalt	40	-
0/31.5 reclaimed asphalt	-	40
0/31.5 crushed aggregate improving the gradation—melaphyre	50	50
0/2 natural aggregate improving the gradation—sand	10	10

MCE_D—fine-grained mineral–cement emulsion mixture; MCE_G—coarse-grained mineral–cement emulsion mixture.

The grading of the mineral mixes was within the field of good grading—Figure 3 [55]. The grading of such a mineral mix was determined without taking into account the cement. The optimal percentage of reclaimed asphalt was adopted based on the analyses [54,56].

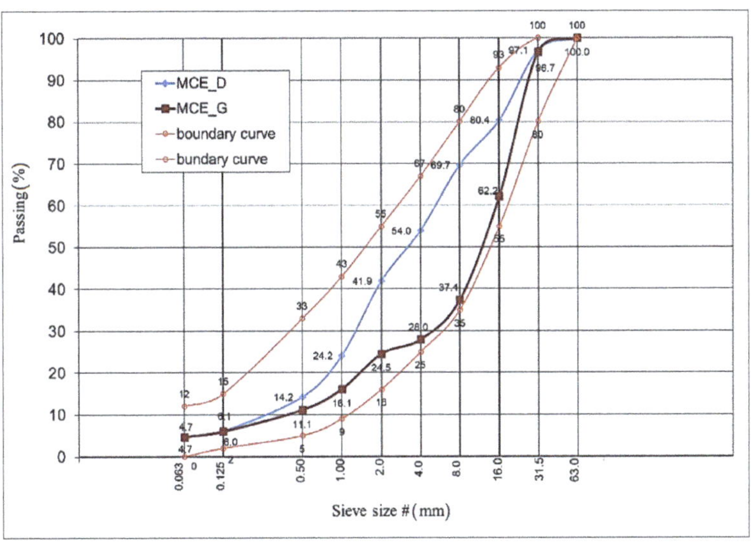

Figure 3. Mineral mixes' grain size distribution and boundary curves.

Both the fine and coarse mixes are grain continuous and comply with the boundary grading curves specified for MCE mixes intended for base layers.

2.3.2. Designing the Optimal Amount of Hydraulic Binder (Cement) and Water

The mutual proportions between the individual components (recycled aggregate, material improving the gradation, hydraulic binder, and asphalt binder) determine the strength and deformation properties of MCE mixtures.

In the design of MCE mixes, based on own experience [57–60] and information recorded in world literature [52,61,62], the percentage of cement was assumed to be 3%. The designed MCE mixes use CEM II class 32.5 Portland cement.

2.3.3. Designation of Optimum Moisture and Maximum Density of the Mix Skeleton

The optimal moisture content of the mix was determined by the Proctor method, based on the standard [63]. The optimal moisture and the maximum bulk density of the mix skeleton were determined.

From the tests [54,56], it was determined that for a fine-grained mix, the optimum humidity is 8.0%, with the maximum density of the mineral skeleton of the mixtures of 2.142 g/cm^3. In turn, for a coarse-grained mixture, the optimal humidity is 7.8%, with the maximum density of the mineral skeleton of the mixes of 2.198 g/cm^3.

Designing the Optimal Amount of Asphalt Emulsion

Having a specific cement content and knowing the optimum humidity of the mixes in question, it was possible to determine the percentage of asphalt emulsion. For the analyzed MCE mixes, the asphalt emulsion 60/40 (AE 60/40) was used in accordance with the requirements of the standard [64]. Additionally, according to the MCE mix designing guide [55], it is recommended that the emulsion meets the following conditions: asphalt type 50/70 or 70/100.

In the design of MCE mixes, using own experience [57–60] and knowledge taken from the world literature [65–67], the percentage of asphalt emulsion 60/40 at 5% was set. Such emulsion content allowed to obtain the total binder content in fine-grained mixtures at the level of 5.1%, and in coarse-grained mixes at the level of 4.9% (taking into account the binder contained in the emulsion and the destructor).

The amount of water added to the MCE resulted from the optimal moisture content of the mineral–cement mixes and the amount of water from the emulsion.

2.3.4. The Composition of MCE Mixes

With the screening data prepared for all mineral materials in the MCE mix and after determining the appropriate content of cement, asphalt emulsion and water, the final design of the composition of the recycled fine-grained and coarse-grained MCE mixes was prepared. These mixes were called reference mixes. The percentages of individual components are presented in the Tables 3 and 4.

Table 3. Composition of the fine-grained mix.

Components	Mineral Mixture (MM)	Mineral–Cement Emulsion Mixture (MCE_Ref)
	(%)	(%)
0/10 reclaimed asphalt	40	34.4
0/31.5 crushed aggregate improving the gradation	50	43.0
0/2 natural aggregate improving the gradation	10	8.6
Cement	-	3.0
Asphalt emulsion 60/40	-	5.0
Water	-	6.0

Table 4. Composition of the coarse-grained mix.

Components	Mineral Mixture (MM)	Mineral–Cement Emulsion Mixture (MCE_Ref)
	(%)	(%)
0/31.5 reclaimed asphalt	40	34.5
0/31.5 crushed aggregate improving the gradation	50	43.1
0/2 natural aggregate improving the gradation	10	8.6
Cement	-	3.0
Asphalt emulsion 60/40	-	5.0
Water	-	5.8

The next step in the research was to determine the composition of MCE mixes, which instead of cement contained an innovative binder. Seven such binders were used, and they differed in their content of three components: cement, lime, and dust.

This binder was designed to reduce stiffness and ensure service life for the MCE mix. The compositions of individual binders are presented in Table 1.

2.4. Research

Mechanical properties for the established compositions of MCE mixes (16 mixes—14 with the innovative binder plus two with cement binder only) were determined. The use of an innovative cement binder in recycled materials resulted in the reduction of cracking in MCE mixes. In the research cycle, the following tests were taken into account:

- Tests of the complex module by the four-point bending test on prismatic materials (4PB-PR) method according to [68] at temperatures of −10, +5, +13, +25, and +40 °C;
- Module tests using the indirect tension test on cylindrical specimens (IT-CY) method according to [68] at temperatures of −10, +5, +13, +25, and +40 °C;
- Tests of shrinkage by the ring method according to [69] at a temperature of +25 °C.

2.4.1. Reduction of Stiffness after Using Dusts Instead of Cement (Stiffness Tests)

Tests of the indirect tensile stiffness modulus, IT-CY, were carried out according to the standard [68]. Four cylindrical (Marshall) samples with a nominal diameter of 101.5 mm and a height of 63.5 mm were used for the analyses for each test and characteristic temperature. The tests were carried out at the following temperatures: −10, +5, +13, +25, and +40 °C. The samples were compacted using 2×75 blows in a Marshall automatic compactor (MULTISERW-MOREK, Brzeznica, Poland). Due to the presence of cement mixes and hydraulic binders in the composition, the curing period of the samples was 28 days. The view of the Marshall samples is shown in Figure 4.

During the tests, detailed registration of the loading vertical force and horizontal displacement was made, which was used to determine the stiffness modulus of the tested material—Equation (1).

$$E_{IT-CY} = \frac{F \cdot (v + 0.27)}{(z \cdot h)} \tag{1}$$

where E_{IT-CY}—stiffness modulus (MPa), F—maximum vertical force (N), v—Poisson's ratio of the material (–), z—amplitude of the obtained deformation (mm), h—average thickness of the sample (mm).

The sample loading scheme is shown in Figure 5.

Figure 4. Marshall samples.

Figure 5. Scheme of the indirect tension test on cylindrical specimens (IT-CY) study.

The tests of the complex module by the bending method of a four-point beam were performed according to the standard [68], and the fatigue life tests according to the standard [70]. The same test samples are used in both cases. Six samples in the form of bars with nominal dimensions of b = 60 mm, h = 50 mm, L_{tot} = 400 mm (according to the equipment's limitation) were used for the analyses, for each test and characteristic temperature. The tests were carried out at the following temperatures: −10, +5, +13, +25, and +40 °C. The samples were compacted with a roller with the use of skid plates according to the standard [71] to obtain the volumetric density determined on the Marshall samples. Then, bars were cut from the compacted slab to the required nominal dimensions. Due to the presence of cement mixtures and hydraulic binders in the composition, the curing period of the samples was 28 days. The view of the sample in the form of a bar is shown in Figure 6.

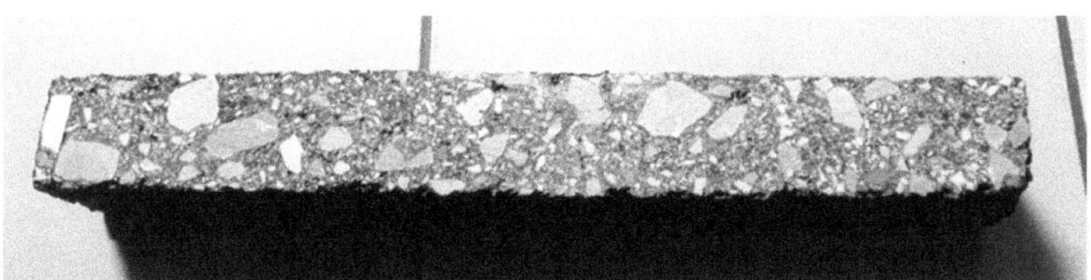

Figure 6. Sample for testing with the four-point bending test on prismatic materials (4PB-PR) method.

During the tests with the method of bending a four-point beam, a detailed registration of the vertical loading force and the vertical displacement in the middle of the beam span was made. The obtained data were used to determine the complex modulus—Equation (2).

$$E_{4PB-PR} = \sqrt{E_1^2 + E_2^2} \tag{2}$$

where

E_{4PB-PR}—dynamic complex modulus of MCE mixture (MPa),
E_1—the real component—Equation (3),
E_2—the imaginary component—Equation (4).

$$E_1 = \gamma \cdot \left(\frac{F}{z} \cdot \cos(\varphi) + \frac{\mu}{10^6} \cdot \omega^2 \right) \tag{3}$$

$$E_2 = \gamma \cdot \left(\frac{F}{z} \cdot \sin(\varphi) \right) \tag{4}$$

where F—load (kN), z—deflection in the middle of the sample's span (mm), φ—phase angle (°), ω—loading frequency (rad/s), μ—mass factor (g), γ—form factor (mm^{-1}).

In the case of testing the complex module, a constant load was applied in the form of a given micro-strain at the level of $\varepsilon = 50 \times 10^{-6}$ m/m with a frequency of 10 Hz. The sample loading scheme is shown in Figure 7.

Figure 7. Scheme of the 4PB-PR test.

2.4.2. Cracking Reduction Caused by Shrinkage and Stiffness (Shrinkage Test)

Shrinkage analyzes of the tested materials were performed using the ring method according to [69] as for concrete conglomerates. For tests, a mixture of calibrated sand with cement binder is prepared, in which shrinkage is generated as a result of setting. The appropriate proportions of the ingredients (sand, binder, water) were determined on the basis of the consistency tests presented in [72].

Table 5 shows the w/s (water to binder) coefficients necessary to obtain slurries and mortars of standard consistency for individual binders. The consistency of the mortars was determined by the drop cone [73] and spreading method [74].

Table 5. Ratios w/s (water to binder).

Binder	*w/s* for the Mortar
3 V	0.60
1 V	0.79
4 C	0.75
5 C	0.68
2 V	0.76
7 C	0.66
6 C	0.62

After the mortars were made, they were placed in the [69] ring. The method of compaction was vibrating for a period of 15 s. The test temperature was 25 °C. The curing periods of the samples were from 14 to 122 days. An example of how the ring is connected (view of the test apparatus) with an example of the test run is shown in Figure 8.

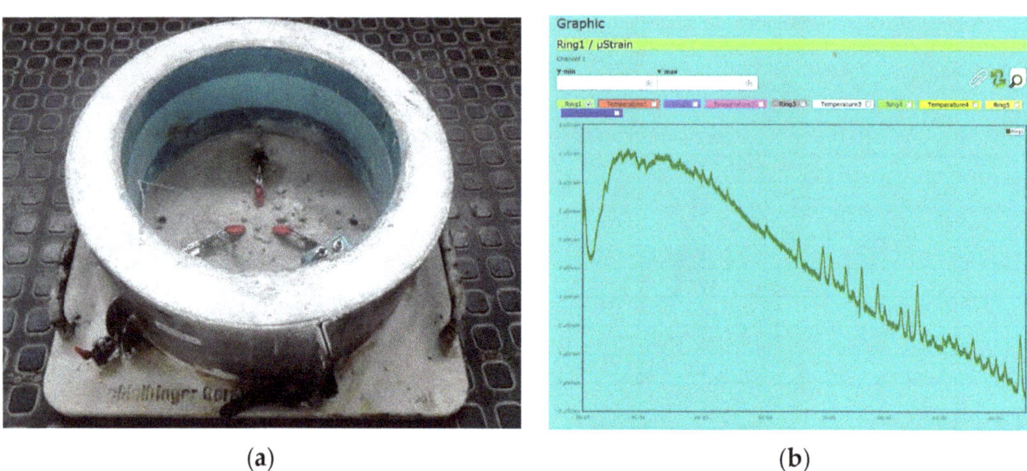

(a) (b)

Figure 8. (a) View of the test apparatus and (b) the test run.

3. Results and Discussion

3.1. Different Methods for Stiffness Comparisons (4PB-PR and IT-CY)

On the basis of the obtained detailed results, the average values of the indirect tensile stiffness modulus, IT-CY, along with the standard deviations of the results were determined for all tested MCE mixes, divided into coarse-grained and fine-grained mixes—Figures 9 and 10.

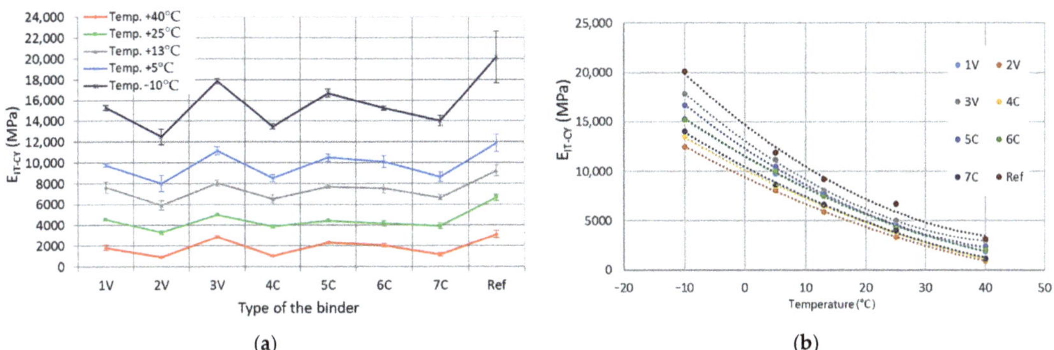

Figure 9. IT-CY stiffness modulus—coarse-grained MCE mix: (**a**) binder dependency, (**b**) temperature dependency.

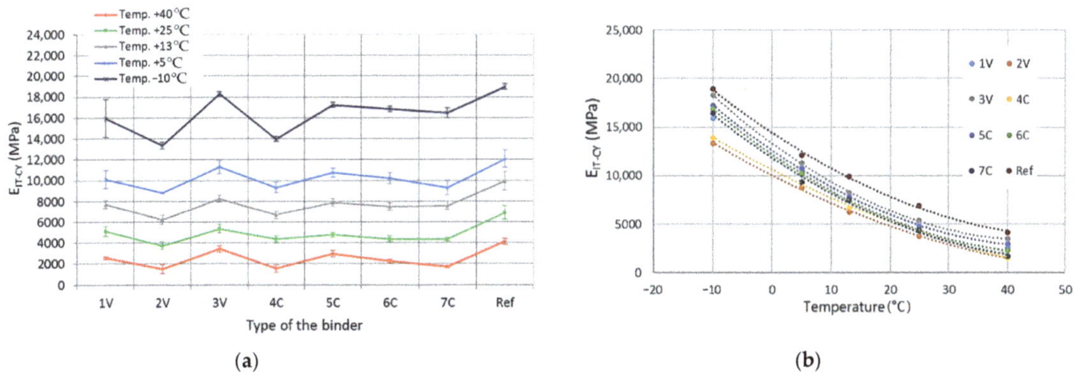

Figure 10. IT-CY stiffness modulus—fine-grained MCE mix: (**a**) binder dependency, (**b**) temperature dependency.

Based on the analyses of the test results, it was found that the test temperature had a significant impact on the IT-CY stiffness modulus of MCE mixes. The stiffness of the mixes decreases with increasing test temperature. Depending on the grain size and binder 1V, 2V, 3V, 4C, 5C, 6C, 7C, and cement (Ref) used, MCE mixes may have a different material specification. The use of innovative binders made it possible to reduce the stiffness of individual MCE mixes compared to that of reference mixes containing classic cement in the entire temperature range—Figure 11.

Depending on the binder used, both for fine-grained and coarse-grained mixtures, the decrease in the value of the modules was from 40% to 80% of the value of the module of the reference mixture with cement only. The values of modulus of fine-grained and coarse-grained mixes for mixes with the content of a given binder were similar.

The greatest reduction in stiffness was achieved by using 2V, 4C, and 7C binders, both in coarse-grained and fine-grained mixes.

Based on the detailed results obtained, the mean values of the complex modulus and fatigue life along with the standard deviations of the results were determined in the four-point beam test for all tested MCE mixes, divided into coarse-grained and fine-grained mixes—Figures 12 and 13.

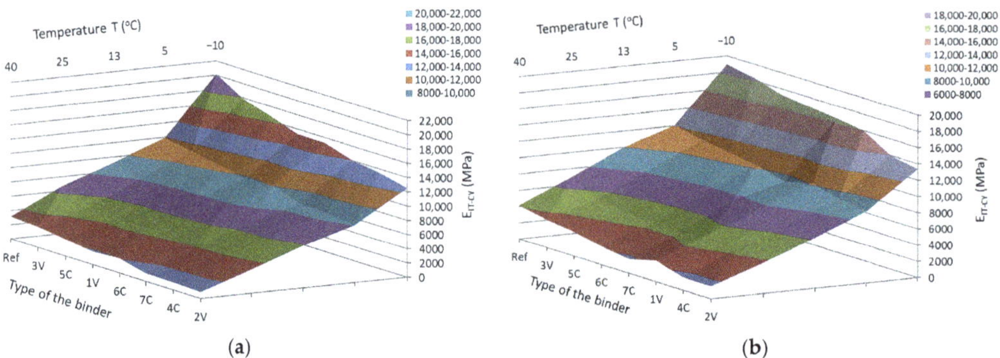

Figure 11. IT-CY module: (**a**) coarse-grained mixtures, (**b**) fine-grained mixtures.

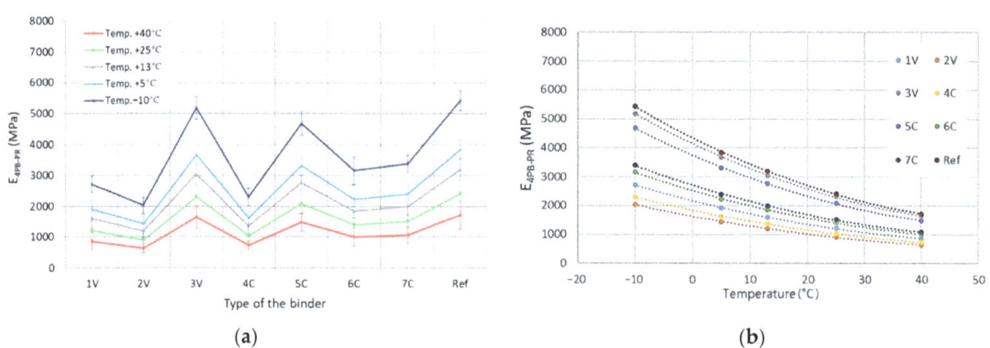

Figure 12. 4PB-PR complex modulus—coarse-grained MCE mix: (**a**) binder dependency, (**b**) temperature dependency.

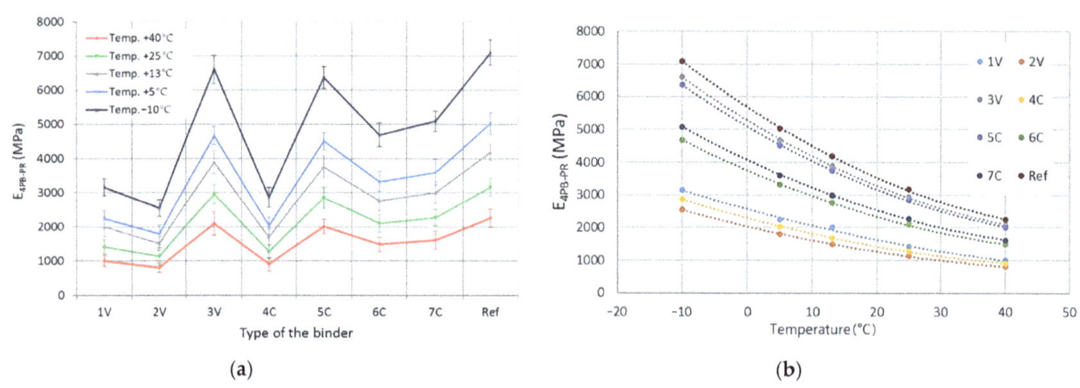

Figure 13. Complex modulus 4PB-PR—fine-grained MCE mix: (**a**) binder dependency, (**b**) temperature dependency.

The obtained test results indicate that with the increase in test temperature, the values of the complex modulus 4PB-PR decrease. Both for coarse-grained and fine-grained mixes, the lowest values of modulus were observed for 2V and 4C mixes. In turn, the highest values of the complex modulus were obtained by the mixes of 3V and 5C. The same mixes achieved the maximum values of the IT-CY stiffness modulus. The use of innovative binders made it possible to reduce the stiffness of individual MCE mixes compared to that of reference mixes containing classic cement in the entire temperature range—Figure 14.

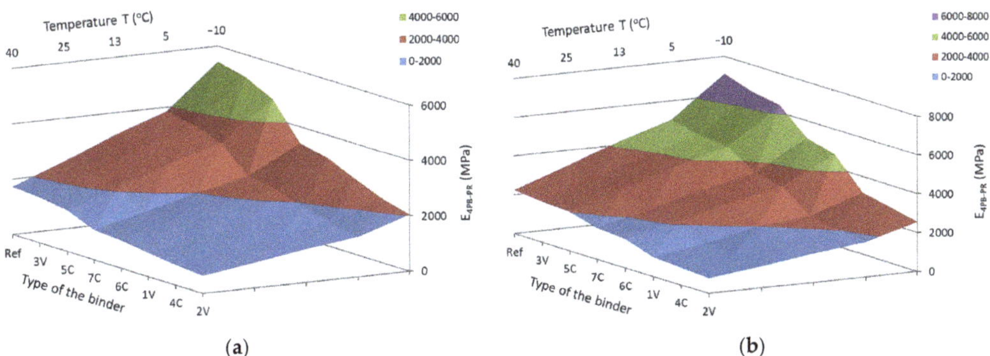

(a)   (b)

Figure 14. 4PB-PR module: (**a**) coarse-grained mixes, (**b**) fine-grained mixes.

3.2. Correlation of the Results of 4PB-PR and IT-CY Tests—Conversion Factors

As a result of the analyses carried out and the observation of similar dependencies between the stiffness of the tested material obtained by the 4PB-PR and IT-CY methods, an attempt was made to obtain a correlation for these two types of modules, Figure 15. Such correlation will allow for the future use of these tests to estimate the durability of the pavement road layer in MCE technology.

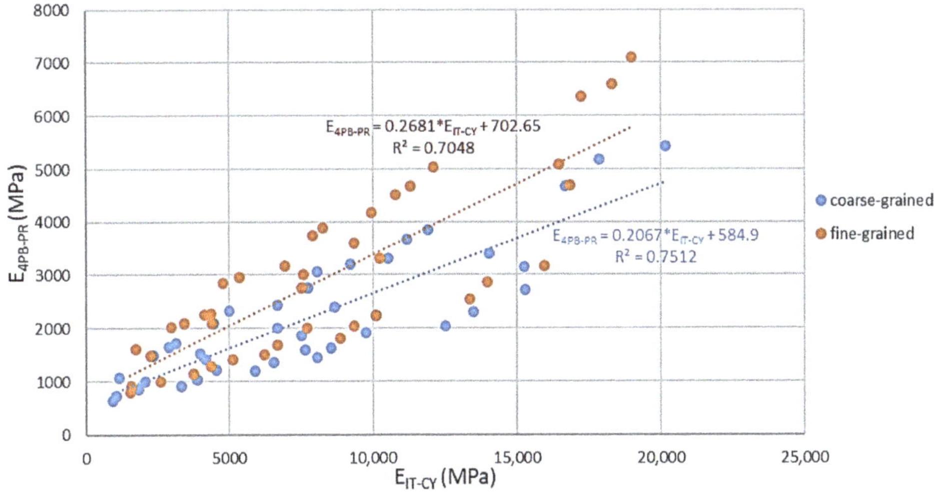

Figure 15. Correlation of modules obtained by 4PB-PR and IT-CY methods.

Correlations in Figure 15 were set for different compositions of MCE mixtures (fine grained and course grained). The results are shown for all tested temperatures with application of different UCPPs (cement dusty by-products) content. In spite of different variables in the tests, obtained correlations can be set as linear functions with a high coefficient of determination, R. On the basis of the conducted analyses, a good correlation was obtained between the stiffness determined according to the IT-CY indirect tensile method and the stiffness of MCE mixes determined by the bending method of a four-point beam, 4PB-PR—Equation (5).

$$E_{4PB-PR} = A \cdot E_{IT-CY} + B \tag{5}$$

where

E_{4PB-PR}—dynamic complex modulus of MCE mixture (MPa),
E_{IT-CY}—stiffness modulus of MCE mixture (MPa),
A—slope factor depending on the grain size of the mix, A = 0.2067–0.2681 (-),
B—directional form factor depending on the grain size of the mix, B = 584.9–702.65 (-).

These dependencies will allow the interchangeable application of the above-mentioned tests, which can be used in the fatigue life prediction of road surfaces, taking into account the same test conditions.

3.3. Shrinkage Analysis

The phenomenon of shrinkage is observed during curing of samples of mixes with cement binders. As a result, it leads to the appearance of cracks in the material—Figure 16.

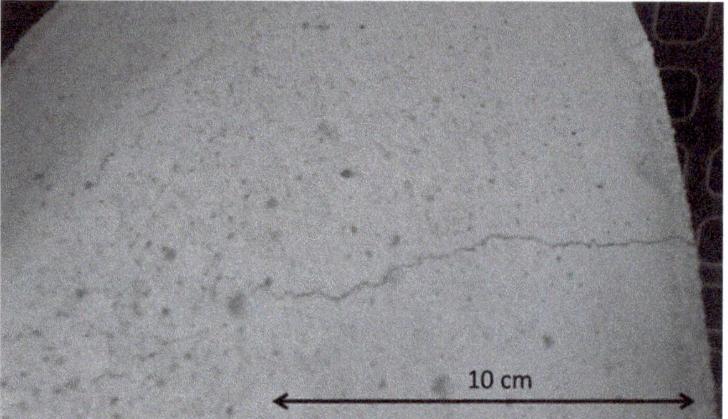

Figure 16. The occurrence of shrinkage cracks.

In order to reduce the occurrence of shrinkage cracks, innovative waste binders were used instead of classic cement. The appropriate stiffness was maintained. In the shrinkage analyses, the cement binder was used as a reference against three innovative binders 1V, 2V, and 7C, which clearly differed in the properties demonstrated in the stiffness tests using the 4PB-PR and IT-CY methods. The tests were carried out at the temperature of 25 °C. The relationship between the curing time of the samples and the micro-contraction deformation is shown in Figure 17.

A significant decrease in destructive strains was observed after the use of innovative binders. At the same time, the curing time to destructive cracks was significantly longer.

As a result of the tests, it was possible to establish the relationship between the stiffness of the tested material and the shrinkage generated during its curing. The relationship between 4PB-PR stiffness and shrinkage is shown in Figure 18, while the relationship between IT-CY and shrinkage is shown in Figure 19. The analyses relate to the curing temperature of 25 °C. The minimum level of stiffness for MCE mixtures is 2000 MPa according to the criteria specified in [58]. This condition was met for the fine-grained mixture with the binder 7C.

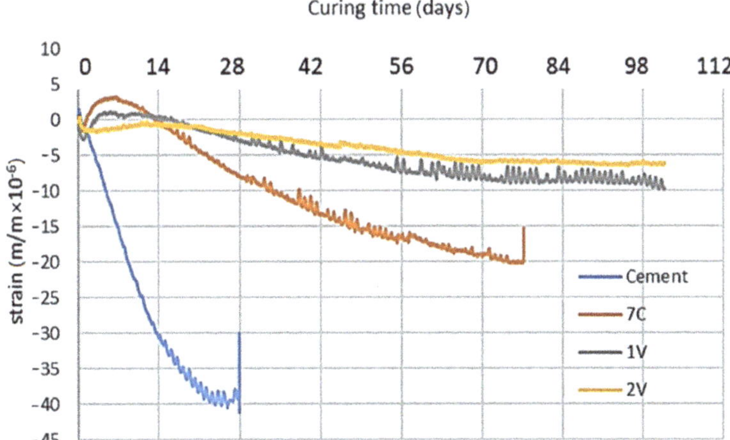

Figure 17. Increase in shrinkage micro-strain during material curing time.

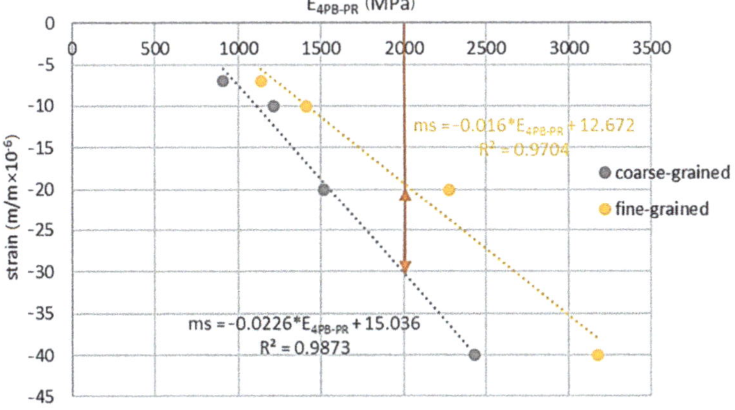

Figure 18. 4PB-PR stiffness vs. shrinkage.

Estimated levels of micro-strain for a minimum stiffness level of 4PB-PR = 2000 MPa are between −30 and −20 micro-strains. For such a range of shrinkage strains, it is possible to assume the IT-CY stiffness level—Figure 19.

As a result of the analyses, it was found that the minimum level of IT-CY stiffness modulus for MCE mixes should be in the range of 5000–6000 MPa. The obtained data made it possible to estimate the plane of stiffness of the tested recycled materials. As the stiffness increases, the probability of contraction cracks occurring in it increases. Micro-strain shrinkage (ms) dependency on IT-CY stiffness modulus and 4PB-PR complex modulus for fine-grained mix is presented in Figure 20. This dependency is shown in the Equation (6) with the coefficient of determination, $R^2 = 0.99$.

$$ms = 19.8525 - 0.0134 \cdot E_{4PB-PR} - 0.0024 \cdot E_{IT-CY} \tag{6}$$

where

ms—shrinkage of MCE mixture (μm),

E_{4PB-PR}—dynamic complex modulus of MCE mixture (MPa),

E_{IT-CY}—stiffness modulus of MCE mixture (MPa).

The use of materials from the central area of the plane in the road surface will allow for adequate durability of the layer with the maximum reduction of shrinkage cracks.

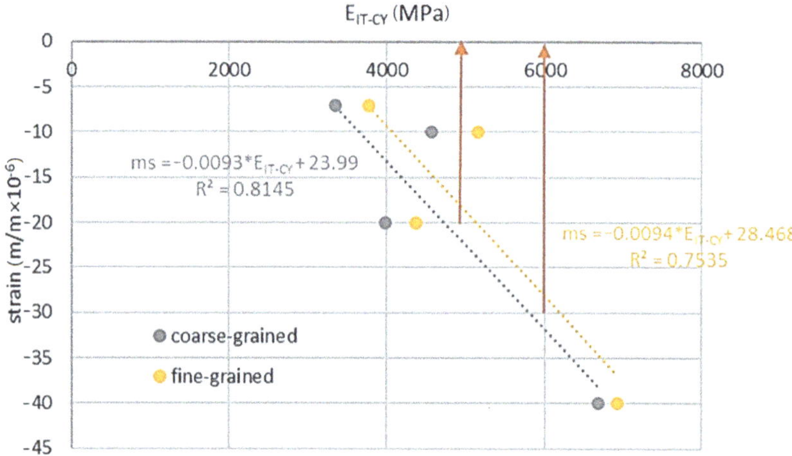

Figure 19. IT-CY stiffness vs. shrinkage.

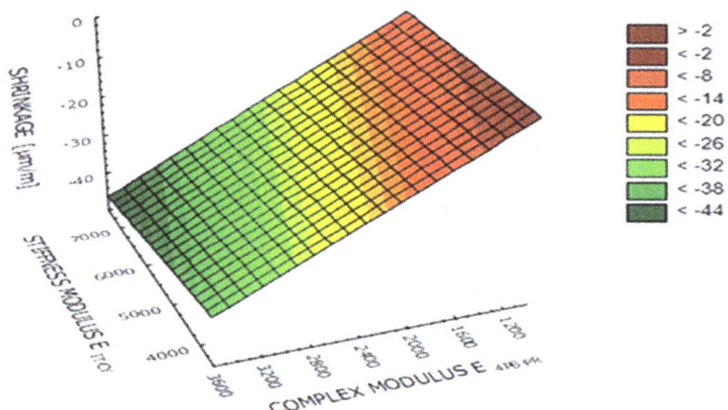

Figure 20. Stiffness plane of MCE mixes.

4. Practical Application of MCE Mixes

As part of the research on the use of recycled materials and the assessment of the actual suitability of waste in the form of cement by-products (UCPPs) in combination with reclaimed asphalt (RAP) for the structural layers of road pavements, an experimental section was made in the arrangement of layers of the flexible pavement structure. The thicknesses of the individual layers are shown in Table 6. The foundation uses MCE with a cement binder and MCE with an innovative type 5C binder.

Table 6. Layout of pavement construction layers used in the experimental section.

Layer Type	Material	Layer Thickness (m)
Wearing/base course layer	SMA (stone matrix asphalt) 0/16 mm	0.08
base layer	MCE and MCE + binder	0.20
	natural subsoil	

The stiffness tests of the executed layers will be correlated with laboratory tests and will be used to estimate the fatigue life of the pavement layers. The results of these studies will be the subject of further publications.

5. Conclusions

The conducted analyses for MCE mixes (16 mixes—14 with the innovative binder plus two with cement binder only) showed the following:

- Coarse-grained and fine-grained mixes containing cement binder marked as MCE_G_Ref and MCE_D_Ref were adopted as reference MCE mixes. Then, instead of cement, innovative road binders were used, containing cement, hydrated lime, and cement dusty by-products (UCPPs). All mixes can be used to make the road surface foundation.
- The analyzes of the IT-CY stiffness modulus showed that the test temperature had a significant impact on the stiffness for MCE mixes. With increasing test temperature, the stiffness of the mixes decreased. The use of innovative binders allowed to reduce the stiffness of individual MCE mixes compared to that of reference mixes containing classic cement in the entire temperature range. The greatest reduction in stiffness was obtained after using 2V, 4C, and 7C binders both in coarse-grained and fine-grained mixes. The reduction in the value of the modules accounted for 40–80% of the value of the modulus of the reference mixture only with cement.
- Based on the analyses of the results of the 4PB-PR complex module, it was found that the value of this module was significantly influenced by the grain size distribution of the MCE mix. Fine-grained mixes were characterized by greater homogeneity and higher complex modulus as compared to coarse-grained mixes. Despite the microscopic cracking phenomenon, the MCE layer will be able to take loads and ensure adequate fatigue life of the entire road surface structure.
- With the increase in temperature, the percentage decrease in the values of the modules obtained by the 4PB-PR method was slightly lower than that determined by the IT-CY method. In the case of both described modules for fine-grained and coarse-grained mixes with the content of all the binders used for the tests, depending on their value on the temperature change, they maintained the same curvilinear course.
- As the stiffness of the material increases, the likelihood of shrinkage cracking increases. The research showed a good correlation between the stiffness test results and the shrinkage strains. The higher the stiffness of the material, the greater the shrinkage deformation. The use of innovative binders allowed to reduce shrinkage cracks and extend the time to maximum deformation. For a minimum accepTable 4PB-PR stiffness level of 2000 MPa, the shrinkage value is between −20 and −30 micro-strains. Further reduction of the material stiffness is not desirable due to its too low fatigue life [58,75]. For the determined level of deformation, it was possible to estimate the minimum stiffness level in the IT-CY method. The IT-CY stiffness modulus of mineral–cement emulsion mixes should be reduced as much as possible but not less than 5000–6000 MPa.
- The developed relationships between the stiffness of 4PB-PR and IT-CY will allow the maximum use of various types of research equipment in the process of designing and making MCE mixes in road surfaces.
- Observation of the experimental section and identification of the stiffness modules of the foundation layer with the MCE will allow the correlation of the results of the

modules of the built-in layer and those determined in the laboratory for the same materials. The results of the work will be the subject of further publications.

- The demonstrated reduction of shrinkage cracks in MCE mixes, through the use of cement dusty by-products (UCPPs), allows for the conclusion that the layers of road pavements containing these binders will have a higher fatigue life compared to that of conventional solutions based on cement binder.
- The use of cement dusty by-products (UCPPs) and reclaimed asphalt (RAP) to make pavement layers can significantly reduce the amount of unprocessed waste that causes environmental degradation. At the same time, the use of recycled products allows reducing the consumption of natural resources in road construction.

Author Contributions: Methodology, Ł.S., J.K., and A.S.; formal analysis, Ł.S., J.K., and A.S.; investigation, Ł.S., J.K., and A.S.; writing—original draft preparation, Ł.S., J.K., and A.S.; writing—review and editing, Ł.S. and J.K.; funding acquisition A.S. All authors have read and agreed to the published version of the manuscript.

Funding: This research was funded by The National Center for Research and Development (Polish NCBR) grant number TECHMATSTRATEG1/349326/9/NCBR/2017.

Data Availability Statement: The data presented in this study are available on request from the corresponding author.

Acknowledgments: The research results were developed as part of the project titled "The innovative technology used the binding agent optimization that provides the long service life of the recycled base layer" (TECHMATSTRATEG1/349326/9/NCBR/2017) within the scientific undertaking of Strategic Research and Development Program titled "Modern Materials Technology" (TECHMATSTRATEG I), which is financed by the National Center for Research and Development (Polish NCBR).

Conflicts of Interest: The authors declare no conflict of interest. The funders had no role in the design of the study; in the collection, analyses, or interpretation of data; in the writing of the manuscript; or in the decision to publish the results.

Abbreviations

The following acronyms are used in this paper:

MCE	mineral–cement emulsion mixture
UCPP	cement dusty by-product
CDW	construction and demolition waste
RAP	reclaimed asphalt pavement
CBR	California bearing ratio
RFA	recycled fine aggregate
RCA	recycled concrete aggregate
WCR	water–cement ratio
CRM	cold recycled mix
CSM	cement stabilized macadam
TS	thermal shrinkage
DS	dry shrinkage
AE 60/40	asphalt emulsion (60% asphalt and 40% water)
4PB-PR	four-point bending test on prismatic specimens
IT-CY	indirect tension test on cylindrical specimens
SMA	stone matrix asphalt

References

1. Zieliński, K. Impact of recycled aggregates on selected physical and mechanical characteristics of cement concrete. *Procedia Eng.* **2017**, *172*, 1291–1296. [CrossRef]
2. Tavira, J.; Jiménez, J.R.; Ledesma, E.F.; López-Uceda, A.; Ayuso, J. Real-scale study of a heavy traffic road built with in situ recycled demolition waste. *J. Clean. Prod.* **2020**, *248*, 119219. [CrossRef]
3. Liu, L.; Li, Z.; Cai, G.; Liu, X.; Yan, S. Humidity field characteristics in road embankment constructed with recycled construction wastes. *J. Clean. Prod.* **2020**, *259*, 120977. [CrossRef]

4. Liu, L.; Li, Z.; Congress, S.S.C.; Liu, X.; Dai, B. Evaluating the influence of moisture on settling velocity of road embankment constructed with recycled construction wastes. *Constr. Build. Mater.* **2020**, *241*, 117988. [CrossRef]

5. Cristelo, N.; Vieira, C.S.; de Lurdes Lopes, M. Geotechnical and geoenvironmental assessment of recycled construction and demolition waste for road embankments. *Procedia Eng.* **2016**, *143*, 51–58. [CrossRef]

6. Soleimanbeigi, A.; Edil, T.B. Compressibility of recycled materials for use as highway embankment fill. *J. Geotech. Geoenvironmental Eng.* **2015**, *141*, 04015011. [CrossRef]

7. Soleimanbeigi, A.; Edil, T.B.; Benson, C.H. Engineering properties of recycled materials for use as embankment fill. In Proceedings of the Geo-Congress 2014 Technical Papers: Geo-Characterization and Modeling for Sustainability, Atlanta, Georgia, 23–26 February 2014; pp. 3645–3657. [CrossRef]

8. Arulrajah, A.; Disfani, M.M.; Horpibulsuk, S.; Suksiripattanapong, C.; Prongmanee, N. Physical properties and shear strength responses of recycled construction and demolition materials in unbound pavement base/subbase applications. *Constr. Build. Mater.* **2014**, *58*, 245–257. [CrossRef]

9. Azam, A.M.; Cameron, D.A. Geotechnical properties of blends of recycled clay masonry and recycled concrete aggregates in unbound pavement construction. *J. Mater. Civil Eng.* **2013**, *25*, 788–798. [CrossRef]

10. Gabr, A.R.; Cameron, D.A. Properties of recycled concrete aggregate for unbound pavement construction. *J. Mater. Civil Eng.* **2012**, *24*, 6. [CrossRef]

11. Jiménez, J.R.; Agrela, F.; Ayuso, J.; López, M. A comparative study of recycled aggregates from concrete and mixed debris as material for unbound road sub-base. *Mater. Construcción* **2011**, *61*, 289–302. [CrossRef]

12. O'Mahony, M.M.; Milligan, G.W.E. *Use of Recycled Materials in Subbase Layers. Transportation Research Record No. 130. Transportation Research Board*; National Research Council: Washington, DC, USA, 1991; pp. 73–80.

13. Vegas, I.; Ibañez, J.A.; Lisbona, A.; de Cortazar, A.S.; Frías, M. Pro-normative research on the use of mixed recycled aggregates in unbound road sections. *Constr. Build. Mater.* **2011**, *25*, 2674–2682. [CrossRef]

14. Zhang, J.; Ding, L.; Li, F.; Peng, J. Recycled aggregates from construction and demolition wastes as alternative filling materials for highway subgrades in China. *J. Clean. Prod.* **2020**, *255*, 120223. [CrossRef]

15. Huber, S.; Henzinger, C.; Heyer, D. Influence of water and frost on the performance of natural and recycled materials used in unpaved roads and road shoulders. *Transp. Geotech.* **2020**, *22*, 100305. [CrossRef]

16. Mehrjardi, G.T.; Azizi, A.; Haji-Azizi, A.; Asdollafardi, G. Evaluating and improving the construction and demolition waste technical properties to use in road construction. *Transp. Geotech.* **2020**, *23*, 100349. [CrossRef]

17. López-Alonso, M.; Martinez-Echevarria, M.J.; Garach, L.; Galán, A.; Ordoñez, J.; Agrela, F. Feasible use of recycled alumina combined with recycled aggregates in road construction. *Constr. Build. Mater.* **2019**, *195*, 249–257. [CrossRef]

18. Norambuena-Contreras, J.; Quilodran, J.; Gonzalez-Torre, I.; Chavez, M.; Borinaga-Treviño, R. Electrical and thermal characterisation of cement-based mortars containing recycled metallic waste. *J. Clean. Prod.* **2018**, *190*, 737–751. [CrossRef]

19. Hagnell, M.K.; Åkermo, M. The economic and mechanical potential of closed loop material usage and recycling of fibre-reinforced composite materials. *J. Clean. Prod.* **2019**, *223*, 957–968. [CrossRef]

20. Ren, J.; Wang, S.; Zang, G. Effects of recycled aggregate composition on the mechanical characteristics and material design of cement stabilized cold recycling mixtures using road milling materials. *Constr. Build. Mater.* **2020**, *244*, 118329. [CrossRef]

21. Nanjegowda, V.H.; Biligiri, K.P. Recyclability of rubber in asphalt roadway systems: A review of applied research and advancement in technology. *Resour. Conserv. Recycl.* **2020**, *155*, 104655. [CrossRef]

22. Wang, Q.-Z.; Wang, N.-N.; Tseng, M.-L.; Huang, Y.-M.; Li, N.-L. Waste tire recycling assessment: Road application potential and carbon emissions reduction analysis of crumb rubber modified asphalt in China. *J. Clean. Prod.* **2020**, *249*, 119411. [CrossRef]

23. Liu, L.; Cai, G.; Zhang, J.; Liu, X.; Liu, K. Evaluation of engineering properties and environmental effect of recycled waste tire-sand/soil in geotechnical engineering: A compressive review. *Renew. Sustain. Energy Rev.* **2020**, *126*, 109831. [CrossRef]

24. Fedrigo, W.; Núñez, W.P.; Kleinert, T.R.; Matuella, M.F.; Ceratti, J.A.P. Strength, shrinkage, erodibility and capillary flow characteristics of cement-treated recycled pavement materials. *Int. J. Pavement Res. Technol.* **2017**, *10*, 393–402. [CrossRef]

25. Kox, S.; Vanroelen, G.; Van Herck, J.; de Krem, H.; Vandoren, B. Experimental evaluation of the high-grade properties of recycled concrete aggregates and their application in concrete road pavement construction. *Case Stud. Constr. Mater.* **2019**, *11*, e00282. [CrossRef]

26. Nwakaire, C.M.; Yap, S.P.; Yuen, C.W.; Onn, C.C.; Koting, S.; Babalghaith, A.M. Laboratory study on recycled concrete aggregate based asphalt mixtures for sustainable flexible pavement surfacing. *J. Clean. Prod.* **2020**, *262*, 121462. [CrossRef]

27. Zhang, L.W.; Sojobi, A.O.; Kodur, V.K.R.; Liew, K.M. Effective utilization and recycling of mixed recycled aggregates for a greener environment. *J. Clean. Prod.* **2019**, *236*, 117600. [CrossRef]

28. Visintin, P.; Xie, T.; Bennett, B. A large-scale life-cycle assessment of recycled aggregate concrete: The influence of functional unit, emissions allocation and carbon dioxide uptake. *J. Clean. Prod.* **2020**, *248*, 119243. [CrossRef]

29. Jiménez, J.R.; Ayuso, J.; Agrela, F.; López, M.; Galvín, A.P. Utilisation of unbound recycled aggregates from selected CDW in unpaved rural roads. *Resour. Conserv. Recycl.* **2012**, *58*, 88–97. [CrossRef]

30. Chomicz-Kowalska, A.; Maciejewski, K. Performance and viscoelastic assessment of high-recycle rate cold foamed bitumen mixtures produced with different penetration binders for rehabilitation of deteriorated pavements. *J. Clean. Prod.* **2020**, *258*, 120517. [CrossRef]

31. Iwański, M.; Mazurek, G.; Buczyński, P. Bitumen Foaming Optimisation Process on the Basis of Rheological Properties. *Materials* **2018**, *11*, 1854. [CrossRef] [PubMed]
32. Iwański, M.; Mazurek, G.; Buczyński, P.; Zapała-Sławeta, J. Multidimensional analysis of foaming process impact on 50/70 bitumen ageing. *Constr. Build. Mater.* **2020**, *266*, 121231. [CrossRef]
33. Birgisdóttir, H.; Bhander, G.; Hauschild, M.Z.; Christensen, T.H. Life cycle assessment of disposal of residues from municipal solid waste incineration: Recycling of bottom ash in road construction or landfilling in Denmark evaluated in the ROAD-RES model. *Waste Manag.* **2007**, *27*, S75–S84. [CrossRef] [PubMed]
34. Loaiza, A.; Colorado, H.A. Marshall stability and flow tests for asphalt concrete containing electric arc furnace dust waste with high ZnO contents from the steel making process. *Constr. Build. Mater.* **2018**, *166*, 769–778. [CrossRef]
35. Bostanci, S.C. Use of waste marble dust and recycled glass for sustainable concrete production. *J. Clean. Prod.* **2020**, *251*, 119785. [CrossRef]
36. Czapik, P.; Zapała-Sławeta, J.; Owsiak, Z.; Stępień, P. Hydration of cement by-pass dust. *Constr. Build. Mater.* **2020**, *231*, 117139. [CrossRef]
37. Owsiak, Z.; Czapik, P. Testing the cement, hydrated lime and cement by-pass dust mixtures hydration. *Roads Bridges Drog. Mosty* **2020**, *19*, 135–147. [CrossRef]
38. Bai, G.; Zhu, C.; Liu, C.; Liu, B. An evaluation of the recycled aggregate characteristics and the recycled aggregate concrete mechanical properties. *Constr. Build. Mater.* **2020**, *240*, 117978. [CrossRef]
39. Chen, T.; Luan, Y.; Zhu, J.; Huang, X.; Ma, S. Mechanical and microstructural characteristics of different interfaces in cold recycled mixture containing cement and asphalt emulsion. *J. Clean. Prod.* **2020**, *258*, 120674. [CrossRef]
40. Chen, T.; Ma, T.; Huang, X.; Ma, S.; Tang, F.; Wu, S. Microstructure of synthetic composite interfaces and verification of mixing order in cold-recycled asphalt emulsion mixture. *J. Clean. Prod.* **2020**. [CrossRef]
41. Mallick, R.B.; Radzicki, M.J.; Zaumanis, M.; Frank, R. Use of system dynamics for proper conservation and recycling of aggregates for sustainable road construction. *Resour. Conserv. Recycl.* **2014**, *86*, 61–73. [CrossRef]
42. Sultan, A.A.M.; Lou, E.; Mativenga, P.T. What should be recycled: An integrated model for product recycling desirability. *J. Clean. Prod.* **2017**, *154*, 51–60. [CrossRef]
43. Bendimerad, A.Z.; Delsaute, B.; Rozière, E.; Staquet, S.; Loukili, A. Advanced techniques for the study of shrinkage-induced cracking of concrete with recycled aggregates at early age. *Constr. Build. Mater.* **2020**, *233*, 117340. [CrossRef]
44. Mazurek, G.; Iwański, M. Optimisation of the innovative hydraulic binder composition for its versatile use in recycled road base layer. In *IOP Conference Series: Materials Science Engineering*; IOP Publishing Ltd.: Bristol, UK, 2019; Volume 603, p. 032044. [CrossRef]
45. Owsiak, Z.; Czapik, P.; Zapała-Sławeta, J. Properties of a Three-Component Mineral Road Binder for Deep-Cold Recycling Technology. *Materials* **2020**, *13*, 3585. [CrossRef]
46. Abate, S.Y.; Song, K.-I.; Song, J.-K.; Lee, B.Y.; Kim, H.-K. Internal curing effect of raw and carbonated recycled aggregate on the properties of high-strength slag-cement mortar. *Constr. Build. Mater.* **2018**, *165*, 64–71. [CrossRef]
47. Wang, J.; Zhang, J.; Cao, D.; Dang, H.; Ding, B. Comparison of recycled aggregate treatment methods on the performance for recycled concrete. *Constr. Build. Mater.* **2020**, *234*, 117366. [CrossRef]
48. Yildirim, S.T.; Meyer, C.; Herfellner, S. Effects of internal curing on the strength, drying shrinkage and freeze–thaw resistance of concrete containing recycled concrete aggregates. *Constr. Build. Mater.* **2015**, *91*, 288–296. [CrossRef]
49. Li, Q.; Wang, Z.; Li, Y.; Shang, J. Cold recycling of lime-fly ash stabilized macadam mixtures as pavement bases and subbases. *Constr. Build. Mater.* **2018**, *169*, 306–314. [CrossRef]
50. Wu, M.; Zhang, Y.; Jia, Y.; She, W.; Liu, G. Study on the role of activators to the autogenous and drying shrinkage of lime-based low carbon cementitious materials. *J. Clean. Prod.* **2020**, *257*, 120522. [CrossRef]
51. Abdollahnejad, Z.; Mastali, M.; Woof, B.; Illikainen, M. High strength fiber reinforced one-part alkali activated slag/fly ash binders with ceramic aggregates: Microscopic analysis, mechanical properties, drying shrinkage, and freeze-thaw resistance. *Constr. Build. Mater.* **2020**, *241*, 118129. [CrossRef]
52. Zhou, J.; Zeng, M.; Chen, Y.; Zhong, M. Evaluation of cement stabilized recycled concrete aggregates treated with waste oil and asphalt emulsion. *Constr. Build. Mater.* **2019**, *199*, 143–153. [CrossRef]
53. Iwański, M.; Chomicz-Kowalska, A.; Buczyński, P.; Mazurek, G. Optimization of the binding agent composition of universal nature in the recycled base layers. Development of nomograms of obtained properties for mortars and grouts in accordance with the assumed experiment plan for the designed mixed binders. In *TECHMATSTRATEG "Modern Material Technologies" Program*; Report No. 2/1/PŚk/2018; Kielce University of Technology: Kielce, Poland, 2018.
54. Szydło, A.; Mackiewicz, P.; Skotnicki, Ł.; Kuźniewski, J. Assessment of the impact of an innovative binder on the physical, mechanical and rheological properties of a recycled base layer made of a mineral-binder mixture with asphalt emulsion. In *TECHMATSTRATEG "Modern Material Technologies" Program*; Report No. 3/2/PWr/2019; Wroclaw University of Science and technology: Wroclaw, Poland, 2019.
55. General Directorate for National Roads and Motorways (GDDKiA). *Instruction for Designing and Embedding of Mineral-Cement-Emulsion Mixtures (MCE)*; General Directorate for National Roads and Motorways: Gdańsk, Poland, 2014.

56. Szydło, A.; Mackiewicz, P.; Skotnicki, Ł.; Kuźniewski, J. *Innovative Technology Using the Optimization of the Binder Intended for the Technology of Deep Cold Recycling of the Pavement Structure Ensuring Its Operational Durability-Report of the Institute of Civil Engineering on Wroclaw University of Science and Technology no. SPR 74*; Wroclaw University: Wroclaw, Poland, 2019.

57. Kuźniewski, J.; Skotnicki, Ł. Influence of the compaction method on mineral-cement emulsion mixture properties. *J. Mater. Civil Eng.* **2016**, *28*, 04016138. [CrossRef]

58. Kuźniewski, J.; Skotnicki, Ł. Properties of mineral-cement emulsion mixtures based on concrete aggregates from recycling. *Case Stud. Constr. Mater.* **2019**, e00309. [CrossRef]

59. Kuźniewski, J.; Skotnicki, Z.Ł.; Szydło, A. Fatigue durability of asphalt-cement mixtures. *Bull. Pol. Acad. Sci. Tech. Sci.* **2015**, *63*, 107–111. [CrossRef]

60. Skotnicki, Z.Ł.; Kuźniewski, J.; Szydło, A. Stiffness identification of foamed asphalt mixtures with cement, evaluated in laboratory and in situ in road pavements. *Materials* **2020**, *13*, 1128. [CrossRef] [PubMed]

61. Du, S. Mechanical properties and shrinkage characteristics of cement stabilized macadam with asphalt emulsion. *Constr. Build. Mater.* **2019**, *203*, 408–416. [CrossRef]

62. Guha, A.H.; Assaf, G.J. Effect of Portland cement as a filler in hot-mix asphalt in hot regions. *J. Build. Eng.* **2020**, *28*, 101036. [CrossRef]

63. *PN-EN 13286-2. Unbound and Hydraulically Bound Mixtures–Part. 2: Test. Methods for the Determination of the Laboratory Reference Density and Water Content–Proctor Compaction*; Polish Committee for Standardization: Warsaw, Poland, 2010.

64. *PN-EN 13808:2013-10. Bitumen and Bituminous Binders–Framework for Specifying Cationic Bituminous Emulsions*; Polish Committee for Standardization: Warsaw, Poland, 2013.

65. Gui-Ping, H.; Wing-Gun, W. Effects of moisture on strength and permanent deformation of foamed asphalt mix incorporating RAP materials. *Constr. Build. Mater.* **2008**, *22*, 125–130. [CrossRef]

66. Yan, J.; Ni, F.; Yang, M.; Li, J. An experimental study on fatigue properties of emulsion and foam cold recycled mixes. *Constr. Build. Mater.* **2010**, *24*, 2151–2156. [CrossRef]

67. Zhang, Z.; Cong, C.; Xi, W.; Li, S. Application research on the performances of pavement structure with foamed asphalt cold recycling mixture. *Constr. Build. Mater.* **2018**, *169*, 396–402. [CrossRef]

68. *PN-EN 12697-26. Bituminous Mixtures-Test. Methods for Hot Mix Asphalt–Part. 26: Stiffness*; Polish Committee for Standardization: Warsaw, Poland, 2012.

69. *ASTM C1581/C1581M. Standard Test. Method for Determining Age at Cracking and Induced Tensile Stress Characteristics of Mortar and Concrete under Restrained Shrinkage*; ASTM International (ASTM): West Conshohocken, PA, USA, 2018.

70. *PN-EN 12697-24. Bituminous Mixtures-Test. Methods for Hot Mix Asphalt–Part. 24: Resistance to Fatigue*; Polish Committee for Standardization: Warsaw, Poland, 2012.

71. *PN-EN 12697-33. Bituminous Mixtures-Test. Methods for Hot Mix Asphalt–Part. 33: Specimen Prepared by Roller Compactor*; Polish Committee for Standardization: Warsaw, Poland, 2019.

72. Owsiak, Z.; Zapała-Sławeta, J.; Czapik, P. Optimization of the binding agent composition of universal nature in the recycled base layers, Test results of an innovative binder. In *TECHMATSTRATEG "Modern Material Technologies" Program*; Report No. 1/1/PŚk/2018; Kielce University of Technology: Kielce, Poland, 2018.

73. *PN-EN 12350-2. Testing Fresh Concrete–Part. 2: Slump Test*; Polish Committee for Standardization: Warsaw, Poland, 2009.

74. *PN-EN 12350-5. Testing Fresh Concrete-Part. 5: Flow Table Test*; Polish Committee for Standardization: Warsaw, Poland, 2009.

75. Szydło, A.; Stilger-Szydło, E.; Krawczyk, B.; Mackiewicz, P.; Skotnicki, Ł.; Kuźniewski, J.; Dobrucki, D. *The Use of Recycled Materials. Task 6, Recycling of Concrete Pavements, Report of Faculty of Civil Engineering on Wroclaw University of Science and Technology no. SPR 22*; Wroclaw University: Wroclaw, Poland, 2018.

 *materials*

Article

Warsaw Glacial Quartz Sand with Different Grain-Size Characteristics and Its Shear Wave Velocity from Various Interpretation Methods of BET

Katarzyna Gabryś [1,*], Emil Soból [2], Wojciech Sas [1], Raimondas Šadzevičius [3] and Rytis Skominas [3]

[1] Water Centre WULS, Warsaw University of Life Sciences, 02-787 Warsaw, Poland; wojciech_sas@sggw.edu.pl

[2] Institute of Civil Engineering, Department of Geotechnics, Warsaw University of Life Sciences, 02-787 Warsaw, Poland; emil_sobol@sggw.edu.pl

[3] Institute of Hydraulic Engineering, Vytautas Magnus University Agriculture Academy, 53361 Kaunas, Lithuania; raimondas.sadzevicius@vdu.lt (R.Š.); rytis.skominas@vdu.lt (R.S.)

* Correspondence: katarzyna_gabrys@sggw.edu.pl; Tel.: +48-2259-35-405

Abstract: After obtaining the value of shear wave velocity (V_S) from the bender elements test (BET), the shear modulus of soils at small strains (G_{max}) can be estimated. Shear wave velocity is an important parameter in the design of geo-structures subjected to static and dynamic loading. While bender elements are increasingly used in both academic and commercial laboratory test systems, there remains a lack of agreement when interpreting the shear wave travel time from these tests. Based on the test data of 12 Warsaw glacial quartz samples of sand, primarily two different approaches were examined for determining V_S. They are both related to the observation of the source and received *BE* signal, namely, the first time of arrival and the peak-to-peak method. These methods were performed through visual analysis of BET data by the authors, so that subjective travel time estimates were produced. Subsequently, automated analysis methods from the GDS Bender Element Analysis Tool (BEAT) were applied. Here, three techniques in the time-domain (TD) were selected, namely, the peak-to-peak, the zero-crossing, and the cross-correlation function. Additionally, a cross-power spectrum calculation of the signals was completed, viewed as a frequency-domain (FD) method. Final comparisons between subjective observational analyses and automated interpretations of BET results showed good agreement. There is compatibility especially between the two methods: the first time of arrival and the cross-correlation, which the authors considered the best interpreting techniques for their soils. Moreover, the laboratory tests were performed on compact, medium, and well-grained sand samples with different curvature coefficient and mean grain size. Investigation of the influence of the grain-size characteristics of quartz sand on shear wave velocity demonstrated that V_S is larger for higher values of the uniformity coefficient, while it is rather independent of the curvature coefficient and the mean grain size.

Keywords: shear wave velocity; sand; bender elements test; grain-size characteristics

Citation: Gabryś, K.; Soból, E.; Sas, W.; Šadzevičius, R.; Skominas, R. Warsaw Glacial Quartz Sand with Different Grain-Size Characteristics and Its Shear Wave Velocity from Various Interpretation Methods of BET. *Materials* **2021**, *14*, 544. https://doi.org/10.3390/ma14030544

Received: 16 December 2020
Accepted: 19 January 2021
Published: 23 January 2021

Publisher's Note: MDPI stays neutral with regard to jurisdictional claims in published maps and institutional affiliations.

Copyright: © 2021 by the authors. Licensee MDPI, Basel, Switzerland. This article is an open access article distributed under the terms and conditions of the Creative Commons Attribution (CC BY) license (https://creativecommons.org/licenses/by/4.0/).

1. Introduction

Shear wave (S-wave) velocity (V_S) is a very important parameter in the field of geotechnical earthquake engineering [1,2]. The magnitude of shear wave velocity is determined from both in situ and laboratory tests, and this is used for computing the low-strain shear modulus (G_{max}) [3]. G_{max} is an essential input parameter for dynamic stability analysis of slopes, embankments, dams, etc. The most versatile and portable method to assess G_{max} in the laboratory is the study of shear wave propagation through bender elements (BE) tests or the ultrasonic method [4]. The other group of methods is based on vibrations such as torsional shear (TS) and resonant column (RC) tests [5].

This paper is dedicated to the bender elements test (BET) as a non-destructive test that has gained popularity in the laboratory determination of the low-strain shear modulus, due to its simplicity in use, low cost, and aforementioned non-destructive operation [6]. Bender elements are made of two piezoelectric ceramic sheets with a central shim of usually ferrous nickel alloys to enhance its strength. Bender elements operate as an electromechanical transducer. When a small voltage is applied to one element, it will bend due to the polarization that has induced across its plates, and thus it will act as a transmitter element. On the contrary, when an element bends, a voltage is generated and so it will act as a receiver one [7]. As the strain induced in soil due to bender element movement is in the elastic strain range of soils behavior ($<10^{-3}\%$), based on the theory of elastic wave propagation [8], the low-strain shear modulus can be derived from Equation (1):

$$G_{max} = \rho \cdot V_S^2 \tag{1}$$

where ρ is soil density and V_S is shear wave velocity. The transmitter and receiver elements are inserted at the two opposite ends of soil specimens and the time lag(t), between the input and output wave signals, is measured by triggering and propagating a wave impulse through the soil specimen at a certain frequency and voltage [9]. Shear wave velocity is then computed using Equation (2):

$$V_S = \frac{L_{tt}}{t} \tag{2}$$

where L_{tt} is the tip-to-tip distance between transmitter and receiver [10].

Research methods with the use of piezoelements, although known and applied for several decades [11], have not received any formal standards. They were developed according to the criteria created for the needs of subsequent research centers and may not necessarily be universally adopted. The multitude of design solutions for apparatus does not help in unification. In addition to the few solutions available commercially, there are many devices constructed mainly in research centers. Apart from the same main principle of operation, these apparatuses differ quite significantly in terms of detailed hardware solutions and software. To systematize this issue and try to estimate the impact of the methodology on results under the auspices of the Japanese Technical Committee TC-29, a parallel study was conducted and published the results provide insight into the scale of the problem [12].

Two aspects of the methodology of BET are of fundamental importance: (1) interpretation of the wave's arrival time and (2) selection of the wave's frequency. First of all, adequately detecting the time travel (t) is one of the most important obstacles in data interpreting of BET. Many studies were made considering this issue and many testing and interpretation methods have been proposed so far. To diverse methodologies created over the years, there can be included the simplest methods built on the immediate observation of the wave traced and measurements of the time interval between starting points. For interpretation of the received signals, more elaborated techniques, supported by signal processing and spectrum analysis tools. can be also applied [13]. All the methods created a deal with "appropriate" criteria to select the arrival time. The initial classification of such methods appears in Arulnathan et al. [14]. Viana Da Fonseca et al. [13] updated this classification more recently including developed methods and presenting a combined framework taking advantage of both time and frequency domain interpretations. Time domain methods are direct measurements based on plots of electrical signals versus time, whereas the frequency domain methods involve analyzing the spectral breakdown of the signals and comparing phase shifts of the components [15]. Several researchers, like Greening and Nash [16], prefer the frequency domain methods because they potentially allow the automation of signal processing and avoid the difficulties associated with picking the first arrival. On the other hand, they are either unreliable or require considerable user intervention to provide a reasonable result [17]. It was found that there is no specialized technique with an adequate level of accuracy and reproducibility to be adopted as a standard. The time domain and frequency domain analysis applied in this work to estimate the travel time is presented in detail later in the article. Experimental Results and Discussion.

Considering the determination of the exact travel time of shear wave between the transmitter and receiver, it is also necessary to raise the issue of the wave's shape transmitted through the soil sample. Most early studies using BET generated a single square-wave pulse [13]. The problem with the square wave is that it is composed of a wide spectrum of frequencies [18]. From the received signal of the square wave alone, it is uncertain whether shear wave arrival is at the point of first deflection, the reversal point, or some other point. To reduce the degree of subjectivity in the interpretation, and to avoid the difficulty in interpreting the square wave response, Viaggiani and Atkinson [15] suggested using a sine pulse as the input signal. Sine-wave pulses have become more popular, as these have shown to primarily give more reliable time measurements [19].

The second analyzed methodological aspect is the choice of the wave frequency used in the study. Different frequencies give different results, and not all devices suitable for testing have a choice of frequencies. Unfortunately, due to the apparatus diversity, but also the diversity of the soil material, again, it is not possible to adopt a uniform approach. There is one very important and widely commented guideline in the literature, related to the useful frequency range [20], described in detail further in the manuscript (part 3). Nevertheless, it is worth mentioning that it is related to the useable frequency range [21]. It is a condition that the travel distance (L_{TT}) to the wavelength (λ) should not be less than 2, in some cases and some types of soil 3 (reaching according to some authors even range.

$2 < L_{TT}/\lambda < 9$ [21]). It is all due to a strong near-field effect, which can distribute the received signal in a manner preventing correct interpretation. The recommendation of TC-29 [12] in this aspect is to perform research in a broader scope of frequency spectrum and analyze the results in terms of their consistency. The authors' examples illustrating this issue are shown in part 3 of the article.

Among the factors affecting the accuracy of BET are also sample geometry and its size [21]. Arroyo [22] stated that the sample size could produce (a) effects introduced by end rebounds which provoke interferences and signal overlap, and (b) effects due to the cylindrical boundary that produce and interfered with signal where each frequency travels at a different velocity especially when wavelengths are comparable with the size of the specimen. Rio [23] in his thesis stated that the best results are obtained for slenderness ratios greater than 2. Therefore, specimens with small diameters are more affected by reflections from lateral boundaries [21].

To conduct BET, BE transducers are plugged into both ends of the specimen [24]. Historically there has been concern over whether the installation of BEs may cause some degree of disturbance to the examined samples. Boonyatee et al. [25] in their study investigated the effects of BE installation on V_S measurements. The effects were inspected by comparing tV_S obtained before and after the receiver, BE is penetrated into the soil sample. The penetration tests, performed by varying the rate of penetration, size of the sample, and consolidation pressure, revealed that the installation of piezoelectric transducers generates almost no disturbance.

Wave velocity depends on various soil parameters, namely, confining stress, void ratio, moisture content, etc. In recent years, there has been growing interest in understanding the potential influence of grain-size characteristics on V_S or G_{max}, such as, e.g., the effect of particle size [26], the effect of gradation [27], and the effect of fines [28].

Cho et al. [29] conducted notable studies on the effect of grain shape, reporting a summary from the literature about natural and crushed sands and quantified shape parameters for these sands. A database of shape parameters was created using the reference shape chart proposed by Krumbein and Sloss [30]. One of the significant findings of their study [28] was that a decrease in the roundness of sand grains leads to a decrease in shear wave velocity or stiffness. It is worth noting that analyzed data cover a range of sands with different size distributions. A careful examination of the database showed that the coefficients of uniformity (C_u) of these sands vary from as low as 1.4 to as high as 5.5.

Quick communication about the fundamental question as to whether shear wave velocity (V_S) of sand is dependent on particle shape is a work by Liu and Yang [31]. The authors proved that a sand specimen with angular particles tends to exhibit higher V_S or G_{max} values than a sand specimen with rounded particles.

One of the particular studies of the effect of particle-size distribution on the stiffness of sand using the resonant column technique has been conducted by Wichtmann and Triantafyllidis [32]. It has been demonstrated for a constant void ratio that in the investigated range (0.1 mm $\leq d_{50} \leq$ 6 mm, $1.5 \leq C_u = \frac{d_{60}}{d_{10}} \leq 8$) the small strain shear modulus (G_{max}) does not depend on the mean grain size (d_{50}), but significantly decreases with an increased coefficient of uniformity (C_u), where d_{60} and d_{10} represent the particle sizes that 60% and 10% of the sand mass are smaller than, respectively. This result is in agreement with that of Iwasaki and Tatsuoka [33], which was also derived from several *RC* tests on the sand. For poorly graded sands ($C_u <$ 1.8, 0.16 mm $\leq d_{50} \leq$ 3.2 mm) without a fines content (i.e., no grains smaller than d = 0.074 mm), the values of $G_{max}(e)$ did not depend on d_{50}. Furthermore, Iwasaki and Tatsuoka [33] could not observe a significant influence on the grain shape. Similar G_{max} values were measured for sands with round, subangular, and angular grains. However, the works of Wichtmann and Triantafyllidis [32] and Liu et al. [27] raise concerns that the variation of V_S reported by Cho et al. [29] may not be a true reflection of the shape effect, but rather be associated with varying gradation.

The experimental work of Patel et al. [34], showing an appreciable size dependence of small-strain stiffness, does not agree with that of Witchmann and Triantafyllidis [32] and Iwasaki and Tatsuoka [33]. It does, however, appear to be consistent with that of Bartake and Singh [35], who performed BE tests on three dry samples of sand with similar gradation and found that the G_{max} value increased as d_{50} of sand decreased. It is worth recalling the laboratory observations of Sharifipour et al. [36], where the BE technique was used to measure shear wave velocity but in glass beads. This work adds further uncertainty viz., for glass beads of three different nominal sizes (1.0, 2.0, and 3.0 mm) the authors obtained an opposite result to Patel et al. [34]. The value of V_S increased with increasing particle size.

Another interesting research work was done by Menq and Stokoe [37] who performed *RC* tests on natural river sand with different d_{50} and C_u values without fines content. For each sand, three different initial densities (loose, medium dense, and dense) were studied. In contrast to Iwasaki and Tatsuoka [33], a slight increase in G_{max} with increasing d_{50} was measured when void ratio (e) and mean effective stress (p) were kept constant. Furthermore, the curves of $G_{max}(e)$ were steeper for the coarse material. Then, the authors observed that G_{max} increases with C_u for constant relative density (D_r). It should also be mentioned a study of Lontou and Nikolopoulou [38], where they showed a slight increase of G_{max} with the mean grain size up to d_{50} = 1.8 mm. Significantly higher values for $d_{50} >$ 1.8 mm may be influenced by the small specimen size (diameter of the specimen equal to 4.8 mm).

However, it should be as well emphasized that the experimental data from BET in the literature always seem to indicate that particle size affects G_{max}, although opposite trends were observed in the G_{max} variation with grain size. On the other hand, still, the *RC* test data always seem to suggest that there is not any particle size effect. In this respect, a question appeared: Does the testing method have any effect on small-strain stiffness? Yang and Gu [26] tried to find the correct answer. For the fine glass beads, the BE measurements of G_{max} are comparable to the *RC* measurements, with differences being less than about 10%. Both tests showed a trend that G_{max} decreases slightly with d_{50}, particularly for glass beads at a loose state.

In a more recent study, Altuhafi et al. [39] introduced a new shape parameter SAGI to collectively account for aspect ratio, convexity, and sphericity of sand grains and showed for a database of natural sand that G_{max}, normalized for size and gradation, increases with SAGI. In the current context, whether shear wave velocity or the associated stiffness depends on the particle size distribution and/or grain shape remains still an open question [29,40,41].

In this paper, an experimental investigation is performed to study some aspects of proper selection and verification of the test conditions for BET methods of shear wave velocity (V_S) determination. Moreover, for laboratory research Warsaw glacial quartz sand with different grain-size distribution was chosen. An extensive array of bender element tests in triaxial apparatus is performed on saturated compacted natural sands. The main test results are presented along with their interpretation and discussion. These results show the complex nature of determining V_S using the BET method. Despite the increasing popularity of BET, considerable uncertainty remains in signal interpretation, and thus in the estimated shear wave velocity and shear stiffness [26]. A good example of large scatter in V_S results evaluated from a single BE test on a specimen of natural clay is shown in the work of Clayton [42]. Similar results were obtained from international parallel bender element tests on uniform Toyoura sand [43]. These observations, together with the contradictory results in the literature for stiffness variation with particle size, underline the need for a careful examination of BET, especially for granular material, in which wave propagation is complex owing to its particulate nature. Therefore, the authors made some efforts to clarify several issues closely related to the reliability of BE measurements:

- the characteristics of received signals in both the time and frequency domains over a wide range of excitation frequencies and wavelengths;
- how changes of particle size alter the characteristics of received signals;
- the performance of different interpretation methods under a variety of combinations of test; and
- conditions (i.e., grain size, excitation frequency, and confining stress).

2. Materials and Methods

2.1. Characterization of Materials Used

To explore the influence of grain-size distribution on the shear wave velocity of saturated compacted sand, several series of experiments were performed on twelve different soil samples. These samples were made of fractionated glacial quartz sand with particle sizes ranging from 100 to 4000 µm. The research material was obtained from the Vistula river valley from the area of Warsaw. To receive the material with different grain size, i.e., medium-grained (coefficient of uniformity, C_u, should be in the range from 6 to 15) and well grained (C_u should be bigger than 15) [44], sand samples with the coefficient of uniformity of 12.0, 14.0, and 16.0 were prepared. Moreover, it was assumed that some of the tested soils, on the one hand, will compact well and be suitable, e.g., construction of embankments (coefficient of curvature, $C_c = \frac{d_{30}^2}{d_{10} \cdot d_{60}}$, should be in the range from 1 to 3). On the other hand, some of the samples will not be susceptible to compaction and therefore of little use for construction purposes (C_c should not be between 1 and 3) [45]. To meet these requirements, sand samples with C_c equal to successively 1.0, 2.0, 3.0, and 4.0 were prepared.

The method of organizing samples for laboratory tests included several phases. First, the sand was sieved through sieves with the following mesh sizes: 4.0, 2.0, 1.0, 0.5, 0.25, 0.125, and 0.063 mm (7 gradations). Then, the grain size distribution curves shown in Figure 1 were mixed from these gradations in the correct proportions to receive the samples planned. Following Eurocode 7 [46], the material was classified as coarse sand (CSa). The sands S1 to S4 (Figure 1a) have the uniformity coefficient of $C_u = 12.0$, the effective particle size of $d_{10} = 0.063$ mm, the equivalent diameter of particles $d_{60} = 0.760$ mm, different mean grain sizes in the range $0.50 \leq d_{50} \leq 0.70$ mm, and different curvature coefficient in the range $1.0 \leq C_c \leq 4.0$. The materials S5 to S8 (Figure 1b) have the uniformity coefficient of $C_u = 14.0$, the effective particle size of $d_{10} = 0.063$ mm, the equivalent diameter of particles $d_{60} = 0.885$ mm, different mean grain sizes in the range $0.62 \leq d_{50} \leq 0.72$ mm, and different curvature coefficient in the range $1.0 \leq C_c \leq 4.0$. The sands S9 to S12 (Figure 1c) have the uniformity coefficient of $C_u = 16.0$, the effective particle size of $d_{10} = 0.063$ mm, the equivalent diameter of particles $d_{60} = 1.0$ mm, different mean grain sizes in the range $0.70 \leq d_{50} \leq 0.90$ mm, and different curvature coefficient in the range $1.0 \leq C_c \leq 4.0$.

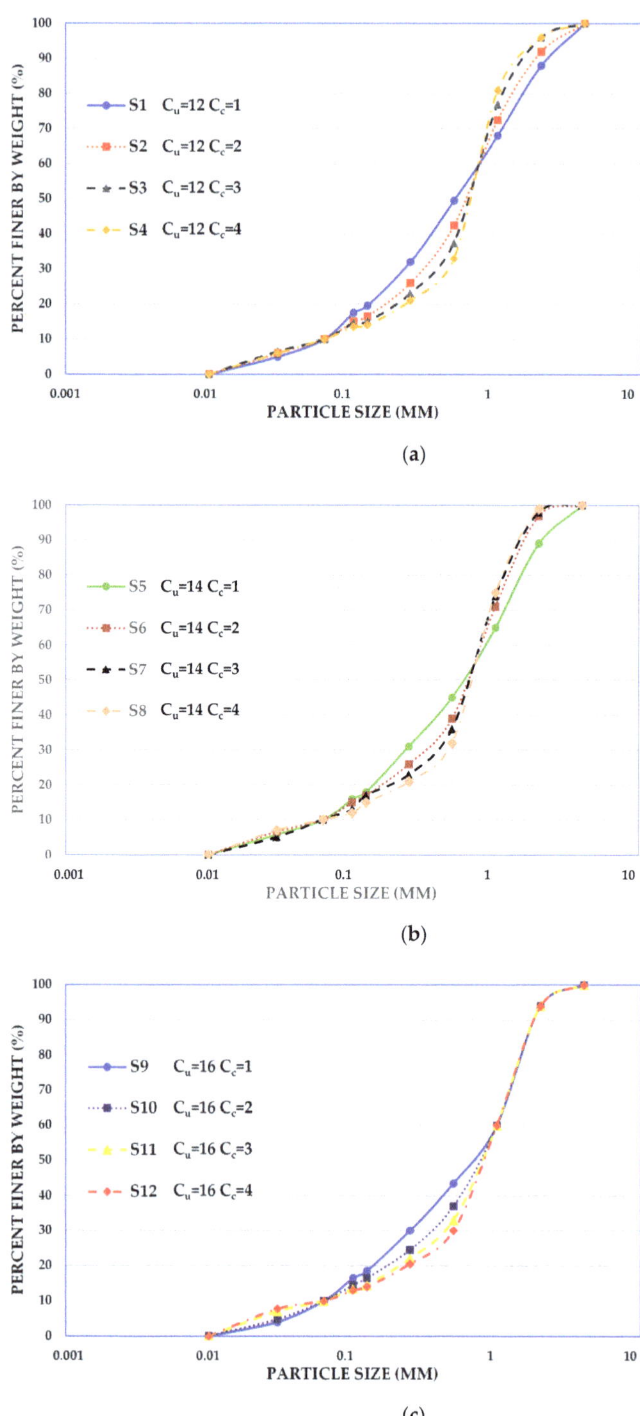

Figure 1. Tested grain size distribution curves for the group of sands with C_u equal to (**a**) 12, (**b**) 14, and (**c**) 16.

The basic properties of the test materials are summarized in Table 1. These are data on equivalent particle sizes: d_{10}, d_{30}, d_{50}, d_{60}, the minimum and the maximum void ratios (e_{min}, e_{max}) (determined according to Polish standard code [47], and the optimum moisture content (m_{opt}) and the maximum dry density (ρ_{dmax}) (from Proctor test).

Table 1. Basic properties of the soils tested in the experimental investigation.

Sample Name	Gradation				e_{min} (-)	e_{max} (-)	m_{opt} (%)	ρ_{dmax} (g/cm^3)
	d_{10} (mm)	d_{30} (mm)	d_{50} (mm)	d_{60} (mm)				
S1	0.063	0.220	0.50	0.760	0.221	0.568	6.7	2.22
S2	0.063	0.310	0.61	0.760	0.227	0.577	7.3	2.20
S3	0.063	0.379	0.65	0.760	0.233	0.596	8.4	2.17
S4	0.063	0.440	0.70	0.760	0.244	0.596	9.1	2.12
S5	0.063	0.240	0.62	0.885	0.221	0.577	6.8	2.22
S6	0.063	0.340	0.70	0.885	0.244	0.596	7.2	2.20
S7	0.063	0.410	0.72	0.885	0.238	0.596	7.6	2.16
S8	0.063	0.480	0.68	0.885	0.233	0.587	7.5	2.16
S9	0.063	0.250	0.70	1	0.227	0.596	7.3	2.21
S10	0.063	0.355	0.78	1	0.233	0.577	7.1	2.21
S11	0.063	0.435	0.82	1	0.221	0.523	6.9	2.21
S12	0.063	0.500	0.90	1	0.238	0.559	7.0	2.19

According to literature-empiric diagrams [30] and after observing the individual particles of representative samples through a microscope (magnification 20 times), the natural glacial quartz sands from Warsaw comprise sub-rounded to particles. Images of the individual particles of representative samples are given in Figure 2. The shape of the particles in terms of roundness and sphericity [28] were both estimated to range from 0.5 to 0.7.

2.2. Experimental Equipment, Specimens Preparation, and Testing Program

In this study, shear wave velocity was directly measured for sand specimens using piezoelectric bender elements (the GDS Bender Element system) installed in the modified triaxial apparatus. The set-up of the apparatus, manufactured by the British company GDS Instruments Ltd. (Hook, Hampshire, UK), is shown in Figure 3. All the samples had a cylindrical shape with the same diameters and heights. The apparatus can accommodate a soil specimen 70 mm in diameter and 140 mm high, with a water-filled cell pressure and an internal linear variable differential transducer of high resolution. Diameter $D = 70$ mm is the maximum achievable diameter for which our specimens can be easily molded. The size and proportions of all examined specimens were in agreement with the suggestions of previously reported works (see Introduction). The modified triaxial apparatus is mounted with vertically placed piezo-element inserts, produced by the aforementioned company GDS Instruments Ltd., on the top cap and the pedestal, respectively, which work as both bender elements and extender elements, similar to the configuration described by Leong et al. [48]. This allows the propagation of S-waves (bender element configuration) and P-waves (extender element configuration) through the body of the specimen. Consequently, measurements of the shear and compression wave velocities (V_S and V_P, respectively) can be carried out based on the obtained wave arrival times with the wave propagation direction along the specimen axis.

The length of the bender element inserted into the soil specimen was optimized (always about 1.5 mm) to avoid compromising the power transmitted or received by the elements. This is achieved by fixing the element further down inside the insert and then filling the remaining volume with flexible material. This allows the element to achieve maximum flexure at its tip, while only protruding into the sample by a reasonable distance. Advantages of this include prolonged life by increased resilience to breakage and easier sample preparation, particularly on very stiff samples where only a small recess for the element is required.

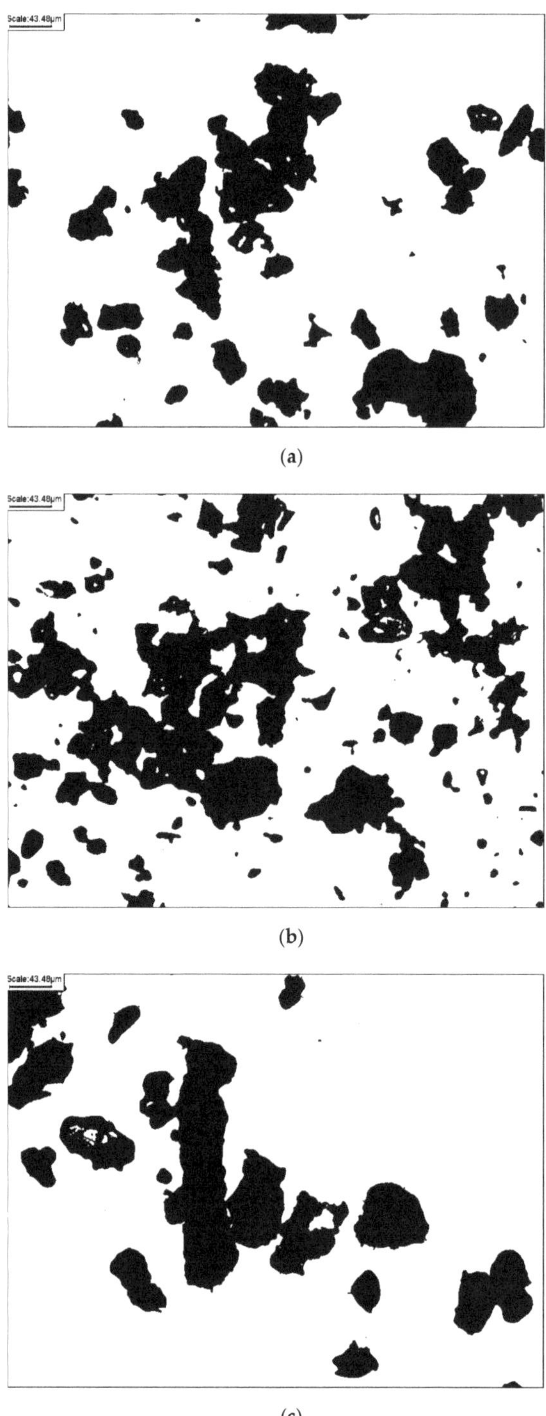

(a)

(b)

(c)

Figure 2. Microscopic images of individual particles of representative samples of this study: (**a**) S2, (**b**) S8, and (**c**) S9.

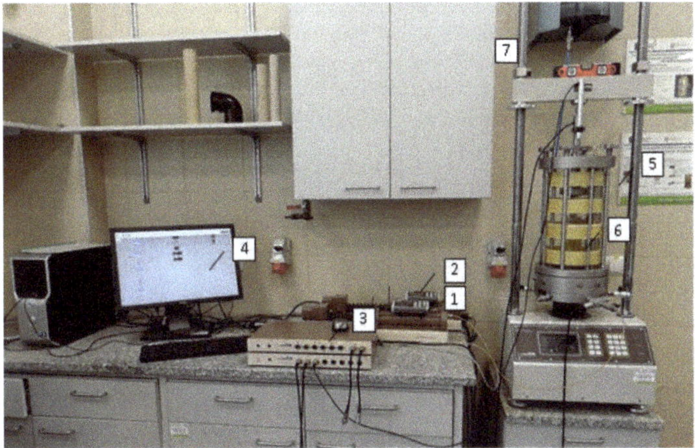

Figure 3. Set-up of dynamic testing system at Water Centre in Warsaw University of Life Sciences: (1) back pressure controller, (2) cell pressure controller, (3) data logger, (4) PC and control software, (5) frame, (6) cell, (7) water tank.

When performing BET, one of its most important aspects is the phase orientation of the elements [24]. The authors always checked the relationship between the received signals concerning the source signal. The desirable orientation was "n-phase". If the orientation was correct, the source and received traces were exactly "in-line".

During BET, shear wave velocity was calculated from the simple measurement of propagation distance (Δs) and propagation time (Δt). Based on several previous works, it is generally accepted that the travel distance is the distance between the tips of two *BEs* [21]. Therefore, in the presented research the travel distance was the tip-to-tip length, which is the height of the specimen minus the length of each *BE*.

Specimens of Warsaw glacial quartz sand were prepared using the moist tamping method by under compaction technique, which is similar in principle to the methods used by several researchers in testing granular soils (see, e.g., in [49]). The optimum moisture content was adopted to receive moist specimens. A predetermined mass of soil was mixed with an appropriate amount of water to obtain a moisture content close to the optimal one. Each soil specimen was prepared in five layers, and it was compacted using a tamper after placing each layer [27]. No obvious segregation was observed during specimen preparation. After sample preparation, the drive head and the load cell were installed. To stabilize the specimens, a suction of 35 kPa was applied to the specimens. The dimensions of the specimens were measured accurately and the initial void ratio was determined. The cell and back pressure were then increased simultaneously to keep a constant isotropic effective stress of 35 kPa.

When a triaxial cell was assembled a specimen was flushed with CO_2 and then with de-aired water. The specimens were saturated with a Skempton's pore-water pressure parameters B-value over 95%. Subsequently, to reach the target, the confining and back-pressure were raised step by step with a difference of 30 kPa. All the specimens were tested in a fully saturated state. After saturation, they were subjected to an isotropic stress path (compression), and the dynamic tests (S-wave and P-wave measurements by bender elements) were conducted at mean effective stress $p' = 45$ kPa for specimens S1–S10, except for specimens S11 and S12. These were tested at progressively increasing mean effective stress equal to $p' = 45, 90$, and 180 kPa. During each consolidation process, the volume changes of the specimen and the corresponding deformation were measured. In the next step, after the completed consolidation, wave velocities measurements were performed under a range of systematically changing excitation frequencies. The frequency range was selected based on the literature review and the authors' own experience. This article focuses primarily on shear wave velocity using one type of signal input: sinusoidal signal. However, for most of the specimens, compression wave velocity was also estimated. Note that specimens S1–S10 were tested in one series, at one confining stress (45 kPa). In contrast, specimens S11 and S12 were subjected to multi-stage consolidation, each one with increasing confining stress. In Table 2, the summary of the test series is presented.

Table 2. Summary of test series.

Test Series	Sample Name	Wave Period (ms); Mean Effective Stress (kPa)	
		$(T_S; p')$	$(T_P; p')$
I	S1	(0.2, 0.25, 0.4, 0.45; 45)	(0.02, 0.05; 45)
II	S2	(0.2, 0.25, 0.4, 0.45; 45)	
III	S3	(0.2, 0.25, 0.4, 0.45; 45)	(0.05; 45)
IV	S4	(0.2, 0.25, 0.4, 0.45; 45)	(0.02, 0.05; 45)
V	S5	(0.2, 0.25, 0.4, 0.45; 45)	(0.02, 0.05; 45)
VI	S6	(0.2, 0.25, 0.4, 0.45; 45)	(0.02, 0.05; 45)
VII	S7	(0.2, 0.25, 0.4, 0.45; 45)	
VIII	S8	(0.2, 0.25, 0.4, 0.45; 45)	(0.02, 0.05; 45)
IX	S9	(0.2, 0.25, 0.4, 0.45; 45)	(0.02, 0.05; 45)
X	S10	(0.2, 0.25, 0.4, 0.45; 45)	
XI -1		(0.06, 0.08, 0.1, 0.2, 0.3, 0.4, 0.5; 45)	
XI - 2		(0.04, 0.06, 0.08, 0.1, 0.2, 0.3, 0.4, 0.5; 90)	(0.03; 90)
XI - 3	S11	(0.06, 0.08, 0.1, 0.2, 0.3, 0.4, 0.5; 180)	
XII - 1		(0.04, 0.06, 0.08, 0.1, 0.2, 0.3, 0.4, 0.5; 90)	(0.02; 90)
XII - 2		(0.04, 0.06, 0.08, 0.1, 0.2, 0.3, 0.4, 0.5; 180)	(0.02, 0.03; 180)
XIII - 1		(0.04, 0.05, 0.06, 0.08, 0.1, 0.2, 0.25, 0.3, 0.4, 0.45, 0.5; 180)	(0.01, 0.02, 0.03, 0.04; 180)
XIV -1		(0.04, 0.05, 0.06, 0.08, 0.1, 0.2, 0.25, 0.3, 0.4, 0.45, 0.5; 45)	(0.03, 0.04, 0.05, 0.06; 45)
XIV- 2		(0.04, 0.05, 0.06, 0.08, 0.1, 0.2, 0.25, 0.3, 0.4, 0.45, 0.5; 90)	(0.03, 0.04, 0.05, 0.06, 0.08; 90)
XIV - 3	S12	(0.04, 0.05, 0.06, 0.08, 0.1, 0.2, 0.25, 0.3, 0.4, 0.45, 0.5; 180)	(0.03, 0.04, 0.05, 0.06, 0.08; 180)
XV - 1		(0.04, 0.05, 0.06, 0.08, 0.1, 0.2, 0.25, 0.3, 0.4, 0.45, 0.5; 90)	(0.08, 0.1; 90)
XV - 2		(0.04, 0.05, 0.06, 0.08, 0.1, 0.2, 0.25, 0.3, 0.4, 0.45, 0.5; 180)	(0.02, 0.03, 0.04, 0.06, 0.08, 0.1; 180)
XVI - 1		(0.04, 0.05, 0.06, 0.08, 0.1, 0.2, 0.25, 0.3, 0.4, 0.45, 0.5; 180)	(0.04, 0.06, 0.08, 0.1; 180)

3. Experimental Results and Discussion

3.1. Synopsis of Experimental Results

3.1.1. Piezo-Elements Signal Analysis

The estimation of the wave velocities from the bender/extender elements was performed primarily by adopting two different approaches related to the observation of the source and received bender/extender element signals, namely, the first time of arrival method (FTA; a different name for start-to-start method) and the peak-to-peak method (PTP). They are both considered as typical techniques applied in the time-domain (TD). The time-domain methods generally determine the travel time directly from the time lag between the transmitted and received signals [24,43,50].

The first above-mentioned method is based on the visual inspection of the received signal and is still quite controversial and subjective due to the complex received signal, wave's reflection, and the near-field effect. Many studies reported that the near-field effect decreases with the increase of frequency or the ratio of wave path length to wavelength ($\frac{L_{tt}}{\lambda}$). Arulnathan et al. [14] reported that the near field effect disappears when this ratio is larger than 1.0. Pennington et al. [51] pointed out that when the $\frac{L_{tt}}{\lambda}$ values range from 2.0 to 10.0, a good signal can be obtained. Wang et al. [52] advocated a ratio greater than or equal to 2.0 to avoid the near field effect. Similarly, a value of 3.33 was recommended by Leong et al. [53] to improve the signal interpretation.

The peak-to-peak method is also widely applied in signal interpretation. In this technique, the time delay between the peak of the transmitted signal and the first major peak of the received signal is regarded as the travel time [54]. As the frequency of the received signal may be slightly different from that of the transmitted signal, and the nature of the soil and size of the sample often affect the shape of the signal which could present more than one peak, great attention should be paid to the calculation of travel time using this method.

Typical plots of the input and output waves from bender and extender element tests for the specimens with code name S11 and S12 are given in Figure 4a,b. A wave was produced by a displacement in the source transducer due to applied excitation voltage equal to 10 V or 14 V. Wave transmission created a displacement in the receiver, which in turn resulted in a voltage that could be measured. The frequency of the input signal was variable and decreased from the value f = 25 kHz to f = 2.0 kHz for S-wave velocity and from f = 100 kHz to f = 10 kHz for P-wave velocity. For each test frequency, five separate source element triggers were applied to the specimen, with the received signal output then stacked in the time domain to remove random signal noise [55].

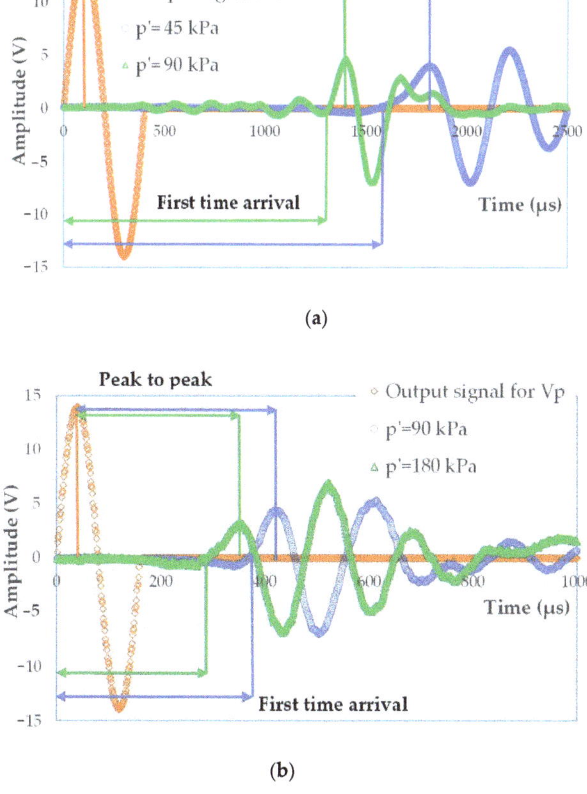

Figure 4. Example of the signal analysis/interpretation for (**a**) the bender element tests for specimen S11, p' = 45 kPa, f = 10 kHz, and (**b**) the extender element tests for specimen S12, p' = 90 kPa, f = 12.5 kHz.

From the data shown in Figure 4a, it can be easily seen that the maximum and first peaks do not coincide for tested sandy soils. These oscillations are derived from ambiguous peaks observed at the beginning of the received signal. This behavior [43] is due to reflected P-waves and does not represent the arrival of an S-wave [55]. The effect of the reflected wave is more pronounced for sandy soils than for any other kind of soil because of long reverberations and low damping. At this place, there is also a phenomenon called the near-field effect observed. The near-field component of the wave causes the transmitted wave to be distorted at its origin point and the wave starts with downward or upward deflection [56]. To minimize the near-field effect, as suggested in the literature [57,58], the higher input frequency and wavelength were used. However, in some examined cases, despite the triggering frequency higher than 2 kHz, the $\frac{L\mu}{\lambda}$ ratio was around 1 or even slightly less. At that time, the near-field effect was not avoided.

A comparison between the two selected different interpretation methods of shear and compression wave velocities is given in Figure 5. The data shown in this figure are clustered around the $45°$ lines, indicating an excellent agreement between the two analyzed methods. For more than half the number of the specimens (around 55% of the test results), higher shear wave velocity values were obtained from *PTP* method. The situation is similar in the case of compression wave velocity values. Here, for 61% of the data, V_P from the *PTP* method is greater than V_P from the *FTA* technique. Nevertheless, the differences between the results from both compared methods, for 99% of the results, are up to 10% for S-wave, and up to 20% for P-wave. The minimum difference between studied techniques in the case of V_S is 0.1 m·s^{-1}, whereas the maximum is -42.1 m·s^{-1}. In the case of V_P, the minimum difference is equal to 2.0 m·s^{-1}, whereas the maximum is -223.0 m·s^{-1}. From Figure 5 can be also seen that the *FTA* method of identification results in a relatively smaller variation as compared with *PTP* method.

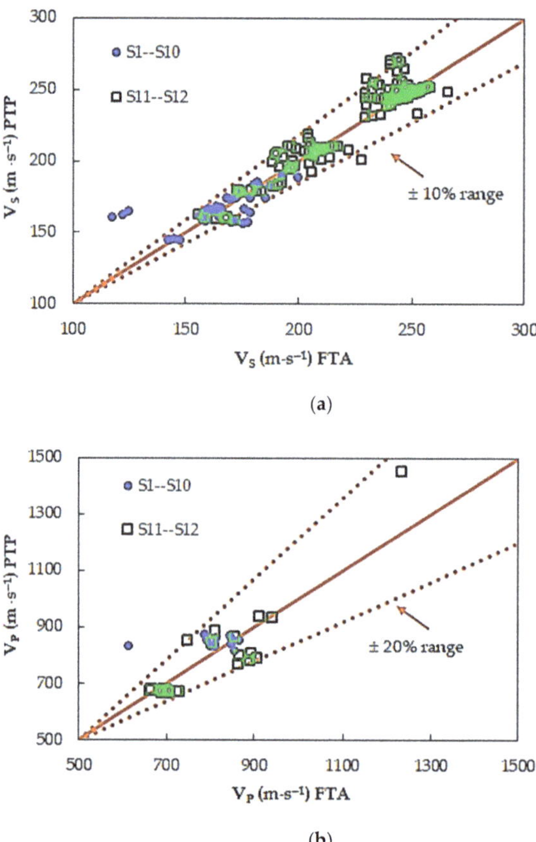

Figure 5. Comparison of (**a**) S-wave and (**b**) P-wave velocities, between the first time of arrival (FTA) and peak-to-peak (PTP) methods of signal analysis.

In Figure 6, shear wave velocities obtained for different source wave frequency for the test specimens S11 and S12 are presented. In this exemplary figure, a clear scatter of the results of V_S calculated based on two selected signal analysis methods depending on the source frequency can be perceived. In the case of the lowest frequencies used ($f < 3.3$ kHz), the significant variation of the results was obtained (from around 4 m·s^{-1} to 30 m·s^{-1} —specimen S11 and from around 11 m·s^{-1} to again around 30 m·s^{-1} —specimen S12). It is most likely related to the aforementioned phenomenon of the near field effect. This was explained, e.g., in the work of Sánchez-Salinero et al. [59]. The authors suggested the lower limit of frequency of 3.3 kHz as the value which corresponds to an approximate propagation distance-to-wavelength ratio equal to two, avoiding data that may include near field effects. In the case of specimen S11, a relatively large discrepancy between the S-wave velocities was noted for $f = 5$ kHz (average 14.5 m·s^{-1}), whereas for specimen S12, the scatter in estimated V_S values for this frequency as well for frequency $f = 10$ kHz was the minimal (average 0.1 m·s^{-1}). Therefore, for the investigated soils, it seems appropriate to choose a frequency closest to 10 kHz as the characteristic frequency. Generally, for all tested Warsaw glacial quartz sands, as the source frequency increases from the characteristic frequency, comparable V_S values can be received, regardless of the travel time determination technique adopted (here, *FTA* and *PTP*).

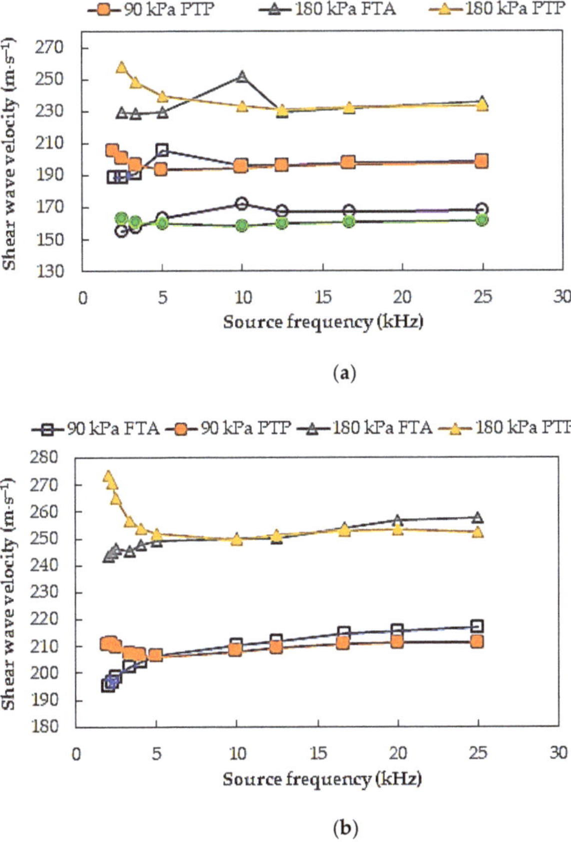

Figure 6. Shear wave velocities for different source frequencies and different mean effective stress, data for the test specimen: (**a**) S11, test series XI, steps 1, 2, 3, and (**b**) S12, test series XV, steps 1, 2.

In Figure 6, the variations in shear wave velocity values for different mean effective stress ($p' = 45, 90$, and 180 kPa) are shown beside. Shear wave velocities increase as the effective stress increases, as expected. On average, this increase for all obtained data is 16–17%.

To verify any possible influence of multi-stage consolidation process on the V_S results, shear wave velocities of specimen S12 from three consecutive consolidation stage, for mean effective stress $p' = 180$ kPa, corresponding to the test series, with numbers respectively XIV, XV, and XVI, are summarized in Figure 7. Based on the analysis of the results, it was found that the differences between the respective V_S values from individual test series of the same soil specimen were on average 4 m·s^{-1} (around 1%). The multi-stage consolidation procedure seems to be in fairly good agreement with the conventional one.

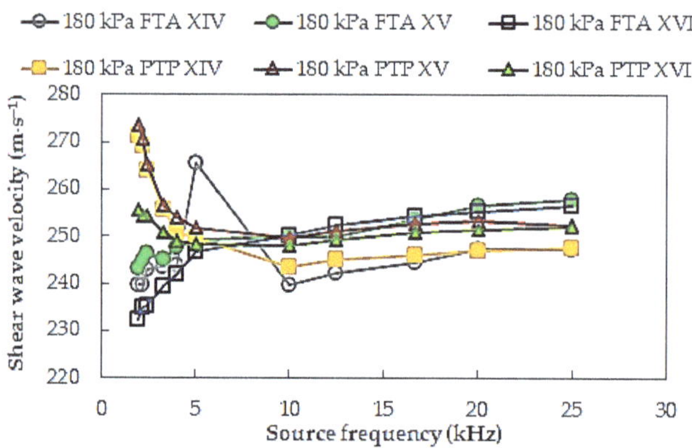

Figure 7. Shear wave velocities from multi-stage consolidation process.

3.1.2. Multi-Method Automated Tool for Travel Time Analyses—GDS BEAT

A lack of agreement when interpreting the S-wave travel time from the bender elements test triggered the development, by the British company GDS Instruments, of a new software tool to automate the interpretation process using several analysis methods recommended in the literature [12,14,60]. The main aim of this tool was to allow travel time estimations to be conducted objectively via a simple user interface, providing both visual and numerical representations of the estimated travel times. Implementation of the tool was completed by creating Add-Ins for Microsoft Excel: The Interactive Analysis tool and the Batch Analysis tool, a decision based on the ubiquitous use of the software. The details of the GDS Bender Element Analysis Tool user interface can be found in the work of Rees et al. [55].

For implementation, variations of the three different approaches were chosen:

- observation of points of interest within the received wave signal via software algorithm (time-domain technique),
- cross-correlation of the source and received signals (time-domain technique), and
- a cross-power spectrum calculation of the signals (frequency-domain method).
- The operation of the program is mainly based on one specific method of numerical analysis of the obtained results, using three factors:
- objective marking of points A, B, C, D (Figure 8) with the help of a software algorithm;
- mutual connection of generating and receiving elements signals; and
- calculation of the signal power curve spectrum for time estimation in the frequency-domain method.

The major first peak (point D) is located by scanning the received signal and determining the maximum, as well as the most positive output. The corresponding time signature defines point D. Next, point B is defined by scanning the wave signal from time zero up to point D and locating the minimum, as well as the most negative output. The time signature corresponding to this minimum thus describes point B. To find point C, it is required to scan the received wave signal between points B and D in such a way as to locate the output closest to zero. The corresponding time on the timeline is the one that is assigned to point C, whereas point A is computed during the interaction of the individual values. Starting at time zero, the mean and standard deviation of 10 consecutive outputs (e.g., n_1–n_{10}) are calculated, followed by 5 consecutive outputs (n_{11}–n_{15}). Subsequently, it is judged whether at least 3 standard deviations are more negative than the calculated mean. If it is true, the time signature of the first of the 5 subsequent outputs (i.e., n_{11}) is used to define point A. If it is false, the iteration proceeds by determining the mean and standard deviation of the next set of 10 consecutive outputs (i.e., n_2–n_{11}) until a "true" condition are reached [55].

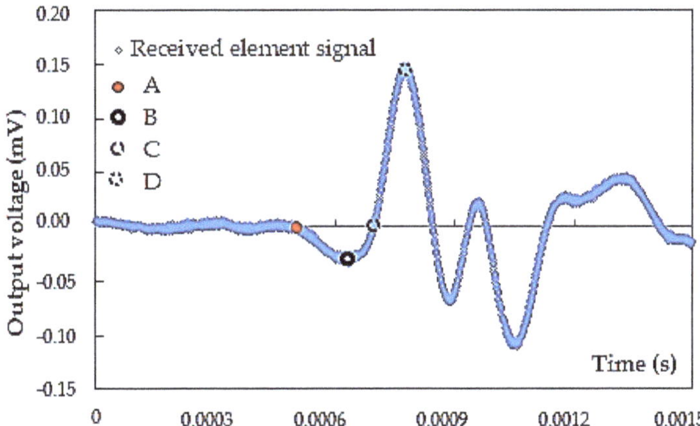

Figure 8. Receiver time charts: (**A**) the first deflection, (**B**) first bump, (**C**) zero-crossing, and (**D**) first major peak. An example of the test results from specimen S1, f = 4 kHz.

It is interesting how within BEAT the cross-correlation and the cross-power spectrum calculation of source and received wave signals work. Cross-correlation values are calculated from the source and received element signals at each data time stamp. The time at which the maximum calculated cross-correlation value occurs is then used as an estimate of the shear wave travel time. The cross-power spectrums, obtained via a Fast Fourier Transform (FFT), are used to create a phase angle versus frequency plot. The slope of this plot is then used to estimate the shear wave travel time, based on a linear best fit across a defined frequency window. Note, GDS BEAT automatically uses a frequency window of 0.8 to 1.2 times the specified source element frequency; however, this can be manually altered by the user when running the Interactive Analysis.

In Figures 9 and 10, a summary of the V_S values obtained by various techniques of travel time determination is presented. The data combined in Figure 9 concern the first ten specimens, namely, from S1 to S10, tested at p' = 45 kPa. Those in Figure 10, however, relate to two specimens, S11 and S12, isotropically consolidated to three mean effective stresses, p' = 45, 90, and 180 kPa, and additionally subjected to multi-stage consolidation. All the results are resumed here, regardless of the value of the $\frac{L_{tt}}{\lambda}$ ratio. By performing the GDS BEAT program, the results concerning the following four automated analysis methods were received: three time-domain techniques, namely, peak-to-peak, zero-crossing, and cross-correlation, and one frequency domain technique, i.e., cross-spectrum. These four methods were subsequently compared with two non-automated subjective analyses, including the first time of arrival method and the peak-to-peak technique (discussed in 3.1.1. Piezo-elements signal analysis). From both Figures 9 and 10, it can be noted that the V_S values are nonuniform for different source frequencies.

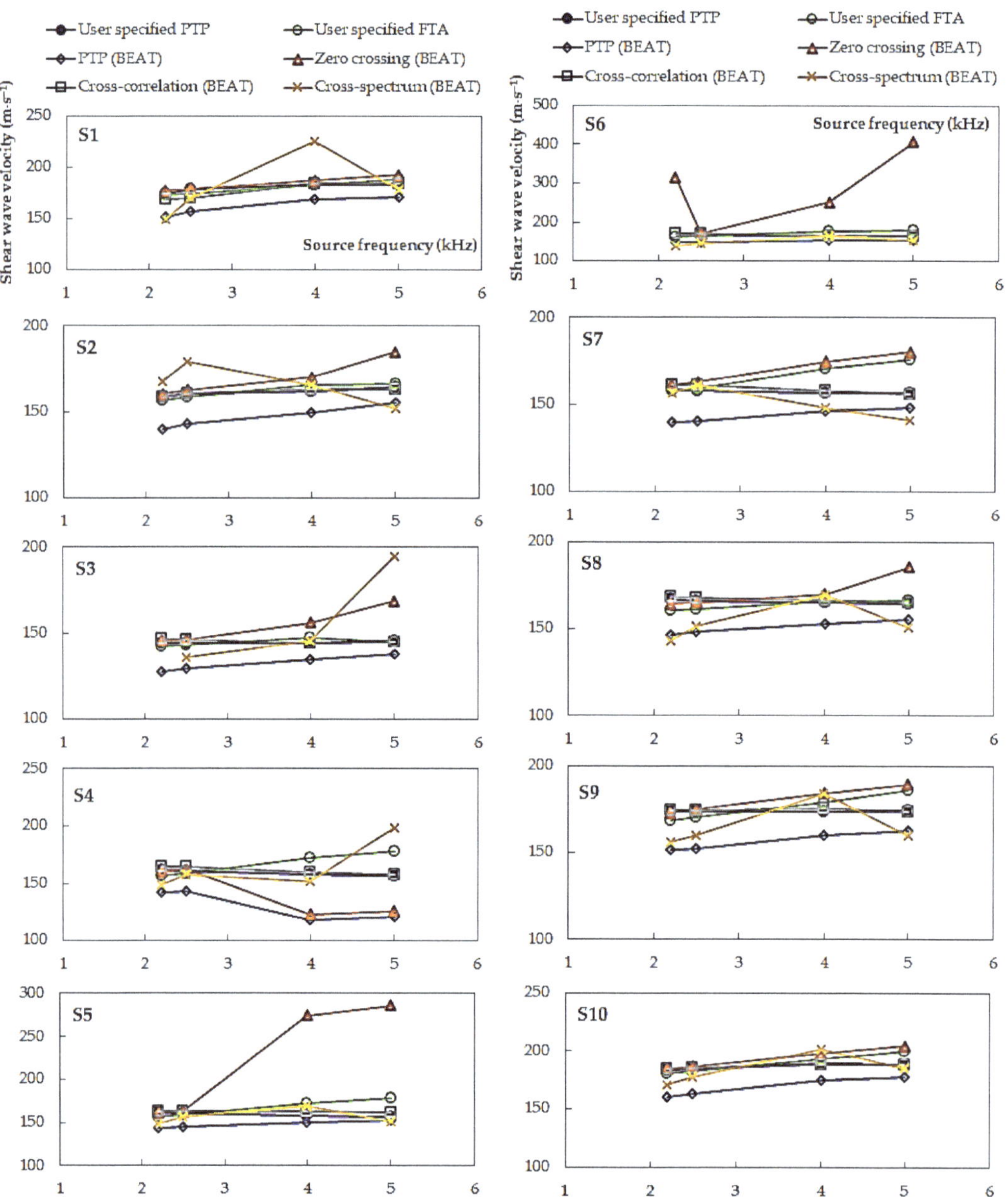

Figure 9. Shear wave velocities calculated from various methods of interpretation of *BE* test results, data for specimens S1–S10.

The values of V_S in Figure 9 indicate significant variation in the cost estimates obtained using the cross-spectrum calculation. The dispersion of the results for this method ranges from 15.8 m·s^{-1} (specimen S6) to 76.2 m·s^{-1} (specimen S1). The scatter observed from the zero-crossing method is also noteworthy, especially in the case of two specimen S5 (ΔV_S = 124.4 m·s^{-1}) and S6 (ΔV_S = 236.4 m·s^{-1}). For the rest of the tested specimens, however, these are the values at the level of around 20.0 m·s^{-1}. Conversely, the difference in the results obtained from the peak-to-peak automated analysis made by BEAT ($\pm$3.7 m·s^{-1})

and the cross-correlation function (± 5.0 m·s^{-1}) is relatively minimal. The V_S results in Figure 9 show also that in the case of Warsaw glacial quartz sand tested at the preset pressure of 45 kPa the smallest dispersion of shear wave velocity (± 16.0 m·s^{-1}) was recorded for the smallest source frequencies, i.e., f = 2.2 kHz and f = 2.5 kHz.

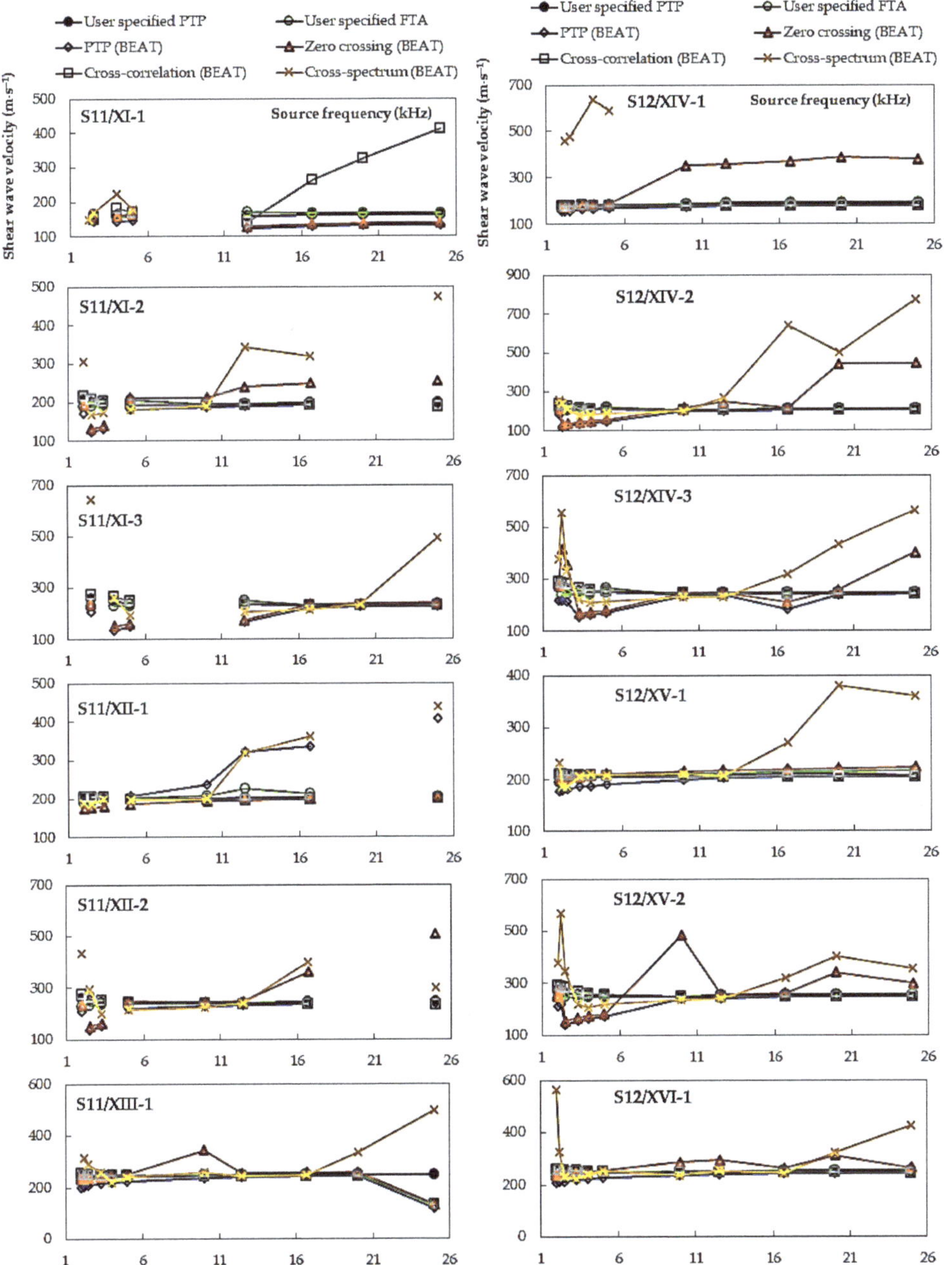

Figure 10. Shear wave velocities calculated from various methods of interpretation of BE test results, data for specimens S11–S12.

Inspection of Figure 10 indicates that the largest scatter of the results was obtained again for the cross-spectrum calculation, ranging from 193.3 m·s^{-1} (specimen S12/XV-2) to 593.2 m·s^{-1} (specimen S12/XIV-2). Analyzing the other methods, the zero-crossing technique is also characterized by a quite large discrepancy in the V_S values, where $\Delta V_{S,min}$ = 23.8 m·s^{-1} and $\Delta V_{S,max}$ = 356.3 m·s^{-1}. It may be suggested that estimates taken from these two above mentioned methods may be unreliable. In particular, the scatter observed from the cross-spectrum analyses has been previously reported following other studies [49]. Furthermore, the authors of the article themselves in their research devoted shear wave velocity of two types of anthropogenic material [42] received quite a large scatter in the submitted data from the frequency domain method. Therefore, it is quite dangerous to identify the arrival time only with this method. A significant decrease in the discrepancy of the results ($\pm$15.0 m·s^{-1}), for specimens S11 and S12, was obtained for the peak-to-peak method specified by the user. The scatter in calculations from the cross-correlation function for these two specimens was also rather small ($\pm$40.0 m·s^{-1}). These values suggest each method is relatively robust, an observation also made for the cross-correlation function after reviewing recent studies comparing analysis methods [15]. In the case of specimens S11 and S12, tested at three preset pressures (45, 90, and 180 kPa), the smallest dispersion of shear wave velocity ($\pm$131.0 m·s^{-1}) was recorded for the source frequencies equal to f = 3.3, 4.0, and 5.0 kHz. The highest scatter of the results were obtained in the case of both the lowest (f = 2 kHz) and the highest (f = 25 kHz) frequencies set in BET.

In Figure 11, the mean values of V_S calculated for all methods of interpretation listed in the article, after applying the frequency criterion, are presented. Therefore, wave velocities from the frequency of 3.3 kHz were used for further discussion of the results, avoiding data that may include near-field effects [59]. For most of the tests of Warsaw glacial quartz sands (for 99% sand specimens), the lowest average values of V_S were obtained from the peak-to-peak methods produced by BEAT. The highest V_S values for 11 specimens were gained from the frequency domain technique, whereas the remaining 9 specimens had their highest values from the zero-crossing method.

In the next order, the statistical analysis of all the data presented in Figure 11 was executed. The use of the frequency criterion did not significantly affect the obtained results. The notable variation in the mean V_S values resulted again in applying the cross-spectrum technique ($\pm$775.6 m·s^{-1}). A small spread of the results was once more provided by the peak-to-peak method specified by the user ($\pm$107.7 m·s^{-1}) and the cross-correlation function ($\pm$108.0 m·s^{-1}). The smallest one, however, was obtained using the first time of arrival method ($\pm$103.8 m·s^{-1}). This technique of travel time identification in BEs testing also gave the lowest standard deviation, amounting to 31.8 m·s^{-1}. After estimating standard error, for the first time of arrival method and the cross-correlation method, the standard error is insignificant: just 1.2 m·s^{-1}. This can mean that these two above analyzed interpretation methods are correct. An analysis of uncertainty was also required, to approve the precision and credibility of this study. The relative uncertainty in the range of 0.1% to 10% is typical for laboratory experiments [42]. Based on the data summarized in Figure 11, lower uncertainty of the results were gained for the time domain techniques, at the level of 16% (the first time of arrival method) and of 18% (the cross-correlation method and the peak-to-peak specified by user).

In this study, the comparison of four time-domain methods, namely, the user-specified peak-to-peak together with the peak-to-peak by BEAT (Figure 12), as well as the user-specified first time of arrival together with the zero-crossing (Figure 13), was completed. For each pair, the same methods are analyzed, but the first one is the subjective analysis made by the authors themselves during BE tests, whereas the second one is the automated analysis by the performance of BEAT. The minimum and maximum differences in the shear wave velocity values are included in the figures below. It is visible that these differences depend on the tested sand specimens. The average difference between the two peak-to-peak techniques was around 22.2 m·s^{-1}, which is 11%. The average difference between the first time of arrival and zero-crossing techniques was around 33.4 m·s^{-1}, which is 14%. The results obtained can be considered as preliminary results suggesting the use of BEAT may decrease subjectivity when interpreting travel times using standard observational techniques, while still allowing accurate estimates of the shear wave velocity values, keeping in mind the type of soil tested.

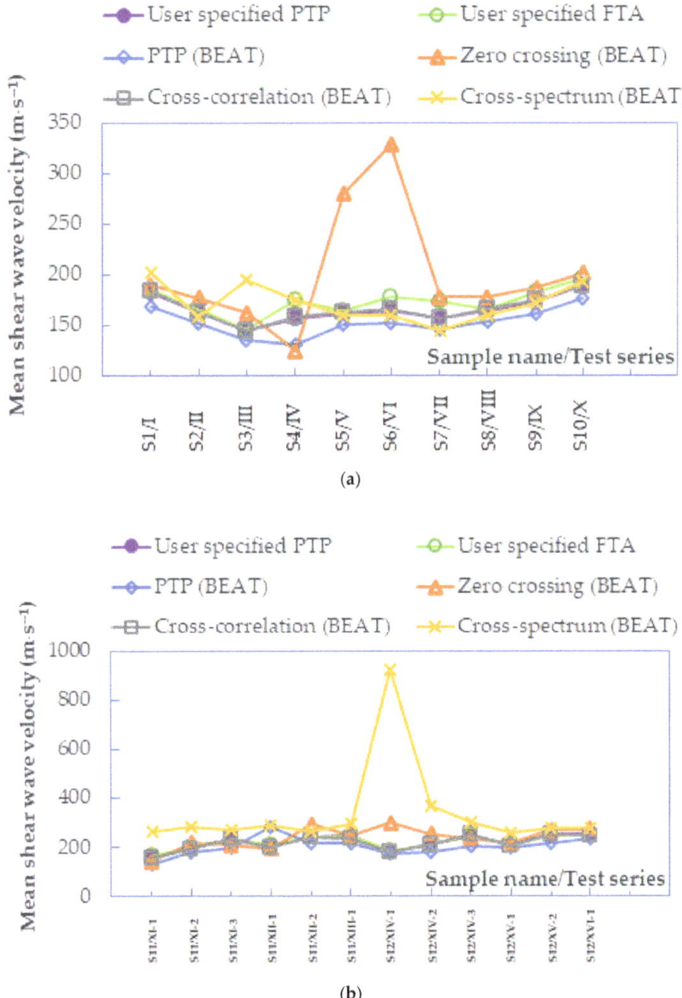

Figure 11. Summary of mean values of shear wave velocities from different interpretation methods when $f < 3.3$ kHz, data for specimens (**a**) S1–S10 and (**b**) S11–S12.

For further analysis, i.e., the influence of grain size characteristics on shear wave velocity of Warsaw glacial quartz sand, the mean values of V_S obtained for $p' = 45$ kPa were chosen. Moreover, such results were selected for which the quality of received signals was found to be satisfactory, also near-field effects and attenuation were found to be reduced. Then, the focus was put only on two interpreting travel times method: on the standard observational technique—the first time of arrival method (FTA)—and on the cross-correlation (CC) function obtained via BEAT analyses. These are the methods that provide the most consistent results for the studied soils.

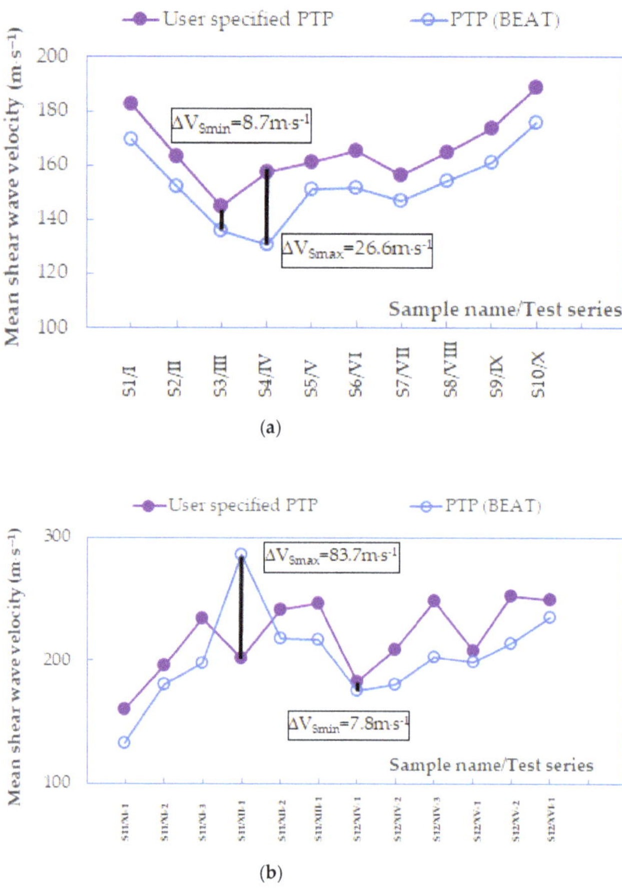

Figure 12. Mean values of shear wave velocities from peak-to-peak methods when $f < 3.3$ kHz, data for specimens (**a**) S1–S10 and (**b**) S11–S12.

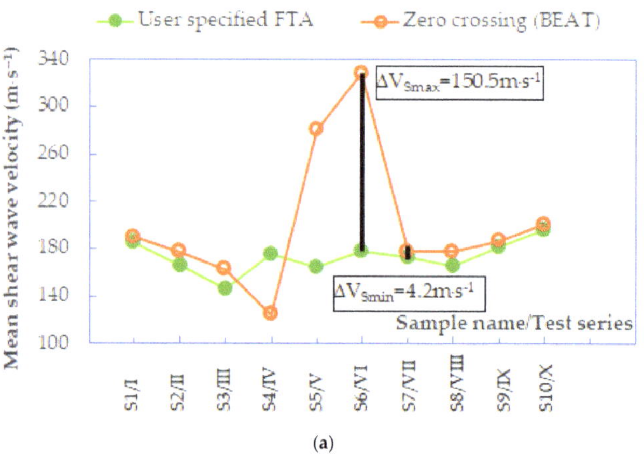

Figure 13. *Cont.*

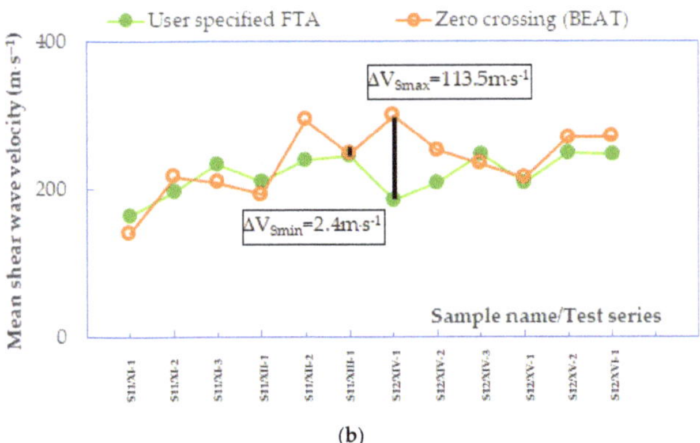

(b)

Figure 13. Mean values of shear wave velocities from the first time of arrival and zero-crossing methods when $f < 3.3$ kHz, data for specimens (**a**) S1–S10 and (**b**) S11–S12.

3.2. Effect of Grain Size Characteristics

In Figure 14, the values of mean shear wave velocity obtained for all tested Warsaw glacial quartz sands at $p' = 45$ kPa, from two selected time-domain methods for determining the wave travel time, versus the coefficient of curvature, are presented. This parameter took values from 1 to 4, whereas the test material was divided into three groups depending on the coefficient of uniformity. No clear dependence of the curvature coefficient (C_C) on the V_S values can be detected from the date in Figure 14. The highest values of V_S have examined sands characterized by $C_C = 2$, regardless of the C_U value. The smallest shear wave velocity was obtained for sands with $C_C = 3$. The greatest scatter of the V_S results characterizes the specimens with $C_C = 4$ ($\Delta V_{S,avg} = 10\%$) and $C_C = 2$ ($\Delta V_{S,avg} = 8\%$), in the case of the *FTA* method. The smallest, however, have the specimens with $C_C = 1$. When considering the results from the CC method, the significant variation in the V_S estimates, i.e., 10%, was gained for sands with $C_C = 1$, while the minimal one for sands with $C_C = 3$ ($\Delta V_{S,avg} = 5\%$).

In Figure 15, it is demonstrated that, in contrast to the curvature coefficient, the shear wave velocity of Warsaw glacial quartz sand is influenced by the uniformity coefficient (C_U). As C_U increased, shear wave velocity increased too. The only doubts can be raised by the results of sands with $C_C = 1$. In the case of the *FTA* method, the greatest scatter of the V_S results characterizes the very well-graded specimens, with $C_U = 16$ ($\Delta V_{S,avg} = 16\%$), while the smallest one characterizes the well-graded sands with $C_U = 14$ ($\Delta V_{S,avg} = 12\%$). The dispersion of the V_S values is more significant for the second considered method. Here, namely, sands with the lowest C_U are characterized by the greatest variability of V_S ($\Delta V_{S,avg} = 18\%$). On the other hand, for the specimens with $C_U = 14$, the scatter of the results is relatively minimal ($\Delta V_{S,avg} = 5\%$).

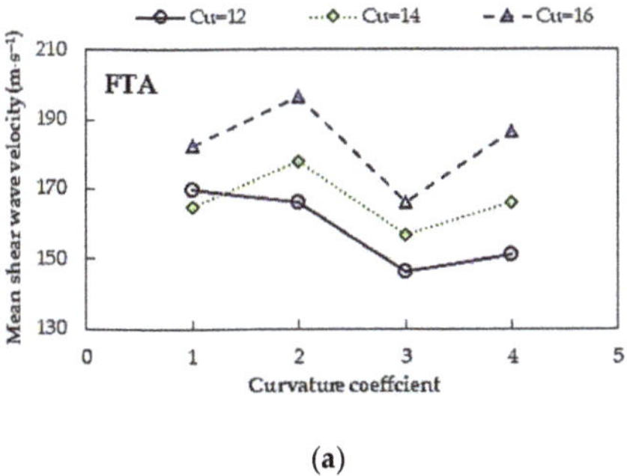

(a)

Figure 14. *Cont.*

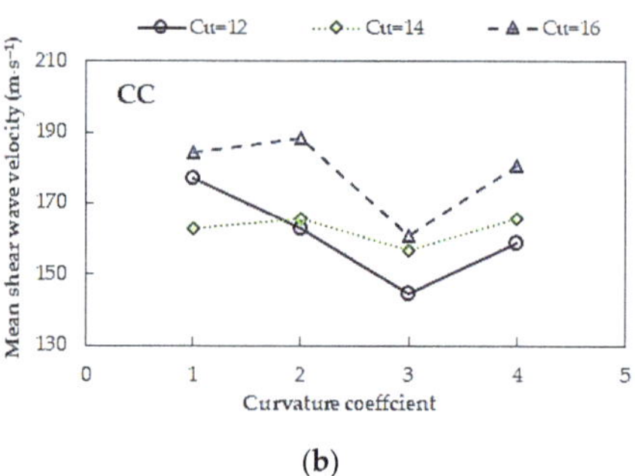

(b)

Figure 14. Effect of curvature coefficient (C_C) on mean shear wave velocity for sands of the same coefficient of uniformity (C_U) from: (**a**) first time of arrival method and (**b**) cross-correlation method.

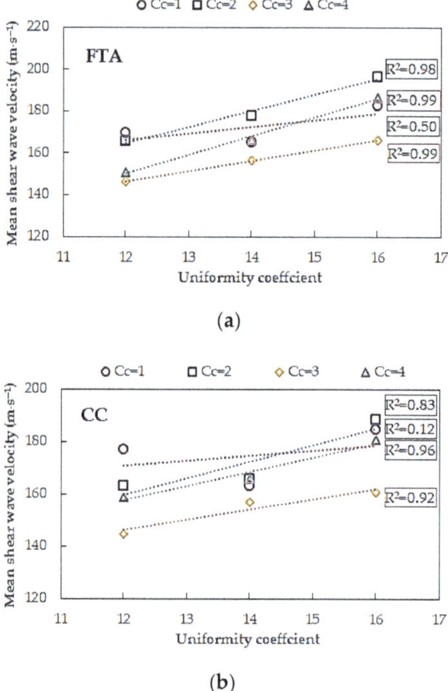

(a)

(b)

Figure 15. Effect of uniformity coefficient (C_U) on mean shear wave velocity for sands of the same coefficient of curvature (C_C) from: (**a**) first time of arrival method and (**b**) cross-correlation method.

In Figure 16, the mean shear wave velocity is plotted versus the mean grain size. The effect of d_{50} on the V_S values is hardly demonstrated in this study. In some of the tested materials, i.e., when $C_U = 12$, V_S decreased around 16% with increasing d_{50}. However, for Warsaw sands with $C_U = 14$ or $C_U = 16$, as d_{50} increased, an average of 9% to 15% increase in the V_S results was noted. This decrease in the V_S values is more visible than the increase.

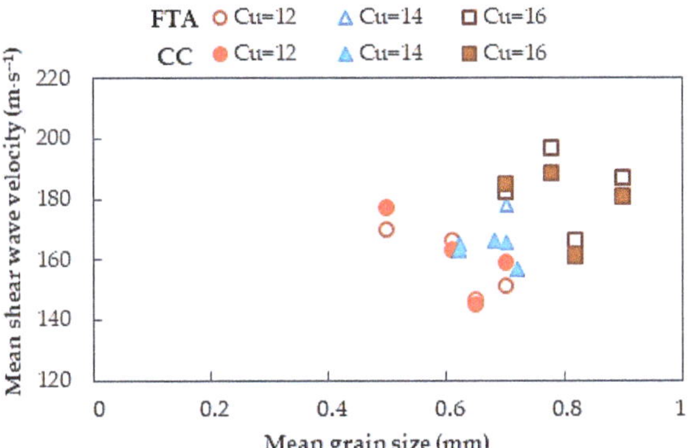

Figure 16. Effect of mean grain size (d_{50}) on mean shear wave velocity for sands from the first time of arrival method and cross-correlation method.

4. Concluding Remarks

The results of 12 clean quartz samples of sands from the Warsaw region presented in the article show, first of all, the complex specificity of performing shear wave velocity determination using the BET method. It is visible that despite the universality of this type of apparatus, special attention should be paid to the methodology of work and the correct selection of the test parameters. As has been shown, both the aspect of selecting the frequency of the test and the method of interpreting the results appropriate for a given geomaterial are very important. Users of the BET method should be aware of all factors that may influence the obtained results.

Concerning the difference in the arrival time identification method for Warsaw glacial quartz sandy soils, the time domain interpretation methods, namely, the first time of arrival (FTA) and the cross-correlation (CC) techniques, provide V_S results that are more consistent compared to the other methods. The scatter of the results for these two methods was undoubtedly smaller, even up to 7 times smaller than for the most questionable method, as it turned out here the cross-spectrum (CS). Quite small values of standard deviation and standard error allow us to conclude that these methods are relatively robust. The latter of these two techniques, i.e., CC, was used in the application of the GDS Bender Element Analysis Tool. The CS method proved to be the interpretative technique that must be used with great caution for the tested sandy samples.

If data at different frequencies are available within the sine input wave, the results of shear wave velocity for analyzed soils by input frequency closest to 10 kHz should be selected. A frequency close to this value 10 kHz is a characteristic frequency for Warsaw glacial quartz sandy soils.

It should be noted as well that BEAT can offer an accurate, objective explanation of BET data via a simple user interface. By using such a tool, a significant reduction in the time needed for the interpretation of the V_S results by different methods, at the same time, deserves a great emphasis. Automation is also a direction that allows objectification and popularization of the interpretation method.

A comprehensive experimental program has been performed using bender elements incorporated in the triaxial apparatus to define the combined effects of grain-size characteristics on the shear wave velocity of Warsaw glacial quartz sands too. As a measure of grain-size characteristics, three parameters were involved: the uniformity coefficient (C_U), the curvature coefficient (C_C), and the mean grain size (d_{50}). Despite the narrow range of variability of the particle size curves, the test results show that shear wave velocity is not affected by both the C_C and d_{50} of the tested material. In contrast, for most of the analyzed cases, the V_S values significantly increased with increasing the uniformity coefficient, with an average increase of 13.5%.

In future work, the laboratory tests will be extended to include Warsaw quartz sands with different grain size distribution curves. For each new material, tests with different pressures and densities are planned. After database expansion, the authors would like to examine some selected from the literature expressions of G_{max} to inspect if they can predict the measured values of G_{max} with the right level of accuracy. Additionally, comparative tests are planned to actually compare the stiffness of Warsaw quartz sands from various laboratory methods: bender elements (BE), resonant column (RC), and torsional shear (TS).

Author Contributions: Conceptualization, K.G. and E.S.; methodology, E.S.; formal analysis, K.G.; investigation, E.S. and K.G.; project administration, R.Š. and R.S.; resources, W.S.; data curation, K.G. and E.S.; writing—original draft preparation, K.G.; writing—review and editing, K.G. and W.S.; visualization, K.G.; supervision, W.S., R.Š., and R.S. All authors have read and agreed to the published version of the manuscript.

Funding: This research received no external funding.

Data Availability Statement: Data available on request due to their size properties. The data presented in this study are available on request from the corresponding author.

Conflicts of Interest: The authors declare no conflict of interest.

References

1. Kulkarni, M.P.; Patel, A.; Singh, D.N. Application of shear wave velocity for characterizing clays from coastal regions. *KSCE J. Civ. Eng.* **2010**, *14*, 307–321. [CrossRef]
2. Karray, M.; Hussien, M.N. Shear wave velocity as a function of cone penetration resistance and grain size for Holocene-age uncemented soils: A new perspective. *Acta Geotech.* **2017**, *12*, 1129–1158. [CrossRef]
3. Zekkos, D.; Sahadewa, A.; Woods, R.D.; Stokoe, K.H. Development of Model for Shear Wave Velocity of 419 Municipal Solid Waste. *J. Geotech. Geoenviron. Eng.* **2014**, *140*, 04013030. [CrossRef]
4. Lee, M.-J.; Choo, H.; Kim, J.; Lee, W. Effect of artificial cementation on cone tip resistance and small strain shear modulus of sand. *Bull. Eng. Geol. Environ.* **2011**, *70*, 193–201. [CrossRef]
5. Sas, W.; Gabryś, K.; Szymański, A. Experimental studies of dynamic properties of Quaternary clayey soils. *Soil Dyn. Earthq. Eng.* **2017**, *95*, 29–39. [CrossRef]
6. Ingale, R.; Patel, A.; Mandal, A. Performance analysis of piezoceramic elements in soil: A review. *Sens. Actuators A Phys.* **2017**, *262*, 46–63. [CrossRef]
7. Shirley, D.J.; Hampton, L.D. Shear-wave measurements in laboratory sediments. *J. Acoust. Soci. Am.* **1978**, *63*, 607–613. [CrossRef]
8. Kramer, S. *Geotechnical Earthquake Engineering*; Prentice-Hall: Upper Saddle River, NJ, USA, 1996.
9. Donovan, J.O.; Marketos, G.; Sullivan, C.O. Novel Methods of Bender Element Test Analysis. In *Geomechanics from Micro to Marco*; Soga, K., Kumar, K., Biscontin, G., Kuo, M., Eds.; Taylor & Francis Group: London, UK, 2015; pp. 311–316.
10. Camacho-Tauta, J.F.; Reyes-Ortiz, O.J.; Alvarez, J.D.J. Comparison between resonant-column and bender element tests on three types of soils. *Dyna* **2013**, *80*, 163–172.
11. Dyvik, R.; Madhus, C. Lab measurements of G_{max} using bender elements. In *Advance in the Art of Testing Soils under Cyclic Conditions*; Koshla, V., Ed.; ASCE: New York, NY, USA, 1985; pp. 186–196.
12. Yamashita, S.; Kawaquchi, T.; Nakata, Y.; Mikami, T.; Fujiwara, T.; Shibuya, S. Interpretation of interpretation parallel test on the measurement of G_{max} using bender element. *Soils Found.* **2009**, *49*, 631–650. [CrossRef]
13. Da Fonseca, A.V.; Ferreira, C.; Fahey, M. A framework interpreting bender element tests, combining time-domain and frequency-domain methods. *Geotech. Test. J.* **2009**, *32*, 1–17.
14. Arulnathan, R.; Boulanger, R.W.; Riemer, M.F. Analysis of bender element tests. *Geotech. Test. J.* **1998**, *21*, 120–131.
15. Viggiani, G.; Atkinson, J.H. Interpretation of bender element tests. *Géotechnique* **1995**, *45*, 149–154. [CrossRef]
16. Greening, P.D.; Nash, D.F.T. Frequency Domain Determination of G_0 Using Bender Elements. *Geotech. Test. J.* **2004**, *27*, 1–7. [CrossRef]
17. Airey, D.; Mohsin, A.K.M. Evaluation of Shear Wave Velocity from Bender Elements Using Cross-Correlation. *Geotech. Test. J.* **2013**, *36*, 506–514. [CrossRef]
18. Jovičic, V.; Coop, M.R.; Simic, M. Objective criteria for determining G_{max} from bender element tests. *Géotechnique* **1996**, *46*, 357–362. [CrossRef]
19. Blewett, J.; Blewett, I.J.; Woodward, P.K. Phase and Amplitude Responses Associated with the Measurement of Shear-Wave Velocity in Sand by Bender Elements. *Can. Geotech. J.* **2000**, *37*, 1348–1357. [CrossRef]
20. Godlewski, T.; Szczepański, T. *Metody Określania Sztywności Gruntów W Badaniach Geotechnicznych (In Polish) Methods for Determining Soil Stiffness in Geotechnical Investigations*; Poradnik ITB: Warsaw, Poland, 2015.
21. Camacho-Tauta, J.F.; Alvarez, J.D.J.; Reyes-Ortiz, O.J. A procedure to calibrate and perform the bender element test. *Dyna* **2012**, *79*, 10–18.

22. Arroyo, M. Pulse Tests in Soils Samples. Ph.D. Thesis, University of Bristol, Bristol, England, 2001.
23. Rio, J. Advances in Laboratory Geophysics Using Bender Elements. Ph.D. Thesis, University of London, London, UK, 2006.
24. Sas, W.; Gabryś, K.; Soból, E.; Szymański, A. Dynamic Characterization of Cohesive Material Based on Wave Velocity Measurements. *Appl. Sci.* **2016**, *6*, 49. [CrossRef]
25. Boonyatee, T.; Chan, K.H.; Mitachi, T. Effect of bender element installation in clay samples. *Géotechnique* **2010**, *60*, 287–291. [CrossRef]
26. Yang, J.; Gu, X.Q. Shear stiffness of granular material at small strains: Does it depend on grain size? *Géotechnique* **2013**, *63*, 165–179. [CrossRef]
27. Liu, X.; Yang, J.; Wang, G.H.; Chen, L.Z. Small-strain shear modulus of volcanic granular soil: An experimental investigation. *Soil Dyn. Earthq. Eng.* **2016**, *86*, 15–24. [CrossRef]
28. Ruan, B.; Miao, Y.; Cheng, K.; Yao, E.-l. Study on the small strain shear modulus of saturated sand-fines mixtures by bender element test. *Eur. J. Environ. Civ. Eng.* **2018**, *25*, 1–11. [CrossRef]
29. Cho, G.-C.; Dodds, J.; Santamarina, C.J. Particle shape effects on packing density, stiffness, and strength: Natural and crushed sands. *J. Geotech. Geoenviron. Eng.* **2006**, *132*, 591–602. [CrossRef]
30. Krumbein, W.C.; Sloss, L.L. *Stratigraphy and Sedimentation*, 2nd ed.; Freeman: San Francisco, CA, USA, 1963.
31. Liu, X.; Yang, J. Shear wave velocity in the sand: Effect of grain shape. *Géotechnique* **2017**, *68*, 1–7. [CrossRef]
32. Wichtmann, T.; Triantafyllidis, T. Influence of the grain-size distribution curve of quartz sand on the small-strain shear modulus G_{max}. *J. Geotech. Geoenviron. Eng.* **2009**, *135*, 1404–1418. [CrossRef]
33. Iwasaki, T.; Tatsuoka, F. Effects of grain size and grading on dynamic shear moduli of sands. *Soils Found.* **1977**, *17*, 19–35. [CrossRef]
34. Patel, A.; Bartake, P.; Singh, D. An Empirical Relationship for Determining Shear Wave Velocity in Granular Materials Accounting for Grain Morphology. *Geotech. Test. J.* **2009**, *32*, 1–10. [CrossRef]
35. Bartake, P.P.; Singh, D.N. Studies on the determination of shear wave velocity in sands. *Geomech. Geoengin.* **2007**, *2*, 41–49. [CrossRef]
36. Sharifipour, M.; Dano, C.; Hicher, P.Y. Wave velocities in assemblies of glass beads using bender-extender elements. In Proceedings of the 17th ASCE Engineering Mechanics Conference, Newark, DE, USA, 13–16 June 2004.
37. Menq, F.Y.; Stokoe, K.H., II. Linear dynamic properties of sandy and gravelly soils from large-scale resonant tests. In *Deformation Characteristics of Geomaterials*; Benedetto, D., Doanh, T., Geoffroy, H., Sauzéat, C., Eds.; Swets & Zeitlinger: Lisse, The Netherlands, 2003; pp. 63–71.
38. Lontou, P.B.; Nikolopoulou, C.P. *Effect of Grain Size on Dynamic Shear Modulus of Sands: An Experimental Investigation (In Greek)*; Department of Civil Engineering, University of Patras: Patras, Greece, 2004.
39. Altuhafi, F.N.; Coop, M.R.; Georgiannou, V.N. Effect of Particle Shape on the Mechanical Behavior of Natural Sands. *J. Geotech. Geoenviron. Eng.* **2016**, *142*, 04016071. [CrossRef]
40. Payan, M.; Khoshghalb, A.; Senetakis, K.; Khalili, N. Effect of particle shape and validity of G_{max} models for sand: A critical review and a new expression. *Comput. Geotech.* **2016**, *72*, 28–41. [CrossRef]
41. Wang, M.; Pande, G.; Kong, L.; Feng, Y. Comparison of Pore-Size Distribution of Soils Obtained by Different Methods. *Int. J. Geomech.* **2016**, *17*, 06016012. [CrossRef]
42. Clayton, C.R.I. Stiffness at small strain: Research and practice. *Géotechnique* **2011**, *61*, 5–38. [CrossRef]
43. Wang, Y.; Benahmed, N.; Ciu, Y.-J.; Tang, A.-M. A novel method for determining the small-strain shear modulus of soil using bender elements technique. *Can. Geotech. J.* **2016**, *54*, 280–289. [CrossRef]
44. PN-EN ISO 14688-2: 2006. Badania Geotechniczne. Oznaczanie I klasyfikowanie Gruntów. Część 2: Zasady Klasyfikowania. Available online: https://sklep.pkn.pl/pn-en-iso-14688-2-2018-05p.html (accessed on 27 November 2019).
45. Szymański, A. *Mechanika Gruntów*; Wydawnictwo SGGW: Warsaw, Poland, 2007.
46. ISO 17892-4:2016 Geotechnical Investigation and Testing—Laboratory Testing of Soil—Part 4: Determination of Particle Size Distribution. Available online: https://sklep.pkn.pl/pn-en-iso-17892-4-2017-01e.html (accessed on 18 January 2018).
47. PN-88/B-04481. Grunty Budowlane. Badania Próbek Gruntu. Available online: https://sklep.pkn.pl/pn-b-04481-1988p.html (accessed on 30 June 1988).
48. Leong, E.C.; Cahyadi, J.; Rahardjo, H. Measuring shear and compression wave velocities of soil using bender-extender elements. *Can. Geotech. J.* **2009**, *46*, 792–812. [CrossRef]
49. Tatsuoka, T.; Iwasaki, T.; Yoshida, S.; Fukushima, S.; Sudo, H. Shear modulus and damping by drained test on clean sand specimens reconstituted by various methods. *Soils Found.* **1979**, *19*, 39–54. [CrossRef]
50. Gabryś, K.; Sas, W.; Soból, E.; Głuchowski, A. Application of bender elements technique in the testing of anthropogenic soil-recycled concrete aggregate and its mixture with rubber chips. *Appl. Sci.* **2017**, *7*, 741. [CrossRef]
51. Pennington, D.S.; Nash, D.F.T.; Lings, M.L. Horizontally Mounted Bender Elements for Measuring Anisotropic Shear Moduli in Triaxial Clay Specimens. *Geotech. Test. J.* **2001**, *24*, 133–144. [CrossRef]
52. Wang, Y.H.; Lo, K.F.; Yan, W.M.; Dong, X.B. Measurement Biases in the Bender Element Test. *J. Geotech. Geoenviron. Eng.* **2007**, *133*, 564–574. [CrossRef]
53. Leong, E.C.; Yeo, S.H.; Rahardjo, H. Measuring Shear Wave Velocity Using Bender Elements. *Geotech. Test. J.* **2005**, *28*, 488–498. [CrossRef]

54. Ogino, T.; Kawaguchi, T.; Yamashita, S.; Kawajiri, S. Measurement Deviations for Shear Wave Velocity of Bender Element Test Using Time Domain, Cross-Correlation, and Frequency Domain Approaches. *Soils Found.* **2015**, *55*, 329–342. [CrossRef]
55. Rees, S.; Le Compte, A.; Snelling, K. A new tool for the automated travel time analyses of bender element tests. In Proceedings of the 18th International Conference on Soil Mechanics and Geotechnical Engineering, Paris, France, 2–6 September 2013; pp. 2843–2846.
56. Lee, J.S.; Santamarina, J.C. Bender elements: Performance and signal interpretation. *J. Geotech. Geoenviron. Eng.* **2005**, *131*, 1063–1070. [CrossRef]
57. Patel, A.; Singh, D.N.; Singh, K.K. Performance analysis of piezo-ceramic elements in soils. *Geotech. Geol. Eng.* **2010**, *28*, 681–694. [CrossRef]
58. Brignoli, E.; Gotii, M.; Stokoe, K. Measurement of shear waves in laboratory specimens by means of piezoelectric transducers. *Geotech. Test. J.* **1996**, *19*, 384–397.
59. Sánchez-Salinero, I.; Roesset, J.M.; Stokoe, K.H., II. *Analytical Studies of Body Wave Propagation and Attenuation. Geotechnical Engineering Report*; GR86-15; University of Texas: Austin, TX, USA, 1986.
60. Styler, M.A.; Howie, J.A. Comparing frequency and time domain interpretations of bender element shear wave velocities. In Proceedings of the GeoCongress 2012, Oakland, CA, USA, 25–29 March 2012.

 *materials*

Article

Determination of the Shear Modulus of Pine Wood with the Arcan Test and Digital Image Correlation

Piotr Bilko *, Aneta Skoratko, Andrzej Rutkiewicz and Leszek Małyszko

Department of Mechanics and Building Structures, Institute of Geodesy and Civil Engineering, Faculty of Geoengineering, University of Warmia and Mazury in Olsztyn, 10-719 Olsztyn, Poland; aneta.skoratko@uwm.edu.pl (A.S.); andrzej.rutkiewicz@uwm.edu.pl (A.R.); leszek.malyszko@uwm.edu.pl (L.M.)
* Correspondence: piotr.bilko@uwm.edu.pl

Abstract: Arcan shear tests with digital image correlation were used to evaluate the shear modulus and shear stress–strain diagrams in the plane defined by two principal axes of the material orthotropy. Two different orientation of the grain direction as compared to the direction of the shear force in specimens were considered: perpendicular and parallel shear. Two different ways were used to obtain the elastic properties based on the digital image correlation (DIC) results from the full-field measurement and from the virtual strain gauges with the linear strains: perpendicular to each other and directed at the angle of $\pi/4$ to the shearing load. In addition, the own continuum structural model for the failure analysis in the experimental tests was used. Constitutive relationships of the model were established in the framework of the mathematical multi-surface elastoplasticity for the plane stress state. The numerical simulations done by the finite element program after implementation of the model demonstrated the failure mechanisms from the experimental tests.

Keywords: digital image correlation; Arcan shear test; wood; orthotropic shear modulus; elastic-plastic material; finite element method

Citation: Bilko, P.; Skoratko, A.; Rutkiewicz, A.; Małyszko, L. Determination of the Shear Modulus of Pine Wood with the Arcan Test and Digital Image Correlation. *Materials* **2021**, *14*, 468. https://doi.org/10.3390/ma14020468

Received: 30 November 2020
Accepted: 13 January 2021
Published: 19 January 2021

Publisher's Note: MDPI stays neutral with regard to jurisdictional claims in published maps and institutional affiliations.

Copyright: © 2021 by the authors. Licensee MDPI, Basel, Switzerland. This article is an open access article distributed under the terms and conditions of the Creative Commons Attribution (CC BY) license (https://creativecommons.org/licenses/by/4.0/).

1. Introduction

Wood is an organic, naturally grown material and is commonly used for creating all kinds of goods and structures in many branches of industry. Softwood, which is mainly used for structural and load-bearing purposes in civil engineering, at a micro-scale level, is built from axial tracheids connected between themselves by a lignin matrix. The tracheids (see Figure 1b, which shows a tracheid in a perpendicular cut) are long, thin cells organized in a way that their length is parallel to the length of the log and are the main source of the wood strength. They are "glued" by lignin at the edges of its cell walls and create the annual rings. The micro-scale built is a basis for understanding the macro behavior and strength of clear wood (i.e., a material considered as without flaws, e.g., resin pores or knots), which is generally high in the longitudinal direction (denoted as L), where the tracheid's generate strength, and low in the two other directions, i.e., radial and tangential (denoted by R and T, respectively), where the lignin matrix has lower mechanical properties (see Figure 1b,c for direction denoting). Therefore, the weakest mechanical properties of wood are those at the direction normal to the fibers during tension (i.e., R and T) or shear along the longitudinal direction (i.e., L), causing a rupture between annual rings. Exposure to these types of stresses easily leads to cracking, which usually forms along the grain direction, choosing the path of least resistance.

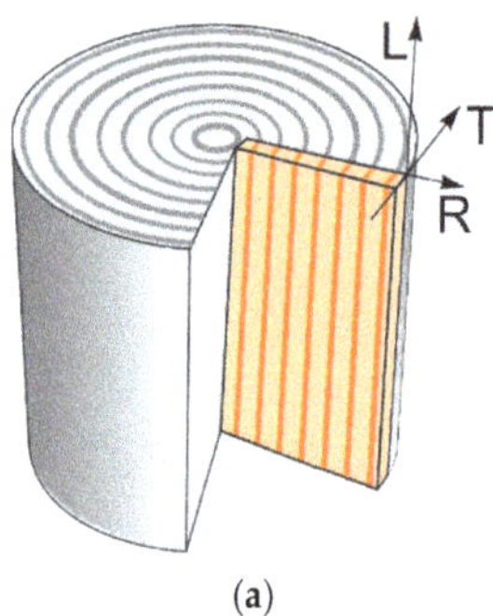

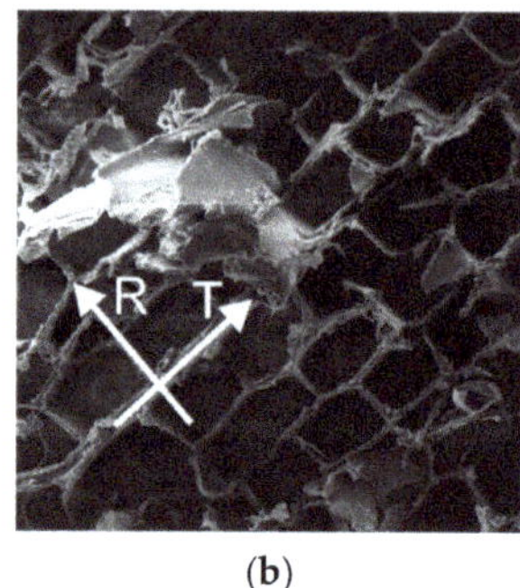

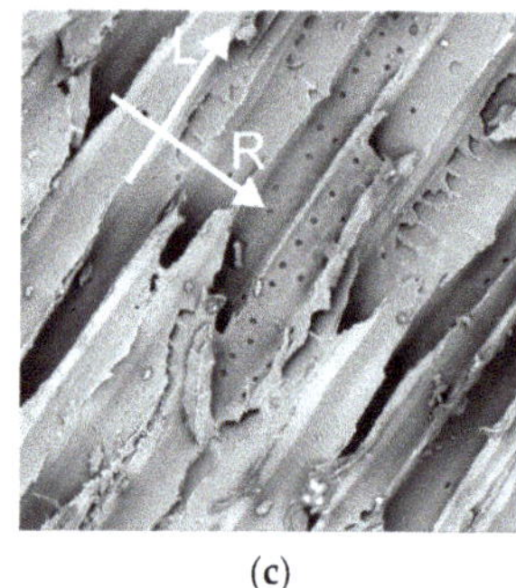

(a) (b) (c)

Figure 1. The wood material axes on a log view (**a**); scanning electron microscope photograph of a RT plane at 746 times magnitude (**b**); and scanning electron microscope photograph of a LR plane at 533 times magnitude (**c**).

The heterogeneity, orthotropy and high variability of naturally grown wooden materials makes both modeling and experimental investigations challenging. The natural origin of wood, being its major advantage, is also a major obstacle in the advancement of wood research. The lack of manufacturing control of the material properties, as is possible in artificial materials, as well as the complicated internal structure of wood from micro- to macro-scale, gives a substantial level of uncertainty in the interpretations of test results and model approaches [1,2]. The clear wood is often treated mechanically as an orthotropic material with three specified material axes (Figure 1a), i.e., the longitudinal (L), radial (R) and tangential (T) ones. Its macroscopic behavior originates from its microscopic structure—fibers—and their directional arrangement (Figure 1b,c). Moreover, in one annual ring, the mechanical characteristics are different due to early (spring) and late (autumn) growth characteristics of this ring. The earlywood grows more quickly and is weaker, which is in opposition to the latewood. In general, it rises the additional issue of material inhomogeneity, which is typical for materials of natural origin (e.g., wood and soil). It is also an issue where is the limit of considering wood as a homogeneous material, which is a common engineering practice. This is important especially while examining the shearing of LR plane of orthotropy with the direction of the shearing load (P) parallel with the longitudinal material axis (L). A crack may lie in the LR plane and may propagate in one of two directions, from which more practical importance has the one along the lower strength path parallel to the grain. This system of propagation, where L is the direction in which the crack propagates (the LR,L system), will predominate as a result of the low strength and stiffness of wood perpendicular to the grain. The opposite of this is the LR,R system, where the crack propagates in the direction R.

In general, the experimental determination of the wood behavior in shear has always been influenced by difficulties in obtaining a pure and uniform shear state. This issue resulted in many different experimental methods: the Arcan test, off-axis tests, Iosipescu test, four-point bending test, etc. [3]. Numerical simulations, performed for tests, usually indicate a combination of normal and shear stresses, making difficult to interpret the pure shear behavior. Among the mentioned methods, the Arcan shear test [4] is considered to create a rather uniform and pure state of shear stress among the critical cross section. The main problem arises with boundary conditions, which are strongly dependent on the type of specimen fixture and the distance to the critical cross section, and can influence the behavior of the specimen. The Arcan test on wood has been studied (see, e.g., [5–9]), where strains are measured using strain gauges [6,9], with video extensometers [5] or not measured at all [7]. Tests with the digital image correlation were used by the authors of [8,10]. The work in [6] is of significant importance because it gives the shear constants in all three material planes at two load-to-material axes directions, which are rarely obtained due to the labor-intensive nature of such tests.

Although shearing tests have been performed using many techniques, the increasement in measuring technology makes it possible today to gather more information and obtain new results. Technology of the digital image correlation (DIC) enables recording and analyzing the whole surface of the specimen, on both sides [11]. The possibilities can be shown in a simple example presented in Figure 2. Measurements using strain gauge T-rosettes enable measuring two values at one "point" (this point is distributed along the strain gauge grid length) (Figure 2a) [7]. The DIC enables measuring the displacements (and further calculate strains) of approximately 2400 points in the observed area. However, the accuracy of the system is still studied.

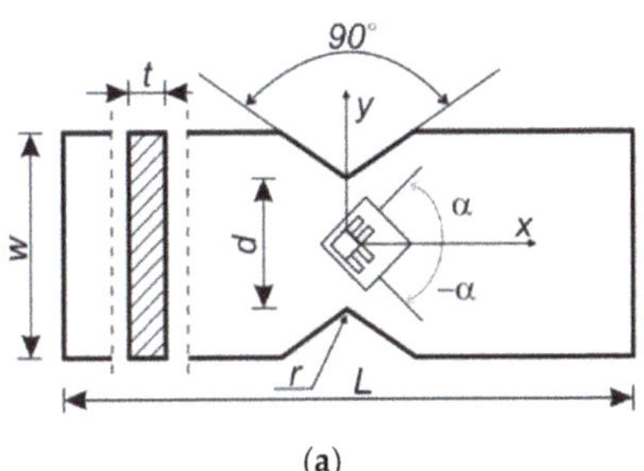

(a)

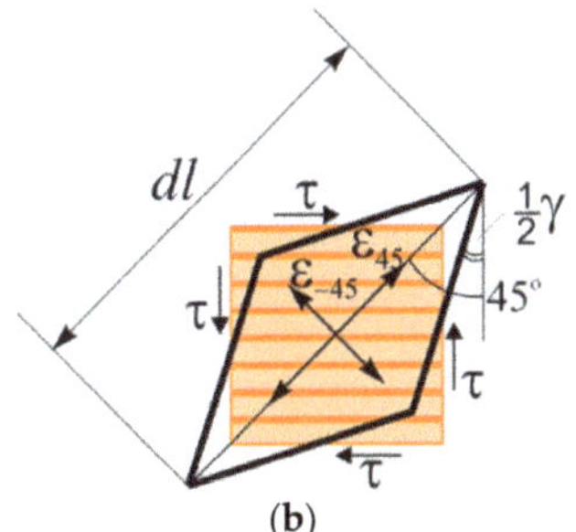

(b)

Figure 2. The standard shear modulus test specimen-strain gauge T-rosette and a way of measurement of the shear angle (**a**); and a state of pure shear (**b**). All dimensions are in millimeters (mm) unless otherwise noted.

In this paper, apart from the experimental studies described in Section 2, the results of numerical simulations of the performed tests are also presented in Section 3. The simulations were carried out using the own orthotropic material model of clear wood, as discussed in [12]. Three basic failure mechanisms in plane stress are distinguished in the model: failure due to tensile, compressive and shear stresses. The composite failure criterion consists of three analytical expressions, each of them being a limit equilibrium condition of the material in a complex stress state.

2. Experimental Studies

2.1. Background Theory

The 2D strains components, the normal strains $\varepsilon_x, \varepsilon_y$ and the shear strain ε_{xy}, are directly calculated in the software of the digital image correlation system [13] from the symmetrical material stretch tensor **U**:

$$\mathbf{U} = \sqrt{\mathbf{F}^T\mathbf{F}} = \begin{pmatrix} 1+\varepsilon_x & \varepsilon_{xy} \\ \varepsilon_{xy} & 1+\varepsilon_y \end{pmatrix}, \tag{1}$$

where **F** is the deformation gradient. The shear angle γ_{xy} without the rigid rotation is calculated as:

$$\gamma_{xy} = \gamma_x + \gamma_y = \arctan\left(\frac{\varepsilon_{xy}}{1+\varepsilon_x}\right) + \arctan\left(\frac{\varepsilon_{yx}}{1+\varepsilon_y}\right), \tag{2}$$

where γ_x and γ_y are the corresponding shear angles of the two sides of the deformed elemental square (see also [7]). Note that the software gives the strains and angles at the certain points referring to an arbitrary coordinate system (x, y).

The constitutive law of linear elasticity for the orthotropic material is determined by nine independent material parameters because of the strain and stress tensors symmetries and the existence of the elasticity energy function. As is the case for the orthotropic material, the formulas of Hooke's law depend on the orientation of the coordinate system in a reference to the principal axes of material symmetry, i.e., the axes of orthotropy. The shear moduli in the frame of reference aligned with the orthotropic axes, so-called the technical moduli, can be calculated independently of other material constants as:

$$G_{LR} = \frac{\tau_{LR}}{\gamma_{LR}}, \ G_{LT} = \frac{\tau_{LT}}{\gamma_{LT}}, \ G_{RT} = \frac{\tau_{RT}}{\gamma_{RT}}, \tag{3}$$

where $\tau_{LR}, \tau_{LT}, \tau_{RT}$ are the shear components of the stress tensor in the planes LR, LT, RT, respectively, and $\gamma_{LR}, \gamma_{LT}, \gamma_{RT}$ are the corresponding shear angles.

To determine the shear angle γ in the chosen plane, the following engineering geometrical considerations is used additionally. Let us consider an infinitely small element in the plane LR, which is in a state of pure shear (Figure 2b). When shearing, the right angles change by the value $\gamma = \gamma_{LR}$. The one diagonal is then lengthened with the strain ε_{45} and the second diagonal shortens with the strain ε_{-45} and:

$$\tan\left(\frac{\pi}{4} - \frac{\gamma}{2}\right) = \frac{dl(1 + \varepsilon_{-45})}{dl(1 + \varepsilon_{45})} \approx \frac{1 - \gamma/2}{1 + \gamma/2}, \tag{4}$$

where dl is the diagonal length of the element. From (4), we get:

$$\gamma = \frac{2(\varepsilon_{45} - \varepsilon_{-45})}{2 + \varepsilon_{45} + \varepsilon_{-45}}. \tag{5}$$

To derive the engineering shear angle γ in the DIC system, the construction of the virtual strain gauges is required in the same way as for two-element strain gauge rosettes, e.g., for a $10 \times 10 \ \mathrm{mm}^2$ square, where $\varepsilon_{45}, \varepsilon_{-45}$ are the linear strains, perpendicular to each other and directed at 45° angle to the shearing load.

2.2. Specimens, Equipment and Methods

The dimensions of the specimen were preliminarily defined by the basic numerical tests on several different configurations of the critical cross-section height and curvature. Nine different shapes were modeled checking stress distribution by means of the finite element method. The dimensions modified were: the shear area height $h \to \{36, 40\}$ mm, the initial diameter $d \to \{4, 8, 12\}$ mm and the inclination angle $\varphi \to \{45°, 60°, 75°, 90°\}$ of the cutting lines (see Figure 3a). The aim was to achieve a pure shear state in the middle section of the specimen; hence, it was sufficient to adopt an isotropic material and perform the simplified analysis only in the elastic range. The most satisfactory results were obtained for the following dimensions: $h = 36$ mm, $d = 8$ mm and $\varphi = 90°$ (all dimensions are shown in Figure 3d). The obtained tangential stress distribution was characterized by low variability with practically zero values of the associated normal stresses. The stress distributions for the middle cross-section are shown in Figure 3e,f.

Storage and processing of wood specimens were performed in normal conditions of 65% relative humidity and temperature of 20 °C. The pre-specimens were firstly cut from 16 mm planks (planks cut from the central part of the log) of pine wood (*Pinus sylvestris* L.) with rectangular dimensions of 100 mm × 150 mm concerning two perpendicular directions of LR plane (in Figure 3b,c, the arrows show the shearing directions with respect to the material axes). Further, the notches were made using a milling-machine with a down spindle and a saw to create the required shape (Figure 3d). The tests were performed immediately after the transportation to the testing facility and prepared for the DIC measurements. Therefore, the surface of the specimens was sprayed using a black aerosol can to create a more stochastic pattern. Since wood has an inhomogeneous surface, there little paint was needed. Further, the specimens were inserted into the Arcan fixtures (Figure 4). The fixture dimensions and shape were designed by the authors and water cut from an 8mm thick stainless-steel plate. The fixture elements were connected by 8 and 12 mm steel screws with steel plates used as distances, while the specimens were tightened by steel tooth plates (Figure 4d).

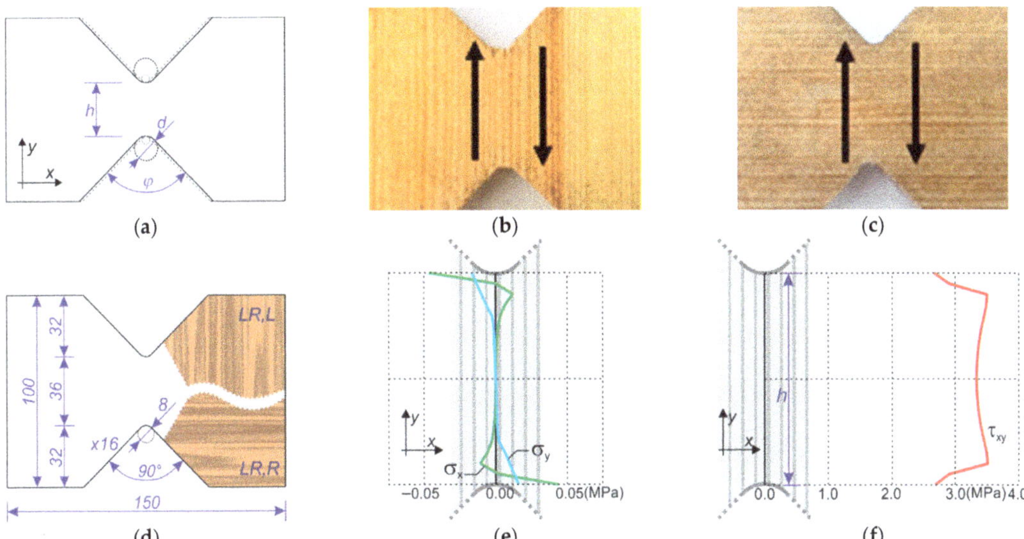

Figure 3. The specimens: (**a**) dimensions subjected to variation; (**b**) LR,L orientation; (**c**) LR,R orientation; (**d**) geometry of specimen used for shear tests; and (**e**,**f**) stress distributions in the middle section for LR,L specimens. All dimensions are in millimeters (mm) unless otherwise noted.

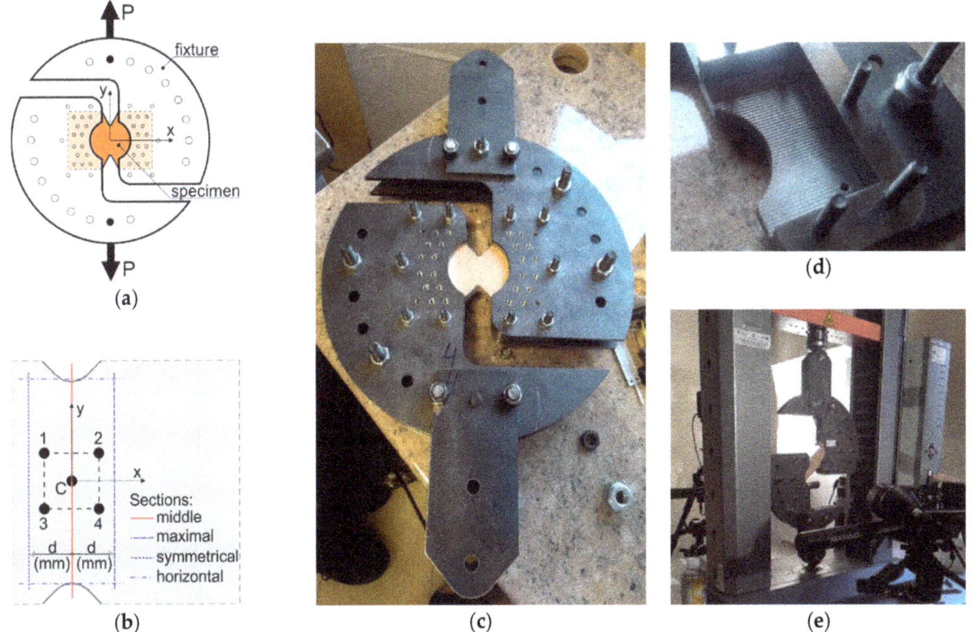

Figure 4. The Arcan fixture: (**a**) scheme; (**b**) the explanation of measuring points/sections; (**c**) physical fixture; (**d**) distances and tooth plates; and (**e**) experimental setup with the DIC system.

The test was performed using the 10 kN nominal force universal testing machine, equipped with the Arcan fixture. A digital image correlation system [13] was used to obtain the full-field displacement distribution and visualize the shear strain uniformity. The testing machine gave information on the forces, while the DIC system on the shear angles. The preparation of the DIC system started with choosing the calibration object, here the manufacturer's CP90/20 was used, which allowed measuring an area between 78×65 mm^2 and 130×105 mm^2 (final area was approximately 85×70 mm^2). Afterwards, the typical system calibration was performed. The software options were chosen as: facet size 19×19 pixels and faced step (distances between facets) 15×15 pixels, as proposed by the manufacturer. The DIC used in this experiment was composed of two 5Mpix cameras (resolution 2448×2050). The starting points for calculations were chosen, as recommended, at areas where the displacements were minimal (typically, the lower left part of the specimen). The experimental setup of the DIC system is shown in Figure 4e.

Twelve specimens were tested. Six of them were oriented so that the shear direction was parallel to the L axis (LR,L-specimens) and the other six with the shear direction parallel to the R axis (LR,R-specimens). Both tests were tracked automatically by displacement with a constant value of 0.35 mm/min and an initial force of 130 and 70 N for the LR,L and LR,R specimens, respectively. The ultimate forces were taken from the testing machine at the moment of failure (LR,L specimens) and at the moment of first horizontal crack (LR,R specimens). The values of ultimate shear angle and shear modulus were calculated for central point of the cross-section (indexed "c") (Point C in Figure 4b and Expression (2)) and for Points 1–4 presented in Figure 4b (indexed as "g") (from the virtual gauges located on the diagonals of the central square of 10×10 mm^2 and Expression (5)). The ultimate shear angle values were taken at the moment of failure.

The calculations of the LR shear moduli were performed according to Formula (3) with the shear angle obtained from (2) for the G^c moduli and with the shear angle obtained from (5) for the apparent G^g moduli. The shear modulus was determined as a secant modulus in the range between 25% and 50% of the maximal external force P^{ult} in each specimen. Such values were chosen due to very small shear angles for the measuring system resolution (initially 10–40% of the maximal force was considered). The expression for calculating the modulus from the experimental results can be written as follows:

$$ G = \frac{\Delta\tau}{\Delta\gamma} = \frac{\tau^{50\%P} - \tau^{25\%P}}{\gamma^{50\%P} - \gamma^{25\%P}}, \tag{6} $$

where the shear angles are taken as from Equation (2) or (5) for $G = G^c$ and $G = G^g$, respectively. The nominal shear stresses τ were computed as a ratio between the force P and nominal cross-section A_{nom}, i.e., $\tau = P / A_{nom}$.

The shear angle maps were generated by the software for stages just before the failure i.e., rupture for the LR,L specimens and first horizontal crack for the LR,R specimens. Three different vertical sections were prepared to provide complete information on the distribution of strains along the cross-section of the LR,L specimens, as shown in Figure 4b: the "middle cross-section" (red continuous line), the "maximal cross section" (blue dashed line) and the "symmetrical cross section" (blue dashed line). For the Specimens LR,R, two sections of different lengths were used: the "middle cross-section" and the horizontal blue dash-dot one.

In addition, for the obtained $\tau - \gamma$ relationship, a linear approximation of results was done by the use of the least square method. The method finds the best fit, in this case the shear moduli itself, which is a tangent of the angle between the linear fit and the horizontal axes. The shear angles were taken from Expression (2) and the fitting range was 25–100% of the maximal external force P^{ult} (rupture in LR,L direction and first crack in the LR,R direction).

2.3. Results

The post-processing of strain values was performed in the DIC software. Maps of the shear angle and charts at the moment just before failure for all six LR,L specimens are shown in Figures 5 and 6, respectively.

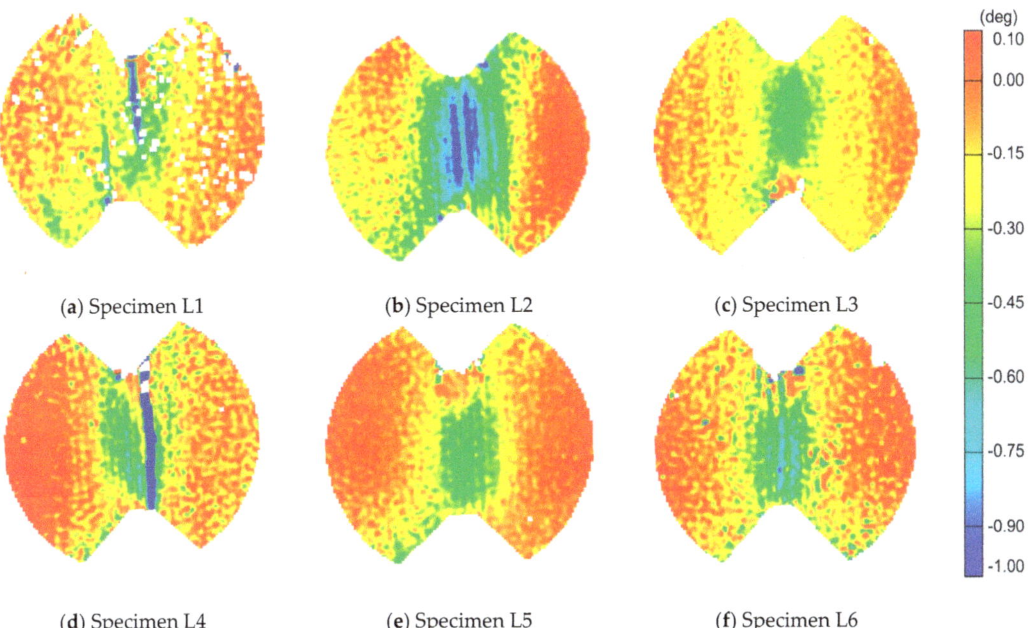

(**a**) Specimen L1   (**b**) Specimen L2   (**c**) Specimen L3

(**d**) Specimen L4   (**e**) Specimen L5   (**f**) Specimen L6

Figure 5. The shear angle maps just before the failure for each LR,L-type specimen: (**a–f**) for each of the samples from L1 to L6 respectively.

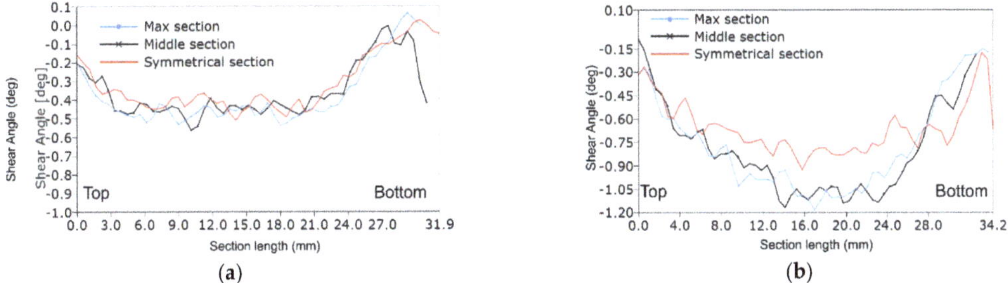

(**a**)   (**b**)

Figure 6. The shear angle plots along critical sections for LR,L-type specimens: (**a**) Specimen L3; and (**b**) Specimen L2.

The LR,L results in the maps of shear angle show different failure sections. The maps for Specimens L1 (Figure 5a) and L4 (Figure 5d) clearly show that the failure arises not in the central cross-section as expected, but next to it. This is generated due to the material inhomogeneity, where most likely a wider strip of earlywood was defining the path of failure. The strains in Specimens L2–L6 are concentrated around the central part of the specimen. In addition, Specimens L3, L5 and L6 present more uniform strain distribution among the others. Figure 6 shows the distribution of the shear angle along vertical sections. In general, the distribution of deformations in the central part is close to parabolic. However, in the case of Specimen L3, there is a more even distribution across the width as compared to sections from the symmetrical to max section (Figure 6a). In Specimen L2 (Figure 6b), the greater values of the angle, and hence the greater stress intensity, are on the right side of the central axis.

In Figures 7 and 8, we can find similar information for LR,R specimens. Figure 7 shows the deformation maps. The failure crack occurs horizontally along a line taken from the edge of the weakened central section (see Specimens R1, R3, R4 and R6 in Figure 7). In Figure 8, an increase in the value of shear strains can be noticed in the vicinity of the initial and end coordinates corresponding to the edges of the specimen (vertical middle section, black solid line). The red line of the results for the horizontal section also shows the highest values in the middle, i.e., near the critical edge.

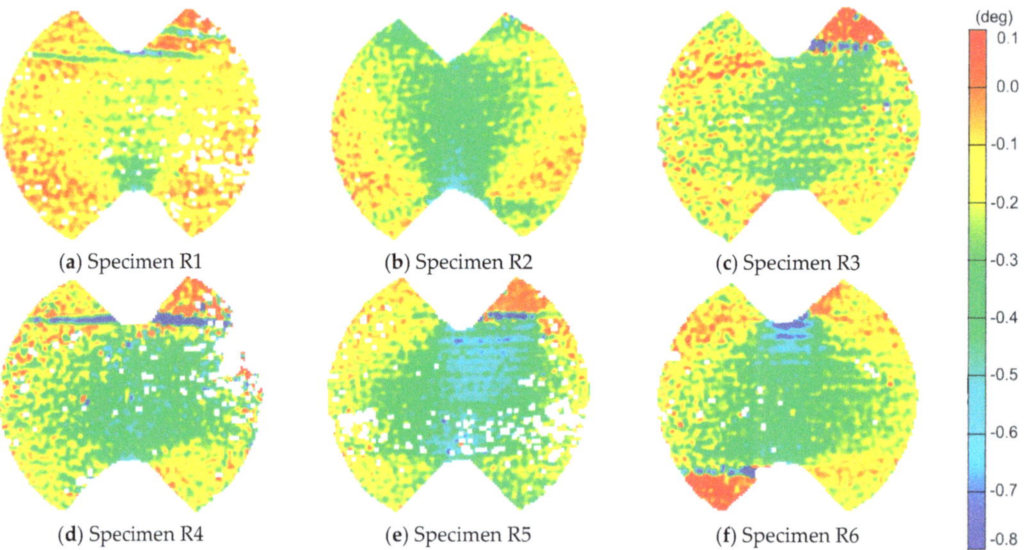

(**a**) Specimen R1   (**b**) Specimen R2   (**c**) Specimen R3

(**d**) Specimen R4   (**e**) Specimen R5   (**f**) Specimen R6

Figure 7. The shear angle maps just before the failure for each LR,R-type specimen: (**a–f**) for each of the samples from R1 to R6 respectively.

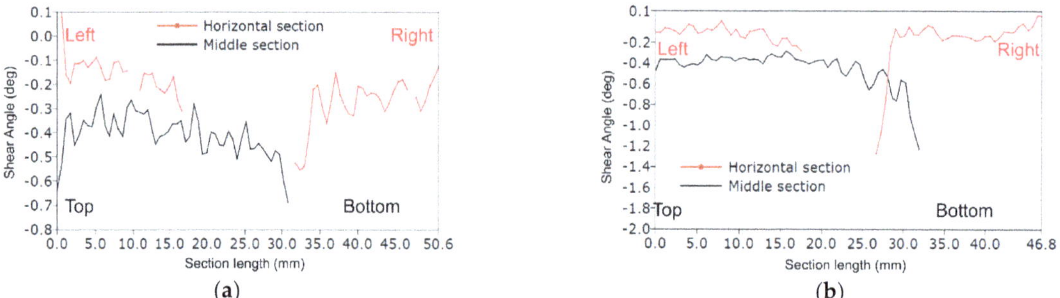

Figure 8. The shear angle plots along critical sections for LR,R-type specimens: (**a**) Specimen R2; and (**b**) Specimen R6.

The obtained results on shear moduli, ultimate strength and strain are shown in Table 1 for the LR,L specimens and in Table 2 for the LR,R specimens. The apparent shear modulus G^g and apparent ultimate shear angle $\gamma^{g,ult}$ from Expression (5) the coefficient of variation (COV) did not exceed 24%, which are quite big but at an acceptable level in wood. The same values calculated by (2) show acceptable COVs for the modulus (G^c) and a moderately high value for the ultimate shear angle ($\gamma^{c,ult}$).

Table 1. The results from the monotonic test of the LR,L specimens.

	P^{ult}	$\tau = P^{ult}/A_{nom}$	$\gamma^{g,ult}$	G^{g}	$\gamma^{c,ult}$	G^{c}
	(N)	(MPa)	(10^{-3})	(MPa)	(10^{-3})	(MPa)
Mean	2448 ± 404	4.4 ± 0.7	11.5 ± 3.6	392 ± 55	8.73 ± 3.9	470 ± 133
SD	492	0.85	4.37	67	4.75	161
COV	34.3	20.1	38	17.1	54.5	34.3

^{ult} corresponds to the maximum external force; ^{g} the values from the virtual gauges; ^{c} the values for central point of the cross-section.

Table 2. The results from the monotonic test of the LR,R specimens.

	P^{ult}	$\tau = P^{ult}/A_{nom}$	$\gamma^{g,ult}$	G^{g}	$\gamma^{c,ult}$	G^{c}
	(N)	(MPa)	(10^{-3})	(MPa)	(10^{-3})	(MPa)
Mean	2904 ± 513	5.16 ± 0.9	6.07 ± 1.1	813 ± 146	5.88 ± 1.2	816 ± 123
SD	673	1.18	1.43	192	1.54	162
COV	23.2	22.8	23.6	23.7	26.2	19.8

^{ult} corresponds to the maximum external force; ^{g} the values from the virtual gauges; ^{c} the values for central point of the cross-section.

The values of shear strength for the LR,L specimens had a mean value of 4.4 ± 0.7 MPa, with a standard deviation of 0.85 MPa and a COV of 19.2%. The ultimate force had a mean value of 2448 ± 404 N, a standard deviation of 492 N and a COV of 20.1%. The apparent shear modulus had a value of 392 ± 55 MPa, a standard deviation of 67 MPa and a COV of 17.1%. In linear elastic orthotropic theory, the LR and RL moduli are considered as equal. The value of the modulus obtained from LR,L specimens is relatively low. The reason lies in the high variability of the measured shear angles: a COVs of 34% from the Expression (5) and 54% for the Expression (2). Hence, they were considered nonrealistic results and disregarded.

Figures 9 and 10 show the stress–strain relationships for LR,L and LR,R specimens, respectively. The shear angles were taken from Expression (2) for the central point. The red lines show a linear fit in the range of results from 0.25 of the ultimate force up to the moment of failure. The LR,L specimens present a brittle failure (Figure 9), while LR,R present a linear behavior up to the first crack. After the crack, the specimen is changing the configuration and the specimen is slightly rotating and deforming, which is why the experimental points seem to lay on each other, due to stress loss after the crack, especially on the specimens shown in Figure 10b,e. The obtained LR,L average modulus of shear is approximately 350 MPa, while that of LR,R is approximately 840 MPa. These results are similar to those shown above in Tables 1 and 2. This confirms the validity of the previously adopted methods.

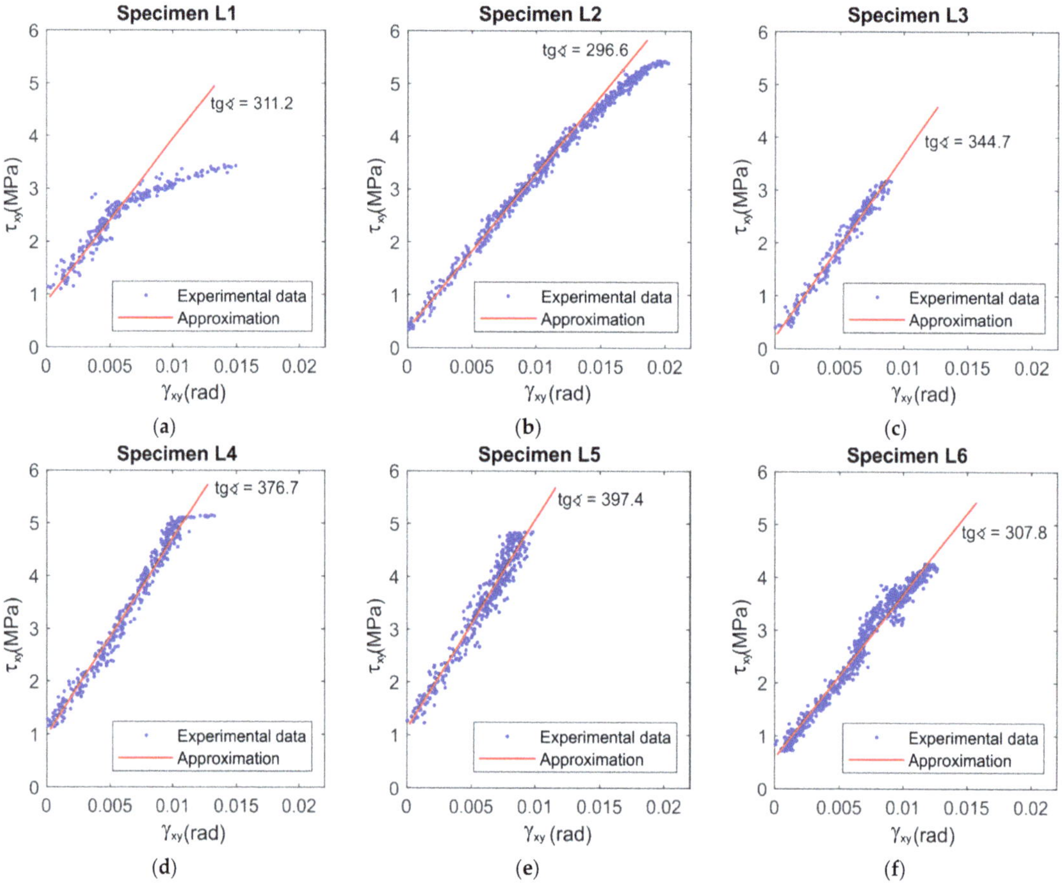

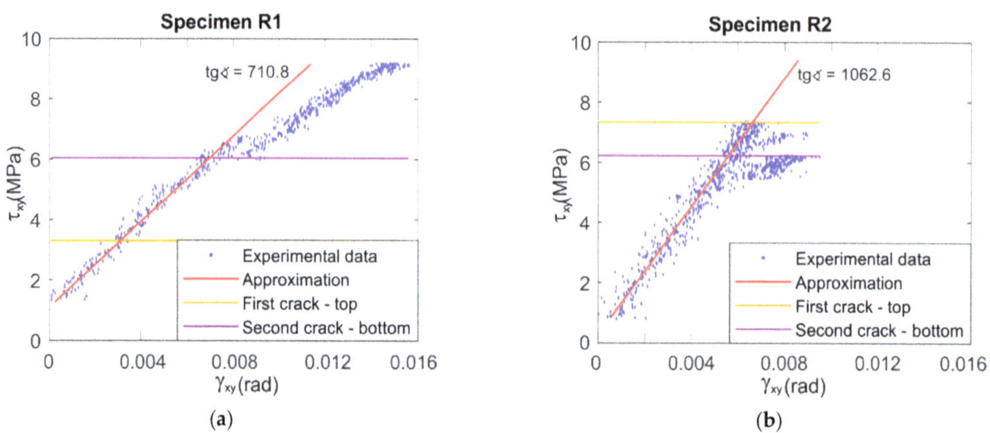

Figure 9. Charts of the relationship of shear stress–shear angle at the central point for LR,L specimens with their linear approximation: (**a**–**f**) for each of the samples from L1 to L6 respectively.

Figure 10. *Cont.*

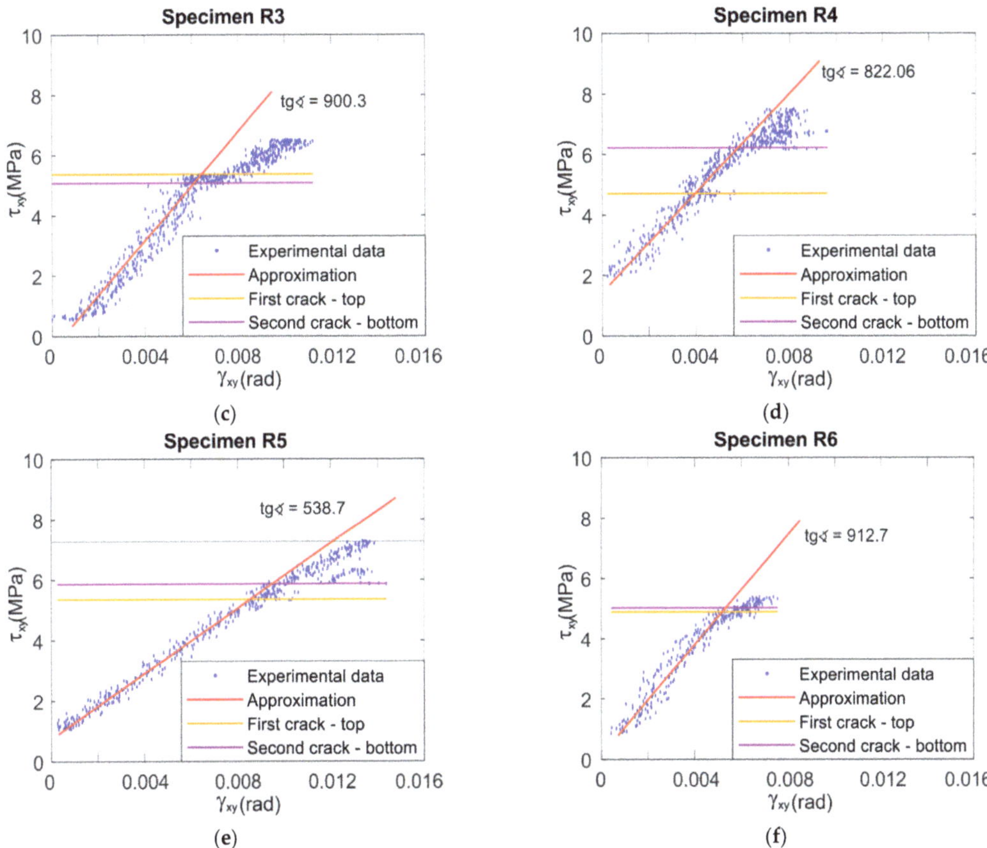

Figure 10. Charts of the relationship of shear stress–shear angle at the central point for LR,R specimens with their linear approximation: (**a–f**) for each of the samples from R1 to R6 respectively.

The results of the LR,L specimens shows a very low average modulus of shear of approximately 400 MPa, while the LR,R modulus is approximately 800 MPa. It may be caused by a relatively great ratio of earlywood-to-latewood width in the used wooden specimens. This ratio depends only on the growth conditions of the tree. The experimental scheme causes that, in the LR,L direction, the earlywood becomes dominant in material behavior as the more susceptible material part, whereas, in the LR,R direction, both material parts, early- and latewood, are working simultaneously.

Another issue lays in system accuracy. Consider the noise of the system; analyzing the chart of the L3 specimen in Figure 6a, successive points from each line can differ even by 25%. These values are calculated using Expression (2), which uses a small area to gather information called facets, i.e., a square defined in pixel size in the software (in this case 19×19 pixels) and corresponding true dimensions of approximately 0.5×0.5 mm^2 (in this particular case). For homogeneous fields with large strain values, such as plastic flows in steel or displacements of parts, this is sufficient. However, it may be more complex to calculate a very inhomogeneous strain field, where early- and latewood particles are mixed, and strains are very low. Therefore, we can consider the maps as a qualitative source of information. However, this does not forbid the quantitative analysis—the noise of the results is relatively high but can be reduced to some level by averaging results, as well as considering using a greater calculating area—of the aforementioned facets, where it is possible to increase such fields to dimensions of even 50×50 or 100×100 pixels to reduce the noise at a cost of calculation time, which was already proved by some experimental works. Moreover, the displacement measuring accuracy can be easily increased by using a greater length of the measuring base. That is why the apparent values of the moduli and ultimate shear angles are reliable—the values are calculated using a 10×10 mm^2 square and the displacement of these points is seen by the system well.

3. Numerical Simulations of Failure Mechanisms in the Tests

The numerical simulations of timber shearing in the Arcan test (Figure 4a [8]) with the failure modes and mechanisms are the purpose of this section after the implementation of the own continuum structural models from [12,14] into the commercial finite element code [15]. The Arcan test is considered to create a rather uniform and pure state of shear stress among the critical cross section. However, an appropriate constitutive model and the analysis by means of the finite element method allow more detailed insight into the sequence in which crack zones develop. The constitutive relationships of the model have been established in the framework of the mathematical elastic–plastic theory of small displacements. The model is based on the three orthotropic failure criteria that were earlier proposed by Geniev and next incorporated into the plasticity condition as the composite yield surface [16,17]. This orthotropic failure criteria can be regard as generalization of the well-known an isotropic maximum principal stress criterion of Rankine extended to the tension and compression anisotropic regimes and the Mohr–Coulomb strength criterion for the shear regime. They have the following forms in the plane state of stresses and in the frame of reference coincided with the axes of the principal stresses:

$$\left(\frac{\cos^2 \varphi}{Y_{t1}} + \frac{\sin^2 \varphi}{Y_{t2}} \right)\sigma_1 + \frac{\sigma_1 \sigma_2}{Y_{t1} Y_{t2}} + \left(\frac{\sin^2 \varphi}{Y_{t1}} + \frac{\cos^2 \varphi}{Y_{t2}} \right)\sigma_2 - 1 = 0, \tag{7}$$

$$\left(\frac{\cos^2 \varphi}{Y_{c1}} + \frac{\sin^2 \varphi}{Y_{c2}} \right)\sigma_1 + \frac{\sigma_1 \sigma_2}{Y_{c1} Y_{c2}} + \left(\frac{\sin^2 \varphi}{Y_{c1}} + \frac{\cos^2 \varphi}{Y_{c2}} \right)\sigma_2 + 1 = 0, \tag{8}$$

$$\sigma_1^2 - 2(1 + 2\mu^2)\,\sigma_1 \sigma_2 + \sigma_2^2 + 2\mu\,(C_{11} + C_{22})(\sigma_1 + \sigma_2) + \\ + 2(C_{11} - C_{22})\,(|\sin 2\varphi| - \mu \cos 2\varphi)(\sigma_1 - \sigma_2) - 4 C_{11} C_{22} = 0, \tag{9}$$

where φ denotes an angle between the axis of the first principal stress and the first axis of orthotropy.

The four uniaxial strength parameters $Y_{\Delta i}$, $i = 1, 2$ appear in the Rankine-type criteria described by Formulas (7) and (8), which are obtained from the two tensile tests $(\Delta = t)$ and two compressive tests $(\Delta = c)$ in the directions of the first and second axes of orthotropy, respectively. In Formula (9), we can find the parameter of internal friction μ and the shear strength parameters C_{11} and C_{22} obtained from the tests with the predetermined shear failure planes which are coincided with the orthotropy axes. Three different characteristic values of the shear stress can be calculated from Criterion (9) for the stress state $\sigma_1 = -\sigma_2$ and the angle $\varphi = 0^0$, 45^0, 90^0. The value of the shear stress for the angle $\varphi = 45^0$ is of the particular interest, because the direction of the shearing then coincides with the orthotropic axes. This shear stress can be helpful in setting the strength parameters in the numerical simulation of the experimental tests, and it is obtained from the following relationship:

$$\tau_{\max} = \frac{\sqrt{(C_{11} - C_{22}) + 4(1 + \mu^2)C_{11}C_{22}} - (C_{11} - C_{22})}{2(1 + \mu^2)}. \tag{10}$$

Contours of the failure criteria in the principal stress state are presented in Figure 11 for different values of the angle φ and in the axes of orthotropy.

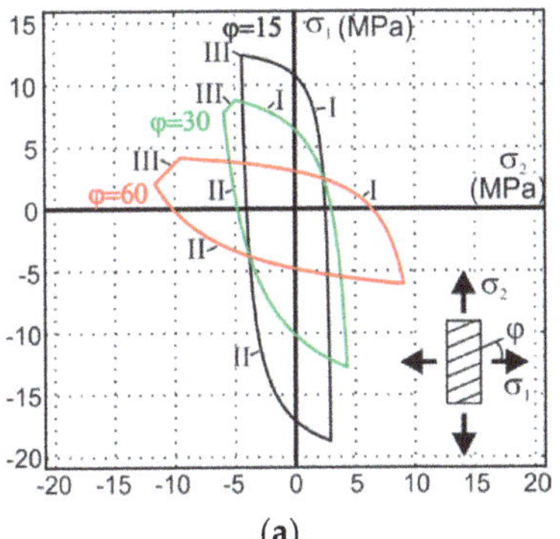

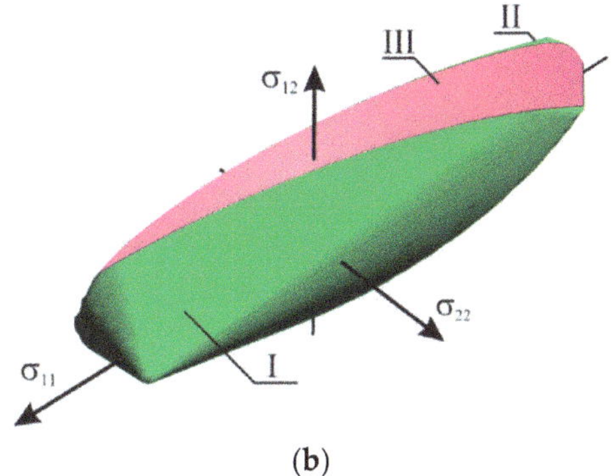

(a) **(b)**

Figure 11. Contours of orthotropic strength criteria: (**a**) in the principal stress space; and (**b**) in the axes of orthotropy (I) and (II) are Rankine-type criterion for tension and compression regime and (III) is the Mohr–Coulomb shear failure criterion.

3.1. Implementation of the Model

The more detailed discussion of the constitutive equations of the similar models and their numerical implementation into the commercial FEM system has been recently presented [18,19]. The plastic part of the strain tensor is defined by a flow rule associated with the yield function given by the plasticity (failure) criterion written in the following form:

$$f_\Delta(\sigma, \alpha_{\Delta,in}) = \frac{1}{2}\sigma \cdot \mathbf{P}_\Delta \cdot \sigma + \mathbf{p}_\Delta \cdot \sigma - (1 + K_\Delta \alpha_{\Delta,in}) = 0 \tag{11}$$

where K_Δ is a given constant plastic parameter and α_{in} is an internal hardening variable, hence the fourth- and second-order symmetric tensor functions $\mathbf{P}_\Delta$ and $\mathbf{p}_\Delta$ are dependent on the strength parameters of Criteria (7–9). The double contraction of the tensors is denoted by one dot. The plastic parameter $K_\Delta = 0$ for the perfect plasticity, $K_\Delta > 0$ for the hardening and $K_\Delta < 0$ for the softening behavior [12]. Since the model consists of three yield surfaces (11), we identify the material parameters by adding the subscript index Δ, where $\Delta = t$ is assigned to the tension Condition (7), $\Delta = c$ to the compression Condition (8) and $\Delta = s$ to the shear Condition (9).

The model was implemented as the user-supplied subroutine into the FE system DIANA [15], in which the nonlinear material behavior is updating over the equilibrium step within a framework of an incremental-iterative algorithm of the finite element method with a return-mapping algorithm and a consistent tangent stiffness operator for the plane stress state. The implementation was a very demanding programming task of the subroutine USRMAT in the FORTRAN language, which is described in detail in [18,19]. The formulation of the model during the implementation was presented based on the assumption that the principal axes of orthotropy coincided with the Cartesian frame of reference for stresses and strains in finite element computations. The tensor functions $\mathbf{P}_\Delta$ and $\mathbf{p}_\Delta$ have then the following matrix representations for tension and compression regimes:

$$\mathbf{p}_t \Rightarrow \begin{bmatrix} \dfrac{1}{Y_{tL}} \\ \dfrac{1}{Y_{tR}} \\ 0 \end{bmatrix}, \ \mathbf{p}_c \Rightarrow \begin{bmatrix} \dfrac{1}{Y_{cL}} \\ \dfrac{1}{Y_{cR}} \\ 0 \end{bmatrix}, \ \mathbf{P}_\Delta \Rightarrow \begin{bmatrix} 0 & \dfrac{-1}{Y_{\Delta L}Y_{\Delta R}} & 0 \\ \dfrac{-1}{Y_{\Delta L}Y_{\Delta R}} & 0 & 0 \\ 0 & 0 & \dfrac{2}{Y_{\Delta L}Y_{\Delta R}} \end{bmatrix} \tag{12}$$

and

$$\mathbf{p}_s \Rightarrow \begin{bmatrix} \dfrac{\mu}{C_{LL}} \\[2mm] \dfrac{\mu}{C_{RR}} \\[2mm] \dfrac{C_{LL}-C_{RR}}{C_{LL}C_{RR}}sign(\tau_{LR}) \end{bmatrix}, \ \mathbf{P}_s \Rightarrow \begin{bmatrix} \dfrac{1}{2C_{LL}C_{RR}} & \dfrac{-(1+2\mu^2)}{2C_{LL}C_{RR}} & 0 \\[3mm] \dfrac{-(1+2\mu^2)}{2C_{LL}C_{RR}} & \dfrac{1}{2C_{LL}C_{RR}} & 0 \\[3mm] 0 & 0 & \dfrac{2(1+\mu^2)}{C_{LL}C_{RR}} \end{bmatrix} \tag{13}$$

for the shear regime. In Formulas (12) and (13), the frame of reference is denoted as (x_L, x_R, x_T) and the shear strength parameter, e.g., C_{LL}, is obtained from the direct shear test in which the normal to the shear plane is predetermined in direction of the first axis of orthotropy.

Several tests confirmed the correctness of the proposed numerical algorithm for the anisotropic continuum. The multi-surface model enables the identification of the relevant macroscopic failure modes. The separated description of the three regimes also allows the modeling of their respective post-failure behavior with modern hardening/softening evolution laws, although an intersection of different yield surfaces defines corners that require special attention in the numerical algorithm.

3.2. FEM Modeling and Results

The finite element mesh was created out of 1321 nodes and 1232 elements. The geometry of FE mesh with boundary conditions is presented in Figure 12a. The type of used elements was the Q8MEM (isoparametric, cuboid eight node elements). The force was inducted by displacement of the upper arm of the fixture, so the analyses were carried out with indirect displacement control. The following material parameters were adopted based on the tests [10]: the Young's moduli $E_{LL} = 13.7$ MPa, $E_{RR} = 1.1$ MPa, the shear modulus $G_{LR} = 820$ MPa and the Poisson's ratio $\nu_{LR} = 0.45$. Other material parameters are presented in Table 3. When assuming the shear strength, the results from the experiments discussed above were considered. However, due to their large spread, it was decided to round the values. It should be noted that in Table 3 the yield strengths are assumed in the numerical computations, which can be different from the experimental strengths.

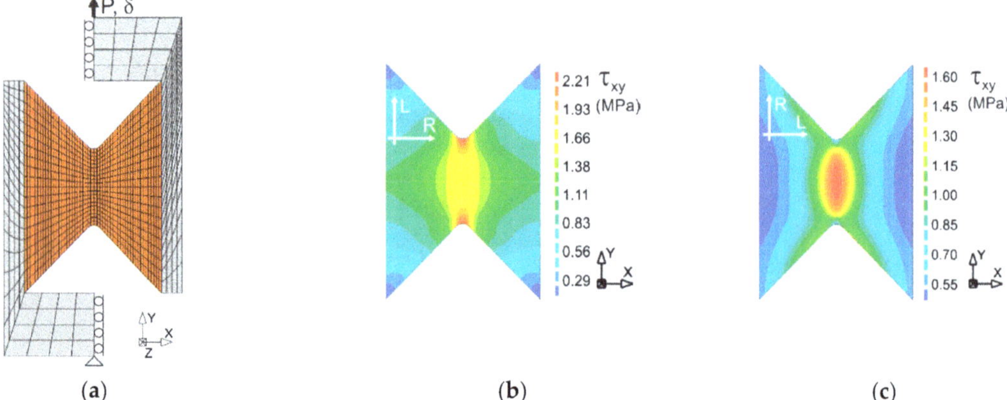

Figure 12. Distributions of the shear stress for different configurations of the specimens: (**a**) the finite element mesh; (**b**) Specimen LR,L; and (**c**) Specimen LR,R.

Table 3. Material parameters for numerical simulations.

Criterion	Parameters		
Compression	Y_{cL} (MPa) 40.8	Y_{cR} (MPa) 7.8	K_c 0.0
Tension	Y_{tL} (MPa) 80.5	Y_{tR} (MPa) 3.8	K_t 0.0
Shear	C_{LL} (MPa) 6.0	C_{RR} (MPa) 4.5	μ 0.1

Figures 12–14 present the results of the numerical simulations of the test for the specimens with different orientations—LR,L and LR,R. Figure 12 presents shear stress distributions in the elastic state for the displacement level of $\delta = 0.1$ mm. This level corresponds to values of the external force P = 1.5 kN for Specimen LR,L (Figure 13a, Line a) and P = 0.75 kN for Specimen LR,R (Figure 13a, Line b). It is seen in Figure 12 that the uniform state of shear stress occurs only in the middle of the specimens and its distributions are different depending on the orientation of the material axes.

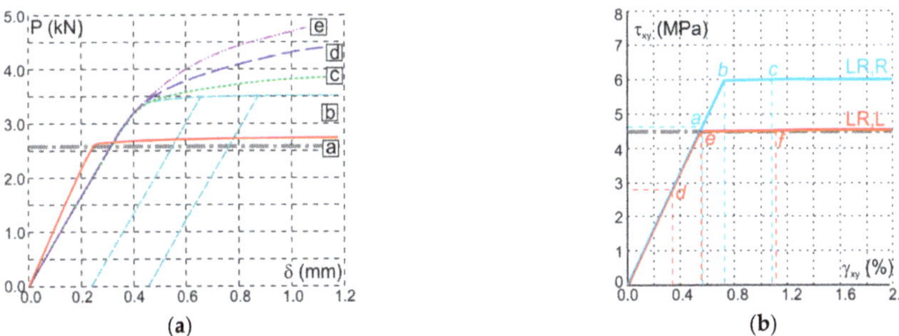

(a)   (b)

Figure 13. Graphs of the responses obtained from the numerical tests-description in the text: (**a**) load vs. displacement graphs; and (**b**) stress–strain curves for the middle point of the specimen.

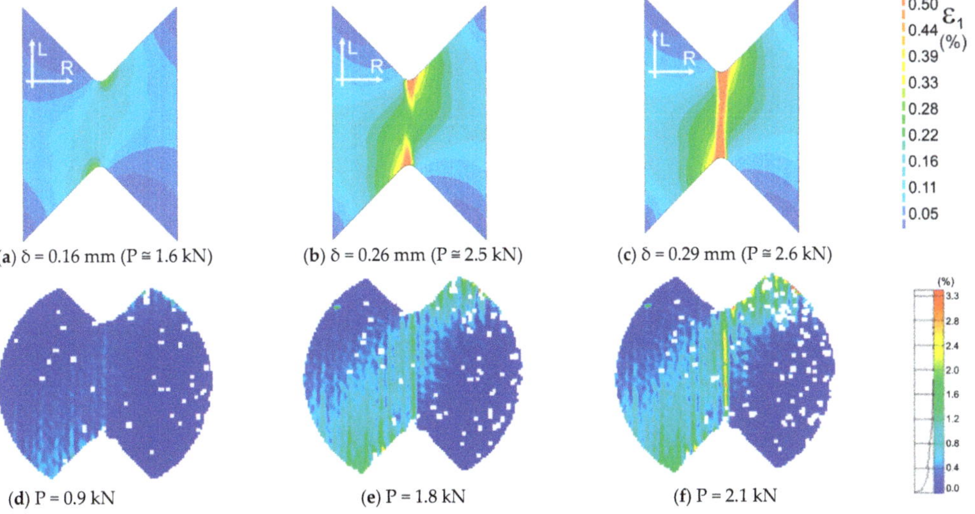

(**a**) $\delta = 0.16$ mm (P $\cong$ 1.6 kN)   (**b**) $\delta = 0.26$ mm (P $\cong$ 2.5 kN)   (**c**) $\delta = 0.29$ mm (P $\cong$ 2.6 kN)

(**d**) P = 0.9 kN   (**e**) P = 1.8 kN   (**f**) P = 2.1 kN

Figure 14. Comparison of the maps of the maximum principal strains ε_1 obtained numerically (**a–c**) and calculated from displacements measured by the DIC method (**d–f**) for LR,L orientation.

Figure 13a shows the relationship between reaction and displacement of the upper support (P − δ) for different values of the plastic parameter. Lines [a] and [b] are for the perfect plasticity. For the comparison, Lines [c]–[e] are also shown in Figure 13a for Specimen LR,R and the hardening plasticity with $K_s = 10$ for Line [c], $K_s = 50$ for Line [d] and $K_s = 200$ for Line [e]. Good agreement was found between numerical and experimental results of the ultimate external force for the perfect plasticity and Specimen LR,L—2.73 (the numerical simulation) and 2.58 kN (the experiment)—and the worse agreement for the Specimen LR,R—3.50 and 2.90 kN, respectively.

The value 2.90 kN of the ultimate external force corresponds to the moment of the first crack appearance. The second crack appears with an average load of 3.30 kN, which is closer to the numerical result. The hardening effect similar to the experimental results is visible in Figure 10 and can be easily controlled by the appropriate selection of K_s parameter. Lines [c]–[e] in Figure 13a are an example of the possibilities offered by the model. The exact fit will be the subject of further research. Figure 13b shows the relationship between shear stress and strain in the central point. Red Line [a] is for Specimen LR,L and perfect plasticity ($K_s = 0$). The obtained maximum strength was 4.50 MPa, which corresponds to the adopted value of the shear strength (C_{RR}). Blue Line [b] is for Specimen LR,R and perfect plasticity ($K_s = 0$). Again, the obtained maximum strength 6.0 MPa corresponds to the adopted value of the shear strength (C_{LL}). The slope in the elastic range is consistent with the adopted value of the modulus. Another path to destruction has been observed. For the test in the LR,L configuration, the first active was the shear criterion, while, in the LR,R configuration, the tensile criterion was activated first. Points marked with letters a and d shown in Figure 13b correspond to the first moments of reaching the failure criterion.

In Figures 14 and 15, we can find maps of the plastic strains at different stages of the tests. The maps in Figure 14a–c correspond to the test moments marked with points d–f in the diagram of Figure 13b, respectively. The maps in Figure 15a–c correspond to the test moments marked with points a–c in the diagram of Figure 13b, respectively. The obtained mechanisms are compatible with those obtained experimentally and shown in Figures 14d–f and 15d–f.

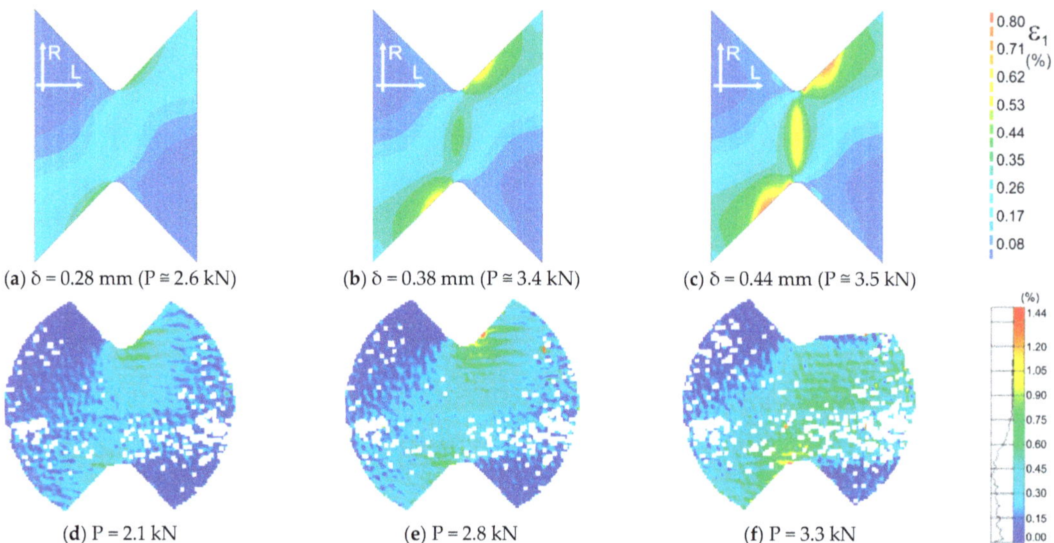

(**a**) δ = 0.28 mm (P ≅ 2.6 kN) (**b**) δ = 0.38 mm (P ≅ 3.4 kN) (**c**) δ = 0.44 mm (P ≅ 3.5 kN)

(**d**) P = 2.1 kN (**e**) P = 2.8 kN (**f**) P = 3.3 kN

Figure 15. Comparison of the maps of the maximum principal strains ε_1 obtained numerically (**a–c**) and calculated from displacements measured by the DIC method (**d–f**) for LR,R orientation.

4. Conclusions

The paper presents experimental investigations of wood shearing in the LR plane for two different directions of loading. Twelve specimens were tested. Forces and shear angles were measured using a testing machine and digital image correlation. Some of the material constants and strengths were determined. Shear angle maps and charts for the critical cross sections are presented.

The usage of the DIC system showed that it is capable of gathering more information on the experiment than typically used measuring techniques such as strain gauges. The graphical presentation in the form of maps showed a great inhomogeneity on the specimen surface in case of the shear angles distribution.

The results on the two different directions of loading show that it may be necessary to reconsider the specimen shape and border conditions of the test to obtain the homogenized material parameters. This issue is depending to a very large ratio between earlywood and latewood among annual rings. This issue may affect the results of specimens of different species with lower early-to-latewood ratios, to a certain extent.

Our own constitutive model for the analysis of wooden structures in biaxial plane stress states, implemented into the finite element code, was used to analyze behavior of wood during shearing in the Arcan test. Experimental determination of the shear behavior has always been influenced by difficulties in obtaining a state of pure and uniform shear in test specimens. Model calibration allows adjustment to experimental results. For different specimen material axis orientations, adequate destruction mechanisms were obtained.

Author Contributions: Conceptualization, L.M.; methodology, A.S. and P.B.; software, A.S. and P.B.; validation, A.S. and P.B.; investigation, A.S., P.B. and A.R.; resources, P.B.; writing—original draft preparation, P.B. and A.R.; writing—review and editing, L.M. and P.B.; visualization, P.B.; supervision, L.M.; and funding acquisition, A.S. All authors have read and agreed to the published version of the manuscript.

Funding: This research received no external funding.

Institutional Review Board Statement: Not applicable.

Informed Consent Statement: Not applicable.

Data Availability Statement: The data presented in this study are available on request from the corresponding author.

Conflicts of Interest: The authors declare no conflict of interest.

References

1. Benvenuti, E.; Orlando, N.; Gebhardt, C.; Kaliske, M. An orthotropic multi-surface damage-plasticity FE-formulation for wood: Part I—Constitutive model. *Comput. Struct.* **2020**, *240*, 106350. [CrossRef]
2. Fakoor, M.; Khezri, M.S. A micromechanical approach for mixed mode I/II failure assessment of cracked highly orthotropic materials such as wood. *Theor. Appl. Fract. Mech.* **2020**, *109*, 102740. [CrossRef]
3. Bru, T.; Olsson, R.; Gutkin, R.; Vyas, G.M. Use of the Iosipescu test for the identification of shear damage evolution laws of an orthotropic composite. *Compos. Struct.* **2017**, *174*, 319–328. [CrossRef]
4. Arcan, M. A New Method for the Analysis of Mechanical Properties of Composite Materials. In Proceedings of the 3rd International Congress on Experimental Mechanics, Los Angeles, CA, USA, 13–18 May 1973.
5. Dahl, K.B.; Malo, K.A. Linear shear properties of spruce softwood. *Wood Sci. Technol.* **2009**, *43*, 499–525. [CrossRef]
6. Huang, D.; Bian, Y.; Zhou, A.; Sheng, B. Experimental study on stress–strain relationships and failure mechanisms of parallel strand bamboo made from Phyllostachys. *Constr. Build. Mater.* **2015**, *77*, 130–138. [CrossRef]
7. Liu, J.Y. New shear strength test for solid wood. *Wood Fiber Sci.* **1984**, *16*, 567–574.
8. Małyszko, L.; Rutkiewicz, A. The shear behaviour of pine wood in the Arcan test with the digital image correlation. *MATEC Web Conf.* **2017**, *117*, 00113. [CrossRef]
9. Xavier, J.C.; Garrido, N.M.; Oliveira, M.; Morais, J.L.; Camanho, P.P.; Pierron, P.P. A comparison between the Iosipescu and off-axis shear test methods for the characterization of Pinus Pinaster Ait. *Compos. Part A Appl. Sci. Manuf.* **2004**, *35*, 827–840. [CrossRef]

10. Małyszko, L.; Rutkiewicz, A.; Kowalska, E. Experimental determination of in-plane tensile moduli of elasticity for pine wood. In *Lightweight Struct. In Civil Eng. Contemporary Problems. Monograph from Scientific Conference of IASS Polish Chapters. XXI LSCE 2015, Rzeszów, 4 December 2015*; Tarczewski, R., Bieniek, Z., Eds.; Agat Studio Graficzne Jerzy Burski: Warsaw, Poland, 2015; pp. 147–152.
11. Jeong, G.Y.; Park, M.J. Evaluate orthotropic properties of wood using digital image correlation. *Constr. Build. Mater.* **2016**, *113*, 864–869. [CrossRef]
12. Małyszko, L. Orthotropic Yield Criteria in a Material Model for Timber Structures. *Mech. Control* **2011**, *30*, 221–228.
13. *GOM Testing—ARAMIS User Manual—Software v6.3 and Higher*; GOM: Braunshweig, Germany, 2016. Available online: https://www.gom.com/metrology-systems/aramis.html (accessed on 19 December 2020).
14. Klovanych, S.; Małyszko, L. Plasticity in Structural Engineering. In *Models for Timber and Masonry Structures*; Wydawnictwo Uniwersytetu Warmińsko-Mazurskiego Olsztyn: Olsztyn, Poland, 2019.
15. DIANA—Finite Element Analysis. User's Manual Release 9.3. Release Notes. TNO DIANA BV, Delftechpark 19a, 2628 XJ Delft, The Netherlands. 2008. Available online: https://dianafea.com/diana-manuals (accessed on 19 December 2020).
16. Geniev, G. On a strength criterion of timber in a plane stress state. Stroit. *Mech. Rascet Sooruz. Mosk.* **1981**, *3*, 15–20.
17. Geniev, G.; Małyszko, L. Selected strength and plasticity problems of anisotropic structural materials. In Proceedings of the International IASS Symposium on Lightweight Structures in Civil Engineering, Warsaw, Poland, 24–28 June 2002; Micro-Publisher Jan B. Obrębski Wydawnictwo Naukowe: Warsaw, Poland, 2002; pp. 309–316.
18. Małyszko, L.; Jemioło, S.; Bilko, P. Implementation tests of orthotropic elastic-plastic models into a commercial finite element code. In *XVI Scientific Conf. of IASS Polish Chapters: Lightweight Structures in Civil Engineering. Contemporary Problems*; Warsaw University of Technology: Warsaw, Poland, 2010; pp. 51–56.
19. Bilko, P.; Małyszko, L. An Orthotropic Elastic-Plastic Constitutive Model for Masonry Walls. *Materials* **2020**, *13*, 4064. [CrossRef] [PubMed]

Article

The Increase in the Elastic Range and Strengthening Control of Quasi Brittle Cement Composites by Low-Module Dispersed Reinforcement: An Assessment of Reinforcement Effects

Dominik Logoń [1,*], Krzysztof Schabowicz [1], Maciej Roskosz [2] and Krzysztof Fryczowski [3]

[1] Faculty of Civil Engineering, Wrocław University of Science and Technology, Wybrzeże Wyspiańskiego 27, 50-370 Wrocław, Poland; krzysztof.schabowicz@pwr.edu.pl

[2] Faculty of Mechanical Engineering and Robotics, AGH University of Science and Technology, aleja Mickiewicza 30, 30-059 Kraków, Poland; mroskosz@agh.edu.pl

[3] Faculty of Energy and Environmental Engineering, Silesian University of Technology, ul. Akademicka 2A, 44-100 Gliwice, Poland; krzysztof.fryczowski@polsl.pl

[*] Correspondence: dominik.logon@pwr.edu.pl

Abstract: This paper presents the possibility of using low-module polypropylene dispersed reinforcement (E = 4.9 GPa) to influence the load-deflection correlation of cement composites. Problems have been indicated regarding the improvement of elastic range by using that type of fibre as compared with a composite without reinforcement. It was demonstrated that it was possible to increase the ability to carry stress in the Hooke's law proportionality range in mortar and paste types of composites reinforced with low-module fibres, i.e., $V_f = 3\%$ (in contrast to concrete composites). The possibility of having good strengthening and deflection control in order to limit the catastrophic destruction process was confirmed. In this paper, we identify the problem of deformation assessment in composites with significant deformation capacity. Determining the effects of reinforcement based on a comparison with a composite without fibres is suggested as a reasonable approach as it enables the comparison of results obtained by various universities with different research conditions.

Keywords: quasi-brittle cement composites; low-module polypropylene fibres; elastic range

Citation: Logoń, D.; Schabowicz, K.; Roskosz, M.; Fryczowski, K. The Increase in the Elastic Range and Strengthening Control of Quasi Brittle Cement Composites by Low-Module Dispersed Reinforcement: An Assessment of Reinforcement Effects. *Materials* **2021**, *14*, 341. https://doi.org/10.3390/ma14020341

Received: 8 December 2020
Accepted: 7 January 2021
Published: 12 January 2021

Publisher's Note: MDPI stays neutral with regard to jurisdictional claims in published maps and institutional affiliations.

Copyright: © 2021 by the authors. Licensee MDPI, Basel, Switzerland. This article is an open access article distributed under the terms and conditions of the Creative Commons Attribution (CC BY) license (https://creativecommons.org/licenses/by/4.0/).

1. Introduction

The development of cement composites results in an increase in compressive strength without significantly improving bending strength [1,2]. The brittleness of cement composites causes rapid destruction, which is particularly disadvantageous in high-strength structures. Researchers have attempted to limit the brittleness of cement composites by trying to increase the flexural or bending strength using various fibre reinforcements [1–39]. Fibre reinforcement requires good rheological properties of the mix (for random dispersion), which determines a higher amount of cement. Short reinforcement controls the effect of multi-cracking, and longer reinforcement improves the toughness and strength [3,20,24]. The best results with respect to flexural strength and toughness have been achieved with a high-strength matrix and fibres [37].

High strength and high Young's modulus reinforcement (steel fibre E = 210 GPa and carbon fibres E = 30–300 GPa) can be applied to increase stress corresponding to first crack appearance [29–35]. Improvement of the elastic range by means of low-module reinforcement is difficult to achieve, and therefore hybrid reinforcement is frequently used, with both high and low extension strength and Young's modulus, and with different lengths [4,17,39]. High-module reinforcement controls the elastic range, while low-module reinforcement controls the deflection range after exceeding f_{cr} [5–12]. Existing publications do not indicate the effectiveness of applying low-module fibres for improving Hooke's law proportionality range, which limits their use in construction materials for controlling the deformation and crack propagating process [13,15,16].

In this paper, we take into account the possibility of improving elastic range by using the application of low-module polypropylene reinforcement. Such fibres are commonly used, and a number of papers have been written with respect to their use. However, the existing papers do not focus on the possibility of using low-module polypropylene reinforcement to improve Hooke's law proportionality range and obtain ESD composites (E, elastic range; S, strengthening control; and D, deflection control). These effects reduce the brittleness of the cement composites and help to avoid the catastrophic destruction process [13–20].

There are different methods of calculating flexural toughness, for example, the JSCE method, the ASTM C 1018 and the EN 14651 standard consistent with RILEM recommendations [40–45]. There is no one universal method to accurately describe and compare reinforcing ESD effects. The existing norms and standards describing these effects seem to be insufficient. This makes the assessment and comparison of the obtained effects difficult. Despite a number of formulas for the calculation of the reinforcement effect, new methods and modifications of the existing ones are still being proposed. Various papers have shown that rheological properties influence mechanical properties, especially with respect to concrete composites, whose rheological properties are much worse than those of mortar or paste composites [18,25].

According to a literature review, previous works have not shown the possibility of significantly increasing stress in the elastic range with low-module polypropylene fibres in structural elements as compared with matrix (unreinforced composite) [24,28,36,38].

In this paper, damaged composites were obtained with significant deflection and flexural strength that equalled or exceeded the strength corresponding to the first crack. Some of the effect had already been presented in our own works [21–27] for small beams of cement composites with synthetic structural polypropylene fibres [28] but not with respect to matrix (unreinforced composite).

In this paper, we focus on the limitation of the catastrophic destruction process by means of ESD effects in structural elements. The main goal of the paper was to improve the elastic range of cement composites by means of low-module fibres, which required the introduction of the maximum volume of dispersed reinforcement. That effect was obtained with the maximum volume of fibres for mortar, $V_f = 3\%$, and for paste, $V_f = 6\%$.

2. Materials and Methods

2.1. Materials Used for Tests

The following materials were used for the preparation of the cement composites: Portland cement (c) CEM I 42.5R (Górażdże cement plant, Górażdże, Poland) silica fume (10% c), fly ash (20% c), superplasticizer (SP, Sika company, Baar, Switzerland) tap water (w), and w/binder = 0.35. The sand used in the research is sold as sand for the production of ordinary concrete. The grain size distribution of the sand was 0–2 mm.

The composites were reinforced with randomly dispersed fibres (Figure 1 and Table 1) and synthetic structural polypropylene fibres (compliance with ASTM C 1116), specific weight 0.91 kg/dm^3, flexural strength $f_t = 620$–758 MPa, E = 4.9 GPa, l = 54 mm, equivalent diameter 0.48 mm, and l/d = 113.

Figure 1. Synthetic structural polypropylene fibres l = 54 mm.

Table 1. Tested specimens.

Symbol	Specimen	Cement:Sand (Volume)	V_f [%]
$MV_f0\%$	mortar	1:4.5	0
$MV_f2\%$	mortar	1:4.5	2
$MV_f3\%$	mortar	1:4.5	3
$ZV_f6\%$	paste	-	6

2.2. Preparation of Specimens for Tests

The specimens $MV_f3\%$ and $ZV_f6\%$ were reinforced with the maximum volume possible to disperse polypropylene fibres. All tested samples were demoulded and notched. Each beam was turned by 90° and cut to the depth of 30 mm (cut width 3 mm).

Components were mixed in the concrete mixer, and then used to mould samples. Beams (150 mm × 150 mm × 600 mm) were cast in slabs, and then cured in water at $20 \pm 2\,°C$. After 180 days of ageing, the beams were prepared for the bending test, Figure 2. Figure 2a presents a sample prepared for the four-point bending test.

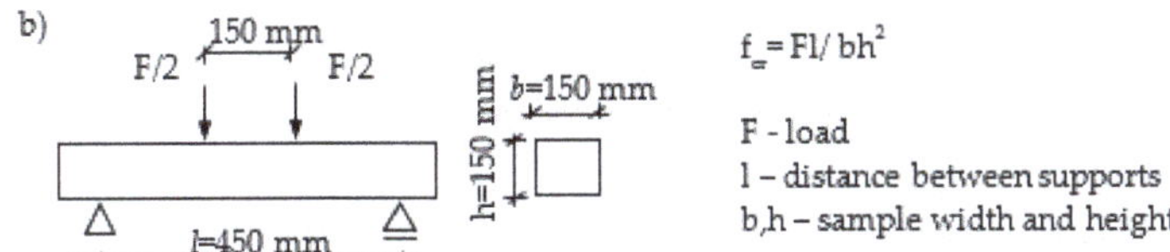

Figure 2. Four-point bending test. (**a**) Specimen before test, paste V_f = 6%; (**b**) Diagram of the test.

2.3. Description of the Test Stand

Four-point bending tests were carried out on the testing machine with closed-loop servo control displacement. The load-deflection curves (Figures 3–9) were obtained according to ASTM C 1018, but the test was based on the measurement of the displacement of crosshead. The following data was obtained:

- Tensile strength at bending f_{max} (MOR, the modulus of rupture), tensile strength at first crack f_{cr} (LOP, the limit of proportionality);
- The characteristic points on the load-deflection curve, f_x (F_x-load, ε_x-deflection, and W_x-energy);
- Energy (work) as proportional to the area under the load-deflection curve up to the characteristic point.

Additionally, deflection was recorded by means of two LVDT sensors located as in Figure 2. During the test, the bending load and deflection of the specimen were measured. The testing procedure corresponded to the requirements of the ASTM C 1018 standard.

The ESD reinforcement effect (i.e., elastic range, strengthening control, deflection control) is presented by characteristic points f_x and areas A_X under he load-deflection curve, Figure 3).

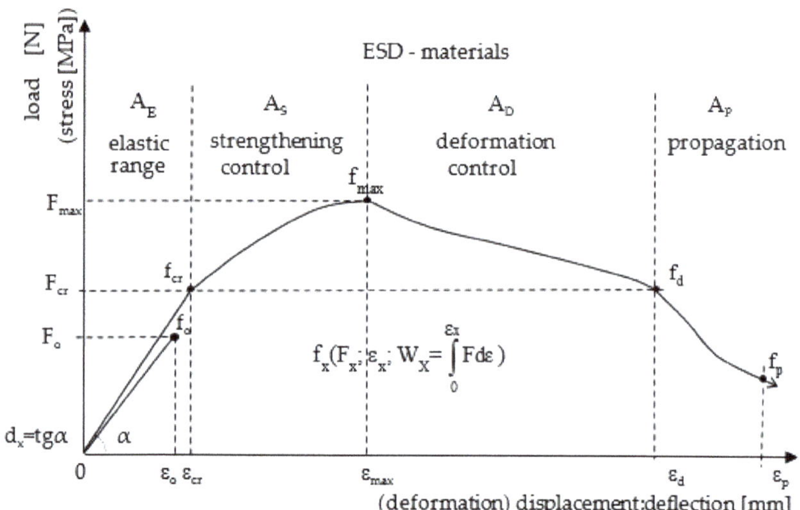

Figure 3. ESD composites depending on load deflection. Areas: A_E, elastic range; A_S, strengthening control; A_D, deflection control; A_P, propagation.

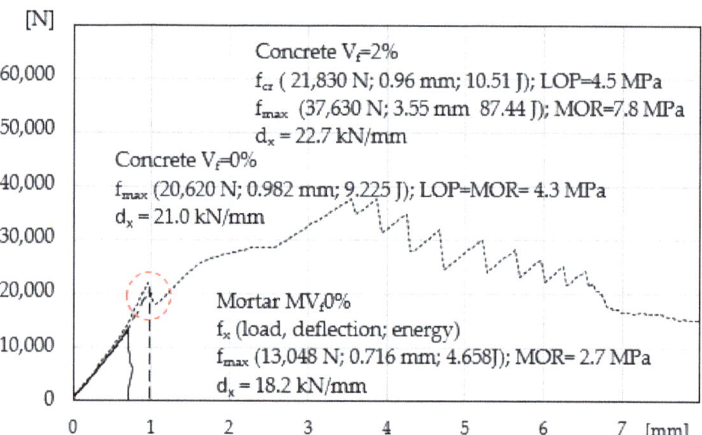

Figure 4. Load-deflection curve in the four-point bending test, mortar MV$_f$0% (matrix) and concrete without fibres V_f = 0% and with V_f = 2% fibres.

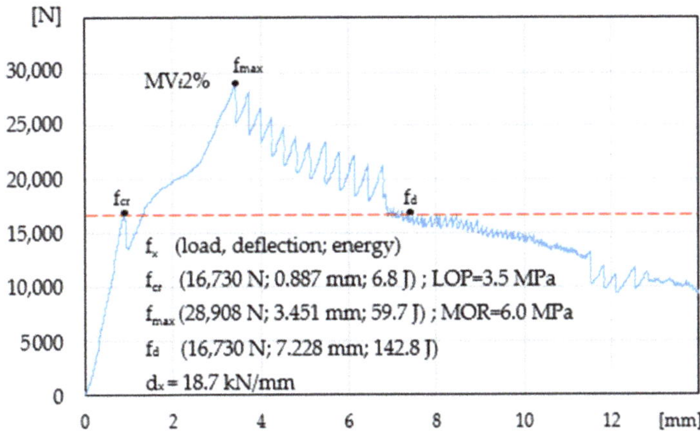

Figure 5. Load-deflection curve in the four-point bending test, mortar MV$_f$2%.

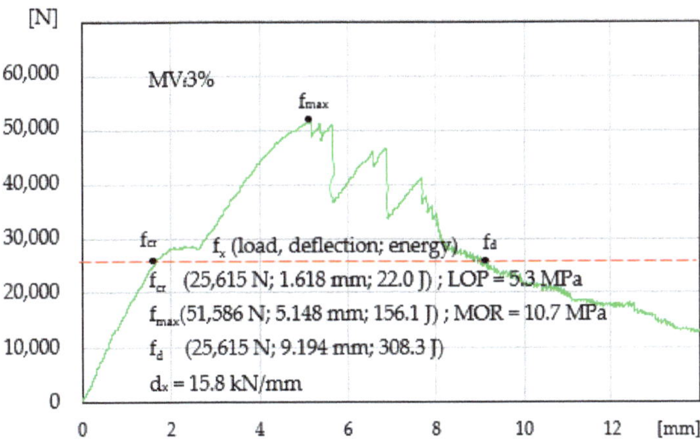

Figure 6. Load-deflection curve in the four-point bending test, mortar MV$_f$3%.

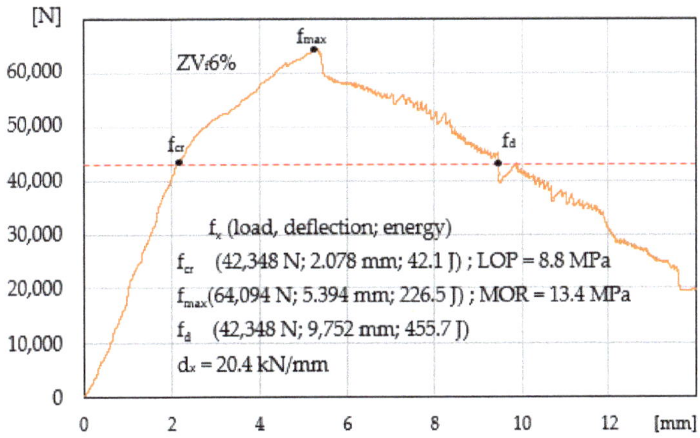

Figure 7. Load-deflection curve in the four-point bending test, paste ZV$_f$6%.

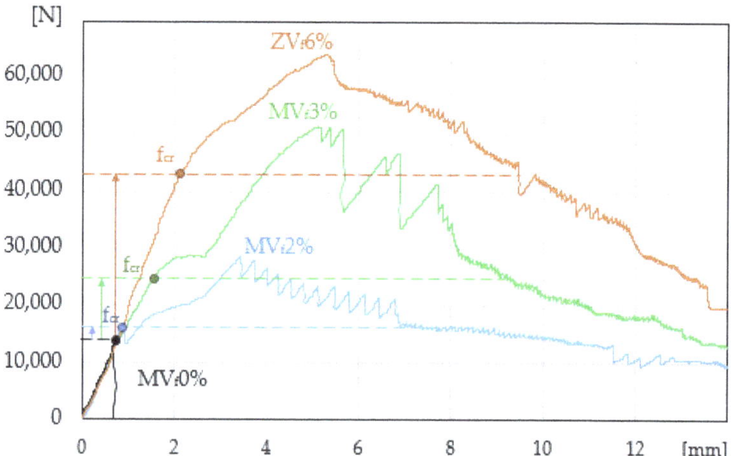

Figure 8. Load-deflection curve in the four-point bending test, comparison of samples.

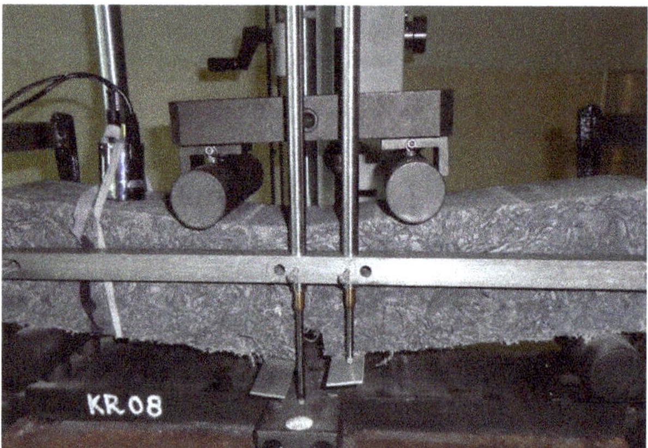

Figure 9. The four-point bending test, paste ZV$_f$6% after the test.

3. Test Results

The load-deflection curve for mortar MV$_f$0% is presented in Figure 4. Additionally, the results presented earlier [24] for concrete without fibres and with the maximum volume content of the same fibres V_f = 2% are included (insignificant increase in the elastic range of concretes with the maximum volume of fibres was obtained). A typical load-deflection correlation for cement composites without reinforcement was obtained, with catastrophic destruction process (deflection as displacement of crosshead (mm)). The figure presents data corresponding to the maximum ability to carry stress $f_{cr} = f_{max}$ = MOR and deformation capacity d_x. The specimen is a reference for the other tested composites.

Figure 5 presents mortar with 2% of fibres. Characteristic points f_{cr}, f_{max}, f_d, have been determined, which enable the identification of the following areas: elastic range A_E = 6.8 J, strengthening control A_S = 52.9 J, deflection control A_D = 83.1 J, and propagation area A_E is not significantly larger than A_E of the specimen without reinforcement. In addition, LOP, MOR, and d_x have been determined.

Mortar with the maximum possible volume of fibres is presented in Figure 6. The obtained results indicate a significant improvement of the properties of ESD composites in the following areas: elastic range A_E = 22.0 J, strengthening control A_S = 134.1 J, deflection control A_D = 130.2 J, and propagation area A_S with 3% of fibres is larger than A_D.

Additionally, LOP and MOR were determined. We found that reinforcement significantly contributed to the increased deformation capacity in the elastic range $d_x = 15.8$.

Figure 7 shows a paste specimen $ZV_f6\%$ with the maximum possible volume of fibres $V_f = 6\%$. The best results were obtained regarding the ability to carry stress in the elastic range $A_E = 184.4$ J, strengthening control area $A_S = 134.1$ J, deflection control area $AD = 271.3$ J, and propagation area A_P. Significant improvement of the ability to carry stress has been achieved for LOP and MOR and a slight improvement of deformation capacity d_x.

The compilation of all the tested specimens (load-deflection curves) is presented in Figure 8. The curves illustrate the scale of obtained ESD effects as compared with the matrix (mortar without fibres). For structural reasons, it is important to improve stress in Hooke's law area and not those corresponding to f_{max}. As the presented curves show, the specimens with the content exceeding $V_f = 2\%$ may show significant ESD effects.

Figure 9 shows specimen $ZV_f6\%$ after the four-point bending tension test. The figure presents two sensors measuring deflection relative to the neutral axis. It should be noted that there are significant differences regarding the displacement of the cut edges in the case of considerable deformations of the specimen. The visible differences result in significant differences in the measurements of ESD composites' deflection.

Table 2 presents a compilation of the results of the four-point bending tension test of the tested specimens (load, deflection, absorbed energy, LOP, MOR, and d_x) with the data that correspond to various characteristic points f_x (f_{cr}, f_{max}, and f_d).

Table 2. Tested specimens, a compilation of data in relation to characteristic points f_x.

Composite	Load (N)	Deflection (mm)	Energy (J)	LOP (MPa)	MOR (MPa)	d_x (kN/mm)
$MV_f0\%$						
$f_{cr} = f_{max}$	13,048	0.716	4.7	2.7	2.7	18.2
$MV_f2\%$						
f_{cr}	16,730	0.887	6.8	3.5		18.7
f_{max}	28,908	3.451	59.7		6.0	
f_d	16,730	7.228	142.8			
$MV_f3\%$						
f_{cr}	25,615	1.618	22.0	5.3		15.8
f_{max}	51,586	5.148	156.1		10.7	
f_d	25,615	9.194	308.3			
$ZV_f6\%$						
f_{cr}	42,348	2.078	42.1	8.8		20.4
f_{max}	64,094	5.394	116.5		13.4	
f_d	42,348	9.752	455.7			

In order to compare the results with results obtained in other research centres, the results are compared with a reference matrix (specimen of mortar without reinforcement $MV_f0\%$). The obtained results indicate multiple changes (improvement/deterioration of properties) as compared with the original composite, Table 3.

During the analysis of the results, significant discrepancies were found regarding the determination of deflection by means of crosshead displacement and by means of reference to the neutral axis of the specimen with respect to Hooke's law proportionality range, point f_{cr}. In the case of a beam made of paste with the maximum content of fibres $ZV_f6\%$, the following deflection was recorded relative to the neutral axis in point f_{cr}, sensor 1, 0.774 mm and sensor 2, 0.879 mm, Figure 9 The average deflection was 0.827 mm, while the corresponding crosshead displacement was significantly larger and equalled 2.078 mm.

Table 3 proposes a method of comparing the results. The reference matrix was mortar without reinforcement (any other composite can be taken as a reference point).

Table 3. The tested specimens, a compilation of the obtained results as compared with a reference matrix.

Composite	Load (N)	Deflection (mm)	Energy (J)	LOP (MPa)	MOR (MPa)	d_x (kN/mm)
matrix $MV_f0\%$ $f_0 = f_{cr} = f_{max}$	13,048	0.716	4.7	2.7	2.7	18.2
$MV_f2\%/MV_f0\%$ $f_{cr/0}$	1.28	1.24	1.45	1.30		1.03
$f_{max/0}$	2.22	4.82	12.70		2.22	
$f_{d/0}$	1.28	10.09	30.38			
$MV_f3\%/MV_f0\%$ $f_{cr/0}$	1.96	2.56	4.68	1.96		0.87
$f_{max/0}$	3.95	7.19	33.21		3.96	
$f_{d/0}$	1.96	12.84	65.60			
$ZV_f6\%/MV_f0\%$ $f_{cr/0}$	3.25	2.90	8.96	3.26		1.12
$f_{max/0}$	4.91	7.53	24.79		4.96	
$f_{d/0}$	3.95	13.62	96.96			

4. Discussion

Increasing the ability to carry stress in the elastic range in ESD cement composites reinforced with low-module fibres is very limited (as compared with high-module reinforcement) and even impossible if the volume of fibres is low. The conducted tests show that with the content of fibres exceeding $V_f = 2\%$ in mortar cement composites there is such a possibility. Previous tests carried out on concrete composites indicated there was no such possibility with $V_f = 2\%$ [24], Figure 4. In the presented tests (Figure 8), we found that it was possible, if a large volume of dispersed reinforcement was introduced at the level of $V_f = 3\%$. It was impossible in concrete composites [13,28], due to the deteriorating (along with the increasing fibre content) rheological properties of the mixes [25].

The best ESD effects were demonstrated in paste mixes (with the best rheological parameters as compared with mortars and concretes), which enabled the addition of the largest volume of fibres $V_f = 6\%$, Figure 8. It should be noted that such a large content of dispersed reinforcement makes it difficult to form the mix, which makes it predestined for use in the prefabrication of cement composites. The use of smaller quantities of the reinforcement (which improved the rheological parameters of the mixes) resulted in decreased ESD effects in those composites. The obtained results indicate the possibility of using this type of reinforcement, for example, in prefabricated thin-walled composites such as building facade cladding panels. ESD effects limit the catastrophic destruction process of these composites (e.g., earthquake or mechanical damage). Paste with the maximum fibres volume was proposed as the ESD composite instead of mortar, because it was possible to introduce twice as many fibres into the paste composite, which resulted in a significant improvement of the elastic range of those composites, which was the main goal of this paper.

In the presented test results, emphasis was placed again on the need to compare the obtained effects f_{cr}, f_{max}, f_d with the parameters of a reference matrix (without reinforcement). The ESD composites should show greater ability to carry stress in the elastic range, strengthening control area, and a considerable range of deflections in the deflection control area. Assigning appropriate symbols to the tested composites enables the identification of their behaviour under load, and it is possible to describe them in detail on the basis of the characteristic points f_x (F_x, load; ε_x, deflection; W_x, energy and areas A_x, Figure 3).

The determination of the absorbed energy by means of ASTM 1018 (I_5, I_{10}, I_{15}) and crack mouth opening displacement (CMOD) did not correlate with f_{cr} f_{max}, and f_d, rendering a description of the behaviour under load in various deflection areas impossible. We found that the values of deflection measured by means of crosshead displacement were considerably higher than those measured in relation to the neutral axis. The measurement of deflection in the elastic range in accordance with [40] is a good method of deflection assessment. We confirmed that the deflection of ESD composites outside the elastic range relative to the neutral axis could not be controlled, Figure 9. Despite the significantly larger deflections measured by means of crosshead displacement as compared with the measurement in relation to the neutral axis, the results should be considered to be a good basis for the assessment of the elastic range, assuming that those diagrams are made based on the relationship between linear force and deflection (the initial deflections and settlement on supports resulting in disproportionately larger deflections are not taken into consideration).

An attempt to standardise the symbols and description of the method of testing various building materials under load seems to be justified. The existing recommendations and guidelines focus on tests and symbols ascribed differently to different materials when describing the behaviour of materials under load, Figure 3.

The testing of materials under load also needs to be standardised in terms of the method of assessing the obtained results. The impact of a number of variables (size of the specimen, cutting, load application speed, humidity, etc.) often makes the results from tests obtained in various universities difficult to compare. Again, we suggest the possibility of assessing the results based on a comparison with composite without reinforcement, and therefore the results could be approximately compared with the results obtained at various universities, limiting the influence of the scale of specimens and testing methods on the obtained results, since the results would be assessed based on multiple improvements/deteriorations as compared with a reference matrix. The possibilities for comparing results are presented in Table 3 in relation to the four-point bending test, which requires further discussion.

5. Conclusions

We have demonstrated that it is possible to improve the ability to carry stress in Hooke's law proportionality range in cement composites reinforced with low-module fibres if a large quantity of dispersed reinforcement exceeds $V_f = 3\%$. That condition cannot be fulfilled in traditional concrete structures due to worse rheological parameters of the mix as compared with mortar or paste composites.

The best ESD effects were demonstrated in the elastic range (and additionally in strengthening and deflection control areas) in paste with the maximum volume of fibres $V_f = 6\%$.

We suggest that there is a need to assess the obtained effects f_{cr}, f_{max}, f_d based on a comparison with the parameters of matrix (specimens without reinforcement), in order to identify quasi-brittle composites as ESD, with increased ability to carry stress in the elastic range, strengthening control area, and a considerable range of deflections in the deflection control area.

The values of deflection measured by means of crosshead displacement were demonstrated to be considerably higher than those measured in relation to the neutral axis. However, it was difficult to assess the ESD effects in relation to the neutral axis and crack mouth opening displacement.

Author Contributions: K.S. correction of results and applications; D.L. analysed the test results and performed editing, prepared the specimens, and analysed the test results. M.R., K.F. and K.S. writing—original draft preparation. M.R., K.F. and K.S. writing—review and editing. All authors have read and agreed to the published version of the manuscript.

Funding: This research received no external funding.

Data Availability Statement: Not applicable.

Conflicts of Interest: The authors declare no conflict of interest.

References

1. Banthia, N.; Sappakittipakorn, M. Toughness enhancement in steel fibre reinforced concrete through fibre hybridization. *Cem. Concr. Res.* **2007**, *37*, 1366–1372. [CrossRef]
2. Banthia, N.; Soleimani, S.M. Flexural response of hybrid fibre reinforced cementitious composites. *ACI Mater. J.* **2005**, *192*, 5.
3. Bentur, A.; Mindess, S. *Fibre Reinforced Cementitious Composites*; Elsevier: London, UK, 1990.
4. Brandt, A.M. Fibre reinforced cement-based (FRC) composites after over 40 years of development in building and civil engineering. *Compos. Struct.* **2008**, *86*, 3–9. [CrossRef]
5. Brandt, A.M. *Cement Based Composites. Materials Mechanical Properties and Performance*, 2nd ed.; Taylor & Francis: Abingdon, UK, 2009.
6. Campione, G.; Mindess, S.; Papia, M. Tensile strength of medium and high strength fibre reinforced concrete: A Comparison of different testing techniques. In Proceedings of the Brittle Matrix Composites 6, Warsaw, Poland, 9–11 October 2000; pp. 83–92.
7. Chen, M.; Zhong, H.; Zhang, M. Behaviour of recycled tyre polymer fibre reinforced concrete under dynamic splitting tension. *Cem. Concr. Compos.* **2020**, *114*, 103764. [CrossRef]
8. Deng, Z. The fracture and fatigue performance in flexure of carbon fibre reinforced concrete. *Cem. Concr. Compos.* **2005**, *27*, 131–140. [CrossRef]
9. Fantilli, A.P.; Jóźwiak-Niedźwiedzka, D. The effect of hydraulic cements on the flexural behavior of wool reinforced mortars. *Acad. J. Civ. Eng.* **2020**, *37*, 287–292, ISSN 2680-1000. [CrossRef]
10. Felekoğlu, B.; Tosun, K.; Baradan, B. Effects of fibre type and matrix structure on the mechanical performance of self-compacting micro-concrete composites. *Cem. Concr. Res.* **2009**, *39*, 1023–1032. [CrossRef]
11. Cascardi, A.; Lerna, M.; Micelli, F.; Aiello, M.A. Discontinuous FRP-Confinement of Masonry Columns. *Front. Built Environ.* **2020**, *5*, nr 147. [CrossRef]
12. Foti, D. Innovative techniques for concrete reinforcement with polymers. *Constr. Build. Mater.* **2016**, *112*, 202–209. [CrossRef]
13. Foti, D. Preliminary analysis of concrete with waste bottles PET fibers. *Constr. Build. Mater.* **2011**, *25*, 1906–1915. [CrossRef]
14. Glinicki, M.A. Testing of macro-fibre reinforced concrete for industrial floors. *Cem. Lime Concr.* **2008**, *4*, 184–195, ISSN 1425-8129.
15. Hsie, M.; Tu, C.; Song, P.S. Mechanical properties of polypropylene hybrid fibre-reinforced concrete. *Mater. Sci. Eng.* **2008**, *494*, 153–157. [CrossRef]
16. Jasieńko, J.; Logoń, D.; Misztal, W. Trass-lime reinforced mortars in strengthening and reconstruction of historical masonry walls. *Constr. Build. Mater.* **2016**, *102*, 884–892. [CrossRef]
17. Kanda, T.; Li, V.C. Interface property and apparent strength of high-strengthhydrophilicfibre in cement matrix. *J. Mater. Civ. Eng.* **1998**, *10*, 5–13. [CrossRef]
18. Kang, S.-T.; Kim, J.-K. The relation between fibre orientation and tensile behavior in an Ultra High Performance Fibre Reinforced Cementitious Composites (UHPFRCC). *Cem. Concr. Res.* **2011**, *41*, 1001–1014. [CrossRef]
19. Karihaloo, L. Optimum design of high-performance steel fibre-reinforced concrete mix. In Proceedings of the Brittle Matrix Composites 6, Warsaw, Poland, 9–11 October 2000; pp. 3–16.
20. Kim, D.J.; Naaman, A.E.; El-Tawil, S. Comparative flexural behavior of four fibre reinforced cementitious composites. *Cem. Concr. Compos.* **2008**, *30*, 917–928.1032B. [CrossRef]
21. Kucharska, L.; Logoń, D. Influence of the composition of matrices in HPFRCC on the effects of their ageing. In Proceedings of the Non-Traditional Cement & Concrete II, Brno, Czech Republic, 14–16 June 2005; pp. 344–353.
22. Kucharska, L.; Brandt, A.M. Pitch-based carbon fibre reinforced cement composites. In *Materials Engineering Conference ASCE. Materials for the New Millenium*; Chong, K.P., Ed.; ASCE: Washington, DC, USA, 1996; Volume 1, pp. 1271–1280.
23. Logoń, D. FSD cement composites as a substitute for continuous reinforcement. In Proceedings of the Eleventh International Symposium on Brittle Matrix Composites BMC-11, Warsaw, Poland, 28–30 September 2015; Brandt, A.M., Ed.; Institute of Fundamental Technological Research: Warsaw, Poland, 2015; pp. 251–260.
24. Logoń, D. Hybrid reinforcement in SRCC concrete. In Proceedings of the Third International Conference on Sustainable Construction Materials and Technologies SCMT3, Kyoto, Japan, 18–21 August 2013; p. e154.
25. Logoń, D. Quality control of cement mixes using cube flow and electrical resistance tests. In Proceedings of the Non-Traditional Cement & Concrete IV, Brno, Czech Republic, 27–30 June 2011; Bílek, V., Keršner, Z., Eds.; Brno University of Technology: Brno, Czech Republic, 2011; pp. 238–247.
26. Logoń, D.; Schabowicz, K.; Wróblewski, K. Assessment of the mechanical properties of ESD pseudoplastic resins for joints in working elements of concrete structures. *Materials* **2020**, *13*, 2426. [CrossRef]
27. Logoń, D. The application of acoustic emission to diagnose the destruction process in FSD cement composites. In Proceedings of the International Symposium on Brittle Matrix Composites BMC-11, Warsaw, Poland, 28–30 September 2015; Brandt, A.M., Ed.; Institute of Fundamental Technological Research: Warsaw, Poland, 2015; pp. 299–308.
28. Logoń, D. Safe cement composites SRCC—The rope effect in HPFRC concrete. In *Brittle Matrix Composites*; BMC 10; Elsevier: Warsaw, Poland, 2012; pp. 253–254.
29. Naaman, E.; Reinhardt, H.W. Proposed classification of HPFRC composites based on their tensile response. *Mat. Struct.* **2006**, *39*, 547–555. [CrossRef]
30. Naaman, A.E. Engineered steel fibres with optimal properties for reinforcement of cement composites. *J. Adv. Concr. Technol.* **2003**, *1*, 241–252. [CrossRef]

31. Naaman, A.E.; Reinhardt, H.W. *High Performance Fibre Reinforced Cement Composites (HPFRCC4)*; Pro 30; RILEM: Paris, France, 2003; p. 546.
32. Nataraja, M.C.; Dhang, N.; Gupta, A.P. Toughness characterization of steel fibre-reinforced concrete by JSCE approach. *Cem. Concr. Res.* **2000**, *30*, 593–597. [CrossRef]
33. Qian, C.X.; Stroeven, P. Development of hybrid polypropylene-steel fibre-reinforced concrete. *Cem. Concr. Res.* **2000**, *30*, 63–69. [CrossRef]
34. Ranachowski, Z.; Schabowicz, K. The contribution of fibre reinforcement system to the overall toughness of cellulose fibre concrete panels. *Constr. Build. Mater.* **2017**, *156*, 1028–1034. [CrossRef]
35. Šimonová, H.; Topolář, L.; Schmid, P.; Keršner, Z.; Rovnaník, P. Effect of carbon nanotubes in metakaolin-based geopolymer mortars on fracture toughness parameters and acoustic emission signals. In Proceedings of the BMC 11 International Symposium on Brittle Matrix Composites, Warsaw, Poland, 28–30 September 2015; pp. 261–288.
36. Sun, Z.; Xu, Q. Microscopic, physical and mechanical analysis of polypropylene fibre reinforced concrete. *Mater. Sci. Eng. A* **2009**, *527*, 198–204. [CrossRef]
37. Taylor, M.; Lydon, F.D.; Barr, B.I.G. Toughness measurements on steel fibre-reinforced high strength concrete. *Cem. Concr. Compos.* **1997**, *19*, 329–340. [CrossRef]
38. Xu, B.; Toutanji, A.H.; Gilbert, J. Impact resistance of poly(vinyl alcohol) fibre reinforced high-performance organic aggregate cementitious material. *Cem. Concr. Res.* **2010**, *40*, 347–351. [CrossRef]
39. Yao, W.; Li, J.; Wu, K. Mechanical properties of hybrid fibre-reinforced concrete at low fibre volume fraction. *Cem. Concr. Res.* **2003**, *33*, 27–30. [CrossRef]
40. ASTM 1018. *Standard Test Method for Flexural Toughness and First Crack Strength of Fibre–Reinforced Concrete*; American Society for Testing and Materials: Philadelphia, PA, USA, 1992; Volume 04.02.
41. EN 14651. *Test Method for Metallic Fibre Concrete. Measuring the Flexural Tensile Strength (Limit of Proportionality (LOP), Residual)*; European Committee for Standarization: Brussels, Belgium, 2005.
42. Japan Concrete Institute Standard JCI-S-003-2007. *Method of Test for Bending Moment–Curvature Curve of Fibre-Reinforced Cementitious Composites*; Japan Concrete Institute: Tokyo, Japan, 2007.
43. Japan Concrete Institute Standard JCI-S-002-2003. *Method of Test for Load-Displacement Curve of Fibre Reinforced Concrete by Use of Notched Beam*; Japan Concrete Institute: Tokyo, Japan, 2003.
44. Japan Concrete Institute Standard JCI-S-001-2003. *Method of Test for Fracture Energy of Concrete by Use of Notched Beam*; Japan Concrete Institute: Tokyo, Japan, 2003.
45. TC Membership. *RILEM TC 162-TGF: Test and Design Methods for Steel Fibre Reinforced Concrete*; Bending Test. Final Recommendation; International Union of Laboratories and Experts in Construction Materials, Systems and Structures: Paris, France, 2002; Volume 35, pp. 579–582.

Article

Diagnostics of Concrete and Steel in Elements of an Historic Reinforced Concrete Structure

Paweł Tworzewski, Wioletta Raczkiewicz *, Przemysław Czapik and Justyna Tworzewska

Faculty of Civil Engineering and Architecture, Kielce University of Technology, Al. Tysiąclecia Państwa Polskiego 7, 25-314 Kielce, Poland; ptworzewski@tu.kielce.pl (P.T.); p.czapik@tu.kielce.pl (P.C.); j.tworzewska@tu.kielce.pl (J.T.)
* Correspondence: wiolar@tu.kielce.pl; Tel.: +48-41-34-24-582

Abstract: Existing buildings, especially historical buildings, require periodic or situational diagnostic tests. If a building is in use, advanced non-destructive or semi-destructive methods should be used. In the diagnosis of reinforced concrete structures, tests allowing to assess the condition of the reinforcement and concrete cover are particularly important. The article presents non-destructive and semi-destructive research methods that are used for such tests, as well as the results of tests performed for selected elements of a historic water tower structure. The assessment of the corrosion risk of the reinforcement was carried out with the use of a semi-destructive galvanostatic pulse method. The protective properties of the concrete cover were checked by the carbonation test and the phase analysis of the concrete. X-ray diffractometry and thermal analysis methods were used for this. In order to determine the position of the reinforcement and to estimate the concrete cover thickness distribution, a ferromagnetic detection system was used. The comprehensive application of several test methods allowed mutual verification of the results and the drawing of reliable conclusions. The results indicated a very poor state of the reinforcement, loss in the depth of cover and sulphate corrosion.

Keywords: reinforced concrete; diagnostic testing; corrosion; carbonation; galvanostatic pulse method; phase composition analysis; X-ray analysis; thermal analysis

Citation: Tworzewski, P.; Raczkiewicz, W.; Czapik, P.; Tworzewska, J. Diagnostics of Concrete and Steel in Elements of an Historic Reinforced Concrete Structure. *Materials* **2021**, *14*, 306. https://doi.org/10.3390/ma14020306

Received: 21 November 2020
Accepted: 5 January 2021
Published: 8 January 2021

Publisher's Note: MDPI stays neutral with regard to jurisdictional claims in published maps and institutional affiliations.

Copyright: © 2021 by the authors. Licensee MDPI, Basel, Switzerland. This article is an open access article distributed under the terms and conditions of the Creative Commons Attribution (CC BY) license (https://creativecommons.org/licenses/by/4.0/).

1. Introduction

Periodic inspections of existing buildings, as well as renovation or modernisation works and even planned demolition works, require appropriate diagnostic tests [1–3]. In standard periodic inspections aimed at assessing the technical condition of the facility, the procedures for carrying out such tests are governed by relevant regulations, e.g., construction law in Poland [4]. However, in other cases, especially when it comes to examinations after a construction failure or due to the desire to change the existing utility function of the facility (despite the existing general recommendations), it is necessary to approach a given case individually and prepare an original program of diagnostic tests. Particular attention should be paid to the research of historic buildings, in which any invasive work interfering with the structure of the elements is often prohibited [2–7]. In such situations, it is usually necessary to use non-destructive or semi-destructive methods [1,3,8].

The significant and rapid development of technology in the field of measurement methods has recently contributed to the use of tools in construction diagnostics that allow destructive tests of elements and specimens as well as measurements with methods that are non-destructive or semi-destructive in their effects on building structures [1,8–11]. Additionally, measurements can be made with ever-greater precision. Due to the multitude of test methods that can be used, it is important to choose the right measurement method appropriate to the situation. If possible, it is worth verifying the obtained results with several methods. This is very important in obtaining reliable results, minimising measurement errors and reaching accurate conclusions.

371

Carefully considered diagnostic tests allow verification of the data contained in historic project documentation or obtained on the basis of an analytical model [2,3,7,12]. Based on appropriately selected and performed diagnostic tests, it is possible to develop reliable expertise indicating the causes of possible damage. Recommendations regarding the scope, order and manner of repair and upgrading work can be made.

However, not only should a well-prepared program of diagnostic tests account for an assessment of the existing condition, it should also provide a forecast for the durability of the structure in the future. This applies in particular to load-bearing structures operated in unfavourable environmental conditions, in which the design assumptions for ensuring adequate durability were incorrect [1,7–16]. In the article [16], the authors pay special attention to the chemical tests of the material taken from the structure (usually small samples of the material obtained while performing semi-destructive tests), which allows assessment of the chemical condition of the elements in the context of protection against corrosion during its future operation. The authors believe that the decisive factor in the possibility of effective and durable repair of a corroded reinforced concrete structure is its chemical condition.

The comprehensive and systematised classification of test methods for the diagnosis of concrete structures can be found in [17].

In the article, the authors also provided information on external and internal factors influencing the degradation of the structure, the degree of their impact and the optimal selection of research methods (destructive, non-destructive and semi-destructive). Diagnostic tests can be divided into tests aimed at estimating the load-bearing capacity of structure elements and strength parameters of materials (very often destructive or semi-destructive tests) and tests aimed at assessing the durability of elements and the entire structure. The authors took into account the usefulness of using various methods of assessing the durability of structures depending on the degradation mechanisms and their impact on durability. Non-destructive methods (NDT) were analysed, i.e., those that do not violate the integrity of the structure in any way, and semi-destructive methods (SDT), i.e., those that require material sampling or that slightly disturb the integrity of the structure. The authors proposed the classification of physicochemical and biological diagnostic methods to assess the durability of RC structures and presented the research equipment [17].

In the case of reinforced concrete structures, diagnostic tests primarily concern the assessment of the state of reinforcement and the concrete cover, which directly determines the durability of the structure [1,10,13–15,18–20].

In reinforced concrete structures, concrete (if it is tight enough, was made according to the appropriate recipe and has the right thickness) is the basic and best protection for the reinforcement [10,11,13,21,22]. The protective role of concrete is related to its alkaline reaction (pH $\approx$ 12.5–13.5), giving the effect of passivation at the interface between concrete and steel (formation of a micro-layer with very low ionic conductivity), which for all practical purposes prevents corrosion of the bars. Unfortunately, because of the negative long-term effects of physical and chemical factors of the external environment as well as to possible mechanical damage, the protective qualities of the concrete cover deteriorate.

Most often the deterioration is the effect of carbon dioxide in the air and changes in temperature and humidity of the environment, which leads to gradual carbonation of concrete and directly affects the loss of the protective properties of the cover in relation to the reinforcement [10,11,21–24].

As a result of carbonation, the pH of the concrete gradually decreases, and the areas of neutralised concrete extend further and further into the cover, eventually reaching the passive layer. In areas where the pH of the concrete drops below ~11.8, the surface of the reinforcing steel is depassivated, which directly leads to the initiation of electrochemical corrosion of the reinforcement.

On the other hand, in the case of seaside facilities located in the coastal zone (port and tourist buildings), exposed either to the direct influence of sea salt or to the so-called salt mist, as well as road structures (bridges, viaducts, flyovers), concrete surfaces, parking lots and open garages, due to the use of de-icing agents containing NaCl, the cause of degradation is the action of chlorides [10,21,23–27]. Sulphates that degrade concrete are also dangerous [28,29].

On the other hand, mechanical damage to the concrete cover (caused by accidents or minor damage accompanying the use of the facility), although visible to the naked eye in the form of cracks and defects, paradoxically often does not indicate an advanced process of structural degradation [30].

Nevertheless, the assessment of the load-bearing capacity of structural elements is a very important factor in the diagnostics of buildings. For this purpose, computational and experimental models are often used [31–33]. It is especially important in the case of unusual load, such as high temperature caused by fire [34]. In recent years, neural networks have been used for diagnostics of structural elements of buildings [35,36].

Usually, diagnostics of reinforced concrete structure elements is carried out in several ways. One of them concerns the measurements of the thickness and continuity of the concrete cover and the location of the reinforcement. The second are electrochemical, non-destructive or semi-destructive tests, which indirectly allow one to determine and forecast the degree of concrete and steel corrosion advancement. Third is the testing of the material, i.e., fragments of concrete cover taken from the structure, on the basis of which the chemical composition of concrete and its microstructure are determined. Computational and experimental models are also used to assess the load-bearing capacity of structural elements.

The applied research methods are quite known in the diagnostics of reinforced concrete structures. However, often the tests are performed selectively—they are limited to one type of method, which provides limited information. The authors would like to point out that only the knowledge and combination of several methods allows for the preparation of an appropriate research plan and a full assessment of the condition of the structure, especially if it concerns historic objects, in the diagnostics of which it is only possible to use non-destructive or semi-destructive methods, as in the described example. Additionally, the authors would like to draw attention to the measurement techniques used. While the ferromagnetic method used in the research can be successfully replaced by other methods (which are described in more detail in this chapter), the galvanostatic pulse method seems to be the most precise and easy to use compared to other commonly used electrochemical methods. X-ray analysis and phase composition analysis are also noteworthy as techniques allowing for very precise recognition of the effects of carbonation.

2. Materials and Methods

The thickness of the concrete cover is one of the key factors determining the durability of reinforced concrete structures susceptible to aggressive external environment, because it determines the speed of penetration of aggressive substances that can damage the passive layer and initiate corrosion of the reinforcement [10,11,13,15,29].

There are many test methods that allow determining the thickness of the cover and locating the reinforcement bars, their diameter and mutual position. Currently, a number of the methods in use are based on changes in electromagnetic waves penetrating through the tested reinforced concrete element medium. The tests can be performed using electromagnetic, radiological, ultrasonic, or radar methods. The precision of the obtained results as well as the effectiveness and efficiency of the research as a function of time depend on the research method used.

One of the most accurate among those mentioned above is the radiological (radiographic) method, which uses ionising radiation (X-rays or gamma rays) [37–40]. Because of this, it is possible to determine the geometrical parameters of the reinforcement with an accuracy of 5%. It is the best method in terms of accuracy that allows locating the reinforcement and its parameters in complex reinforced concrete elements. However, its application requires the use of complex equipment and protection against harmful radiation. The long duration of the research also contributes to the low popularity of this type of research.

A more widely used alternative to radiographic methods are electromagnetic methods [3,37], which use the phenomenon of electromagnetic field changes according to the ferromagnetics (steel) system in a reinforced concrete element [1,41–43]. Modern devices used to perform tests with this method are easy to use (intuitive). The analysis of the scanned reinforcement image is also not complicated [44,45]. Unfortunately, this type of research is not the most accurate and is susceptible to the influence of many factors.

The radar method can also be used to locate reinforcement in concrete. It uses the phenomenon of the penetration and reflection of an electromagnetic wave (in the range of 100 MHz–2 GHz) through a medium (reinforced concrete element) which, depending on the medium material type (concrete or steel), the wave penetrates into depth or is scattered or reflected [46–50]. In the case of the currently used equipment, the method of conducting the research itself is similar to the electromagnetic method, but the interpretation of the results requires more knowledge. Fortunately, the development of computer software greatly facilitates the operation and analysis of the obtained phalograms. As in the case of the previous method, this method is also susceptible to the influence of many factors that may disturb or prevent the analysis of the obtained results.

Impact-echo [51] and impulse response [52] methods, which rely on the analysis of the flow time of a mechanically excited wave in the tested medium (concrete), can also be used to assess the continuity of the reinforcement. The acoustic emission (AE) method is also more and more widely used [53–55]. It consists of the analysis of differences in the acoustic wave propagation speed in reinforced concrete. However, these methods do not allow for measuring the diameter of the bars or for the precise measurement of the reinforcement cover thickness.

In most cases, electromagnetic methods are effective and sufficient. In the tests described in this article, the PS 200 Hilti (Schaan, Liechtenstein) scanner was used to locate the reinforcement and determine the cover parameters. This device has been on the market for a long time. It enables easy and quick measurement and subsequent analysis of the obtained results. The performed measurements allow one to estimate the cover thickness and the diameter of the reinforcement. The ability to perform surface scans and combine them is particularly useful. As mentioned earlier, this is not an accurate method, and any magnetic materials in the concrete, such as tie wire or other ferrous particles, as well as reinforcement corrosion products, will distort the results. However, it is still the fastest method of reinforcement detection and in most cases it is sufficient in the diagnosis of reinforced concrete elements [15].

Some of the mentioned methods allow the detection of defects and discontinuities of the material that are not visible on the surface. The method of acoustic emission allows, among others, the recording of acoustic signals caused by defects in the microstructure of the material during the use of the structure [55]. The advantage of this method is its sensitivity and, as a result, its ability to determine various destructive processes that took place in the tested element at a given time.

On the other hand, defects in the form of scratches and cracks visible on the concrete surface can be diagnosed quite simply with the use of practical handy tools [1]. To measure the width, depth and activity of cracks, one can use feeler gauges or templates (paper or plastic), as well as a Brinell magnifier with a scale (Dioptra Turnov, Czech Republic) (optical instrument with a magnification of 8 to 40 times, with a scale of 0.05 mm or 0.1 mm). Mechanical feeler gauges (drawing gauges, gauges or crack gauges) allow one to measure the width of the crack as well as monitor displacements in one, two, or even three directions. In complicated cases where access to the crack is difficult, the problem can be overcome by the use of microcameras, which are an electronic version of a magnifier with a scale. Endoscopes (rigid—boroscopes—and flexible—fiberoscopes) are often used to inspect hard-to-reach places). Advanced optical measurement systems are increasingly used to measure the cracks and track their development on real objects (DIC-digital image correlation) [56].

More and more often, a scanning electron microscope (Hitachi Group, Dusseldorf, Germany) is also used to analyse the defects in the microstructure of concrete (SEM). SEM tests are very accurate but require special preparation of samples taken from the structure; thus, SEM tests are associated with slight damage to the structure. Therefore, these tests should be classified as semi-destructive tests. On their basis, it is possible to identify the causes of cracks [57,58].

An inventory of cracks allows one to estimate the causes of damage, locate places of potential weakening of elements, identify possible corrosion of the reinforcement and indicate the areas from which material for laboratory tests should be taken.

Electrochemical tests are particularly valuable because of the possibility of conducting them in the field with simultaneous assessment of the interaction of the current steel and concrete properties in real operating conditions. They make it possible to determine areas exposed to corrosion on the surface of elements as well as the degree of reinforcement corrosion advancement and to forecast its progression over time [10,11,16]. These tests are performed using specialised devices that allow the measurement of appropriate electrical quantities, the values of which indicate the course of the corrosion process on the reinforcement [1,10,14]. Currently, the most commonly used methods are mainly the potentiometric method, half–cell potentials (measurement of stationary potential and its gradient) [59], the polarisation method—measurement of corrosion current density [60–63]—and the resistance method [14] (concrete resistance—measurement of concrete cover resistivity). Currently, in the available literature, the most frequently described research is conducted using the half-cell potential method, although the experience of the authors and other researchers shows that this method is less reliable than the polarisation method (e.g., galvanostatic pulse method). In the paper [64], the authors presented the results of chemical tests as well as the measurements by the use of the galvanostatic pulse technique indicating an advanced corrosion process of the reinforcement in the elements of the bridge structure, while the half-cell potential method did not show it. The authors of this paper repeatedly drew similar conclusions from their research [23,29,30].

Among the above-mentioned electrochemical methods, the most reliable, and at the same time that which allows one to conduct measurements in the field in a relatively uncomplicated manner, are the measurements of the corrosion current density.

Polarisation methods, depending on the method of initiating the disturbance, are three in number: the electrochemical impedance spectroscopy method (disorder results from the action of alternating current in a wide frequency range); the linear polarisation resistance method (disorder is generated by applying a linearly changing potential); and the galvanostatic pulse method (the disorder is generated by a current with a certain intensity value).

In the case of diagnostic tests, it is worth using various measurement techniques simultaneously to verify the obtained results. It is particularly important to combine research with electrochemical methods with material tests and appropriate control opencasts.

The research described in the article used the GP-5000 GalvaPulse™ (Force Technology, Brøndby, Denmark) set for measurements using the galvanostatic pulse polarisation method. The equipment allows one not only to measure the corrosion current density (which allows one to estimate the corrosive activity of the reinforcement and forecast its rate), but also to measure the stationary potential of reinforcement and the resistivity of the concrete cover. The scope of the research makes it possible (on the basis of an analysis of the conditions conducive to corrosion) to determine the areas on the structure where the probability of corrosion is relatively higher than elsewhere. The obtained results are analysed in relation to certain criterion values, which in the form of tabulated data are attached to the apparatus (Table 1). Depending on the obtained values of the stationary reinforcement potential and the resistivity of the concrete cover, it is possible to conclude on the value of the corrosion probability of the reinforcement in the tested area. However, these are not always reliable measurements. On the basis of the value of the corrosive current density, its corrosive activity can be estimated and its rate forecast over time. At the same time, it should be remembered that the criteria presented in GP-5000 GalvaPulse™ set cannot be used elsewhere. When performing measurements with other equipment (due to a different type of reference electrode or a different way of polarisation of the reinforcement), the reference criteria are different.

Table 1. Criteria for assessing the degree of reinforcement corrosion risk [65].

Criteria for Assessing the Degree of Reinforcement Corrosion Risk by Use the Galvanostatic Pulse Method				
		Reinforcement corrosion activity		Corrosion pace; [mm·year^{-1}]
Corrosion Current Density	i_{cor} [µA·cm^{-2}]	<0.5	not forecasted corrosion activity	<0.006
		0.5–2.0	Irrelevant corrosion activity	0.006–0.023
		2.0–5.0	Low corrosion activity	0.023–0.058
		5.0–15.0	Moderate corrosion activity	0.058–0.174
		>15.0	high corrosion activity	>0.174
Reinforcement Stationary Potential	E_{st} [mV]	<−350	95% of corrosion probability	
		−350–−200	50% of corrosion probability	
		>−200	5% of corrosion probability	
Concrete Cover Resistivity	Θ [kΩ·cm]	≤10	high corrosion probability	
		10–20	medium corrosion probability	
		≥20	small corrosion probability	

A reinforcement covered in concrete is protected against corrosion by the steel passivation mechanism. However, it is only effective when the rebars are well-covered by concrete. Concrete may be affected by the environment and change its properties overtime. The carbonation process is considered to be very unfavourable. The effect of carbon dioxide that penetrates the concrete dissolves in the pore solution, resulting in the formation of carbonic acid. This acid reacts with the cement paste alkaline components, which results in the formation of carbonates and a lowering of concrete pH. When the pH drops below a certain level (pH < 11.8), the steel depassivates and in the presence of water and oxygen begins to corrode [21]. Carbonation, however, is a slow process. By detecting it at an early stage, it can be countered. For this reason, test methods that allow for quick and easy diagnosis of the carbonation process in concrete are being developed. Each of them has its own advantages and disadvantages. The advantages and limitations of many methods have been discussed by Qiu [22].

The most common methods for determining carbonation are colourimetric tests. They consist in spraying a solution on the tested concrete surface. Solution in contact with the concrete, depending on its pH, causes its colour to change according to its interaction with the surface. The most popular indicator used for this purpose is phenolphthalein. The study of carbonation with the use of phenolphthalein indicator is recommended by EN 13295:2004 and RILEM CPC-18. In this study, phenolphthalein, in contact with non-carbonated concrete with a high pH, causes its colouration to a characteristic, intense magenta colour. As a result of carbonation, the concrete pH may drop below 9. In this case, phenolphthalein produces no colouration of concrete. Based on this, carbonated and uncarbonated concrete can be distinguished. The determination of the concrete surface condition itself is unsatisfactory for diagnostic purposes. In tests, it is usually important to check the extent of carbonation in the concrete, which requires slight structural damage to collect a sample. This sample can then be used for other tests. However, due to the necessity to take samples from the structure, it is not a test used for continuous control of its condition. The limitations of this method also result from the phenolphthalein specificity, which can evaporate from the concrete surface in the period between the application and the measurement. As a result, the phenolphthalein concentration can change, which may affect the concrete colouration [22]. Time plays an important role in this type of test because of the individual phases of the cement matrix of concrete colour at different rates [66]. Phenolphthalein changes the concrete surface colour at a pH range from 8.5 up to 9.5. Meanwhile, depassivation of steel occurs when the pH drops to 11.5–11.8 [10,21]. For this reason, in colourimetric tests other indicators are also used, e.g., timoloftaleine, alizarin yellow R, alizarin yellow GG, or T^{F}PLPt [22,67]. Thymolphthalein changes colour at a pH of 9.3–10.5, while T^{F}PLPt can be used for pH tests in the range 11.0–13.5. The disadvantage of tests with phenolphthalein and thymolphthalein is the colouration of concrete to one colour only. As a result, they perform well in determining the depth of carbonation, but it is difficult to use them to determine its degree of advancement in particular areas. For this reason, colourimetric tests are currently being developed that also allow the identification of the pH distribution on the tested concrete surface. This type of test is the rainbow-test, which allows one to perform pH tests in the 5 to 13 range [10,22,68,69]. In this test, each change in pH by 2 is associated with a significant change in the obtained concrete colour. Concrete that is considered non-carbonated has the violet colour characteristic for pH = 11. Similarly, when a T^{F}PLPt indicator is used, the different colours are obtained depending on the pH of the concrete being tested [67]. In this case, non-carbonated concrete turns greenish or cream-coloured. In the case of this indicator, it is possible to perform more accurate analyses of the pH distribution on the tested surfaces using a camera with appropriate band-pass filters.

Compared to other tests, the colourimetric test's advantage is the speed and the ease of obtaining information about the pH lowering, which is a factor directly responsible for steel depassivation. More complicated but more precise tests of concrete pH concern the tests of the pH of the pore solution [10]. For this purpose, the tested concrete must be crushed to a particulate in the range of 0–0.2 mm, and then a model liquid must be obtained from such material by preparing a water extract or by vacuum or pressure extraction methods. The easiest way is to prepare a water extract by mixing ground concrete with distilled water in a 1:1 ratio and storing it for 24 h. After this period, the solid phase is filtered off and the pH of the liquid is measured with a pH-meter. The solution obtained in this way, however, does not fully correspond to the liquid that occurs naturally in the pores of concrete, as the pores are under a pressure other than atmospheric. A more accurate representation of the pore liquid is obtained by performing a vacuum extraction. For this purpose, after preparing a water extract of ground concrete, it is concentrated under vacuum. The best material for research is obtained using pressure extraction. During this, the test liquid is obtained by squeezing it from the concrete sample in a special press, applying a pressure of 450–500 MPa. In this way, small amounts of test liquid are obtained. By testing the pore liquid, one can measure the pH more accurately, but it is more difficult to determine the depth of carbonation, which requires many tests on concrete samples taken from greater and greater depth.

Computer X-ray tomography is a modern test that allows one to determine the carbonation front progress (XCT) [70,71]. Using this method, it is possible to test samples that do not require significant processing, and it is not necessary (e.g., as in colourimetric methods) to expose the tested surfaces. Carbonation measurement with this method is based on the fact that as a result of carbonation and the formation of calcium carbonate, the density of the concrete increases and its porosity decreases, which means that it absorbs more radiation [21,70]. As a result of this test, it is possible to determine the depth of both carbonated and uncarbonated concrete as well as its porosity, and to distinguish cracks in the concrete. In the case of reinforced concrete, it can also determine the location of the bars and determine whether they are not subject to corrosion as evidenced by a layer of rust, with a lower density and cracks around the bars.

Another test utilising X-rays that is used in testing the durability of concrete is X-ray diffraction (XRD). It can be used to determine the phase composition of concrete. Carbonated cement paste in concrete is characterised by reduced compactness and/or evident absence of portlandite. In this pastes in the place of portlandite various calcium carbonates are formed [28,66,72–75]. As a result of carbonation on the surface of the portlandite, amorphous calcium carbonates are formed, which over time turn into vaterite, from which aragonite can be formed. Ultimately, as a result of polymorphic transformations, calcium carbonate is produced in the form of calcite. In this form, calcium carbonate is often present in concrete as a component of aggregate, but it will not be present in the form of vaterite and aragonite. Thus, the detection of one of the metastable polymorphic forms of calcium carbonate in the course of research may indicate the ongoing process of carbonation. Such tests can be performed on small fragments of crushed concrete or cement paste matrix separated from it, that had been taken while performing other tests. For testing purposes it is also worth collecting drill cuttings, taken from different depths of concrete, in order to determine the depth to which the cement paste has been carbonated (using the Profile Grinder kit; (Force Technology, Brøndby, Denmark). However, testing of samples in which the cement paste is mixed with aggregate is very difficult, because the diffraction pattern is usually dominated by the aggregate peaks and it is difficult to identify the peaks originating from the cement matrix phases.

Fourier transformation infrared spectroscopy (FTIR) is also used to study the changes in the phase composition of the cement paste undergoing carbonation [22,72]. It can also be used to study the transformations of portlandite and the C–S–H phase as well as the $CaCO_3$ polymorphism. In addition to being able to determine the amount of the resulting carbonation products, one can also study its depth. In order to test the carbonation depth, the peak for C–O bonds corresponding to a wavelength of 1415 cm^{-1} is measured. This peak decreases until a constant amplitude is attained after reaching the depth at which the carbonation occurs. The results of the molecular structure studies undertaken with FTIR testing provide more accurate information about the partially carbonated zone than the results of the phenolphthalein indicator test.

Infrared spectroscopy studies can be combined with other types of research, thus obtaining more accurate information about the composition of the carbonated paste. They can be complemented with XRD analysis [72,73], but can also be combined with thermal analysis [76]. Thermal analysis by the thermogravimetric (TG) method is a popular method that allows one to quantify the cement paste carbonation degree. It is based on the measurement of the sample weight loss during its heating. By taking samples from different depths of concrete, it is possible to determine the exact content in them of $Ca(OH)_2$ and $CaCO_3$, and thus it can be stated to what extent the former has transformed into the latter. Using this technique, it is also possible to study $CaCO_3$ formed as a result of carbonation of the C–S–H phase, because crystallisation degree and thermal stability of the carbonates formed in this way is reduced [22]. However, it is not possible to distinguish polymorphic forms of calcium carbonate in concrete by thermal analysis. For this reason, this analysis is often supplemented with the phase composition analysis using the XRD or FTIR method.

Microstructural studies using a scanning electron microscope (SEM) or a transmission electron microscope (TEM) are often performed as complementary studies to phase analyses [22,28,72,74,77]. By observing the microstructure, changes can be observed on the surfaces of portlandite crystals as a result of the carbonation process [71] in the form of visible carbonation products (Figure 1). Thus, it is possible to estimate qualitatively to what extent the concrete is carbonated. Because of the differences in the porosity of the carbonated and non-carbonated paste [21,75,77], when conducting research on specimens at low magnifications, the depth of carbonation can also be determined. In such a study, Rimmelé et al. [78] were able to distinguish four successive zones of cement paste: non-carbonated core inside the paste, dissolution front, carbonation front and carbonated zone. These tests, as with those performed with the computed tomography method, allow for the location and assessment of the condition of the reinforcement. For SEM tests in this area (as opposed to XCT tests), samples should be specially prepared—they should be made into a polished section or thin section.

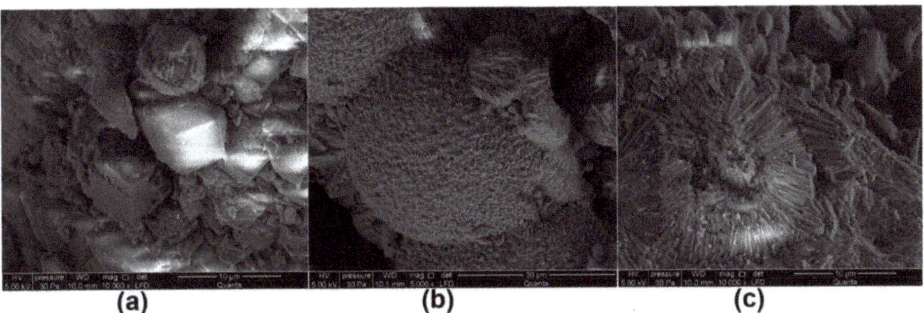

Figure 1. Calcium carbonates in carbonated drips on concrete described in [28]: (**a**) calcite, (**b**) vaterite, (**c**) aragonite.

As a result of the analysis of the phase composition using the XRD, FTIR and DTG methods as well as the observation of the microstructure, it is possible to detect other substances potentially dangerous for reinforced concrete. Because of the possible corrosion of the reinforcement, attention is drawn to the presence of Friedel's salts and other compounds containing chlorides. The presence of chlorides may cause the degradation of the cement paste, reducing its passivation properties, as well as having a significant impact on the corrosion of the reinforcement.

In order to study the progress of concrete carbonation, other spectroscopic methods can also be used: Raman, nuclear magnetic resonance and electrochemical impedance [22,79]. Carbonation and corrosion of the reinforcement can also be monitored using piezo-impedance transducers [80] or by measuring the absorption of gamma radiation [81].

3. Assessment of Selected Elements of the Historic Water Tower Structure

The following part of the article contains a description of the research aimed at assessing the condition of reinforcement and concrete cover in selected reinforced concrete elements of a historic water tower structure. The galvanostatic pulse method was used to assess the risk of reinforcement corrosion in concrete and forecast its rate over time in the structure elements, as well as X-ray diffraction and thermal diffraction analysis of the concrete cover. The ferromagnetic detection system was used to locate reinforcement in the structure elements.

3.1. The Research Methods and Used Materials

The water tower located in Zabrze (Silesia, Poland) was designed by an architect, August Kind, and Friedrich Loose, a building advisor. It was erected in 1909. The height of the building with respect to the ground level is about 45 m. The building has a unique construction (Figure 2a). It is based on an octagonal plan (the diameter of the base is 23 m) with 9 brick pillars—8 in the corners and 1 central. The pillars support the tank located at a height of about 28 m topped with a mansard roof. Between the pillars, in addition to one underground storey, there are three storeys accommodating a residential and office area.

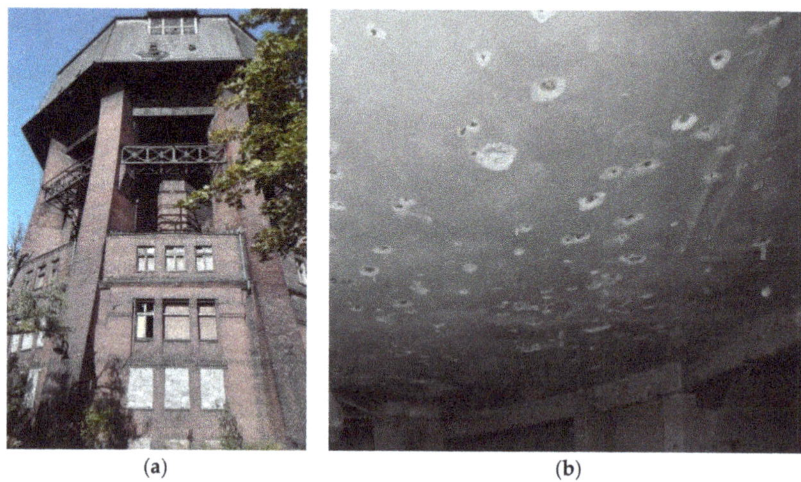

(**a**) (**b**)

Figure 2. The picture of a historic water tower: (**a**) picture of the object, (**b**) pop outs in the concrete cover.

At the end of 2018, an assessment of the tower's structure was carried out as part of the reconstruction, extension and modernisation strategy scheduled in the project to revitalise it for social, educational, scientific and cultural purposes. During the inventory, numerous concrete cavities were observed on all reinforced concrete elements of the supporting structure in the residential and office areas (Figure 2b). For this reason, it was decided to assess the state of reinforcement and concrete cover based on carefully selected tests.

Because of the historic character of the building, only non-destructive and semi-destructive testing were possible. A building inventory was developed and the structural system of the building was identified. Two elements were chosen for detailed diagnostic tests: a ceiling slab above the second floor and a load-bearing column (Figure 3).

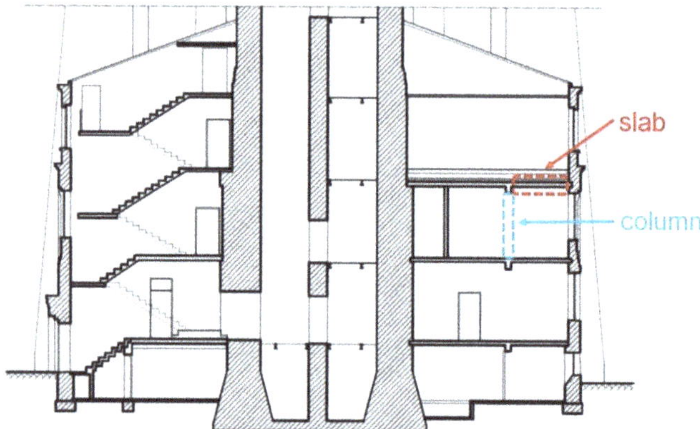

Figure 3. The vertical cross-section of the water tower lower part with measurement locations—the plan sketch.

The testing was carried out in two stages. In Stage I, in situ tests were performed to determine the rebar location and layout, rebar corrosion degree and the carbonation depth in the concrete cover. Samples were taken for laboratory tests. Stage II included laboratory tests of the concrete cover. The Hilti PS200 scanner was used to locate reinforcing bars in the concrete (Figure 4a). Software added to the device helped determine the thickness of the concrete cover. The device operates on the principle of electromagnetic induction [10,82]. The measurements helped to assess deviations in the concrete cover thickness in the tested areas. Thickness deviations can seriously affect the protective properties of the concrete [15].

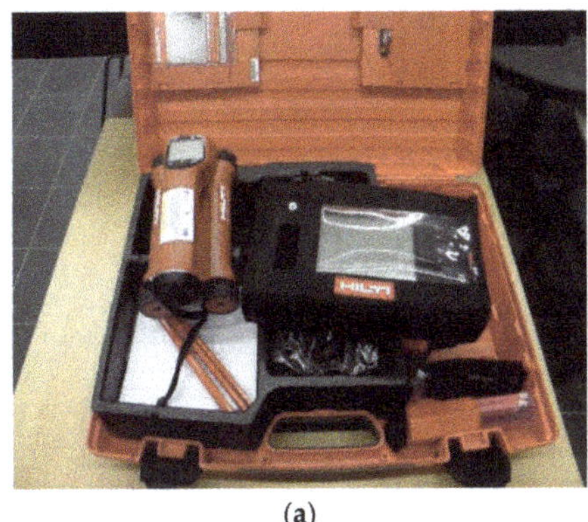 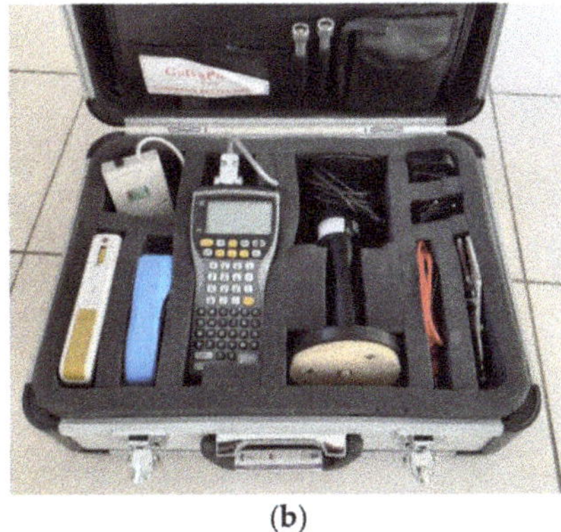

(a)   (b)

Figure 4. Measuring devices: (**a**) PS 200 Hilti scanner, (**b**) GP-5000 GalvaPulse™ kit.

The degree of reinforcement corrosion was evaluated via the electrochemical galvanostatic pulse method (GPM) [13,23,65]. This measurement method is an alternative to the more commonly used half-cell potential measurements. The commonly utilised potential mapping technique, measuring the half-cell potential on a concrete surface, sometimes has led to misinterpretation. For this reason the GPM has been introduced as a more advanced technique. Both electrochemical methods assume that corrosion of reinforcement in concrete is an electrochemical process. A steel bar serves as an electrode and the alkaline liquid filling in the concrete pores as an electrolyte. Local anode and cathode microcells formed on the surface of the bar generate the flow of electric charges through the bar, while the liquid filling the concrete pores carries the ions. Measurements are taken along the bar on the concrete surface after connecting the device to the bar. It is therefore necessary to expose the rebar along a few centimetres of its length. In the galvanostatic pulse method, the short-time anodic current pulse is impressed galvanostatically, which leads to the polarisation of the reinforcement and allows one to measure certain electrical quantities, i.e., the reinforcement's stationary potential (E_{st}), corrosion current density (i_{cor}) and concrete cover resistivity (Θ). The obtained values, after comparing them with the criterion values (Table 1), allow assessing the reinforcement corrosion degree in concrete. In these tests, the GP-5000 GalvaPulse™ set was used to make measurements (Figure 4b). The main elements of the set include the control and recording device (PSION minicomputer), silver-chloride reference electrode (Ag/AgCl) and calibration device. The advantage of using the GalvaPulse™ set is the relatively short measurement time—usually no more than thirty seconds at any given point.

Depth of carbonation into the concrete cover at the sites of exposed rebar was determined via the carbonation test (1% phenolphthalein solution test) [83].

Stage II included laboratory tests of the concrete cover samples extracted from the structure. X-ray diffractometry (XRD) and thermal analysis (DTA-TG) methods were used [84,85]. The phase composition of the samples was identified by X-ray diffraction in an Empyrean diffractometer (PANalytical, Almelo, The Netherlands) equipped with a Cu lamp and a X'Celerator silicon detector. The analyses were carried out in the range from 5 to 75°θ, and the ICDD PDF-2 database was used to interpret the diffractograms. The thermal analysis was performed in STD Q600analyser (TAinstruments, Hüllhorst, Germany). The measurements were carried out in a nitrogen atmosphere increasing the temperature to 1000 °C at a rate of 10 °C/min.

3.2. Tests Results

3.2.1. Reinforcement Location and Concrete Cover Thickness Measurement

The rebar position and layout in the elements were determined with the Ferroscan detector. For this purpose, a 3×3 reference grid with a cell size of 0.15×0.15 m was attached to the concrete surface in the selected site on the slab (Figure 2). This was followed by running the scanner along the grid rows and columns ("Imagescan" mode). The location of the rebars helped select the site for concrete cover removal for testing. The scans and measurements carried out on the exposed rebars found that the main reinforcement of the slab was composed of square bars of 7.5 mm side-spaced about 8–15 cm apart. The software (PS 200 Software, release 5.4.2.1) included in the device, after entering the data from the initial rebar diameter measurements, measured the thickness of the main rebar coverage at twenty points. On the basis of the obtained concrete cover thickness values, the minimum, maximum and average values as well as standard deviation and the variability index were determined. Obtaining a large number of such scans of the structure and the statistical analysis of the results allows to find places with the smallest cover thickness and the greatest dispersion of values. Corrosion of the reinforcement in these places is most likely due to the fastest destruction of the concrete cover. The results are compiled in Table 2. The graphical representation of the rebar detection results is shown in Figure 5a.

Table 2. Concrete cover thickness for the slab and column.

Average Value c [mm]	Maximum Value c_{max} [mm]	Minimum Value c_{min} [mm]	Standard Deviation s [mm]	Coefficient of Variation V [%]
		Slab		
21.3	26	18	2.5	12
		Column		
38	50	38	8	21

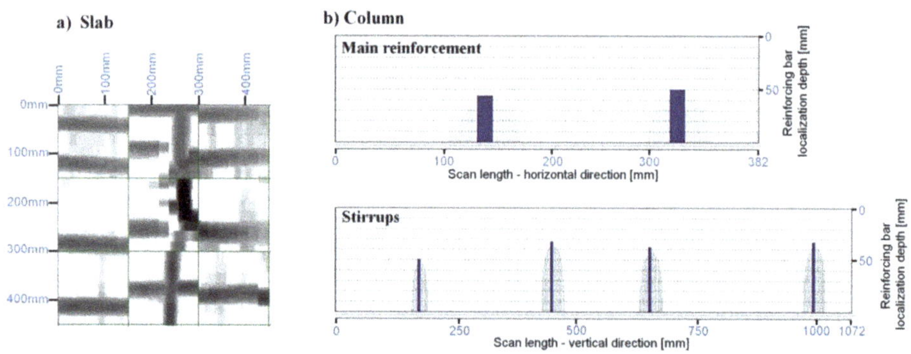

Figure 5. Image of reinforcing bars location in concrete elements, obtained with the use of Ferroscan device.

The rebar diameter measurements found that four square bars with 14 mm cross-section were used as the main reinforcement of the column. Because of the small width of the column, scanning was performed using the "Quickscan" method. The results of concrete cover thickness measurements are presented in Table 2. The graphical representation of the rebar detection results is shown in Figure 5b.

3.2.2. Degree of Reinforcing Bar Corrosion

Because of the visible poor condition of the concrete and the concrete cover carbonation confirmed by in situ tests, it was necessary to assess the state of reinforcement in the structure. The degree of reinforcement corrosion was assessed in the selected areas of the floor slab and column (Figure 3). Four bars were accessed in the floor slab (Figure 6a). Six points for measurements were planned to be set on the concrete surface along each bar (at ~10 cm spacing), that is to say, a total of twenty-four measurement points were established. Unfortunately, for an unknown reason, all rebar no. 1 measuring points specified by coordinates (1.1)–1.6) failed to provide results useful for analysis. Ultimately, the results from eighteen measurement points were analysed. In the column, two main reinforcing bars were accessed at a height of about 1.0 m from the floor level (Figure 6b). Along these bars, four points on each of them were determined at a spacing of ~10 cm, which gave a total of eight measurement points.

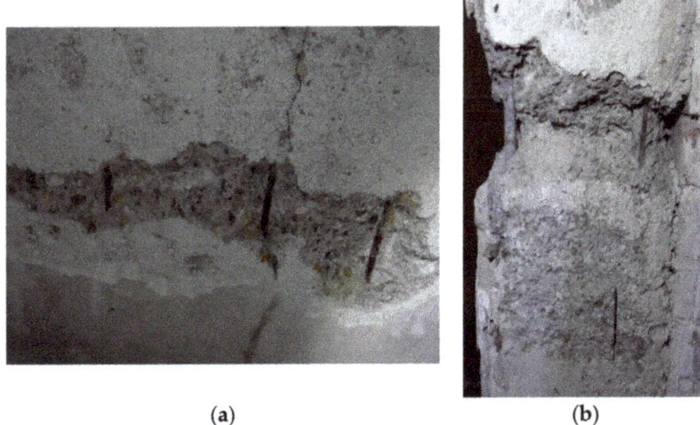

Figure 6. (**a**) Floor slab with exposed reinforcing bars; (**b**) column with exposed section of the bar under test.

The measurement results of all three parameters (E_{st}, Θ, i_{cor}) from the GP-5000 GalvaPulse™ device, measured at individual points, are summarised in Table 3 (slab) and Table 4 (column). The values of the predicted corrosion rate [$\mu m \cdot year^{-1}$] and maps of the distribution of values of all measured parameters were generated in the PsiWin program (included in the GP-5000 GalvaPulse™ set), Figure 6a for the floor slab and Figure 6b for the load-bearing column.

Table 3. Results from the measurements of the reinforcement stationary potential, corrosion current density and resistivity of concrete cover in the floor slab.

Meas. Points	Values			Meas. Points	Values			Meas. Points	Values		
	E_{st}	i_{cor}	Θ		E_{st}	i_{cor}	Θ		E_{st}	i_{cor}	Θ
(2, 1)	−376.06	3.53	4.90	(3, 1)	−307.06	12.23	2.10	(4, 1)	−324.14	15.71	2.00
(2, 2)	−242.30	2.98	4.00	(3, 2)	−327.65	8.82	4.70	(4, 2)	−362.29	12.02	2.60
(2, 3)	−256.27	3.65	3.10	(3, 3)	−346.85	6.07	4.10	(4, 3)	−384.76	6.83	5.10
(2, 4)	−304.25	2.81	6.40	(3, 4)	−575.50	3.66	6.40	(4, 4)	−422.67	6.50	6.30
(2, 5)	−421.74	4.23	2.30	(3, 5)	−596.33	3.44	6.60	(4, 5)	−508.80	5.84	6.80
(2, 6)	−453.80	4.92	8.00	(3, 6)	−533.84	13.19	5.90	(4, 6)	−535.95	10.74	2.00

Table 4. Results from the measurements of the reinforcement corrosion potential, corrosion current density and resistivity of concrete cover in the load-bearing column, storey II.

Meas. Points	Values			Meas. Points	Values		
	E_{st}	i_{cor}	Θ		E_{st}	i_{cor}	Θ
(1, 1)	−18372	0.19	27.50	(2, 1)	−316.19	17.53	4.30
(1, 2)	−173.89	1.55	13.30	(2, 2)	−568.48	6.71	9.90
(1, 3)	−169.44	2.47	14.20	(2, 3)	−467.14	8.19	6.80
(1, 4)	−142.76	2.59	8.80	(2, 4)	−437.42	7.28	6.60

The results were analysed based on the criteria presented in Table 1, which allowed determining the probability of rebar corrosion in the areas that were examined as well as assessing the corrosion degree and estimating the rebar corrosion rate.

In the floor slab, the analysis of the stationary potential of the reinforcement showed that the probability of corrosion in the selected area of the slab was very high. At ten measurement points, it exceeded E_{st} = −350 mV, which indicates a 95% probability of corrosion. At the remaining eight points, a 50% probability of reinforcement corrosion was obtained. The results of concrete cover resistivity also indicate a high probability of reinforcement corrosion, because at all eighteen points the values were lower than Θ = 10 kΩ·cm (Figure 7). Measurements of corrosion current density indicated the ongoing corrosion process of the bars that were tested, although the intensity of this process varied. Low corrosion activity (i_{cor} = 2–5 µA·cm^{-2}) was obtained for rebar no. 2 and two points of rebar no. 3. Moderate corrosion activity (i_{cor} = 5–15 µA·cm^{-2}) was observed at four points of rebar no. 3 and five points of rebar no. 4. High corrosion activity was found only at one point of rebar no. 4, where the corrosion current density exceeded i_{cor} = 15–A·cm^{-2}. At the same time, the results allowed estimating the corrosion rate in the tested rebars within the 33–57 µm/year range in rebar no. 2, 40–153 µm/year in rebar no. 3, and 68–182 µm/year in rebar no. 4, indicating a very poor condition of the floor slab reinforcement.

The same analysis was carried out for two reinforcing bars in the load-bearing column. A low 5% probability of corrosion (the values of the rebar stationary potential did not exceed E_{st} = −200 mV) was obtained at all four points of the rebar no. 1. But in rebar no. 2, the reinforcement stationary potential was much higher. At one point, rebar no. 2 indicated a 50% probability of corrosion, and at the other 3 points the corrosion probability was 95% (E_{st} < −350 mV). The values of concrete cover resistivity in rebar no. 1 indicated the ambiguous probability of corrosion, from small (at one point, Θ > 20 kΩ·cm) through medium (at two points, 10 < Θ < 20 kΩ·cm) to large (at one point, Θ < 10 kΩ·cm). In rebar no. 2, the occurrence of corrosion probability at all points was defined as large (Θ < 10 kΩ·cm). Measurements of corrosion current density of the column reinforcement were consistent with the measurements of the corrosion potential and resistivity of the concrete cover. At two points determining the corrosion activity in rebar no. 1, the measured values did not exceed i_{cor} = 2 µA·cm^{-2}, which indicated that the activity was not predicted or negligible, and at the remaining two points, the values slightly exceeded i_{cor} = 2 µA·cm^{-2}, which should be read as low corrosion activity. The rate of corrosion development in this rebar was estimated at 2.2–30.1 µm/year. However, at the points determining the corrosion activity in the rebar no. 2, the values of the corrosion current density were much higher, indicating a moderate (at three points, i_{cor} = 6.71–8.19 µA·cm^{-2}) and even high corrosion activity in this rebar (at one point, i_{cor} = 17.53 µA·cm^{-2}). The rate of corrosion development in rebar no. 2 can be estimated as 203.3 µm/year.

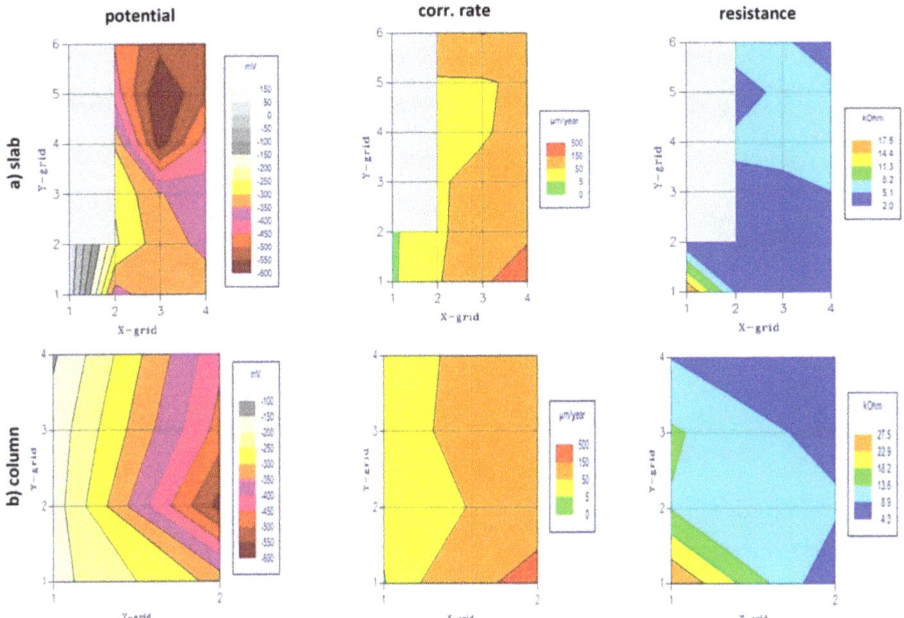

Figure 7. Measurement results of galvanostatic pulse method.

3.2.3. Laboratory Testing of Concrete Cover

Field assessment of depth of carbonation into the concrete surface (carbonation test with a phenolphthalein solution) of the removed concrete cover pieces revealed complete carbonation of the concrete cover (no concrete colouration, indicating pH < 8.5).

Laboratory tests of the concrete samples show that as a result of the carbonation process, the paste underwent significant changes. Three polymorphic forms of calcium carbonate ($CaCO_3$): calcite, aragonite and vaterite are visible in the X-ray diffraction pattern in Figure 8a [28,86]. Analysis of the used aggregate (Figure 8b) reveals the presence of only a trace of calcite. Despite intense indications from $CaCO_3$, portlandite peaks are still clearly visible in the sample. Analysis of the paste sample confirms the presence of a significant amount of aggregate dust fractions, which is represented by the intense peaks of dolomite and quartz appearing on the X-ray patterns. The presence of dolomite dust contributes to noticeable loss of sample mass above 600 °C. This is also associated with the presence of carbonation products. Gypsum and sylvite found in the paste indicate that the concrete may have been exposed to sulphate and chloride attack. The aggregate tested consists of dolomite, which is accompanied by quartz mixed with pyrite (Figure 8b).

The results of the X-ray analysis of the aggregate are supported by the results of the thermal analysis. Analysis of the DTA curve of the aggregate (Figure 9) shows a double endothermic effect characteristic of dolomites [87,88], whose first maximum (728 °C) is associated with the thermal decomposition of dolomite, and the second (783 °C) with decarbonation of the resulting calcite. At the same time, the DTA curve shows a very low endothermic effect at 573 °C associated with the polymorphic transformation of low-temperature α-quartz into high-temperature β-quartz.

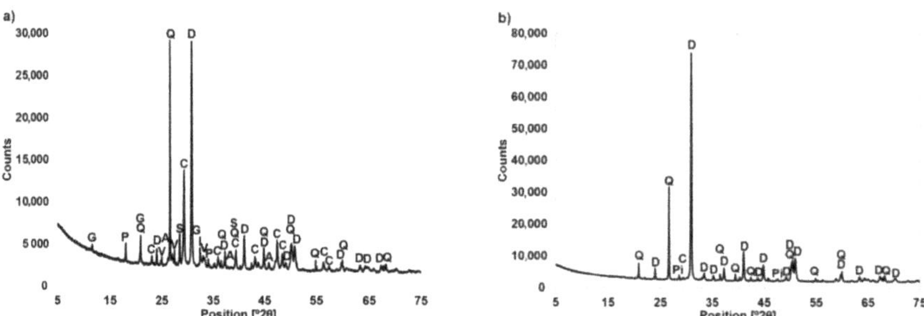

Figure 8. Phase analysis (XRD) of (**a**) cement paste, (**b**) aggregate recovered from the concrete. Notation: A—Aragonite, C—calcite, D—Dolomite, G—Gypsums, I—Ilmenite, Q—Quartz, P—Portlandite, Pi—Pyrite, S—Sylvite, V—Vaterite.

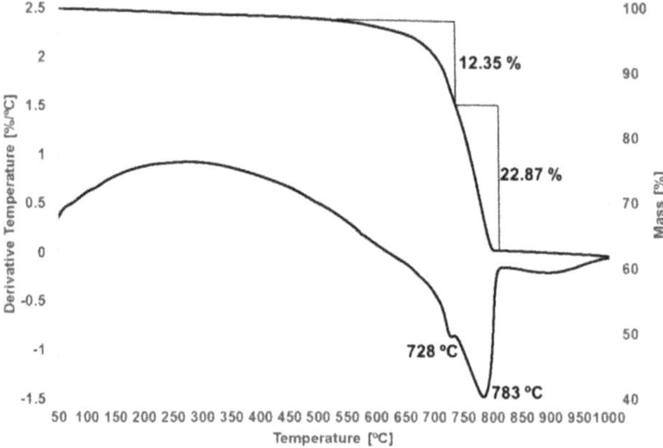

Figure 9. Thermal analysis of aggregate.

The phase composition analysis allows for precise recognition of effects carbonation. Based on the analysis, the possibility of the presence of calcium carbonate in concrete as a result of its introduction with carbonate aggregate was excluded. It has also been established that $CaCO_3$ is found in the 110-year old concrete both in stable calcite form as well as vaterite and aragonite. As a result of X-ray analysis, it was found that the degradation of concrete caused by carbonation could additionally be caused by the effect of sulphates.

4. Conclusions

Diagnostics of historic objects is usually associated with limitations resulting from the possibility of using only non-destructive or semi-destructive methods and the possibility of taking a small number of samples for material testing. However, the tests carried out on selected elements of the supporting structure of the historic 110-year old water tower indicate that they are sufficient for the initial assessment of the condition of reinforcement and concrete cover of the structure. Based on the local visual assessment supplemented with measurements of the concrete cover thickness, the locations of the remaining tests were easily determined. In situ concrete pH measurements allowed one to determine whether concrete still fulfils its protective role in relation to the reinforcement. However, only a full and thorough analysis of laboratory tests allowed the determination of the causes of corrosion. To obtain reliable results, tests should be carried out with several mutually verifying methods, such as electrochemical assessment tests of the reinforcement corrosion level, supplemented with laboratory tests of the concrete cover.

Performed research and analysis allowed the following conclusions to be drawn:

1. Non-destructive testing of the rebar location and layout performed using the Ferroscan gave accurate information on the location of reinforcing bars, their diameters and their cover thickness, all of which were positively verified after small pieces of concrete were removed.
2. Concrete cover distress was due to sulphate attack attributed to the presence of gypsum in the paste.
3. Non-destructive corrosion degree measurements of the reinforcing bars in the floor slab and in the load-bearing column determined the probability of corrosion in the examined areas, estimated corrosion activity in individual rebars, and predicted its pace over time. The results indicate a very poor state of the reinforcement in both the floor slab (high corrosion probability and locally high or moderate corrosion activity in the reinforcement) and column (high probability of corrosion and high corrosion activity in one of the two bars).
4. The carbonation test for in situ concrete showed total depassivation across the cover depth.
5. Material tests confirmed the information on the degree of concrete carbonation in the tested elements. Concrete carbonation was accelerated by the destruction of the concrete structure as a result of sulphate corrosion.

The described tests still need to be supplemented with diagnostic tests to assess the condition of the structure, mainly in terms of its load-bearing capacity. These tests should include, among others, determining the strength parameters of concrete and steel, crack morphology and displacement measurements. Because of the historical nature of the object, non-destructive and semi-destructive sclerometric, acoustic, pull-off methods can be used in the research.

Author Contributions: Conceptualisation, P.T. and W.R.; methodology, P.T., W.R., P.C.; formal analysis, P.T., W.R., P.C. and J.T.; writing—original draft preparation, P.T., W.R., P.C. and J.T.; writing—review and editing, P.T., W.R., P.C. All authors have read and agreed to the published version of the manuscript.

Funding: This research was funded by grant number 02.0.06.00/2.01.01.01. 0007; MNSP. BKWB. 16.001 "Analysis of limit states, durability and diagnostics of structures and methods and tools for quality assurance in construction" [Kielce University of Technology, Kielce, Poland]. Revitalization project of the water tower located in Zabrze at ul. Zamoyskiego 2 for social, educational, scientific and cultural purposes co-financed by the European Regional Development Fund under the operational program of the Silesian Voivodeship for 2014-2020, implemented by the Coal Mining Museum in Zabrze.

Data Availability Statement: No new data were created or analyzed in this study. Data sharing is not applicable to this article.

Conflicts of Interest: The authors declare no conflict of interest.

References

1. Raczkiewicz, W. *Building Diagnostics. Selected Methods of Materials As Well As Elements and Structures Test*; Kielce University of Technology: Kielce, Poland, 2019. (In Polish)
2. Szmygina, B. *Adaptation of Historic Buildings to Contemporary Utility Functions*; Lublin Scientific Society, Lublin University of Technology: Warsaw, Poland, 2009. (In Polish)
3. Stawiski, B.; Kania, T. Building diagnostics versus effectiveness of repairs. *MATEC Web Conf.* **2018**, *174*, 03005. [CrossRef]
4. Act of 7 July 1994 (as Amended); Journal of Laws 1994 No. 89 Item 414. Available online: https://isap.sejm.gov.pl/isap.nsf/DocDetails.xsp?id=WDU19940890414. (accessed on 8 January 2021).
5. Ślusarek, J.; Szymanowska-Gwiżdż, A.; Krause, P. Damage to historical balconies in view of building physics. In Proceedings of the 3rd World Multidisciplinary Civil Engineering, Architecture, Urban Planning Symposium (WMCAUS), Prague, Czech, 18–22 June 2018; Volume 471. [CrossRef]
6. Orłowski, Z.; Szklennik, N. The scope of building modernization—As a result of the diagnostic analysis of the object. *Civil Environ. Eng.* **2011**, *2*, 353–360. (In Polish)
7. Chybiński, M.; Kurzawa, Z.; Polus, Ł. Problems with Buildings Lacking Basic Design Documentation. *Procedia Eng.* **2017**, *195*, 24–31. [CrossRef]

8. Schabowicz, K. Non-destructive testing of materials in civil engineering. *Materials* **2019**, *12*, 3237. [CrossRef]
9. Schabowicz, K.; Sterniuk, A.; Kwiecińska, A.; Cerba, P. Comparative Analysis of Selected Non-Destructive Methods of Specific Diagnosis. *Mater. Sci. Eng.* **2018**, *365*, 032063. [CrossRef]
10. Zybura, A.; Jaśniok, M.; Jaśniok, T. *Diagnostics of Reinforced Concrete Structures. Tests on Reinforcement Corrosion and Concrete Protective Properties*; PWN: Warsaw, Poland, 2011.
11. Drobiec, Ł.; Jasiński, R.; Piekarczyk, A. *Diagnostics of Reinforced Concrete Structures. Methodology, Field Tests, Laboratory Tests of Concrete and Steel*; PWN: Warsaw, Poland, 2010. (In Polish)
12. Zhou, Z.; Jin, H. Evaluation of the protection of historical buildings. In Proceedings of the Procedia Engineering, International Conference on Electric Technology and Civil Engineering (ICETCE), Lushan, China, 22–24 April 2011. [CrossRef]
13. Raczkiewicz, W.; Wójcicki, A. Evaluation of effectiveness of concrete coat as a steel bars protection in the structure—Galvanostatic pulse method. In Proceedings of the 26th International Conference on Metallurgy and Materials, Brno, Czech, 24–26 May 2017; pp. 1425–1431.
14. Brodnan, M.; Koteš, P.; Bahleda, F.; Šebök, M.; Kučera, M.; Kubissa, W. Using non-destructive methods for measurement of reinforcement corrosion in practice. *Prot. Against Corros.* **2017**, *3*, 55–58. [CrossRef]
15. Tworzewski, P. Impact of concrete cover thickness deviations on the expected durability of reinforced concrete structures. *Constr. Rev.* **2017**, *11*, 52–55. (In Polish)
16. Hulimka, J.; Kałuża, M. Basic Chemical Tests of Concrete during the Assessment of Structure Suitability—Discussion on Selected Industrial Structures. *Appl. Sci.* **2020**, *10*, 358. [CrossRef]
17. Hoła, J.; Bień, J.; Sadowski, Ł.; Schabowicz, K. Non-destructive and semi-destructive diagnostics of concrete structures in assessment of their durability. *Bull. Pol. Acad. Sci. Tech. Sci.* **2015**, *63*, 87–96. [CrossRef]
18. Otieno, M.; Ikotun, J.; Ballim, Y. Experimental investigations on the influence of cover depth and concrete quality on time to cover cracking due to carbonation-induced corrosion of steel in RC structures in an urban, inland environment. *Constr. Build. Mater.* **2019**, *198*, 172–181. [CrossRef]
19. Anterrieu, O.; Giroux, B.; Gloaguen, E.; Carde, C. Non-destructive data assimilation as a tool to diagnose corrosion rate in reinforced concrete structures. *J. Build. Eng.* **2019**, *23*, 193–206. [CrossRef]
20. Filipek, R.; Pasierb, P. New, nondestructing diagnostic method of corrosion for reinforced concrete structures. *Corros. Prot.* **2017**, *12*. [CrossRef]
21. Kurdowski, W. *Cement and Concrete Chemistry*; Springer: Berlin/Heidelberg, Germany, 2014.
22. Qiu, Q. A state-of-the-art review on the carbonation process in cementitious materials: Fundamentals and characterization techniques. *Constr. Build. Mater.* **2020**, *247*. [CrossRef]
23. Raczkiewicz, W.; Wójcicki, A. Selected aspects of forecasting the level of reinforcing steel corrosion in concrete by the electro-chemical method. *Weld. Rev.* **2017**, *11*, 28–33.
24. Jaśniok, M.; Jaśniok, T. Measurements on Corrosion Rate of Reinforcing Steel under various Environmental Conditions, Using an Insulator to Delimit the Polarized Area. *Procedia Eng.* **2017**, *193*, 431–438. [CrossRef]
25. Raczkiewicz, W. Influence of the Air-Entraining Agent in the Concrete Coating on the Reinforcement Corrosion Process in Case of Simultaneous Action of Chlorides and Frost. *Adv. Mater. Sci.* **2018**, *1*, 13–19. [CrossRef]
26. Bogas, J.A.; Real, S. A Review on the Carbonation and Chloride Penetration Resistance of Structural Lightweight Aggregate Concrete. *Materials* **2019**, *12*, 3456. [CrossRef]
27. Melchers, R. Long-Term Durability of Marine Reinforced Concrete Structures. *Mar. Sci. Eng.* **2020**, *8*, 290. [CrossRef]
28. Czapik, P.; Owsiak, Z. Chemical corrosion of external stairs—Case study. In Proceedings of the MATBUD'2018—8th Scientific-Technical Conference on Material Problems in Civil Engineering, Cracow, Poland, 25–27 June 2018; Volume 163, pp. 1–7. [CrossRef]
29. Bacharz, K.; Raczkiewicz, W.; Bacharz, M.; Grzmil, W. Manufacturing Errors of Concrete Cover as a Reason of Reinforcement Corrosion in a Precast Element—Case Study. *Coatings* **2019**, *9*, 702. [CrossRef]
30. Tworzewski, P.; Raczkiewicz, W.; Grzmil, W.; Czapik, P. Condition assessment of selected reinforced concrete structural elements of the bus station in Kielce. *MATEC Web Conf.* **2019**, *284*. [CrossRef]
31. Bednarz, L.; Górski, A.; Jasieńko, J.; Rusiński, E. Simulations and analyses of arched brick structures. *Autom. Constr.* **2011**, *20*, 741–754. [CrossRef]
32. Lourenço, P.B. Computations on historic masonry structures. *Prog. Struct. Eng. Mater.* **2002**, *4*, 301–319. [CrossRef]
33. Navrátil, J.; Drahorád, M.; Ševčík, P. Assessment of Load-Bearing Capacity of Bridges. *Solid State Phenom.* **2017**, *259*, 113–118. [CrossRef]
34. Ferreira, D.R.; Araújo, A.; Fonseca, E.M.; Piloto, P.A.; Pinto, J. Behaviour of non-loadbearing tabique wall subjected to fire—Experimental and numerical analysis. *J. Build. Eng.* **2017**, *9*, 164–176. [CrossRef]
35. Nguyen, Q.H.; Ly, H.B.; Tran, V.Q.; Nguyen, T.A.; Phan, V.H.; Le, T.T.; Pham, B.T. A Novel Hybrid Model Based on a Feedforward Neural Network and One Step Secant Algorithm for Prediction of Load-Bearing Capacity of Rectangular Concrete-Filled Steel Tube Columns. *Molecules* **2020**, *25*, 5348. [CrossRef]
36. Nikoo, M.; Sadowski, Ł.; Nikoo, M. Prediction of the Corrosion Current Density in Reinforced Concrete Using a Self-Organizing Feature Map. *Coatings* **2017**, *7*, 160. [CrossRef]

37. Runkiewicz, L. The use of non-destructive methods to assess the technical condition of large-panel buildings. *Weld. Rev.* **2014**, *86*, 51–59. (In Polish)

38. Drobiec, Ł.; Jasiński, R.; Piekarczyk, A. Ways of locating reinforcing steel in reinforced concrete structures. Radiological method, Part II. *Build. Rev.* **2007**, *12*, 31–37. (In Polish)

39. Drobiec, Ł.; Jasiński, R.; Piekarczyk, A. Ways of locating reinforcing steel in reinforced concrete structures. Radiological method, Part I. *Build. Rev.* **2007**, *11*, 34–39.

40. Redmer, B.; Weise, F.; Ewert, U.; Likhatchev, A. Location of Reinforcement in Structures by Different Methods of Gamma-Radiography, International Symposium (NDT-CE 2003). Non-Destructive Testing in Civil Engineering. 2003. Available online: https://www.ndt.net/article/ndtce03/papers/v020/v020.htm. (accessed on 8 January 2021).

41. De Alcantara, N.P.; Costa, D.C.; Guedes, D.S.; Sartori, R.V.; Bastos, P.S.S. A Non-Destructive Testing Based on Electromagnetic Measurements and Neural Networks for the Inspection of Concrete Structures. *Adv. Mater. Res.* **2011**, *301*, 597–602. [CrossRef]

42. Helal, J.; Sofi, M.; Mendis, P. Non-destructive testing of concrete: A review of methods. *Electron. J. Struct. Eng.* **2015**, *14.1*, 97–105.

43. Wiwatrojanagul, P.; Sahamitmongkol, R.; Tangtermsirikul, S. A method to detect lap splice in reinforced concrete using a combination of cover meter and GPR. *Constr. Build. Mater.* **2018**, *173*, 481–494. [CrossRef]

44. Salman, A.A. Non-Destructive Test of Concrete Structures Using: FERROSCAN. *Eng. Technol. J.* **2011**, *29*, 2933–2941.

45. Mishra, M.; Grande, C. *Probabilistic NDT Data Fusion of Ferroscan Test Data Using Bayesian Inference*; Structural Analysis of Historical Constructions: Leuven, Belgium, 2016. [CrossRef]

46. Shaw, M.R.; Millard, S.G.; Molyneaux, T.C.K.; Taylor, M.J.; Bungey, J.H. Location of steel reinforcement in concrete using ground penetrating radar and neural networks. *NDT E Int.* **2005**, *38*, 203–212. [CrossRef]

47. Lachowicz, J.; Rucka, M. Application of GPR method in diagnostics of reinforced concrete structures. *Diagnostics* **2015**, *16*, 31–36.

48. Tosti, F.; Ferrante, C. Using Ground Penetrating Radar Methods to Investigate Reinforced Concrete Structures. *Surv. Geophys.* **2019**, *41*, 485–530. [CrossRef]

49. Rucka, M.; Wojtczak, E.; Zielińska, M. Interpolation methods in GPR tomographic imaging of linear and volume anomalies for cultural heritage diagnostics. *Measurement* **2020**, *154*, 107494. [CrossRef]

50. Lachowicz, J.; Rucka, M. Diagnostics of pillars in St. Mary's Church (Gdańsk, Poland) using the GPR method. *Int. J. Arch. Heritage* **2019**, *13*, 1223–1233. [CrossRef]

51. Hlaváč, Z.; Anton, O.; Garbacz, A. Hlav Detection of Steel Bars in Concrete by Impact-Echo. *Trans. Transp. Sci.* **2009**, *2*, 122–127. [CrossRef]

52. Moczko, A.; Rybak, J. Impulse response– nowoczesna metoda nieniszczącej defektoskopii konstrukcji betonowych. *Bud. Technol. Archit.* **2010**, *1*, 46–50.

53. Logoń, D.; Schabowicz, K. The Recognition of the Micro-Events in Cement Composites and the Identification of the Destruction Process Using Acoustic Emission and Sound Spectrum. *Materials* **2020**, *13*, 2988. [CrossRef] [PubMed]

54. Trąmpczyński, W.; Goszczyńska, B.; Bacharz, M. Acoustic Emission for Determining Early Age Concrete Damage as an Important Indicator of Concrete Quality/Condition before Loading. *Materials* **2020**, *13*, 3523. [CrossRef] [PubMed]

55. Świt, G. Acoustic Emission Method for Locating and Identifying Active Destructive Processes in Operating Facilities. *Appl. Sci.* **2018**, *8*, 1295. [CrossRef]

56. Tworzewski, P.; Goszczyńska, B. An Application of an Optical Measuring System to Reinforced Concrete Beams Analysis. In Proceedings of the 2016 Prognostics and System Health Management Conference (PHM-Chengdu), Chengdu, China, 19–21 October 2016. [CrossRef]

57. Raczkiewicz, W.; Wójcicki, A.; Grzmil, W.; Zapała-Sławeta, J. Impact of Environment Conditions on the Degradation Process of Selected Reinforced Concrete Elements. *Mater. Sci. Eng.* **2019**, *471*, 032048. [CrossRef]

58. Czapik, P. Microstructure and Degradation of Mortar Containing Waste Glass Aggregate as Evaluated by Various Microscopic Techniques. *Materials* **2020**, *13*, 2186. [CrossRef] [PubMed]

59. Zhu, X.E.; Dai, M.X. A Discuss on Basing Half-Cell Potential Method for Estimating Steel Corrosion Rate in Concrete. *Appl. Mech. Mater.* **2012**, *166*, 1926–1930. [CrossRef]

60. Frølund, T.; Jensen, F.; Bäßler, R. Determination of reinforcement corrosion rate by means of the galvanostatic pulse technique. In Proceedings of the First International Conference on Bridge Maintenance, Safety and Management IABMAS 2002, Barcelona, Spain, 14–17 July 2002.

61. Pacheco, A.R.; Schokker, A.J.; Volz, J.S.; Hamilton, H.R. Linear Polarization Resistance Tests on Corrosion Protection Degree of Post-Tensioning Grouts. *ACI Mater. J.* **2011**, *108*, 365–370.

62. Jaśniok, M.; Jaśniok, T. Evaluation of Maximum and Minimum Corrosion Rate of Steel Rebars in Concrete Structures, Based on Laboratory Measurements on Drilled Cores. *Procedia Eng.* **2017**, *193*, 486–493. [CrossRef]

63. Rengaraju, S.; Neelakantan, L.; Pillai, R.G. Investigation on the polarization resistance of steel embedded in highly resistive cementitious systems—An attempt and challenges. *Electrochim. Acta* **2019**, *308*, 131–141. [CrossRef]

64. Liu, J.; Jiang, Z.; Zhao, Y.; Zhou, H.; Wang, X.; Zhou, H.J.; Xing, F.; Li, S.; Zhu, J.-H.; Liu, W. Chloride distribution and steel corrosion in a concrete bridge after a long-term exposure to natural marine environment. *Materials* **2020**, *13*, 3900. [CrossRef]

65. GalvaPulse. Available online: http://www.germann.org/TestSystems/GalvaPulse/GalvaPulse.pdf (accessed on 20 March 2014).

66. Steiner, S.; Lothenbach, B.; Proske, T.; Borgschulte, A.; Winnefeld, F. Effect of relative humidity on the carbonation rate of portlandite, calcium silicate hydrates and ettringite. *Cem. Concr. Res.* **2020**, *135*, 106116. [CrossRef]

67. Liu, E.; Ghandehari, M.; Brückner, C.; Khalil, G.; Worlinsky, J.; Jin, W.; Sidelev, A.; Hyland, M.A. Mapping high pH levels in hydrated calcium silicates. *Cem. Concr. Res.* **2017**, *95*, 232–239. [CrossRef]
68. Jasieńko, J.; Moczko, M.; Moczko, A.; Dżugar, R. Testing the mechanical and physical properties of concrete in the bottom perimeter ring of the dome of the Centennial Hall in Wrocław. *Conserv. News* **2010**, *27*, 21–34.
69. Runkiewicz, L.; Hoła, J. Technical diagnostics of reinforced concrete structures. *Eng. Constr.* **2018**, *7*, 397–405. (In Polish)
70. Han, J.; Sun, W.; Pan, G.; Caihui, W. Monitoring the Evolution of Accelerated Carbonation of Hardened Cement Pastes by X-Ray Computed Tomography. *J. Mater. Civ. Eng.* **2013**, *25*, 347–354. [CrossRef]
71. Šavija, B.; Luković, M. Carbonation of cement paste: Understanding, challenges, and opportunities. *Constr. Build. Mater.* **2016**, *117*, 285–301. [CrossRef]
72. Tomasin, P.; Mondin, G.; Zuena, M.; El Habra, N.; Nodari, L.; Moretto, L.M. Calcium alkoxides for stone consolidation: Investigating the carbonation process. *Powder Technol.* **2019**, *344*, 260–269. [CrossRef]
73. Tracz, T.; Zdeb, T. Effect of Hydration and Carbonation Progress on the Porosity and Permeability of Cement Pastes. *Materials* **2019**, *12*, 192. [CrossRef]
74. Chen, Y.; Peng, L.; Zhiwu, Y. Effects of Environmental Factors on Concrete Carbonation Depth and Compressive Strength. *Materials* **2018**, *11*, 2167. [CrossRef]
75. Grzmil, W.; Owsiak, Z. The Influence of carbonation of self-compacting concrete with granulated blast furnace slag addition on its chosen properties. *Cem. Lime Concr.* **2013**, *18*, 137–143.
76. Stepkowska, E.T. Simultaneous IR/TG study of calcium carbonate in two aged cement pastes. *J. Therm. Anal. Calorim.* **2006**, *84*, 175–180. [CrossRef]
77. Czapik, P.; Wolniewicz, M. Microstructure of cement paste containing large amount of silica fume. Monografie Technologii Betonu. Proceedings of 10th Dni Betonu Conference, Wisła, Poland, 8–10 November 2018; Kijowski, P., Ed.; Polish Cement Association: Krakow, Poland, 2018; pp. 571–583.
78. Rimmele, G.; Barlet-Gouédard, V.; Porcherie, O.; Goffé, B.; Brunet, F. Heterogeneous porosity distribution in Portland cement exposed to CO2-rich fluids. *Cem. Concr. Res.* **2008**, *38*, 1038–1048. [CrossRef]
79. Dong, B.-Q.; Qiu, Q.-W.; Xiang, J.-Q.; Huang, C.-J.; Xing, F.; Han, N.; Lu, Y.-Y. Electrochemical impedance measurement and modeling analysis of the carbonation behavior for cementititous materials. *Constr. Build. Mater.* **2014**, *54*, 558–565. [CrossRef]
80. Talakokula, V.; Bhalla, S.; Ball, R.J.; Bowen, C.R.; Pesce, G.L.; Kurchania, R.; Bhattacharjee, B.; Gupta, A.; Paine, K. Diagnosis of carbonation induced corrosion initiation and progression in reinforced concrete structures using piezo-impedance transducers. *Sens. Actuators A Phys.* **2016**, *242*, 79–91. [CrossRef]
81. Villain, G.; Thiery, M.; Platret, G. Measurement methods of carbonation profiles in concrete: Thermogravimetry, chemical analysis and gammadensimetry. *Cem. Concr. Res.* **2007**, *37*, 1182–1192. [CrossRef]
82. Urbanowicz, D.; Warzocha, M. The use of ferromagnetic devices in the diagnostics of building structures. *Constr. Rev.* **2015**, *5*, 32–35. (In Polish)
83. PN-EN 13295:2005. Products and Systems for the Protection and Repair of Concrete Structures. Test Methods. Determination of Carbonation Depth in Hardened Concrete by Phenolphthalein Method. Polish Committee for Standardization: Warsaw, Poland, 2005.
84. Krzywobłocka-Laurów, R. *Instruction 357/98. Concrete Phase Composition Tests*; Building Research Institute: Warsaw, Poland, 1998.
85. Awoyera, P.; Britto, B.F. Foamed concrete incorporating mineral admixtures and pulverized ceramics: Effect of phase change and mineralogy on strength characteristics. *Constr. Build. Mater.* **2020**, *234*, 117434. [CrossRef]
86. Liu, S.; Dou, Z.; Zhang, S.; Zhang, H.; Guan, X.; Feng, C.; Zhang, J. Effect of sodium hydroxide on the carbonation behavior of β-dicalcium silicate. *Constr. Build. Mater.* **2017**, *150*, 591–594. [CrossRef]
87. Pelovski, Y.; Dombalov, I.; Petkova, V. Mechano-chemical Activation of Dolomite. *J. Therm. Anal. Calorim.* **2001**, *64*, 1257–1263. [CrossRef]
88. Chernykh, T.N.; Nosov, A.V.; Kramar, L.Y. Dolomite magnesium oxychloride cement properties control method during its production. *IOP Conf. Ser. Mater. Sci. Eng.* **2015**, *71*, 012045. [CrossRef]

materials

MDPI

Article

Mechanical Performance and Environmental Assessment of Sustainable Concrete Reinforced with Recycled End-of-Life Tyre Fibres

Magdalena Pawelska-Mazur [1,*] and Maria Kaszynska [2,*]

[1]   Faculty of Civil and Environmental Engineering, Gdansk University of Technology, ul. Narutowicza 11/12, 80-233 Gdansk, Poland
[2]   Faculty of Civil and Environmental Engineering, West Pomeranian University of Technology in Szczecin, Al. Piastów 17, 70-310 Szczecin, Poland
*   Correspondence: mmazur@pg.edu.pl (M.P.-M.); Maria.Kaszynska@zut.edu.pl (M.K.)

Abstract: The presented research's main objective was to develop the solution to the global problem of using steel waste obtained during rubber recovery during the tire recycling. A detailed comparative analysis of mechanical and physical features of the concrete composite with the addition of recycled steel fibres (RSF) in relation to the steel fibre concrete commonly used for industrial floors was conducted. A study was carried out using micro-computed tomography and the scanning electron microscope to determine the fibres' characteristics, incl. the EDS spectrum. In order to designate the full performance of the physical and mechanical features of the novel composite, a wide range of tests was performed with particular emphasis on the determination of the tensile strength of the composite. This parameter appointed by tensile strength testing for splitting, residual tensile strength test (3-point test), and a wedge splitting test (WST), demonstrated the increase of tensile strength (vs unmodified concrete) by 43%, 30%, and 70% relevantly to the method. The indication of the reinforced composite's fracture characteristics using the digital image correlation (DIC) method allowed to illustrate the map of deformation of the samples during WST. The novel composite was tested in reference to the circular economy concept and showed 31.3% lower energy consumption and 30.8% lower CO_2 emissions than a commonly used fibre concrete.

Keywords: fibre-reinforced concrete; recycled steel fibres; micro-computed tomography; scanning electron microscopy; tensile strength

Citation: Pawelska-Mazur, M.; Kaszynska, M. Mechanical Performance and Environmental Assessment of Sustainable Concrete Reinforced with Recycled End-of-Life Tyre Fibres. *Materials* **2021**, *14*, 256. https://doi.org/10.3390/ma14020256

Received: 15 December 2020
Accepted: 31 December 2020
Published: 6 January 2021

Publisher's Note: MDPI stays neutral with regard to jurisdictional claims in published maps and institutional affiliations.

Copyright: © 2021 by the authors. Licensee MDPI, Basel, Switzerland. This article is an open access article distributed under the terms and conditions of the Creative Commons Attribution (CC BY) license (https://creativecommons.org/licenses/by/4.0/).

1. Introduction

Due to the increasing EU car production, the problem of end-of-life tyre utilisation in an eco-friendly way is still growing. It is estimated that around 3.3 million tons of used tyres per year in Europe require recycling. In EC countries, restrictive regulations have been introduced to support solutions for the recovery, recycling or reuse of tyres towards the reduction of their harmful impact on the environment. In accordance with the Landfill Directive (1999/31/EC), the disposal of end-of-life tyres to landfills is prohibited. The directive introduced a ban on the storage of all used tyres since 2003, and since 2006 this ban included additionally crumbled tyres. Another End-of-Life Vehicle Directive (2000/53/EC) ordered the removal of tyres from vehicles before scrapping, and the Waste Incineration directive obliged cement plants using tyres as fuel to achieve lower limits for the content of pollutants in waste gases.

Used tyres (end-of-life tyres ELT) are recycled in two ways: towards obtaining energy or material. Novel solutions for raw materials or innovative products resulting from ELT are desired. However, still, up to 64% of used tyres are subject to energy recycling, i.e., they are burned mostly in cement kilns [1].

Due to the growing awareness of the need for sustainable development in the construction sector, new research has been conducted on the use of raw materials derived from

391

tyre recycling processes. Tyre shredding in an ambient method is the most commonly used mode of ELT recycling. Pre-cut tyres are shredded in special granulators and rolling mills. Pneumatic separation is used to remove the textile fibres, and the steel fibres are removed by means of an electromagnet. Tyres are 100% recyclable. All tyre components, i.e., rubber, metals, and textiles, can be recovered. However, only rubber is widely reused in various commercial products, as evidenced by numerous studies published in the literature [2–15]. Other derived raw materials like steel and synthetic fibres have not yet found any wide application as a full-value raw material or a full-fledged component of other products. Tyres contain 16–25% of steel by mass. The separated recycled steel fibres (RSF) are treated as waste in the recovery of rubber. Their commercial value refers to the scrap price, and they are pressed into briquettes as separated steel fibres and sold to steel mills.

The use of steel fibres is much more difficult compared to the rubber raw material due to their heterogeneity, irregular shape, and frequent rubber contamination. In the majority of tyre recovery plants using the ambient grinding method, steel collected from separators is contaminated with rubber by up to 10%. Together with irregular geometric characteristics, these are the key features that determine the suitability of recycled fibres for their further high-quality applications. Fibres from various tyre recycling plants were considered for further research, but only those obtained from the recycling process invented by KAHL Holding GmbH showed the lowest degree of pollution. It was noticed that the latest KAHL technology focused on the high effectiveness of rubber recovery, allowing obtaining steel fibres with no greater than 2% impurity (Figure 1a).

Research on the methods of the valuable ELT steel application in the construction sector was conducted at numerous institutes. The majority of the studies focused on fibre concrete as one of the possible areas for the use of steel fibres from recycled tyres. According to numerous publications [16–18], the addition of uniformly dispersed reinforcement in the form of steel fibres improves the cooperation of fibres with the concrete matrix when transferring loads, reduces the tendency to form micro-shrinkage pattern in concrete and hinders the propagation of scratches when loading concrete. Steel fibres added to the concrete matrix, considered as a discrete reinforcement of concrete, prevent damage to the element after a crack.

Strengthening the concrete matrix with heterogeneous RSF fibres made of high-quality steel can improve the concrete properties. However, in most cases, the researchers had encountered a problem of high rubber contamination and developed a pretreatment method to clean fibres before their use in concrete [19–21]. The use of almost pure RSF fibres allows their effective use in concrete without any pretreatment processes.

The main objective of the research was to develop the method of valuable steel tyre cord application in the construction sector. This paper demonstrates the results of studies conducted in order to prove the feasibility of replacing industrially produced fibres (ISF) with RSF fibres in the concrete composite commonly used for industrial floors.

2. Materials

2.1. Components of Concrete Mix

In order to determine the full characteristics of the designed composite with RSF fibres, 3 different concrete mixes were designed based on the components previously selected within preliminary tests and recommendations for the design of industrial concrete floors.

CEM II/B-S 42.5N-NA low-alkali slag Portland cement (Gorazdze cement plant—Heidelberg Cement Group, Gorażdże, Poland) with an average specific surface area of $4110 \text{ cm}^2/\text{g}$ was used as the most commonly used cement in Poland for industrial floors. A PCE polymer superplasticiser (Sika, Warsaw, Poland) was used added at an amount of 1%.

Glacial origin aggregates found in Pomerania (Dech-Pol aggregates plant, Gdynia, Poland) were used to make all concrete mixtures. Natural washed aggregates with fractions 0/2, 2/8, and 8/16 were used. A sand point equal to 43% was assumed. For gravel, the percentage advantage of natural aggregate fraction 2/8 (60%) over the 8/16 (40%) mm fraction was adopted, as it is a solution reducing the cost of fibre concrete and supporting environmental protection due to the prevalence of fine aggregates occurrence in northern Poland.

2.2. Characteristics of Recycled Steel Fibres (RSF) and Industrial Steel Fibres (ISF)

The use of almost pure fibres RSF without any pretreatment was essential in research. RSF fibres (Figure 1a) are characterised by different shapes, lengths and diameters. Therefore, for a more detailed analysis of the geometric characteristics of RSF, an in-depth study was carried out using the scanning electron microscope methods including the Energy Dispersive X-ray Spectroscopy (EDS) spectrum (Jeol Ltd., Tokyo, Japan) and the micro-computed tomography (μCT) (Bruker, Kontich, Belgium).

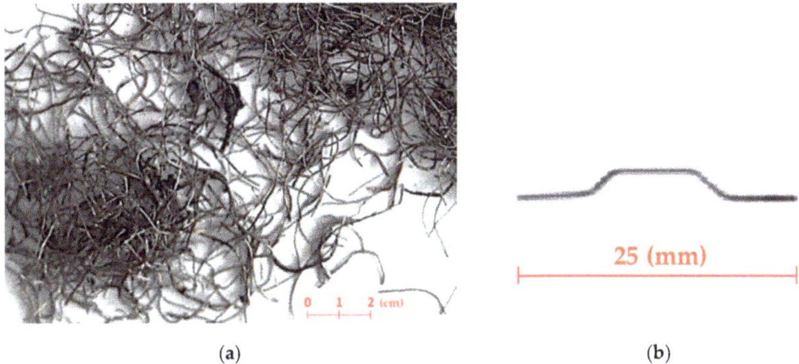

(a)   (b)

Figure 1. (**a**) Recycled steel fibres (RSF) derived in KAHL technology (ambient grinding method). (**b**) Industrial produced steel fibres (ISF).

The structural and chemical analysis of RSF fibres was performed using a Field Emission Gun (FEG) scanning electron microscope (Jeol Ltd., Tokyo, Japan), the operation of which is based on Energy Dispersive X-ray Spectroscopy (EDS) analysis. Use of a scanning microscope (model JSM-7900F, Jeol Ltd., Tokyo, Japan) combines the highest quality of imaging with chemical and structural analysis in the nanometer scale. EDS analysis allowed for the surface and volume identification of chemical elements included in the composition of tested RSF fibres.

SEM images (Figure 2) taken by use of a scanning electron microscope show that the surface of RSF fibres can be contaminated with rubber to various degrees. In addition, surface distortions are visible as a result of mechanical damages during use or the recycling process.

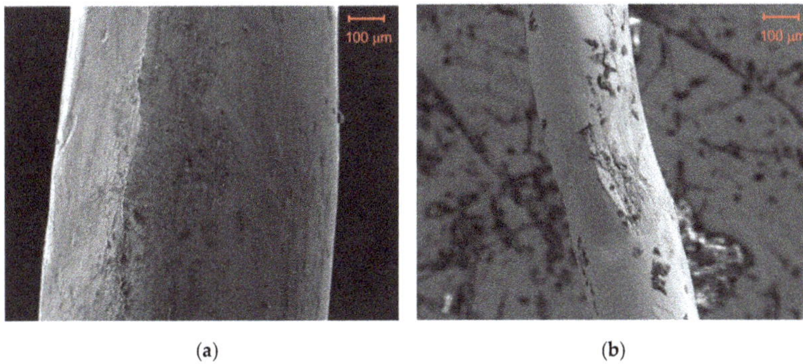

(a)   (b)

Figure 2. SEM image by use of a scanning electron microscope image of RSF fibres with a diameter of: (**a**) 600 μm and (**b**) 200 μm.

The EDS analysis showed that the chemical composition of some RSF fibres contains percentages of copper and zinc (Figure 3). The presence of both elements in approximately 2:1 weight ratio confirms that the steel cord fibres are covered with a thin layer of brass. Covering the fibres with a soft alloy coating is beneficial because of the further increase in the adhesion of fibres to the cement matrix.

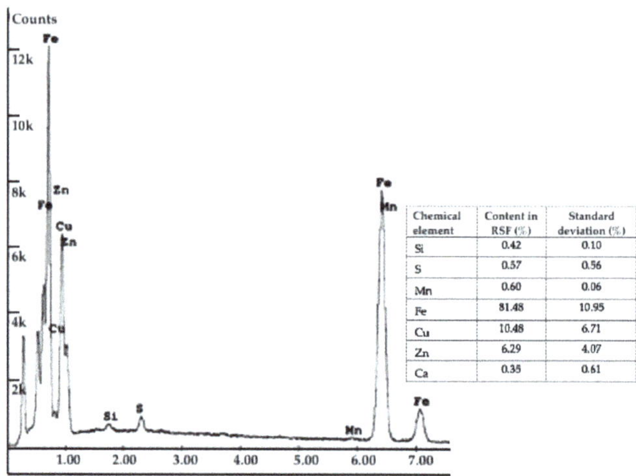

Chemical element	Content in RSF (%)	Standard deviation (%)
Si	0.42	0.10
S	0.57	0.56
Mn	0.60	0.06
Fe	81.48	10.95
Cu	10.48	6.71
Zn	6.29	4.07
Ca	0.35	0.61

Figure 3. EDS spectrum for RSF recognised under X-ray analysis.

The 3D Skyscan 1173 X-ray microtomography (Bruker, Kontich, Belgium) was used to perform a comprehensive analysis of the characteristics and distribution of fibres in the concrete. X-ray microtomography (μCT) is a 3D imaging technique that uses X-rays to create cross sections of a physical object to recreate a virtual model (3D model) without destroying the sample. The following parameters were determined as part of the research:

- the percentage distribution of the fibre length,
- percentage distribution of the diameters of the fibres,
- percentage distribution of fibre slenderness,
- percentage distribution of the fibre orientation.

The research was carried out on scans of 75 mm side samples made of RSFRC concrete. The maximum size of the test samples was narrowed down to a cube with a side of 70 mm to eliminate boundary disturbances. As part of the analysis of the characteristics of fibres, the following parameters were determined: relative frequency of the fibre length and relative frequency of the fibre diameter.

The study has proved that about 77% of RSF fibres have a length in the range of 5 to 30 mm (Figure 4). When analysing the fibre length distribution, it should be emphasised that it concerns real lengths. RSF fibres, due to their irregular shape, have a much smaller effective length l_{ef} of anchoring the fibres. At the same time, it was found that about 90% of RSF fibres have a diameter in the range of 0.1 to 0.4 mm (Figure 5). Another calculation of the average percentage slenderness distribution of RSF fibres indicated that the slenderness of the fibres is in the range 10–150, and the most (~27%) has the slenderness in the range 30–60 (Figure 6). The geometrical characteristics of the steel fibres are illustrated in Figure 1a with a visible variable length and irregular shape of the fibres. RSF fibres are made of steel with a tensile strength of over 2200 MPa [22], which is twice as high as the commonly used fibres for concrete. Due to the low fibre contamination, a density of 7800 kg/m^3 was assumed for RSF. The mechanical and geometrical features of RSF are presented in Table 1.

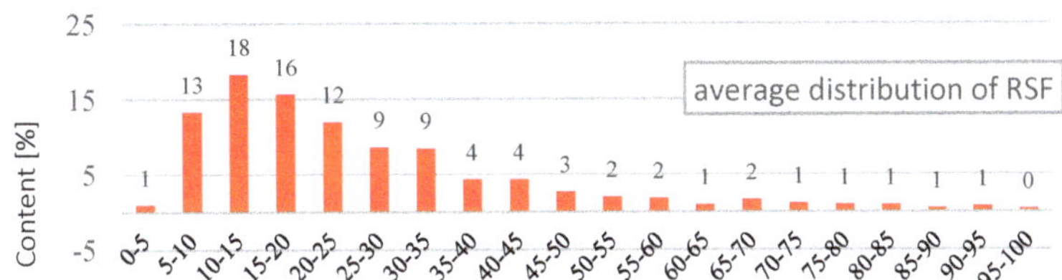

Figure 4. The average percentage of RSF length distribution.

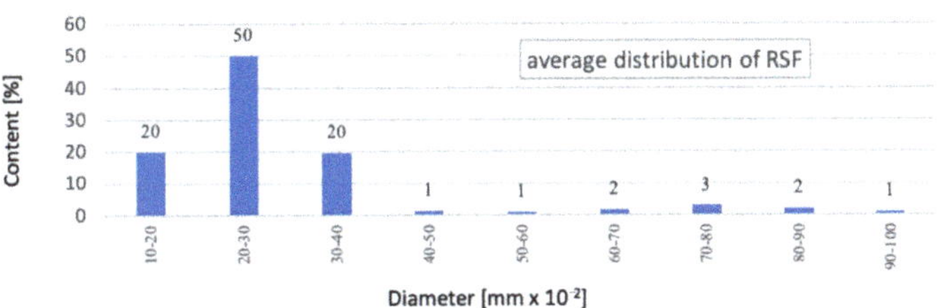

Figure 5. The average percentage of RSF diameter distribution.

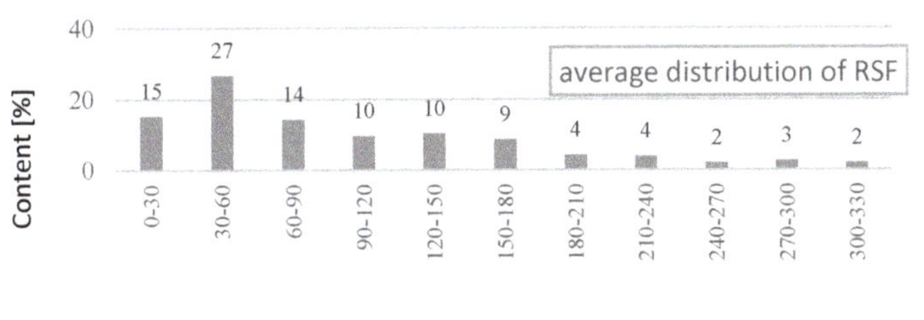

Figure 6. Average percentage RSF slenderness (*l/d*) distribution.

Table 1. Mechanical and geometrical features of RSF.

Characteristics of RSF	Value
RSF length (mm) (77%)	5–30
RSF diameter (mm) (90%)	0.1–0.4
RSF slenderness	10–150
RSF figure	irregular
RSF tensile strength (MPa) [22]	2200
RSF density (kg/m^3)	7800

For the comparative tests, industrially produced steel fibres (ISF) (Figure 1b) with a hook anchorage, 0.5 mm in cross section, 25 mm long and specified by the manufacturer (Baubach Metall E.S. GmbH, Effelder, Germany) as WLS-25/0.5/H were used. According to the technical characteristics declared by the manufacturer (Table 2), the tensile strength of the fibres is at least 1100 N/mm^2. The above-mentioned fibres, in their length and cross section, are the closest to those obtained in the ambient recycling by the KAHL processing.

Table 2. Mechanical and geometrical features of ISF (according to the producer's declaration).

Characteristics of ISF	Value
ISF length (mm) (77%)	25
ISF diameter (mm) (90%)	0.5
ISF slenderness	50
ISF Steel type	Group I EN 14889-1:2006
ISF tensile strength (MPa)	1100

2.3. Composition of Concrete Mix

Concerning the determination of the physical and mechanical characteristics of the designed composite, a wide range of tests was carried out on unreinforced concrete (NC), reinforced with ISF fibres (SFRC) and RSF fibres (RSFRC). The composition of the concrete mix was determined in accordance with the recommendations for the design of industrial concrete floors and based on the conclusions of preliminary tests.

The recipe of each mix (Table 3) was initially appointed analytically, based on the method of three equations, assuming that the amount of cement in the mix is 320 kg/m^3 and the factor $w/c = 0.5$. The additional design assumptions were as follows; S3 consistency class of concrete mix, concrete class of min. C25/30; adoption of RSF fibres with a content of 50 kg/m^3 and a comparatively ISF fibres with a content of 25 kg/m^3. The application of double content of RSF fibres versus ISF fibres was conducted due to previous study on geometrical characteristics of the recycled fibres, where it was demonstrated that only about 60% of RSF fibres improve the parameters of the composite.

Table 3. Composition of concrete mixes without fibres (NC), with ISF fibres (SFRC), and with RSF fibres (RSFRC).

Concrete Mix Composition	NC	SFRC	RSFRC
cement CEM II/B-S 42.5N-NA (kg/m^3)	320	320	320
sand 0–2 mm (kg/m^3)	457	456	454
gravel 2–8 mm (kg/m^3)	808	805	801
gravel 8–16 mm (kg/m^3)	687	684	681
water (kg/m^3)	160	160	160
ISF fibres WLS-25/0.5/H (kg/m^3)	-	25	-
RSF fibres (kg/m^3)	-	-	50
superplasticiser (kg/m^3)	3.2	3.2	3.2

3. Methodology

3.1. Assigning the Rheological Properties of the Concrete Mix

The study was initiated by assigning the rheological properties of the concrete mix. The research was conducted on the determination of the consistency, the air content and density of concrete mixes which contained fibres (ISF and RSF) or were plain. The consistency of the concrete mix was tested by the Abrams cone slump method in accordance with the EN 12350-2 standard, the air content was determined using the pressure method based on the Boyle–Mariott law in accordance with the EN 12350-7 code and the concrete mix density was tested in accordance with the EN 12350-6 standard.

3.2. Tests on Physical and Mechanical Features of Concrete Composites

The next stage of the research was concerned with the determination of the parameters of concrete composites. It contained the analysis of the distribution of ISF and RSF fibres in studied concrete mixes. The imagines of the distribution of ISF and RSF fibres in the samples of concrete composites were taken, and the orientation of their position in concrete was determined in relation to the Cartesian system. The research was carried out on 3D scans of 75 mm side samples made of RSFRC and SFRC concrete (Figure 7). The experimental procedure followed the research on the analysis of the RSF characteristics (Section 2.2), where 3D Skyscan 1173 X-ray microtomography was used.

Figure 7. Determination of the distribution of ISF and RSF fibres in the samples of concrete composites using 3D Skyscan 1173 X-ray microtomography.

In order to designate the full characteristics of the physical and mechanical features of the proposed new composite, a wide range of tests was performed, starting with the compressive strength test. That test was carried out on 6 cubic samples with a side of 150 mm for each of the 3 series, in accordance with the EN 12390-3 standard, using CONTROLS testing machine (Controls S.p.A., Liscate, Italy) enabling electronic recording of the results.

Assigning other mechanical properties, particular emphasis was put on the determination of the tensile strength of the composite. This parameter was appointed by three different methods: tensile strength testing for splitting (Brazilian method), residual tensile strength test (3-point test) and testing of tensile strength by a wedge splitting (WST method). The last two methods allowed us to determine the dependence of *strength-CMOD* for the tested samples.

The Brazilian method of splitting a concrete sample during compression is the most commonly used method of testing concrete tensile strength. According to the EN 12390-6 standard, tests were performed on cubic samples with a side of 150 mm, 6 samples for each of the 3 series after 28 days of maturation, using CONTROLS testing machine.

In the case of composites reinforced with steel fibres, the determination of the residual tensile strength is the key test used in the calculation of structural elements made of fibre concrete according to the recommendations of RILEM TC 162-TDF. The addition of steel fibres is intended to transfer stresses in the tension zone after the first crack has appeared. Tensile strength was measured in accordance with EN 14651 code. The method is performing a 3-point test (Figure 8) in order to determine the proportionality limit and residual strength on beam-shaped samples with dimensions: $150 \times 150 \times 700$ mm (3 samples for each concrete mix: NC, SFRC and RSFRC). The samples were prismatically incised in the middle of the span to the height of 25 mm in order to force a crack in the beam axis. The span of beam supports was 500 mm.

Figure 8. Test stand for determination of the residual tensile strength (3-point bending test).

The wedge splitting test (WST) is an alternative method of determining the σ-CMOD relationship. The test was developed in 1986 by H.N. Linsbauer and E.K. Tschegga [23]. Since then, it has been widely used to determine the mechanical properties of fractures of brittle and quasi-brittle materials. The optical method of digital image correlation (DIC) was used to determine the map of sample deformation during stretching by splitting with a wedge. The applied DIC system was equipped with a camera enabling obtaining two-dimensional results. As part of this work, the WST tests were carried out on cubic samples with a side of 71 mm made of three different concrete formulas: without fibres—NC, concrete reinforced with ISF fibres—SFRC, and RSF—RSFRC fibres. The samples were incised in the upper plane to a depth of 10 mm and a width of 5 mm in the centre of the sample, where the relationship between the crack width (CMOD) and the applied force F was then tested. The samples were loaded with a controlled movement speed of 0.001 mm/min in a testing machine INSTRON 5569 (Instron, Electromechanical & Industrial Products Group, Norwood, MA, USA) (Figure 9). This type of control allowed us to obtain a constant decrease in strength in the period after reaching the maximum stress corresponding to the applied force F_{max}.

Figure 9. Test stand for testing tensile cracking characteristics when splitting concrete sample with a wedge (WSP) using digital image correlation (DIC) method.

Due to the intended use of the designed composites in industrial concrete floors, additional tests were performed on the impact of RSF fibre addition on adhesion and abrasion of concrete. The adhesion of concrete layers was determined by the pull-off test in accordance with EN 1542 code. The test was performed for 7 samples from each of the 3 series of concrete. The abrasion resistance was determined using a Boehme dial, over which a test specimen with dimensions $71 \times 71 \times 71$ mm^3 was fixed, according to the procedure described in EN 13892-3. Abrasion was determined by the loss of height and volume of the sample determined on the basis of the weight loss of the tested sample.

4. Results

4.1. Analysis of Fibre Distribution in a Cement Matrix

Concrete reinforced with fibre is considered an anisotropic material. Using 3D X-ray micro-computed tomography, images of the distribution of ISF and RSF fibres in the samples of concrete composites were taken, and the orientation of their position in concrete was determined in relation to the Cartesian system. The results of the comparison of the percentage distribution of the RSF and ISF fibres in relation to the Z-axis (Figure 10) prove the advantage of the fibres arranged in a perpendicular direction to the concreting direction of the samples. Approximately 73% RSF and 74% ISF are laid at an angle of Θ in the range of 45° to 90°. The observed phenomena, although contradicting the assumption of anisotropy of fibre-reinforced concrete, in the case of floor applications, are advantageous due to the predominance of tensile forces acting perpendicularly to the direction of concreting.

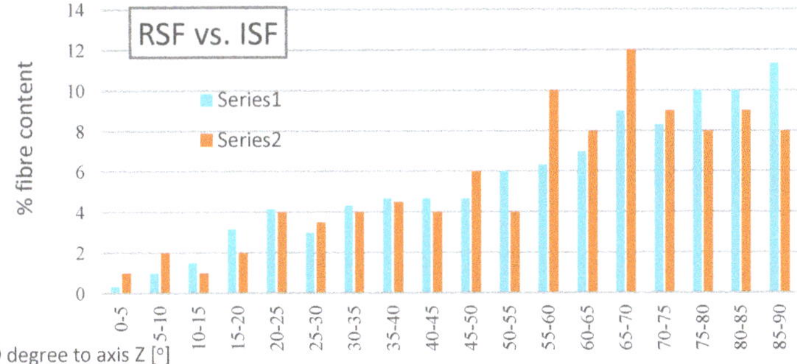

Figure 10. The average percentage fibre distribution in samples containing RSF fibres (Series 1) and ISF fibres (Series 2) in relation to their location, angle Θ, relative to the Z-axis.

The analysis of the fibre inclination angle Φ in the XY plane carried out for the same samples showed an even distribution for both types of fibres. The percentage distribution of the position angle of the fibres Φ in the XY plane for 5° did not show a greater share than 4.2% for both ISF fibres and RSF fibres (Figure 11).

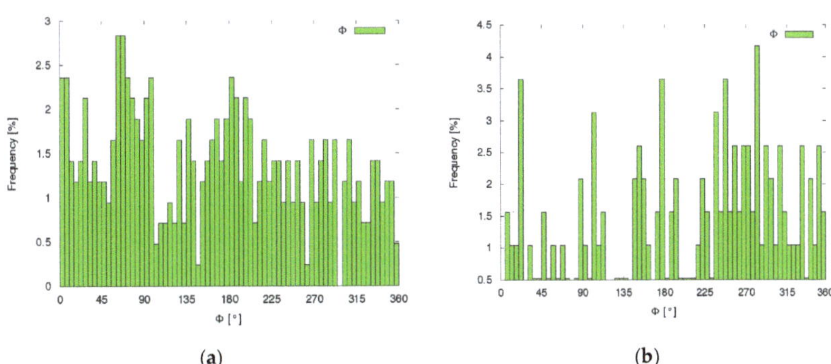

Figure 11. The percentage distribution of fibres in relation to their position in the XY plane (Φ angle): (a) RSF fibres and (b) ISF fibres.

3D images of samples (Figure 12) show an even distribution of ISF and RSF fibres in concrete. An illustrated lack of clusters of fibres, i.e., "balls", is a significant finding. It proves no additional requirements are needed for mixing the components during the production of the proposed fibre concrete.

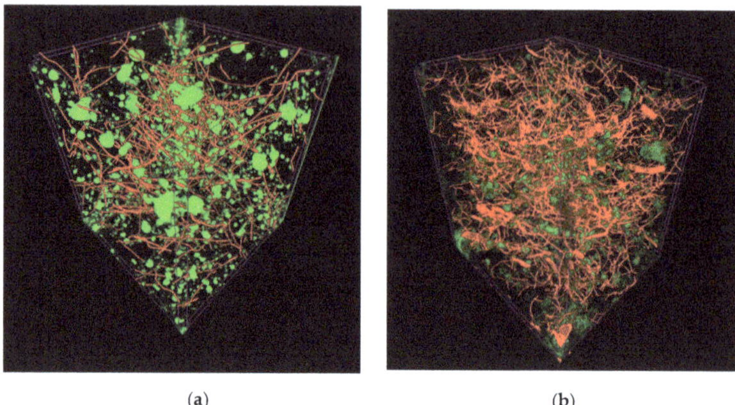

(a)   (b)

Figure 12. 3D computer micro-tomography image of SFRC (**a**) and RSFRC (**b**); fibre distribution (red): SFRC—0.32% and RSFRC 0.64%; and air pores (green): SFRC—2.76% and RSFRC 1.73%.

4.2. Tests on the Rheological Properties of the Concrete Mix

Studies have shown that dosing of ISF-dispersed reinforcement in the amount of 25 kg/m^3 caused the Abrams Cone slump by 6 cm, and the addition of RSF fibres in the amount of 50 kg/m^3 resulted in the decrease by another 2 cm in comparison to mixtures without fibres (Figure 13). These results correspond to a lowering of the consistency class for both mixtures modified with steel fibres from S4 (for NC) to S3 (for SFRC and RSFRC).

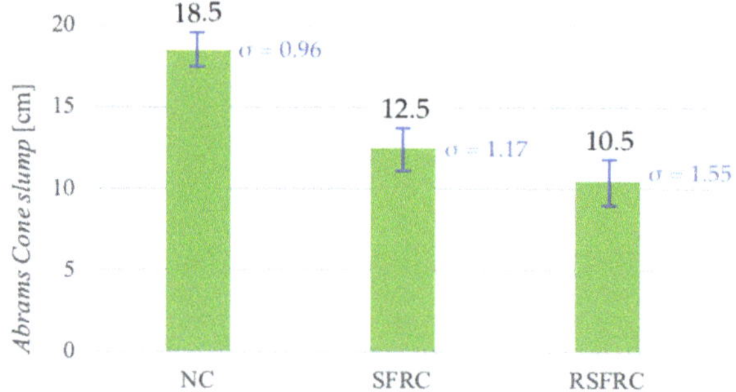

Figure 13. Comparison of the consistency test results with the cone slump test for mixtures without fibres (NC) and with steel fibres (SFRC and RSFRC).

During testing, efforts were made to avoid sedimentation of the ingredients while maintaining relatively low air content. The air content test results (Figure 14) show the increase in air content at concrete mixes modified with still fibres: up to 2.15% for the SFRC mix and up to 1.90% for the RSFRC mix versus 1.70% for unmodified concrete mixes (NC).

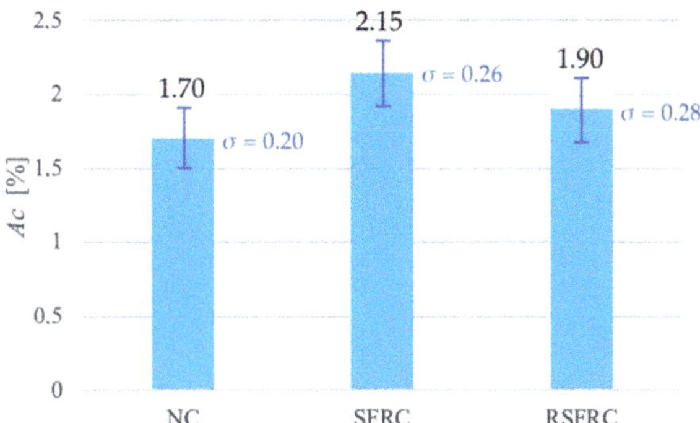

Figure 14. Comparison of the air content (*Ac*) for mixtures without fibres (NC) and with steel fibres (SFRC and RSFRC).

4.3. Compressive Strength of Concrete

The results of compressive strength tests are presented in Table 4 and Figure 15.

Table 4. Compression strength results after 28 days.

Compression Strength	NC	SFRC	RSFRC
Compression force F_c (kN]	966	1096	1179
medium compression strength f_{cm} (MPa)	42.9	48.7	52.4
Standard deviation (f_{cm}) (MPa)	2.27	1.29	2.48
characteristic compression strength f_{ck} (MPa)	39.9	44.7	48.4
Class of Concrete	C30/37	C35/45	C35/45

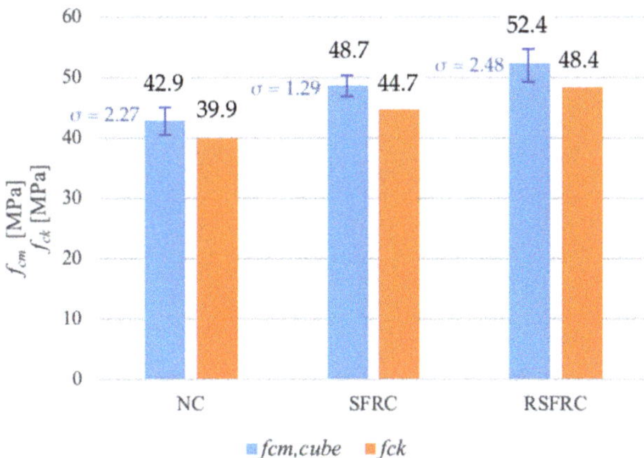

Figure 15. Comparison of medium (f_{cm}) and characteristic (f_{ck}) compression strength of tested concretes without fibres (NC) and with steel fibres (SFRC and RSFRC).

4.4. Tensile Strength of Concrete

The tensile strength of the composite was appointed by three different methods: tensile strength testing for splitting (Brazilian method), residual tensile strength test (3-point test) and testing of tensile strength by a wedge splitting (WST method). Results of these tests are presented in Table 5.

In the case of tensile strength at splitting, the average strength of ISF fibre modified concrete is 31% higher, and in the case of RSF fibre modified concrete, it is 43% higher than the average strength of reference concrete without fibres.

The 3-point bending test allowed to determine the dependence of the applied force on the deflection (*F-deflection*) and the tensile stresses on the opening width of crack (σ-*CMOD*) for composites without fibres and modified with ISF and RSF fibres with a crack width up to 4 mm. To better illustrate the behaviour of samples in the early deformation phase, Figure 16 compares plots of *F-deflection* relationships for non-fibre (NC) and fibre (SFRC and RSFRC) composites obtained during residual strength testing at a deflection of 0.4 mm, which shows differences in the behaviour of the samples before scratching.

Table 5. Comparison of tensile strength for concretes without fibres (NC) and with fibres (SFRC and RSFRC).

-	NC	SFRC	RSFRC
Tensile strength testing for splitting	-	-	-
$f_{ct,sp}$ (MPa)	2.71	3.54	3.87
Increase to NC (%)	-	31	43
Standard deviation ($f_{ct,sp}$) (MPa)	0.28	0.25	0.24
Residual tensile strength			
$LOP = f_L$ (MPa)	3.59	4.82	4.67
Increase to NC (%)	-	34	30
Standard deviation (f_L) (MPa)	0.18	0.07	0.38
f_{R1k} (MPa) at $CMOD_1 = 0.5$ mm	-	2.29	2.45
f_{R2k} (MPa) at $CMOD_2 = 1.5$ mm	-	1.15	1.66
f_{R3k} (MPa) at $CMOD_3 = 2.5$ mm	-	0.91	1.26
f_{R4k} (MPa) at $CMOD_4 = 3.5$ mm	-	0.72	0.96
$f_{R3k}/f_{R1k} > 0.5$	-	0.4 < 0.5	0.51 > 0.5
$f_{R1k}/f_L > 0.4$	-	0.47 > 0.4	0.53 > 0.4
Tensile strength by WST method			
σ_{NT} (MPa)	1.79	2.63	3.04
Increase to NC (%)	-	47	70

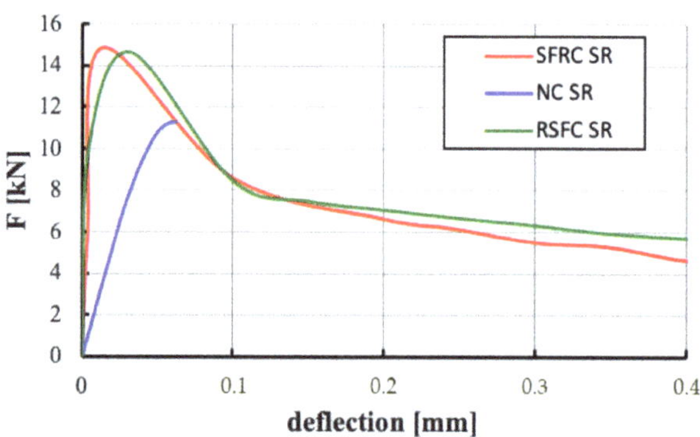

Figure 16. Comparison of diagrams of *F-deflection* relationships for non-fibre (NC) and fibre (SFRC and RSFRC) concretes obtained during the residual strength test with deflection up to 0.4 mm (3-point test).

The σ-*CMOD* relationship diagram (Figure 17) illustrates the convergence of results for samples containing ISF fibres and RSF fibres. According to the results obtained (Table 5), RSFRC composites demonstrate a higher residual flexural strength f_{R1k} corresponding to a crack opening equal to $CMOD_1 = 0.5$ ($f_{R1k} = 2.45$ MPa) versus $f_{R1k} = 2.29$ MPa for SFRC composites. A similar relationship applies to the residual strength f_{R3k} corresponding to a crack opening equal to $CMOD_3 = 2.5$. For RSFRC samples, the residual strength f_{R3k} is 1.26 MPa, and for SFRC samples the residual strength f_{R3k} is significantly lower, it is 0.91 MPa.

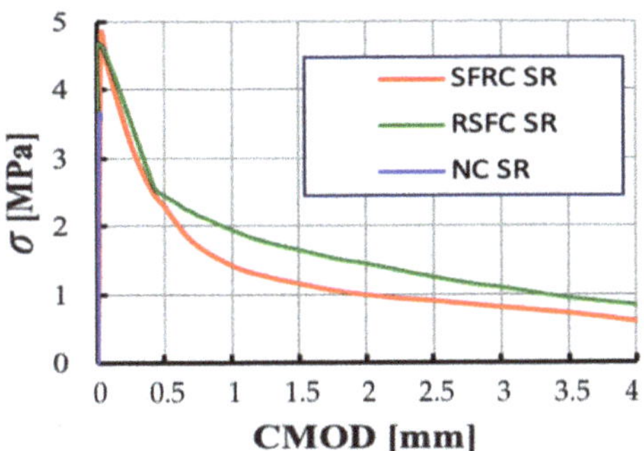

Figure 17. Comparison of diagrams of σ-*CMOD* relationships for concrete without fibres (NC) and with fibres (SFRC and RSFRC) obtained during the residual strength test for CMOD up to 4 mm (3-point test).

The tensile strength test by splitting referring to the WST method showed the strength σ_{NT} of NC concrete constituted about 60–70% of the tensile strength by splitting of concrete reinforced with steel fibres (SRFC and RSFCR). The values of the corresponding maximum tensile stress σ_{NT} for concrete: NC, SRFC and RSFCR are shown in Table 5. The diagrams of the σ-*CMOD* relationship (Figure 18) demonstrate a very high convergence of results for samples reinforced with diffuse ISF and RSF fibres.

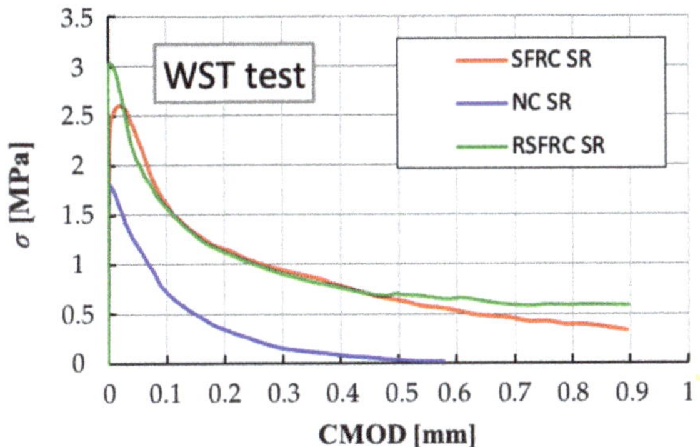

Figure 18. Comparison of diagrams of σ-*CMOD* relationships for concrete without fibres (NC) and with fibres (SFRC and RSFRC) at WST test.

The optical method of Digital Image Correlation (DIC) was used to determine the map of sample deformation during stretching by wedge splitting test (WST). By analysing the places of deformation concentration, the sample cracking zone was located. Comparison of deformation propagation map images in the elastic zone (for a point corresponding to the application of ~85% force F_{max} behind the vertex (85% after F) indicates much faster crack penetration in concrete without fibres. In addition, it was observed that the crack location zone was more curved in the case of concrete with ISF fibres than RSF fibres (Figure 19). The colour scale indicates the intensity of the horizontal deformation of the RSFRC sample.

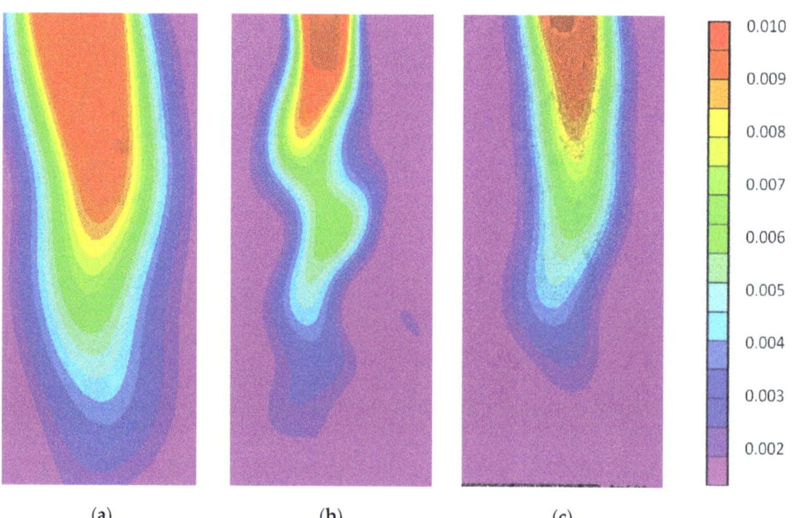

(a)   (b)   (c)

Figure 19. Images of deformation propagation map in the elastic zone obtained using Digital Image Correlation (DIC) for samples: (**a**) NC, (**b**) SFRC and (**c**) RSFRC.

The above observation of the WST test confirmed the picture showing the scratches formed in the ISF sample (Figure 20).

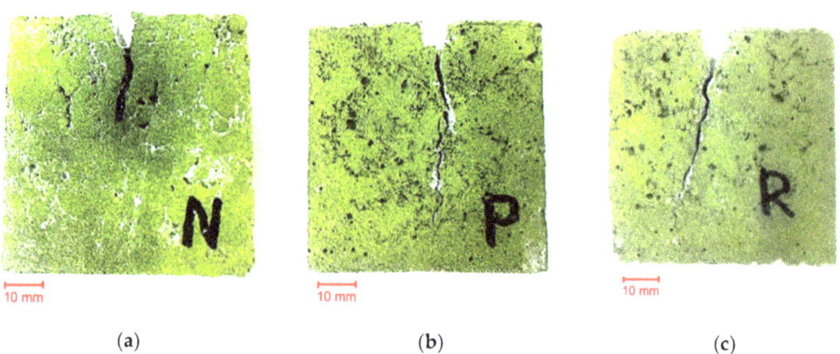

(a)   (b)   (c)

Figure 20. Pictures were taken with a digital camera illustrating the scratches formed in the samples as a result of the WSP test: (**a**) NC), (**b**) SFRC and (**c**) RSFRC.

4.5. Adhesion and Abrasion of Concrete Tests

The results of a pull-off peel test are demonstrated at Table 6, while the results of Boehme's abrasion resistance tests are shown at Table 7.

Table 6. Results of pull-off peel tests.

Adhesion Strength-Pull-Off Test	NC	SFRC	RSFRC
f_h [MPa]	2.50	2.45	2.49
Standard deviation (f_h)	0.41	0.37	0.46

Table 7. Results of Boehme's abrasion resistance tests.

Boehme's Abrasion Resistance	NC	SFRC	RSFRC
ΔV (cm^3/50 cm^2]	8.48	7.58	7.47
Standard deviation (ΔV)	1.04	0.56	0.38
Δl (mm]	1.70	1.53	1.47
Standard deviation (Δl)	0.17	0.12	0.10
Boehme's abrasion resistance class	A9	A9	A9

4.6. Comparative Analysis of Energy Consumption and CO$_2$ Emissions of the Production of Concrete Reinforced with ISF and RSF Fibres

To determine the energy consumption and CO$_2$ emission parameters for individual components of fibre concrete, indicators published by the Central Statistical Office for the Polish Economy [24] (Table 8) and indicators derived from scientific studies were adopted [24].

Table 8. Determination of energy consumption and CO$_2$ emissions for 1 m^3 of fibre concrete with ISF fibres.

Concrete Component	Energy Consumption Factor (MJ/kg]	CO$_2$ Emission (kg CO$_2$/kg]	Amount in 1 m^3 (kg]	Energy Consumption (MJ]	CO$_2$ Emission (kg CO$_2$]
cement (kg]	3 *	0.3 **	320	960	96
water (dm^3]	0.05 **	0 **	160	8	0
sand 0/2 (kg]	0.1 **	0.007 **	805	80.5	5.635
gravel 2/8 (kg]	0.1 **	0.007 **	684	68.1	4.788
gravel 8/16 (kg]	0.1 **	0.007 **	454	45.6	3.192
fibre ISF (kg]	21.0 ***	1.95 **	25	525	48.75
		SUMA		**1687.5** (MJ]	**158.4** (kg CO$_2$]

(*) calculation based on indicators published by the Central Statistical Office [24] on the assumption that CEM II/B-S 42.5N-NA contains 70%; (**) indicators derived from publications [25]; (***) calculation based on indicators published by the Central Statistical Office [24] assuming the following processes in the production of steel fibres: production of iron pig iron, steel from electric furnaces, cold rolling of steel and wire drawing. In addition, 0.5 MJ/kg wire cutting energy was adopted.

5. Discussion

5.1. Impact of RSF Fibre Addition on the Rheological Properties of the Concrete Mix

The addition of steel fibres significantly affects all rheological properties of the fresh concrete mix. Fibre parameters such as type, length, shape and their content in the cement matrix impact on workability, consistency and air content. The results obtained so far relating to concretes modified with steel fibres prove that workability and consistency deteriorate with the increase in steel fibre content. The above phenomenon is explained by the fact that the spatially condensed fibre system in the mixture hinders the free movement of aggregate grains. In addition, part of the cement paste directly surrounds fibres instead of sand grains.

Research conducted on fibre concrete in the 1970s [16] showed that the consistency of concrete modified with RSF fibres depends on two parameters: fibre content in the concrete mix and their slenderness *l/d*. The results obtained on consistency tests are consistent in the abovementioned statement through lowering the consistency class for both mixtures modified with steel fibres from S4 to S3.

The air content in the concrete mix is closely related to the compacting process and the occurring phenomenon of removing air from the cement paste. In case of fibre concrete mixes, the reduction of air bubbles during compacting is more difficult due to higher aeration of the mix, and the limitation of components displacement due to the fibre addition. If the compacting process is too long, a risk appears of disturbing the spatial distribution of the fibres and, consequently, their falling to the bottom of the mould. The test results (Figure 14) confirmed the effect of fibres on the increase in air content: up to 2.15% for the SFRC mix and up to 1.90% for the RSFRC mix. Increased air content in fibre reinforced concrete was also observed when testing hardened concrete samples with an X-ray micro-computed tomography (Figure 9), where SFRC concrete contained 2.76% and RSFRC concrete only 1.73% of air voids. It has been shown that the effect of fibres on loosening the mix structure and hardened concrete is greater in the case of homogeneous fibres (ISF) than hybrid fibres (RSF).

5.2. Impact of RSF Fibre Addition on Strength Parameters of the Designed Composite

According to the literature [26], experimental tests of compressive strength show no significant impact of dispersed reinforcement with a content of up to 1% of concrete volume. It was observed that an addition of ISF fibres in an amount exceeding 1.5% effect an increase in compressive strength up to 15%. Other sources [27,28] indicate an increase of 10 to 30% in the compressive strength of fibre concrete already with the addition of ISF fibres in the amount exceeding 0.5% of the concrete volume. Due to the loosening of the fibre concrete structure at the presence of fibres, experimental tests may also show a slight decrease in strength.

Test results cited in the literature on fibre concrete made from recycled tyres showed no significant effect on compressive strength [29,30] or a slight increase in strength (by 9% at a fibre content of 0.41% and by 17% at a volumetric content fibre 0.46%).

Own research showed an increase in compressive strength (Table 4 and Figure 15) by 13.5% for concrete modified with ISF fibres (at 0.32% fibre volume content) and by 22% at RSF (at 0.64% fibre volume content). The high compressive strength obtained in the test for both concretes modified with steel fibres is explained by the inclusion of fibres in the control of static stress in the three-axis stress state occurring in compressed cubes. When the concrete expands in a direction perpendicular to the compressive force, the fibres can bridge emerging scratches until the steel fibres are torn out of the cement matrix, which results in an increase in compressive strength.

When determining the strength parameters, special emphasis was placed on the tensile strength of the new composite. Low tensile strength in elements of ordinary concrete has an effect on the formation of scratches. In published results of world studies [26,27] it has been shown that the addition of steel fibres increases the tensile strength of concrete by 10 to 30% with a spatial arrangement of fibres and from 30 to 70% with directed fibre distribution.

During that study, three different tests on the tensile strength of the composite: splitting test (Brazilian method), residual tensile strength test (3-point test) and a wedge splitting test (WST method) have confirmed the increase in tensile strength when modifying concrete with RSF fibres (Table 5).

Analysis of the graphs of the *F-deflection* relationships at the bending test of fibre concrete elements (Figure 16) demonstrated that the formation of the first crack does not lead to the sudden destruction of the element made of RSFRC concrete. It is assumed that the development of a critical crack results in tensile stresses to be absorbed by adjacent fibres. Thus, subsequent scratches lead to further deformation of the element but not to complete destruction.

The results of experimental studies also showed a significant impact of sample size on concrete tensile strength when determined by various methods: bending, splitting and direct stretching. It was found that the value of concrete tensile strength at bending is greater than determined in the splitting test. The assumption of linear elasticity and flat sections of concrete when determining the tensile strength at bending causes an increase in the calculated value of tensile strength compared to the actual value.

Analysis of the results of their own research confirms the above observation. The determined value of concrete tensile strength at bending f_L obtained during the 3-point test is higher by 21% for RSFRC and by 36% for SFRC than the strength $f_{ct,sp}$ determined in the Brazilian splitting test. Similarly, the effect of sample size, a different method of force application, and a calculation procedure based only on the effect of the horizontal component when calculating the size may explain the much lower tensile strength value obtained by splitting according to the WST method than in the case of the Brazilian method. Despite the differences in values of the tensile strengths, an increase in tensile strength for RSFRC concrete from 30% (tensile strength determined in the bending test) to 70% (tensile strength determined in the splitting test during the WST test) relative to NC concrete was observed for all measuring methods (Figure 21).

Additional analysis of the σ-CMOD charts obtained during the 3-point test and the WST test (Figure 22) indicates a great convergence of the nature of the behaviour of the samples with the RSF and ISF fibres in the elastic zone, which indicates the possibility of replacing ISF fibres with recycled RSF fibres without reducing the strength parameters fibre concrete.

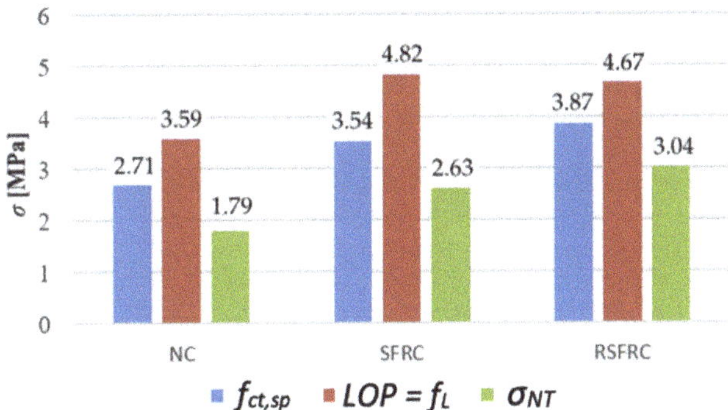

Figure 21. Comparison of tensile strength of tested concretes without fibres (NC) and with steel fibres (SFRC and RSFRC) appointed by Brazilian method, 3-point test and WST.

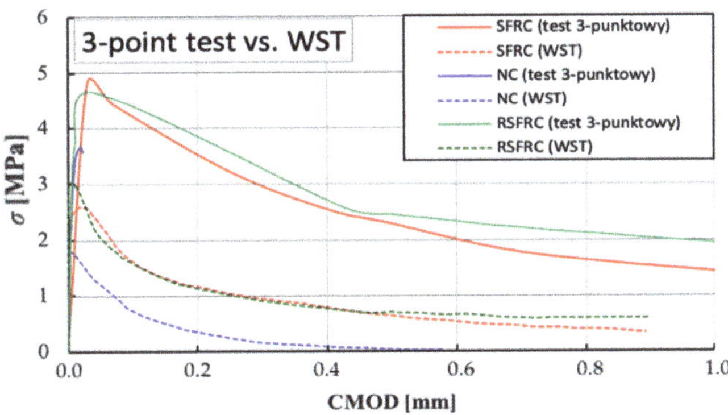

Figure 22. Comparison of σ-CMOD diagrams determined in the bending tests (3-point test) versus wedge splitting test (WST).

5.3. Impact of RSF Fibre Addition on Adhesion and Abrasion of Concrete

In the case of steel fibre-modified composites, in the case of a pull-off peel test, the dispersion of results is quite large due to the small peel surface and the heterogeneous distribution of the fibres in its area. No apparent effect of steel fibres on the adhesion in concrete was observed. The average surface tensile strength for a composite without NC fibres and for a composite modified with ISF fibres is 2.50 MPa and 2.45 MPa, respectively.

The highest peel strength (2.49 MPa) was observed for the composite modified with RSF fibres. The test results confirm that the designed RSFRC concrete meets the requirement for industrial floors, where it is recommended that the value of the surface tensile strength tested by the pull-off test be above 1.5 MPa (Table 6).

The results of tests on concrete abrasion resistance showed a significant influence of steel fibres on composite abrasion resistance. The test carried out on the Boehme disk showed a volume loss for the SFRC samples smaller by 11%, and for the RSFRC samples smaller by 15% compared to those obtained for NC concrete (Table 7).

5.4. Environmental Impact of the Production of Concrete Reinforced with ISF and RSF Fibres

Comparing the energy consumption and CO_2 emission for the concrete recipe with RSF fibres (Table 9) to ISF fibres (Table 8) shows 31.3% lower energy consumption and 30.8% lower CO_2 emission than for concrete with ISF fibres. It is clearly demonstrated that industrially produced steel fibres, due to complicated and energy-intensive production processes, are the second component (after cement) significantly affecting the natural environment. Replacing industrially produced steel fibres with waste fibres obtained in the recovery of rubber from used tyres may impact on the reduction of energy consumption and greenhouse gas emissions. Considering the extension of reused components in tyre recycling process, the proposed solution fully responds to the Circular Economy concept.

Table 9. Determination of energy consumption and CO_2 emissions for 1 m^3 of fibre concrete with RSF fibres.

Concrete Component	Energy Consumption Factor (MJ/kg]	CO_2 Emission (kg CO_2/kg]	Amount in 1 m^3 (kg]	Energy Consumption (MJ]	CO_2 Emission (kg CO_2]
cement (kg]	3 *	0.3 **	320	960	96
water (dm^3]	0.05 **	0 **	160	8	0
sand 0/2 (kg]	0.1 **	0.007 **	801	80.1	5.607
gravel 2/8 (kg]	0.1 **	0.007 **	681	68.1	4.767
gravel 8/16 (kg]	0.1 **	0.007 **	454	45.4	3.178
fibre RSF (kg]	0 ****	0 ****	50	0	0
			SUMA	**1161.6** (MJ]	**109.55** (kg CO_2]

(*)–(**) according to Table 8. (****) zero energy consumption and zero CO_2 emission value were assumed for RSF fibres constituting waste during rubber recovery.

6. Conclusions

In this study, the possibility of replacing industrial fibres in concrete with RSF fibres treated as waste from tyre recycling processes was demonstrated. To estimate the efficiency of RSF fibre concrete, a wide range of strength tests was performed, emphasising the determination of the tensile strength of the composite. The research and analysis showed that RSF fibres have parameters comparable to those of industrial steel fibres used to reinforce concrete industrial floors.

The following conclusions can be drawn from the research described in this paper.

- The geometric examination of the fibres confirmed that the fibres from the recycling of tyres used for the tests are characterised by varying lengths, diameter, slenderness and shape. Such a mixture can be classified as hybrid fibres, which show higher efficiency in concrete than fibres of the equal length. However, considering the geometrical characteristics of the fibres, is was shown that only about 60% of the RSF fibres in concrete improve the parameters of the composite. This indicates the application of double content of RSF fibres (50 kg/m^3) versus ISF fibres (25 kg/m^3) in proposed concrete mix.

- The research showed that steel fibres' addition significantly affects all rheological and mechanical properties of the concrete: its workability (lower), consistency (reduction by one class), air content in the mix, and strength parameters. The addition of steel fibres can enhance concrete performance, especially the compressive strength (by 13.5% for composite modified with ISF fibres and by 22% for concrete modified with RSF fibres). Tensile strength tests carried out by three methods: Brazilian splitting, bending (3-point test), and WST splitting confirmed the increase in tensile strength when modifying concrete with RSF fibres, respectively, by 43%, 30% and 70% in comparison to the average strength of reference concrete without fibres. Moreover, RSF fibres significantly improved the abrasion resistance of the composite (by 42%).
- The calculation of environmental parameters of concrete with RSF fibres showed significantly lower energy consumption (by 31.3%) and lower CO2 emission (by 30.8%) than concrete with ISF fibres due to the energy-consuming production processes of industrial fibres.

The main limitation for the substitution of ISF fibres with fibres derived from the recycling of tyres in concrete was recognised in improper distribution of irregular RSF fibres and the threat of fibre clusters, i.e., "balls". However, the 3D images of samples had demonstrated the lack of clusters of RSF fibres and therefore proved that no additional requirements are needed for mixing the components during the production of proposed fibre concrete.

The above-mentioned conclusions prove the possibility of replacing industrial fibres with RSF fibres without reducing fibre concrete's strength parameters. In addition, due to the growing public awareness of the need to protect the environment and the climate consequences associated with excessive greenhouse gas emissions, all solutions to reduce CO_2 emissions and waste recycling have great potential for rapid commercialisation. Therefore, further investigations are recommended to explore the possibility of utilising RSF as an eco-friendly material to mitigate the challenges of the sustainable construction industry.

Author Contributions: Conceptualisation, M.P.-M. and M.K.; methodology, M.P.-M.; validation, M.P.-M.; formal analysis, M.P.-M. and M.K.; investigation, M.P.-M. and M.K.; writing-original draft preparation, M.P.-M. and M.K.; writing—review and editing, M.P.-M. and M.K.; supervision, M.P.-M. and M.K.; All authors have read and agreed to the published version of the manuscript.

Funding: This research received no external funding.

Institutional Review Board Statement: Not applicable.

Informed Consent Statement: Not applicable.

Data Availability Statement: Not applicable.

Conflicts of Interest: The authors declare no conflict of interest.

References

1. Etrma Statistics—2017. Available online: https://www.etrma.org/wp-content/uploads/2019/11/ELT-Management-Figures-2017-vf.xlsx.pdf (accessed on 15 April 2019).
2. Ganjian, E.; Khorami, M.; Maghsoudi, A.A. Scrap−tyre−rubber replacement for aggregate and filler in concrete. *Constr. Build. Mater.* **2009**, *23*, 1828–1836. [CrossRef]
3. Jing, L. A New Composite Slab Using Crushed Waste Tires as Fine Aggregate in Self-Compacting Lightweight Aggregate Concrete. *Materials* **2020**, *13*, 2551. [CrossRef]
4. Gesoglu, M.; Güneyisi, E.; Hansu, O.; İpek, S.; Asaad, D.S. Influence of waste rubber utilization on the fracture and steel–concrete bond strength properties of concrete. *Constr. Build. Mater.* **2015**, *101*, 1113–1121. [CrossRef]
5. Guo, Y.C. Evaluation of properties and performance of rubber-modified concrete for recycling of waste scrap tire. *J. Clean. Prod.* **2017**, *148*, 681–689. [CrossRef]
6. Gupta, T.; Chaudhary, S.; Sharma, R.K. Assessment of mechanical and durability properties of concrete containing waste rubber tire as fine aggregate. *Constr. Build. Mater.* **2014**, *73*, 562–574. [CrossRef]
7. Ho, A.C. Effects of rubber aggregates from grinded used tyres on the concrete resistance to cracking. *J. Clean. Prod.* **2012**, *23*, 209–215. [CrossRef]

8. Xie, J.-h.; Guo, Y.-c.; Liu, L.-s.; Xie, Z.h. Compressive and flexural behaviours of a new steel-fibre-reinforced recycled aggregate concrete with crumb rubber. *Constr. Build. Mater.* **2015**, *79*, 263–272. [CrossRef]

9. Lijuan, L.; Shenghua, R.; Lan, Z. Mechanical properties and constitutive equations of concrete containing a low volume of tire rubber particles. *Constr. Build. Mater.* **2014**, *70*, 291–308.

10. Silva, F.M. Investigation on the properties of concrete tactile paving blocks made with recycled tire rubber. *Constr. Build. Mater.* **2015**, *91*, 71–79. [CrossRef]

11. Busic, R.; Bensic, M.; Milicevic, I.; Strukar, K. Prediction Models for the Mechanical Properties of Self-Compacting Concrete with Recycled Rubber and Silica Fume. *Materials* **2020**, *13*, 1821. [CrossRef]

12. Son, K.S.; Hajirasouliha, I.; Pilakoutas, K. Strength and deformability of waste tyre rubber-filled reinforced concrete columns. *Constr. Build. Mater.* **2011**, *25*, 218–226. [CrossRef]

13. Noaman, A.T.; Abu bakar, B.H.; Akil, H.M. Experimental investigation on compression toughness of rubberized steel fibre concrete. *Constr. Build. Mater.* **2016**, *115*, 163–170. [CrossRef]

14. Thomas, B.S.; Gupta, R.C. Long term behaviour of cement concrete containing discarded tire rubber. *J. Clean. Prod.* **2015**, *102*, 78–87. [CrossRef]

15. Turatsinze, A.; Bonnet, S.; Granju, J.-L. Mechanical characterisation of cement–based mortar incorporating rubber aggregates from recycled worn tyres. *Build. Environ.* **2005**, *40*, 221–226. [CrossRef]

16. Naaman, A.E. Tensile Strain-Hardening FRC Composites: Historical Evolution since the 1960. In *Advances in Construction Materials*; Grosse, C.U., Ed.; Springer: Berlin/Heidelberg, Germany, 2007; pp. 181–202. [CrossRef]

17. Labib, W.; Eden, N. An investigation into the use of fibres in concrete industrial ground-floor slabs. In *Proceedings International Postgraduate Research Conference in the Built and Human Environment*; University of Salford: Salford, UK, 2006; pp. 466–477.

18. Naaman, A.E. High Performance Fiber Reinforced Composites: Classification and Applications. In Proceedings of the International Workshop "Cement Based Materials and Civil Infrastructure", Karachi, Pakistan, 10–11 December 2007; pp. 389–401.

19. Tlemat, H.; Pilakoutas, K.; Neocleous, K. Demonstrating Steel Fibres from Waste Tyres as Reinforcement in Concrete: Material Characterization. In Proceedings of the First International Conference on Innovative Materials and Technologies for Construction and Restoration, Lecce, Italy, 6–9 June 2004; Volume 1, pp. 172–185.

20. Tlemat, H.; Pilakoutas, K.; Neocleous, K. Design Issues for Concrete Reinforced with Steel Fibres Recovered from Waste Tyres. *J. Mater. Civ. Eng. ASCE* **2006**, *18*, 677–685.

21. Tlemat, H.; Pilakoutas, K.; Neocleous, K. Stress Strain Characteristic of SFRC using Recycled Fibre. *Mater. Struct.* **2006**, *39*, 365–377. [CrossRef]

22. Aiello, M.A. Use of steel fibers recovered from waste tyres as reinforcement in concrete: Pull-out behaviour, compressive and flexural strength. *Waste Manag.* **2009**, *29*, 1960–1970. [CrossRef]

23. Linsberger, H.N.; Tschegg, E.K. Fracture energy determination of concrete with cube specimens. *Zem. Beton* **1986**, *31*, 38–40.

24. Central Statistical Office for the Polish Economy. ISBN 978-83-89641-04-5, Katowice, 2011. Available online: http://stat.gov.pl/cps/rde/xbcr/gus/oz_wskazniki_zrownowazonego_rozwoju_Polski_us_kat.pdf (accessed on 15 April 2019).

25. Casanas, V. Energy as Indicator of Environmental Impact on Building Systems. Available online: http://wsb14barcelona.org/programme/pdf_poster/P-029.pdf (accessed on 15 April 2019).

26. Glinicki, M.A. Beton ze Zbrojeniem Strukturalnym. In *XXV Ogólnopolskie Warsztaty Pracy Projektanta Konstrukcji*; 2010; pp. 279–308. Available online: http://www.ippt.pan.pl/Repository/o70.pdf (accessed on 15 April 2019).

27. Jamroży, Z. *Beton i Jego Technologie*; Wydawnictwo Naukowe PWN: Warsaw, Poland, 2005; pp. 415–417.

28. Zych, T. Współczesny fibrobeton—Możliwość kształtowania elementów konstrukcyjnych i form architektonicznych. *Archit. Czas. Tech.* **2010**, *18*, 371–386.

29. Martinelli, E.; Caggiano, A.; Xargay, H. An experimental study on the post- cracking behaviour of Hybrid Industrial/Recycled Steel Fibre-Reinforced Concrete. *Constr. Build. Mater.* **2015**, *94*, 290–298. [CrossRef]

30. Centonze, G. Concrete Reinforced with Recycled Steel Fibers from End of Life Tires: Mix-Design and Application. *Key Eng. Mater.* **2016**, *711*, 224–231. [CrossRef]

 *materials*

Article

Numerical Analysis of Shear and Particle Crushing Characteristics in Ring Shear System Using the PFC^{2D}

Sueng-Won Jeong [1], Kabuyaya Kighuta [2], Dong-Eun Lee [3] and Sung-Sik Park [2,*]

[1] Geologic Environment Division, Korea Institute of Geoscience and Mineral Resources, Daejeon 34132, Korea; swjeong@kigam.re.kr
[2] Department of Civil Engineering, Kyungpook National University, 80 Daehakro, Bukgu, Daegu 41566, Korea; dankabuyaya@knu.ac.kr
[3] Department of Architectural Engineering, Kyungpook National University, 80 Daehakro, Bukgu, Daegu 41566, Korea; dolee@knu.ac.kr
* Correspondence: sungpark@knu.ac.kr; Tel.: +82-53-950-7544

Abstract: The shear and particle crushing characteristics of the failure plane (or shear surface) in catastrophic mass movements are examined with a ring shear apparatus, which is generally employed owing to its suitability for large deformations. Based on results of previous experiments on waste materials from abandoned mine deposits, we employed a simple numerical model based on ring shear testing using the particle flow code (PFC^{2D}). We examined drainage, normal stress, and shear velocity dependent shear characteristics of landslide materials. For shear velocities of 0.1 and 100 mm/s and normal stress (NS) of 25 kPa, the numerical results are in good agreement with those obtained from experimental results. The difference between the experimental and numerical results of the residual shear stress was approximately 0.4 kPa for NS equal to 25 kPa and 0.9 kPa for NS equal to 100 kPa for both drained and undrained condition. In addition, we examined particle crushing effect during shearing using the frictional work concept in PFC. We calculated the work done by friction at both peak and residual shear stresses, and then used the results as crushing criteria in the numerical analysis. The frictional work at peak and the residual shear stresses was ranged from 303 kPa·s to 2579 kPa·s for given drainage and normal stress conditions. These results showed that clump particles were partially crushed at peak shear stress, and further particle crushing with respect to the production of finer in shearing was recorded at residual shear stress at the shearing plane.

Keywords: residual shear stress; particle crushing; ring shear test; particle flow code (PFC^{2D}); frictional work

Citation: Jeong, S.-W.; Kighuta, K.; Lee, D.-E.; Park, S.-S. Numerical Analysis of Shear and Particle Crushing Characteristics in Ring Shear System Using the PFC^{2D}. *Materials* **2021**, *14*, 229. https://doi.org/10.3390/ma14010229

Received: 6 December 2020
Accepted: 31 December 2020
Published: 5 January 2021

Publisher's Note: MDPI stays neutral with regard to jurisdictional claims in published maps and institutional affiliations.

Copyright: © 2021 by the authors. Licensee MDPI, Basel, Switzerland. This article is an open access article distributed under the terms and conditions of the Creative Commons Attribution (CC BY) license (https://creativecommons.org/licenses/by/4.0/).

1. Introduction

Erosion and rainfall-induced mass movements could result in significant life loss and property damage in urban areas. After slope failure initiation, spreading mass movements are strongly related to the frictional characteristics of the movement stages. In particular, evaluation of residual shear stress is crucial when a significant propagation of mass movements is expected after the onset of slope failure. For the safety of ecosystems, rapid downward moving masses involving soil, rock, water or their combinations have been intensively studied [1–5]. Evaluating failure and post-failure processes of rapid landslides requires in-depth knowledge of various scientific disciplines, such as geomorphology, geomechanics, hydraulics, and rheology, to predict, prevent, and stabilize the mass movements. Moreover, shear stress is an important mechanical parameters necessary to understand the landslide mobilization. Various shear tests have been conducted to investigate shearing characteristics of geomaterials [6–18]. The shear strength of granular materials is still challenging to determine using both experimental and numerical methods.

Bagherzadeh-Khalkhali and Mirghasemi [6] have investigated the direct shear strength of coarse-grained soils using experimental and numerical analysis under different normal

411

stresses. They found that the characteristics of coarse-grained soil varied from strain-hardening to softening at the shearing duration as the normal stress increased, and the internal friction angle decreased with stress level increase. Cabalar et al. [7] have performed triaxial and cyclic tests to assess the strength of different sands, considering the effects of the particle shapes. They found that crush stone sands extracted from the northern region of Cyprus show a significantly higher strength than sands obtained from Gaziantep because of their different shapes of sand particles. They have also claimed that the degrees of sphericity and roundness of the particles increase the strength of sand and can reduce its volumetric strain. The above-mentioned experimental tests exhibit limitations in evaluating the overall shear characteristics of diverse mass movements. The most extensively used shear tests are the direct shear, triaxial shear, and ring shear tests; each of these has advantages and limitations. In contrast to other shear tests, the ring shear test can measure shear stress for large displacement [18]. In addition, it is performed with advanced equipment capable of controlling the consolidation, drainage, and shearing speed under static and dynamic conditions. Numerical analysis is often used as a way to overcome the limitations of laboratory experiments.

Lobo-Guerrero and Vallejo [19] proposed a discrete element method to simulate the evolution of sugar particle crushing subjected to ring shear testing; they used the Particle Flow Code in two dimensions (PFC^{2D}). The periodic movement of particles in the ring shear apparatus was simulated with two parallel periodic boundaries and two saw-toothed standard boundaries; these were made with several edges to model the shearing surface. They found that the residual coefficient of the sugar materials was maintained constant in spite of the particles being crushed. However, erroneous stress computations can occur when combining periodic space boundaries and standard boundaries because the periodic space is not compatible with the standard boundary in PFC [20]. Moreover, the roughened shearing surface with multiples edges negatively affects the particle response and the stress calculation, especially for particles distributed in the corners. In addition, neither the periodic space nor the standard boundary method take into account the interactions between machine components and particles.

A more accurate and efficient evaluation of the ring shear characteristics of highly mobile landslide materials using the PFC^{2D} is needed. To simulate the ring shear test, we used a general boundary, which is crucial and useful for simulating the interactions between granular materials and machine components. We modeled the particle crushing using the clump method and the frictional work technique from PFC^{2D}. The numerical results validated the results obtained from the ring shear experiments. The shear stress was calculated considering four different shearing velocities (0.01, 0.1, 1, and 100 mm/s), drainage condition, and normal stresses varying from 20 to 150 kPa. The discrete element method in PFC^{2D} requires contact models involving micromechanical properties of the granular materials. Material mechanical properties obtained experimentally are taken as macromechanical properties and are computed using the micromechanical properties by trial-and-error in PFC^{2D} [20]. The obtained microproperties can be utilized to design the shear behaviors of the waste materials in landslide hazards using PFC.

2. Materials and Methods

2.1. Materials

The waste materials were sourced from Busan Metropolitan city, Korea. They are taken from Imgi mine deposits, where the landslide occurs due to intense rainfall [21–23]. The landslide materials are mainly contained sub-graded and angular grains composed of pyrite, kaolinite, sericite, pyrophyllite, and quartz. The sample used for laboratory shear ring testing was composed of 35% gravel, 63% sand, and 2% other fine materials (i.e., fine particles that are more than 50% of soil passes 0.075 mm sieve). Thus, the soil sample can be considered as coarse-grained sediments. Porosity, the ratio of volume of voids to the total volume of the soil, is approximately 40%. These waste materials are categorized as gravelly sandy soils. Their mean diameter, effective grain size, and the uniformity coefficient are

1.5 mm, 0.3 mm, and 5 mm, respectively. The materials used are very similar to typical landslide materials encountered in Korea. Table 1 summarizes the geotechnical properties of the materials used. This work focuses more on the numerical analysis; more details about sample preparation and material properties can be found in [21].

Table 1. Physical properties of waste materials.

Specific Gravity	Water Content (%)	Total Unit Weight (t/m^3)	Dry Unit Weight (t/m^3)	Liquid Limit (%)	Porosity (%)	USCS
2.63	6.9	1.7	1.59	24.5	39.5	SM

2.2. Experimental Program

The ring shear test is suitable for investigating the shear characteristics of landslides because it offers several advantages and permits the measurement of shear at large displacement; it can also be used to investigate the mechanical characteristics of sliding surfaces due to large shear displacements [14]. We performed laboratory ring shear testing with a ring shear apparatus designed at the Korea Institute of Geoscience and Mineral Resources (KIGAM). This machine can quantitatively simulate the consolidation, drainage, and shear velocity in static and dynamic loading conditions. The outer and inner diameters and the height of the shear box of the KIGAM ring shear apparatus were 250, 110, and 75 mm, respectively. The shear box consists of an upper and a lower boxes, as shown in Figure 1. During the ring shear test, the upper box is fixed and the lower one rotates. The shear surface is clearly visible after testing (Figure 1b). Landslides may occur in a diverse shape and size. Normal stress can be considered based on the soil thickness where the shear surface observes. Shear velocity is important in determining the shear strength with respect to the landslide movement rate. There are numerous types of landslides, which are ranged from very slow to very rapid speed. Drainage is one of important conditions in the landslide initiation, because it is directly related to the generation of pore water pressure in shear surface (or landslide movement). Drained condition can be applied for no pore water pressure condition; thus, it can be used to reproduce very slow landslides, such as a creep motion of clay-rich landslides (e.g., a few centimeter per year). In the experimental program, the valve located in the ring shear box is open; thus, the water can freely move during shearing. No pore water pressure occurs. However, undrained condition is specifically used for a relatively rapid landslide occurrence (e.g., higher than 1.8 m/hr). The same boundary conditions are used in the numerical analysis, as detailed in the next section. We experimentally measured the normal stresses, vertical displacement from a linear variable differential transformer, pore pressure, and torques. The parameters considered in the experimental tests are: normal stress, drainage condition, and shear velocity, as presented in Table 2. Details on the laboratory experiments are found in [21].

Table 2. Experimental parameters.

Test Condition	Velocity (mm/s)	Normal Stress (kPa)
Drained	0.01	20
	0.1	40
	1	60
	100	80
Undrained		100
		150

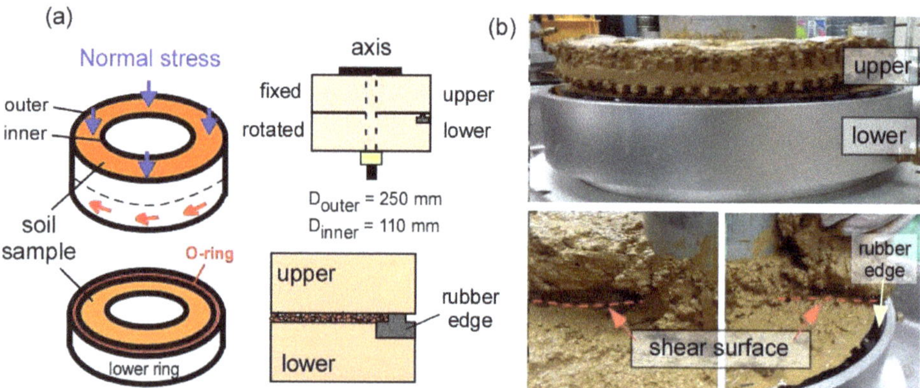

Figure 1. Schematical illustration of the ring shear box and shear surface after testing: (**a**) configuration of ring shear box and (**b**) observation of shear surface after testing.

3. Numerical Model

3.1. Discrete Element Method Description in PFC^{2D} and the Clump Method

The particle flow code (PFC^{2D}) developed by Cundall is a discrete-element-method (DEM)-based software designed to simulate the movement and interaction of stressed granular assemblies. Cundall and his colleagues [24–26] are among the frontier researchers to apply the discrete element method to the movement of granular assemblies. The granular assembly consists of different particles that displace independent of one another, and the interaction between particles occurs only at contact points or interfaces. The PFC assumptions are as follows: the particles are considered rigid bodies; a soft-contact approach characterizes the particles at contact points, where they are permitted to overlap; the contact between particles can be defined by bonds; the shape of particles is either circular or spherical, with unit thickness; the overlap magnitude is related to the force of contact by the force–displacement law [20]; the overlap is small compared to the particle size. It denotes the relative contact displacement in the normal direction. The overlap equation is given by:

$$U^n = \begin{cases} R^{[A]} + R^{[B]} - d & \text{(particle − particle contact)} \\ R^{[b]} - d & \text{(patricle − boundary contact)} \end{cases} \tag{1}$$

where $R^{[A]}$ and $R^{[B]}$ are the radii of particles in contact. $R^{[b]}$ and d are the radius of a particle in contact with a wall (boundary) and the distance between particles centers, respectively.

In addition, the calculation process involves applying alternatively a force–displacement formulation at contact points and the Newton's second law to the rigid bodies. Thus, the motion of each rigid body due to contact and forces applied on it is determined by Newton's second law; further, the update of contact forces computed from the relative motion at each contact is governed by the force–displacement law. This law is applied for both particle–particle and particle–wall (i.e., model boundary) contacts. The computational scheme is a time stepping algorithm that consists of applying repeatedly the law of motion to each particle, a force-displacement law to each contact, and a constant updating of wall positions as shown in Figure 2. A detailed description of the DEM in PFC can be found in [27].

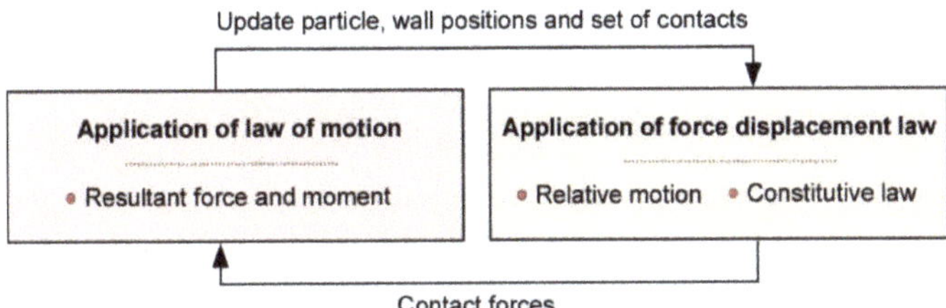

Figure 2. Discrete Element Method (DEM) computational scheme in PFC^{2D}.

In the past decades, the particles generated for any granular assembly were simply circular or spherical; however, with the modern DEM, one can create a general particle shape using two or more circular or spherical particles [28–30]. The process of creating a particle of any shape is termed the clump or clustering method [27]. The particle created in the granular assembly may be a two-, three-, or four-particle clump, depending on the number of the particles forming it. For example, to simulate a granular assembly containing triangular or square particles or grains of a more natural shape, one only needs to combine few predefined simple particles to create the intended particle shape. The creation of triangular or square-shaped particles is illustrated in Figure 3a,b. The contact model is defined only between clump particles. The contact stiffness model, slip and separation model, and bonding model are the three different contact models provided by the PFC. The bonding models are classified into two types: contact bond models can simply produce a force and parallel bond models that can produce both a force and a moment. Herein, we employed the parallel bond as a cementation material between clump particles because it provides efficiently rotational movement of the particles in the granular system.

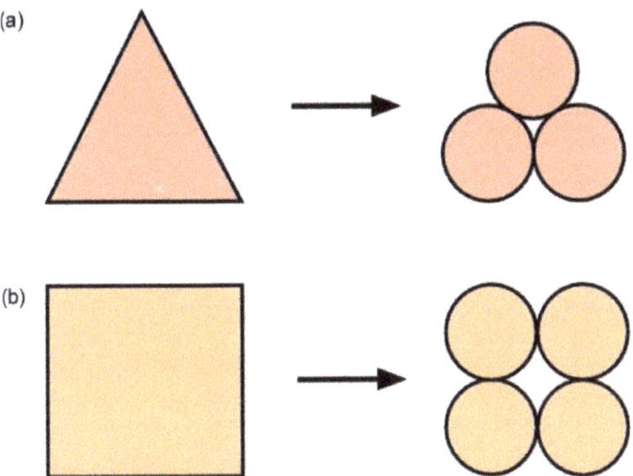

Figure 3. Illustration of clump formation: (**a**) Triangular particle made from nine overlapping circular particles and (**b**) Square particle made from 15 overlapping circular particles.

3.2. Simulation Procedure

As it is extremely difficult to measure micromechanical properties of soil and rock materials in laboratory experimentation, in which only the macromechanical properties of materials can be measured, the micromechanical properties of synthetic materials in PFC^{2D} can be used to obtain the macromechanical properties of granular materials by the trial-and-error method [20]. Although the DEM simulation cannot take into account as many particles as used in an experimental sample, it does guarantee a good approximation [20]. PFC^{2D} version 4.0 supports up to 100,000 particles for one granular assembly. In this study, to reduce the computational time, 6830 particles were used to simulate the ring shear test.

We performed ring shear simulation to investigate the normal stress and shear velocity effects on the shear stress. The clump method, as demonstrated in the particle flow code PFC^{2D}, was used to generate the granular systems (Figure 4). The clump particles were created with five circular particles (Figure 4a) and random size distribution in the assembly. The particle cementation material that bonded the clump particles was set to parallel bond (Figure 4b,c). Since there is no fluid connection at the contact between touching particles in PFC^{2D}, both drained and undrained conditions were simulated using the parallel bond between clump-particles and the calibration process may be used to obtain the macro-properties [20]. The micromechanical properties of the Lac du Bonnet material [20] were used to simulate the waste materials. Table 3 presents the materials properties of the clump particle assembly system. Different normal stresses were installed in the assembly using the initial stress installation procedure. After normal stress installation, a ring-shaped boundary was created using a general wall mechanism (Figure 4d). The top section of the ring shear box was assumed to simulate the 3D ring shear experiment. The ring shear box was rotated by applying to the outer boundary rotational velocities (i.e., 0.01, 0.1, 1, and 100 mm/s).

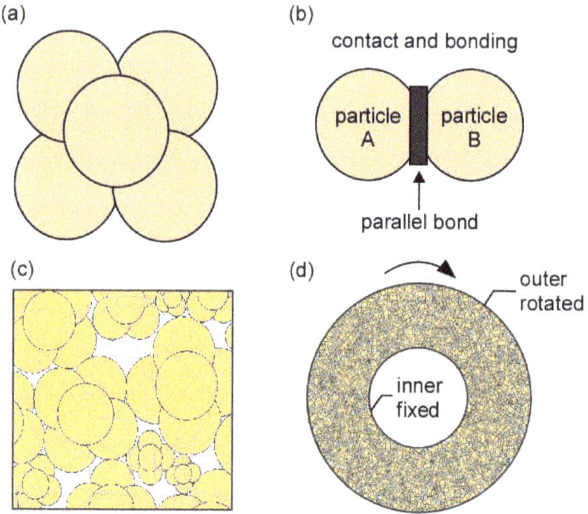

Figure 4. Clump particles and assembly: (**a**) clump particle, (**b**) parallel bond idealization between particles, (**c**) parallel bond connection between clump particles and (**d**) clump particles in ring shear box.

Table 3. Synthetic material properties.

Clump Particle	Cementing Material (Parallel Bond)
Bulk density 1700 kg/m^3 R_{max}/R_{min} [1] = 5.0 Modulus of elasticity = 6.1 MPa Normal to shear stiffness ratio = 2.5 Friction coefficient = 0.5	Bond-radius = 1 Modulus of elasticity = 6.1 MPa Normal to shear stiffness ratio = 2.5 Normal strength = Shear strength = mean ± std.dev = 162 ± 37 MPa

[1] R_{max} = clump particle maximum radius, R_{min} = clump particle minimum radius, and std.dev = standard deviation.

Our results showed that the shear stress increases when the normal stress and shear velocity increase. The shear angular velocity creates a centrifugal force on the particles. The materials are tested under normal stress of 20, 40, 60, 80, 100, and 150 kPa. The material was generated in the ring-shaped vessel in order to produce an isotropic and well-connected granular assembly at a specified normal stress. To perform an accurate ring shear simulation, the clump particle assembly was created using with the material genesis procedure [20].

4. Results and Discussions

We employed a series of numerical models to investigate the effects of the normal stress and shear velocity on the shear stress. The microproperties designed in PFC enable the simulation of the macromechanical properties obtained from laboratory experiments using a trial-and-error procedure. The three main macroproperties considered in this study are the modulus of elasticity, peak stress, and residual stress. After the materials were generated according to the material genesis procedure, using the micromechanical properties listed in Table 3, we repeatedly conducted numerical ring shear tests. Then, the resulting numerical values were directly compared to the experimental results by matching macromechanical properties. To reproduce the relevant behaviors of the waste materials, we determined the appropriate microproperties by a calibration process in which the response of the synthetic material is compared directly with the measured response of the waste materials. The results obtained experimentally were compared to those obtained by numerical simulations. Based on the research findings, the followings are highlighted: (a) shearing time, (b) shear velocity, (c) normal stress, and (d) crushing phenomenon during shearing.

4.1. Shear Stress and Shearing Time

We compared the experimental and numerical results for shear stress characteristics during a period of 300 s in the ring shear system for given drainage and normal stress conditions. To examine the effect of drainage and normal stress on the shear stress, we plotted shear stress vs. shearing time curves at a shear velocity of 0.1 mm/s; the normal stress was constant during each test (Figure 5). There is a clear peak value in shear stress–time relationships regardless of drainage condition. For a normal stress of 25 kPa, the experimental and numerical evaluations revealed a slope difference of 0.81 and 0.28 kPa/s in drained and undrained conditions, respectively; while for a normal stress of 100 kPa, the slope difference between the experimental and numerical curves was 0.56 and 0.39 kPa/s in drained and undrained condition, respectively. In addition, under the undrained condition, numerical and experimental results showed the similar peak stress values of 10.1 and 13.7 kPa at normal stress of 25 and 100 kPa, respectively; a peak stress difference of 0.3 kPa was obtained under the drained condition for both 25 and 100 kPa normal stresses. These results show that the experimental and numerical results for both drained and undrained condition are in good agreement.

A sudden drop of the shear stress appeared for both the experimental and numerical curves after the peak shear stress was found. Regardless of the normal stress level and drainage conditions, the materials evaluated here presented a strain-softening behavior (Figure 5). For a normal stress of 25 kPa, the calculated and experimental times at which the peak stress was reached differed by 1.1 s for the undrained condition, and 8.5 s for the drained condition. For a normal stress of 100 kPa, the calculated and measured time needed to reach the peak stress differed by 5.7 s for the undrained condition, and 0.9 s for the drained condition. These differences might be due to the difference in the time step scheme used for the calculation of the shear stress in PFC^{2D} [20].

Furthermore, Figure 5 illustrates the residual shear stress induced by the resistance of the clump particles after the drop in peak shear stress. This resistance is due to inter-particle friction and inter-locking effect between clump particles. Thus, numerical analysis is an efficient way to explain particle rearrangement with respect to the reduction in shear strength. Stabilization is reached for 150–300 s for both drained and undrained conditions with various normal stresses. The shape of the clump particles is also crucial to create some resistance after the drop in peak stress. We assumed that the residual shear stress was the shear stress measured during the stabilization period that followed the sudden drop in peak shear stress. In the granular assembly, a progressive clump particle crushing occurred after the drop in peak shear stress.

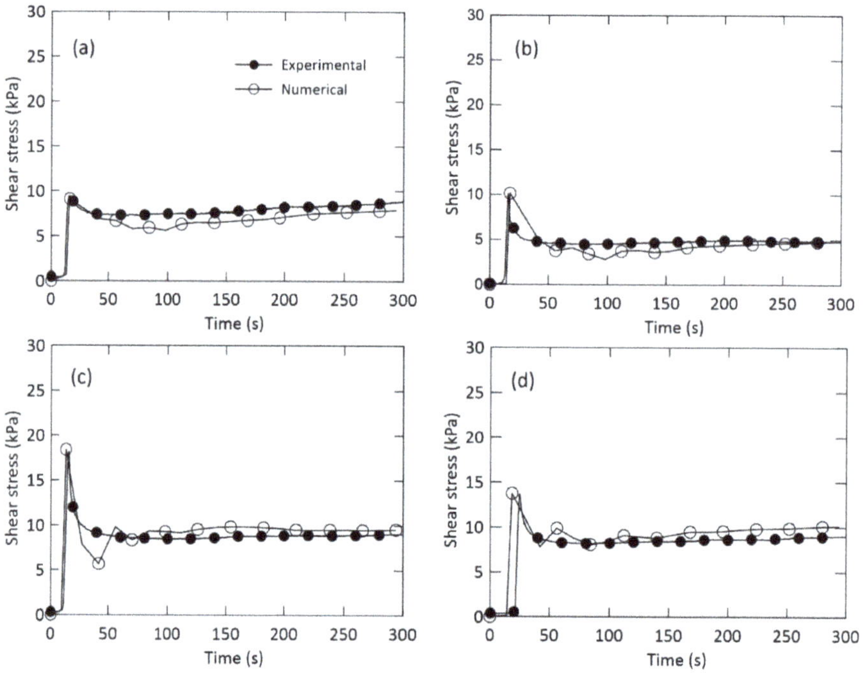

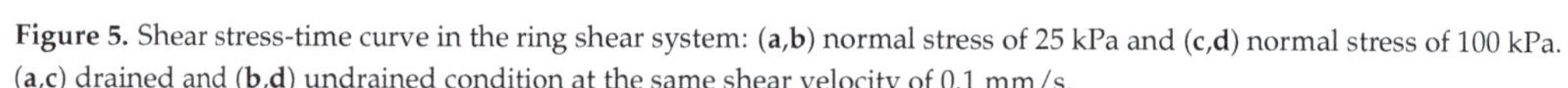

Figure 5. Shear stress-time curve in the ring shear system: (**a**,**b**) normal stress of 25 kPa and (**c**,**d**) normal stress of 100 kPa. (**a**,**c**) drained and (**b**,**d**) undrained condition at the same shear velocity of 0.1 mm/s.

For a normal stress of 25 kPa, the difference in residual shear stress between the experimental and numerical analyses was 0.4 kPa for both the drained and undrained conditions (Figure 5a,b). For a normal stress of 100 kPa, the residual shear stress obtained experimentally and numerically differed by 0.7 kPa and 0.9 kPa for drained and undrained conditions, respectively (Figure 5c,d). As a result, the larger the normal stress, the larger the difference. These results show that the residual shear stress values we obtained experimentally and numerically were in good agreement.

4.2. Shear Stress and Shear Velocity

The effect of shear velocity on the shear stress is far more specific than those of drainage and normal stresses. We examined the shear characteristics of the waste materials as a function of shear velocity with respect to the peak and residual shear stress values. Figure 6 shows the influence of shear velocity on the peak and residual shear stress under different drainage and normal stresses. In general, the shear stress increased with an increase of shear velocity for all given conditions (Table 4). For a normal stress of 25 kPa under the drained condition, for shear velocities of 0.01, 0.1, 1, and 100 mm/s, the difference in peak shear stress between the experimental and numerical analyses was 0.1, 0.3, 0.1, and 0.7 kPa, respectively; the residual shear stresses differed by 1, 0.4, 0.5, and 1.7 kPa, respectively. For a normal stress of 100 kPa under the drained condition, at shear velocity of 0.01 mm/s, the experimental value of the peak shear stress was in a similar range compared with that obtained from numerical analysis; at shear velocities of 0.1, 1, and 100 mm/s, the difference in peak shear stress between the experimental and numerical evaluations was 0.3, 0.2, and 0.3 kPa, respectively. It seems that shear stress is not strongly affected by low shear speed (i.e., 0.01 mm/s) in ring shear apparatus used. Moreover, the residual shear stresses differed by 0.4, 0.7, 1, and 2.3 kPa at shear velocities of 0.01, 0.1, 1, and 100 mm/s, respectively (Figure 6a,b).

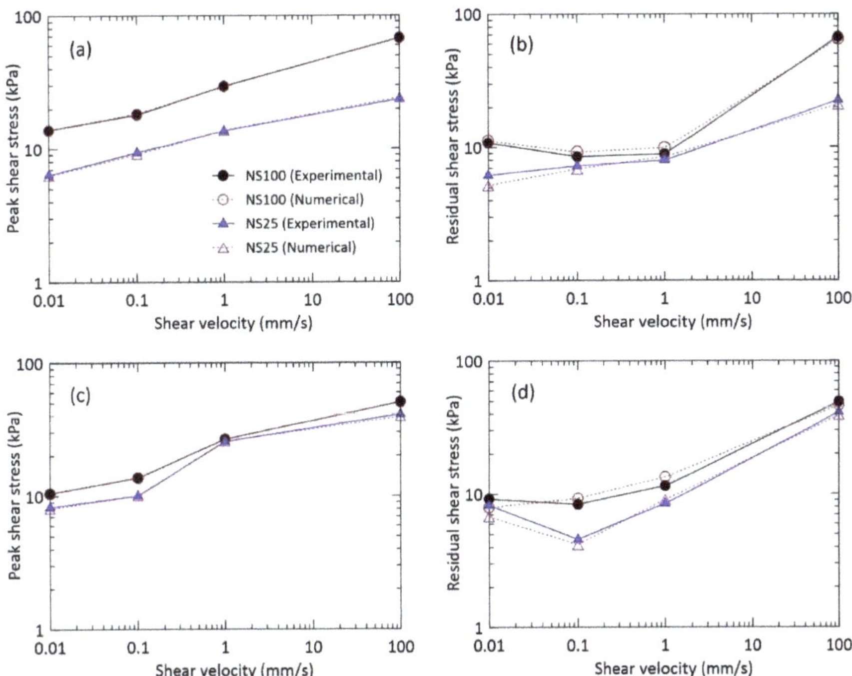

Figure 6. Peak and residual shear stress as a function of shear velocity: (**a**,**b**) drained and (**c**,**d**) undrained conditions.

Table 4. Comparison of peak and residual shear stresses as a function of normal stress.

	NS25–τ_p	NS25–τ_r	NS100–τ_p	NS100–τ_r
Drained	$\tau = 12.8 \cdot V^{0.14}$	$\tau = 10.3 \cdot V^{0.14}$	$\tau = 29.5 \cdot V^{0.17}$	$\tau = 17.3 \cdot V^{0.21}$
Undrained	$\tau = 19.1 \cdot V^{0.18}$	$\tau = 12.1 \cdot V^{0.21}$	$\tau = 23.3 \cdot V^{0.17}$	$\tau = 16.2 \cdot V^{0.19}$

Note: NS = normal stress, τ_p = peak shear stress, τ_r = residual shear stress, and V = shear velocity (mm/s).

For a normal stress of 25 kPa under the undrained condition, at a shear velocity of 0.1 mm/s, the numerical peak shear stress was similar to that obtained experimentally; at the shear velocities of 0.01, 1, and 100 mm/s, the difference in peak shear stress values obtained experimentally and numerically was 0.2, 0.1, and 1.8 kPa, respectively; the residual shear stresses differed by 1.5, 0.4, 0.6, and 2.2 kPa, respectively. For a normal stress of 100 kPa under the undrained conditions, at shear velocities of 0.01, 0.1, and 1 mm/s, similar peak shear stresses were obtained by both numerical and experimental evaluations; at a shear velocity of 100 mm/s, the difference in peak shear stress between experiment and numerical analysis was 0.1 kPa. The residual shear stresses differed by 1.23, 0.9, 1.9 and 3 kPa at shear velocities of 0.01, 0.1, 1, and 100 mm/s, respectively (Figure 6c,d). These results show that the shear stress increases with a shear velocity increase. Similar results were obtained by Fukuoka et al. [31].

4.3. Shear Stress and Normal Stress

Figure 7 presents the influence of normal stress on shear stress under the drained and undrained conditions, obtained by both numerical and experimental analysis. The shearing velocity (0.1 mm/s) was employed to examine the influence of the normal stress on the shear stress. For a normal stress of 20, 40, 60, 80, 100, and 150 kPa under the drained condition, the values obtained for the difference in peak shear stress between the experimental and numerical analysis were 0.2, 0.1, 0.1, 0.1, 0.3, and 0.2 kPa, respectively. At normal stress of 80 kPa, the experimental residual shear stress was similar to that obtained by numerical analysis; for normal stress of 20, 40, 60, 100, and 150 kPa, the residual shear stresses differed by 0.8, 0.1, 0.1, 0.7, and 0.1 kPa, respectively.

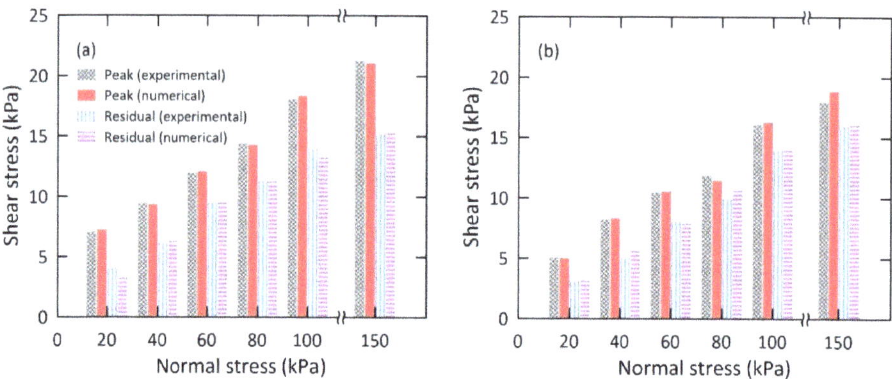

Figure 7. Shear stress vs. normal stress: (**a**) drained and (**b**) undrained condition.

Under the undrained condition of the soil, the difference in peak shear stress between the experimental and numerical analysis was 0.1, 0.1, 0.1, 0.4, 0.2, and 0.9 kPa for normal stresses of 20, 40, 60, 80, 100, and 150 kPa, respectively. For the normal stress of 20, 40, 60, 80, and 150 kPa, the difference between the experimental and numerical residual shear stress was 0.1, 0.6, 0.1, 0.7, and 0.1 kPa, respectively. These results show that the shear stress increases as the normal stress increases. Similar results were found by several other researchers [31]. These experimental and numerical results are in good agreement. In particular, the residual shear stresses obtained from experimental and numerical analysis are very similar one another; however, the peak shear stress under the drained condition is gradually increasing with normal stress and has almost three times higher than the counterpart. It may be due to the fact that there is more strong interaction between particles under the drained condition and results in high shear resistance. Under the undrained condition, water may pay an important role in the crushing and breakage process of granular material. It seems that the lubrication effect of fine particles may occur under the undrained condition. As previously mentioned, after the sudden and abrupt drop of the peak shear stress, the post-failure of the shear characteristics was examined using the residual shear stress. This phase is mainly characterized by the particle crushing mechanism, which is discussed in the next section.

4.4. Particle Crushing Characteristics

Obtaining information about the micromechanics of particle crushing in laboratory experiments is very difficult. However, this obstacle can be overcome by simulating the particle crushing using the discrete element method. The various shapes of particles may be created with two or more single particles using the clump logic; the generated particles are considered as rigid bodies. In the discrete element method, crushing of the granular material is defined as breakage of one or more particles off the clump particle. Thus, one or more criteria are required for implementing the particle crushing. The clump particle can be partially or completely broken when the crushing criteria are satisfied. For this purpose, a user-defined function using FISH language in PFC^{2D} was implemented.

Crushing mechanisms occur from two different mechanisms: abrasion and overstressing. Abrasion occurs through friction, when a particle rubs on another particle and is progressively abraded or broken. Overstressing occurs when a crack is generated in the clump particle, which is broken into two or more smaller particles as the crack enlarges [19,32,33]; this crack is created by excessive application of forces including compressive, tensile, and diametrical forces. In this work, we used the abrasion mechanism to simulate the crushing of clump particles.

Previous studies have simulated crushing of particles in a system based on a single particle crushing experimental data. In many situations, particles crushing in a granular system may often occur simultaneously. In this study, particle crushing was modeled using the energy dissipated by frictional sliding at contact points between particles. This energy is termed frictional work [20]. As previously mentioned, the residual shear stress was maintained constant owing to the frictional resistance between clump particles. The crushing of particles occurs when the required quantity of frictional work is produced between particles; this frictional work quantity was used to evaluate the particle crushing process in our granular assembly. Figure 8 presents the frictional work and development of shear zone in ring shear tests. For the laboratory experiment, the frictional energy was assumed to be the area under the shear stress-time relationship curves after reaching the peak shear stress (Figure 8a). A progressive development of shear zone is related to the reduction in shear stress in landslides. It can be illustrated in ring shear box (Figure 8b). Compared to the initial state of shearing, the shear zone is getting larger and larger during shearing. Shearing may create the finer particles in shear zone and result in the reduction in shear strength (i.e., strain softening behavior) during shearing. According to the previous research findings, large particles can be concentrated in the center of ring shear box, small particles can be accumulated mostly at the lower part of ring shear box due to the vertical movement occurred in shearing [2,15,31]. In addition, for fine-grained sediments, the shear surface is very thin (e.g., less than 1 mm thick), but for coarse-grained sediments, the shear surface is larger with shearing time [2,31,34].

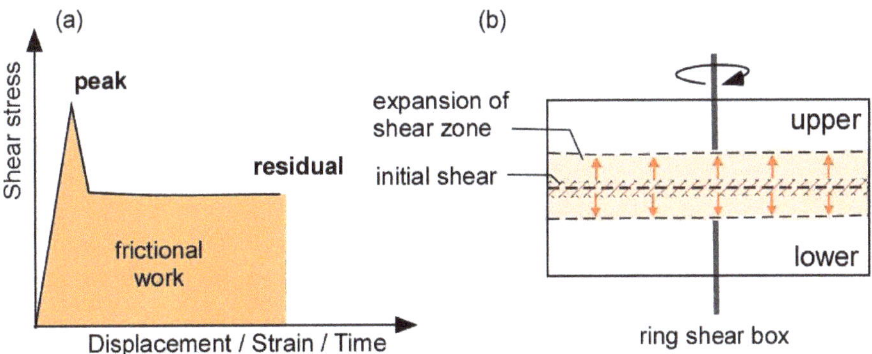

Figure 8. Frictional energy and shear zone in ring shear test: (**a**) shear stress-strain relationship to determine the peak and residual shear stress and (**b**) shear zone in ring shear box.

For the DEM simulation, the crushing of clump particles was allowed until the required frictional energy was reached in the granular assembly system, depending on the shearing velocity. The frictional work, W_f, is computed as [27]:

$$W_f = \sum_{N_c}\left[(F_i^s)(\Delta D_i^s)^{slip}\right] \tag{2}$$

where N_c, F_i^s and $\left(\Delta D_i^s\right)^{slip}$. are the number of contacts, average shear force, and increment of the slip displacement, respectively, at the contact for the current time step. The increment of the slip displacement produced over a time step Δt is given by:

$$\Delta D_i^s = V_i^s \Delta t \tag{3}$$

where V_i^s is the relative shear motion at contact, and is calculated as:

$$V^s = \left(\dot{x}_i^{[\varnothing^2]} - \dot{x}_i^{[\varnothing^1]}\right) t_i - \omega_3^{[\varnothing^2]}\left|x_k^{[C]} - x_k^{[\varnothing^2]}\right| - \omega_3^{[\varnothing^1]}\left|x_k^{[C]} - x_k^{[\varnothing^1]}\right| \tag{4}$$

where $\dot{x}_i^{[\varnothing^j]}$ and $\omega_3^{[\varnothing^j]}$ are the translational and rotational velocity of the entity $\varnothing^j$, respectively. These are expressed as:

$$\left\{\varnothing^1,\ \varnothing^2\right\} = \left\{ \begin{array}{l} \{A,\ B\}\ (\text{particle}\ -\ \text{particle contact}) \\ \{b,\ w\}\ (\text{particle}\ -\ \text{boundary contact}) \end{array} \right. \tag{5}$$

and $t_i = \{-n_2,\ n_1\}$ (n_1 and n_2 are the unit normal vectors).

Frictional energy is an important mechanical property of granular materials. In this study, the frictional work–shearing time relationship is examined. Figure 9 presents the variation in frictional work computed from the measured shear stress and time response at a constant velocity (i.e., 0.1 mm/s) for different drainage and normal stress conditions; D-NS25 and UD-NS25 denote the drained and undrained condition for normal stress of 25 kPa, respectively. The frictional work increases linearly with time. The frictional work obtained at peak shear stress was 303, 190, 220, and 155 kPa·s for D-NS100, UD-NS100, D-NS25, and UD-NS25, respectively. The frictional work at residual shear stress was 2579, 2418, 2280, and 1387 kPa·s for D-NS100, UD-NS100, D-NS25, and UD-NS25, respectively. Interestingly, it can be seen that the frictional energy is much more sensitive under the normal stress conditions than under the drainage conditions. Compared to the others, there is a large difference in the frictional energy at the end of testing (i.e., 300 s) for the normal stress of 25 kPa under the undrained condition. The total shearing time (i.e., 300 s) can be one of the limitations in this study. Nevertheless, it is considered to be sufficient to understand the shear and crushing characteristics of landslide materials with respect to the initiation of slope failure. The condition of normal stress of 25 kPa indicates a relatively shallow soil thickness. It has approximately 1 meter thick in the field. Water moves freely through soil matrix in the drained conditions; however, in the undrained condition, water captured in or surrounding shear zone due to water infiltration during or after heavy rainfall event may result in a sudden reduction in shear strength and cause high mobilization of landslide materials.

The clump particle crushing was permitted after the granular assembly systems reached the corresponding frictional work at peak and residual shear stresses. Figure 10 shows a progressive occurrence of clump particle crushing at peak and residual shear stress. The blue arrows in Figure 10 indicate where progressive crushing mechanism occurred in the ring boundary; the clump particle crushing occurs mainly on the shearing surface at the outer ring boundary. The results showed that the clump particles were partially crushed at peak shear stress, i.e., one particle has been separated from the original clump particle (Figure 10a). The shearing area with the peak value is approximately 5%–7% of the total (see Figure 5). More particle crushing occurred in the residual shear stress state (Figure 10b) due to substantial friction work, i.e., a progressive crushing continued occurring. For a given shear displacement (or time), the number of particles is approximately 10–16 at peak shear stress (Figure 10a), but it is approximately 15–19 at residual shear stress (Figure 10b) when we select a specific part and look closely in the upper and lower parts of ring shear box. The crushing or abrasion of the clump particle made of several particles at peak shear stress progressively abraded as the frictional work increased; the crushing occurred mainly in the vicinity of the shearing boundary because of not only the shear stress, but also to compressive stress and the diametrical and centrifugal forces created in the shear ring granular assembly system. The compressive stress is mainly due to the applied normal stress, whereas the diametrical and centrifugal forces are produced between clump particles by the translational and rotational velocities in the granular assembly system. Other researchers have also found that the particle crushing occurred principally on the shearing surface [35].

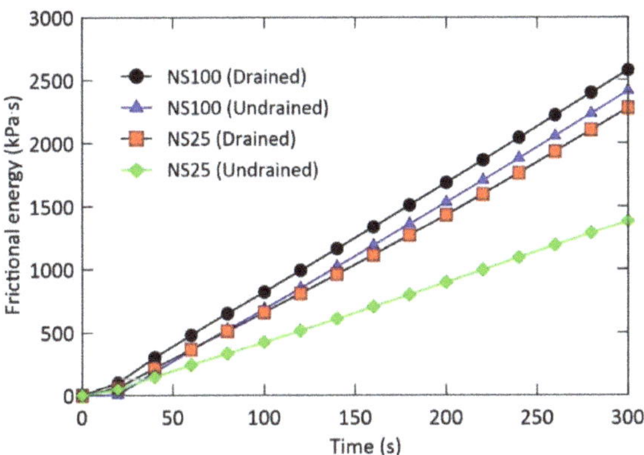

Figure 9. Frictional energy and shearing time dependent on drainage and normal stress condition.

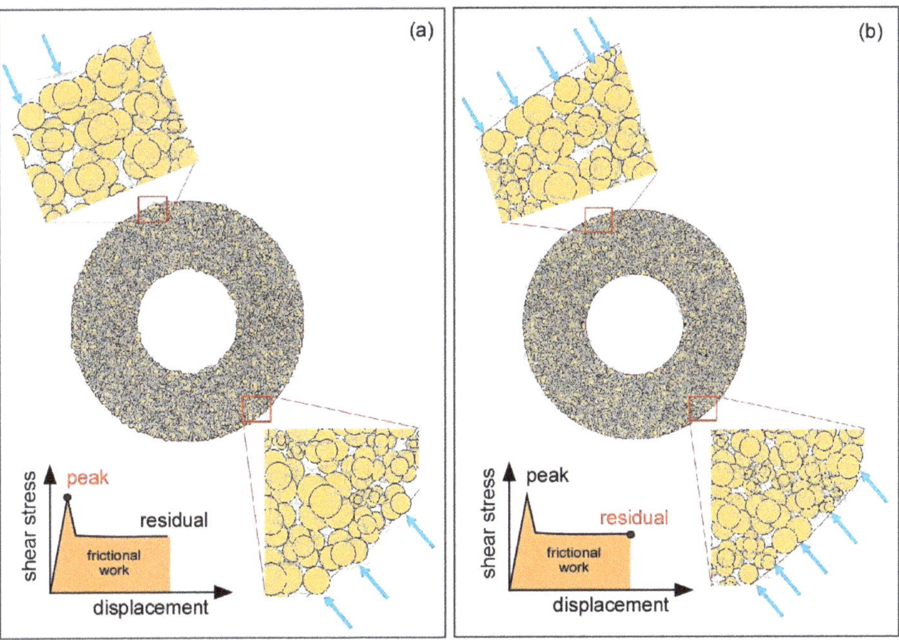

Figure 10. Clump particle crushing at: (**a**) peak shear stress and (**b**) residual shear stress. Arrows indicate a progressive occurrence of clump particle crushing.

5. Conclusions

In this study, a simple ring shear numerical model was developed to investigate the shear and particle crushing characteristics of granular materials. The peak and residual shear stresses are strongly affected by variations of the normal stress, shear velocity, and drainage conditions. As expected, the shear stress increases with an increase in normal stress and shear velocity, regardless of the drainage conditions. In general, there is a very good agreement between experimental and numerical results. For both drained and undrained conditions, the shear stress reaches a peak value rapidly and then undergoes a sharp drop followed by a period of variations before stabilizing (i.e., a typical strain-softening behavior). This may be related to the occurrence of clump particle rearrangement and crushing during shearing. Further, the differences in peak and residual shear stresses are much larger under the drained condition rather than under the undrained condition. In other words, it can be expected that under low normal stress and undrained condition the soil can be mobilized much more easily than in the opposite case.

The particle crushing phenomenon in ring shear test is analyzed using DEM because it affects directly the shear stress. Using the frictional work concept in PFC^{2D}, a new FISH language was implemented in PFC to simulate the clump particle crushing at both peak and residual shear stresses. As the friction work increases monotonically, the frictional work at residual shear stress is greater than that obtained at the peak shear stress. Therefore, the clump particles were partially crushed at peak shear stress, i.e., one particle has been separated from the original clump particle made of five particles; further, more crushing occurred (and was visualized) during the residual shear stress state owing to substantial frictional work. This explains the progressive crushing mechanism in the simulation of the ring shear test using DEM. The crushing (or abrasion) of the clump particle made of four particles at peak shear stress progressed as the frictional work increased. The clump particle crushing mainly occurred in the vicinity of the outer ring boundary due to limited shearing. In future studies, the shear and crushing characteristics over a long period of shearing time should be investigated through three-dimensional analysis.

Author Contributions: Conceptualization, S.-W.J. and S.-S.P.; methodology, K.K. and S.-W.J.; software, K.K. and S.-S.P.; validation, S.-W.J. and D.-E.L.; formal analysis, K.K., S.-W.J. and S.-S.P.; investigation, S.-W.J. and D.-E.L.; resources, S.-W.J. and K.K.; data curation, S.-S.P. and K.K.; writing—original draft preparation, K.K., S.-W.J. and S.-S.P.; writing—review and editing, K.K., S.-S.P., S.-W.J. and D.-E.L.; visualization, K.K. and S.-S.P.; supervision, S.-W.J. and S.-S.P.; project administration, S.-S.P., S.-W.J. and D.-E.L.; funding acquisition, S.-S.P., S.-W.J. and D.-E.L. All authors have read and agreed to the published version of the manuscript.

Funding: This work was supported by a National Research Foundation of Korea (NRF) grant funded by the Korean government (MSIT) (No. NRF-2018R1A5A1025137) and Korea Institute of Geoscience and Mineral Resources (KIGAM) research project (No. 20-3412-1).

Institutional Review Board Statement: Not applicable.

Informed Consent Statement: Not applicable.

Data Availability Statement: Data sharing not applicable to this article as no datasets were generated or analysed during the current study.

Conflicts of Interest: The authors declare no conflict of interest.

References

1. Okada, Y.; Sassa, K.; Fukuoka, H. Undrained Shear Behavior of Sands Subjected to Large Shear Displacement and Estimation of Excess Pore-Pressure Generation from Drained Ring Shear Tests. *Can. Geotech. J.* **2005**, *42*, 787–803. [CrossRef]
2. Sassa, K.; Fukuoka, H.; Wang, G.; Ishikawa, N. Undrained Dynamic-Loading Ring-Shear Apparatus and Its Application to Landslide Dynamics. *Landslides* **2004**, *1*, 7–19. [CrossRef]
3. Wang, G.; Suemine, A.; Schulz, W.H. Shear-Rate-Dependent Strength Control on the Dynamics of Rainfall-Triggered Landslides, Tokushima Prefacture, Japan. *Earth Surf. Process. Landf.* **2010**, *35*, 407–416.
4. Fukuoka, H.; Sassa, K.; Wang, G. Influence of Shear Speed and Normal Stress on the Shear Behavior and Shear Zone Structure of Granular Materials in Naturally Drained Ring Shear Tests. *Landslides* **2007**, *4*, 63–74. [CrossRef]

5. Li, D.; Yin, K.; Glade, T.; Leo, C. Effect of Over-Consolidation and Shear Rate on the Residual Strength of Soils of Silty Sand in the Three Gorges Reservoir. *Sci. Rep.* **2017**, *7*, 5503. [CrossRef]

6. Bagherzadeh-Khalkhali, A.; Mirghasemi, A.A. Numerical and Experimental Direct Shear Tests for Coarse-Grained Soils. *Particuology* **2009**, *7*, 83–91. [CrossRef]

7. Cabalar, A.F.; Dulundu, K.; Tuncay, K. Strength of Various Sands in Triaxial and Cyclic Direct Shear Tests. *Eng. Geol.* **2013**, *156*, 92–102. [CrossRef]

8. Cresswell, A.W.; Powrie, W. Triaxial Tests on an Unbonded Locked Sand. *Géotechnique* **2004**, *54*, 107–115. [CrossRef]

9. Feda, J. Notes on the Effect of Grain Crushing on the Granular Soil Behavior. *Eng. Geol.* **2002**, *63*, 93–98. [CrossRef]

10. Edincliler, A.; Cabalar, A.F.; Cagatay, A.; Cevik, A. Triaxial Compression Behavior of Sand–Tire Wastes Using Neural Networks. *Neural Comput. Appl.* **2012**, *21*, 441–452. [CrossRef]

11. Charles, J.A.; Watts, K.S. The Influence of Confining Pressure on the Shear Strength of Compacted Rockfill. *Géotechnique* **1980**, *30*, 353–367. [CrossRef]

12. Tika, T.E.; Vaughan, P.R.; Lemos, L.J.L.J. Fast shearing of pre-existing shear zones in soil. *Géotechnique* **1996**, *2*, 197–233. [CrossRef]

13. Iverson, N.R.; Mann, J.E.; Iverson, R.M. Effect of Soil Aggregates on Debris-Flow Mobilization: Results from Ring-Shear Experiments. *Eng. Geol.* **2010**, *114*, 84–92. [CrossRef]

14. Jeong, S.W.; Park, S.S. Effect of the Surface Roughness on the Shear Strength of Granular Materials in Ring Shear Tests. *Appl. Sci.* **2019**, *9*, 2977. [CrossRef]

15. Jeong, S.W.; Park, S.S.; Fukuoka, H. Shear and Viscous Characteristics of Gravels in Ring Shear Tests. *Geosci. J.* **2018**, *22*, 11–17. [CrossRef]

16. Wang, S.; Luna, R.; Zhao, H. Cyclic and Post-cyclic Shear Behavior of Low-Plasticity Silt with Varying Clay Content. *Soil Dyn. Earthq. Eng.* **2015**, *75*, 112–120. [CrossRef]

17. Jensen, R.P.; Bosscher, P.J.; Plesha, M.E.; Edil, T.B. DEM Simulation of Granular Media-Structure Interface: Effects of Surface Roughness and Particle Shape. *Int. J. Numer. Anal. Methods Geomech.* **1999**, *23*, 531–547. [CrossRef]

18. Sadrekarimi, A.; Olson, S.M. A New Ring Shear Device to Measure the Large Displacement Shearing Behavior of Sands. *Geotech. Test. J.* **2009**, *12*, 197–208.

19. Lobo-Guerrero, S.; Vallejo, L.E. Modeling Granular Crushing in Ring Shear Tests: Experimental and Numerical Analyses. *Soils Found.* **2006**, *46*, 147–157. [CrossRef]

20. Itasca Consulting Group Inc. *PFC-Particle Flow Code, Ver. 4.00*; Itasca: Minneapolis, MN, USA, 2008.

21. Jeong, S.W.; Park, S.S.; Fukuoka, H. Shear Behavior of Waste Materials in Drained and Undrained Ring Shear Tests. *Geosci. J.* **2014**, *18*, 459–468. [CrossRef]

22. Jeong, S.W. Geotechnical and Rheological Characteristics of Waste Rock Deposits Influencing Potential Debris Flow Occurrence at the Aban-Doned Imgi Mine, Korea. *Environ. Earth Sci.* **2015**, *73*, 8299–8310. [CrossRef]

23. Jeong, S.W.; Wu, Y.H.; Cho, Y.C.; Ji, S.W. Flow Behavior and Mobility of Contaminated Waste Rock Materials in the Abandoned Imgi Mine in Korea. *Geomorphology* **2018**, *301*, 79–91. [CrossRef]

24. Cundall, P.A.; Strack, O.D.L. A Discrete Numerical Model for Granular Assemblies. *Géotechnique* **1979**, *29*, 47–65. [CrossRef]

25. Cundall, P.A. Formulation of a Three-Dimensional Distinct Element Method-Part I. a Scheme to Detect and Represent Contacts in a System Composed of Many Polyhedral Blocks. *Int. J. Rock Mech. Mining Sci. Geomech. Abstr.* **1988**, *25*, 107–116. [CrossRef]

26. Hart, R.; Cundall, P.A.; Lemos, J. Formulation of a Three-Dimensional Distinct Element Method-Part II. Mechanical Calculations for Motion and Interaction of a System Composed of Many Polyhedral Blocks. *Int. J. Rock Mech. Mining Sci. Geomech. Abstr.* **1988**, *25*, 117–125. [CrossRef]

27. Itasca Consulting Group Inc. *UDEC. Universal Distinct Element Code, Version 4.0*; Itasca: Minneapolis, MN, USA, 2004.

28. Nguyen, D.H.; Azema, E.; Sornay, P.; Radjai, F. Rheology of Granular Materials Composed of Crushable Particles. *Eur. Phys. J.E.* **2018**, *41*, 50. [CrossRef]

29. Nguyen, D.H.; Azema, E.; Sornay, P.; Radjai, F. Bonded-Cell Model for Particle Fracture. *Phys. Rev.* **2015**, *91*, 022203.

30. Artoni, R.; Neveu, A.; Descantes, Y.; Richard, P. Effect of Contact Location on the Crushing Strength of Aggregates. *J. Mech. Phys. Solids.* **2019**, *122*, 406–417. [CrossRef]

31. Fukuoka, H.; Sassa, K.; Wang, G.; Sasaki, R. Observation of Shear Zone Development in Ring-Shear Apparatus with a Transparent Shear Box. *Landslides* **2006**, *3*, 239–251. [CrossRef]

32. Trent, B.C.; Margolin, L.G. A Numerical Laboratory for Granular Solids. *Eng. Comput.* **1992**, *9*, 191–197. [CrossRef]

33. Einav, I. Breakage Mechanics—Part I: Theory. *J. Mech. Phys. Solids.* **2007**, *55*, 1274–1297. [CrossRef]

34. Park, S.S.; Jeong, S.W.; Yoon, J.H.; Chae, B.G. Ring Shear Characteristics of Two Different Soils. *J. Korean Geotech. Soc.* **2013**, *5*, 39–52. (In Korean) [CrossRef]

35. Tsoungui, O.; Vallet, D.; Charmet, J. Numerical Model of Crushing of Grains inside Two-Dimensional Granular Materials. *Powder Technol.* **1999**, *105*, 190–198. [CrossRef]

Article

Numerical–Experimental Analysis of Polyethylene Pipe Deformation at Different Load Values

Adam Gnatowski [1,*]**, Agnieszka Kijo-Kleczkowska** [2]**, Mateusz Chyra** [1] **and Dariusz Kwiatkowski** [1]

[1] Department of Technology and Automation, Faculty of Mechanical Engineering and Computer Science, Czestochowa University of Technology, Armii Krajowej 21, 42-201 Czestochowa, Poland; mateuszchyra@wp.pl (M.C.); kwiatkowski@ipp.pcz.pl (D.K.)

[2] Department of Thermal Machinery, Faculty of Mechanical Engineering and Computer Science, Czestochowa University of Technology, Armii Krajowej 21, 42-201 Czestochowa, Poland; kijo@imc.pcz.pl

* Correspondence: gnatowski@ipp.pcz.pl

Abstract: Polymer pipes are used in the construction of underground gas, water, and sewage networks. During exploitation, various external forces work on the pipeline, which cause its deformation. In the paper, numerical analysis and experimental investigations of polyethylene pipe deformation at different external load values (500, 1000, 1500, and 2000 N) were performed. The authors measured strains of the lower and upper surface of the pipe during its loading moment using resistance strain gauges, which were located on the pipe at equal intervals. The results obtained from computer simulation and experimental studies were comparable. An innovative element of the research presented in the article is recognition of the impact of the proposed values of the load of polyethylene pipe on the change in its deformation.

Keywords: polyethylene pipe; mechanical properties of polyethylene; resistance strain; computer simulation

Citation: Gnatowski, A.; Kijo-Kleczkowska, A.; Chyra, M.; Kwiatkowski, D. Numerical–Experimental Analysis of Polyethylene Pipe Deformation at Different Load Values. *Materials* **2021**, *14*, 160. https://doi.org/10.3390/ma14010160

Received: 2 December 2020
Accepted: 28 December 2020
Published: 31 December 2020

Publisher's Note: MDPI stays neutral with regard to jurisdictional claims in published maps and institutional affiliations.

Copyright: © 2020 by the authors. Licensee MDPI, Basel, Switzerland. This article is an open access article distributed under the terms and conditions of the Creative Commons Attribution (CC BY) license (https://creativecommons.org/licenses/by/4.0/).

1. Introduction

Polyethylene plastics are used in various branches of global industry, mainly in extrusion and injection technology in the form of pipes, foils, and various types of packaging. The chemical, physical, mechanical and aesthetic properties of polymer materials depend on the conditions of use: temperature, load time, type of deformation, atmospheric conditions, UV radiation, design solutions, soil parameters in which the pipeline works, and external forces, e.g., car traffic [1–10].

Polymer pipes are used in the construction of underground gas, water and sewage networks. This is due to their low weight compared to, e.g., steel pipes, which makes their transport and assembly much easier. Furthermore, polymer pipes are characterized by chemical inertness and very good mechanical properties [11–13].

Polymer pipe is a flexible material; therefore, it can be deformed at different external loads. The pipeline exploited in the soil react to the loads by deformation of its surfaces and by change in its cross-section. The value of the pipe deformation depends on the value of vertical force acting on the pipe and on the type and degree of soil compaction in which the pipeline is used [14–16]. The pipeline should be properly backfilled in the soil, and it must be on an even, uniform surface, free of large and sharp stones. During exploitation, the pipe is subjected to various loads (for example, the weight of the ground, buildings, road and rail traffic, embankments, and other objects).

At high loads, failure can occur due to the pipe wall breakage. A particularly dangerous case, from the point of view of the mechanical strength of the pipe, is pipeline installation at construction sites. At this stage, there is no working pressure in the pipe, which can cause it to stiffen. Furthermore, excavators, earth-filled lorries, move on such sites. High loading of pipelines can be caused by excessive soil layers directly above the

427

pipe. This may lead to significant pipe deformation and, consequently, to pipeline damage even before its exploitation [17–19].

For this reason, it is essential to conduct experimental studies to prevent pipeline failure in real-life conditions. In regard to polyethylene material use, its thermal parameters are also very important [20–23].

In the paper, numerical analysis and experimental investigations of polyethylene pipe deformation at various external load values were performed. The studies were carried out on a specially designed test stand. The lateral strains on the lower and upper surface of the pipe were measured at the following pipe loadings: 500, 1000, 1500, and 2000 N.

2. Materials and Methods

In this paper, numerical analysis and experimental investigations of pipe deformation under the influence of various external load values were performed.

2.1. Numerical Analysis of the Influence of External Load on the Deformation of Polyethylene Pipe

Computer simulation was carried out for a high-density polyethylene pipe (HD-PE) loaded by evenly distributed soil and an external force on the central part of the pipe surface. The numerical analysis was performed by using the ADINA System 9.3.4 (ADINA R & D, Inc., 71 Elton Avenue, Watertown, MA 02472, USA) program. The spatial model of the system includes a soil block with approximate dimensions of 400 mm × 3100 mm × 400 mm in which the analyzed pipe is located. Computer simulation was carried out for the pipe with the following approximate dimensions: outer diameter of 40 mm, wall thickness of 3.7 mm, and length of 2300 mm. The pipe was placed in soil at a depth of 315 mm. In the central part of the model's upper surface, an evenly distributed load was placed over an area of 320,000 mm^2. Four load cases, 500, 1000, 1500, and 2000 N, were considered. The following boundary conditions were adopted in the tests:

-   Restraining of the system's lower surface;
-   Restraining of the system's side surfaces;
-   No restraint of the pipe.

A diagram of the pipe–soil model is shown in Figure 1.

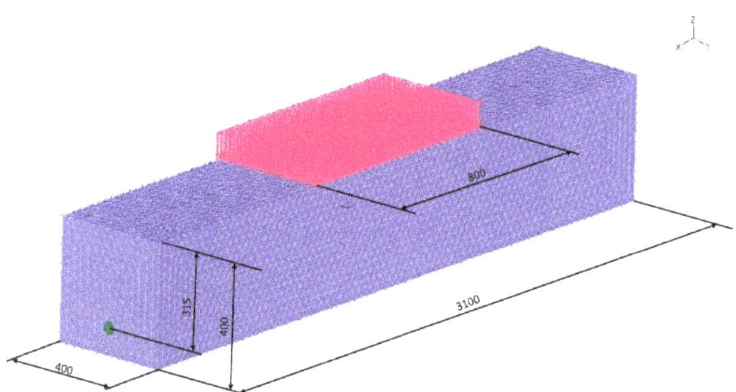

Figure 1. Diagram of the numerical model of the pipe–soil system. (Units in mm)

The dimensioned pipe model with the marked section plane is shown in Figure 2.

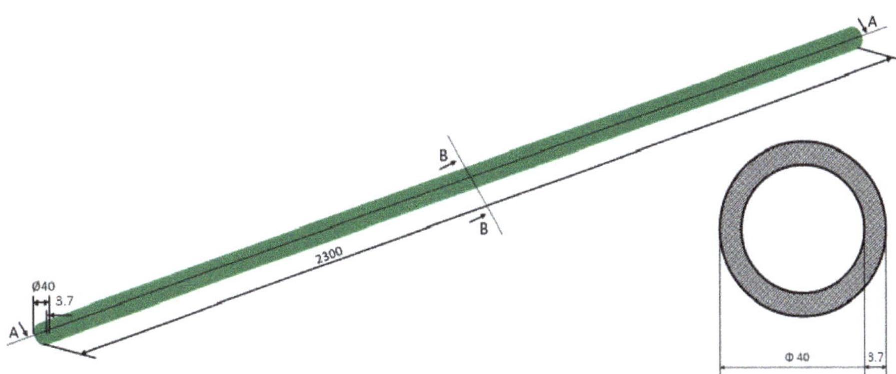

Figure 2. Model of the analyzed pipe. (Units in mm).

The tested model of the pipe–soil system consists of 237,766 finite elements (75,599 for the pipe; 162,167 for the soil) and 377,615 nodes. The soil was modeled with the use of the elastic–ideally plastic Coulomb–Mohr model, which is one of the most frequently used models in soil numerical descriptions [24–27]. The elastoplastic models describe the state of deformation and soil load in the zones subject to the limit state. In the Coulomb–Mohr model, the limit state is the same as the plastic surface. In the stress space, the yield surface for the Coulomb–Mohr model is defined by the following relationship [24]:

$$\frac{1}{2}(\sigma_1 - \sigma_3) + \frac{1}{2}(\sigma_1 + \sigma_3)\sin\Theta - c\,\cos\Theta = 0 \tag{1}$$

where:

σ_1, σ_2, σ_3—main stresses, MPa; θ—internal friction angle of soil, °; c—cohesion, MPa.

In the spaces σ_1, σ_2, and σ_3, the plastic area is limited by the side surfaces of the pyramid with the base of the hexagon (Figure 3), whose side lengths and angles between them change with the value of the internal friction angle.

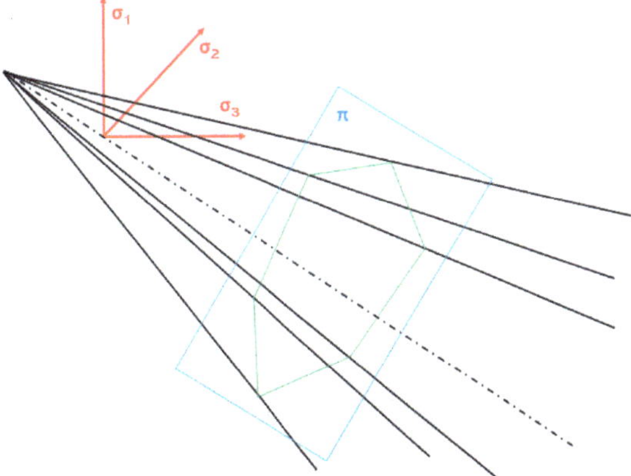

Figure 3. Plasticity area in the Coulomb–Mohr model.

The following geotechnical parameters were used in the computer simulation [28]:
- Compressibility modulus p = 20 MPa;
- Poisson's ratio v = 0.32;

- Volumetric weight w = 0.000018 N/mm^3;
- Cohesion c = 0.017 MPa;
- Internal friction angle θ = 17°.

The pipe was built on the basis of the elastic–isotropic material model. This model requires the definition of data such as Young's modulus and Poisson's ratio. In the discussed case, the following material parameters were adopted [29]:

- Young's modulus E = 1000 MPa;
- Poisson's ratio v = 0.46.

The tensile strength of the pipe material carried out according to the standard PN-EN ISO 527-2: 2012 [30] was determined (Figure 4). The samples were tested using an electromechanical tensile testing machine type ZWICK100 (ZwickRoell, August- Nagel-Straße 11, 89079 Ulm, Germany), with a measuring range of 0–100 kN.

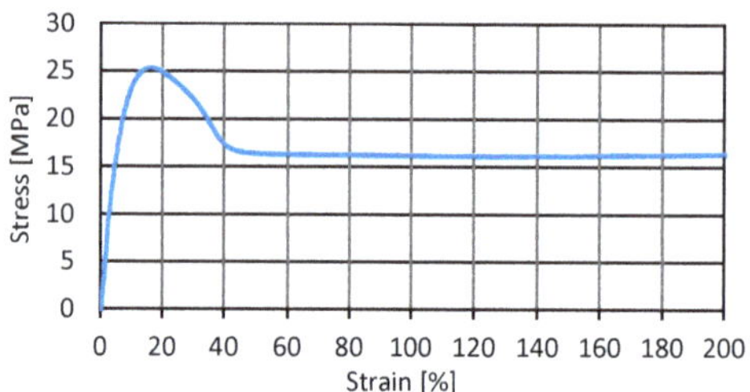

Figure 4. Diagram of the relationship between tensile strength and elongation of a high-density polyethylene pipe (HD-PE).

2.2. Experimental Research of the Influence of External Load on the Deformation of Polyethylene Pipe

The experimental research was carried out on a test stand made of oriented strand board (the approximate dimensions of the box were 3100 mm × 400 mm). A plastic peephole was installed in the central part of the test stand's side wall. The stand was used to perform an experiment of simulating the actual conditions that prevail during pipe deformation caused by the load. The test pipe was placed in the box on a sand bed. The same type of sand was used for backfilling of the pipe. The loads of 500, 1000, 1500, 1500, and 2000 N were applied to the upper surface of the backfill in the central part of the box. The value of deformation was recorded at the moment of pipe load. A scheme of the box is shown in Figure 5.

Experimental investigations were carried out using a high-density polyethylene (HD-PE) pipe (pipe dimensions: length of 2300 mm, external diameter of 40 mm, and pipe wall thickness of 3.7 mm).

On the lower and upper surface of the pipe, the strain gauges were placed at equal intervals of 120 mm with numbers from 1 to 20 (the sensor marked as 10 was located in the central part of the pipe). In this research, electrical and hose strain gauges by Microtechna (manufacturer of the strain gauges) were used, which were glued to the pipe walls. The sensors were used to record changes in the lateral strains of the pipe at different external loads. The strain gauges were made of one piece of wire which was glued to the paper or foil in a hose manner.

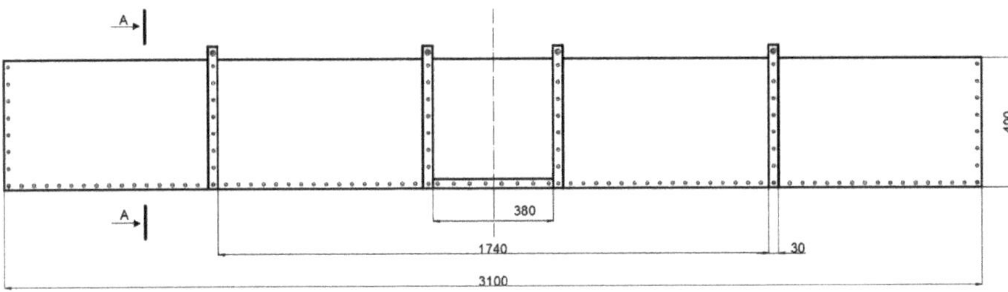

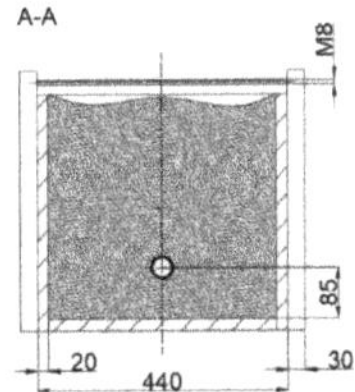

Figure 5. Schematic of the test stand. (Units in mm).

In the resistance strain gauges used in the experiment, the strains were measured based on the relationship between electrical resistance and the length of the wire—Equation (2):

$$R = \delta \frac{L_d}{A} \tag{2}$$

where:

R—the electrical resistance of the wire, Ω;

δ—specific resistance of the wire, Ω;

L_d—wire length, mm;

A—wire cross-sectional area, mm.

The relative increment of strain gauge resistance is described in Equation (3):

$$\frac{\Delta R}{R_1} = \frac{\Delta \delta}{\delta} + \frac{\Delta L_d}{L_d} - \frac{\Delta A}{A} \tag{3}$$

where:

$\frac{\Delta \delta}{\delta}$—relative increment of specific resistance;

$\frac{\Delta L_d}{L_d}$—relative strain of the wire;

$\frac{\Delta A}{A}$—relative change in the cross-section of the wire.

In order to determine $\Delta A/A$, a square with ABCD sides was determined on the cross-section of the wire. At the load applied, the lengths of the square sides with values of $(1 + \varepsilon_x)$ and $(1 + \varepsilon_y)$ were deformed. The square cross-sectional area was initially equal to $A_k = 1$, and after deformation, it was as described in Equation (4):

$$A'_k = (1 + \varepsilon_x) - (1 + \varepsilon_y) \tag{4}$$

where:

ε_x—strain of the cross-section in the x-direction;

ε_y—strain of the cross-section in the y-direction.

The relative change in the wire cross-section is described in Equation (5):

$$\frac{\Delta A}{A} = \frac{A'_k - A_k}{A_k} = \frac{(1 + \varepsilon_x)(1 + \varepsilon_y) - 1}{1} = \varepsilon_x + \varepsilon_y + \varepsilon_x \varepsilon_y \tag{5}$$

Excluding the product $\varepsilon_x \varepsilon_y$ as an infinitesimally small quantity, and taking into account the fact that during stretching, the strain gauge wire is in a unidirectional state of stress, that is, $\varepsilon_x = \varepsilon_y = -v\varepsilon$, the relative change in the wire cross-section can be obtained using Equation (6):

$$\frac{\Delta A}{A} = -2v\varepsilon \tag{6}$$

The relative increase in strain gauge resistance can be written as Equation (7):

$$\frac{\Delta R}{R} = \left(\frac{\left(\frac{\Delta \delta}{\delta} \right)}{\varepsilon} + 1 + 2v \right) \varepsilon \tag{7}$$

The relative strain is described in Equation (8):

$$\varepsilon = \frac{1}{\left(1 + 2v + \frac{\Delta \delta / \delta}{\varepsilon} \right)} \frac{\Delta R}{R} \tag{8}$$

The value of the denominator in Equation (8) is the constant k, called the strain gauge constant, according to Equation (9) [31]:

$$k = 1 + 2v + \frac{\Delta \delta / \delta}{\varepsilon} \tag{9}$$

The relative strain can be written as Equation (10):

$$\varepsilon = \frac{1}{k} \left(\frac{\Delta R}{R} \right) \tag{10}$$

where:

k—strain gauge constant: 2.15;

ΔR—relative increase in electrical resistance;

R—the wire's electrical resistance.

The value of the constant k in in Equation (9) depends on the sensor wire material, and it ranges between 1.6 and 3.6. For strain gauges used in this experiment, the constant k was 2.15 (value of k is provided by the manufacturer of the strain gauges).

3. Results and Discussion

Below are the results of the numerical simulation and experimental tests of polyethylene pipe loading.

3.1. Numerical Analysis of the Influence of External Load on the Deformation of Polyethylene Pipe

The results of the numerical analysis, illustrating the distribution of longitudinal strains of the tested pipe for different values of external load, are presented in Figure 6. In order to accurately illustrate the distribution of deformations, two sections were made on the models: transverse in the central part of the pipe and longitudinal along the pipe's entire length.

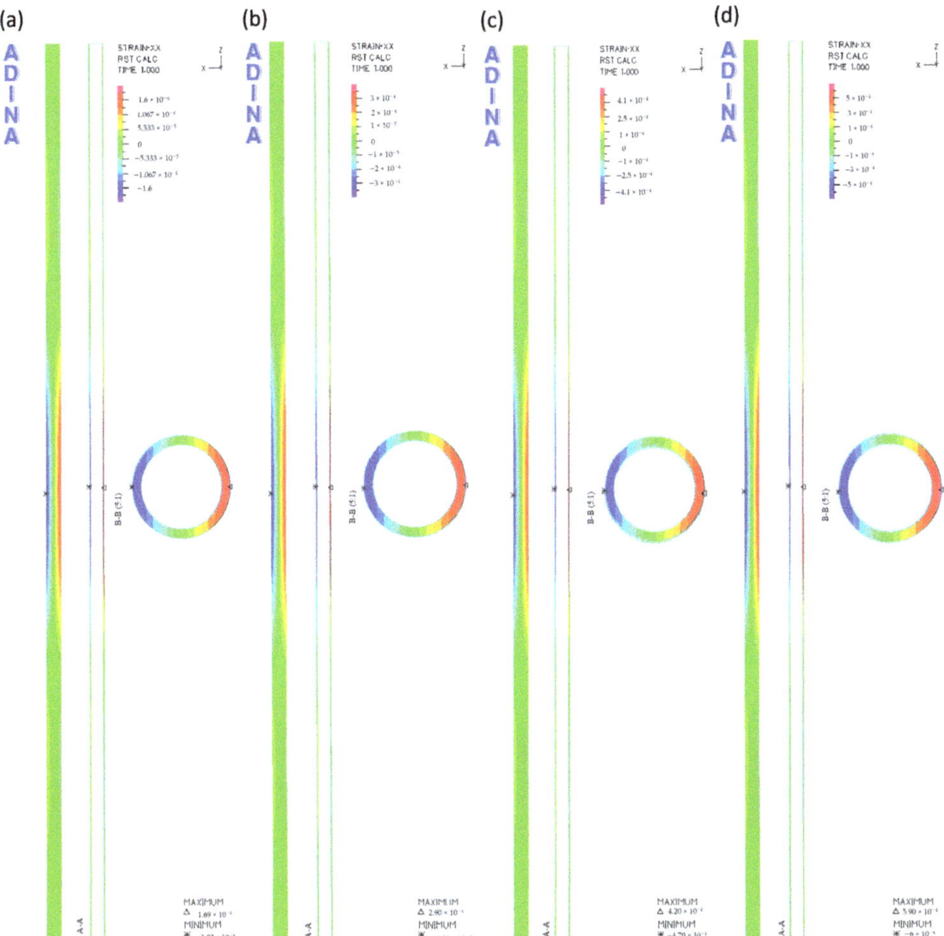

Figure 6. The pipe longitudinal deformations ε_x at the following loads: (**a**) 500 N, (**b**) 1000 N, (**c**) 1500 N, (**d**) 2000 N.

The nature of the longitudinal deformations of the pipe is the same for each load case. Considering the top surface of the pipe, the following deformations were noted: negative in the middle part and positive at the ends of the pipe model. On the pipe's lower surface, the following deformations were noted: positive in the middle part and negative at the ends of the pipe model.

Figure 7 shows the values of the largest determined deformations of the tested pipes, depending on the external load.

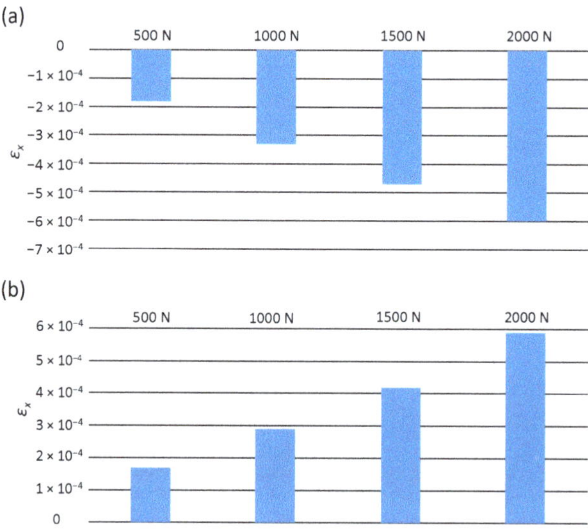

Figure 7. The maximum values of longitudinal deformations ε_x: (**a**) on the top surface of the pipes, (**b**) on the bottom surface of the pipes.

Figure 7 shows that an increase in deformations is directly proportional to the external load increasing. For example, at the load of 500 N, the following values of maximum deformation occur: -1.8×10^{-4} on the upper surface, and 1.7×10^{-4} on the pipe's lower part. At the load of 2000 N, this value is approximately 3.3 times greater.

Figure 8 presents the results obtained from the numerical analysis, illustrating the distribution of the longitudinal strains of the pipe at the considered values of the external load. The model also includes the cross-section in the central part of the pipe and the longitudinal section along its entire length.

The distribution of transverse deformations of the tested pipe is of the same nature in each analyzed load case. On the lower and upper surface of the pipe, in its middle part, the deformation values are negative, while at the pipe ends, the values are positive.

Figure 9 shows the values of the maximum deformations of the tested pipe, depending on the external load.

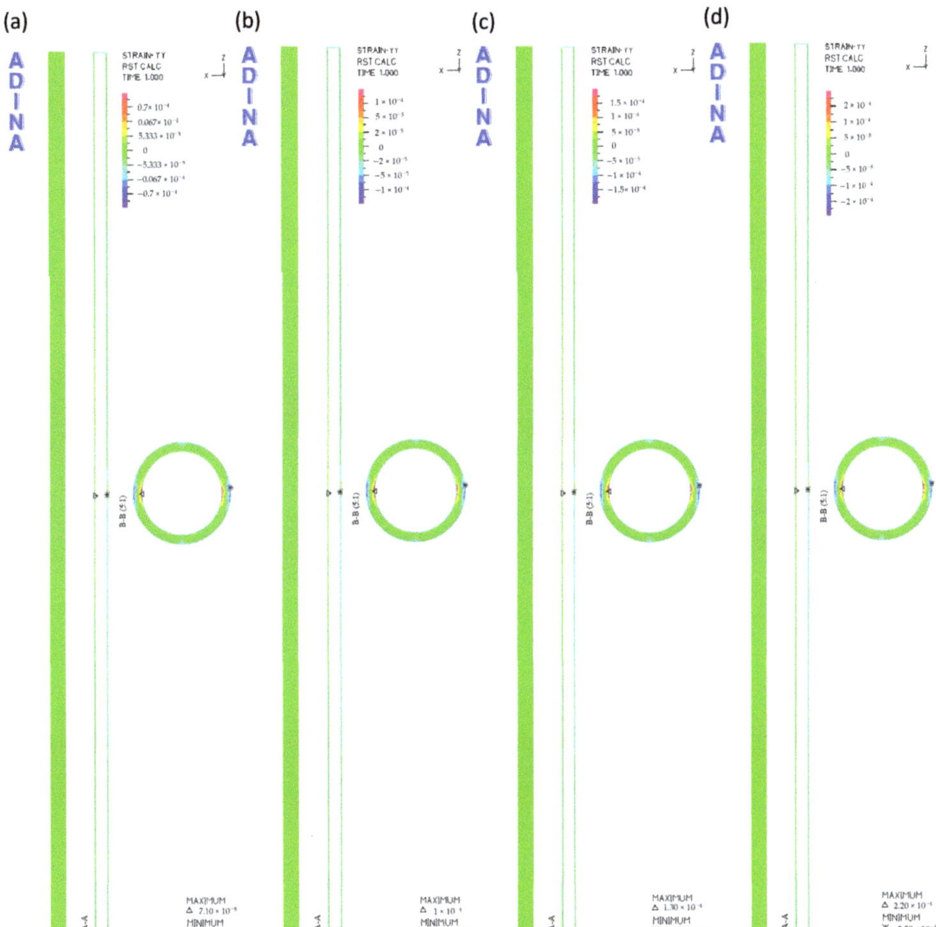

Figure 8. The pipe transverse deformations ε_y at the following loads: (**a**) 500 N, (**b**) 1000 N, (**c**) 1500 N, (**d**) 2000 N.

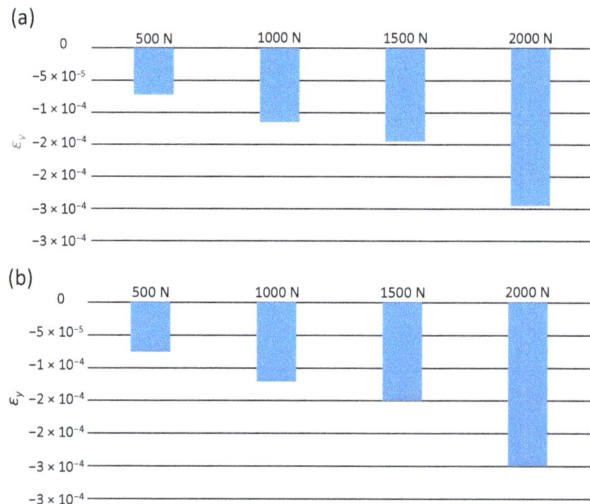

Figure 9. The maximum values of transverse deformations ε_y: (**a**) on the upper surface of the pipes, (**b**) on the lower surface of the pipes.

The highest values of negative deformations in each load variant occur in the central part of the pipe. Along with the increase in the value of the soil load, the lateral deformation increased. It can be concluded that, as in the case of longitudinal deformations, the smallest values of pipes transverse deformations occur at the load of 500 N. On the upper surface of the pipe, the value is equal to -7.3×10^{-5}, while in the lower part, it is equal to -7.5×10^{-5}. In the case of loading of 2000 N, an increase in the deformation value of -2.8×10^{-4} at the top was observed, while an increase in the deformation value of -3×10^{-4} at the bottom of the pipe in relation to the load of 500 N was noted. The values of longitudinal and transverse deformations are small, falling within the yield point of polyethylene [32,33]. Therefore, in the operating conditions of the pipelines, such soil loading would not cause the pipe to break or disturb the transport of the medium.

3.2. Comparative Analysis of Numerical Simulation and Experimental Research of the Influence of the External Load on Polyethylene Pipe Deformation

Figure 10 present the results of experimental tests of the pipe longitudinal strain measurements, using electric resistance strain gauges as shown. The horizontal axis of the charts shows the numbers of successive strain gauges attached to the pipe's surface, while the ordinate axis shows the values of the recorded strains. For comparison, the graphs also show the longitudinal deformations of the pipe obtained during the numerical analysis.

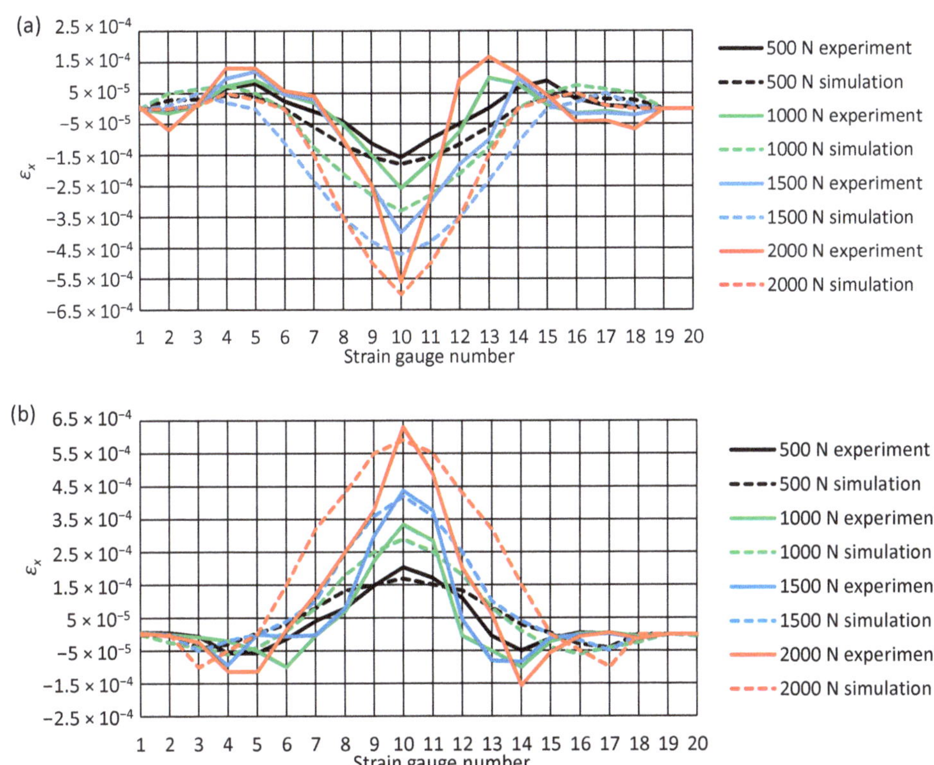

Figure 10. Longitudinal deformations of the pipe ε_x at different external loads: (**a**) top of the pipe, (**b**) bottom of the pipe.

The results of experimental tests of longitudinal deformation of the pipe at the considered loads confirmed the results obtained during the numerical analysis, and they have the same trend. The highest negative deformation appeared to occur on the pipe's upper surface, while the maximum positive deformation was recorded in its lower part.

Figure 11 shows the results of the measurement of the transverse deformations of the pipe obtained during the experimental tests. As in the case of the pipe longitudinal deformation analysis, the numbers of successive strain gauges are marked on the abscissa axis of the graphs, while the values of the registered deformations of the pipe are located on the ordinate axis. The diagrams also show the results obtained during the numerical analysis.

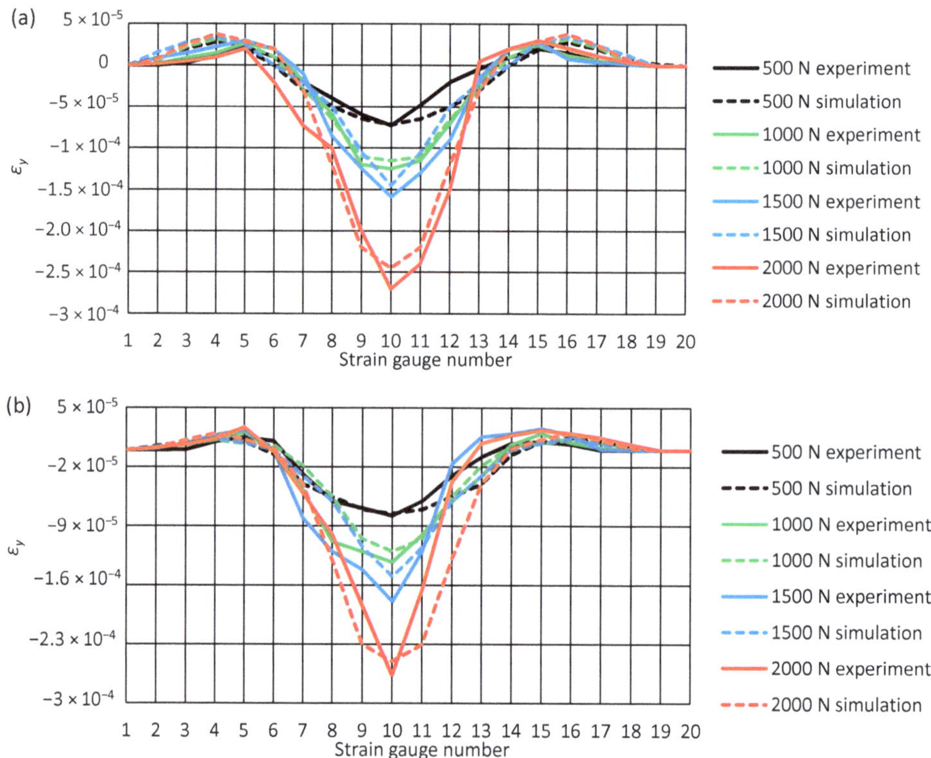

Figure 11. Transverse deformations of the pipe ε_y at different external loads: (**a**) top of the pipe, (**b**) bottom of the pipe.

As in the case of longitudinal deformations of the pipe, the results of the experimental tests also agree with the results obtained from the computer simulation. The nature of deformations is the same in each analyzed case. On the lower and upper surface of the pipe, the greatest negative values of deformation always occur in the center of the pipe. In turn, deformations with a positive sign are located at the pipe ends. In each load variant, the positive deformations are slightly greater on the pipe's upper surface.

Table 1 summarizes the differences in the deformation values of ε_x and ε_y observed in the results obtained from the simulation and the experiment for the central part of the pipe (where the pipe's highest deformation values occur).

Table 1. Comparison of the differences in the values of longitudinal deformations ε_x and transverse deformations ε_y observed in the results of computer simulation and experimental tests.

	500 N	1000 N	1500 N	2000 N
		ε_x		
top of the pipe	2.11×10^{-5}	7.43×10^{-5}	7.26×10^{-5}	4.34×10^{-5}
down the pipe	3.43×10^{-5}	4.23×10^{-5}	1.78×10^{-5}	4.31×10^{-5}
		ε_y		
top of the pipe	1.13×10^{-5}	1.06×10^{-5}	1.43×10^{-5}	2.56×10^{-5}
down the pipe	2.35×10^{-5}	1.28×10^{-5}	2.98×10^{-5}	1.66×10^{-5}

The largest recorded difference in deformation ε_x between the results obtained from the computer simulation and the results of experimental tests concerns the upper surface of the pipe for a load of 1000 N. The maximum value of deformation obtained from the experiment is 7.43×10^{-5} smaller than the value obtained from the simulation. The greatest difference in the value of the maximum deformation ε_y was recorded for the lower surface of the pipe at the load of 1500 N. The difference in values between the experimental test and the numerical analysis is 2.98×10^{-5}.

The slight differences in the obtained results confirm the high accuracy of the computer simulation and the experimental research.

4. Conclusions

It is necessary to conduct research on changes in pipe properties as a result of degradation processes. The results of such studies may contribute to the prediction of failure-free operation of pipelines, as well as earlier planning of their repairs or replacements, which is reflected in the reduction of downtime in the supply or receipt of utilities, both from households and production companies.

Conducting experimental tests and numerical simulations allowed for a comparative analysis between them and recognition of the impact of the proposed values of polyethylene pipes' load on the change in their deformation. The high consistency of the results of computer simulations with the results of experimental tests obtained in the work indicates the appropriate application of the models in the problem under consideration (soil modeling—elastic–ideally plastic Coulomb–Mohr model; pipe modeling—elastic–isotropic model).

The authors plan to conduct further research on the change in the mechanical properties of polyethylene pipes after aging, corresponding to operation of a few years (2, 5 and 10 years).

Author Contributions: Conceptualization, A.G., A.K.-K.; Methodology, A.G., A.K.-K., D.K.; Data curation, A.G., A.K.-K., D.K., M.C.; Formal analysis, A.G., A.K.-K., D.K.; Investigation, A.G., A.K.-K., M.C.; Methodology, A.G., A.K.-K., Resources, A.G., M.C.; Supervision, A.G., A.K.-K.; Validation, A.G., A.K.-K.; Visualization, A.G., A.K.-K., M.C.; Writing—original draft, A.G., A.K.-K., M.C.; Writing—review & editing, A.G., A.K.-K.; All authors have read and agreed to the published version of the manuscript.

Funding: This research received no external funding.

Institutional Review Board Statement: Not applicable.

Informed Consent Statement: Not applicable.

Data Availability Statement: Not applicable.

Conflicts of Interest: The authors declare no conflict of interest.

References

1. Deblieck, R.A.C.; van Beek, D.J.M.; McCarthy, M.; Mindermann, P.; Remerie, K.; Langer, B.; Lach, R.; Grellmann, W. A simple intrinsic measure for rapid crack propagation in bimodal polyethylene pipe grades validated by elastic–Plastic fracture mechanics analysis of data from instrumented Charpy impact test. *Polym. Eng. Sci.* **2017**, *57*, 13–21. [CrossRef]
2. Kliszewicz, B. Numerical 3D analysis of buried flexible pipeline. *Eur. Sci. J.* **2013**, *9*, 93–101.
3. Hubert, L.; David, L.; Séguéla, R.; Vigier, G.; Degoulet, C.; Germain, Y. Physical and mechanical properties of polyethylene for pipes in relation to molecular architecture. I. Microstructure and crystallisation kinetics. *Polymer* **2001**, *42*, 8425–8434. [CrossRef]
4. Kadhim, L.F. Mechanical properties of high density polyethylene/chromium trioxide under ultraviolet rays. *Int. J. Appl. Eng.* **2017**, *10*, 2517–2526.
5. Kamweru, P.K.; Ndiritu, F.G.; Kinyanjui, T.; Muthui, Z.W.; Ngumbu, R.G.; Odhiambo, P.M. UV absorbtion and dynamic mechanical analysis of polyethylene films. *Int. J. Phys. Sci.* **2014**, *9*, 545–555.
6. Madhu, G.; Bhunia, H.; Bajpai, P.K.; Chaudhary, V. Mechanical and morphological properties of high density polyethylene and polylactide blends. *J. Polym. Eng.* **2014**, *34*, 813–821. [CrossRef]
7. Amjadi, M.; Fatemi, A. Tensile behavior of high-density polyethylene including the effects of processing technique, thickness, temperature, and strain rate. *Polymers* **2020**, *12*, 1857. [CrossRef]
8. Spalding, M.A.; Chatterjee, A. *Handbook of Industrial Polyethylene and Technology: Definitive Guide to Manufacturing, Properties, Processing, Applications and Markets*; Wiley: Hoboken, NJ, USA, 2017.

9. Xu, M.-m.; Huang, G.-y.; Feng, S.-s.; McShane, G.J.; Stronge, W.J. Static and dynamic properties of semi-crystalline polyethylene. *Polymers* **2016**, *8*, 77. [CrossRef]
10. Elleuch, R.; Taktak, W. Viscoelastic behavior of HDPE polymer using tensile and compressive loading. *J. Mater. Eng. Perform.* **2006**, *15*, 111–116. [CrossRef]
11. Haager, M.; Pinter, G.; Lang, R.W. Ranking of PE-HD Pipe grades by fatigue crack growth performance. In Proceedings of the Plast. Pipes XIII, Washington, DC, USA, 2–5 October 2006.
12. Pyo, S.; Woo, J.; Park, J.; Kim, M.; Choi, S. Measurement of rapid crack propagation in pressure pipes: A static S4 approach. *Polym. Test.* **2012**, *31*, 439–443. [CrossRef]
13. Bilgin, Ö.; Stewart, H.E.; O'Rourke, T.D. Thermal and mechanical properties of polyethylene pipes. *J. Mater. Civ. Eng.* **2007**, *19*, 1043–1052. [CrossRef]
14. Maess, M.; Wagner, N.; Gaul, L. Dispersion curves of fluid filled elastic pipes by standard FE models and eigenpath analysis. *J. Sound Vib.* **2006**, *296*, 264–276. [CrossRef]
15. Bilgin, Ö. Modeling viscoelastic behavior of polyethylene pipe stresses. *J. Mater. Civ. Eng.* **2014**, *26*, 676–683. [CrossRef]
16. Grellmann, W.; Langer, B. *Deformation and Fracture Behaviour of Polymer Materials*; Springer International Publishing: Berlin/Heidelberg, Germany, 2017.
17. Frank, A.; Pinter, G. Evaluation of the applicability of the cracked round bar test as standardized PE-pipe ranking tool. *Polym. Test.* **2014**, *33*, 161–171. [CrossRef]
18. Kliszewicz, B. Veryfication of numerical model of pipeline—Soil system on the basis of laboratory testing. *J. Civ. Eng. Environ. Archit.* **2014**, *31*, 115–126. (In Polish)
19. Djebli, A.; Bendouba, M.; Aid, A.; Talha, A.; Benseddiq, N.; Benguediab, M. Experimental analysis and damage modeling of high-density polyethylene under fatigue loading. *Acta Mech. Solida Sin.* **2016**, *29*, 133–144. [CrossRef]
20. Weon, J.-i. Effects of thermal ageing on mechanical and thermal behaviors of linear low density polyethylene pipe. *Polym. Degrad. Stab.* **2010**, *95*, 14–20. [CrossRef]
21. Gnatowski, A.; Kijo-Kleczkowska, A. Selected physical properties and structure of materials based on modified polyamide 6. *Int. J. Numer. Methods Heat Fluid Flow* **2020**, *30*, 1577–1588. [CrossRef]
22. Gnatowski, A.; Kijo-Kleczkowska, A.; Gołębski, R.; Mirek, K. Analysis of polymeric materials properties changes after addition of reinforcing fibers. *Int. J. Numer. Methods Heat Fluid Flow* **2020**, *30*, 2833–2843. [CrossRef]
23. Merah, N.; Saghir, F.; Khan, Z.; Bazoune, A. Effect of temperature on tensile properties of HDPE pipe material. *Plast. Rubber Compos.* **2006**, *35*, 226–230. [CrossRef]
24. Cudny, M.; Binder, K. Criteria of soil shear strength in geotechnics. *Mar. Eng. Geotech.* **2005**, *6*, 456–465. (In Polish)
25. Grosel, S.; Pachnicz, M.; Różański, A.; Sobótka, M.; Stefaniuk, D. Influence of bedding and backfill soil type on deformation of buried sewage pipeline. *Studia Geotechnica et Mechanica* **2018**, *40*, 313–320. [CrossRef]
26. van den Bogert, P.A.J.; van Eijs, R.M.H.E. Why Mohr-circle analyses may underestimate the risk of fault reactivation in depleting reservoirs. *Int. J. Rock Mech. Min. Sci.* **2020**, *136*, 104502. [CrossRef]
27. Inn Woo, S.; Seo, H.; Kim, J. Critical-state-based Mohr-Coulomb plasticity model for sands. *Comput. Geotech.* **2017**, *92*, 179–185. [CrossRef]
28. *Geotechnical Design*; PN-EN 1997-1:2008; Polski Komitet Normalizacyjny: Warsaw, Poland, 2008. (In Polish)
29. Madaj, A.; Węgrzynowski, M.; Janusz, L. Long term observations of a 3-span highway corrugated steel box bridge on Gniezno Bypass. *Arch. Inst. Civ. Eng.* **2012**, *12*, 197–203. (In Polish)
30. PN-EN ISO 527-2. *Plastics—Determination of Mechanical Properties under Static Stretching*; Polski Komitet Normalizacyjny: Warsaw, Poland, 2012. (In Polish)
31. La Mantia, F.P.; Morreale, M.; Botta, L.; Mistretta, M.C.; Ceraulo, M.; Scaffaro, R. Degradation of polymer blends: A brief review. *Polym. Degrad. Stab.* **2017**, *145*, 79–92. [CrossRef]
32. Gnatowski, A.; Chyra, M.; Baranowski, W. Analysis of thermomechanical properties and morphology of polyethylene pipes after aging by UV radiation. *Polimery* **2014**, *59*, 308–313. [CrossRef]
33. Yayla, P.; Bilgin, Y. Squeeze-off of polyethylene pressure pipes: Experimental analysis. *Polym. Test.* **2007**, *26*, 132–141. [CrossRef]

 *materials*

Article

Mechanical Behavior of GFRP Connection Using FRTP Rivets

Takayoshi Matsui [1,*], **Yoshiyuki Matsushita** [2] and **Yukihiro Matsumoto** [1]

[1] Department of Architecture and Civil Engineering, Toyohashi University of Technology, Aichi, Toyohashi 441-8580, Japan; y-matsum@ace.tut.ac.jp

[2] IO INDUSTRY CO., Ltd., Kosai, Shizuoka 431-0302, Japan; matsushitay@io-industry.co.jp

* Correspondence: matsui.takayoshi.us@tut.jp; Tel.: +81-532-44-6845

Abstract: In recent years, the application of fiber-reinforced plastics (FRPs) as structural members has been promoted. Metallic bolts and rivets are often used for the connection of FRP structures, but there are some problems caused by corrosion and stress concentration at the bearing position. Fiber-reinforced thermoplastics (FRTPs) have attracted attention in composite material fields because they can be remolded by heating and manufactured with excellent speed compared with thermosetting plastics. In this paper, we propose and evaluate the connection method using rivets produced of FRTPs for FRP members. It was confirmed through material tests that an FRTP rivet provides stable tensile, shear, and bending strength. Then, it was clarified that non-clearance connection could be achieved by the proposed connection method, so initial sliding was not observed, and connection strength linearly increased as the number of FRTP rivets increased through the double-lapped tensile shear tests. Furthermore, the joint strength of the beam using FRTP rivets could be calculated with high accuracy using the method for bolt joints in steel structures through a four-point beam bending test.

Keywords: GFRP; FRTP; rivet; connection

1. Introduction

Citation: Matsui, T.; Matsushita, Y.; Matsumoto, Y. Mechanical Behavior of GFRP Connection Using FRTP Rivets. *Materials* **2021**, *14*, 7. https://dx.doi.org/doi:10.3390/ma14010007

Received: 28 November 2020
Accepted: 18 December 2020
Published: 22 December 2020

Publisher's Note: MDPI stays neutral with regard to jurisdictional claims in published maps and institutional affiliations.

Copyright: © 2020 by the authors. Licensee MDPI, Basel, Switzerland. This article is an open access article distributed under the terms and conditions of the Creative Commons Attribution (CC BY) license (https://creativecommons.org/licenses/by/4.0/).

Fiber-reinforced plastics (FRPs) are used for repairing and reinforcing structures because they have a high strength-to-weight ratio and corrosion resistance. In recent years, the application of FRPs as structural members, e.g., in pedestrian bridges, buildings, and large roofs, has been promoted [1]. Bai and Keller [2] introduced an FRP pedestrian bridge built in 2005 and investigated its dynamic response behavior. They suggested that the connection method affects structural dynamic behavior. Votsis et al. [3] investigated the structural behavior of a novel type of FRP bridge—the Aberfeldy footbridges—and they evaluated their dynamic properties by experiment and the finite element method to clarify the long-term performance of FRP bridges. Evernden and Mottram [4] introduced FRP buildings and their manufacturing process to develop FRP buildings in the United Kingdom. Yang et al. [5] proposed new space frame structures with a grass fiber-reinforced plastic (GFRP) connection method, and the method of structural design and modeling for finite element analysis was clarified. Matsumoto and Yonemaru [6] investigated mechanical performance of a CFRP roof truss member under long-term loading conditions, and it was confirmed that properties were not varied even if specimens were exposed outside. In order to use FRPs as structural members, it is necessary to study the connection method. Several methods have been proposed for connecting FRP members, such as mechanical joints, adhesively bonded joints, and composites of these. Coelho and Mottram [7] summarized the bolted connection and its strength, and they suggested that the connection of pultruded FRP should be established by considering the material characteristics, such as orthotropic material properties and the estimation of several failure modes. Ueda et al. [8] proposed a new connection method using a carbon fiber-reinforced thermoplastics (CFRTP) rod to joint CFRP plates like a rivet. They clarified the shear strength of CFRTP rod with 5.2 mm

441

diameter as 2.5–3.6 kN, and demonstrated that the specific joint strength can be effectively increased. Ascione et al. [9] proposed adhesively bonded GFRP beam-column connections with an improved connection by angle member and stiffener, and experimentally investigated the connection strength. They concluded that the fully bonded connection could provide rigid and higher connection strength with cohesive failure. However, strength variation and stability was not evaluated depending on bonding condition and material imperfection. In the case of mechanical joints, metallic bolts and rivets are often used, but the corrosion resistance of FRP member is not fully utilized because these joints are degraded by corrosion. Furthermore, a bolted connection lacks initial stiffness because frictional resistance cannot be performed. Connection strength may also decrease due to unavoidable non-uniform clearances between base member and bolt/rivet shank because an FRP member cannot redistribute bearing stress by plastic deformation. Marra et al. [10] clarified un-uniform load distribution in multi-bolt joints because of bolt-hole clearance and bolt position by finite element analysis, and they evaluated the stress distribution coefficients. However, an improved method was not suggested. Matsumoto et al. [11] reported that the diameter ratio of the bolt and bolt hole greatly affects bearing strength, and bearing strength can be improved by decreasing the diameter of the bolt hole to close a clearance.

On the other hand, fiber-reinforced thermoplastics (FRTPs), which are FRPs that use thermoplastic resins, have attracted attention in composite material fields. FRTPs can be remolded by heating, so they have the possibility of secondary processing, recycling, and reuse in addition to FRPs. Moreover, FRTPs could reduce costs because they can be manufactured with excellent speed by injection molding or press molding compared with thermosetting plastics. Yeong et al. [12] demonstrated the manufacture of a CFRTP and GFRTP coupon, and they tested them under tension, bending, and indentation loads. They suggested that the manufacturing process using FRTP could reduce molding costs, and showed comparable strength and elastic modulus to conventional CFRP and GFRP. To develop and apply FRTPs to engineering fields, research for the evaluation of mechanical characteristics by Doan [13] and Cao [14] was carried out, and manufacturing processes were also evaluated in recent years by Fan [15] and Rodonò [16].

On the basis of this background, we studied the connection methods to take advantage of the features of FRP materials and structures. In this paper, we propose a connection method using FRTP rivets to provide a solution for problems caused by corrosion and clearance, and evaluate the connection strength and mechanical behavior through material tests and structural experiments.

2. Fiber-Reinforced Thermoplastic (FRTP) Rivet and Its Connection Method

Figure 1 shows the geometry and image of the FRTP rivets used in this study. The FRTP rivets were made of polyamide 6 (PA6) with 50 wt.% glass fiber (Durethan BKV50HEF 900,116 DUS022, Tokyo, Japan) by injection molding. The length of the glass fiber was approximately 0.35 mm. Table 1 shows the properties of the FRTP material.

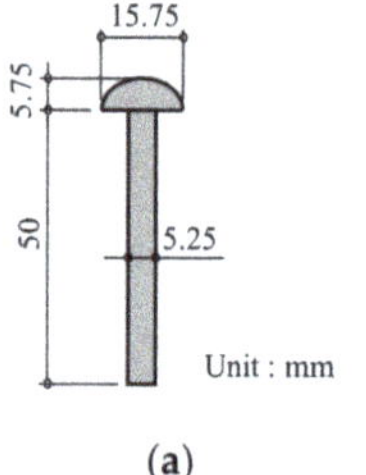

(a)

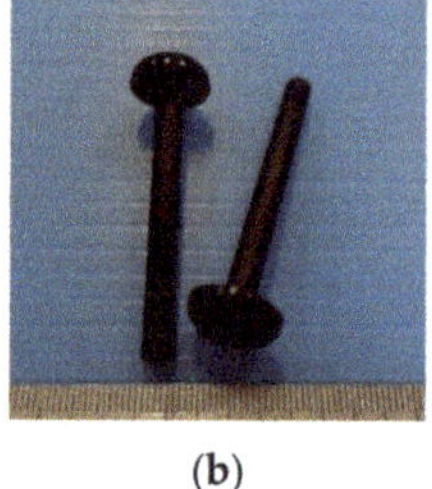

(b)

Figure 1. Fiber-reinforced thermoplastic (FRTP) rivet. (a) Geometry of FRTP rivet; (b) image of FRTP rivets.

Table 1. Properties of FRTP material.

Property	Dry Condition	Wet Condition	Testing Method
Tensile Strength (MPa)	205	128	ISO527-1,-2
Elastic Modules (GPa)	16.2	9.5	ISO527-1,-2
Bending Strength (MPa)	320	203	ISO178A
Melting Point (°C)		220	ISO11357-1,-3
Density		1.57	ISO1183

We evaluated the connection method using tapping screws that do not require holes to insert a tapping screw for FRP members [17]. The initial stiffness of the connection increased, and stress concentration was reduced because there was no clearance between FRP base member and tapping screw. However, the connection using the tapping screw lacked pull-out and fatigue strength because it did not contain a nut at the drilling side. To improve the non-clearance connection, the connection method using FRTP rivets and tentative tapping screws was proposed to provide higher initial connection stiffness and minimize preparation for mechanical connection. Figure 2 shows the connection method using FRTP rivets. First, a tentative connection by tapping screws to produce holes and fix FRP members was performed (Figure 2a). The nominal outer diameter of the tapping screw was 5.5 mm. Second, tapping screws were ejected while fixing FRP members, and FRTP rivets were inserted (Figure 2b). Third, the rivet head was thermoformed using a heating die (Figure 2c). Processes (b) and (c) were applied to all FRTP rivets one by one. Lastly, the FRTP connection could be made as shown in Figure 2d. In addition, a tentative connection by tapping screws does not necessarily have to be drilled for all rivet positions. In this case, other rivet holes could be drilled with a drilling machine in the same way as in bolt-hole preparation.

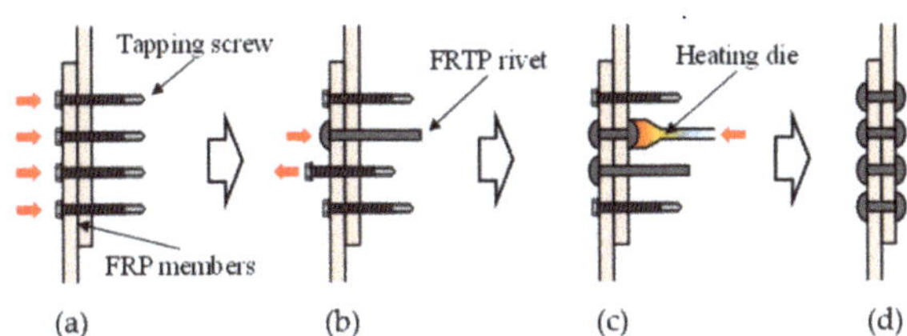

Figure 2. Connection method. (**a**) Tentative connection by tapping screw to produce holes and fix fiber-reinforced plastic (FRP) members; (**b**) ejecting tapping screw and inserting FRTP rivets; (**c**) thermoforming rivet head; (**d**) completed FRTP connection.

3. Material Tests

3.1. Tensile Test

Figure 3 shows the material tensile test method and setup. Total number of specimens for material tensile test was 5. Tensile force was applied to the FRTP rivet through the bottom steel pipe and top testing frame because it was difficult to directly apply tensile force to the FRTP rivet. The FRTP rivet was fixed to the steel pipe using epoxy adhesive with 30 mm length for chucking to the tensile testing machine. The shank of the FRTP rivet was then sandwiched between two half-notched steel plates and fixed with another steel plate and bolts. Then, it was hooked on a test frame combined with square steel pipe.

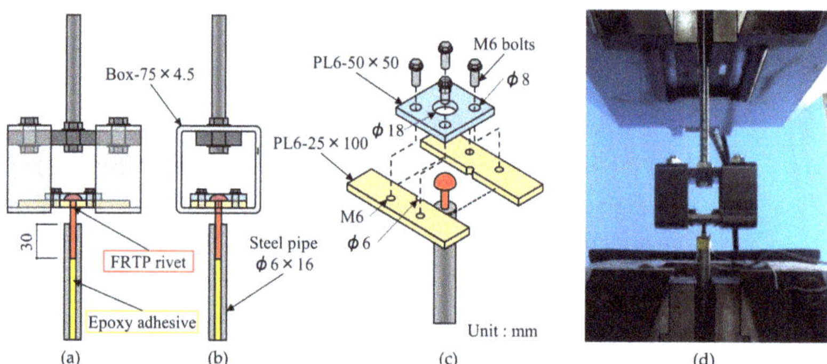

Figure 3. Material tensile test method: (**a**) Front view; (**b**) side view; (**c**) detail of testing parts; (**d**) experiment setup.

Figure 4 shows the failure mode of the FRTP rivet under tensile loading. Material tensile breaking occurred at the corner between rivet head and shank in all specimens. Figure 5 shows the tensile forces, average value, and coefficient of variation obtained by the material tensile test. The tensile strength of the right vertical axis was obtained by dividing the tensile force of the left vertical axis by the cross-section area of the rivet shank with an outer diameter of 5.25 mm. The tensile strength of the FRTP rivet was evaluated to be approximately 69% of the nominal tensile strength of the material at wet condition, as shown in Table 1. This is because stress concentration occurred due to rapid cross-sectional changing at the corner between rivet head and shank. Stable tensile strength of the FRTP rivet is demonstrated because of the small coefficient of variation.

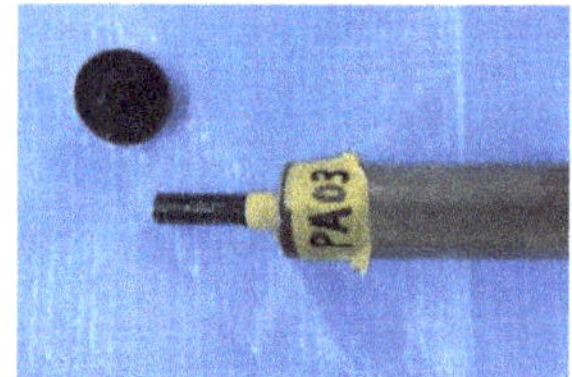

Figure 4. Failure mode of FRTP rivet under tensile loading.

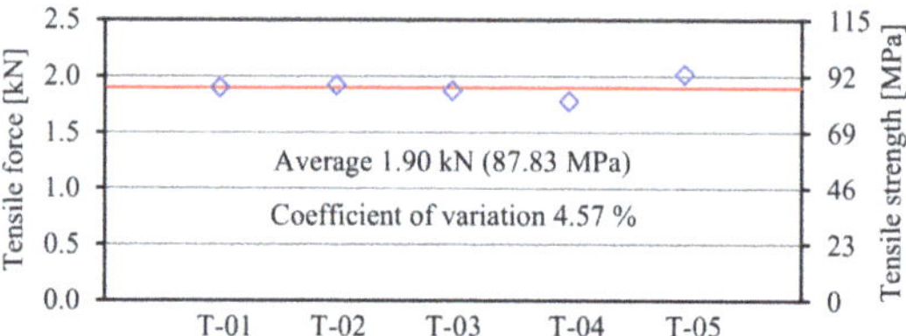

Figure 5. Material tensile test results.

3.2. Tensile Shear Test

Figure 6 shows the material tensile shear test method and setup. Total number of specimens for material tensile shear test was 10. Shear force was applied to the FRTP rivet through steel plates. The top center plate and cover plates were fixed by high-strength bolts. The FRTP rivet was inserted into a 6 mm hole of cover plates and bottom center plate to apply shear load without bending.

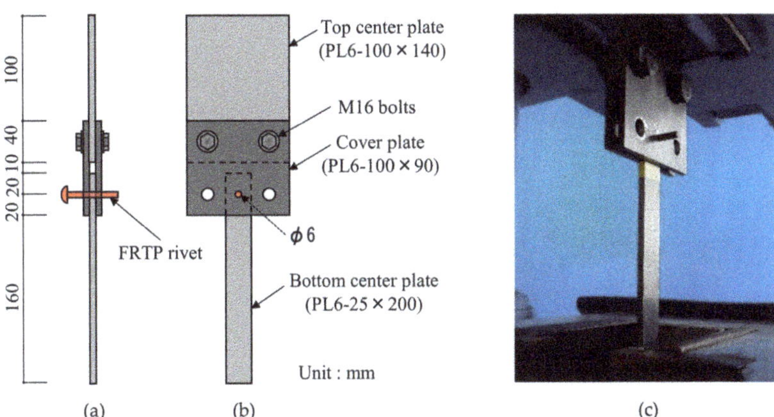

Figure 6. Material shear test method: (**a**) Front view; (**b**) side view; (**c**) experiment setup.

Figure 7 shows the failure mode of FRTP rivet under shear loading. Material shear breaking occurred at the boundaries between steel plates in all specimens. Figure 8 shows the shear forces, average value, and coefficient of variation obtained by the material tensile shear test. The shear strength of the right vertical axis was obtained by dividing the shear force of the left vertical axis by twice the cross-section area of the rivet shank, because shear force was applied to two cross-sections of the rivet shank. The stable shear strength of the FRTP rivet is demonstrated because of the small coefficient of variation.

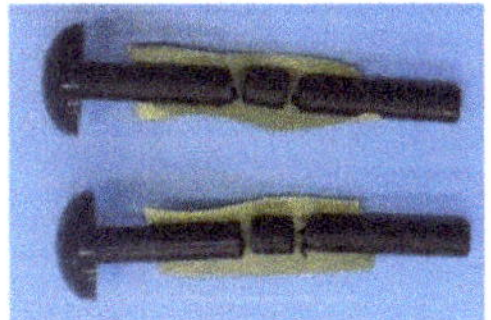

Figure 7. Failure mode of FRTP rivet under shear loading.

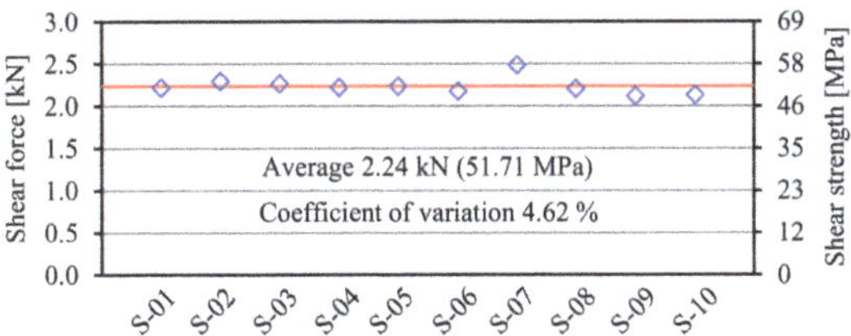

Figure 8. Material tensile shear test results.

3.3. Bending Test

Figure 9 shows the material bending test method and setup. Total number of specimens for the material bending test was 10. Bending force was applied to the FRTP rivet by using the longitudinally extended test frame of the material shear test by 9 mm thickness spacers inserted between cover plates and top center plate.

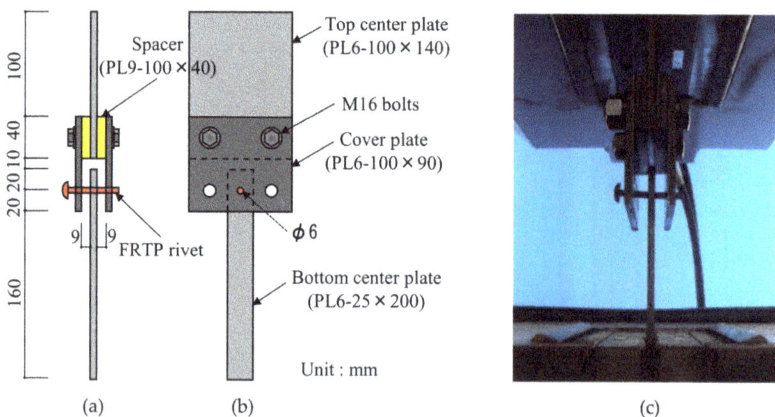

Figure 9. Material bending test method: (**a**) Front view; (**b**) side view; (**c**) experiment setup.

Figure 10 shows the failure mode of the FRTP rivet under bending loading. Material bending breaking occurred at the center of rivet shank all specimens. Figure 11 shows bending forces, average value, and coefficient of variation obtained by the material bending test. The bending strength of right vertical axis σ_R was calculated by the bending force of left vertical axis F_R from the following equation.

$$\sigma_R = \frac{F_R l_B}{2Z_R} \tag{1}$$

where l_B is the distance between steel plates (l_B = 9 mm), and Z_R is the section modulus of the rivet shank. The bending strength of the FRTP rivet was between the nominal bending strengths of material at dry and wet conditions. In addition, a small coefficient of variation was observed, so the stable bending strength of the FRTP rivet could be demonstrated.

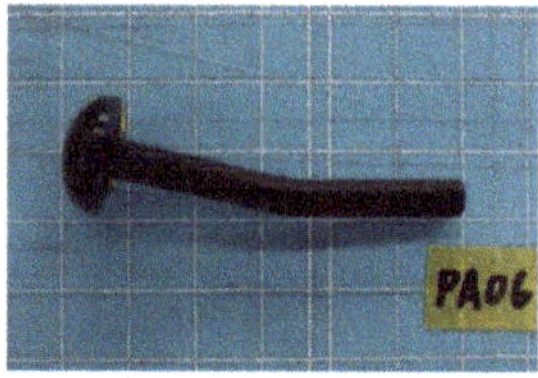

Figure 10. Failure mode of FRTP rivet under bending.

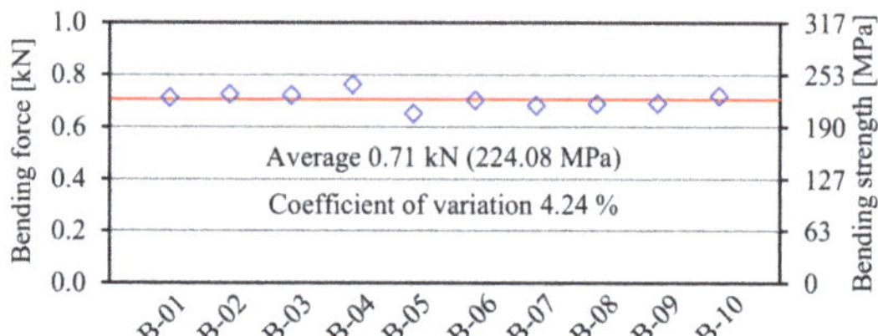

Figure 11. Material bending test results.

4. Connection Strength under Tensile Shear Loading

This section discusses the connection strength using single or multiple FRTP rivets for GFRP plates through double-lapped tensile shear tests. Figure 12 shows the connection test specimens and setup. Double-lapped GFRP plates with 40 mm width connected by

FRTP rivets were used. GFRP plates were made by pultrusion using unsaturated polyester resin and glass fiber (PLALLOY^{TM} by AGC Matex Co., Ltd., Kanagawa, Japan); the glass roving layer (approx. 4.5 mm) was sandwiched between the continuous strand mat layers (approximately 0.25 mm). Table 2 shows the mechanical properties of the GFRP plate. The number of specimens for the connection test was 5 for each connection type depending on the number of used rivets: 1, 4, 8, and 12. The rivets were arranged low at intervals of 20 mm. Specimen names were D1, D4, D8, and D12 according to the number of rivets used.

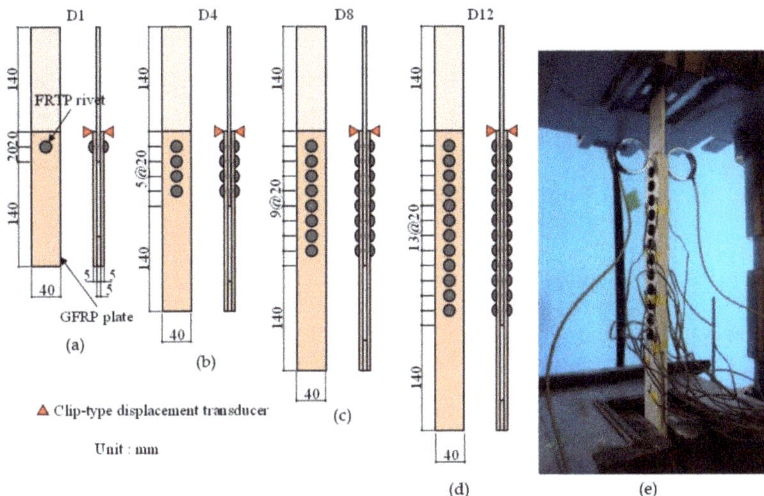

Figure 12. Connection test specimens: (**a**) Single rivet; (**b**) four rivets; (**c**) eight rivets: (**d**) 12 rivets; (**e**) experiment setup (12-rivet specimen).

Table 2. Mechanical properties of grass fiber-reinforced plastic (GFRP).

Property	Value	Testing Method
Fiber content (wt.%)	53	JIS K 7165
Longitudinal tensile strength (MPa)	411	JIS K 7165
Longitudinal elastic modulus (GPa)	28	JIS K 7052

All specimens were monotonically tested using a tensile test machine. Two clip-type displacement transducers were mounted on both sides of the specimen to measure the relative displacement of the connection, as shown in Figure 12.

As a result of the connection test, the shear failure of the FRTP rivets was obtained in all specimens, as shown in Figure 13a. Figure 14a shows the maximal loading, average values, and standard deviation obtained by the connection test. Average strength obtained by the D1 specimens was 2.72 kN, which was 21% higher than the average of the material test result. This is because a slight bending moment was applied due to the clearance between cover plate and rivet in the material shear test, but only shear force could be applied in the connection test by the non-clearance connection method shown in Figure 2. Figure 13b shows a close-up image of around the FRTP rivet of the broken specimen. Since the nominal outer diameter of the tapping screw was 5.5 mm, the clearance between FRTP rivet and GFRP plate was theoretically 0.25 mm. In fact, Figure 13b confirms that the clearance was very small. The connection using FRTP rivets could provide quite stable strength because of the small coefficient of variation regardless of the number of rivets. Figure 14b shows the maximal loading per unit rivet. Maximal loading per unit rivet was calculated by the maximal loading shown in Figure 14a divided by the number of rivets used for the connection. Even if the number of rivets increased, the connection strength per unit rivet was almost the same. Moreover, the connection strength per unit rivets was higher than the shear strength of material test was. Therefore, connection strength using FRTP rivets linearly increased as the number of rivets increased. Figure 15 shows the load-relative displacement relations. Initial sliding was not observed in all specimens using multiple rivets because non-clearance could be achieved, as shown in Figure 13b.

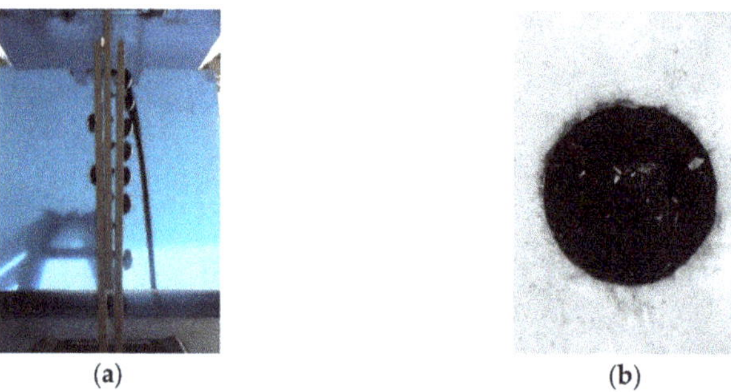

(a)   (b)

Figure 13. Typical failure mode. (**a**) Failure mode of eight-rivet specimen; (**b**) close-up image around FRTP rivet.

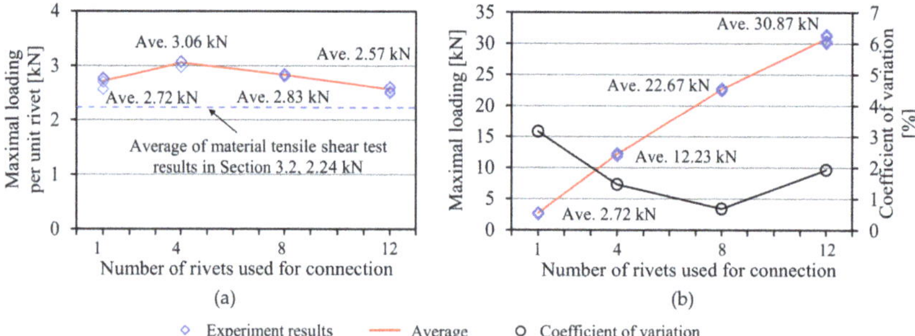

Figure 14. Experiment result. (**a**) Maximal loading; (**b**) maximal loading per unit rivet.

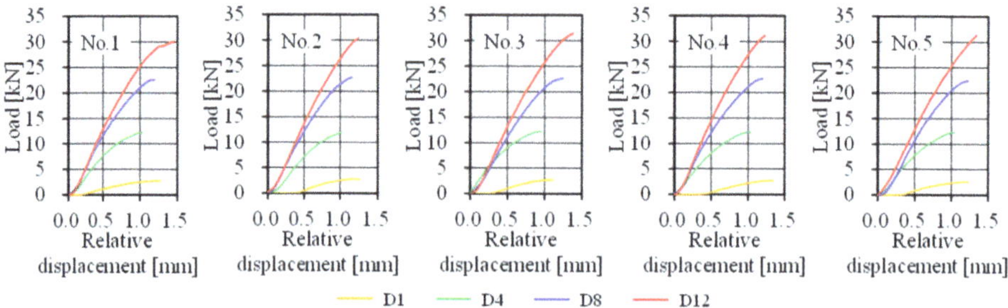

Figure 15. Load-relative displacement relations.

5. Mechanical Behavior of Beam Joint under Bending

This section discusses the strength of the beam joint using FRTP rivets for a GFRP beam through a four-point bending test. Figures 16 and 17 show bending test specimen and setup, respectively. The specimen was produced with longitudinally jointed two H-shaped GFRP beams with a 5 mm gap using FRTP rivets through GFRP splice plates. The H-shaped GFRP beam member consisted of two pultruded GFRP channel-shaped members that were adhesively bonded back-to-back by epoxy adhesive. The mechanical properties of the GFRP channel-shaped member are shown in Table 2. The GFRP splicing plates were molded by vacuum-assisted resin transfer molding using two-directional glass woven fabric (ERW580-554A) and epoxy resin.

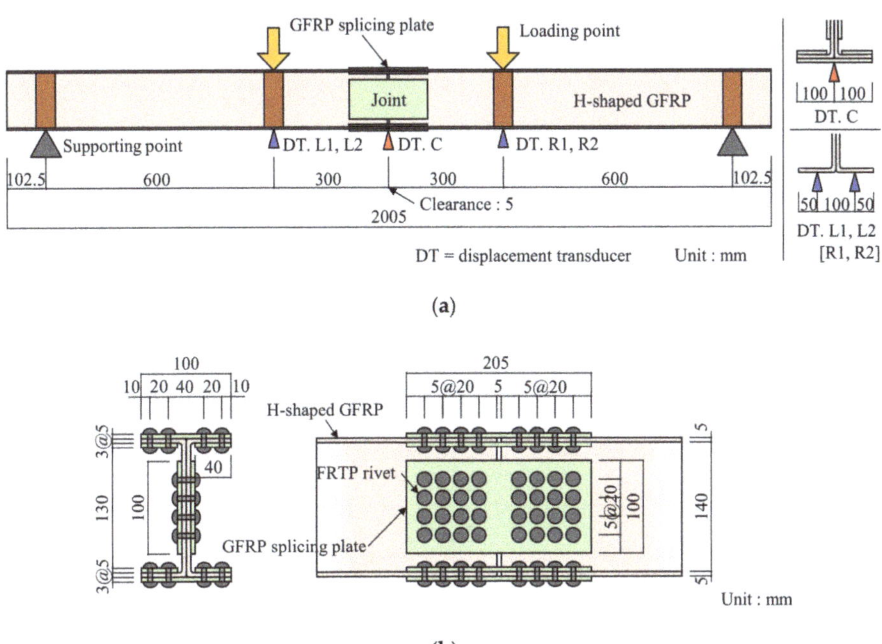

(a)

(b)

Figure 16. Bending test specimen. (**a**) Specimen overview; (**b**) joint detail.

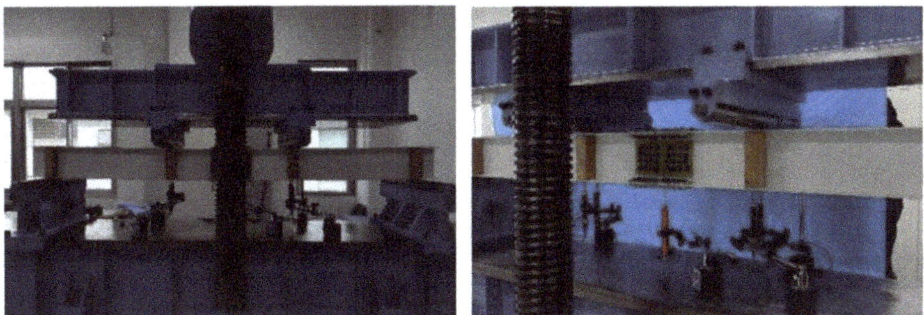

Figure 17. Experiment setup.

Figure 16b shows the joint details. The number of rivets in width, height, and longitudinal direction were four, four, and eight, respectively. Thus, the total number of rivets used for joining was 96. The total number of rivets was determined so that the rivet fracture preceded the beam member fracture and buckling to evaluate the strength of the FRTP rivet connection. The bending test was performed on only one specimen, but bending test behaviors were expected to be stable because connection strength using FRTP rivets was stable, as shown in Section 4.

Maximal loading at the joint was calculated by multiplying rivet strength and distance from the neutral axis or the center point of rotation with the same method as that for bolt joints in steel structures, as shown in Figure 18. Bending moment resisted by flange rivets M_f was calculated by the following equation.

$$M_f = n_{Rf} \cdot F_R \cdot h \tag{2}$$

where n_{Rf} is the number of FRTP rivets in upper/lower flange of one side of beam member F_R is the shear force of the FRTP rivets, which was obtained from D1 specimens shown in Figure 12 of the connection test; and h is distance between the center of the thickness of the upper and lower flanges. In this study, $n_{Rf} = 16$, $F_R = 2.72$ kN, and $h = 145$ mm. Bending moment resisted by web rivets M_w was calculated by the following equation.

$$M_w = \sum r_i \cdot F_{Ri} = \sum r_i \cdot \frac{h_i}{h/2} \cdot F_R \tag{3}$$

where r_i is distance from the center of gravity of web joint to one rivet; and F_{Ri} is the shear force resisted by the rivet, which was calculated by multiplying the shear force of FRTP rivets F_R by the ratio of the vertical distance of rivet h_i to the vertical distance of center of flange $h/2$ based on the beam neutral axis. Therefore, maximal loading at the joint P_R was calculated using the bending moments obtained from the Equations (2) and (3) by the following equation.

$$P_R = 2\frac{M_f + M_w}{l} \tag{4}$$

where l is distance from loading point to supporting point of the specimen ($l = 600$ mm). As a result of the calculation, maximal loading at the joint could be estimated as 22.35 kN.

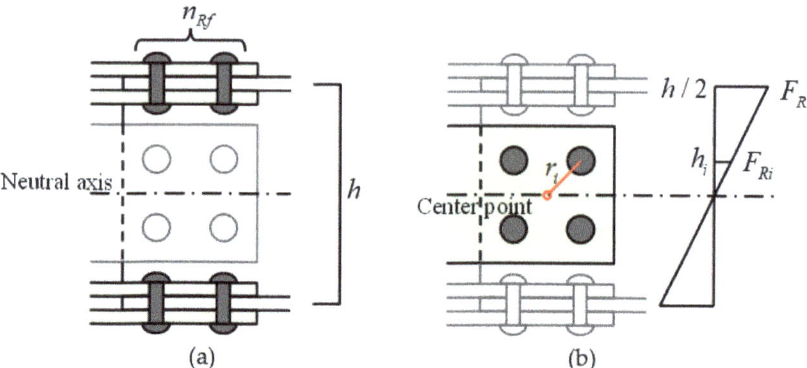

Figure 18. Calculation of maximal loading at the joint. (**a**) Flange; (**b**) web.

The specimen was monotonically tested using a compression test machine. Displacement transducers were mounted onto the middle of the span and the loading points of the specimen to measure the displacement of the specimen, as shown in Figure 16a.

Figure 19 shows the failure mode of the bending test specimen. Rivets were almost broken, with shear failure mode on the right side of the web and the bottom flange, and the failure position was the surface between web and splice plate. This failure mode was as expected because shearing force acting on the rivets reached the maximum at the top and bottom of the beam. Figure 20 shows the load-displacement relations. The load rapidly decreased due to the rivets breaking after reaching maximal loading. The maximal loading obtained by bending test was 22.61 kN, which had a 1.2% error from theoretical value P_R. Thus, it is demonstrated that joint strength using FRTP rivets can be calculated with high accuracy from the shear strength of a rivet and the distance from the center using the method for bolt joints in steel structures. Maximal displacement at the middle of span obtained by bending test was 22.26 mm, and the average of the displacements at the loading points of the specimen obtained by bending test was 17.60 mm.

Figure 19. Failure mode.

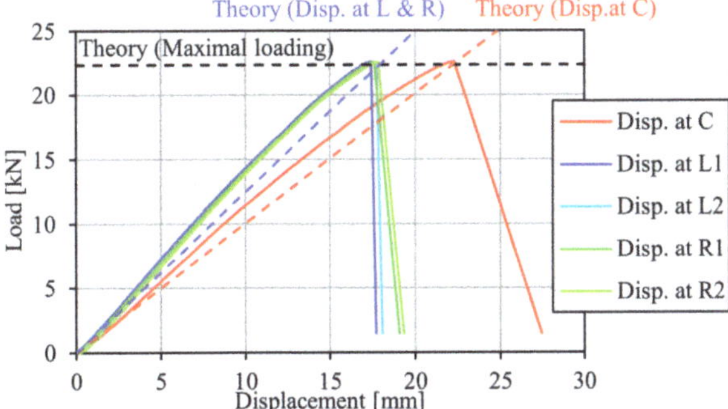

Figure 20. Load-displacement relations.

In this study, we considered three displacement components to evaluate deformation, i.e., the bending and shear deformations obtained from fundamental beam theory, and rotation at the joint as shown in Figure 21. Displacements due to bending, δ_{BC} and δ_{BL} were calculated by the following equations. Subscript C, middle of span; subscript L, loading points.

$$\delta_{BC} = \frac{23Pl^3}{48EI} \tag{5}$$

$$\delta_{BL} = \frac{5Pl^3}{12EI} \tag{6}$$

where P is the applied load, E is the longitudinal elastic modulus of the H-shaped GFRP beam shown in Table 2, and I is the moment of inertia of the H-shaped GFRP beam. Displacements due to shear stress δ_{SC} and δ_{SL} were calculated by the following equation.

$$\delta_{SC} = \delta_{SL} = \frac{\kappa Pl}{2GA} \tag{7}$$

where κ is the shear correction factor, which is the ratio of the web cross-sectional area to the total cross-sectional area of the H-shaped GFRP beam; G is the shear modulus of the H-shaped GFRP beam; and A is the cross-sectional area of the H-shaped GFRP beam. Figure 21 shows a model in which the joint of the beam was an elastic hinge to obtain the displacement due to rotation at the joint. Shear force resisted by upper/lower flange rivets Q was calculated from bending moment applied at the joint of beam M by the following equation.

$$Q = \frac{M}{h} = \frac{Pl}{2h} \tag{8}$$

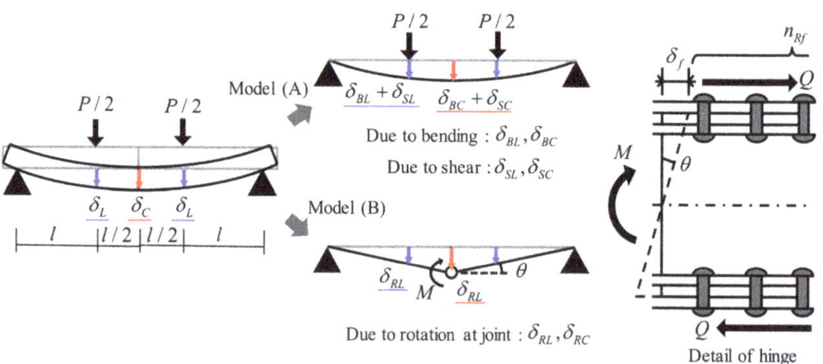

Figure 21. Displacement components of FRTP rivet connection.

Longitudinal displacement of flange plate by rivet deformation of δ_f was calculated by the following equation.

$$\delta_f = \frac{Q}{n_{Rf}K_R} = \frac{Pl}{2n_{Rf}hK_R} \tag{9}$$

where K_R is the shear stiffness of the FRTP rivet; and $K_R = 4.12$ kN/mm, which was obtained from the D1 specimens shown in Figure 12. Angle of rotation at the joint, θ was calculated by the following equation.

$$\theta = \frac{2\delta_f}{h} = \frac{Pl}{n_{Rf}h^2K_R} \tag{10}$$

Displacements due to rotation at the joint, δ_{RC} and δ_{RL} were calculated by multiplying the angle of beam rotation by the distance from the support point as per the following equations.

$$\delta_{RC} = \theta \cdot \frac{3}{2}l = \frac{3Pl^2}{2n_{Rf}h^2K_R} \tag{11}$$

$$\delta_{RL} = \theta \cdot l = \frac{Pl^2}{n_{Rf}h^2K_R} \tag{12}$$

On this basis, displacement at the middle of span and the loading point of the specimen was calculated from the sum of the deformation components as per the following equations.

$$\delta_C = \delta_{BC} + \delta_{SC} + \delta_{RC} \tag{13}$$

$$\delta_L = \delta_{BL} + \delta_{SL} + \delta_{RL} \tag{14}$$

Theoretical displacements obtained by the Equations (13) and (14) are also represented in Figure 18. They were in good agreement with the displacements obtained by bending test at maximal loading; errors were 1.1% at the middle of span and 2.9% at the loading points. Therefore, displacement at maximal loading could be accurately evaluated by considering displacement due to rotation at the joint in addition to displacement due to bending and shear. Displacement during loading could not be perfectly simulated because the obtained displacement from the bending test increased nonlinearly with the load. However, theoretical displacement at the middle of span could be simulated with 15% error the experimental displacement.

6. Conclusions

This paper proposed a connection method of FRP members using FRTP rivets and evaluated the fundamental properties of FRTP rivets through material tests. Connection behavior was also evaluated through the double-lapped tensile shear tests and a four-point bending test of the beam. As a result, the following conclusions were reached.

(1) FRTP rivets provide stable tensile, shear, and bending strength within 4.7% of the coefficient of variation.

(2) Connection strength using FRTP rivets linearly increases as the number of rivets increases.

(3) Initial sliding was not observed at the FRTP rivet connection because non-clearance could be achieved by the proposed connection method.

(4) Beam joint strength using FRTP rivets could be calculated with high accuracy within 1.2% error from the shear strength of a rivet like a bolt joint in steel structure.

(5) Beam displacement using FRTP rivets at maximal loading could be accurately evaluated within 2.9% error by considering displacement due to rotation at the joint in addition to displacement due to bending and shear deflection.

On the basis of this research, GFRP members can be connected with stable loading capacity, and deflections can be evaluated with high accuracy by using FRTP rivet even if bolt holes are not prepared. The connection method in this paper could also reduce construction time and improve the material machining process because onsite drilling can be applied by tapping screws.

Author Contributions: Conceptualization, Y.M. (Yukihiro Matsumoto); methodology, Y.M. (Yukihiro Matsumoto); validation, T.M.; formal analysis, T.M., Y.M. (Yoshiyuki Matsushita) and Y.M. (Yukihiro Matsumoto); investigation, T.M. and Y.M. (Yukihiro Matsumoto); writing—original draft preparation, T.M.; writing—review and editing, Y.M. (Yukihiro Matsumoto) and Y.M. (Yoshiyuki Matsushita); supervision, Y.M. (Yukihiro Matsumoto); project administration, Y.M. (Yukihiro Matsumoto). All authors have read and agreed to the published version of the manuscripts.

Funding: This research was funded by the Natio Taisyun Section and Technology Foundation.

Conflicts of Interest: The authors declare no conflict of interest.

References

1. Zoghi, M. *The International Handbook of FRP Composites in Civil Engineering*; CRC Press: Boca Raton, FL, USA, 2013.
2. Bai, Y.; Keller, T. Modal parameter identification for a GFRP pedestrian bridge. *Compos. Struct.* **2008**, *82*, 90–100. [CrossRef]
3. Votsis, R.A.; Stratford, T.J.; Chryssanthopoulos, M.K.; Tantele, E.A. Dynamic assessment of a FRP suspension footbridge through field testing and finite element modelling. *Steel Compos. Struct.* **2017**, *23*, 205–215. [CrossRef]
4. Evernden, M.C.; Mottram, J.T. A case for houses to be constructed of fibre reinforced polymer components. *Proc. ICE Constr. Mater.* **2012**, *165*, 3–13. [CrossRef]
5. Yang, X.; Bai, Y.; Ding, F. Structural performance of a large-scale space frame assembled using pultruded GFRP composites. *Compos. Struct.* **2015**, *133*, 986–996. [CrossRef]
6. Matsumoto, Y.; Yonemaru, K. Exposure test and long-term characteristics for CFRP structural members. In Proceeding of the 4th Asia-Pacific Conference on FRP in Structures, Melbourne, Australia, 11–13 December 2013.
7. Coelho, A.M.G.; Mottram, J.T. A review of the behaviour and analysis of bolted connections and joints in pultruded fibre reinforced polymers. *Mater. Des.* **2015**, *74*, 86–107. [CrossRef]
8. Ueda, M.; Ui, N.; Ohtani, A. Lightweight and anti-corrosive fiber reinforced thermoplastic rivet. *Compos. Struct.* **2018**, *188*, 356–362. [CrossRef]
9. Ascione, F.; Lamberti, M.; Razaqpur, A.G.; Spadea, S. Strength and stiffness of adhesively bonded GFRP beam-column moment resisting connections. *Compos. Struct.* **2017**, *160*, 1248–1257. [CrossRef]
10. Feo, L.; Marra, G.; Mosallam, A.S. Stress analysis of multi-bolted joints for FRP pultruded composite structures. *Compos. Struct.* **2012**, *94*, 3769–3780. [CrossRef]
11. Matsumoto, Y.; Yamada, S.; Komiya, I. Nonlinear Failure Behavior and Bearing Strength of Bolted Joints in Fiber Reinforced Polymer Plates. In Proceeding of the SAMPE Tech, North Charleston, SC, USA, 22–25 October 2012.
12. Goh, G.D.; Dikshit, V.; Nagalingam, A.P.; Goh, G.L.; Agarwala, S.; Sing, S.L.; Wei, J.; Yeong, W.Y. Characterization of mechanical properties and fracture mode of additively manufactured carbon fiber and glass fiber reinforced thermoplastics. *Mater. Des.* **2018**, *137*, 79–89. [CrossRef]

13. Doan, H.G.M.; Mertiny, P. Creep Testing of Thermoplastic Fiber-Reinforced Polymer Composite Tubular Coupons. *Materials* **2020**, *13*, 4637. [CrossRef] [PubMed]
14. Cao, Z.; Guo, D.; Fu, H.; Han, Z. Mechanical Simulation of Thermoplastic Composite Fiber Variable-Angle Laminates. *Materials* **2020**, *13*, 3374. [CrossRef] [PubMed]
15. Fan, C.; Shan, Z.; Zou, G.; Zhan, L.; Yan, D. Performance of Short Fiber Interlayered Reinforcement Thermoplastic Resin in Additive Manufacturing. *Materials* **2020**, *13*, 2868. [CrossRef] [PubMed]
16. Rodonò, G.; Sapienza, V.; Recca, G.; Carbone, D.C. A Novel Composite Material for Foldable Building Envelopes. *Sustainability* **2019**, *11*, 4684. [CrossRef]
17. Duong, N.N.; Nhut, P.V.; Satake, C.; Matsumoto, Y. Study on Mechanical Behavior of Self-Tapping Screws Connection Using Washers in Single-Lapped Glass Fiber Reinforced Plastic Plates by Experiment and Finite Element Analysis. *Int. Conf. Build. Mater. Constr.* **2018**, *C0027*. [CrossRef]

 *materials*

Article

A Proposal of a Method for Ready-Mixed Concrete Quality Assessment Based on Statistical-Fuzzy Approach

Izabela Skrzypczak [1], Wanda Kokoszka [1], Joanna Zięba [1], Agnieszka Leśniak [2,*], Dariusz Bajno [3] and Lukasz Bednarz [4]

[1]   Faculty of Civil and Environmental Engineering and Architecture, Rzeszow University of Technology, Powstancόw Warszawy 12, 35-082 Rzeszow, Poland; izas@prz.edu.pl (I.S.); wandak@prz.edu.pl (W.K.); j.zieba@prz.edu.pl (J.Z.)

[2]   Faculty of Civil Engineering, Cracow University of Technology, Warszawska 24, 31-155 Kraków, Poland

[3]   Faculty of Civil and Environmental Engineering and Architecture, UTP University of Science and Technology, Al. Prof. S. Kaliskiego 7, 85-796 Bydgoszcz, Poland; dariusz.bajno@utp.edu.pl

[4]   Faculty of Civil Engineering, Wroclaw University of Science and Technology, Wybrzeże Wyspiańskiego 27, 50-370 Wroclaw, Poland; lukasz.bednarz@pwr.edu.pl

*   Correspondence: alesniak@l7.pk.edu.pl

Received: 10 November 2020; Accepted: 8 December 2020; Published: 12 December 2020

Abstract: Control of technical parameters obtained by ready-mixed concrete may be carried out at different stages of the development of concrete properties and by different participants involved in the construction investment process. According to the European Standard EN 206 "Concrete–Specification, performance, production and conformity", mandatory control of concrete conformity is conducted by the producer during production. As shown by the subject literature, statistical criteria set out in the standard, including the method for concrete quality assessment based on the concept of concrete family, continue to evoke discussions and raise doubts. This justifies seeking alternative methods for concrete quality assessment. This paper presents a novel approach to quality control and classification of concrete based on combining statistical and fuzzy theories as a means of representation of two types of uncertainty: random uncertainty and information uncertainty. In concrete production, a typical situation when fuzzy uncertainty can be taken into consideration is the conformity control of concrete compressive strength, which is conducted to confirm the declared concrete class. The proposed procedure for quality assessment of a concrete batch is based on defining the membership function for the considered concrete classes and establishing the degree of belonging to the considered concrete class. It was found that concrete classification set out by the standard includes too many concrete classes of overlapping probability density distributions, and the proposed solution was to limit the scope of compressive strength to every second class so as to ensure the efficacy of conformity assessment conducted for concrete classes and concrete families. The proposed procedures can lead to two types of decisions: non-fuzzy (crisp) or fuzzy, which point out to possible solutions and their corresponding preferences. The suggested procedure for quality assessment allows to classify a concrete batch in a fuzzy way with the degree of certainty less than or equal to 1. The results obtained confirm the possibility of employing the proposed method for quality assessment in the production process of ready-mixed concrete.

Keywords: ready-mixed concrete; construction material; quality assessment; conformity criteria; statistical-fuzzy method

---

1. Introduction

The construction industry is an economic sector characterised by high changeability and diversity. Individual character of the facilities constructed is expressed in their unique qualities, such as form, shape and purpose, and influenced by such factors as environmental conditions (the facility's surroundings), completion time, technologies applied and building materials used. Much of the work related to facility construction involves optimisation of project completion time [1], optimisation of costs [2], energy efficiency [3], which also includes finding optimal technologies [4] and appropriate building materials for the particular project [5]. Execution of construction works within the scheduled time, within the framework of estimated costs and at the assumed quality level is the determinant of success for the investor, the designer and the contractor. The existence of relationship between costs, completion time and project quality, as depicted in the form of project management triangle, is considered to be self-evident [6]. The subject literature provides numerous definitions and interpretations of the term "quality". Considering the concept of quality in the construction industry, it can be defined as meeting the requirements of the designer, the contractor, the owner and the regulatory agencies [7]. The quality of the facilities constructed is directly influenced by the applied quality control procedures for the execution of construction works at construction sites, and procedures related to the production of building materials in permanent production facilities. As shown in [8], these procedures greatly vary, since measures related to quality control of the execution of construction works can be approached in a relatively flexible manner while remaining within the aforementioned provisions, whereas quality control of the construction materials supplied to the market is strictly regulated.

Concrete is a building material widely used in construction [9], while ready-mixed concrete (RMC) is the principal construction material for civil engineering infrastructure [10]. Currently, the world produces 4.4 billion tons of concrete annually, but that number is expected to rise to over 5.5 billion tons by 2050, according to the Chatham House report [11]. Construction concrete produced under quality control guidelines constitutes about 70% of total concrete production [12]. Since the properties of concrete are shaped from the moment of mixing in a process influenced by many factors, assessment of its quality (parameters) can be carried out at different times: during production, during delivery and before/after construction, and importantly, quality assessment can be performed by different participants of the investment process: the producer, the contractor, and the investor. Achieving the desired quality of concrete involves not only conformity assurance, but also appropriate design of concrete mix and selection of suitable ingredients [13–21], proper manufacturing [22–24], development of innovative research methods that aid concrete design aimed at obtaining appropriate properties and durability [25–27] and development of methods for analysing obtained assessment results both during production and in existing constructions [28,29].

The traditional approach to the quality assessment of ready-mixed concrete is through experiments [30,31], which, however, proves to be both time and resource consuming. The proposed statistical-fuzzy-approach-based method for quality assessment can overcome these limitations. The suggested method may be employed in adaptive neuro-fuzzy inference systems and applied to predict the 28-day compressive strength of concrete for concrete mix design by reducing i.a., the number and scope of trials. The application of the proposed procedure combined with the use of artificial neural networks (ANN) ensures the reliable assessment of concrete compressive strength.

According to the European Standard EN-206 "Concrete–Specification, performance, production and conformity" [32], ready-mixed concrete delivered on the construction site as concrete mix is subject to mandatory control for compliance with the criteria set out by EN-206. The assessment is performed by the producer during production. Other procedures for concrete quality control are mostly optional. It should be underscored that conformity control carried out according to the recommended criteria cannot be regarded as statistical control until objective conclusions are drawn in line with the principles of mathematical statistics. Statistical sample method can raise doubts as to the accuracy of estimation of the concrete property being assessed and the classification of the considered concrete batch. With a small sample size ($n = 3$), as is the case in concrete quality control, qualification errors are not uncommon. Statistical quality control arrangements are a result of a "strategic game" between the producer and the consumer, whereas the standard conformity criteria represent a compromise between the quality, economy and safety requirements. Recommended measures for standard conformity control set out in EN-206 continue to evoke discussions, and the research conducted in the field reveals inadequacies [33–38]. These inadequacies concern the analyses of concrete batches conducted before quality assessment and refer to deficiency during continuous production. In view of the above, it would seem justified to seek alternative methods for the quality control of concrete.

In engineering practice, the recognition of the material's compliance with the specification is decided based on the adopted plan for statistical quality control. It is a standard approach based on binary criteria (met/unmet). This is of particular significance in the case of doubts concerning the quality of material (in this case, ready-mixed concrete) already built into the existing structure, where the material quality is especially tightly linked with the structure's safety and reliability. For instance, in the case of prestressed structures, both understated and overstated concrete class has a key influence on the fulfilment of the serviceability limit state condition. Accidental understatement of the concrete class may result in the demolition of a structural component or, in extreme cases, an entire structure (e.g., a bridge).

The present paper aims to propose a novel approach to concrete classification based on combining statistical and fuzzy theories as a means of representation of two types of uncertainty: random uncertainty and information uncertainty. In the field of application of statistical decision procedures–in statistical quality control—there are cases of imprecise definition of quality requirements and imprecise assessment of products subject to quality control. Such state of affairs can be caused by various factors of linguistic, economical and statistical nature. Transition from traditional ("hard") models, with fixed data, relations and limits, to "soft" models that allow some degree of imprecision is made possible by the fuzzy set theory introduced by Zadeh [39].

For the discussed issue of quality control of ready-mixed concrete, a typical situation when fuzzy uncertainty can be taken into consideration is the conformity control of concrete compressive strength, carried out to confirm the declared concrete class. Concrete class is equated with concrete compressive strength (f_{ck}) and constitutes the basis for evaluating the quality of the concrete produced. The proposed procedure for quality assessment of the concrete produced allows for making effective decisions of two types: non-fuzzy (crisp) or fuzzy, which point out to possible solutions and their corresponding preferences.

Quality Control of Ready-Mixed Concrete According to EN-206

Quality control of ready-mixed concrete is carried out with the appliance of standard statistical control procedures set out in the European Standard EN-206 "Concrete–Specification, performance, production and conformity" [32]. Conformity control involves applying two conformity criteria:

(1) Individual assessment result criterion f_{ci}—applied irrespective of the production status (initial or continuous)

$$f_{ci} \geq (f_{ck} - 4)\,\text{N/mm}^2; \tag{1}$$

(2) Mean assessment result criterion f_{cm}—applied in three methods depending on the production status:

– Initial production

$$f_{cm} \geq (f_{ck} + 4)\,\text{N/mm}^2 \quad \text{method A;} \tag{2}$$

– Continuous production

$$f_{cm} \geq (f_{ck} + 1.48 \cdot \sigma)\,\text{N/mm}^2 \quad \text{method B;} \tag{3}$$

– The concept of control chart—method C.

Specific details regarding particular methods can be found in [32]. Conformity is confirmed when both criteria are satisfied.

Conformity control of concrete compressive strength is carried out on concretes of specific composition or concrete families. The majority of concrete manufacturers assess the conformity of the concrete produced in accordance with the criteria for initial production, as these criteria are easier to apply and do not require taking into consideration the impact of coefficient of variation/standard deviation of compressive strength. With high heterogeneity of the concrete produced, conformity criteria for continuous production are more rigorous than for initial production, and therefore, most manufacturers apply the conformity criteria (Method A) recommended for $n = 3$.

The conformity criterion for mean compressive strength value and for sample of size $n = 3$, as set out in EN 206 [32], was established according to the following Equations (4) to (7):

$$f_{cm} \geq f_{ck} + k_1 \tag{4}$$

$$f_{cm} \geq f_{ck} + \left(\frac{k_1}{\sigma}\right) \cdot \sigma \tag{5}$$

$$f_{cm} \geq f_{ck} + \lambda' \cdot \sigma \tag{6}$$

where

$$\frac{k_1}{\sigma} = \lambda' \tag{7}$$

and

$k_1 = 4$—test coefficient value set out by the standard [28],
σ—standard deviation for population.

As proposed by Taerwe [34] and set out in EN 206 [32], the values of λ' for correlated results are given as follows (see Table 1):

Table 1. λ' values for correlated results of mixed size samples [34].

Number (n) of Results	Value λ'
3	2.67
15	1.48

For initial production, the standard conformity criterion was established for constant standard deviation of $4/2.67 = 1.5$ MPa, irrespective of mean compressive strength value.

Applying the conformity criteria set out in EN 206 [28] for a sample of size $n = 3$ (Method A) without providing the standard deviation value may contribute to the deterioration in concrete quality and, in consequence, lead to an excessive recipient risk [35–37].

This is confirmed by the results of random simulations and the analysis of conformity criteria for a sample of size $n = 3$, performed by means of operating characteristic (OC) curves [35]. On the basis of these operations, the following conclusions can be formulated (Figure 1):

- The concrete acceptance probability is not always a compromise between the producer risk and the customer risk. Applying the standard conformity criteria may lead to an excessive customer risk, especially in the case of an assumption of log-normal distribution of compressive strength.
- Applying the standard conformity criteria may lead the producer to adopting strategies involving higher production costs, as it can unnecessarily require higher mean values of production with higher standard deviations. These criteria are not recommended for production with small deviation and may be a reason for concealing the results for samples of understated compressive strength.
- Applying the standard conformity criteria may produce too high values of the consumer risk.

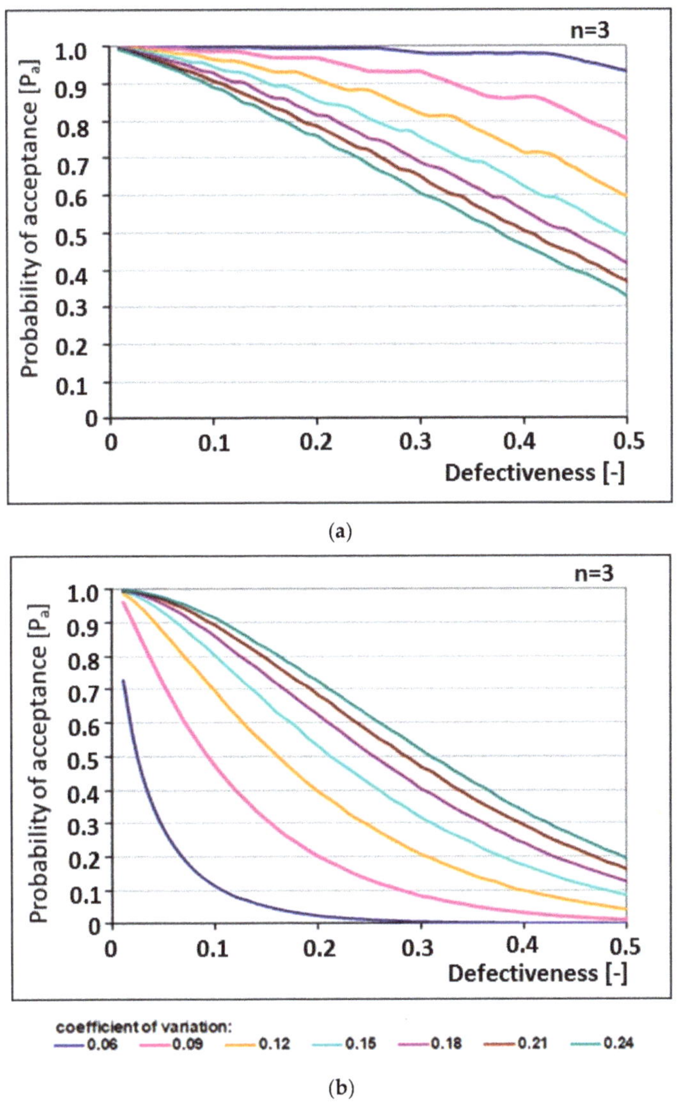

Figure 1. OC curves for conformity criteria for samples of sizes *n* = 3 and normal distribution of concrete compressive strength: for a criterion for (**a**) individual results and (**b**) mean value.

Statistical-fuzzy methods of conformity control could be applied as tools supporting initial production. Assessing the concrete class by determining the degree of certainty of concrete belonging to the class intended at the design stage could be an effective tool in decision-making in view of uncertainties related to concrete classification. The place of the proposed method in the conformity control process is presented in Figure 2.

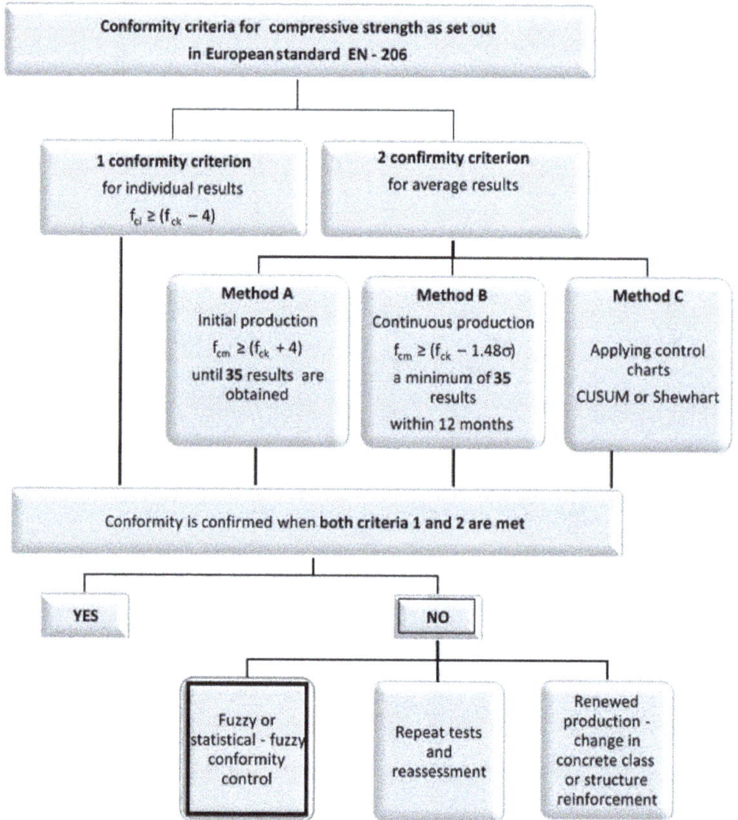

Figure 2. Conformity control of concrete compressive strength according to EN 206 [32], where f_{cm} is the mean compressive strength of concrete, f_{ck} is the characteristic compressive strength of concrete, and σ is the estimate for the standard deviation of a population.

2. Materials and Methods

2.1. Conformity Control of Concrete Compressive Strength in Consideration of Measurement Uncertainty

Conformity criteria set out in EN 206 [32] and other conformity criteria given in technical specification of products all assume that the assessment results obtained are free of measurement uncertainty—which is not true. Each of these values is burdened with measurement "errors" of type I and II. An assessment result is an approximation of the value measured and should be presented along with measurement uncertainty.

As required by ISO/IEC 17,025 [40], it is necessary for all accredited laboratories to specify measurement uncertainty. Every assessment result is, therefore, not a value but an interval, and should be presented with measurement uncertainty taken into account. When relating the assessment result to the conformity criteria set out in standard [32], it is not particular results but intervals that are subject to analysis. Such an analysis was carried out for the purpose of the present paper.

The analysis concerned a population of assessment results for concrete of identical composition, produced by the same concrete batching plant. The concrete analysed was assumed to be of class C20/25 and was characterised by high defectiveness. For the purpose of the analysis, the same criteria were adopted for initial production and overlapping assessment results. The population of results analysed is presented in Figure 3.

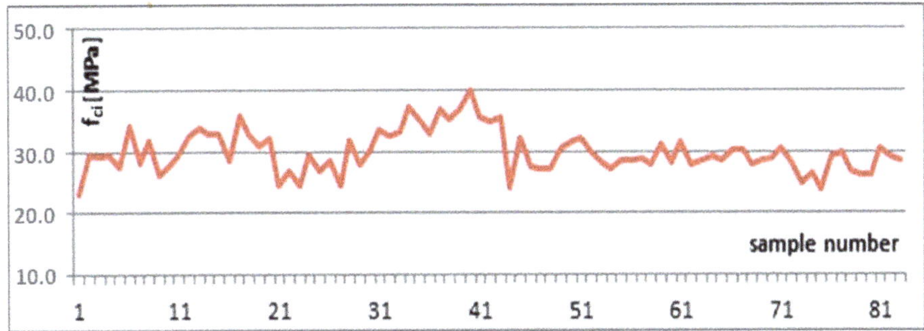

Figure 3. Compressive strength assessment results.

In the case analysed, the conformity criterion concerning particular values did not present a hazard for concrete classification in terms of its compliance with the standard (Table 2). All of the results obtained were higher than required to meet this criterion. For conformity control, the criterion related to the mean value was decisive in approving the concrete batch assessed.

Table 2. Fragment of the table presenting the conformity assessment of the population of results analysed.

Number Sample	Compressive Strength	Criterion 1	Assessment	Criterion 2	Assessment	Compressive Strength + Uncertainty	Criterion 2 + Uncertainty	Assessment
[-]	f_{ci} [MPa]	f_{ci} [MPa]	[-]	f_{cm} [MPa]	[-]	f_{ci} [MPa]	$f_{ci\,min}$ [MPa]	[-]
1	23.1	23.1	met	-		24.2	-	
2	29.7	29.7	met	-		30.8	-	
3	29.1	29.1	met	27.3	unmet	30.2	28.4	unmet
4	29.5	29.5	met	29.4	met	30.6	30.5	met
5	27.5	27.5	met	28.7	unmet	28.6	29.8	met
6	34.3	34.3	met	30.4	met	35.4	31.5	met
7	28.1	28.1	met	30.0	met	29.2	31.1	met
8	31.9	31.9	met	31.4	met	33.0	32.5	met
9	26.1	26.1	met	28.7	unmet	27.2	29.8	met
10	28.0	28.0	met	28.7	unmet	29.1	29.8	met
11	29.4	29.4	met	27.8	unmet	30.5	28.9	unmet
12	32.6	32.6	met	30.0	met	33.7	31.1	met
13	33.8	33.8	met	31.9	met	34.9	33.0	met
14	33.0	33.0	met	33.1	met	34.1	34.2	met
15	32.8	32.8	met	33.2	met	33.9	34.3	met

The population of results analysed was encumbered with an 8-percent error bias. In the case analysed, 42 percent of assessment results did not meet the standard conformity criteria for initial production. As the compressive strength assessment and sampling were conducted by an accredited laboratory, it was possible to establish the value of measurement uncertainty for defining compressive strength. Measurement uncertainty was estimated at 1.1 MPa. With this assumption, bounds of the result intervals were calculated and compared with the standard conformity criteria related to the mean value. With measurement uncertainty taken into account, the number of results that did not meet the standard conformity criteria decreased to 19 percent. In the example presented, the measurement uncertainty of the results obtained is low in relation to the compressive strength values obtained. Even with such a low level of measurement uncertainty, taking it into account in conformity analysis allows for reducing the number of non-compliant results by over 50 percent.

Having analysed the same results according to the criteria for continuous production, it can be observed that about 52 percent of the results do not meet the standard conformity criteria [19]. With measurement uncertainty taken into consideration, the number of non-compliant results is reduced to about 38 percent. This confirms that in the case of high variability of the quality of concrete (standard deviation of the population of results amounting to 3.5 MPa), it is inadvisable to conduct quality control according to the criteria for continuous production.

2.2. Alternative Conformity Criteria for Concrete Compressive Strength

Formulating the statistical conformity criteria for concrete compressive strength remains a complicated issue due to the difficulties related to the insufficiency of statistical methods for small size samples ($n < 15$) and initial production, particularly for samples of size less than or equal to 6.

Employing statistical-fuzzy methods to verify the conformity of a concrete batch might increase the effectiveness of the quality assessment of the concrete produced. Fuzzy functions might be applied on the basis of expertise or marginal distribution parameters (mean and standard deviation) for the considered concrete class and adjacent concrete classes [18,20].

While assessing the quality of the concrete produced, the results of the verification of compliance of concrete compressive strength might be considered as random events, whereas the conformity criteria can be regarded as fuzzy limit values. Conformity criteria for compressive strength, which constitute the basis for the assessment of concrete quality, might be represented as a probability for a random event to be found in a region with fuzzy limits (after Zadeh [39]) or a fuzzy number of known membership function corresponding to the probability that the event belongs to a certain interval [37].

The compressive strength (f_c) of concrete that complies with the conformity criterion can be represented as a fuzzy set (8):

$$T = [f_{cm},\ \mu(f_{cm})]\ \big|\, f_{cm} \in T\ ,\ \mu(f_{cm}) : T \to [0,1] \tag{8}$$

where $\mu_{f_c}(f_{cm})$ is a membership function that assigns each element of compressive strength set $f_{cm} \in T$ a degree of belonging to fuzzy set f_c in interval [0, 1].

Classification of the considered concrete batch into a specific class generally depends on the fulfilment of the condition related to mean compressive strength in sample, f_{cm} (Figure 1b). Sporadically, the condition concerning particular test results f_{ci} is the decisive condition for the fulfilment of the conformity criteria (Figure 1a) [34,35,37,38,41]. Since statistical conformity criteria are found to be insufficient, statistical-fuzzy methods can be applied to define class membership functions, and both standards and expertise can be taken into consideration in the quality control of the concrete produced.

Standard conformity criteria for concrete compressive strength can be given in Equations (9) and (10):

- For method A and sample of size $n = 15$, (9):

$$f_{cm} \ge f_{ck} + 4\ \to T \tag{9}$$

where f_{cm} is the mean compressive strength of concrete, and f_{ck} is the characteristic compressive strength of concrete.

- For method B and sample of size $n \ge 15$, (10):

$$f_{cm} \ge f_{ck} + 1.48\sigma\ \to T \tag{10}$$

where f_{cm} is the mean compressive strength of concrete, f_{ck} is the characteristic compressive strength of concrete, σ-estimate for the standard deviation of a population.

In Equations (6) and (7), the test characteristic T is a fuzzy value of membership function $\mu_T(t)$ that can be determined for specific concrete classes on the basis of a statistical-fuzzy experiment.

In order to determine the membership function for the considered concrete classes (three adjacent concrete classes), statistical-fuzzy method (three-phase method) was applied [42,43]. The method proposed elaborates on the concept by Woliński [43].

The statistical-fuzzy conformity control procedure of concrete compressive strength consists of two stages. The first stage is to determine marginal distribution parameters, and for that purpose, random variables x and y were defined. The variable x represents the point of division of the values of test characteristics T for the considered concrete class and lower. The variable y represents the point of division of test characteristics for the considered concrete class and higher. It is assumed that the pair (x, y) is a two-dimensional, normal random variable, for which marginal distributions $p_x(t)$ and $p_y(t)$ of random variables $x \rightarrow N(m_x, \sigma_x)$ and $y \rightarrow N(m_y, \sigma_y)$ may be determined. Marginal distribution parameters were determined by means of Monte Carlo simulation methods and the following calculation algorithm [37,44]:

1. Generate N groups of random numbers of size $n = 3$ from normal distribution;
2. Randomly select concrete class—Concrete of three adjacent classes C_{i-1}, C_i, C_{i+1} (identica probability of 1/3);
3. Randomly select standard deviation from 2, 3, 4, 5, 6 MPa with 1/5 probability;
4. Repeat (1) and (2) n-times to obtain $f_{ci}, \dots, f_{cn}$;
5. Randomly select defectiveness w from normal distribution;
6. Calculate mean compressive strength of adjacent concrete classes from Equation (11):

$$f_{cm(C_{i-1},C_i)} = \frac{m_{C_{i-1}} + m_{C_i}}{2} \text{ and } f_{cm(C_i,C_{i+1})} = \frac{m_{C_i} + m_{C_{i+1}}}{2} \tag{11}$$

7. Calculate standard deviation from Equation (12):

$$s_{(C_{i-1},C_i)} = \frac{1}{n}\sqrt{(s_{C_{i-1}})^2 + s_{C_i}^2} \text{ and } s_{(C_i,C_{i-1})} = \frac{1}{n}\sqrt{(s_{C_1})^2 + s_{C_{i-1}}^2} \tag{12}$$

8. Determine the characteristic compressive strength for the considered and lower concrete classes from Equation (13):

$$f_{ck(C_{i-1},C_i)} = m_{(C_{i-1},C_i)} - t(w)s_{(C_{i-1},C_i)} \tag{13}$$

and for the considered and higher concrete classes from Equation (14):

$$f_{ck(C_i,C_{i+1})} = m_{(C_i,C_{i+1})} - t(w)s_{(C_i,C_{i+1})} \tag{14}$$

9. Calculate mean compressive strength of the considered and lower concrete classes from Equation (15):

$$f_{cm(C_{i-1},C_i)} = f_{ck(C_{i-1},C_i)} + 4 \tag{15}$$

and of the considered and higher concrete classes from Equation (16):

$$f_{cm(C_{i-1},C_i)} = f_{ck(C_i,C_{i+1})} + 4 \tag{16}$$

10. Create a table for the probability distribution function of random vector (ξ, η) and determine the histogram of marginal distributions by summing rows and columns. The first marginal distribution is the sum of rows and the classification by the considered and lower concrete classes. The second marginal distribution is the sum of columns and the classification by the considered and higher concrete classes.

The obtained graphs of marginal distribution probability functions $p_\xi(x_n)$ and $p_\eta(x_n)$ (marginal distribution parameters) are the basis for determining membership functions of test characteristics for specific concrete classes, i.e., the second stage of calculations.

The calculations were performed in accordance with the adopted algorithm. The membership function of the test characteristic T_i for the considered i-class of concrete and higher can be represented by Equation (17):

$$\mu_{Ci}(f_{cm}) = \int_{-\infty}^{f_{cm}} p_\eta(f_{cm})df_{cm} = F\left(\frac{f_{cm} - m_\eta}{\sigma_\eta}\right) \tag{17}$$

whereas the membership function of the test characteristic F_i for the considered i-class of concrete and higher can be expressed by the following Equation (18):

$$\mu_{Ci-1}(f_{cm}) = \int_{f_{cm}}^{+\infty} p_\xi(f_{cm})df_{cm} = 1 - F\left(\frac{f_{cm} - m_\xi}{\sigma_\xi}\right) \tag{18}$$

The fuzzy membership function for the considered i-class of concrete f_{ci} can be calculated from Equation (19) or (20):

$$\mu_{Ci+1}(f_{cm}) = 1 - \int_{f_{cm}}^{+\infty} p_\xi(f_{cm})df_{cm} - \int_{-\infty}^{f_{cm}} p_\eta f(f_{cm}) \tag{19}$$

$$\mu_{Ci+1}(f_{cm}) = 1 - \left[1 - F\left(\frac{f_{cm} - m_\xi}{\sigma_\xi}\right)\right] - F\left(\frac{f_{cm} - m_\eta}{\sigma_\eta}\right) \tag{20}$$

Eventually, Equation (20) can be written the following Equation (21):

$$\mu_{Ci+1}(f_{cm}) = F\left(\frac{f_{cm} - m_\xi}{\sigma_\xi}\right) - F\left(\frac{f_{cm} - m_\eta}{\sigma_\eta}\right) \tag{21}$$

where $F(z)$ is a Laplace function given by Equation (22):

$$F(z) = \frac{1}{\sqrt{2\pi}} \int_{-\infty}^{z} exp(-0.5z^2)dz \tag{22}$$

Having calculated membership functions for different concrete classes (considered concrete class and adjacent concrete classes) and mean compressive strength for the sample of size n, one may determine the degree of concrete belonging to a specific concrete class. Based on the $\mu_K(f_{cm})$ value, the considered concrete batch can be recognized as a specific concrete class. Such recognition might be more or less accurate, depending on the economic requirements and the impact of classification on the quality assessment of the concrete produced.

2.3. Example of Application of the Statistical-Fuzzy Conformity Criteria for Concrete of Class C20/25

The procedure of statistical-fuzzy conformity control (Section 2.2) was carried out for concrete of class C20/25. By generating 100,000 groups of random numbers of size $n = 3$, consistent with normal distribution, marginal distribution density functions and fuzzy membership functions were estimated for concrete class C25/30 and every second adjacent concrete class, C16/20 and C25/30.

The analysis was carried out for concrete of class C20/25 with the following resulting parameters of marginal distribution of random variable x $\rightarrow$N(m_x,σ_x), i.e., the point of division for concrete of classes C16/20 and C20/25, $m_x = 26.5$ MPa, and $\sigma_x = 4.48$ MPa, respectively. The parameters of marginal distribution of random variable y $\rightarrow$ N(m_y,σ_y), the point of division for concrete of classes C20/25 and C25/30, were estimated as $m_y = 39.8$ MPa and $\sigma_y = 5.46$ MPa, respectively (Figure 4).

The density functions overlap, indicating that the number of classes proposed by the standard is too high, which makes it difficult to classify a concrete batch to a specific class. Irrespective of mean compressive strength value, the membership function graph (green curve) for the considered concrete class C20/25 does not reach value of 1.0, which allows for concluding that the recommended concrete class division is too dense. The above analysis was carried out for concrete class C20/25 and every second adjacent concrete class (Figure 5).

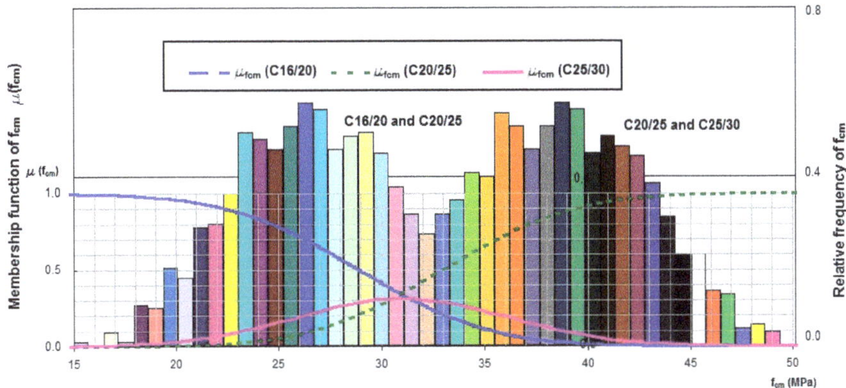

Figure 4. Marginal distribution and membership functions for C20/25 and every adjacent concrete class: C16/20 and C25/30.

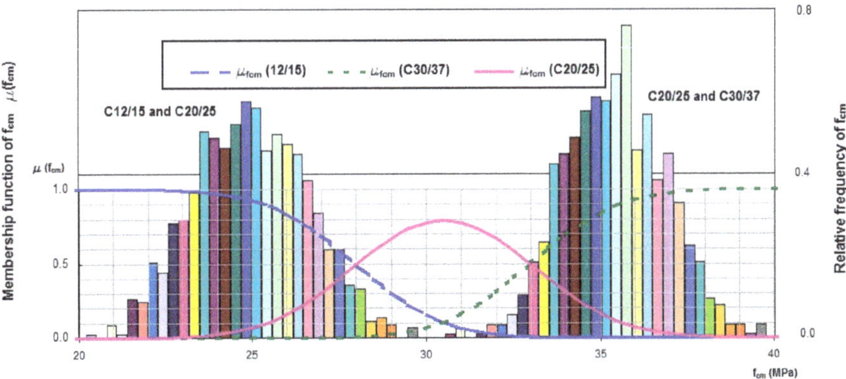

Figure 5. Marginal distribution and membership functions for C20/25 and every second adjacent concrete class: C12/15 and C30/37.

Marginal distribution graphs for the considered concrete class C20/25 and every second adjacent class, C12/15 and C30/37, also overlap, but the maximum abscissa value of the membership function for the considered concrete class C20/25 amounts to 0.83. By performing subsequent calculations, membership functions for separate concrete classes would be obtained, marginal distributions would not overlap, and the membership function graph (green curve) for the concrete class C20/25, for specified values of mean compressive strength, would reach the value of 1.0.

In accordance with Figure 5, an assessment of a concrete batch was carried out for the statistical-fuzzy conformity criterion developed following the algorithm described above. The concrete batch was assessed based on a sample of size $n = 3$ of concrete class C20/25. Mean compressive strength is 30.5 MPa. On the basis of the membership functions determined (Figure 5), it can be concluded that the concrete batches for which mean compressive strength from the sample test amounts to 30.5 MPa can be classified as class C20/25 with a 0.8 degree of certainty. Concrete batches of mean compressive strength from interval (28.0; 30.8) MPa can be classified as class C20/25 or C12/15 with a degree of certainty from 0.5 to 0.8, respectively. Concrete batches of mean compressive strength from interval (30.8; 33.0) MPa can be classified as class C30/37 with a degree of certainty from 0.8 to 0.5.

3. Results and Discussion

The applied statistical-fuzzy methods of concrete classification showed that the concrete classification recommended by the standards includes too many concrete classes of overlapping density distributions (see Figures 4 and 5). Irrespective of mean compressive strength value, the membership function graph plotted for the considered concrete class C16/20 does not reach value of 1.0, which allows for concluding that the recommended concrete class division is too dense. The standards recommended by EN 206 [32] are "too vague" and may lead to understating or overstating concrete class and to concealing the results of understated compressive strength.

Furthermore, when applying the concept of concrete family, standard conformity criteria can conceal the results of understated compressive strength. With the use of the concept of concrete family, small concrete production plants are able to assess the conformity of a larger number of concrete mixes with the benefit for both manufacturer and recipient. Theoretically, the manufacturer can improve the quality of concrete and detect changes in concrete production more quickly, so that the recipient could be informed of the quality of the finished product. What raises doubts is combining the results for different concrete classes of the same family. The results are combined and tested collectively, and as a result, "bad" results (low compressive strength) can be masked by "good" results (high compressive strength). With regard to the concrete family, it is necessary to apply: a single cement type, a single concrete class, aggregate of similar characteristics (granulation, mineralogical composition, geological origin), concretes with or without additions, all consistencies, concretes of limited range of compressive strength.

EN 206 [32] standard does not specify the range of compressive strength. When considering the concrete family composed of four concrete classes: C8/10, C20/25, C25/30 and C30/37, it may be concluded that combining all four classes, i.e., a wide range of classes, is not an appropriate practice. Low values of compressive strength are masked by high values of compressive strength of referential concrete (Figure 6) through transformation and application of the proportionality principle-based method in compliance with CEN CR 13,901 report [45].

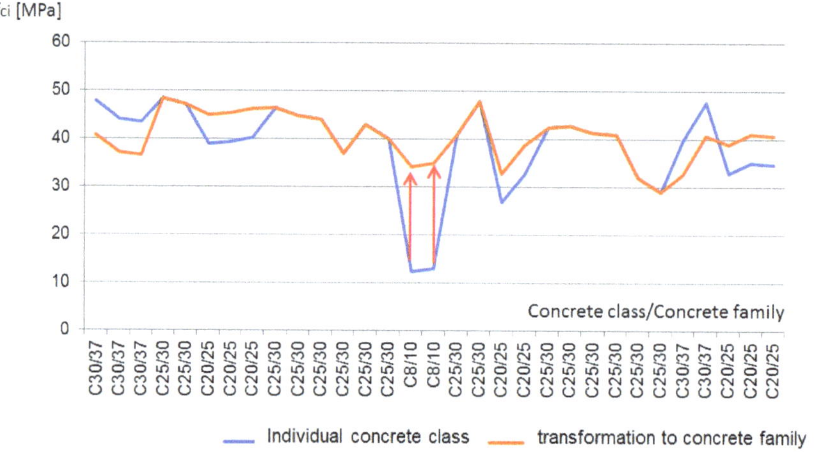

Figure 6. Real results of compressive strength assessment and values for particular classes transformed in relation to the referential concrete in the concrete family.

Therefore, concretes of a limited range of compressive strength should be applied with regard to the concrete family. In accordance with the statistical-fuzzy analysis carried out, it is recommended to limit the range of compressive strength to three adjacent classes so as to ensure the effectiveness of the conformity control performed for the concrete family.

The statistical-fuzzy methods proposed can be applied in cases of non-compliance with the concrete class intended by the design. The decision of either demolition or reinforcement of a structure may be preceded by the fuzzy concrete classification analysis, whose results may impact both the designer's and investor's decisions related to the state of the structure analysed [46].

Taking into account the compressive strength measurement uncertainty broadens the range of acceptability of assessment results obtained. It is in the interest of each party of the construction process for a reliable assessment of concrete conformity to be performed.

In the analysis carried out during conformity assessment, it is important to consider that each result obtained is encumbered with uncertainty, thus disregarding uncertainty completely is not an appropriate approach. The only case when it is possible to disregard measurement uncertainty in assessment is a situation when all of the results obtained meet the conformity criteria. In other instances, i.e., when the product is disqualified on the basis of the results obtained without measurement uncertainty taken into account, such an approach is unadvisable, as it may lead to a falsely negative result for a product that, in fact, meets the standard conformity criteria.

4. Conclusions

Taking into account the compressive strength measurement uncertainty broadens the range of acceptability of assessment results obtained. It is in the interest of each party in the construction process that a reliable assessment of concrete conformity be performed.

In the analysis carried out during conformity assessment, it is important to consider that each result obtained is encumbered with uncertainty, and thus, disregarding uncertainty completely is not an appropriate approach. The only case when it is possible to disregard measurement uncertainty in assessment is a situation when all of the results obtained meet the conformity criteria. In other instances, i.e., when the product is disqualified on the basis of the results obtained without measurement uncertainty being taken into account, such an approach is unadvisable, as it may lead to a falsely negative result for a product that, in fact, meets the standard conformity criteria.

Standard conformity criteria and procedures for assessing the compressive strength of concrete and verifying the concrete's compliance with the requirements set for designed concrete classes frequently lead to inappropriate production-related decisions and strategies. Doubts regarding the assessment and classification of the compressive strength of concrete are, therefore, the reason for seeking new methods based on statistical-fuzzy procedures supporting the quality control of the concrete produced. Statistical-fuzzy methods are, therefore, proposed as an alternative in the quality assessment of ready-mixed concrete:

- The proposed concept of quality assessment allows for minimising the risk of wrong classification of a concrete batch, i.e., overstating or understating the concrete class.
- Employing non-standard methods of conformity control of concrete compressive strength may become a useful tool in the investment-related (technology-related) decision-making process.
- The analyses carried out reveal that the statistical-fuzzy conformity control can play an arbitrary role in the quality assessment of the concrete produced.
- Statistical-fuzzy and fuzzy methods allow to take into account the opposing requirements of safety, quality and economy. Taking these requirements into consideration is made possible by determining a degree of membership lower than 1 for the considered concrete class.
- The alternative method of concrete quality assessment is easy to apply; however, it requires a complex calculation procedure, which significantly limits its universal use in the production process. Widespread application of this method would require implementing specialised utility software developed based on specific algorithms.
- The advantages of the statistical-fuzzy approach are particularly observable when employing the concept of concrete families. It allows to minimise the uncertainty connected to the transformation relation between the results for compressive strength of each concrete family member.

- Based on this approach, a risk matrix may be developed for a construction facility in order to verify the assigned reliability class specified in the construction design.
- Statistical-fuzzy methods are fully compatible with the concept of sustainable construction. Accidental understating of the concrete class results in the rejection of a concrete batch by the recipient. An unsuitable concrete mix is then considered as construction waste, which contradicts the principles of rational use of construction materials and mineral resources.

Author Contributions: Conceptualization, I.S., W.K., J.Z. and A.L.; methodology, I.S. and A.L.; software, I.S.; validation, I.S.; formal analysis, I.S. and A.L.; investigation, I.S., W.K. and J.Z.; resources, I.S., W.K. and J.Z.; data curation, I.S., W.K. and J.Z.; writing—original draft preparation, I.S. and W.K.; writing—review and editing, I.S., W.K., A.L., D.B. and L.B.; visualization, I.S. and W.K.; supervision, I.S.; project administration, I.S. and A.L. All authors have read and agreed to the published version of the manuscript.

Funding: This research received no external funding.

Conflicts of Interest: The authors declare no conflict of interest.

References

1. Krzemiński, M. Optimization of Work Schedules Executed Using the Flow Shop Model, Assuming MultiTasking Performed by Work Crews. *Arch. Civ. Eng.* **2017**, *63*, 3–19. [CrossRef]
2. Leśniak, A.; Zima, K. Cost Calculation of Construction Projects including Sustainability Factors Using the Case Based Reasoning (CBR) Method. *Sustainability* **2020**, *10*, 1608. [CrossRef]
3. Sztubecka, M.; Skiba, M.; Mrówczyńska, M.; Bazan-Krzywoszańska, A. An Innovative Decision Support System to Improve the Energy Efficiency of Buildings in Urban Areas. *Remote Sens.* **2020**, *12*, 259. [CrossRef]
4. Nazarko, L. Technology Assessment in Construction Sector as a Strategy towards Sustainability. *Procedia Eng.* **2015**, *122*, 290–295. [CrossRef]
5. Lee, D.; Lee, D.; Lee, M.; Kim, M.; Kim, T. Analytic Hierarchy Process-Based Construction Material Selection for Performance Improvement of Building Construction: The Case of a Concrete System Form. *Materials* **2020**, *13*, 1738. [CrossRef]
6. Project Management Institute. *A Guide to the Project Management Body of Knowledge (PMBOK Guaid)*; Project Management Institute: Philadelphia, PA, USA, 2000.
7. Arditi, D.; Gunaydin, H.M. Total Quality Management in the Construction Process. *Int. J. Proj. Manag.* **1997**, *15*, 235–243. [CrossRef]
8. Kowalski, D. Quality Assurance of Works and Materials in Construction Projects. *Inżynieria Morska I Geotech.* **2013**, *5*, 362–365.
9. Gursel, A.P.; Masanet, E.; Horvath, A.; Stadel, A. Life-Cycle Inventory Analysis of Concrete Production: A Critical Review. *Cem. Concr. Compos.* **2014**, *51*, 38–48. [CrossRef]
10. Poon, C.S.; Yu, A.T.W.; Jaillon, L. Reducing Building Waste at Construction Sites in Hong Kong. *Constr. Manag. Econ.* **2004**, *22*, 461–470. [CrossRef]
11. Hilburg, J. Concrete Production Produces Eight Percent of the World's Carbon Dioxide Emissions. Architecture, International, News, Sustainability. Available online: https://www.archpaper.com/2019/01/concrete-production-eight-percent-co2-emissions/ (accessed on 20 August 2020).
12. Ready-Mixed Concrete Industry Statistics. Available online: https://mediatheque.snpb.org/userfiles/file/Statistics%20Bound%20Volume%2030_08_2019%20-%20R4.pdf (accessed on 16 July 2020).
13. Gorzelanczyk, T.; Pachnicz, M.; Rozanski, A.; Schabowicz, K. Identification of Microstructural Anisotropy of Cellulose Cement Boards by Means of Nanoindentation. *Constr. Build. Mater.* **2020**, *57*, 119515. [CrossRef]
14. Chen, S.T.T.; Wang, W.C.; Wang, H.Y. Mechanical Properties and Ultrasonic Velocity of Lightweight Aggregate Concrete Containing Mineral Powder Materials. *Constr. Build. Mater.* **2020**, *258*, 119550. [CrossRef]
15. Widianto, A.K.; Wiranegara, J.L.; Hardjito, D. Consistency of Fly Ash Quality for Making High Volume Fly Ash Concrete. *J. Teknol.* **2017**, *79*, 13–20.
16. Mazur, W.; Drobiec, Ł.; Jasiński, R. Research of Light Concrete Precast Lintels. *Procedia Eng.* **2016**, *161*, 611–617. [CrossRef]
17. Debieb, F.; Courard, L.; Kenai, S.; Degeimbre, R. Mechanical and Durability Properties of Concrete Using contaminated Recycled Aggregates. *Cem. Concr. Compos.* **2010**, *32*, 421–426. [CrossRef]

18. Dobiszewska, M.; Beycioglu, A. Investigating the Influence of Waste Basalt Powder on Selected Properties of Cement Paste and Mortar. *Mater. Sci. Eng.* **2017**, *245*, 022027. [CrossRef]

19. Sonebi, M.; Ammar, Y.; Diederich, P. Sustainability of Cement, Concrete and Cement Replacement Materials in Construction. In *Sustainability of Construction Materials*; Woodhead Publishing: Cambridge, UK, 2016; pp. 371–396.

20. Beycioglu, A.; Basyigit, C. Rule-Based Mamdani-Type Fuzzy Logic Approach to Estimate Compressive Strength of Lightweight Pumice Concrete. *Acta Phys. Pol.* **2015**, *128*, 424. [CrossRef]

21. Skrzypczak, I.; Kokoszka, W.; Buda-Ożóg, L.; Janusz Kogut, J.; Słowik, M. Environmental Aspects and Renewable Energy Sources in the Production of Construction Aggregate. In *E3S Web of Conferences*; EDP Sciences: Ulis, France, 2017; p. 00160.

22. Al Ajmani, H.; Suleiman, F.; Abuzayed, I.; Tamini, A. Evaluation of Concrete Strength Made with Recycled Aggregate. *Buildings* **2019**, *9*, 56. [CrossRef]

23. Czarnecki, L.; Hager, T.; Tracz, T. Material Problems in Civil Engineering: Ideas-Driving Forces-Research Arena. *Procedia Eng.* **2015**, *108*, 3–12. [CrossRef]

24. Dobiszewska, M. Waste Materials Used in Making Mortar and Concrete. *J. Mater. Educ.* **2017**, *39*, 133–156.

25. Patil, S.V.; Rao, K.B.; Nayak, G. Quality Improvement of Recycled Aggregate Concrete using six sigma DMAIC methodology. *Int. J. Math. Eng. Manag. Sci.* **2020**, *5*, 1409–1419.

26. Wang, D.; Liu, G.; Li, K.; Wang, T.; Shrestha, A.; Martek, I.; Tao, X. Layout Optimization Model for the Production Planning of Precast Concrete Building Components. *Sustainability* **2018**, *10*, 1807. [CrossRef]

27. Domagała, L. Durability of Structural Lightweight Concrete with Sintered Fly Ash Aggregate. *Materials* **2020**, *13*, 4565. [CrossRef] [PubMed]

28. Słoński, M.; Schabowicz, K.; Krawczyk, E. Detection of Flaws in Concrete Using Ultrasonic Tomography and Convolutional Neural Networks. *Materials* **2020**, *13*, 1557. [CrossRef] [PubMed]

29. Schabowicz, K. Non-Destructive Testing of Materials in Civil Engineering. *Materials* **2019**, *12*, 3237. [CrossRef] [PubMed]

30. Abousnina, R.; Manalo, A.; Ferdous, W.; Lokuge, W.; Benabed, B.; Al-Jabri, K.S. Characteristics, Strength Development and Microstructure of Cement Mortar Containing Oil-ContaminatedS. *Constr. Build. Mater.* **2020**, *252*, 119155. [CrossRef]

31. Ferdous, W.; Manalo, A.; Khennane, A.; Kayali, O. Geopolymer Concrete-Filled Pultruded Composite Beams–Concrete Mix Design and Application. *Cem. Concr. Compos.* **2015**, *58*, 1–13. [CrossRef]

32. *EN 206: 2016 Concrete. Specification, Performance, Production and Conformity*; Polish Committee for Standardization: Warsaw, Poland, 2014.

33. Skrzypczak, I.; Kokoszka, W. Sustainable Methods for Assessing Conformity of Concrete Strength. *J. Civ. Eng. Environ. Archit.* **2015**, *62*, 403–408. (In Polish)

34. Taerwe, L. Evaluation of Compound Compliance Criteria for Concrete Strength. *Mater. Struct.* **1988**, *21*, 13–20. [CrossRef]

35. Skrzypczak, I.; Woliński, S. Influence of Distribution Type on the Probability of Acceptance of Concrete Strength. *Arch. Civ. Eng.* **2007**, *53*, 479–495.

36. Casspeele, R.; Taerwe, L. Conformity Control of Concrete Based on the "Concrete Family" Concept. *Beton Stahlbetonbau* **2008**, *103*, 50–56. [CrossRef]

37. Skrzypczak, I.; Buda-Ożóg, L.; Pytlowany, T. Fuzzy Method of Conformity Control for Compressive Strength of Concrete on the Basis of Computational Numerical Analysis. *Meccanica* **2016**, *51*, 383–389. [CrossRef]

38. Holicky, M.; Vorlicek, M. Fractile Estimation and Sampling Inspection in Building. *Acta Polytech.* **1992**, *32*, 87–96.

39. Zadeh, L.A. Fuzzy Sets and Information Granularity. *Adv. Fuzzy Set Theory Appl.* **1979**, *11*, 3–18.

40. ISO/IEC. *17025:2017 General Requirements for the Competence of Testing and Calibration Laboratories*; Technical Committee: ISO/CASCO Committee on Conformity Assessment: Geneva, Switzerland, 2017.

41. Caltarino, J.M.R. Statistical Criteria for Acceptance of Materials Performance of Concrete Standards. In Proceedings of the ENV 206:1993 and EN 206, 12th ERMCO Congress Lisbon, Lisbon, Portugal, 23–26 June 1998; Volume 1.

42. Li, H.; Yen, V.C. *Fuzzy Sets and Fuzzy Decision-Making*; CRC Press: Boca Raton, FL, USA, 1995.

43. Woliński, S. *Statistical and Fuzzy Compatibility Criteria of Compressive Strength of Concrete*; Cracow University of Technology: Cracow, Poland, 1999.

44. Brandt, S. *Data Analysis*; PWN: Warsaw, Poland, 1998. (In Polish)
45. CEN. *CR 13901:2000 The Use of the Concept of Concrete Families for the Production and Conformity Control of Concrete*; CEN Technical Report; National Standards Authority of Ireland: Dublin, Ireland, 2000.
46. Neshat, M.; Adeli, A.; Sepidnam, G.; Sargolzaei, M. Comparative Study on Fuzzy Inference System for Prediction of Concrete Compressive Strength. *Int. J. Phys. Sci.* **2012**, *7*, 440–456.

Publisher's Note: MDPI stays neutral with regard to jurisdictional claims in published maps and institutional affiliations.

 © 2020 by the authors. Licensee MDPI, Basel, Switzerland. This article is an open access article distributed under the terms and conditions of the Creative Commons Attribution (CC BY) license (http://creativecommons.org/licenses/by/4.0/).

 *materials*

Article

^{1}H NMR Spin-Lattice Relaxometry of Cement Pastes with Polycarboxylate Superplasticizers

Min Pang [1,2], Zhenping Sun [1,2,*], Qi Li [2] and Yanliang Ji [1,2]

[1] Key Laboratory of Advanced Civil Engineering Materials, Ministry of Education, Tongji University, Shanghai 201804, China; pangmin@tongji.edu.cn (M.P.); yanliangji@tongji.edu.cn (Y.J.)

[2] School of Materials Science and Engineering, Tongji University, Shanghai 201804, China; liqi_article@126.com

* Correspondence: szhp@tongji.edu.cn

Received: 30 October 2020; Accepted: 8 December 2020; Published: 10 December 2020

Abstract: ^{1}H spin-lattice relaxometry (T_1, longitudinal) of cement pastes with 0 to 0.18 wt % polycarboxylate superplasticizers (PCEs) at intervals of 0.06 wt % from 10 min to 1210 min was investigated. Results showed that the main peak in T_1 relaxometry of cement pastes was shorter and lower along with the hydration times. PCEs delayed and lowered this main peak in T_1 relaxometry of cement pastes at 10 min, 605 min and 1210 min, which was highly correlated to its dosages. In contrast, PCEs increased the total signal intensity of T_1 of cement pastes at these three times, which still correlated to its dosages. Both changes of the main peak in T_1 relaxometry and the total signal intensity of T_1 revealed interferences on evaporable water during cement hydration by dispersion mechanisms of PCEs. The time-dependent evolution of weighted average T_1 of cement pastes with different PCEs between 10 min and 1210 min was found regular to the four-stage hydration mechanism of tricalcium silicate.

Keywords: nuclear magnetic resonance; spin-lattice relaxometry; proton; hydration kinetics; superplasticizer

---

1. Introduction

Since Bloch [1] and Purcell [2] awarded the Nobel Prize for successful monitoring the magnetic situation of protons in water and parafilm using low-field nuclear magnetic resonance(NMR) instruments, ^{1}H NMR (proton NMR) has been used as an effective technology in cement and concrete research for a long while. ^{1}H NMR could detect the nuclear spin-lattice relaxometry (T_1, longitudinal) or spin–spin relaxometry (T_2, transverse) of ^{1}H nuclei. T_1 and T_2 depend on fluctuations in magnetic dipole–dipole interactions caused by the relative motion of pairs of spins [3]. Relaxometry (relaxation time) may extend due to the relative motion of spins in a fluid as water or oil, theoretical studies have been done by Korb [4,5] previously, and recent studies on cement hydrates by McDonald [6–9].

There are many studies about water transport and distributions in cement-based materials based on T_2 relaxometry [10–13]. T_2 relaxometry has also been used for detecting porosity of oil-well cement pastes [14], C_3S hydrated pastes [15,16], white cement mortars [17], lime plaster–brick systems [18], complex wall materials [19], carbonated cement pastes [20]. Besides macropores in cement-based materials, nanopores (or gel pores) in calcium silicate hydrate (C-S-H) gel could also be demonstrated by T_2 relaxometry [21]. Given that C-S-H gel being one poorly crystalline, which is quasi-amorphous and contains nanopores with water [22], some attempts have been made with water in C-S-H gel pores [23,24]. There are some other applications of T_2 relaxometry in alkali-activated binders [25], lime concrete [26], cement pastes with superabsorbent polymers [27], cellulose ethers [28], woods [29], MgO-based cement [30]. However, the paramagnetic species (mainly Fe^{3+}) in cement pastes could influence T_2 relaxometry [6]. It has been reported that T_2 relaxometry would result in a reduced volume because of its access to the paramagnetic species in a single channel based on the crystalline structure of ettringite [31].

Compared to T_2 relaxometry, there are few applications of T_1 relaxometry in cement-based materials. It has been found that T_1 relaxometry of white cement pastes with w/c = 0.3, 0.4, 0.6, 0.7 could show a significant increase after drying at 105 °C while the freeze–thaw cycling of 25 times could not make obvious changes [32]. Many theoretical studies of T_1 relaxometry in cement-based materials have been done by one joint research group in Yugoslavia and Canada from 1978 to 1996. They have measured the T_1 relaxometry of absorbed water in cement pastes and C_3S pastes during the hardening process [33]. A method has been proposed to determine the specific surface of cement hydrates based on T_1 relaxometry ($1/T_1$) [34]. T_1 relaxometry of Portland cement and white cement, as well as white cement with 5 wt % MSF salts (sulfonated-melamine-formaldehyde), has been monitored, respectively [35–37]. They have also compared signal differences between the synthesized white cement at 40 MHz and at 200 MHz [38]. The fractal geometry of C-S-H gel, quantitative changes of water and $Ca(OH)_2$ in white cement pastes have been monitored successively [39,40]. T_1 relaxometry of calcium aluminate cement, T_1 relaxometry of self-stressed expansive cement, T_1-weighted line shape of white cement pastes have been explored by some coworkers with this group afterward [41–43]. Moreover, some researchers have tried to correlate signals of T_1 relaxometry to microcracks in cement pastes [44]. Other researchers have studied the dynamics of liquid water in pores of cement-based materials based on T_1 relaxometry [45].

As it has been found that 5 wt % MSF salts could change T_1 relaxometry of white cement pastes, which have acted as the superplasticizers [37], could polycarboxylate superplasticizers (PCEs) change T_1 relaxometry of Portland cement pastes? In this research, effects of the synthesized PCEs on the hydration process of Portland cement pastes were initially explored by ^{1}H NMR technique based on T_1 relaxometry and its total signal intensity at three selected time points of 10 min, 605 min, 1210 min, as well as the evolution process of weighted average T_1 from 10 min to 1210 min. We are certain that this study would enrich the applications of T_1 relaxometry in cement-based materials. PCEs are popular chemicals for ultra-high performance concrete (UHPC). Hence, this study could also provide useful data to enrich understandings of the effects of PCEs to cement pastes in UHPC.

2. Materials and Methods

2.1. Materials

The P.II.52.5 cement used in this experiment was purchased from Jiangnan Onoda Cement Co. LTD., Jiangsu, China. The fineness of cement was 315 m^2/kg, and its chemical composition as supplied by its manufacturer is shown in Table 1. The PCEs used in this experiment were synthesized from these materials. Methyl allyl polyethenoxy ether (TPEG) was purchased from Yangzi Aoke Chemical Company(Nanjing, China). Maleic anhydride (MA, analytical-grade), acrylic acid methyl ester (AAME, analytical-grade) and acrylic acid (AA, analytical-grade) were purchased from Shanghai Guoyao Chemical Company (Shanghai, China).

Table 1. Chemical compositions of cement (wt %).

	SiO_2	CaO	Al_2O_3	Fe_2O_3	MgO	Na_2O	K_2O	TiO_2	SO_3	Loss on Ignition
Cement	21.1	64.3	5.3	2.6	1.7	0.2	0.35	0.3	1.7	2.45

2.2. Preparation of PCEs

PCEs were synthesized via aqueous free radical copolymerization at the molar ratios of TPEG: MA: AAME: AA at 6:9:12:6. The average polymerization degree of TPEG was 35, and the Mn of TPEG was 2400 g/mol, which were provided by its manufacturer. The equipment for the synthesized process of PCEs was as same as the previous investigation [46]. The initiator in the synthesized process was $(NH_4)_2S_2O_8$ (ammonium persulfate, AP), which was 3 wt % to total monomers and purchased from

Shanghai Guoyao Chemical Company (Shanghai, China). The temperature in the synthesis process was 80 °C. The adding time of monomers and the soaking time in the synthesized process were 1 h and 0.5 h, respectively. The synthesized formula is shown in Figure 1.

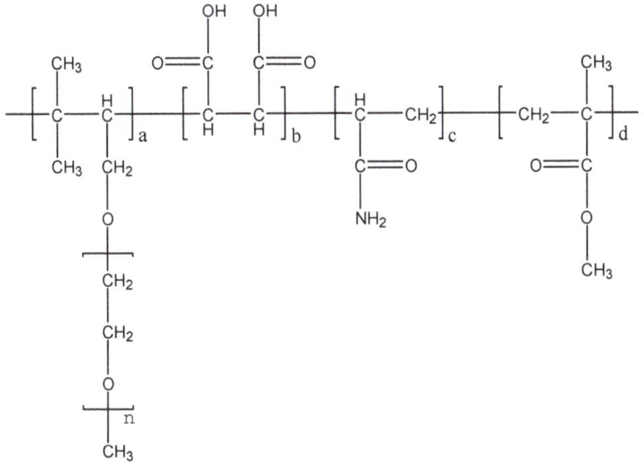

Figure 1. The synthesized formula of polycarboxylate superplasticizers (PCEs).

2.3. NMR Equipment and Theory

The low-field NMR instrument in this experiment was PQ-001 NMR (Niumag Electric Corporation, Shanghai, China). This instrument had a constant magnetic field of 0.49 T, a proton resonance frequency of 21 MHz, a permanent magnet of 32 °C. T_1 relaxometry was detected with the inversion recovery (IR). The IR sequence (π-τ-$\pi/2$-acq) was applied to measure the T_1, where t was the waiting time, acq was the received signal. After the NMR equipment debugged, the prepared samples of different cement pastes with PCEs were poured into an NMR tube with a height between 17 mm to 20 mm and then sealed by a PTFE film.

Commonly, the T_1 relaxometry used in the cement-based materials was based on the fast-exchange model, which was originated from the standard model. Based on the assumption that the molecular exchanging between two phases faster than individual proton relaxation times, the integrated relaxation rate could be described as Equation (1) in [47]. In Equation (1), $T_{1,2}^{bulk}$ and $T_{1,2}^{surf}$ were the proton relaxation times in the bulk and at the surface, f_{surf} and f_{bulk} were the volume fractions of the surface and bulk phases. The correlation between f_{surf} and f_{bulk} could be described as Equation (2). The surface relaxation rate was much higher than the bulk relaxation rate, so Equation (1) was simplified to Equation (3), which had been successfully used [6]. In Equation (3), $\rho_{1,2}$ was the corresponding surface relaxivity, S and V were the pore surface and the pore volume.

$$\frac{1}{T_{1,2}} = \frac{f_{bulk}}{T_{1,2}^{bulk}} + \frac{f_{surf}}{T_{1,2}^{surf}} \tag{1}$$

$$f_{surf} + f_{bulk} = 1 \tag{2}$$

$$\frac{1}{T_{1,2}} = \rho_{1,2}\frac{S}{V} + \frac{1}{T_{1,2}^{bulk}} \approx \rho_{1,2}\frac{S}{V} \tag{3}$$

2.4. Methods

The water-to-cement ratio (w/c) was 0.28, with the detailed proportions listed in Table 2. In cement pastes with PCEs, the spectroscopic observation of the liquid water phase was achieved by exploiting

T_1 relaxometry from the initial time point at 10 min, with intervals of 30 min, to the final time point at 1210 min. The curves of T_1 relaxometry were obtained by fitting the magnetization recovery curves with a log-normal distribution of relaxation times. The setting time of cement pastes with PCEs is shown in Figure 2, which was according to GB/T 1346–2011. It could be seen that the setting time of cement pastes were delayed by PCEs.

Table 2. Mix proportion of cement pastes with PCE (wt %).

Sample	Cement (g)	Water (g)	PCE (g)
S00	1	0.28	0
S06	1	0.28	0.06
S12	1	0.28	0.12
S18	1	0.28	0.18

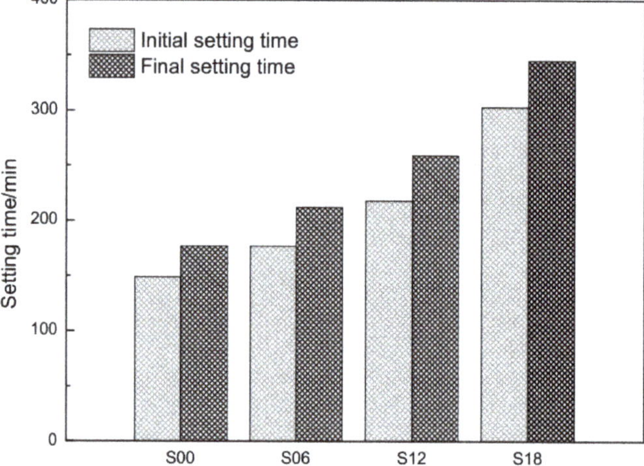

Figure 2. Effects of the synthesized PCEs on the setting time of cement pastes.

3. Results and Discussion

T_1 relaxometry of cement pastes with PCEs at 10 min, 605 min and 1210 min are shown in Figures 3–5, respectively. One can see that there is a main peak in every curve of T_1 relaxometry. To be specific, for T_1 relaxometry at 10 min in Figure 3, the main peak of plain cement pastes (S00) is shorter than those of cement pastes with PCEs. Among T_1 relaxometry of cement pastes with PCEs (S06, S12, S18), the main peak seems to stand at the same time. Furthermore, the main peak of plain cement pastes is obviously lower than those of cement pastes with PCEs. The main peak of S06 (0.06 wt % PCEs) is remarkably lower than those of S12 (0.12 wt % PCEs) and S18(0.18 wt % PCEs), while those of S12 and S18 are nearly the same.

Based on principles of the fast-exchange model [47], any shift of peaks in T_1 relaxometry means changes in the motion trail of a proton (or water). T_1 relaxometry of plain cement pastes (S00) represents the "normal" motion trails of protons in cement pastes at 10 min. Consequently, the prolonged main peaks of cement pastes with PCEs (S06, S12, S18) reveal that the "normal" motion trails of the protons have already been hindered by the dispersed cement grains due to PCEs.

According to [40,43], this main peak represents the quantity of evaporable water in cement pastes. The lowest main peak of plain cement pastes (S00) equals the smallest quantity of evaporable water left in this sample. The higher main peaks of the cement pastes with PCEs (S06, S12, S18) mean that more evaporable water was left in those samples. In other words, the hydration process in plain cement pastes (S00) exhausted more evaporable water than cement pastes with PCEs.

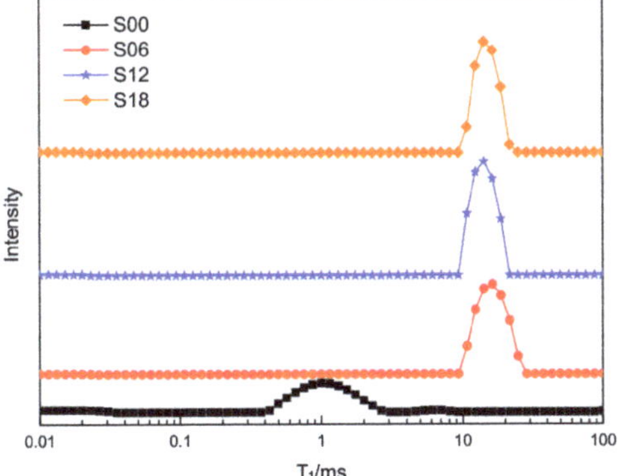

Figure 3. T$_1$ relaxometry of different cement pastes at 10 min.

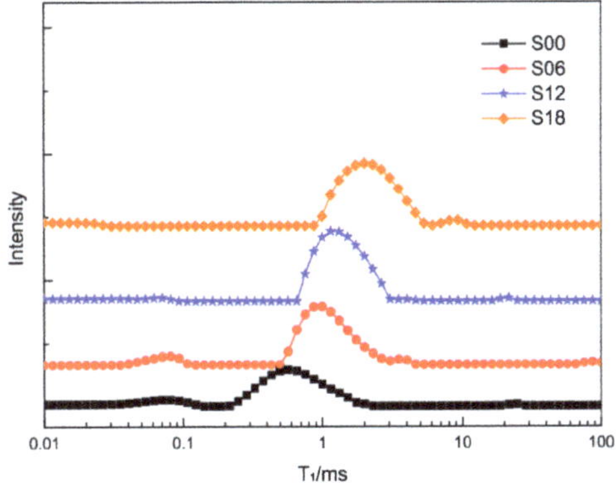

Figure 4. T$_1$ relaxometry of different cement pastes at 605 min.

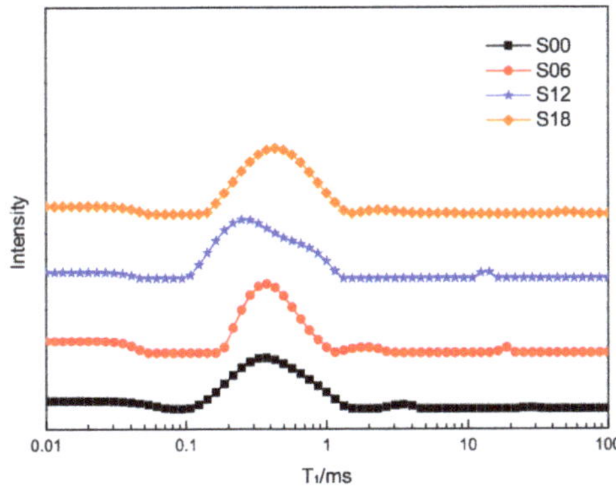

Figure 5. T$_1$ relaxometry of different cement pastes at 1210 min.

It can be seen that the main peaks of cement pastes with PCEs (S06, S12, S18) in Figure 4 are all lower and shorter than those in Figure 3. The deepening hydration process is becoming an extensive consumer of evaporable water. Meanwhile, the conglomerating cement hydrates are becoming strong inhibitors of the "normal" motion trail of protons. All of the main peaks of cement pastes (S00, S06, S12, S18) in Figure 5 are shorter than those in Figure 4. As shown by the setting time of cement pastes with PCEs in Figure 2, the hardened cement pastes at 1210 min have controlled the "normal" motion trail of the protons. Larger area ratios of the main peak in Figure 5 may refer to the bleeding situation in some subregions in hardening cement pastes, which has been revealed by T_1 relaxometry [48].

The total signal intensity of T_1 is proportional to the quantity of evaporable water in cement pastes [49]. It can be found in Figure 6 that the total signal intensity was decreasingly lower from 10 min to 1210 min, which implies the downside of evaporable water in cement pastes. The orders of evaporable water left in cement pastes at three times are opposite to the dosages of PCEs.

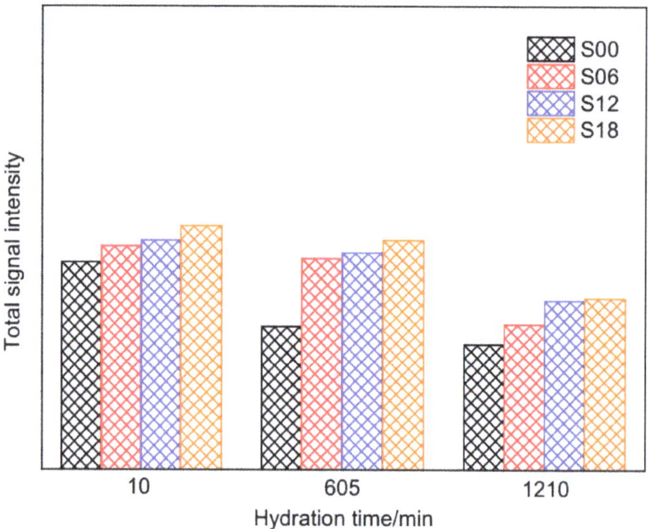

Figure 6. The total signal intensity of cement pastes at different hydration times.

The time-dependent evolution of weighted average T_1 is shown in Figure 7. It is shown that the time evolution of the weighted average T_1 of cement pastes with PCEs has decreased successively. The time-dependent evolution of T_1 has been described according to the four-stage hydration mechanism of tricalcium silicate (3CaO.SiO$_2$, C$_3$S), which occupies 50 wt % to 70 wt % in clinkers [50].

The first stage of C$_3$S hydration is the initial period, which is within 15 min. The initial period in Figure 7 is only 5 min. During the initial period, the original hydration products form gelatinous coatings surrounding cement grains. The entire proton magnetization relaxes with a common T_1, which is due to the fast exchange between water spins in the various environments. The fast exchange may maintain the apparent relaxation homogeneity because of the fluidity of the gelatinous coating and the permeability of the gel–liquid interface [35]. Therefore, the weighted average T_1 of each sample (S00, S06, S12, S18) seems unchanged from 10 min to 15 min in Figure 7. There are still some differences among the initial values of weighted average T_1 at 10 min. The order of these initial values is against the dosages of PCEs. The order of weighted average T_1 at 10 min is the feedback to gelatinous coatings thickness of cement hydrates, which has means that the coatings of hydrates in plain cement pastes were the thickest.

The second stage of C$_3$S hydration is the slow-reaction (dormant) period, which is from 15 min to 120 min. During the dormant period, the weighted average T_1 of each sample (S00, S06, S12, S18) decreased slowly (Figure 7). The fast exchange of water was hindered by the newly formed hydrates coatings. Therefore, the decline of weighted average T_1 occurred, which was shown by comparisons of proton magnetization fractions and relaxation times of H_2O, $Ca(OH)_2$ and C-S-H gel [40]. The third stage of C_3S hydration is the accelerated period, which is from 120 min to 1210 min. During the accelerated period, the weighted average T_1 of each sample (S00, S06, S12, S18) has decreased sharply. The fast exchange of water was heavily hindered by thickening hydrates coatings. The weighted average T_1 of each sample may be regarded as the reverse translation of resistance to the fast exchange of water.

Why does the weighted average T_1 of S18 (most PCEs) have the biggest value during the whole timeline? Based on T_2 relaxometry of cement pastes with PCEs [51], dispersion mechanism of PCEs in OPC [52], the effects of PCEs on C_3A/gypsum [53], the bleeding condition of fresh cement pastes caused by PCEs [54], this answer would come from two aspects: (1) retardation of PCEs on cement hydration; (2) dispersion effects of PCEs on cement grains.

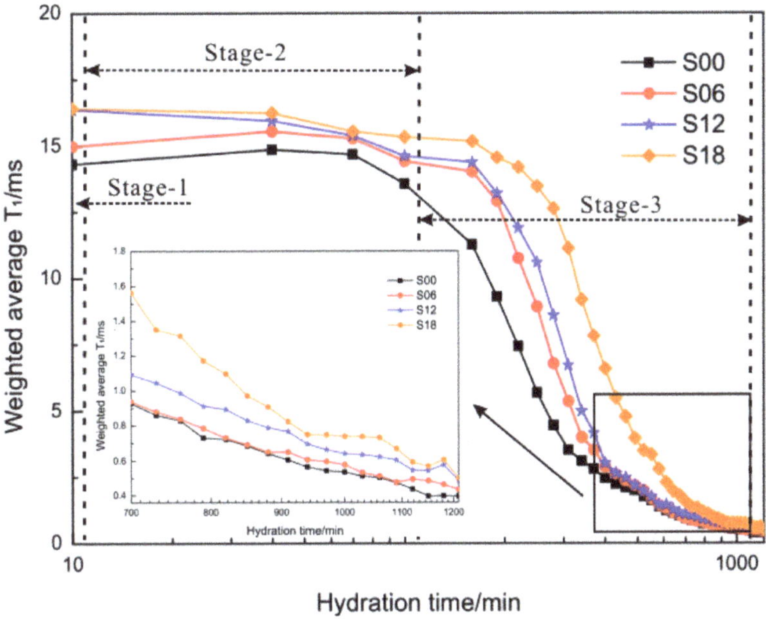

Figure 7. Time-dependent evolution of weighted average T_1 of different cement pastes within 1210 min.

4. Conclusions

(1) The main peak in the T_1 relaxometry of cement pastes at the hydration times of 10 min, 605 min and 1210 min was delayed by polycarboxylate superplasticizers (PCEs). The delayed intensity correlated to the dosage of PCEs. The main peak in T_1 relaxometry of cement pastes became shorter along with the hydration times from 10 min to 1210 min;

(2) The height of the main peak in T_1 relaxometry of cement pastes at these three times was decreased by PCEs. In addition to the larger area ratios of the main peak in T_1 relaxometry of cement pastes at the hydration time of 1210 min due to bleeding, the decreased intensity correlated to the dosage of PCEs;

(3) The main peak in T_1 relaxometry of cement pastes represented the quantity of evaporable water in cement pastes. The delaying situation and the decreasing situation of the main peak was due to the dispersion mechanism and the retardation mechanism of PCEs on cement grains;

(4) The total signal intensity of T_1 of cement pastes at these three times was increased by PCEs. The increasing intensity correlated to the dosage of PCEs. The total signal intensity of T_1 of cement pastes became smaller during the hydration process. As this intensity was proportional

to the quantity of evaporable water, its changes mirrored disturbances of PCEs to situations of evaporable water in the hydration process;

(5)   The time-dependent evolution of weighted average T_1 of cement pastes from 10 min to 1210 min was elevated by PCEs. The elevated intensity correlated to the dosage of PCEs. The curves of weighted average T_1 of cement pastes were well followed by the four-stage hydration mechanism of tricalcium silicate.

Author Contributions: M.P. and Z.S. conceived the project, M.P., Q.L. and Y.J. performed the experiments, M.P., Q.L. and Y.J. carried out the equipment, M.P. wrote the manuscript with the supervision of Z.S. All authors have read and agreed to the published version of the manuscript.

Funding: The authors want to acknowledge the financial support provided by the China National Key R&D Program during the 13th Five-year Plan Period (2016YFC0701004), and Shanghai "Alliance Plan" Project in 2019 (LM201947), the Science and Technology Commission of Shanghai Municipality (19DZ1201404, 19DZ1202702) and the Gansu science and technology-funded project (19YF3GA004). The project is also supported by the Key Laboratory of Advanced Civil Engineering Materials (Tongji University), Ministry of Education.

Conflicts of Interest: The authors declare no conflict of interest.

References

1.   Bloch, F.; Hansen, W.W.; Packard, M.E. Nuclear induction. *Phys. Rev.* **1946**, *69*, 127. [CrossRef]

2.   Purcell, E.M.; Torrey, H.C.; Pound, R.V. Resonance absorption by nuclear magnetic moments in a solid. *Phys. Rev.* **1946**, *69*, 37. [CrossRef]

3.   Faux, D.A.; McDonald, P.J. Nuclear-magnetic-resonance relaxation rates for fluid confined to closed, channel, or planar pores. *Phys. Rev. E.* **2018**, *98*, 063110. [CrossRef]

4.   Korb, J.P. Nuclear magnetic relaxation of liquids in porous media. *New J. Phys.* **2011**, *13*, 035016. [CrossRef]

5.   Monteilhet, L.; Korb, J.P.; Mitchell, J.; McDonald, P.J. Observation of exchange of micropore water in cement pastes by two-dimensional T_2-T_2 nuclear magnetic resonance relaxometry. *Phys. Rev. E* **2006**, *74*, 061404. [CrossRef] [PubMed]

6.   Valori, A.; Rodin, V.; McDonald, P.J. On the interpretation of ^{1}H 2-dimensional NMR relaxation exchange spectra in cements: Is there exchange between pores with two characteristic sizes or Fe^{3+} concentrations? *Cem. Concr. Res.* **2010**, *40*, 1375–1377. [CrossRef]

7.   Gajewicz-Jaromin, A.M.; McDonald, P.J.; Muller, A.C.A.; Scrivener, K.L. Influence of curing temperature on cement paste microstructure measured by ^{1}H NMR relaxometry. *Cem. Concr. Res.* **2019**, *122*, 147–156. [CrossRef]

8.   McDonald, P.J.; Istok, O.; Janota, M.; Gajewicz-Jaromin, A.M.; Faux, D.A. Sorption, anomalous water transport and dynamic porosity in cement paste: A spatially localized ^{1}H NMR relaxation study and a proposed mechanism. *Cem. Concr. Res.* **2020**, *133*, 106045. [CrossRef]

9.   Holthausen, R.S.; McDonald, P.J. On the quantifcation of solid phases in hydrated cement paste by ^{1}H nuclear magnetic resonance relaxometry. *Cem. Concr. Res.* **2020**, *135*, 106095. [CrossRef]

10.   Olaru, A.M.; Blümich, B.; Adams, A. Water transport in cement-in-polymer dispersions at variable temperature studied by magnetic resonance imaging. *Cem. Concr. Res.* **2013**, *44*, 55–68. [CrossRef]

11.   Brocken, H.J.P.; Spiekman, M.E.; Pel, L.; Kopinga, K.; Larbi, J.A. Water extraction out of mortar during brick laying: An NMR study. *Mater. Struct.* **1998**, *31*, 49–57. [CrossRef]

12.   Wyrzykowski, M.; Gajewicz-Jaromin, A.M.; McDonald, P.J.; Dunstan, D.J.; Scrivener, K.L.; Lura, P. Water redistribution-microdiffusion in cement paste under mechanical loading evidenced by ^{1}H NMR. *J. Phys. Chem. C* **2019**, *123*, 16153–16163. [CrossRef]

13.   Wyrzykowski, M.; McDonald, P.J.; Scrivener, K.L.; Lura, P. Water redistribution within the microstructure of cementitious materials due to temperature changes studied with ^{1}H NMR. *J. Phys. Chem. C* **2017**, *121*, 27950–27962. [CrossRef]

14.   Saoût, G.L.; Lécolier, E.; Rivereau, A.; Zanni, H. Micropore size analysis in oil-well cement by proton nuclear relaxation. *Magn. Reson. Imaging* **2005**, *23*, 371–373. [CrossRef]

15.   Plassais, A.; Pomiès, M.P.; Lequeux, N.; Boch, P.; Korb, J.P. Micropore size analysis in hydrated cement paste by NMR. *Magn. Reson. Imaging* **2001**, *19*, 493–495. [CrossRef]

16. Plassais, A.; Pomiès, M.P.; Lequeux, N.; Boch, P.; Korb, J.P.; Petit, D.; Barberon, F. Micropore size analysis by NMR in hydrated cement. *Magn. Reson. Imaging* **2003**, *21*, 369–371. [CrossRef]

17. Zhou, C.S.; Ren, F.Z.; Zeng, Q.; Xiao, L.Z.; Wang, W. Pore-size resolved water vapor adsorption kinetics of white cement mortars as viewed from proton NMR relaxation. *Cem. Concr. Res.* **2018**, *105*, 31–43. [CrossRef]

18. Nunes, C.; Pel, L.; Kunecký, J.; Slížková, Z. The influence of the pore structure on the moisture transport in lime plaster-brick systems as studied by NMR. *Constr. Build. Mater.* **2017**, *142*, 395–409. [CrossRef]

19. Schönfelder, W.; Dietrich, J.; Märten, A.; Koping, K.; Stallmach, F. NMR studies of pore formation and water diffusion in self-hardening cut-off wall materials. *Cem. Concr. Res.* **2007**, *37*, 902–908. [CrossRef]

20. Cano-Barrita, P.F.d.J.; Balcom, B.J.; Castellanos, F. Carbonation front in cement paste detected by T_2 NMR measurements using a low field unilateral magnet. *Mater. Struct.* **2017**, *50*, 150. [CrossRef]

21. Bortolotti, V.; Brizi, L.; Brown, R.J.S.; Fantazzini, P.; Mariani, M. Nano and sub-nano multiscale porosity formation and other features revealed by ^{1}H NMR relaxometry during cement hydration. *Langmuir* **2014**, *30*, 10871–10877. [CrossRef] [PubMed]

22. Youssef, M.; Pellenq, R.J.M.; Yildiz, B. Glassy nature of water in an ultraconfining disordered material: The case of calcium-silicate-hydrate. *J. Am. Chem. Soc.* **2011**, *133*, 2499–2510. [CrossRef] [PubMed]

23. Mcdonald, P.J.; Rodin, V.; Valori, A. Characterisation of intra- and inter-C–S–H gel pore water in white cement based on an analysis of NMR signal amplitudes as a function of water content. *Cem. Concr. Res.* **2010**, *40*, 1656–1663. [CrossRef]

24. Fischer, N.; Haerdtl, R.; McDonald, P.J. Observation of the redistribution of nanoscale water filled porosity in cement based materials during wetting. *Cem. Concr. Res.* **2015**, *40*, 148–155. [CrossRef]

25. Cong, X.Y.; Zhou, W.; Geng, X.R.; Elchalakani, M. Low field NMR relaxation as a probe to study the effect of activators and retarders on the alkali-activated GGBFS setting process. *Cem. Concr. Compos.* **2019**, *104*, 103399. [CrossRef]

26. Faure, P.; Peter, U.; Lesueur, D.; Coussot, P. Water transfers within hemp lime concrete followed by NMR. *Cem. Concr. Res.* **2012**, *42*, 1468–1474. [CrossRef]

27. Yang, J.B.; Sun, Z.P.; Zhao, Y.H.; Ji, Y.L.; Li, B.Y. The water absorption-release of superabsorbent polymers in fresh cement paste: An NMR study. *J. Adv. Concr. Technol.* **2020**, *18*, 139–145. [CrossRef]

28. Patural, L.; Korb, J.P.; Govin, A.; Grosseau, P.; Ruot, B.; Devès, O. Nuclear magnetic relaxation dispersion investigations of water retention mechanism by cellulose ethers in mortars. *Cem. Concr. Res.* **2012**, *42*, 1371–1378. [CrossRef]

29. Cheumani, Y.A.M.; Ndikontar, M.; De Jéso, B.; Sèbe, G. Probing of wood–cement interactions during hydration of wood–cement composites by proton low-field NMR relaxomet. *J. Mater. Sci.* **2011**, *46*, 1167–1175. [CrossRef]

30. Martini, F.; Borsacchi, S.; Geppi, M.; Tonelli, M.; Ridi, F.; Calucci, L. Monitoring the hydration of MgO-based cement and its mixtures with portland cement by ^{1}H NMR relaxometry. *Microporous Mesoporous Mater.* **2018**, *269*, 26–30. [CrossRef]

31. Dalas, F.; Korb, J.P.; Pourchet, S.; Nonat, A.; Rinaldi, D.; Mosquet, M. Surface relaxivity of cement hydrates. *J. Phys. Chem. C* **2014**, *118*, 8387–8396. [CrossRef]

32. Gran, H.C.; Hansen, E.W. Effects of drying and freeze/thaw cycling probed by ^{1}H-NMR. *Cem. Concr. Res.* **1997**, *27*, 1319–1331. [CrossRef]

33. Blinc, R.; Burgar, M.; Lahajnar, G.; Rožmarin, M.; Rutar, V.; Kocuvan, I.; Uršič, J. NMR relaxation study of adsorbed water in cement and C_3S pastes. *J. Am. Ceram. Soc.* **1978**, *61*, 35–37. [CrossRef]

34. Barbič, L.; Kocuvan, I.; Blinc, R.; Lahajnar, G.; Merljak, P.; Zupančič, I. The determination of surface development in cement pastes by nuclear magnetic resonance. *J. Am. Ceram. Soc.* **1982**, *65*, 25–31. [CrossRef]

35. Schreiner, L.J.; Mactavish, J.C.; Miljkovic, L.; Pintar, M.M.; Blinc, R.; Lahajnar, G.; Lasic, D.; Reeves, L.W. NMR line shape-spin-lattice relaxation correlation study of portland cement hydration. *J. Am. Ceram. Soc.* **1985**, *68*, 10–16. [CrossRef]

36. Mactavish, J.C.; Miljkovic, L.; Pintar, M.M.; Blinc, R.; Lahajnar, G. Hydration of white cement by spin grouping NMR. *Cem. Concr. Res.* **1985**, *15*, 367–377. [CrossRef]

37. Miljkovic, L.; Mactavish, J.C.; Jian, J.; Pintar, M.M.; Blinc, R.; Lahajnar, G. NMR study of sluggish hydration of superplasticized white cement. *Cem. Concr. Res.* **1986**, *16*, 864–870. [CrossRef]

38. Lasic, D.D.; Corbett, J.M.; Jian, J.; Mactavish, J.C.; Pintar, M.M.; Blinc, R.; Lahajnar, G. NMR spin grouping in hydrating cement at 200 MHz. *Cem. Concr. Res.* **1988**, *18*, 649–653. [CrossRef]

39. Blinc, R.; Lahajnar, G.; Žumer, S.; Pintar, M.M. NMR study of the time evolution of the fractal geometry of cement gels. *Phys. Rev. B* **1988**, *38*, 2873–2875. [CrossRef]
40. Mactavish, J.C.; Miljkovic, L.; Peemoeller, H.; Corbett, J.M.; Jian, J.; Lasic, D.D.; Blinc, R.; Lahajnar, G.; Milia, F.; Pintar, M.M. Nuclear magnetic resonance study of hydration of synthetic white cement: Continuous quantitative monitoring of water and Ca(OH)$_2$ during hydration. *Adv. Cem. Res.* **1996**, *32*, 155–161. [CrossRef]
41. Kosmač, T.; Lahajnar, G.; Sepe, A. Proton NMR relaxation study of calcium aluminate hydration reactions. *Cem. Concr. Res.* **1993**, *23*, 1–6. [CrossRef]
42. Dolinšek, J.; Apih, T.; Lahajnar, G.; Blinc, R.; Papavassiliou, G.; Pintar, M.M. Two-dimensional nuclear resonance study of a hydrated porous medium: And application to white cement. *J. Appl. Phys.* **1998**, *87*, 3535–3540. [CrossRef]
43. Apih, T.; Lahajnar, G.; Sepe, A.; Blinc, R.; Milia, F.; Cvelbar, R.; Emri, I.; Gusev, B.V.; Titova, L.A. Proton spin–lattice relaxation study of the hydration of self-stressed expansive cement. *Cem. Concr. Res.* **2001**, *31*, 263–269. [CrossRef]
44. Tritt-Goc, J.; Kościelski, S.; Piślewski, N. The hardening of portland cement observed by ^{1}H spin-lattice relaxation and single-point imaging. *Appl. Magn. Reson.* **2000**, *18*, 155–164. [CrossRef]
45. Nestle, N.; Zimmermann, C.; Dakkouri, M.; Kärger, J. Transient high concentrations of chain anions in hydrating cement—Indications from proton spin relaxation measurements. *J. Phys. D Appl. Phys.* **2002**, *35*, 166–171. [CrossRef]
46. Sun, Z.P.; Yang, H.J.; Shui, L.L.; Liu, Y.; Yang, X.; Ji, Y.L.; Hu, K.Y.; Luo, Q. Preparation of polycarboxylate-based grinding aid and its influence on cement properties under laboratory condition. *Constr. Build. Mater.* **2016**, *127*, 363–368. [CrossRef]
47. Korb, J.P. NMR and nuclear spin relaxation of cement and concrete materials. *Curr. Opin. Colloid Interface Sci.* **2009**, *14*, 192–202. [CrossRef]
48. Ji, Y.L.; Pel, L.; Sun, Z.P. The microstructure development during bleeding of cement paste: An NMR study. *Cem. Concr. Res.* **2019**, *125*, 105866. [CrossRef]
49. She, A.M.; Yao, W.; Wei, Y.Q. In-situ monitoring of hydration kinetics of cement pastes by low-field NMR. *J. Wuhan Univ. Technol.-Mater. Sci. Ed.* **2010**, *25*, 692–695. [CrossRef]
50. Odler, I. Chapter-6: Hydration, Setting and Hardening of Portland Cement. In *Lea's Chemistry of Cement and Concrete*, 4th ed.; Hewlett, P., Ed.; Elsevier Science & Technology Books: London, UK, 2004; pp. 241–297.
51. Yu, Y.; Sun, Z.P.; Pang, M.; Yang, P.Q. Probing development of microstructure of early cement paste using ^{1}H low-field NMR. *J. Wuhan Univ. Technol.-Mater. Sci. Ed.* **2013**, *28*, 963–967. [CrossRef]
52. Shui, L.L.; Sun, Z.P.; Yang, H.J.; Yang, X.; Ji, Y.L.; Luo, Q. Experimental evidence for a possible dispersion mechanism of polycarboxylate-type superplasticisers. *Adv. Cem. Res.* **2016**, *28*, 287–297. [CrossRef]
53. Hu, K.Y.; Sun, Z.P. Influence of polycarboxylate Superplasticizers with different functional units on the early hydration of C_3A-gypsum. *Materials* **2019**, *12*, 1132. [CrossRef] [PubMed]
54. Ji, Y.L.; Sun, Z.P.; Yang, J.B.; Pel, L.; Raja, A.J.; Ge, H.S. NMR study on bleeding properties of the fresh cement pastes mixed with polycarboxylate (PCE) superplasticizers. *Constr. Build. Mater.* **2020**, *240*, 117938. [CrossRef]

Publisher's Note: MDPI stays neutral with regard to jurisdictional claims in published maps and institutional affiliations.

© 2020 by the authors. Licensee MDPI, Basel, Switzerland. This article is an open access article distributed under the terms and conditions of the Creative Commons Attribution (CC BY) license (http://creativecommons.org/licenses/by/4.0/).

Article

Comparison of Sandstone Damage Measurements Based on Non-Destructive Testing

Duohao Yin and Qianjun Xu *

State Key Laboratory of Hydroscience and Engineering, Tsinghua University, Beijing 100084, China;
ydh16@mails.tsinghua.edu.cn
* Correspondence: qxu@mail.tsinghua.edu.cn

Received: 21 October 2020; Accepted: 13 November 2020; Published: 16 November 2020

Abstract: Non-destructive testing (NDT) methods are an important means to detect and assess rock damage. To better understand the accuracy of NDT methods for measuring damage in sandstone, this study compared three NDT methods, including ultrasonic testing, electrical impedance spectroscopy (EIS) testing, computed tomography (CT) scan testing, and a destructive test method, elastic modulus testing. Sandstone specimens were subjected to different levels of damage through cyclic loading and different damage variables derived from five different measured parameters—longitudinal wave (P-wave) velocity, first wave amplitude attenuation, resistivity, effective bearing area and the elastic modulus—were compared. The results show that the NDT methods all reflect the damage levels for sandstone accurately. The damage variable derived from the P-wave velocity is more consistent with the other damage variables, and the amplitude attenuation is more sensitive to damage. The damage variable derived from the effective bearing area is smaller than that derived from the other NDT measurement parameters. Resistivity provides a more stable measure of damage, and damage derived from the acoustic parameters is less stable. By developing P-wave velocity-to-resistivity models based on theoretical and empirical relationships, it was found that differences between these two damage parameters can be explained by differences between the mechanisms through which they respond to porosity, since the resistivity reflect pore structure, while the P-wave velocity reflects the extent of the continuous medium within the sandstone.

Keywords: non-destructive testing; P-wave velocity; amplitude attenuation; resistivity; CT scan; sandstone; damage variable

---

1. Introduction

Sandstone is a natural building material and also a common lithology found in the rock surrounding underground engineering sites such as tunnels and underground powerhouses [1]. Sandstone is generally characterized by low strength and high permeability, making it a weak link in the surrounding rock [2]. In underground rock mass engineering, various disturbances will cause damage to the sandstone. Accumulation of sufficient damage will then lead to rock mass failure and affect the stability of the surrounding rock during construction and operation. Rock damage can be measured using a damage test, which provides parameters from which damage variables can be calculated that reflect the damage and can be used to study its evolution. Mechanical parameters such as the elastic modulus and plastic dissipation energy are commonly used for defining damage variables. In addition, damage variables can also be derived from non-destructive testing (NDT) parameters, such as the ultrasonic velocity, wave amplitude attenuation, resistivity and effective bearing area based on computed tomography (CT) scans [3]. These NDT parameters can be obtained faster and more easily than elastic modulus. However, due to the limitations of the measurement methods and their accuracy, these parameters have not been widely used in rock damage mechanics research [4].

This motivates the comparative study of non-destructive measurement parameters that can be used to define the damage variable.

Acoustic testing, including ultrasonic and acoustic emissions testing, are common non-destructive testing (NDT) methods for rocks [5]. Alemu found through testing that the strength of the rock is positively correlated with the wave velocity. When cracks appear in the rock, the strain increases and the wave velocity decreases until the rock is considered to be failed [6]. Heap found that an increase in the peak stress used for the cyclic loading significantly reduced the dynamic elastic modulus of the basalt obtained from ultrasonic tests [7]. As an ultrasonic wave propagates inside a rock, the wave amplitude attenuates at any internal defects. The amplitude attenuation of waves travelling through rock is very sensitive to rock fracture, making it useful for the definition of damage variables [8]. Yim believed the attenuation of ultrasonic waves in rocks is closely related to the properties of internal defects and the porosity of the rock [9]. The simplest measure of ultrasonic wave attenuation is the difference in amplitude between the transmitted wave and the first received wave. Muller suggested that the first wave amplitude attenuation can be taken to reflect rock damage, but the measurement of this parameter is challenging [10].

In addition to ultrasonic parameters, electrical resistivity can be used to define rock damage. Ranade tested the response of resistivity to defects in cement under tension and used resistivity to study the damage evolution [11]. Kahraman found that resistivity is more sensitive to porosity than acoustic velocities, but the direct current (DC) electrical resistivity measurement limits the accuracy of rock resistivity [12]. At present, electrical impedance spectroscopy (EIS) based on alternating current provides the most accurate measurements of rock resistivity. Zisser used the EIS test to study the anisotropy of permeability in dense sandstone [13]. Our previous paper found that the resistivity obtained from the EIS has a high accuracy and can be used to define damage variables in sandstone [14].

From a microstructural perspective, rock damage reflects a change in the effective bearing area. However, current rock measurement techniques struggle to accurately capture the effective bearing area and it can only be measured using expensive meso-testing methods such as CT scanning [15]. Yin observed the damage of granite after ultrasound-assisted rock breaking using CT scan tests [16]. Landis established the damage degree of rock by counting instances of defects extracted from CT images [17].

The aforementioned NDT methods all have their unique advantages. Longitudinal wave (P-wave) velocity and CT scan testing are commonly used in rock damage studies, while resistivity and wave amplitude attenuation are less frequently adopted due to limitations in measurement method and accuracy. Shah compared P-wave velocity and amplitude attenuation that are measured in ultrasonic tests and concluded that they can both be used to reflect rock damage [18]. Yang used two loading methods to cause damage to sandstone, and then defined damage variables using the P-wave velocity and compared them to damage variables that were defined using the elastic modulus [19]. However, systematic comparative studies among different NDT methods are still lacking.

To verify the accuracy of damage variables derived from NDT parameters, this study compared damage variables derived from P-wave velocity, amplitude attenuation, resistivity and the effective bearing area calculated from a CT scan with the damage variables obtained from elastic modulus for sandstone. The comparison considered four areas of difference: differences between the parameters used to define the damage variables; consistency between the different damage variables and the stability of each of the damage tests; differences between microscopic and macroscopic measures of the damage; and differences between the mechanisms that influence measurements made using the acoustic and electrical tests.

2. Materials and Methods

2.1. Sandstone Specimens and Pretreatment

Following ISRM standards and the laboratory's measuring instrument conditions, a representative homogeneous and coarse-grained arkose sandstone core was selected for this study and processed into twelve 75 mm long, cylindrical specimens with diameters of 37.5 mm. The arkose sandstone was taken from Hunan Province, China, and the specimens were drilled from an adjacent location on the same rock block. The two end faces of the specimen were ground to make them parallel to each other and perpendicular to the axis of the cylinder. The physical and mechanical properties of the specimens were measured and the porosity was found to be 14.9%, the dry density was 2313.08 kg/m^3, and the strength was 18.59 MPa. The oxides of the sandstone specimens were found using X-ray fluorescence spectroscopy (XRF), and the minerals composition was obtained by X-ray diffraction (XRD) testing. As shown in Figure 1b, the major mineral compositions of the specimens are quartz and feldspar, and the major clay composition is kaolinite, accounting for 8.49%.

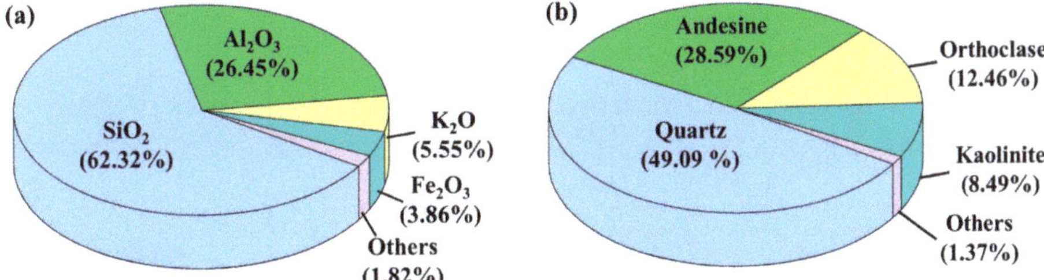

Figure 1. (**a**) Major oxides and (**b**) major minerals composition of rock (wt. %).

Prior to the test, the specimens were placed in a drying oven, heated at 105 °C for 12 h and then cooled naturally, after which the specimens were saturated in a vacuum saturator for 22 h. Uniaxial cyclic loads with different loading paths were applied to each of the twelve specimens so that they were each subjected to different damage, as shown in Figure 2.

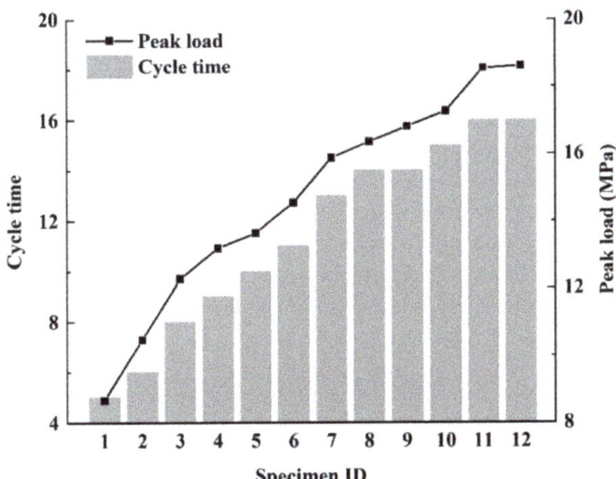

Figure 2. Cyclic loading path for each sandstone specimen.

As shown in Figure 3, stress and strain were measured simultaneously for the specimens during the uniaxial cyclic loading. The portion of the unloading section that accounts for 30%–40% of the strength is assumed to be the straight line section of the stress-strain curve, marked in red in Figure 3, and its secant elastic modulus is taken to be the elastic modulus for the sandstone specimen. The elastic modulus for the first loading cycle in which the peak load exceeds 40% of the strength is taken as the elastic modulus of the specimen before damage, and the elastic modulus of the last loading cycle is taken as the elastic modulus after damage. The equivalent strain assumption allows these to be used to obtain the sandstone damage variable for sandstone derived from the elastic modulus [20]:

$$D = 1 - \frac{E_D}{E_0} \tag{1}$$

where D is the damage variable; E_0 and E_D are the elastic modulus (GPa) for the initial state and for the damaged state, respectively.

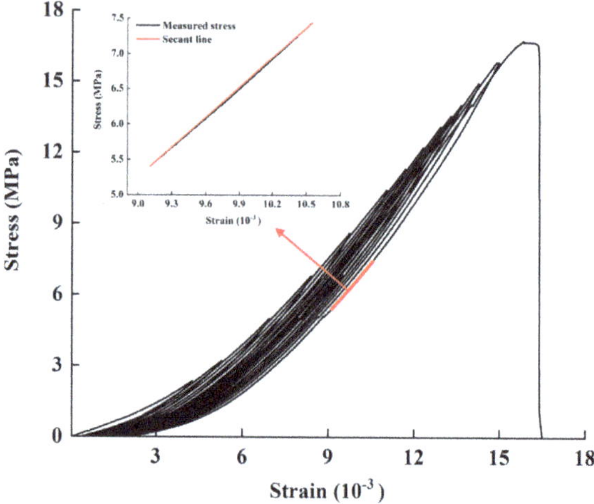

Figure 3. Stress-strain curve measured for the specimen No. 11 during cyclic loading.

2.2. Non-Destructive Testing

2.2.1. Ultrasonic Test

Ultrasonic parameters are largely determined by the internal structure of the rock and reflect the damage properties of the rock. A 33522A ultrasonic tester (Agilent, Santa Rosa, CA, USA) was used to measure the P-wave velocity and first wave amplitude attenuation for the sandstone specimens in the initial state, and as the damage state evolved following the cyclic loading, as shown in Figure 4. After pretreatment, the saturated sandstone specimens were tested with an ultrasonic wave frequency of 50 kHz. The acoustic test was calibrated using a standard aluminum block and an appropriate amount of vaseline was smeared between the ultrasound probe and the specimens to ensure close contact. The specimens were tested in a stress-free state at room temperature, and the ultrasonic probes were fixed to both ends of the specimens.

Figure 4. Ultrasonic testing setup.

The transmitted and received waves can be captured using an oscilloscope as shown in Figure 5, where the voltage represents the wave amplitude. The P-wave velocity and first wave amplitude attenuation can be calculated for the specimens from the captured waveform. The P-wave velocity is equal to the length of the specimen divided by the time interval between the first maximum in the transmitted wave spectrum and the first maximum in the received wave spectrum. The first wave amplitude attenuation is the ratio of the amplitude of the first received wave to the amplitude of the transmitted wave. The damage variable was calculated from the measured parameters using the following equations [21]:

$$D = 1 - \frac{V_D{}^2}{V_0{}^2} \tag{2}$$

$$D = 1 - \frac{F_D}{F_0} \tag{3}$$

where D is the damage variable; V_0 and V_D are the P-wave velocity (m/s) for the initial state and for the evolving damaged state, respectively; F_0 and F_D are the first wave amplitude attenuation for the initial state and for the evolving damaged state, respectively.

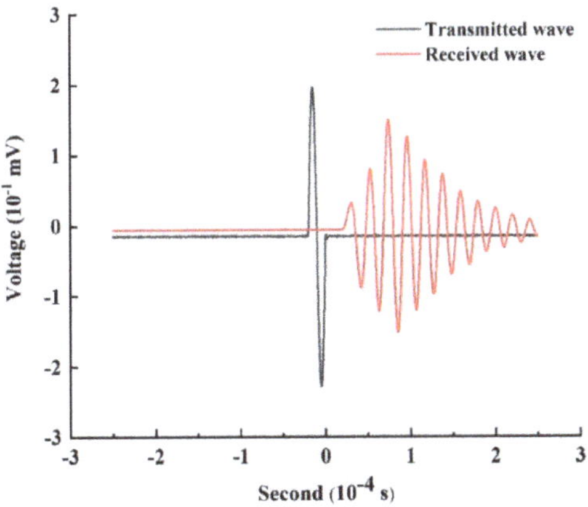

Figure 5. Transmitted and received waves.

2.2.2. EIS Testing

The sandstone specimens were saturated with sodium chloride solution and subjected to EIS testing before and after cyclic loading using an Agilent 4294A precision impedance analyzer, as shown in Figure 6b. The modulus and phase angle of the electrical impedance were measured for frequencies from 100 Hz to 50 MHz. Each specimen was tested three times and the mean of these was recorded. Please refer to our previous paper for a detailed description of the test [14].

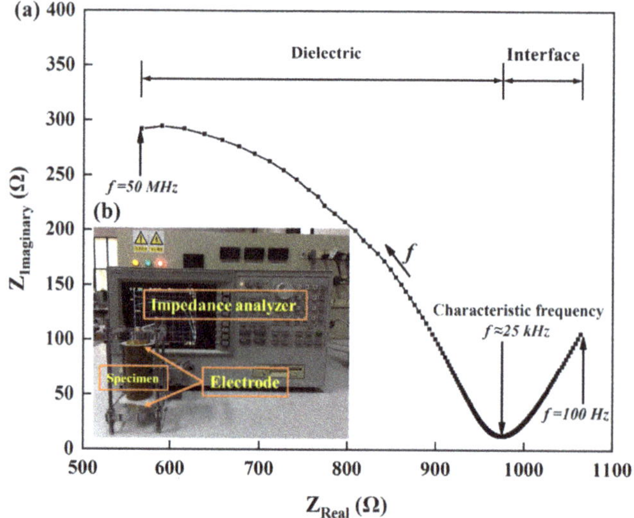

Figure 6. (a) Nyquist diagram and (b) EIS testing.

As shown in Figure 6a, The EIS can be visualized by displaying the modulus and phase angle of the impedance as a Nyquist plot on the complex plane [22]. In the Nyquist diagram, frequency increases from right to left and the EIS is composed of arcs, corresponding to high-frequency impedance, and straight lines, corresponding to low-frequency impedance. For the saturated sandstone specimens, the high-frequency part of the diagram reflects the electrical characteristics of the combined system of rock and pore fluid, and the low-frequency part reflects differences between the electrical properties of the specimen and of the electrode. The frequency corresponding to the intersection of the high- and low-frequency parts of the diagram (around 25 kHz) is called the characteristic frequency and the corresponding resistance is called the characteristic resistance [23,24].

The characteristic resistance was converted to resistivity using the law of resistance, and the difference between the resistivity for the initial and damaged states was used as a measure of the damage of the sandstone specimens. Since current intensity obeys the equivalence assumption in damage mechanics, the damage variable derived from resistivity can be expressed using the following equation [11]:

$$D = 1 - \frac{\rho_D}{\rho_0} \tag{4}$$

where D is the damage variable; ρ_0 and ρ_D (Ω.m) are the resistivity for the initial state and after damage, respectively.

2.2.3. CT Scan Test

Meso-measurement methods such as CT scanning provide an important means of examining rock damage. During the damage process, new pores and cracks are formed in the rock and existing, previously isolated, pore spaces can become connected via the newly created cracks. Materials with different densities attenuate X-ray energy differently and so CT scanning can distinguish pores in the rock [25].

Three representative sandstone specimens (No. 6–8) were selected for CT scan testing before and after damage using a d2 industrial CT system (Diondo, Hattingen, Germany). Each specimen was subjected to a different cyclic load, and the ratio of peak load to strength were 78%, 85% and 88% for specimens 6, 7 and 8, respectively. A schematic illustration of the CT system is shown in Figure 7. Excluding the two ends, the scanning height is 70 mm, and a series of scanned cross-sections of the sandstone specimens were obtained at intervals of 0.5 mm.

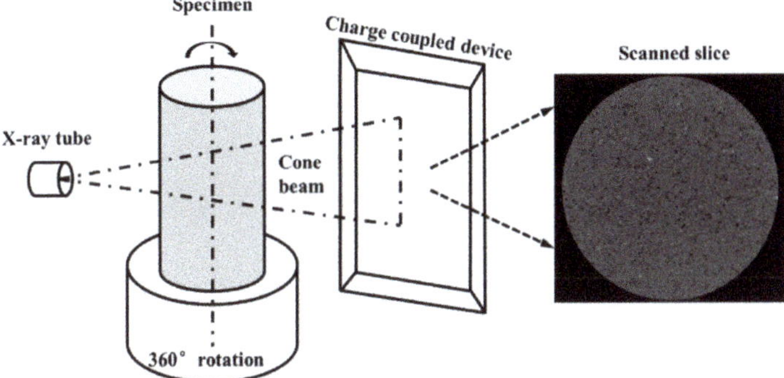

Figure 7. Schematic illustration of the CT system.

CT images that show the pore structure before and after damage can therefore be used to characterize the damage of sandstone specimens. For each specimen, the section for which the pore area covered the greatest proportion of the section after damage was selected first, and then the corresponding section from the CT scan image before damage was selected. The effective bearing area was calculated for each specimen by removing the pore area from the total area before and after the damage separately, allowing the damage variable to be calculated as follows [26]:

$$D = 1 - \frac{A_{Pore}}{A_0} \tag{5}$$

where D is the damage variable; A_{Pore} and A_0 are the pore area and the total area (m^2) of the scanned slices, respectively.

3. Results

3.1. NDT Test Results

An ultrasonic test, an EIS test, and a CT scan test were performed for each specimen before and after damage respectively, then the corresponding resulting parameters were obtained as follows. The P-wave velocities for the twelve sandstone specimens before and after damage under different peak loads and loading cycles are shown in Figure 8. For the initial state, the distribution of P-wave velocities measured for the sandstone specimens is concentrated at around 3000 m/s, with a range of 2900–3100 m/s. For a single specimen, damage leads to a decrease in the P-wave velocity. However, the reduction in wave velocity varies between specimens with different levels of damage. The wave velocities for the sandstone specimens in a damaged state range from 2900 to 2100 m/s, and the decrease, relative to the velocity for the initial state, becomes increasingly significant as the damage increases. In particular, when cracks appeared in the sandstone, the wave velocity decreased by more than 30%, compared with the velocity before damage.

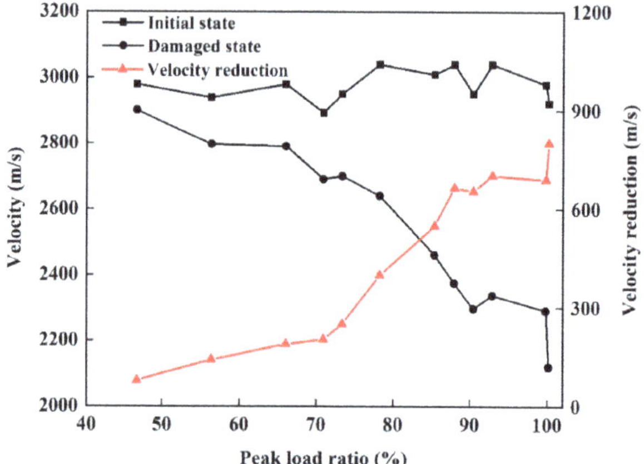

Figure 8. Wave velocities for sandstone specimens before and after damage.

The measured wave amplitude attenuation increased after damage for all twelve sandstone specimens, as shown in Figure 9. The distribution of wave amplitude attenuations for specimens before damage is narrow and is centered on 0.6. When the damage is small, the difference between the wave amplitude attenuation before and after damage is also small. Similar to the P-wave velocity, as the degree of damage increases, the decrease in the wave amplitude ratio, relative to the initial state, becomes greater. When the specimen cracks after cyclic loading, the wave amplitude ratio falls below 0.2. This indicates that the first wave amplitude attenuation is more sensitive to damage than the P-wave velocity when the level of damage approaches failure.

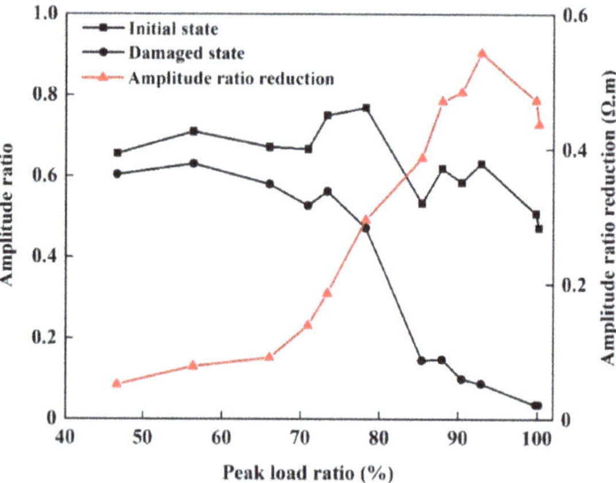

Figure 9. First wave amplitude attenuation for sandstone specimens before and after damage.

Figure 10 shows the resistivity of the twelve sandstone specimens before and after damage under different peak loads and loading cycles. As the loading cycle and peak load increased, the damage accumulated and the resistivity tended to decrease. However, the rate of decrease in resistivity was not uniform. As the damage increased, the rate of decrease in resistivity also increased significantly.

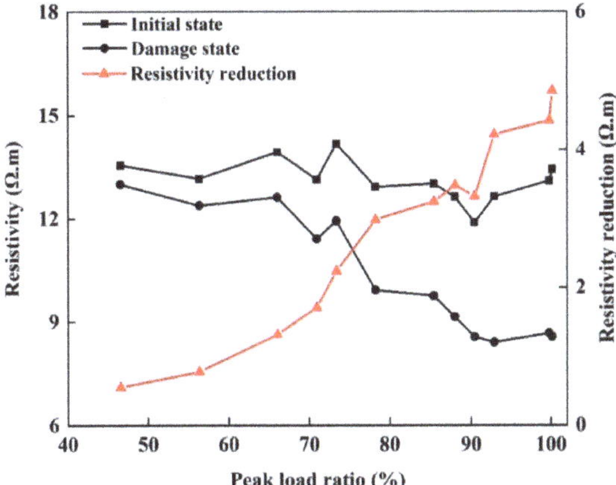

Figure 10. Sandstone resistivity before and after damage under different loading conditions.

In the initial state, the resistivity of sandstone was concentrated around 13 Ω.m, with individual values in the range 12–14 Ω.m. With increasing peak load and loading cycles, the resistivity of sandstone in the damage state showed varying rates of decline and was distributed in the range 8.5–13 Ω.m. When the ratio of peak load to strength was small (after few loading cycles), the resistivity was not much lower than that before damage. As the ratio of peak load to strength and the number of loading cycles increased, the decrease in resistivity became increasingly obvious. Once penetrating cracks had developed, the resistivity of the sandstone specimen dropped sharply (by about 40%) compared to that before the damage.

In CT scan slice images, the different gray levels of the pixels represent different densities of substance. The pores have relatively smaller grayscale values, corresponding to darker pixels, while the matrix has relatively larger grayscale values, corresponding to lighter pixels. Therefore, the pores within the specimen can be identified by selecting a specific threshold value of grayscale. Due to the different cyclic loads applied to specimens No. 6–8, the specimens presented significantly different pore structures after damage. Threshold segmentation was performed on the scanned slices and the pore condition before and after damage was counted separately. Figure 11 shows that the pores and cracks in the damaged specimens increase significantly as the degree of damage increases.

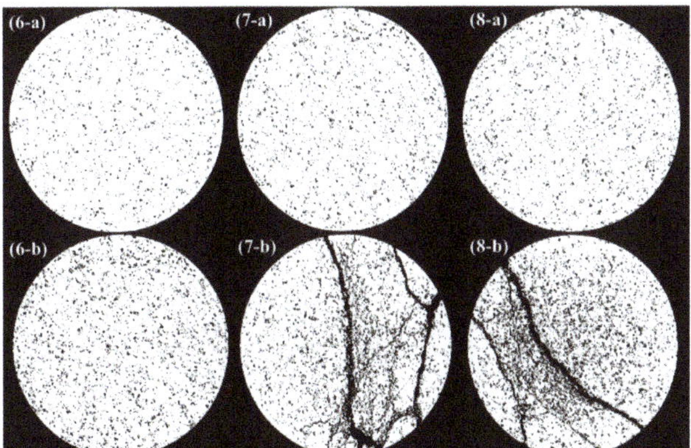

Figure 11. CT scan slice image after being segmented (6–8 represent the specimen ID, while a and b indicate the initial state and damage state respectively).

The change in porosity at the same section before and after damage was obtained by counting the pore area in the grayscale image. In the initial state, the porosity in the section obtained from CT scan is around 12%, with a small dispersion. Specimen No. 6 was not broken after damage, so the change in porosity was relatively small, increasing from 11.8% to 15.5%. The remaining two specimens were both fractured after damage, with a large number of new pores distributed along the cracks and a significant change in porosity. The porosity of specimen No. 7 increased from 12.76% to 19.07% and that of specimen No. 8 increased from 12.74% to 19.67%.

3.2. Damage Variables for Sandstone

The above results show that the four NDT parameters: P-wave velocity, first wave amplitude attenuation, resistivity, and the effective bearing area obtained from the CT scan test can all reflect damage for sandstone. The damage variables obtained from ultrasonic testing are calculated from the P-wave velocity and first wave amplitude attenuation before and after damage. The resistivity obtained from the EIS test, the effective bearing area obtained from the CT scan test, and the elastic modulus obtained from the cycle loading test can also be used to derive damage variables for all twelve sandstone specimens.

Figure 12 shows the differences between the damage variables for sandstone that were derived from the five different measured parameters. All five damage variables increased monotonically as the peak load and cycle times increased. As the values of the damage variables derived from these parameters differ significantly, the consistency between the responses of the different parameters to damage was further discussed. The P-wave velocity is most consistent with the elastic modulus, with a correlation coefficient of 0.991, followed by the resistivity and wave amplitude attenuation, both with a correlation coefficient of 0.885. Among the NDT parameters, the correlation coefficient between P-wave velocity and resistivity was the highest at 0.991 and that between P-wave velocity and amplitude attenuation was 0.958. The correlation coefficient between resistivity and amplitude attenuation is relatively low at 0.939. The high correlation coefficients show that the four macroscopic damage parameters, P-wave velocity, first wave amplitude attenuation, resistivity and elastic modulus, responded consistently to damage.

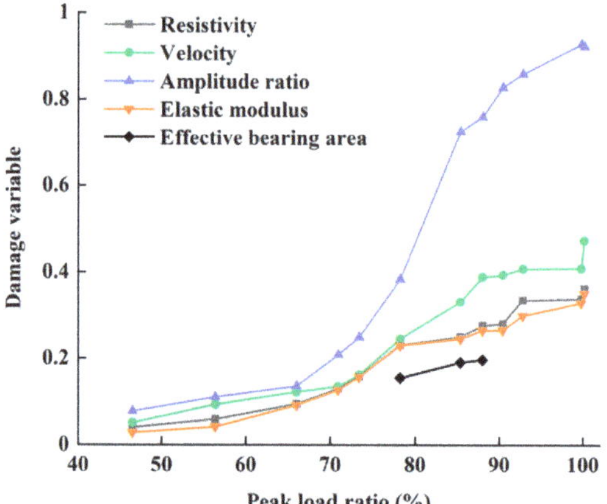

Figure 12. Damage variables derived from different measurement parameters. Note that three specimens were tested using a CT scan.

The damage variables derived from the P-wave velocity and wave amplitude attenuation that were measured in the ultrasonic test were greater than the damage variables that were derived from the other three measurement parameters. The damage variables derived from the first wave amplitude attenuation were greater than those derived from the P-wave velocity. This shows that ultrasonic testing is more sensitive to damage than other tests for sandstone, and that wave amplitude attenuation is the most sensitive parameter to the damage. The measured resistivity and acoustic parameters are NDT parameters. Although the damage variable derived from resistivity is smaller than those derived from the ultrasonic test measurements, it is slightly greater than the damage variable that is derived from the elastic modulus. This suggests that acoustic and electrical damage measurement parameters, in addition to being non-destructive, are superior to the elastic modulus tests in terms of their sensitivity to damage.

CT scan testing allows us to find the effective bearing area for a sandstone specimen, which can be used to define microscopic damage variables. Figure 12 shows the damage variables derived from the effective bearing area for specimens No. 6–8 and compares these with the macroscopic damage variables. The damage variables derived from the CT scan test increase with the degree of damage, in agreement with the macroscopic damage variables. However, the CT scan test is limited by resolution and is insensitive to smaller pores, with the result that the damage variable derived from the effective bearing area is smaller than the damage variables derived from the other damage parameters.

4. Discussion

4.1. Stability of Damage Measurements

The ultrasonic and EIS tests were implemented three times for each specimen, before and after damage. The spread of the three measurements for each parameter can be used to assess the stability of the NDT parameters. Three sandstone specimens that were subjected to different degrees of damage were selected, and three values for the damage variable were calculated from the three repeated measurements for the P-wave velocity, wave amplitude attenuation and resistivity. The results are plotted as a box line diagram in Figure 13.

The results show that damage variables derived from resistivity and amplitude attenuation are less stable when the stress level is low. As the damage accumulates, the stability of the damage variables derived from resistivity and amplitude attenuation increases. The damage variables that are derived from the wave amplitude attenuation and from the resistivity become similarly stable when a specimen is broken. However, the damage variable derived from the P-wave velocity becomes less stable as the level of damage increases, so that when the specimen fails, the damage variable derived from resistivity is the most stable, followed by that derived from wave amplitude attenuation, and the damage variable derived from the P-wave velocity is the least stable.

The reason for these different stabilities is that the three parameters rely on different mechanisms to detect damage. In sandstone, both resistivity and wave amplitude attenuation reflect damage detected via the propagation of electrical current and ultrasonic waves through the pore structure, while the P-wave velocity captures damage detected by the propagation of ultrasonic waves around pores as the waves follow the continuous medium. When the degree of damage is low, the specimen has fewer pores and so electrical current and ultrasonic waves do not propagate along a specific pore channel, which results in the derived damage variables being relatively unstable. Fewer pores means that the medium is more continuous for ultrasonic wave propagation, and so the damage variable derived from ultrasonic wave velocity is relatively stable. In contrast, when the degree of damage is high, continuous pore channels form inside the specimen and the damage variables derived from resistivity and amplitude attenuation are more stable. For high levels of damage, the proportion of space occupied by a continuous medium is reduced, so the damage variables derived from the ultrasonic wave velocity are less stable.

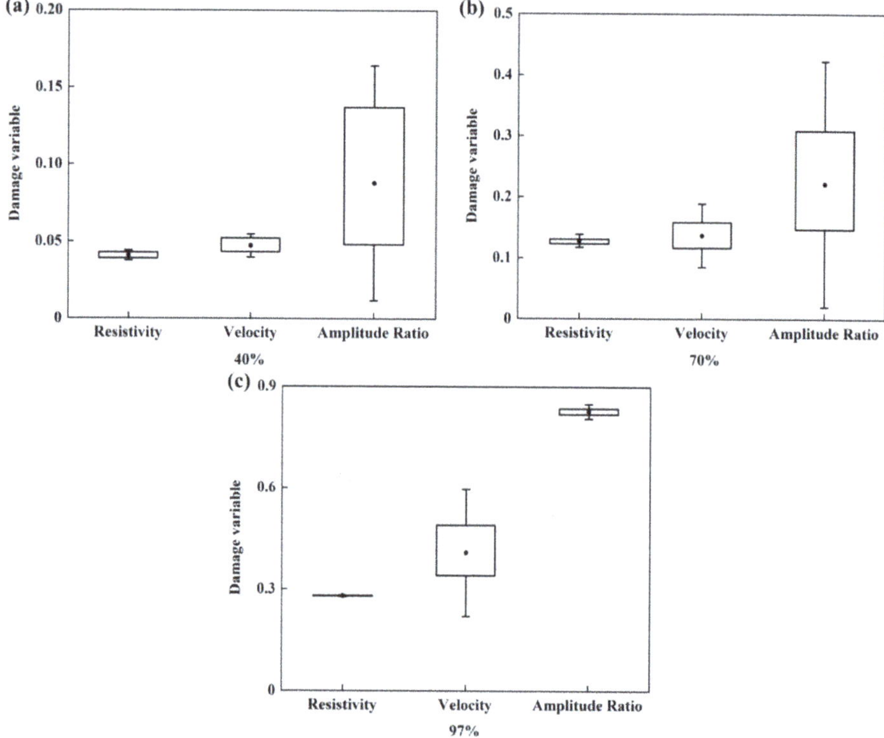

Figure 13. Stability of the damage variables derived from the NDT parameters. (**a**–**c**) correspond to three separate sandstone specimens for which the peak load reached 40%, 70% and 97% of the strength during cyclic loading, respectively.

4.2. Mechanisms Behind Damage Measurements

Ultrasonic testing and EIS testing are NDT methods that measure damage by detecting changes in the internal structure of the specimen. According to the results in the previous section, the difference between the propagation mechanisms for ultrasonic waves and electrical current leads to a difference between the derived damage variables and between the stability of damage testing. Kassab believes that the P-wave velocity measurement can be interpreted as a measure of damage based on the propagation of ultrasonic waves around pores and thus primarily reflects the continuous medium of the rock [27]. On the other hand, Wang thought the resistivity measurement can be interpreted based on the propagation of current in the pore [28]. Both measures are associated with the porosity of the rock, but there are differences between the mechanisms that control the two measurements [29]. To investigate this, we established the relationships between P-wave velocity and porosity, and between resistivity and porosity, using theoretical and empirical models. Using porosity as a common factor, a relationship between P-wave velocity and resistivity was found, and validated using the results from our experiments.

Many models describe the relationship between P-wave velocity and rock porosity, including the Gassmann model, which is based on theoretical assumptions, and the Raymer model, which is based on empirical relationships. The Gassmann model is a theoretical model based on the close relationship between the P-wave velocity and the dynamic elastic modulus of rock [30]. The Gassmann formula for calculating the P-wave velocity is as follows:

$$V_p^2 \rho = \left(K_d + \frac{4}{3}G_d\right) + \frac{\left(1 - \frac{K_d}{K_s}\right)^2}{\left(1 - \varphi - \frac{K_d}{K_s}\right)\frac{1}{K_s} + \frac{\varphi}{K_f}} \tag{6}$$

where, V_p is the P-wave velocity (m/s); ρ is the bulk density (kg/m^3); φ is the porosity; G_d is the shear modulus for rock (GPa); K_d, K_s, and K_f are the moduli for rock, solid fraction of rock, and liquid fraction of rock (i.e., pore fluid), respectively. K_d and G_d are calculated using Krief's formula, which includes the elastic and shear moduli for the rock solid fraction, K_s and G_s, respectively, the porosity φ, and the Krief index, κ ($\kappa = 3$) [31]. Check Table 1 for the values of the above parameters.

$$K_d = K_s(1 - \varphi)^{\kappa/(1-\varphi)} \tag{7}$$

$$G_d = G_s(1 - \varphi)^{\kappa/(1-\varphi)} \tag{8}$$

Table 1. Values used for the material parameters in Equations (6)–(11) (see text for units).

ρ_s	ρ_f	K_s	K_f	G_s	v_s	v_f
1000	0.29	25	2.25	8	3.7	1.4

Raymer's formula is an empirical formula for calculating acoustic velocity for materials with different porosities, based on measured acoustic logging data [32]. The formula is calculated as follows, where v_s and v_f are the P-wave velocities for the rock solid fraction and pore fluid, respectively:

$$v_p = (1 - \varphi)^2 v_s + \varphi v_f \tag{9}$$

The Hashin-Shtrikman (HS) bounds are the theoretical upper and lower boundaries for the elastic and shear moduli and can therefore be used to calculate upper and lower boundaries for the P-wave velocity for rocks [33]. Figure 14 shows the relationship between the P-wave velocity and porosity, calculated using the above models for porosities ranging from 0% to 45%, with a scatterplot of the experimental data superimposed, where porosities are measured separately for all 12 specimens after damage. The model results generally fall within the HS boundaries, and the P-wave velocity calculated by the models decreases with the increase in porosity. The measured data are in good agreement with the Raymer model based on empirical relationships, which is consistent with the results in the previous literature [34].

The relationship between resistivity and porosity can be described by a self-similar model, which is based on self-consistent effective medium theory, or by the Archie model, which is based on empirical relationships. In the self-similar model, resistivity is continuously iterated and its relationship with porosity is then given by the following expression [35]:

$$\rho = \left(\frac{\rho - \rho_s}{\rho_s - \rho_f}\right)^m \rho_f \varphi^{-m} \tag{10}$$

where ρ, ρ_s and ρ_f are the resistivity (Ω.m) of the rock, matrix and pore solution, respectively; φ is the porosity; and m is the cementation exponent, which equals 2.2 for the moderately cemented sandstone specimens in this study [14].

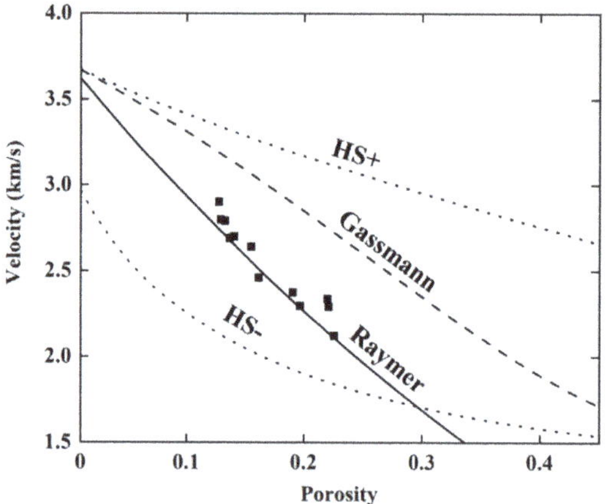

Figure 14. Different P-wave velocity-to-porosity models.

The Archie model is an empirical equation based on measured resistivity logging data [36]. The equation for the Archie model assumes that the resistivities for pure and saturated sandstone are proportional to the resistivity for the pore solution and includes tortuosity factor, *a*, which is equal to 0.8:

$$\rho = a\rho_f\varphi^{-m} \tag{11}$$

If the effect of particle geometry is assumed negligible, then the HS boundary can also provide upper and lower boundaries for the resistivity [37]. Figure 15 shows the experimental data alongside the relationships between resistivity and porosity calculated using the different models. The results show that the resistivity of the sandstone decreases with increasing porosity and the experimental data are in good agreement with the Archie model [30]. When the porosity is low, the experimental data are distributed near the Archie model curve. When porosity is high, the experimental data are scattered between the curves for the Archie model and for the self-similar model.

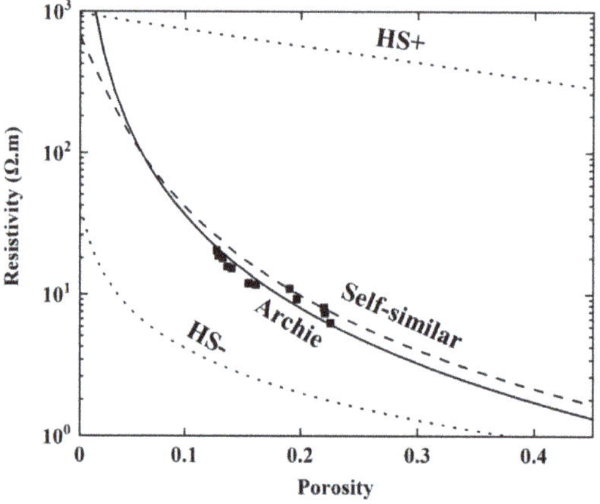

Figure 15. Different resistivity-to-porosity models.

The relationships between the P-wave velocity and porosity with the relationships between resistivity and porosity were combined to establish the relationship between the P-wave velocity and resistivity, using the porosity as a common variable. Figure 16 shows that the experimental data were generally consistent with the Raymer-Archie model, which was constructed from empirical data. This demonstrates the accuracy of the test data and proves that the differences between damage measured parameters of sandstone is caused by the differences in reflecting changes to pore condition between damage measuring methods.

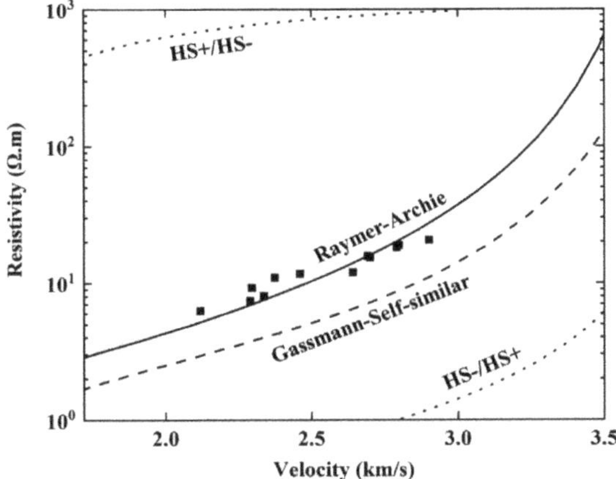

Figure 16. Different models of the relationship between the P-wave velocity and resistivity.

5. Conclusions

Ultrasonic tests, EIS tests and CT scan tests were carried out on sandstone specimens before and after subjecting the specimens to different levels of damage, and the unloading elastic moduli were obtained from the resulting stress-strain curves. Four NDT parameters: P-wave velocity, first wave amplitude attenuation, resistivity, and CT-based effective bearing area, and one destructive test parameter, the elastic modulus, were compared in terms of derived damage variables for sandstone.

The results show that the elastic modulus result in damage variables that agree closely with those derived from NDT damage measurement parameters, justifying the use of NDT test parameters to determine sandstone damage variables. The P-wave velocity agrees more closely than the wave amplitude attenuation with the other damage measurement parameters but is only moderately stable. The wave amplitude attenuation is more sensitive than the other measurement parameters to sandstone damage, and its stability increases significantly as the level of damage increases. Damage variables calculated from the CT scan tests capture microscopic properties, but the finite resolution of the CT scans means that the derived damage variables have lower values than those derived from macroscopic measurement methods.

The resistivity and wave amplitude attenuation reflect the condition of the pore structure, while the P-wave velocity reflects the condition of the continuous medium inside the sandstone. The relationship between the P-wave velocity and resistivity was modeled by considering the relationship of both to porosity. The experimental data were consistent with the Raymer-Archie model, verifying that the different responses of the measured parameters to porosity is what drives the differences between the different damage measurement parameters.

Author Contributions: Q.X. designed the research; D.Y. performed the research and wrote the paper. All authors have read and agreed to the published version of the manuscript.

Funding: This research was funded by the National Key Research and Development Program of China (2017YFC0804602), National Natural Science Foundation of China (51839007 & 51879141) and State Key Laboratory of Hydroscience and Engineering (2019-KY-03).

Conflicts of Interest: The authors confirm that there are no known conflicts of interest associated with this publication.

References

1. Bernardi, D.; DeJong, J.T.; Montoya, B.M.; Martinez, B.C. Bio-bricks: Biologically cemented sandstone bricks. *Constr. Build. Mater.* **2014**, *55*, 462–469. [CrossRef]
2. Dvorkin, J.; Brevik, I. Diagnosing high-porosity sandstones: Strength and permeability from porosity and velocity. *Geophysics* **1999**, *64*, 795–799. [CrossRef]
3. Wang, Y.H.; Liu, Y.F.; Ma, H.T. Changing regularity of rock damage variable and resistivity under loading condition. *Saf. Sci.* **2012**, *50*, 718–722. [CrossRef]
4. Shah, A.A.; Ribakov, Y. Non-destructive evaluation of concrete in damaged and undamaged states. *Mater. Des.* **2009**, *30*, 3504–3511. [CrossRef]
5. Selleck, S.F.; Landis, E.N.; Peterson, M.L.; Shah, S.P.; Achenbach, J.D. Ultrasonic investigation of concrete with distributed damage. *Aci. Mater. J.* **1998**, *95*, 27–36.
6. Alemu, B.L.; Aker, E.; Soldal, M.; Johnsen, Ø.; Aagaard, P. Effect of sub-core scale heterogeneities on acoustic and electrical properties of a reservoir rock: A CO_2 flooding experiment of brine saturated sandstone in a computed tomography scanner. *Geophys. Prospect.* **2013**, *61*, 235–250. [CrossRef]
7. Heap, M.J.; Vinciguerra, S.; Meredith, P.G. The evolution of elastic moduli with increasing crack damage during cyclic stressing of a basalt from Mt. Etna volcano. *Tectonophysics* **2009**, *471*, 153–160. [CrossRef]
8. Antonaci, P.; Bruno, C.L.E.; Gliozzi, A.S.; Scalerandi, M. Monitoring evolution of compressive damage in concrete with linear and nonlinear ultrasonic methods. *Cem. Concr. Res.* **2010**, *40*, 1106–1113. [CrossRef]
9. Yim, H.J.; Kwak, H.G.; Kim, J.H. Wave attenuation measurement technique for nondestructive evaluation of concrete. *Nondestruct. Test. Eva.* **2012**, *27*, 81–94. [CrossRef]
10. Müller, T.M.; Gurevich, B.; Lebedev, M. Seismic wave attenuation and dispersion resulting from wave-induced flow in porous rocks—A review. *Geophysics* **2010**, *75*, 147–164. [CrossRef]
11. Ranade, R.; Zhang, J.; Lynch, J.P.; Li, V.C. Influence of micro-cracking on the composite resistivity of Engineered Cementitious Composites. *Cem. Concr. Res.* **2014**, *58*, 1–12. [CrossRef]
12. Kahraman, S.; Alber, M. Predicting the physico-mechanical properties of rocks from electrical impedance spectroscopy measurements. *Int. J. Rock Mech. Min.* **2006**, *43*, 543–553. [CrossRef]
13. Zisser, N.; Nover, G. Anisotropy of permeability and complex resistivity of tight sandstones subjected to hydrostatic pressure. *J. Appl. Geophys.* **2009**, *68*, 356–370. [CrossRef]
14. Yin, D.H.; Xu, Q.J. Defining the damage variable of sandstone from electrical impedance spectroscopy measurements. In Proceedings of the 14th International Congress on Rock Mechanics and Rock Engineering, Foz do Iguacu, Brazil, 13–18 September 2019; pp. 1053–1060.
15. Wang, Y.; Li, X.; Zhang, B.; Wu, Y. Meso-damage cracking characteristics analysis for rock and soil aggregate with CT test. *Sci. China Technol. Sci.* **2014**, *57*, 1361–1371. [CrossRef]
16. Yin, S.; Zhao, D.; Zhai, G. Investigation into the characteristics of rock damage caused by ultrasonic vibration. *Int. J. Rock Mech. Min.* **2016**, *84*, 159–164. [CrossRef]
17. Landis, E.N.; Zhang, T.; Nagy, E.N.; Nagy, G.; Franklin, W.R. Cracking, damage and fracture in four dimensions. *Mater. Struct* **2007**, *40*, 357–364. [CrossRef]
18. Shah, A.A.; Hirose, S. Nonlinear ultrasonic investigation of concrete damaged under uniaxial compression step loading. *J. Mater. Civil. Eng.* **2010**, *22*, 476–484. [CrossRef]
19. Yang, S.; Zhang, N.; Feng, X.; Kan, J.; Pan, D.; Qian, D. Experimental investigation of sandstone under cyclic loading: Damage assessment using ultrasonic wave velocities and changes in elastic modulus. *Shock Vib.* **2018**, *2018*, 1–13. [CrossRef]
20. Alves, M.l.; Yu, J.; Jones, N. On the elastic modulus degradation in continuum damage mechanics. *Comput. Struct.* **2000**, *76*, 703–712. [CrossRef]
21. Lemaitre, J.; Dufailly, J. Damage measurements. *Eng. Fract. Mech.* **1987**, *28*, 643–661. [CrossRef]
22. Kahraman, S.; Alber, M. Electrical impedance spectroscopy measurements to estimate the uniaxial compressive strength of a fault breccia. *Bull. Mater. Sci.* **2014**, *37*, 1543–1550. [CrossRef]
23. Gomaa, M.M.; Alikaj, P. Effect of electrode contact impedance on A.C. electrical properties of a wet hematite sample. *Mar. Geophys. Res.* **2010**, *30*, 265–276. [CrossRef]
24. Lizarazo-Marriaga, J.; Higuera, C.; Claisse, P. Measuring the effect of the ITZ on the transport related properties of mortar using electrochemical impedance. *Constr. Build. Mater.* **2014**, *52*, 9–16. [CrossRef]

25. Ma, T.; Yang, C.; Chen, P.; Wang, X.; Guo, Y. On the damage constitutive model for hydrated shale using CT scanning technology. *J. Nat. Gas. Sci. Eng.* **2016**, *28*, 204–214. [CrossRef]

26. Ma, T.; Chen, P. Study of meso-damage characteristics of shale hydration based on CT scanning technology. *Petrol. Explor. Dev.* **2014**, *41*, 249–256. [CrossRef]

27. Kassab, M.A.; Weller, A. Anisotropy of permeability, P-wave velocity and electrical resistivity of Upper Cretaceous carbonate samples from Tushka Area, Western Desert, Egypt. *Egypt. J. Petrol.* **2019**, *28*, 189–196. [CrossRef]

28. Wang, H.; Sun, S.Z.; Yang, H.; Gao, H.; Xiao, Y.; Hu, H. The influence of pore structure on P- & S-wave velocities in complex carbonate reservoirs with secondary storage space. *Pet. Sci.* **2011**, *8*, 394–405.

29. Kassab, M.A.; Weller, A. Study on P-wave and S-wave velocity in dry and wet sandstones of Tushka region, Egypt. *Egypt. J. Petrol.* **2015**, *24*, 1–11. [CrossRef]

30. Carcione, J.M.; Ursin, B.; Nordskag, J.I. Cross-property relations between electrical conductivity and the seismic velocity of rocks. *Geophysics* **2007**, *72*, E193–E204. [CrossRef]

31. Krief, M.; Garat, J.; Stellingwerff, J.; Ventre, J. A petrophysical interpretation using the velocities of P and S waves (Full-Waveform Sonic). *Log. Anal.* **1990**, *31*, 355–369.

32. Raymer, L.L.; Hunt, E.R.; Gardner, J.S. An Improved Sonic Transit Time-To-Porosity Transform. In Proceedings of the SPWLA 21st Annual Logging Symposium, Lafayette, Louisiana, 8–11 July 1980; p. 13.

33. Werthmüller, D.; Ziolkowski, A.; Wright, D. Background resistivity model from seismic velocities. *Geophysics* **2013**, *78*, E213–E223. [CrossRef]

34. Grana, D. Probabilistic approach to rock physics modeling. *Geophysics* **2014**, *79*, D123–D143. [CrossRef]

35. Sen, P.N.; Scala, C.; Cohen, M.H. A self-similar model for sedimentary rocks with application to the dielectric constant of fused glass beads. *Geophysics* **1981**, *46*, 781–795. [CrossRef]

36. Archie, G.E. The electrical resistivity log as an aid in determining some reservoir characteristics. *Trans. AIME* **1942**, *146*, 54–62. [CrossRef]

37. Hashin, Z.; Shtrikman, S. A variational approach to the theory of the elastic behaviour of multiphase materials. *J. Mech. Phys. Solids* **1963**, *11*, 127–140. [CrossRef]

Publisher's Note: MDPI stays neutral with regard to jurisdictional claims in published maps and institutional affiliations.

© 2020 by the authors. Licensee MDPI, Basel, Switzerland. This article is an open access article distributed under the terms and conditions of the Creative Commons Attribution (CC BY) license (http://creativecommons.org/licenses/by/4.0/).

MDPI

St. Alban-Anlage 66

4052 Basel

Switzerland

Tel. +41 61 683 77 34

Fax +41 61 302 89 18

www.mdpi.com

Materials Editorial Office

E-mail: materials@mdpi.com

www.mdpi.com/journal/materials